DSP INTEGRATED CIRCUITS

Lars Wanhammar
Linköping University

ACADEMIC PRESS
San Diego London Boston New York Sydney Tokyo Toronto

Academic Press Series in Engineering

Series Editor
J. David Irwin
Auburn University

Designed to bring together interdependent topics in electrical engineering, mechanical engineering, computer engineering, and manufacturing, the Academic Press Series in Engineering provides state-of-the-art handbooks, textbooks, and professional reference books for researchers, students, and engineers. This series provides readers with a comprehensive group of books essential for success in modern industry. A particular emphasis is given to the applications of cutting-edge research. Engineers, researchers, and students alike will find the Academic Press Series in Engineering to be an indispensable part of their design toolkit.

Published books in the series:
Industrial Controls and Manufacturing, 1999, E. Kamen
DSP Integrated Circuits, 1999, L. Wanhammar
Time Domain Electromagnetics, 1999, S.M. Rao
Single and Multi-Chip Microcontroller Interfacing, 1999, G.J. Lipovski

This book is printed on acid-free paper. ∞

Academic Press
A division of Harcourt Brace & Company
525 B Street, Suite 1900, San Diego, CA 92101-4495, USA
http://www.apnet.com

Academic Press
24–28 Oval Road, London NW1 7DX, UK
http://www.hbuk.co.uk/ap/

Library of Congress Cataloging-in-Publication: 98-22149
ISBN: 0-12-734530-2

Printed in the United States of America
99 00 01 02 03 IP 9 8 7 6 5 4 3 2 1

CONTENTS

PREFACE

The book *DSP Integrated Circuits* is used as a textbook for the course "Application-Specific Integrated Circuits for Digital Signal Processing" given at Linköping University. This text is intended to fill a gap in the market for textbooks on design of digital signal processing systems using VLSI technologies. The intent is to present a comprehensive approach to system design and implementation of DSP systems using advanced VLSI technologies. We also try to present a coherent paradigm for the whole design process, i.e., a top-down design approach is stressed throughout the book. The emphasis is on DSP algorithms, scheduling, resource allocation assignment and cir-cuit architectures. We derive an efficient implementation strategy that is based on asynchronous bit-serial processing elements that can be matched to the DSP algorithm and the application requirements. The aim is to minimize power consumption and chip area, but equally important is the use of a structured design methodology that allows an error-free design to be completed according to the project schedule. The presentation necessarily represents a personal view, since there is no unique global view to which everyone in the field agrees.

The textbook presents the design process in a top-down manner. Throughout the text, three case studies are presented. The three examples are selected in order to demonstrate different characteristics of common DSP applications. The first case study involves the design of an interpolator based on lattice wave digital filters. The major design problems are the complicated scheduling of the operations and the resource allocation. The second case study is the design of an FFT processor. The major problem here is the partitioning of the algorithm into appropriate processes that can be mapped onto the processing elements. The third case study is the design of a two-dimensional discrete cosine transform for high-definition TV (HDTV). The major problems here are the high I/O data rate and the high arithmetic work load.

The textbook is aimed for engineers and scientists involved in digital signal processing, real-time systems, including computer-aided design, application-specific integrated circuit design, and VLSI technology. The textbook provides the necessary background in DSP that is needed in order to appreciate the case studies. Of course, it is beneficial if the student has a prior basic understanding of digital signal processing techniques.

I would like to acknowledge my sincere gratitude to Magnus Hörlin, Håkan Johansson, Johan Melander, Erik Nordhamn, Kent Palmkvist, Tony Platt, Magnus Karlsson, Mikael Karlsson Rudberg, Björn Sikström, Mark Vesterbacka, and Torbjörn Widhe for generously providing assistance during the development of the material presented in this book as well as carefully reading the innumerable versions of the manuscript.

Lars Wanhammar, Linköping

1

DSP INTEGRATED CIRCUITS

1.1 INTRODUCTION

Rapid advancements in electronics, particularly in manufacturing techniques for integrated circuits, have already had, and will undoubtedly continue to have, a major impact on both industry and society as a whole. In this book we will discuss various approaches to designing integrated circuits for *digital signal processing* (*DSP*) applications. Modern DSP systems are often well suited to VLSI implementation. Indeed, they are often technically feasible or economically viable only if implemented using VLSI technologies. The large investment necessary to design a new integrated circuit can only be justified when the number of circuits to be manufactured is large, or when the necessary performance requirements are so high that they cannot be met with any other technology. In practice, we often find that both arguments are valid, particularly in communication and consumer applications. Advances in integrated circuit technology also open new areas for DSP techniques, such as intelligent sensors, robot vision, and automation, while simultaneously providing a basis for continuing advancements in traditional signal processing areas, such as speech, music, radar, sonar, audio, video, and communications.

Integrated circuit technology has had a profound effect on the cost, performance, and reliability of electronic circuits. Manufacturing cost is almost independent of the complexity of the system. The cost per integrated circuit (unit cost) for large-volume applications using large chips is dominated by the cost of the chip, while for small and medium size chips the package cost tends to dominate. The whole system cost for small-volume applications is often dominated by the development cost. Unfortunately, the development cost is often difficult to estimate accurately. Increase in system complexity and integration of the manufacturing and design processes tend to increase development costs and cause long design times. However, these adverse effects can be mitigated by extensive use of computer-aided design tools and the use of efficient design methodologies. Today, *computer-aided design* (*CAD*) and *computer-aided manufacturing* (*CAM*) are used extensively in almost all aspects of electronic engineering. To explore VLSI technology optimally it is necessary that the design team cover all aspects of the design, specification, DSP algorithm, system and circuit architecture, logic, and integrated circuit design. Hence, changes in classical design methodologies and in the organization of design teams may be necessary. We will therefore discuss the most common design methodologies used for the design of DSP systems. We will also present a novel methodology and apply it to some common DSP subsystems.

The problem of designing special-purpose DSP systems is an interesting research topic, but, more important, it has significant industrial and commercial relevance. Many DSP systems (for example, mobile phones) are produced in very large numbers and require high-performance circuits with respect to throughput and power consumption. Therefore, the design of DSP integrated circuits is a challenging topic for both system and VLSI designers. DSP integrated circuits are also of economic importance to the chip manufacturers.

1.2 DIGITAL SIGNAL PROCESSING

Signal processing is fundamental to information processing and includes various methods for extracting information obtained either from nature itself or from man-made machines. Generally, the aim of signal processing is to reduce the information content in a signal to facilitate a decision about what information the signal carries. In other instances the aim is to retain the information and to transform the signal into a form that is more suitable for transmission or storage. The DSP systems of interest here are the so-called *hard real-time systems*, where computations must be completed within a given time limit (the sample period). An unacceptable error occurs if the time limit is exceeded.

Modern signal processing is mainly concerned with digital techniques, but also with analog and sampled-data (discrete-time) techniques, which are needed in the interfaces between digital systems and the outside analog world [9, 11]. Sampled-data systems are generally implemented using *switched capacitor* (*SC*) [10] or *switched current* (*SI*) circuits. Most A/D and D/A converters are today based on SC circuit techniques. An important advantage of SC circuits is that they can easily be integrated with digital CMOS circuits on the same chip. Recently, analog circuits such as anti-aliasing filters have also become possible to implement on the same chip. A fully integrated system-on-a-chip is therefore feasible by using a suitable combination of circuit techniques. This will affect both performance and cost of DSP systems.

Generally, complex signal processing systems are synthesized using subsystems that perform the basic DSP operations. Typical operations are frequency selective and adaptive filtering, time-frequency transformation, and sample rate change. In Chapters 3 and 4, we will review some of the most common signal processing functions used in such subsystems. The aim is to provide a background for three typical DSP subsystems that will be used as case studies throughout the book.

1.3 STANDARD DIGITAL SIGNAL PROCESSORS

In principle, any DSP algorithm can be implemented by programming a standard, general-purpose digital signal processor [1]. The design process involves mainly coding the DSP algorithm either using a high-level language (for example, the *C* language) or directly in assembly language. Some high-level design tools allow the user to describe the algorithm as a block diagram via a graphic user interface. The tool automatically combines optimized source codes for the blocks, which are stored in a library, with code that calls the blocks according to the block diagram. Finally, the source code is compiled into object code that can be executed by the processor. This approach allows rapid prototyping, and the achieved performance in terms of execution speed and code size is reasonably good since the codes for the blocks are optimized. However, the performance may become poor if the blocks are too simple since the code interfacing the blocks is relatively inefficient.

Generally, the implementation process, which is illustrated in Figure 1.1, begins with the derivation of an executable high-level description that is subsequently transformed in one or several steps into object code. The representations (languages) used for these transformations are general and flexible so that they can be used for a large set of problems. Further, they are highly standardized.

The key idea, from the hardware designer's point of view, is that the hardware structure (digital signal processor) can be standardized by using a low-level language (instruction set) as interface between the DSP algorithm and the hardware. The digital signal processor can thereby be used for a wide range of applications.

This approach puts an emphasis on short design times and low cost due to the wide applicability of the hardware. Unfortunately, it is not always cost-effective, and often the performance requirements in terms of throughput, power consumption, size, etc. cannot be met. The main reason is mismatch between the capabilities of a standard digital signal processor and the signal processing requirements. The standard processor is designed to be flexible in order to accommodate a wide range of DSP algorithms while most algorithms use only a small fraction of the instructions provided. The flexibility provided by a user-programmable chip is not needed in many applications. Besides, this flexibility does not come without cost.

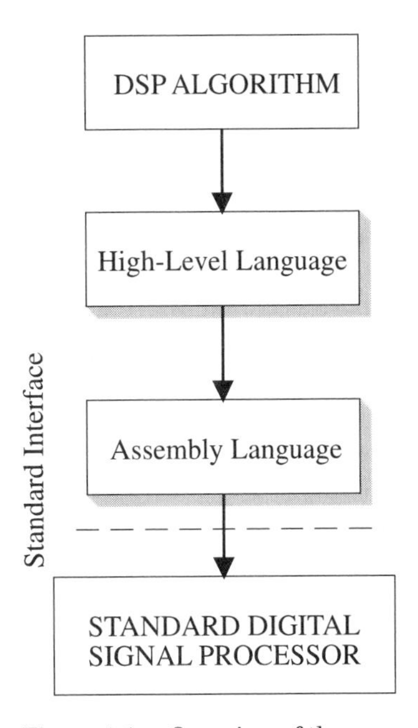

Figure 1.1 Overview of the implementation process using standard signal processors

It should be stressed that if a standard digital signal processor approach can meet the requirements, it is often the best approach. It allows the system to be modified by reprogramming in order to remove errors, and it provides the option of introducing new features that may extend the lifetime of the product. A new design always involves a significant risk that the system will not work properly or that it takes too long to develop and manufacture, so that the market window is lost. A standard digital signal processor approach is therefore an economically attractive approach for some types of DSP applications.

Early standard digital signal processors were based on the Harvard architecture that has two buses and separate memories for data and instructions. Generally, standard digital signal processors are provided with MACs—multiplier-accumulators—in order to perform sum-of-product computations efficiently. The high performance of these processors is achieved by using a high degree of parallelism. Typically, a multiply-and-add, data fetch, instruction fetch and decode, and memory pointer increment or decrement can be done simultaneously. Typical drawbacks are the limited on-chip memory size and the relatively low I/O bandwidth. The architectures used in modern standard digital signal processors will be further discussed in Chapter 8.

Early signal processors used fixed-point arithmetic and often had too short internal data word length (16 bits) and too small on-chip memory to be really efficient. Recent processors use floating-point arithmetic which is much more expensive than fixed-point arithmetic in terms of power consumption, execution time, and chip area. In fact, these processors are not exclusively aimed at DSP applications. Applications that typically require floating-point arithmetic are 3D-graphics, multimedia, and mechanical CAD applications. Fixed-point arithmetic is better suited for DSP applications than floating-point arithmetic since good DSP algorithms require high accuracy (long mantissa), but not the large dynamic signal range provided by floating-point arithmetic. Further, problems due to nonlinearities (rounding of products) are less severe in fixed-point arithmetic. Hence, we conclude that the current generation of standard signal processors is not efficient for many DSP applications.

1.4 APPLICATION-SPECIFIC ICs FOR DSP

The development effort for a large integrated circuit typically ranges between 1 and 10 man-years, depending on the uniqueness of the function, performance constraints, and the availability and performance of design tools. The combined advances in system design capability and VLSI technology have made it possible to economically design unique integrated circuits for use in dedicated applications, so-called *application-specific integrated circuits* (*ASICs*) [14]. This option makes new innovative system solutions practical.

The possibility of incorporating a whole signal processing system into one chip has a multitude of effects. It will dramatically increase the processing capacity and simultaneously reduce the size of the system, the power consumption, and the pin-restriction problem, which may be severe when a system has to be implemented using several chips. Reliability will also increase when the number of pins and the working temperature of the chips are reduced. Although VLSI technology solves or circumvents many problems inherent in older technologies, new limits and drawbacks surface. The main problems originate from the facts that the systems to be designed tend to be very complex and are often implemented in the most advanced VLSI process available. The latter has the adverse effect that the system often must be designed by using untested building blocks and incomplete and unproved CAD tools. Because of the innovative and dynamic nature of DSP techniques, the design team often lacks experience, since a similar system may not have been designed before. These factors make it difficult to estimate accurately the time it will take for the whole design process up to the manufacture of working chips.

Characteristic for DSP is the short step from basic research and innovation to practical applications. Therefore, a strong incentive exists to keep trade and design secrets from the competitors. This is to some extent possible, at least for a reasonably long time (months), if they are put into an application-specific integrated circuit. The cumulative effect is that the total system cost tends to be low and the performance gain provides an incentive to develop application-specific integrated circuits, even for low-volume applications.

1.4.1 ASIC Digital Signal Processors

In order to overcome some of the drawbacks discussed previously, considerable effort has been invested in developing CAD tools for the design of specialized digital signal processors. Generally, these processors are designed (preprogrammed) to execute only a fixed or limited set of algorithms, and cannot be reprogrammed after manufacturing. Typically only some parameters in the algorithms can be set by the user. These signal processors are called *application-specific* signal processors. A signal processor that can only execute a single algorithm is sometimes referred to as an algorithm-specific signal processor. Typically these ASIC processors are used in applications where a standard processor cannot meet the performance requirements (e.g., throughput, power consumption, chip area). High-throughput applications are found in, for example, *high-definition TV* (*HDTV*) and communication systems. Low power requirement is stringent in battery-powered applications. In high-volume applications the lower unit cost, due to the smaller chip area, may be another significant advantage.

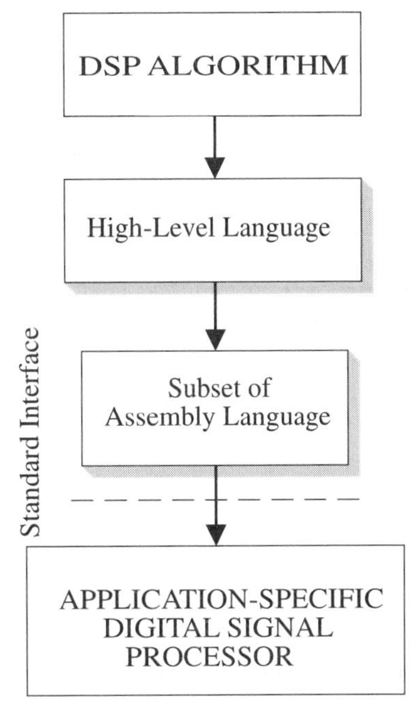

Figure 1.2 Overview of the implementation process using the ASIC digital signal processor approach

The performance in terms of throughput, power consumption, and chip area depends strongly on the architecture and the implemented instruction set. As illustrated in Figure 1.2, the processor can be matched to the algorithm by implementing only those instructions that actually are used and by providing several specialized data paths so that the required throughput is met. Several co-operating processors are often required in high-throughput applications.

A major factor contributing to the overall performance of ASIC signal processors is that the data word length can be adjusted to the requirements. The amount of on-chip memory can therefore be minimized. This is important since it is expensive in terms of chip area to implement large on-chip memories. Note that the use of external memories may result in reduced throughput since the practical data rates are much lower than for internal memories.

A significant performance improvement in terms of throughput, power consumption, and chip area over the standard processor approach is obtained at the cost of a slightly larger design effort. Large efforts are therefore being directed toward automatic design of ASIC signal processors. Major drawbacks of this approach are the inefficiency in terms of chip area and power consumption for applications with small computational workloads, and its inability to meet the throughput requirements in applications with high work loads.

1.4.2 Direct Mapping Techniques

Characteristic for direct mapping techniques is that the DSP algorithm is mapped directly onto a hardware structure without any intermediate representation. The direct mapping approach is particularly suitable for implementing systems with a fixed function, for example, digital filters. This approach allows a perfect match between the DSP algorithm, circuit architecture, and the system requirements. However, algorithms with many data-dependent branching operations may be unsuited to this method. Such algorithms are more easily implemented using the two approaches just discussed. Fortunately, such branching operations are rarely used in DSP algorithms.

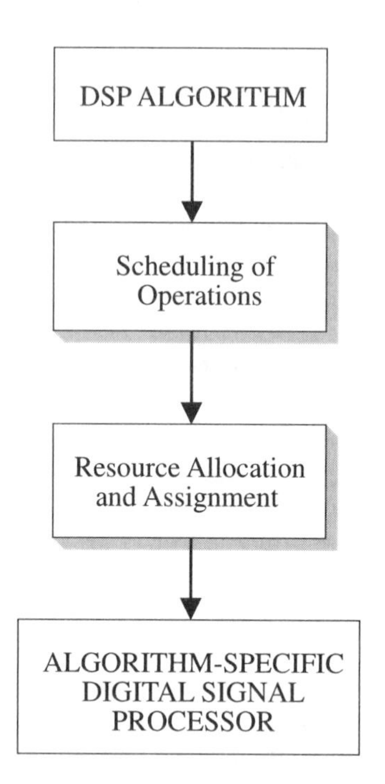

Figure 1.3 The major design steps in the direct mapping approach

Ideally, the design is done sequentially in a top-down manner, as illustrated in Figure 1.3 [5, 12]. In practice, however, several design iterations involving bottom-up evaluation must be carried out in order to arrive at an acceptable solution. The starting point for the design process is the DSP algorithm. The following three design steps are done after the algorithm has been frozen.

- Execution times are assigned to the arithmetic and logic operations in the algorithm. The execution of these operations is then scheduled so that the algorithm can be executed within the given sample period. Generally, several operations must be executed simultaneously. Operations that are not explicitly expressed in the algorithm (for example, address calculations of memory accesses, indices) are also scheduled.
- Computational resources (i.e., processing elements and memories) are allocated and assigned according to the schedule.
- The processing elements (PEs) and memories are connected by appropriate communication channels, and control units that generate the required control signals are provided. The control signals are also derived from the schedule.

This powerful and flexible approach will be developed in detail in subsequent chapters. It is suitable for a wide range of applications, ranging from systems with small work loads and stringent power consumption requirements to systems with high work loads. The former may be found in battery-powered applications (for example, mobile phones), while the latter are typical for many video applications because of their high sample rates. This approach yields very high performance at the cost of a somewhat larger design effort compared to the two approaches discussed earlier.

1.5 DSP SYSTEMS

Generally, a *system* provides an end-user with a complete service. For example, a CD player with amplifier and loudspeakers is a system with three components. The components in a system are often incorrectly referred to as systems or subsystems, although they do not prove a service to the end-user. Figure 1.4 shows an overview of a typical DSP system.

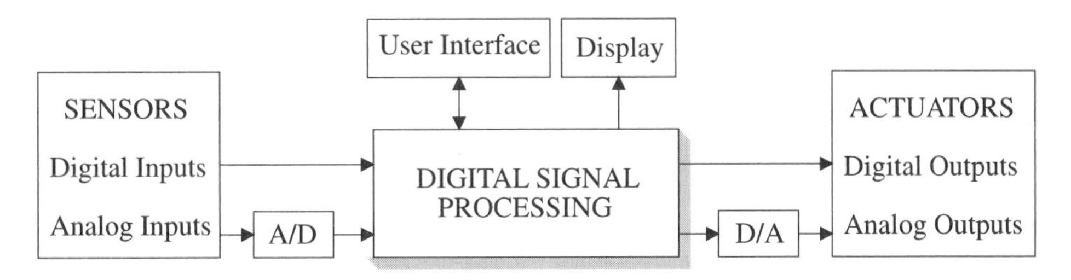

Figure 1.4 Typical DSP system

Generally, the system receives both analog and digital inputs from different sources. The system may also produce both analog and digital outputs. The outputs are often displayed, for example, as an image on a monitor or as sound through a loudspeaker. The outputs may also be used to control actuators that affect the system itself, for example, to change the azimuth angle of the antenna in a tracking radar. The system operator interacts with the system via a user interface to change system parameters such as search mode or frequency range. Keyboards are used as input devices in many applications.

Most systems are today multifunctional, i.e., they appear to simultaneously perform several functions. For example, a radar system may simultaneously perform searching, tracking, communication, and control tasks. Such systems are in practice realized with several subsystems that operate in parallel or sequentially. Often these subsystems are designed to perform only a single function and are referred to as *fixed-function subsystems*.

1.5.1 Facets

Several different representations, called *views* or *facets*, are needed to describe various aspects of the system to be designed (e.g., logic, test, physical, and layout). The aim of a particular view is to clearly represent a certain aspect of the system that is of interest in a particular design stage while other aspects may, or may not, be modeled. Hence, care should be taken so that the use of a specific view is not extended beyond its intended scope.

A *behavioral description* is an input–output description that defines the required action of a system in response to prescribed inputs. The description of the behavior may not include directions about the means of implementation or performance measures such as speed of operation, size, and power dissipation unless they directly affect the application.

A *functional description* defines the manner in which the system is operated to perform its function. Of main interest in the functional view are the signal processing aspects of the DSP system. Furthermore, input and output data rates and buffer sizes are important issues in the functional view.

Figure 1.5 shows a functional view of a typical DSP subsystem using a dataflow model. The complete functional description contains, of course, additional information such as requirements and functional or behavioral descriptions of the blocks. The subsystem in Figure 1.5 is an encoder for video telephony and conferencing. The input is a digital video signal in YCrCb format which in the first block is partitioned into macroblocks of 16 × 16 pixels, each consisting of an 8 × 8 luminance block and two 8 × 8 chrominance blocks. For each macroblock, the motion estimate unit searches the previous frame store for the 16 × 16 macroblock that most closely matches the current macroblock. This macroblock is then subtracted from the current macroblock to obtain a difference macroblock, which in the next block is transformed into the frequency domain using the *discrete cosine transform* (*DCT*). The frequency components are then quantized according to the number of bits that are available for coding. The run length unit replaces sequences with zero-valued frequency components with shorter representations and the quantized values are transformed back by the inverse DCT block. Finally, the entropy encoder converts the remaining frequency components and motion vectors into a variable-length code. The data buffer is needed to maintain a constant-output bit rate. The typical bit rate is 384 kbit/s or more, and the frame rate is in the range of 15 to 30 frames/s.

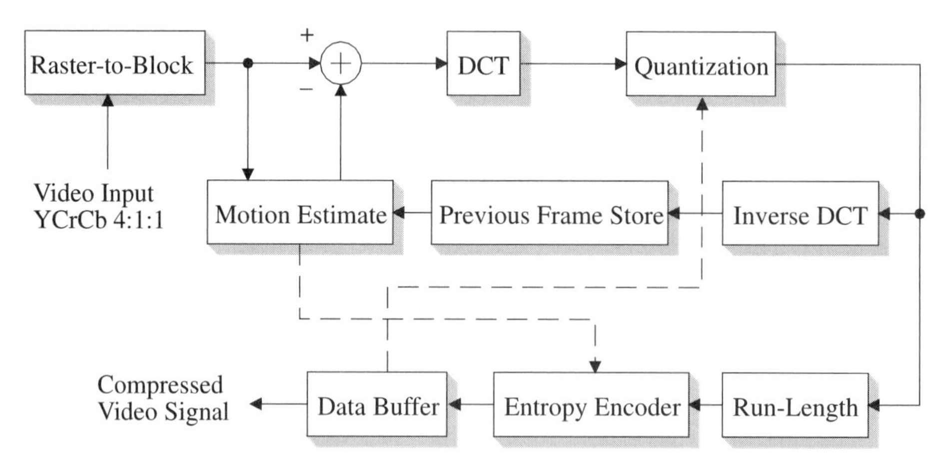

Figure 1.5 Functional view of CCITT H.261 video encoder

The JPEG and MPEG-1 and MPEG-2 standards use similar techniques for coding of video, but the bit rate for the latter is in the range of 3 to 10 Mbit/s. Key components in these systems, or subsystems, from a computation work load point of view, are the DCT and inverse DCT units. We will later discuss the design of these units in more detail.

A *physical view* of a DSP system is shown in Figure 1.6. The hardware organization is of primary concern in the physical view. Typically, the DSP processing is performed by a signal processor, while the user interface and other simple tasks are handled by the host processor. The host processor is usually implemented using a standard computer. Special I/O processors, as illustrated in Figure 1.6, are often required to handle the high input–output data rates. The available processing time and complexities of these three types of tasks vary considerably.

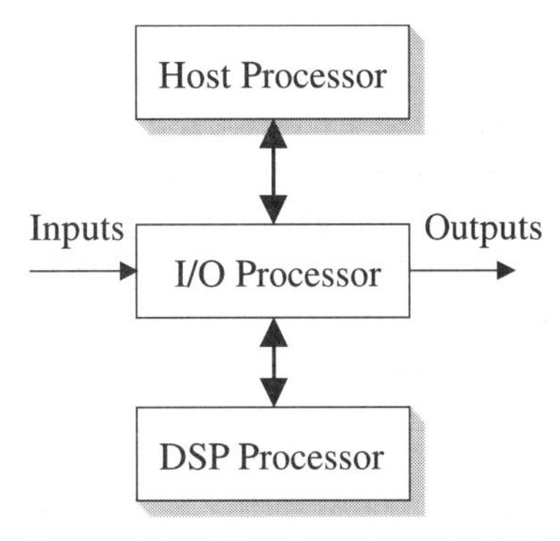

Figure 1.6 Physical view of a DSP system

A common view, the so-called *onionskin view*, used to describe a system is illustrated in Figure 1.7. At the center are the low-level hardware components; the outermost layer usually represents the user interface. Several intermediate layers (coats) may exist between the top and bottom layers. In Figure 1.7 only a few such levels are depicted. The idea is to reduce the design complexity of the system by using a hierarchy of architectures. The components are usually referred to as *virtual machines*. Each virtual machine provides the basic functions that are needed to realize the virtual machine in the next higher layer. The onionskin view represents a pure *hierarchy of virtual machines*.

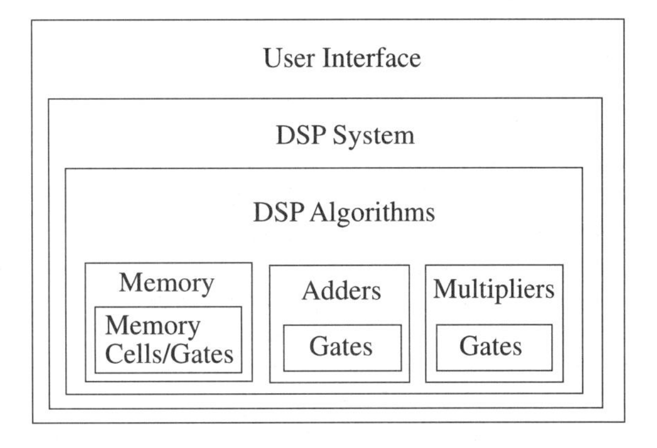

Figure 1.7 Onionskin view of a DSP system

Virtual machines can be implemented in either software or hardware. A pure hardware implementation may be required to obtain sufficiently high throughput for the basic DSP algorithms, while a software implementation is usually preferred for more flexible and irregular algorithms. In other cases, the virtual machines may be implemented as a combination of software and hardware. It is advantageous if the trade-off between software and hardware implementation of the virtual machines can be delayed until all layers in the system have been specified. This allows various design trade-offs to be directly evaluated and compared to the performance requirements.

Typical DSP systems have a hierarchical structure that works with different time frames. For example, the basic signal processing functions in a radar may work with a sample rate of about 10 MHz while the pulse repetition frequency is about 1 kHz. The target data base and user interface may work with an equivalent sample rate of only 10 Hz. Different implementation approaches may therefore be selected depending on the work load and the sample rate. For example, a direct mapping approach or ASIC signal processors may be appropriate for the basic signal processing, while standard signal processor may be used for the complex and irregular functions found in the data base, user interface, etc.

Yet another view is the *architectural description* that is used to describe how a number of objects (components) are interconnected. An architectural description is sometimes referred to as a structural description. In general, a structural description does not describe the functionality of the circuit, but it may include information about actual or estimated performance. Thus, two systems exhibiting the same behavior could be provided by different structures. Note that different structures exist at different levels of the design hierarchy and that behavioral and structural descriptions may appear in the same view.

EXAMPLE 1.1

A behavioral description of an XNOR gate is

$$F = \overline{A \oplus B}$$

Propose two structural descriptions, or architectures, using different types of components.

Figure 1.8 shows a structural description at the logic abstraction level of an XNOR gate that uses behavioral descriptions of the components: inverters, AND gates, and OR gates. Figure 1.9 shows yet another structural description of an XNOR gate with transistors as basic components. Hence, several different structures are possible.

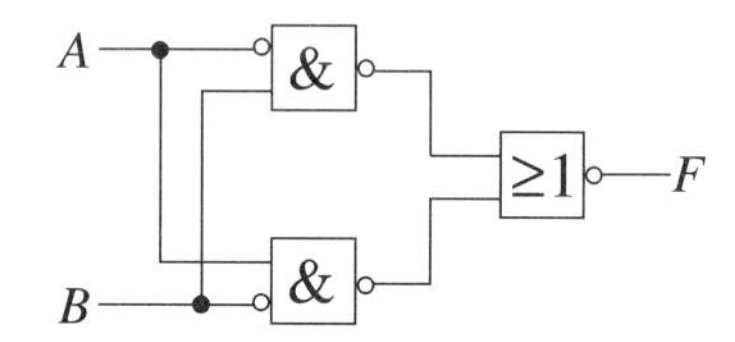

Figure 1.8 Structural description of an XNOR gate

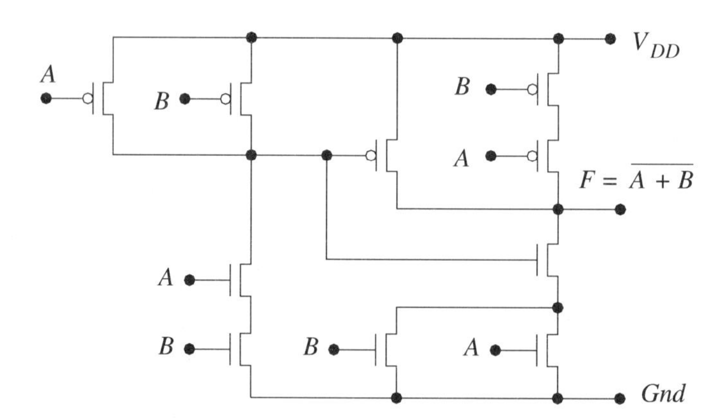

Figure 1.9 Structural description of a CMOS XNOR gate

1.6 DSP SYSTEM DESIGN

For the things we have to learn before we can do them, we learn by doing.

—Aristotle

In the system design phase, a behavioral view of the system is mapped onto an appropriate hardware structure by a sequence of mappings. The starting point for the system design is the *system specification*. Here we assume that the system specification has been preceded by a thorough investigation of the intended market, for example, volume, price, consumer preferences and technical and commercial competition. The specification should also include costs and other constraints due to marketing and maintenance of the DSP system. It is important to consider all costs incurred during the entire life of the system.

The design of a complex DSP system can be partitioned into two major phases: *system design* and *integrated circuit design*, as illustrated in Figure 1.10. These two phases are followed by a manufacturing and testing phase. The design of testable integrated circuits is a very important topic and should be considered early on in the system design phase. However, circuit manufacturing [8] and testing [16] issues will not be discussed in this book.

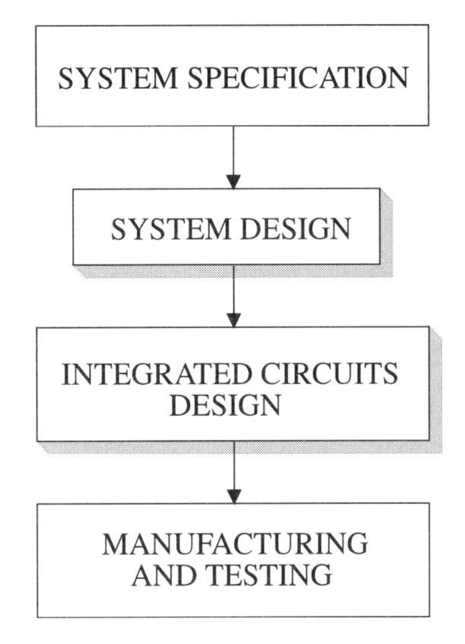

Figure 1.10 The main phases in the design of a DSP system

A *design methodology* is the overall strategy to organize and solve the design tasks at the different steps of the design process. It is not possible to invent a comprehensive design methodology that applies well in all cases, but all efficient methodologies have some common features.

Generally, the design process is viewed as the development of a sequence of models of the system, where each model version is more refined than the previous. The process continues until all design issues are resolved.

It is necessary due to the high complexity of the design problem to follow a *structured design* approach that reduces the complexity. Structured design methods, which will be further discussed in this chapter, are primarily used to

- Guarantee that the performance goals are met and
- Attain a short and predictable design time.

The overall performance goals are typically expressed in terms of subgoals such as acceptable physical size of the system, chip area of the individual integrated circuits, power consumption, and number of pins, to name a few. An important goal is to attain a short and predictable design time so that the product can be placed on the market within the intended time frame. This implies that the risk of ending up with a nonworking integrated circuit due to design errors, erroneous interface, unsatisfactory throughput, etc. must be minimized by using a good design method. Other factors that have a major impact on the design process are the design tools and the layout style used in the integrated circuit design phase. The degree of automation of the design process varies widely, from fully automatic to hand-crafted design [14].

1.6.1 Specification And Design Problem Capture

Besides a description of the tasks that the system shall perform, the system specification should also include requirements on physical size, power consumption, and maximum life cycle costs, etc. Furthermore, a time schedule for the successful completion of the design and target dates for production and market introduction are important issues.

A *specification* has two main parts, as illustrated in Figure 1.11.

- A behavioral description that specifies *what* is to be designed and
- A verification or validation part that describes *how* the design should be verified (validated).

Verification involves a formal process of proving the equivalence of two different types of representations under all specified conditions. Verification of a whole system is rarely done in practice because of the large complexity involved. However, small circuits and modules as well as simple communication protocols represent practical problem sizes.

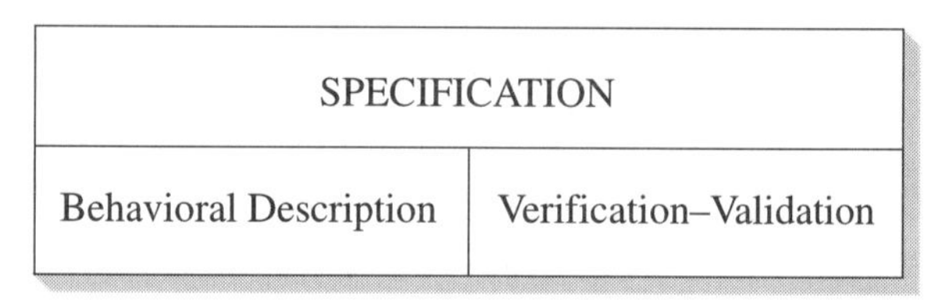

Figure 1.11 Specification

Validation is an informal and less rigorous correctness check. Validation is usually done by simulating the circuit with a finite set of input stimuli to assert that the circuit operate correctly.

A correct and complete specification of the system to be designed is crucial [4]. Recent experience shows that of all custom-designed VLSI circuits that do not work properly, up to 70% of the circuits work without logical or electrical errors, but they do not work in their intended environment. Their specifications, particularly the specifications of the working environments, are erroneous or incomplete. Specification and design problem capture are difficult and not well understood problems.

1.6.2 Partitioning Techniques

Generally, the system design phase consists of a sequence of partitioning steps wherein the system is partitioned into a set of subsystems that are so simple that they can be implemented without difficulty. Partitioning can be performed using different strategies, but most strategies involve a hierarchical partitioning of the system with the aim of reducing design complexity.

Data-Flow Approach

One approach is to partition the system along the data-flow in the system. If the data-flow graph is drawn with data flowing from left to right we can define *vertical* and *horizontal* partitioning as illustrated in Figure 1.12. The former partitions the system into parts that pass data in a sequential manner while the latter partitions the system into parts where data flow in parallel paths [5].

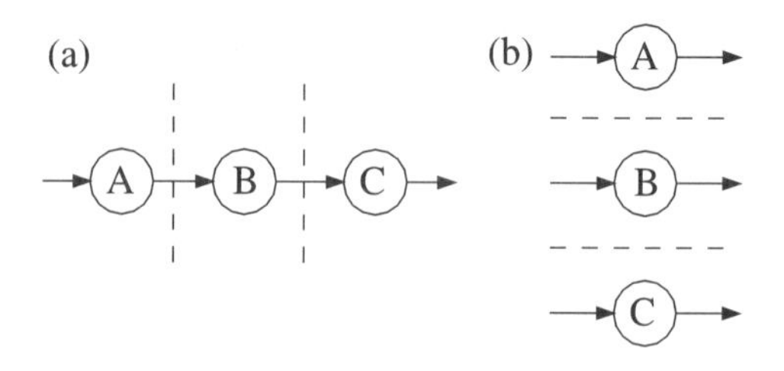

Figure 1.12 (a) Vertical and (b) horizontal partitioning.

The vertical partitioning leads to a sequential system. Such systems can be pipelined so that the subsystems (processors) execute concurrently and pass data sequentially. The horizontal partitioning leads to a set of subsystems working in parallel. The subsystems can be autonomous and need not be synchronized since they do not interchange data. In practice it may not be possible to partition a system in a purely vertical or a purely horizontal style. For example, systems with feedback loops can not be partitioned in this way.

Top-Down Approach

In the *top-down design approach*, the whole system is successively partitioned into a hierarchy of subsystems. On the top level a behavioral description is used. This description is partitioned into a structural description with behavioral descriptions of the components. This process of decomposition is then repeated for the components until sufficiently simple components are obtained. The end result is a functional description of the system. The subsystems are assumed to be implemented by the corresponding hierarchy of virtual machines. Of course, the design becomes easier if these hierarchies are made similar or identical. Figure 1.13 illustrates the top-down approach[1] using a structural decomposition. The design process (partitioning) will essentially continue downward with *stepwise refinement* of the subsystem descriptions [13]. It is advantageous if the partitioning is done so that the complexities at all hierarchical levels are about the same.

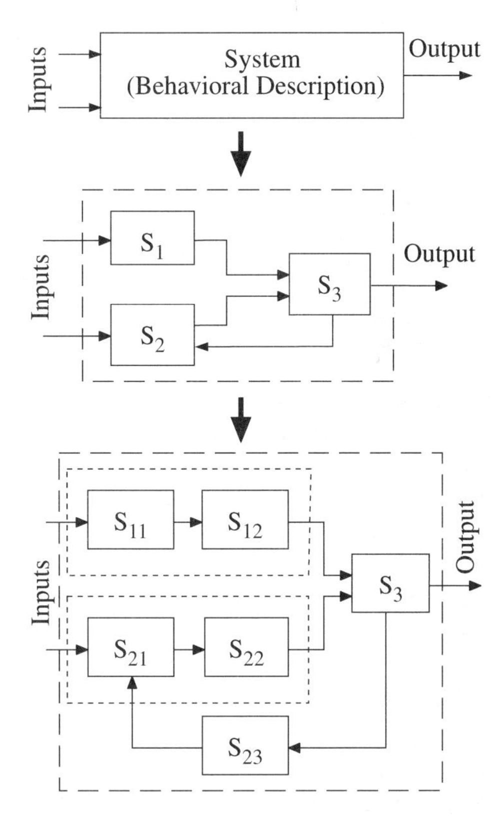

Figure 1.13 The top-down approach

In the top-down approach we stepwise develop the final system by realizing and validating each design level in software. By first building the DSP system in software, the performance can be more accurately estimated. Correctness of the design as well as of the specification can be verified or validated before making a commitment to a particular technology and investing in expensive hardware design. The subsystems are in each design iteration described by using more and more details so that they become closer and closer to their intended implementation. An advantage of this approach is that the system is developed from a global specification and that the successive design models can be checked for

[1] The organization of this book essentially follows a top-down style.

their correctness since they are described using an executable language. The top-down approach guarantees that larger and more important questions are answered before smaller ones.

As mentioned before, and illustrated in Figure 1.14, a typical system design begins with the development of a prototype (non–real-time) of the whole DSP system using either a conventional language, such as *C*, or, preferably, a hardware description language such as VHDL. The latter will be described in brief in section 1.6.6.

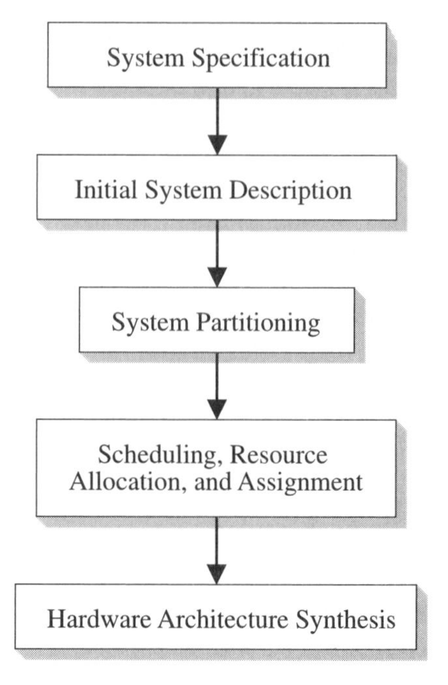

Figure 1.14 Top-down design strategy

After the validation of this initial (often sequential) description of the DSP system, it can be used as the basic system description. Subsequently, the system is hierarchically decomposed into a set of subsystems that at the lowest level implement well-known functions. This is one of the most important tasks in the system design phase—to partition the whole system into a set of realizable subsystems and to determine their individual specifications—because the partitioning will have a major effect on the system performance and cost. Typically, the new system description, which has explicit descriptions of the subsystems, is first derived without regard to time. However, it is advantageous to use at this stage, for example, VHDL, instead of a conventional sequential computer language since such languages do not have mechanisms for describing time and parallel execution of the subsystems.

Generally, a sequential execution of the subsystems cannot meet the real-time requirements of the application. In the next design step, called the scheduling phase, the sequential description is therefore transformed into a parallel description where the subsystems are executed concurrently. In this step, synchronization and timing signals must be introduced between the subsystems.

If a satisfactory solution cannot be found at a certain design level, the design process has to be restarted at a higher level to ensure a correct design. Indeed, the whole design process is in practice an iterative process. Often the whole system design can be split into several parallel design paths, one branch for each main block. The different parts of the system can therefore often be designed by independent design teams.

The next design step involves the mapping of the algorithms that realize the subsystems onto suitable software–hardware structures. This design step can be performed using the strategies discussed in sections 1.3 and 1.4.

In the direct mapping approach, discussed in section 1.4.2, the operations are scheduled to meet the throughput requirements and at the same time minimize the implementation cost. Scheduling techniques for this purpose will be discussed in detail in Chapter 7. Further, in this design step a sufficient amount of resources (i.e., processing elements, memories, and communication channels) must be allocated to implement the system. Another important problem in this design phase is to assign each operation to a resource on which it will be executed. The next design step involves synthesis of a suitable architecture with the appropriate number of processing elements, memories, and communication channels or selection of a standard (ASIC) signal processor. In the former case, the amount of resources as well as the control structure can be derived from the schedule. The implementation cost depends strongly on the chosen target architecture. This design step will be discussed in Chapters 8 and 9.

The last step in the system design phase involves logic design of the functional blocks in the circuit architecture [1, 14]. The result of the system design phase is a complete description of the system and subsystems down to the transistor level.

Figure 1.15 shows a typical sequence of design steps for a digital filter. The passband, stopband, and sample frequencies and the corresponding attenuations are given by the filter specification. In the first step, a transfer function meeting the specification is determined. In the next step, the filter is realized using a suitable algorithm. Included in the specification are requirements for sample rate, dynamic signal range, etc.

The arithmetic operations in the algorithm are then scheduled so that the sample rate constraint is satisfied. Generally, several operations have to be performed simultaneously. The scheduling step is followed by mapping the operations onto a suitable software–hardware architecture. We will later present methods to synthesize optimal circuit architectures.

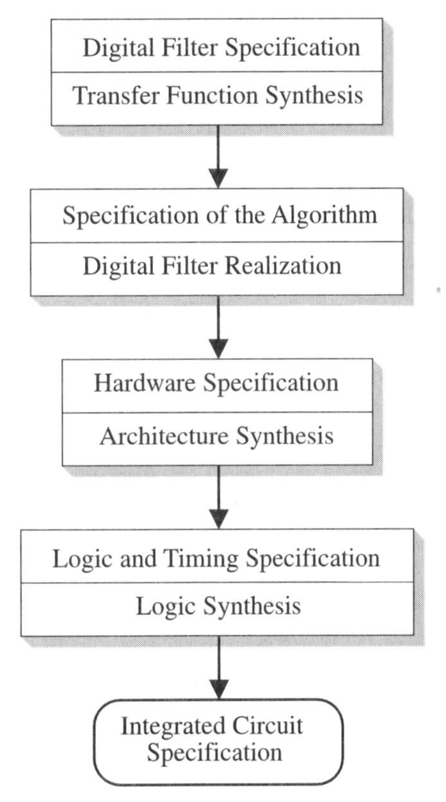

Figure 1.15 Idealistic view of the design phases for a digital filter

The final design step involves the logic design of the architectural components, i.e., processing elements, memories, communication channels, and control units. Communication issues play an important role, since it is expensive in terms of time and power consumption to move information from one point to another on the chip. The final result of the subsystem design phase is a circuit description in terms of basic building blocks: gates, full-adders, flip-flops, RAMs, etc. This description is then used as a specification in the circuit design phase.

The use of a top-down design methodology also forces the designer to carefully define the module interfaces, i.e., use abstractions. In return, the well-defined periphery of a module and its internal function suffice to describe the module at the next higher level in the design hierarchy. This allows internal details to be hidden, so that they do not obstruct the analysis at the next higher level.

The hierarchy of abstractions can, and should, also be used to reduce the volume of design data and to provide suitable representations to speed up the operation of the computer-aided design tools. Note that it may be necessary to store, retrieve, display, and process several hundred megabytes of data if a nonhierarchical approach is used.

The top-down approach relies on the designer's experience since the partitioning must lead to realizable subsystems. From the manager's point of view, it is easy to monitor the progress of the project and check it against the time schedule.

Bottom-Up Approach

The classical approach is the so-called *bottom-up approach* that starts by successively assembling well-known building blocks into more complex blocks until the whole system is realized. Emphasis is placed on realizability of the basic building blocks while communication issues and the overall system performance are less well handled. The probability of getting a nonworking system due to design errors is reduced by using tested building blocks, but the probability is high that the performance requirements are not met. The success of this approach depends to a large extent on the experience of the design team.

Edge-In Approach

Often, a variation of the top-down approach, the so-called *edge-in approach*, is adopted. In this approach the system is successively partitioned into parts, starting from the inputs and outputs, and working inwards. Figure 1.16 shown a typical result of this approach. The process continues until the whole system and its parts have been partitioned into well-known blocks [5]. The edge-in approach tends to put emphasis on the interfaces and communications between the blocks, and it inherently provides good control of the overall performance of the system.

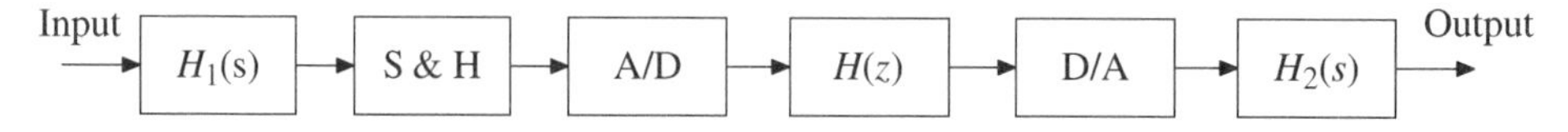

Figure 1.16 Example of edge-in decomposition of a DSP system

Critical Part Approach

Yet another approach is the so-called *critical part approach* that starts by successively designing the critical subsystems in descending order. Obviously, this approach puts emphasis on the feasibility of the design, that is, to meet the performance goals. Development costs may be reduced, as in the bottom-up approach, by using previously designed and proven building blocks.

Meet-In-The-Middle Approach

The aim of a structured design methodology is not only to cope with the high design complexity, but also to increase design efficiency and the probability of an error-free design. As mentioned earlier, the complexity is reduced by imposing a hierarchy of abstractions upon the design. In this way, the system is systematically decomposed into regular and modular blocks. In practice, however, a *meet-in-the-middle* approach is often used. In this approach, which is illustrated in Figure 1.17, the specification-synthesis process is carried out in essentially a top-down fashion, but the actual design of the building blocks is performed in a bottom-up fashion. The design process is therefore divided into two almost independent parts that meet in the middle. The circuit design phase can be shortened by using efficient circuit design tools or even automatic logic synthesis tools. Often, some of the building blocks are already available in a circuit library.

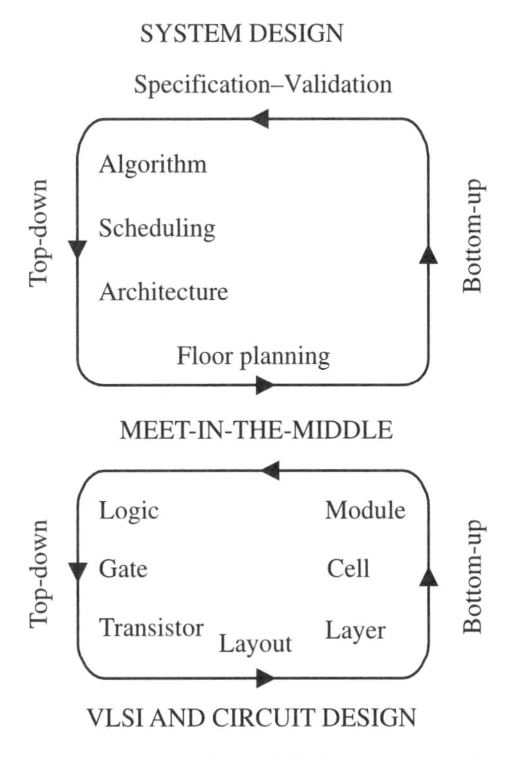

Figure 1.17 Meet-in-the-middle design approach

1.6.3 Design Transformations

Note that Figure 1.15 not only shows how the subsystem design process is partitioned into several phases or *levels of abstraction*, it also shows that each level has a *specification* and a *synthesis phase*. In fact, the whole design process consists of alternating specification and synthesis operations. The result of the synthesis at a particular level, i.e., an implementation at that level, acts as a specification for the next lower level.

At each level in the design process, the representation of the system is refined and transformed into a more detailed, lower-level representation, as illustrated in Figure 1.18. The transformation of a representation from one design level to a lower level is called *synthesis*. Generally, the downward transition between two levels is a one-to-many mapping, since the synthesis process adds information to the design.

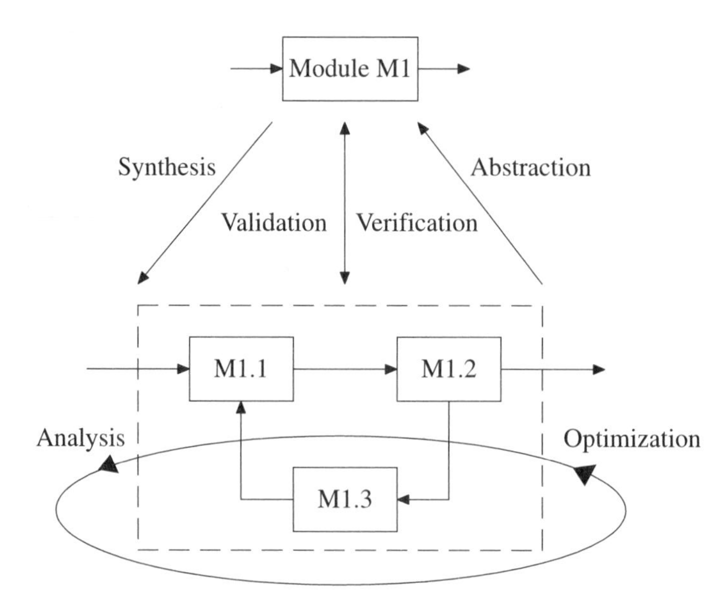

Figure 1.18 Design operations involving two levels of abstraction

Current design tools support automatic synthesis at the lower levels of the design process (for example, where gates and simple logic functions are synthesized) since these mappings only involve one-to-few mappings. The situation is different at the higher design levels where the underlying principles are less understood. Most system design approaches and other high-level synthesis procedures are therefore often based on ad hoc solutions.

The inverse operation to synthesis is called *abstraction*. An abstraction hides details at a lower level. The idea of abstraction is crucial to understanding and handling complex problems.

Another class of transformations that is used to transform representations into equivalent representations at the same abstraction level is called *optimization* transformations. A typical example of optimization is compaction of the layout of wires and transistors on the chip surface to reduce the required chip area. Optimization transformations may be either combinatorial (for example, cell placement) or parametric (for example, resizing transistors in order to improve performance). *Analysis* operations are used to support decision making necessary in the synthesis and optimization transformations. *Verification* and *validation* operations are needed to assure that a representation meets its specification. These operations typically are performed between two levels of abstraction.

1.6.4 Complexity Issues

The design of a complex signal processing system involves several stages of specification, synthesis, optimization, analysis, and verification. The essential aspect is management of the complexity of the design process. In fact, VLSI design is sometimes defined as a design problem where design complexity dominates all other issues. Reducing the design complexity is also necessary in order to reduce the amount of design data, which otherwise would become unreasonably large. Large amounts of data could be stored on a hard drive, but the processing time would become excessive.

There are no direct methods or theories to cope with complexity in a problem as such. The only remedy is to avoid the complexity by introducing some kind of order. Fortunately, this can be done in many ways.

The *complexity* of a system can be measured in terms of the number of interactions between its parts. More formally we have

$$<O, F, R> \tag{1.1}$$

where O is a set of objects with their functional description, F, and their interrelations, R.

The potential complexity grows very fast when the number of parts is increased. Therefore, complexity can be restrained or even reduced if the system is designed so that it is partitioned into groups of low- or noninteracting parts. Complexity is reduced if a set of parts that have a high degree of mutual interaction (coupling) is collected into a module that has few external interactions. The reduction in complexity achieved by grouping several parts into a larger object (module) that can be described by a simpler representation, describing only external communication and without explicit references to any internal interactions, is called *abstraction*.

The idea of hiding internal features of an abstracted module is fundamental to building large systems. For example, in integrated circuits the internal details are removed by the low-impedance drivers and the high-impedance input circuits. The circuits can therefore be interconnected without electrical interaction if some simple fan-out rules are followed. Notice the similarity with current trends in computer software where data and procedures are encapsulated into objects (C++).

Whenever a *hierarchy* of information exists, the information can be subdivided so that the observer can examine the constituent parts and their interrelation at a level with less detail with the aim of controlling the information being handled. Subdivision often implies some tree structures of relationships, where at the lowest levels of the hierarchy the greatest detail is evident. This hierarchy is illustrated in Figure 1.19 with a module that has two different classes of cells: *leaf cells* and *composition cells* [8, 14]. The leaf cells contain low-level objects (e.g., transistors or gates), while the composition cells represent higher-level objects (e.g., full-adders, multipliers, RAM, and ROM).

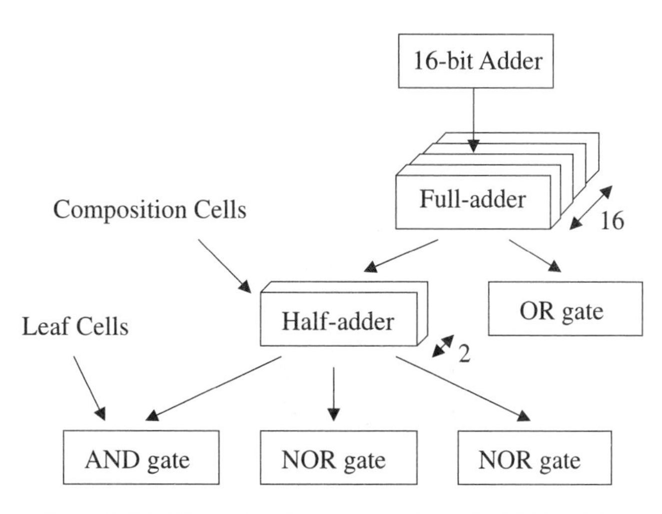

Figure 1.19 Hierarchical representation of a 16-bit adder

Hierarchical abstraction is the iterative replacement of groups of modules. Note that using hierarchy alone does not reduce the complexity. A design that can be completely described by using modules (abstractions) is *modular* and will have low complexity. If the design has only a few types of modules, the complexity is even lower. Such designs have a high degree of *regularity*. A regularity factor can be defined as the ratio of the total number of modules to the number of different modules. *Standardization* that restricts the design domain can be applied at all levels of the design to simplify modules and increase the regularity, thereby reducing the complexity. It is widely believed that the adoption of highly *structured design methodologies* making extensive use of the ideas just discussed, is a necessity for a successful design of complex systems.

1.6.5 The Divide-And-Conquer Approach

A well-known approach to derive low-complexity algorithms is the divide-and-conquer approach which is based on the fact that many large problems can be decomposed into several smaller problems that are easier to solve than the larger one. The decomposition is done recursively until the remaining problems are so small that they can be solved directly. Finally, the solutions are combined to obtain a solution for the original problem [7]. The approach is described by the pseudo-code shown in Box 1.1.

```
function Solve(P);
var
...
begin
  if size(P) ≤ MinSize
    then Solve := Direct_Solution(P)
  else
    begin
      Decompose(P, P1, P2, ..., Pb);
      for i := 1 to b do
        Si := Solve(Pi);
      Solve := Combine(S1, S2, ..., Sb)
    end;
end;
```

Box 1.1. The divide-and-conquer algorithm

The amount of time required at each step is

$$T(n) = \begin{cases} a & \text{for } n \le \text{MinSize} \\ bT\left(\frac{n}{c}\right) + d \cdot n & \text{for } n > \text{MinSize} \end{cases} \tag{1.2}$$

where n is the size of the problem, a is the time required to solve the minimum-size problem, b is the number of subproblems in each stage, n/c is the size of the subproblems, and $d \cdot n$ is the linear amount of time required for decomposition and combination of the problems. MinSize is the size of a problem that is so small that it can be solved immediately.

When comparing different alternative design algorithms it is of interest to determine how rapidly the execution time grows as the problem size increases. We will compare different algorithms by using the notation $O(f(n))$. A function $g(n)$ is a member of $O(f(n))$ if

$$g(n) \in O(f(n)) \text{ if } \lim_{n \to \infty} \frac{g(n)}{f(n)} = \text{const} < \infty \tag{1.3}$$

If the limit exists and const $< \infty$, then function $g(n)$ grows no faster than function $f(n)$. Usually, both $g(n)$ and $f(n) \to \infty$ when $n \to \infty$. Hence, the limit becomes undetermined, but can often be determined using L'Hôpital's rule:

$$\lim_{n \to \infty} \frac{g(n)}{f(n)} = \lim_{n \to \infty} \frac{g'(n)}{f'(n)} \tag{1.4}$$

It can be shown (see Problem 3.5) that divide-and-conquer algorithms have the time-complexity:

$$T(n) \in = \begin{cases} O(n) & \text{for } b < c \\ O(n \log_c(n)) & \text{for } b = c \\ O(n \log_c(b)) & \text{for } b > c \end{cases} \tag{1.5}$$

Thus, recursively dividing a problem, using a linear amount of time, into two problems ($b = 2$) of size $n/2$ ($c = 2$) results in an algorithm with time-complexity of $O(n \log_2(n))$. The *fast Fourier transform* (*FFT*), which is discussed in Chapter 3, is an example of this type of algorithm.

If the number of subproblems were $b = 3$, 4, or 8, then the required execution time would be $O(n^{\log_2(3)})$, $O(n^2)$, or $O(n^3)$, respectively. The execution time grows very rapidly for the last three types of algorithms.

1.6.6 VHDL

Many errors resulting from a faulty specification can be avoided if design ideas can be tested and verified in an early phase of the design process. The use of a *hardware description language* (*HDL*) is one approach used to avoid expensive redesigns.

One such language, *VHDL*[2] has become a de facto standard [2, 3, 6]. The language structure (but not the semantics) is inspired by Ada. VHDL provides most of the features of a high-level programming language plus facilities for the treatment of time.

The main use of VHDL is for documentation and simulation, but a VHDL description of the design problem may also serve as a specification or as the input to automatic logic synthesis tools. It is also possible to use it to validate, by

---

2. The VHDL hardware description language is a result of the VHSIC (Very High Speed Integrated Circuit) program started in 1983 by the U.S. Department of Defense. MIL-STD-454L. VHDL has been standardized by IEEE Std. 1076-1987.

simulation, that the design problem is correctly captured. VHDL descriptions are in practice also used as a communication medium between different design teams and the manufacturer. In fact, some large customers—for example, the U.S. Department of Defense—require a VHDL description of new circuits. The idea is that the VHDL description can later be used in a redesign of the circuit using a more advanced technology.

The VHDL language supports three main styles: behavioral, structural, and data-flow descriptions. In all three styles, the basic unit is the design entity.

Design Entity A module is viewed as a *design entity,* which can be as simple as a NAND gate or as complicated as a digital filter. The description of a design entity in all three styles is divided into the *interface description* and one or more *architectural bodies.* The use of design libraries is encouraged. The interface description (*port map*) specifies the entity's connections with the external environment, whereas the architectural body describes its function which can be described in the three styles just mentioned.

Behavioral Description A pure behavioral description in the architectural body is used for simulating functionality. However, it does not provide any direct correspondence between the behavior and the real hardware.

Structural Description A structure is described by *component* declarations and signal connections in terms of port maps. Components can be described as being composed of lower-level components. Structural descriptions as well as data-flow descriptions can be used for the synthesis of actual hardware.

Data-Flow Description The data-flow description is typically used to describe the system as the flow of data between different units—for example, memories and processing elements. Timing properties are taken into account by describing signal waveforms. Functions to be performed are isolated in block declarations. The activation of blocks is controlled by guarded statements. All signal assignments transpire concurrently. The data-flow description is suitable for description and simulation of signal-flow graphs.

We illustrate some of the basic concepts used in VHDL by the code for a full-adder and a test bench that can be used to validate the code.

EXAMPLE 1.2

Box 1.2 shows the VHDL code that describes a full-adder. The full-adder is realized by using two half-adders and an OR gate.

First we declare the two entities *Half_Adder* and *OR_gate* and their architectural bodies in a behavioral style.

```
-- ENTITY DECLARATIONS
entity Half_Adder is
    port(X, Y: in Bit; Sum, Carry: out Bit);
end Half_Adder;
```

```
entity OR_gate is
   port(In1, In2: in Bit; Out1: out Bit);
end OR_gate;

-- ARCHITECTURAL BODIES
architecture Behavior_desc of Half_Adder is
begin
   process
      begin
         Sum <= X or Y after 5 ns;
         Carry <= X and Y after 5 ns;
         wait on X, Y;
      end process;
end Behavior_desc;
architecture Behavior_desc of OR_gate is
begin
   process
      begin
         Out1 <= In1 or In2 after 5 ns;
         wait on In1, In2;
end process;
end Behavior_desc;

-- ENTITY DECLARATION
entity Full_Adder is
   port(A, B, Carry_in: in Bit; Sum, Carry_out: out Bit);
end Full_Adder;

-- ARCHITECTURAL BODY
architecture Structure of Half_Adder is
-- Signal declarations
signal Temp_sum, Temp_carry1, Temp_carry2: Bit;

-- Local declarations
component HA
   port(X, Y: in Bit; Sum, Carry: out Bit);
end component HA;
component OG
port(In1, In2: in Bit; Out1: out Bit);
end component OG;
for U0: HA use entity Half_Adder(Behavior_desc);
for U1: HA use entity Half_Adder(Behavior_desc);
for U2: OG use entity OR_gate(Behavior_desc);
begin-- Connect the ports of the components
U0: HA
   port(X => A, Y => B, Sum => Temp_sum, Carry => Temp_carry1);
U1: HA
   port(X => Temp_sum, Y => Carry_in, Sum => Sum, Carry => Temp_carry2);
U2: OG
   port(In1 => Temp_carry1, In2 => Temp_carry2, Out1 => Carry_out);
end Structure;
```

Box 1.2. VHDL description of a half-adder and a full-adder

Note that there is a special assignment operator used to propagate signals (<=). The statement **wait on** *X*, *Y*; suspends the logic process until at least one of the signals, *X* or *Y*, is changed. Next we declare the design entity *Full_Adder* and its port map. A structural style is used in the architectural body for the full-adder. The description is in terms of the previously defined components.

EXAMPLE 1.3

Box 1.3 shows the VHDL code for a test bench for testing the full-adder described in Example 1.2.

```
-- ENTITY DECLARATION
entity Test_gen is
   port(A, B, Carry_in: in Bit; Sum, Carry_out: out Bit);
end Test_gen;

-- ARCHITECTURAL BODY
architecture Behavior_desc of Test_gen is
begin
A <= '1', '0' after 20 ns, '0' after 40 ns, '0' after 60 ns, '0' after 80 ns, '1' after
      100 ns,
      '1', after 120 ns, '1' after 140 ns, '1' after 160 ns, '0' after 180 ns ;
B <= '1', '0' after 20 ns, '0' after 40 ns, '1' after 60 ns, '1' after 80 ns, '0' after
      100 ns,
      '0', after 120 ns, '1' after 140 ns, '1' after 160 ns, '0' after 180 ns ;
Carry_in <= '1', '0' after 20 ns, '1' after 40 ns, '0' after 60 ns, '1' after 80 ns,
      '0' after 100 ns, '1' after 120 ns, '0' after 140 ns, '1' after 160 ns,
      '0' after 180 ns ;
end Behavior_desc;

-- Dummy entity for the test bench
entity Test_bench is
end Test_bench;

-- ARCHITECTURAL BODY
architecture Behavior_desc of Test_bench is
signal x, y, z, u, v: Bit;
component Generator
   port(A, B, Carry_in: in Bit; Sum, Carry_out: out Bit);
end component;
component Adder
   port(A, B, Carry_in: in Bit; Sum, Carry_out: out Bit);
end component;

for S0: Generator use entity Test_gen(Behavior_desc);
for S1: Adder use entity Full_Adder(Behavior_desc);
```

```
begin   -- Connect the ports of the components
S0: Generator
   port(x, y, z, u, v);
S1: Adder
   port(x, y, z, u, v);
end Behavior_desc;
```

Box 1.3. VHDL description of a test bench for the full-adder

1.7 INTEGRATED CIRCUIT DESIGN

Integrated circuit technology has made it possible to produce chips with several millions of transistors. However, complex circuits are difficult to design and call for special design methodologies. For example, debugging a flawed VLSI chip is both difficult and time consuming, since the *turnaround time* for design changes ranges from several weeks to many months. Long design times may lead to lost opportunities of marketing the chip ahead of the competition and recouping the investment. This forces batching of design changes and the use of design methods that enforce perfect designs. Thus, the *correctness of the design* is of paramount importance for a successful project.

Generally, the degree of flexibility in the circuit design is very high. The designer specifies the logic and circuit realization, physical placement, and even the details of the individual gates and transistors. Despite this flexibility there are limits in the ability, at one level of the design, to compensate for shortcomings at higher levels. These limits come from inherent constraints in the technology (size, power, and throughput), including the need to limit the addition of new complexity at lower levels of the design. Performance is therefore an issue that must be addressed at all levels of the design. However, experience indicates that design decisions and optimizations at the higher levels in the design process are more important than low-level optimizations.

The VLSI circuit design phase can begin when the system design phase has been completed. In this phase the transistors and their interconnections are laid out on the chip surface. The integrated circuit design phase can be more or less sophisticated, from fully automatic chip generation, down to hand-crafted design of every transistor. We stress that the system solution should be frozen and not allowed to be changed during the layout phase. This is because changes in the design may introduce errors. We will later discuss various design approaches in more detail.

VLSI circuit and system design processes are in many ways similar to traditional software design, but with the more stringent requirement of a "first-time-right" design. Typically, the software design starts with an initial description that is gradually refined until the desired goal is reached. CASE tools that are commonly used for the design of large and complex software may also be used for the system design phase. The basic incremental design philosophy is quite general, but there is one important difference—as the VLSI and system design move downward in the hierarchy, qualitative changes occur in the models. For example, an electrical-level model may not be obtained just by an incremental expansion of a behavioral model of the corresponding logic module. Conversely, in software all of the models belong to the same domain. The transformations needed to derive the VLSI implementation are therefore not straightforward.

Experience gained from both software and hardware projects indicates that it is important that all members of a design team share a common view of, not only the design goals, but also of the design process [12]. Without such a common view severe communication and coordination difficulties may arise in the design process.

1.7.1 System Design Methodology

Integrated circuit design, also called VLSI design, is the process of transforming the structural description of the logic circuitry into a detailed physical layout. This process is complex and error prone. It is therefore mandatory to adhere to a highly structured design methodology for the integrated circuit design phase.

1.7.2 Technical Feasibility

One of the first steps in the design of a DSP system is a feasibility study of the whole project. The question is, can the project, given economic and technical constraints, be finished within the available time frame? If the project is feasible, the next step is to determine a more detailed plan for the whole project. This section deals primarily with the technical feasibility of the VLSI circuit. The economic and time aspects depend largely on the previous experience of the design team, the available CAD tools, and the circuit library. The technical feasibility study of the system generally involves many different issues:

SYSTEM-RELATED

- Partitioning into cabinets, boards, and circuits
- Mixed digital and analog circuits on the same chip
- Clock frequencies
- Power dissipation and cooling
- Circuit area and packaging
- I/O interface

CIRCUIT-RELATED

External

- Interchip propagation delay
- Data transfer frequencies
- Input protection
- Loads that have to be driven, including expected PCB runs
- Available input drivers
- Drive capacity for output buffers
- Restrictions on pin-outs

Internal

- Clock frequencies
- Data transfer frequencies and distances
- Critical timing paths
- Noncritical timing paths
- Power dissipation and cooling

Drive capacity for internal buffers
Circuit area, yield, and packaging
Temperature and voltage effects
Maximum and minimum temperatures, voltages, etc.
Process technology

DESIGN-EFFORT–RELATED
CAD tools
Layout style
Regularity and modularity of the circuits
Module generators
Cell library

The aim is to determine if the system can be built using the given approach. If not, another approach has to be tried or the system requirement has to be relaxed. If a good feasible solution has been found, it can serve as a starting point for the subsequent design process. The design process involves technical trade-offs between system performance, costs, and design effort.

1.7.3 System Partitioning

It is usually advantageous to implement the whole system on a single chip, since external communication is comparatively slow and expensive in terms of power consumption. Since DSP algorithms are characterized by high-speed communication between different parts of the algorithm, it is often difficult to partition the system into several chips that communicate at low speed.

The achievable throughput in a multichip DSP system will ultimately be limited by the I/O bandwidth as the feature size of integrated circuit technologies are scaled down. Scaling down devices does not significantly increase bandwidth of off-chip communication; it mainly increases the internal speed of the circuit. A future solution may be to use new I/O techniques—optoelectronics, for example. Hence, the high computational capacity in highly parallel architectures may be difficult to exploit because of the mundane problem of getting data on and off the chip. Therefore, system solutions that have high computational complexity per input sample will undoubtedly benefit from advances expected in integrated circuit technologies.

A practical approach to system partitioning is to begin by estimating system size in terms of transistor devices needed for memories and logic. The number of chips required can then be estimated by observing the following constraints:

- Assume that the largest economical chip area is, for example, 100 mm^2. Allocate the parts of the system to different chips, taking device density into account.
- The system must be partitioned such that the number of I/O-pins fits the packages, and the power dissipation in each package is not too large. Note that there are rarely any pin-out problems in systems with bit-serial communication.
- The partitioning must lead to reasonable chip-to-chip communication times.

After a good partitioning has been obtained, it can be refined using computer-based methods. Many such partitioning methods have been proposed. The starting point is a system description consisting of the basic building blocks, or modules, with their sizes and interconnections. The aim is to find a partition that satisfies the interconnection constraints and minimizes the number of chips and their sizes. There are several ways to formulate a suitable cost function [5].

Generally, partitioning algorithms can be divided into constructive and iterative improvement methods. Constructive methods are often fast, but the quality of the results is generally not good since the algorithms are "greedy." Solutions quite far from the global minimum are often obtained. Iterative methods start with an initial solution, possibly obtained by a constructive algorithm, and improve the quality of the solution by modifying it incrementally. Fortunately, in practice most systems have only a few natural partitions, which can be found relatively easily by the human designer.

The simplest form of iterative improvement is to select at random a pair of modules belonging to different partitions, and then interchange them. In most algorithms of this type, the interchange of modules is accepted only if the cost function is decreased. If the cost increases, the change is rejected. These algorithms usually lead to fairly good results, but the algorithms discussed next generally lead to even better ones.

The *Kernighan–Lin algorithm*, also called the group migration method, is also based on an interchange of two modules belonging to different partitions. The algorithm achieves its goal by using a scoring function that reflects the cost before and after the two modules have been interchanged.

The problem with greedy algorithms, such as the random interchange method, is that they are often trapped in a local minimum. In the simulated annealing approach (see Chapter 7), changes that increase the cost are accepted with a certain probability. It can be shown that this algorithm will reach the global minimum under certain conditions.

References

[1] Anceau F.: *The Architecture of Microprocessors,* Addison-Wesley, Wokingham England, 1986.

[2] Armstrong J.R.: *Chip-Level Modeling with VHDL,* Prentice Hall, NJ, 1989.

[3] Armstrong J.R. and Gray F.G.: *Structured Logic Design with VHDL,* Prentice Hall, NJ, 1993.

[4] Birtwistle G. and Subrahmanyan: *VLSI Specification, Verification and Synthesis,* Kluwer Academic Pub., Boston, 1988.

[5] Bowen B.A. and Brown W.R.: *System Design, Volume II* of *VLSI Systems Design for Digital Signal Processing,* Prentice Hall, NJ, 1985.

[6] Coelho D.: *The VHDL Handbook,* Kluwer Academic Pub., Boston, 1989.

[7] Cormen T.H., Leiserson C.E., and Rivest R.L.: *Introduction to Algorithms,* MIT Press, Cambridge, MA, McGraw-Hill, New York, 1993.

[8] Dillinger T.E.: VLSI Engineering, Prentice Hall, NJ, 1988.

[9] Geiger R.L., Allen P.E., and Strader II N.R.: *VLSI Design Techniques for Analog and Digital Circuits,* McGraw-Hill, New York, 1989.

[10] Gregorian R. and Temes G.C.: *Analog MOS Integrated Circuits for Signal Processing,* J. Wiley & Sons, New York, 1986.

[11] Haskard M.R. and May I.C.: *Analog VLSI Design—nMOS and CMOS,* Prentice Hall, NJ, 1988.

[12] Lawson H.W.: Philosophies for Engineering Computer-Based Systems, *IEEE Computer*, Vol. 23, No. 12, pp. 52–63, Dec. 1990.
[13] Leung S.S., Fisher P.D., and Shanblatt M.A.: A Conceptual Framework for ASIC Design, *Proc. IEEE*, Vol. 76, No. 7, pp. 741–755, July 1988.
[14] Preas B. and Lorenzetti M. (eds.): *Physical Design Automation of VLSI Systems*, Benjamin/Cummings, Menlo Park, CA, 1988.
[15] Wanhammar L., Afghahi M., and Sikström B.: On Mapping of Algorithms onto Hardware, *IEEE Intern. Symp. on Circuits and Systems*, Espoo, Finland, pp. 1967–1970, June 1988.
[16] Wilkins B.R.: *Testing Digital Circuits: An Introduction*, Van Nostrand Reinhold, Wokingham, England, 1986.

Problems

1.1 (a) Describe briefly a systematic partitioning technique for the design of a complex DSP system. Also give a motivation for the chosen partitioning technique.
(b) Describe the different types of transformations between two adjacent levels of abstraction in the design process. Also describe different types of transformations within a design level.
(c) Define the following concepts and provide an example for each concept:

Abstraction	Hierarchical abstraction
Modularity	Regularity
Architecture	Standard

1.2 (a) What are the basic components of a specification? Provide an example.
(b) Give an example and a counterexample of a behavioral description.
(c) Give an example and a counterexample of an abstraction.

1.3 Describe the main features of a structured design methodology.

1.4 Discuss advantages and disadvantages of an ASIC implementation compared to a software implementation of a DSP system.

1.5 Discuss possibilities to reduce various types of complexity in a system based on the standard and ASIC processor approaches.

1.6 Show that $\ln(n) \in O(n^{\alpha})$ where $\alpha > 0$. That is, $\ln(n)$ grows more slowly than any power of n.

1.7 Show that $n^k \in O(a^n)$ where $k > 0$. That is, powers of n grow more slowly than any exponential function, a^n, where $a > 0$.

1.8 Compare the execution time for algorithms with the complexity of: $O(n)$, $O(n \log(n))$, $O(n^2)$, and $O(n^3)$ for different problem sizes—for example, $n = 10, 100, 1000, 10000$, and 100000.

1.9 (a) Discuss how different design views are supported by VHDL.
(b) Discuss how VHDL supports different methods to reduce the design complexity.

1.10 Derive behavioral, data-flow, and structural descriptions of a full-adder.

2

VLSI CIRCUIT TECHNOLOGIES

2.1 INTRODUCTION

There are two classes of silicon devices: bipolar and unipolar transistors. They differ in that both majority and minority carriers participate in the bipolar transistor action while only minority carriers participate in the unipolar transistor. The main bipolar circuit techniques are ECL/CML (emitter-coupled logic/current mode logic), I_2L (integrated injection logic), and TTL (transistor-transistor logic). These circuit techniques are for various reasons not suitable for large integrated circuits.

There are several types of unipolar devices, e.g., MOS (metal oxide semiconductor) transistor, JFET (junction field-effect transistor), and MESFET (metal-semiconductor field-effect transistor). In practice only MOS transistors; are used for VLSI circuits. The main unipolar circuit technique is CMOS (complementary metal oxide semiconductor) circuits, but numerous alternatives exist [2, 23–26]. CMOS circuits are often augmented with a few bipolar devices, so-called BiCMOS, to achieve higher speed in critical parts.

Gallium arsenide (*GaAs*)–based VLSI circuits have recently become available. They are important because of their high speed and compatibility with optical components, for example, lasers. GaAs-based circuits are therefore interesting candidates for many DSP applications. However, there is a widespread consensus that no other technology will effectively compete with CMOS and BiCMOS for several years to come.

2.2 MOS TRANSISTORS

There are two basic types of MOS devices: *n-channel* and *p-channel* transistors. Figure 2.1 illustrates the structure of an MOS transistor [10, 12, 14, 15, 21]. An n-channel (*nMOS*) transistor has two islands of n-type diffusion embedded in a substrate of p-type. A thin layer of silicon dioxide (SiO_2) is formed on top of the surface between these islands. The *gate* is formed by depositing a conducting material on the top of this layer. The gate was made of metal in now outdated technologies, a fact that explains the name of the device (metal oxide semiconductor transistor).

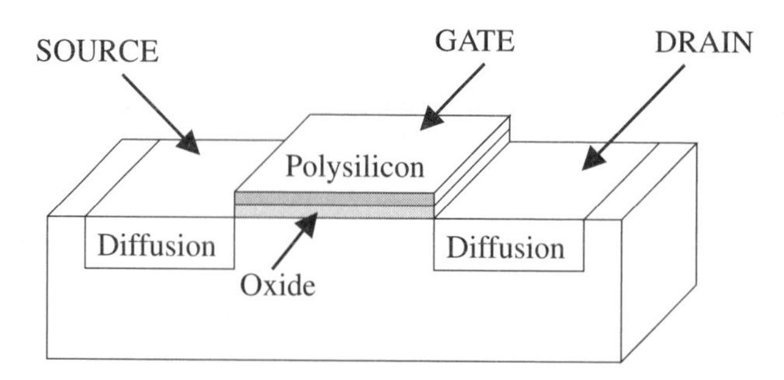

Figure 2.1 The cross section of an MOS transistor

Modern MOS processes use polysilicon for the gate. In the *nMOS* transistor the *channel* is formed under the gate when proper voltages are applied to the terminals. The free charges in an nMOS device are electrons. The p-channel (*pMOS*) transistor is similar to the nMOS type except that the substrate is made of n-type silicon and the diffused islands are made of p-type silicon. The free charges in a pMOS device are holes.

A practical circuit has both n- and p-devices that are isolated from each other by so-called *wells*. Some CMOS processes have both an n-well and a p-well for the p- and n-devices, respectively. Other processes use only a p- or n-type of well in which n- (p-) channel devices are created while p- (n-) channel devices are created directly in the substrate.

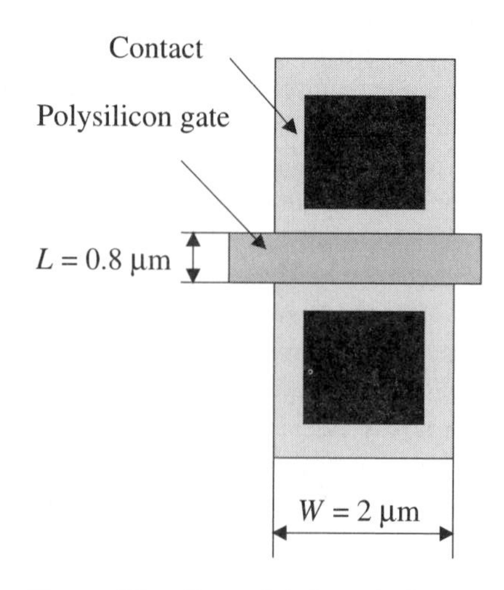

Figure 2.2 Geometric layout of an MOS transistor

The dimensions are successively being reduced with the progress of the process technology [14, 15, 19, 25]. It is common practice to characterize a CMOS process by the minimum feature size. As of 1998, the minimum feature size is in the range 0.25 to 0.35 μm. Figure 2.2 shows the geometric layout of a minimum-size transistor with a channel length of 0.8 μm. Notice that there is a difference between the drawn dimension (masks) and the effective dimension that depends on the process [7, 16].

The drawn width of the transistor is 2.0 μm. For Minimum-size transistors in a typical 0.8 μm CMOS technology we have

$$W_{neff} = W_{drawn} - 2 \cdot 0.58\ \mu m = 0.84\ \mu m$$
$$L_{neff} = L_{drawn} = 0.8\ \mu m$$

and

$$W_{peff} = W_{drawn} - 2 \cdot 0.33\ \mu m = 1.34\ \mu m$$
$$L_{peff} = L_{drawn} + 2 \cdot 0.08\ \mu m = 0.96\ \mu m$$

The MOS transistor is a four-terminal device: *source*, *gate*, *drain*, and *substrate* (well). The source of a transistor is defined so that the charges in the channel move from the source toward the drain. For example, for an n-channel transistor the moving charges are electrons. Here, the source is the terminal that has the lowest potential. The terminal with the highest potential is the source for a p-channel transistor since the moving charges are holes.

Schematic circuit symbols are shown for n- and p-channel transistors in Figure 2.3. The type of substrate is often indicated by an arrow, pointing away from the transistor (right) for an n-doped substrate (i.e., for a p-channel transistor) and toward the transistor (left) for an n-channel transistor. Generally, the substrate (well) of the nMOS device is grounded and the substrate (well) of the pMOS device is connected to the positive power supply rail in a digital circuit.

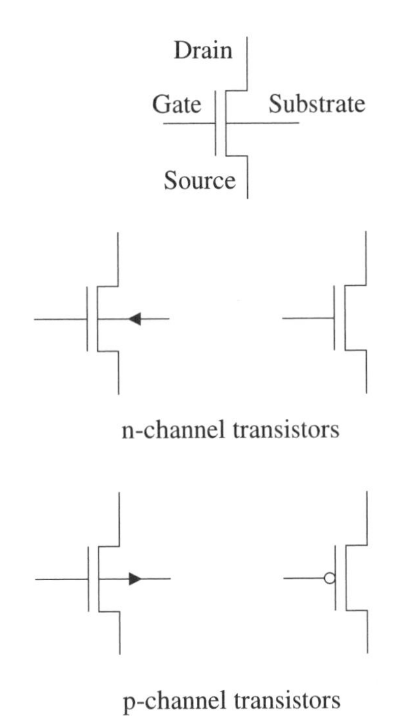

Figure 2.3 Symbols used for MOS transistors

Four different types of devices are possible using MOS technology: n-channel or p-channel transistors that can be either of *enhancement mode* or *depletion mode* type. An enhancement mode transistor does not conduct for zero gate-source voltage while a depletion mode transistor does conduct.

The n-channel enhancement mode transistor will conduct and effectively short-circuit the drain and source if a sufficiently large gate-source voltage is applied, but it will not conduct if the gate-source voltage is less than the threshold voltage. A simple model for the transistor is a voltage-controlled switch that is on if the input voltage is high, and off if it is low. The p-channel transistor will conduct if the gate-source voltage is larger than the threshold voltage.

2.2.1 A Simple Transistor Model

Transistors are characterized by a plot of drain current, I_D, versus the drain-source voltage, V_{DS} for different values of gate-source voltage, V_{GS} [2, 19, 22–25]. All voltages are referenced with respect to the source voltage. The source and substrate are assumed to be connected. Characteristic curves are shown in Figures. 2.4 and 2.5 for a n- and p-channel transistors, respectively.

The two diagrams show typical device characteristics for a 0.8-µm CMOS technology. Large parameter variations occur in practice among devices on different dies while the variations are somewhat smaller between devices on the same die. The diagrams show that drain-source current flows only when the magnitude of the gate-source voltage exceeds a minimum value, called the *threshold voltage*, V_T, i.e., $|V_{GS}| > |V_T|$. This is more clearly illustrated in Figure 2.6 which depicts the I_{Dn}–V_{GS} characteristics for a fixed V_{DS}.

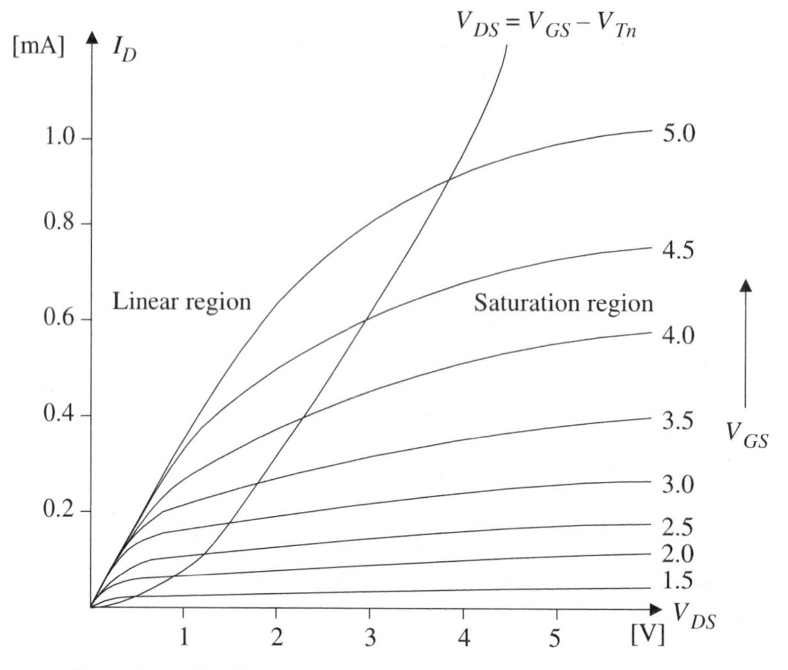

Figure 2.4 I_D–V_{DS} characteristics for an n-channel transistor

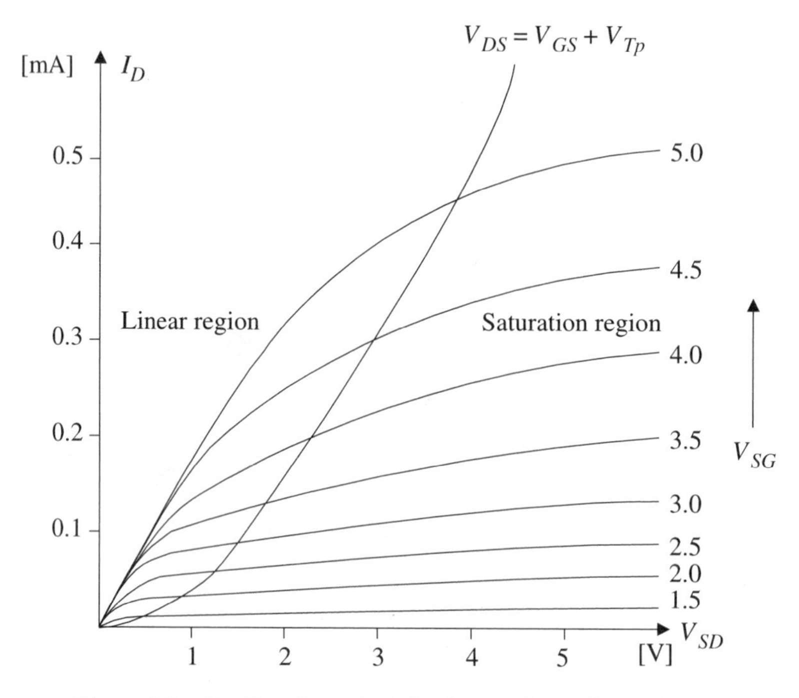

Figure 2.5 I_D–V_{SD} characteristics for a p-channel transistor

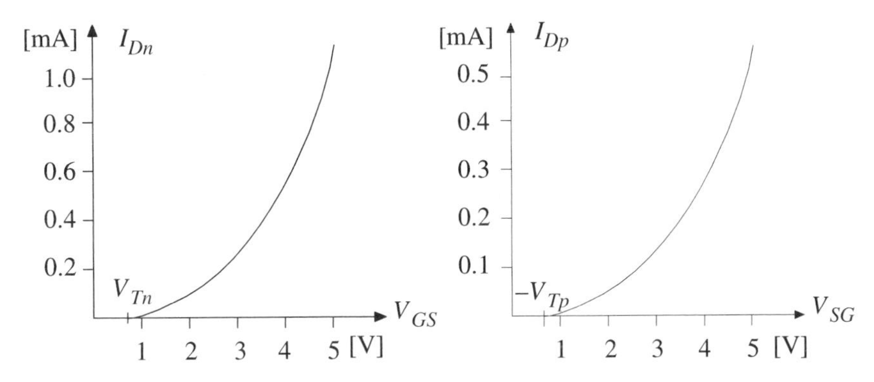

Figure 2.6 I_D–V_{GS} characteristics for saturated n-channel and p-channel transistors

The threshold voltage is adjusted by implanting ions into the substrate in the region below the gate. Typical values for a 5V CMOS process are $V_{Tn} \approx 0.84$ V for an nMOS transistor and $V_{Tp} \approx -0.73$ V for a pMOS transistor. The magnitude of the threshold voltage increases with the source-bulk voltage. The bulk is either the substrate or a well in which the transistor is embedded. Typically, $V_{Tn} \approx 0.84$ V at $V_{DS} = 0$ and $V_{Tn} \approx 1.7$ V at $V_{DS} = 5$ V.

The diagrams shown in Figures. 2.4 and 2.5 can be divided into two regions. The region to the left of this curve is called the linear region. The region where I_D remains practically constant (almost independent of V_{DS}) is called the saturation region.

The drain current[1], I_D, for an nMOS transistor can be modeled by:

$$I_{Dn} = \begin{cases} 0 & V_{GS} - V_{Tn} < 0 \quad \text{(cutoff)} \quad (2.1a) \\ \beta_n \left(V_{GS} - V_{Tn} - \frac{V_{DS}}{2}\right)V_{DS} & V_{GS} - V_{Tn} > V_{DS} \quad \text{(linear)} \quad (2.1b) \\ \frac{\beta_n}{2}\left(V_{GS} - V_{Tn}\right)^2 & V_{GS} - V_{Tn} < V_{DS} \quad \text{(saturated)} \quad (2.1c) \end{cases}$$

where $\beta_n = \mu_n \varepsilon W/T_{ox}L$, μ_n is the average mobility of the charge carriers (electrons for n-channel and holes for p-channel devices), ε is the permittivity of the SiO_2, and T_{ox} is the thickness of the gate oxide.

If the drain-source voltage is increased beyond $V_{DSsat} = V_{GS} - V_{Tn}$ the effective length of the channel is reduced. This effect is referred to as *channel-length modulation*. In order to account for the increase in I_{Dn}, Equation (2.1c) is multiplied by a factor $[1 + \alpha\,(V_{DS} - V_{DSsat})]$ where α is the channel-length modulation factor which typically is in the range 0.02 to 0.05. It can often be neglected in simple analysis of digital circuits but usually not in analysis of analog circuits.

---

1. Shockley's model.

The corresponding model for the drain current for a p-channel transistor is

$$I_{Dp} = \begin{cases} 0 & V_{SG} - V_{Tp} < 0 \quad \text{(cutoff)} \quad (2.2a) \\ \beta_p (V_{SG} - V_{Tp} - \frac{V_{SD}}{2}) V_{SD} & V_{SG} - V_{Tp} > V_{SD} \quad \text{(linear)} \quad (2.2b) \\ \frac{\beta_p}{2} (V_{SG} - V_{Tnp})^2 & V_{SG} - V_{Tp} < V_{SD} \quad \text{(saturated)} \quad (2.2c) \end{cases}$$

where $\beta_p = \mu_p \varepsilon W / T_{ox} L$. We have $V_{DS} = -V_{SD}$ and $V_{GS} = -V_{SG}$. Note that both I_{Dp} and V_{Tp} are negative for p-channel devices. The magnitude of the threshold voltage, V_{Tn}, for the n-channel transistor is about the same as for the p-channel transistor while the polarities of the currents and voltages of an n-channel transistor are the opposite of those for a p-channel transistor. The mobility of the charges in n-channel devices is about two to three times larger than in p-channel devices. Thus, the current flowing through the n-channel transistor is about two to three times larger than in the corresponding p-channel transistor.

Equations (2.1) and (2.2) correspond to a simple transistor model, the so-called LEVEL 1 model, that is used in SPICE[2]. The model is suitable for elementary analysis of circuits, while more accurate models are required for more accurate analysis and for devices with submicron channel lengths. Typical values for a 0.8-μm CMOS process are $\mu_n \approx 4.62\ 10^{-2}\ m^2/Vs$, $\mu_p \approx 1.6\ 10^{-2}\ m^2/Vs$, $\varepsilon = 3.9\ \varepsilon_0$ for SiO_2, and $\varepsilon_0 \approx 8.85$ pF/m. Typically T_{ox} is in the range 100 Å to 200 Å, [1 Å = 10^{-10} m] and $\alpha \approx 0.02$ to 0.05. We get with, $T_{ox} \approx 155$ Å, $\beta_n \approx 103\ (W/L)\ \mu A/V^2$ and $\beta_p \approx 36\ (W/L)\ \mu A/V^2$. For minimum-size devices we have: $\beta_{nmin} \approx 108\ \mu A/V^2$ and $\beta_{pmin} \approx 50 \mu A/V^2$.

2.3 MOS LOGIC

Many different logic MOS circuit techniques have evolved since the 1970s [2, 5–7, 14, 15, 19, 20, 23–25]. These techniques can broadly be partitioned into two main groups:

- Static logic circuits,
- Dynamic logic circuits.

Complex static logic functions can be implemented by placing logic circuits between clocked memory elements. In principle, the logic circuits are combinational and memoryless. Such logic circuits are therefore referred to as static logic circuits since their operation is independent of time. The operation of these circuits depends only on the clocking of the memory elements. They can therefore operate at very low frequencies or even be stopped. In some applications this property is exploited to put the hardware into a standby mode where the circuitry works at a much lower clock frequency to reduce power consumption. In other applications entire functional units may be shut down to conserve power.

2. SPICE stands for Simulation Program with Integrated Circuit Emphasis. University of California at Berkeley.

Dynamic logic circuits are based on the temporary storage of information as charges in stray and gate capacitances. The charges must therefore be periodically restored. This is done by transferring the charges between different storing capacitances and at the same time performing the logic functions. An external control signal (i.e. a clock) is required to control the transfer of information in dynamic circuits. Advantages of dynamic logic circuits are higher clock frequencies, often lower power consumption, and smaller chip area. The main drawback is that the noise margin is only about V_T. Hence, special care has to be taken to reduce the noise.

In MOS logic circuits the output node of a logic circuit is connected via a conducting transistor network to a logic signal source, except during switching between the two states. This source can be either V_{DD}, *Gnd*, or an input signal. In the case of V_{DD} or *Gnd* the output signal level is restored. To transfer the correct logic value sufficiently fast to the next stage, the depth of the network must be small.

2.3.1 nMOS Logic

Generally, a *logic function* is implemented using a circuit that implements the so-called *switching function*. We will use the following notation:

$F(A, B, \dots)$ = logic function to be implemented as a circuit

$S(A, B, \dots)$ = switching function implemented as a transistor network

Networks that can be successively described by serial and/or parallel connections between switches and subnetworks can be analyzed by a form of Boolean algebra[3]—switching algebra;.

In nMOS logic, a switch network of nMOS transistors is connected between ground and the output, as shown in Figure 2.7 [14]. If the network is conducting, a path is established between *Gnd* and the output, and the logic output value is 0. On the other hand, if the switch network is open-circuited, the output is pulled high through the load (weakly conducting transistor). The output will be a logic 1. The logic function performed by this circuit is the inverse of that performed by the switching network.

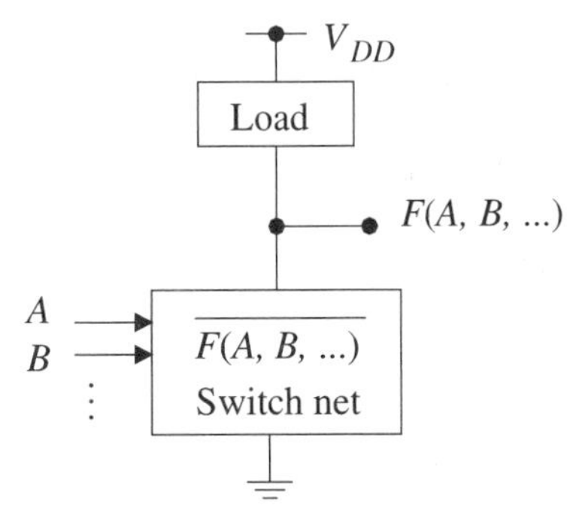

Figure 2.7 nMOS logic circuit

Since only low voltages are transferred by the switch network, only nMOS transistors are needed. The switching function for the n-transistor network needed for implementing a logic function is

$S_n(A, B, \dots) = \overline{F(A, B, \dots)}$ = switching function for the n-transistor network

$F(A, B, \dots)$ = logic function of the nMOS circuit

3. Introduced by C.E. Shannon in his M.Sc. Thesis.

Figure 2.8 shows an nMOS inverter with an n-channel depletion mode transistor, T_1 and an n-channel enhancement mode transistor T_2. The inverter itself has an intrinsic stray capacitance and is typically used to drive other gates that have capacitive inputs. It is therefore appropriate to model the inverter's typical load by a capacitor which the inverter charges or discharges. The depletion mode transistor is called a pull-up device. It is often convenient to model this transistor with a (nonlinear) resistor. The transistor, T_2, acts as a pull-down device. The pull-down transistor will be off if the input to the inverter is lower than the threshold voltage, V_{Tn}. Therefore, the output of the inverter will be charged to V_{DD} by the conducting pull-up transistor. Only a negligible leakage current will flow through the transistors. The power dissipation in this state is negligible.

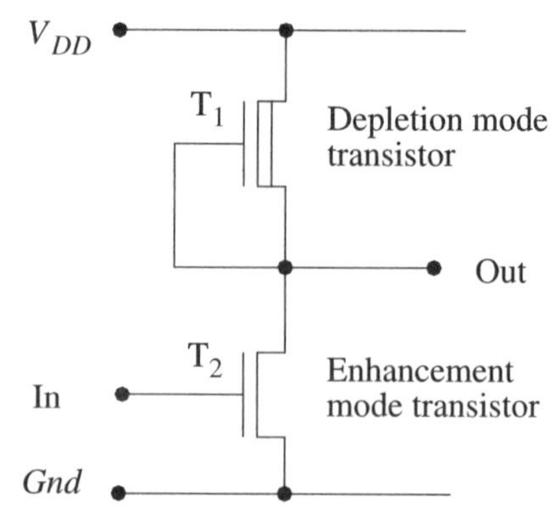

Figure 2.8 nMOS inverter

The pull-down transistor will conduct if the input to the inverter is larger than V_T. In this state, a current will flow through both of the transistors and the inverter will dissipate a significant amount of power. The output voltage, when the output is low, is determined by the ratio of the effective drain-source resistances:

$$V_{\text{out low}} = \frac{R_2}{R_1 + R_2} = V_{DD}$$

nMOS logic is therefore called *ratioed logic*. The output voltage should ideally be zero for a perfect inverter. However, to be able to drive another inverter or gate, the output has only to be sufficiently low, that is, less than the threshold voltage of the following stage. For example, assume that $V_{DD} = 5$ V and $V_T = 1$ V. This implies that $R_1 >> 4\ R_2$ and the effective load capacitance, CL, will discharge through R_2 at least four times faster during the pull-down phase than during the pull-up phase. This asymmetry of switching times is a major problem in ratioed logic.

Now, R_1 must be made small in order to obtain fast switching circuits. This leads to an even smaller value for R_2 in order to satisfy the ratio criterion just discussed. Small effective drain-source resistances will result in large power dissipation due to the increased current. Further, small drain-source resistances require a large silicon area for the transistors. In practice, a trade-off must therefore be made between speed, power dissipation, and transistor sizes.

A two-input NAND gate is shown in Figure 2.9. Note that the low output voltage depends on the relative strength of the pull-up and pull-down branches. The latter consists in a NAND-gate of two transistors in series. The two transistors T_2 and T_3 must therefore be made very wide in order

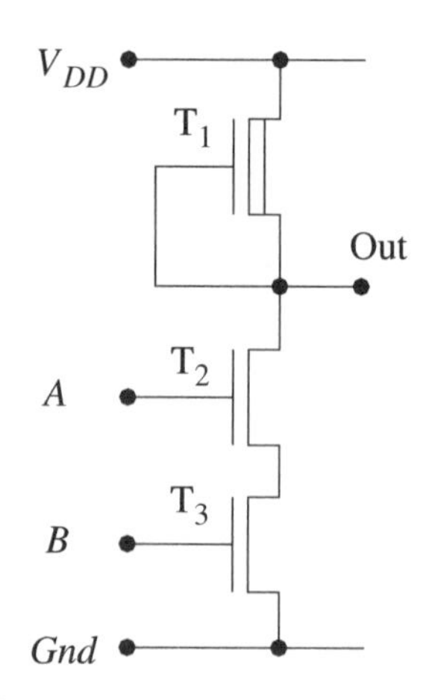

Figure 2.9 nMOS NAND gate

to get a sufficiently low output voltage. It is therefore preferable to use NOR gates instead of NAND gates, since the n-devices, in the former case, are connected in parallel.

The main drawbacks with nMOS logic circuits are the static power dissipation and the fact that a logic 0 at the output is obtained as a ratio between two impedances. The ratio is critical to the function of the gate and requires careful design. The load can be implemented using either a resistor or an nMOS transistor. The nMOS transistor can be of the enhancement type, operating in either the saturated or nonsaturated region, or of the depletion type. The best overall performance for nMOS circuits is obtained by using depletion mode load transistors, but then some extra steps in the fabrication process are needed.

2.3.2 CMOS Logic Circuits

In complementary MOS logic circuits, CMOS[4], two switch networks are connected in series between V_{DD} and ground as shown in Figure 2.10. The networks are the complement of each other. The intermediate node is the output node. Each of the two networks transfers only one type of logic signal. Hence, the upper and lower networks need only to have pMOS and nMOS transistors, respectively. The switching function for the two transistor networks needed to implement a logic function $F(A, B, \dots)$ are

$$S_n(A, B, \dots) = \overline{F(A, B, \dots)} \quad = \text{switching function for the n- transistor network}$$
$$S_p(A, B, \dots) = F(\bar{A}, \bar{B}, \dots) \quad = \text{switching function for the p- transistor network}$$

Note that in the expression for S_p, the input signals are inverted, because of the pMOS transistors inverting property. One, and only one, of the two networks S_n and S_p is conducting, i.e., S_n and S_p are each other's inverse. This guarantees that no current can flow between V_{DD} and ground in the static state, i.e., the gate dissipates practically no static power. This is an important advantage compared to nMOS circuits. However, a significant amount of power is dissipated when the circuit switches and charging and discharging the stray and load capacitances take place.

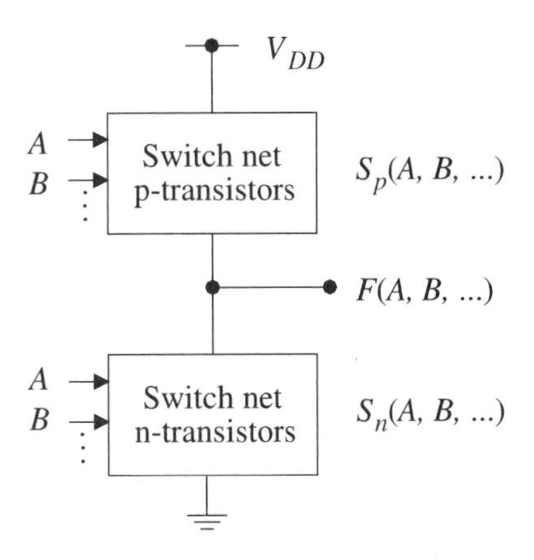

Figure 2.10 CMOS logic circuit

A CMOS inverter is shown in Figure 2.11 [2, 5, 7, 15, 19, 23–25]. The n-channel transistor, T_2 will conduct and the p-channel transistor, T_1 will be off if the input is high, and vice versa. In either case, only one of the transistors conducts. The input signal to the CMOS inverter acts like a control signal that connects the output either to V_{DD}, through the p-channel transistor, or to ground through the n-channel transistor. This means that the output voltage levels are

[4] CMOS logic was invented in 1963 by F. M. Wanlass.

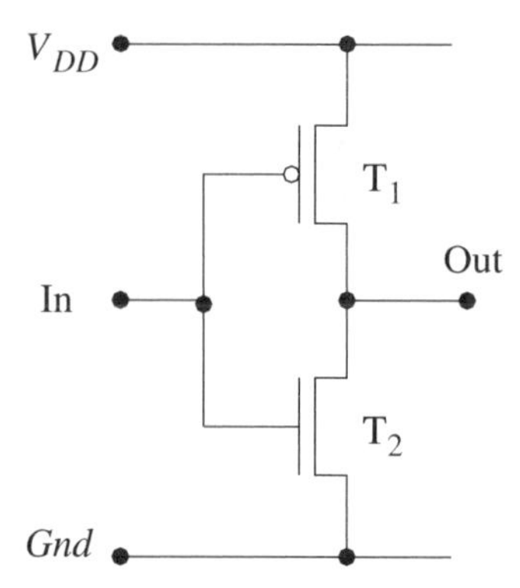

Figure 2.11 CMOS inverter

independent of the relative sizes of the p- and n-channel devices. CMOS logic is therefore referred to as *ratioless logic*. This implies that CMOS circuits produce a larger output voltage swing, resulting in a higher noise immunity than the corresponding nMOS circuits. The p-transistor is often made wider than the corresponding n-channel transistor to get symmetric switching times.

Figure 2.12 shows a two-input CMOS NAND gate. It requires more transistors than the corresponding nMOS gate, but it is more robust and easier to design, since CMOS logic is ratioless. Generally, it is difficult to relate the number of transistors and the required chip area. The number of contacts between the different layers used for wiring also plays an important role in determining the area.

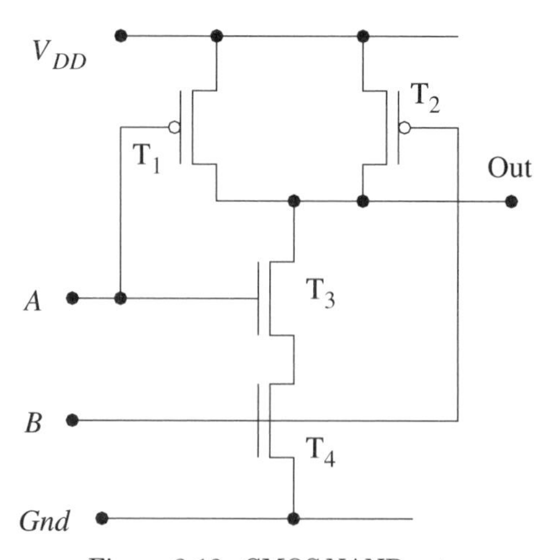

Figure 2.12 CMOS NAND gate

The main advantages of CMOS over nMOS circuits are low power dissipation, high noise margin, high speed, robustness, and ease of design. Disadvantages of CMOS compared to nMOS are slightly lower device density and a more complex manufacturing process. However, these drawbacks tend to be small for modern VLSI processes and circuit techniques.

2.3.3 Propagation Delay in CMOS Circuits

The *propagation delay*, τ_p, in a digital circuit is defined:

$$\tau_p = \tau_{out} - \tau_{in} \tag{2.3}$$

where τ_{out} and τ_{in} are the time instances when the output and input voltage cross the $V_{DD}/2$ level, respectively [2, 7, 23–25].

Speed and power dissipation of digital circuits can be estimated using a simple model that neglects the short-circuit current flowing during a transition. This assumption is reasonable if the slope of the input signal is large. Ideally, the input signal is a step function. The current through the n-channel transistor during the discharge phase is approximated by a pulse as shown in Fig. 2.13. The peak current is equal to the saturation current of the n-channel transistor with $V_{GSn} = V_{DD}$.

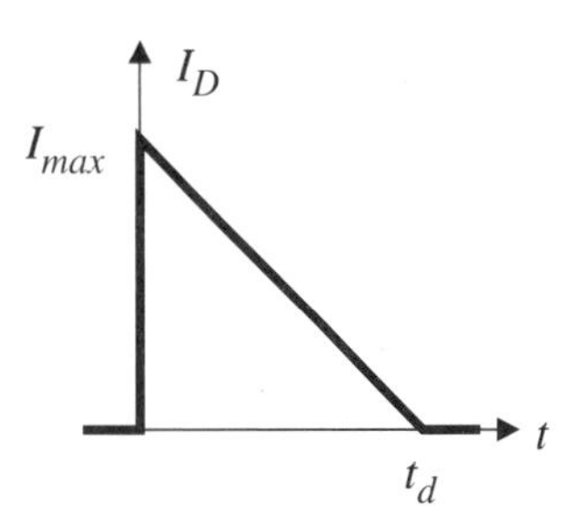

Figure 2.13 Approx. of I_D during charge or discharge

The time required to remove the charge stored on C_L is determined by

$$Q = C_L V_{DD} = \int_0^{t_d} I_D(t)dt = \frac{I_{max} t_d}{2}$$
$$= \frac{\beta_n}{2}(V_{DD} - V_{Tn})^2 \frac{t_d}{2}$$

Solving for t_d we get

$$t_d = \frac{4 C_L V_{DD}}{\beta_n (V_{DD} - V_{Tn})^2} \tag{2.4}$$

Similarly, during the charging phase, the current is approximated by a pulse of duration t_c. The peak current is equal to the saturation current of the p-channel transistor with $V_{SGp} = V_{DD}$. The time required to charge C_L is determined by

$$Q = C_L V_{DD} = \int_0^{t_c} -I_D(t)dt = \frac{-I_{max} t_c}{2} = \frac{\beta_p}{2}(V_{DD} - V_{Tp})^2 \frac{t_c}{2}$$

and

$$t_c = \frac{4 C_L V_{DD}}{\beta_p (V_{DD} - V_{Tp})^2} \tag{2.5}$$

The load capacitance comes from stray capacitances associated with the inverter itself, the load driven by the inverter, and the wiring. The self-load capacitance for a minimum-size inverter is about 10 fF while the input capacitance to a minimum-size inverter is about 8 fF.

The propagation delays are (see Problem 2.6)

$$\tau_{pHL} = -t_d \ln(0.5)$$
$$\tau_{pLH} = -t_c \ln(0.5)$$
$$\tau_p = 0.5\,(\tau_{pHL} + \tau_{pLH})$$

A more accurate expression than Equations (2.4) and (2.5) is

$$t_c = \frac{4 C_L V_{DD}}{\beta_p (V_{DD} - V_{Tp})^{\alpha}} \quad \text{where } \alpha \approx 1.55 \tag{2.6}$$

EXAMPLE 2.1

Estimate the propagation delay for a CMOS inverter that is loaded with five identical inverters connected in parallel. The wiring corresponds to a load of about 5 fF. Compare the result with a SPICE simulation. Use the following values:

$$\mu_n = 4.62\ 10^{-2}\ \text{m}^2/\text{Vs}\ ,\ \mu_p = 1.6\ 10^{-2}\ \text{m}^2/\text{Vs}\ ,\ T_{ox} \approx 155\ \text{Å}$$

$$W_{ndrawn} = 2.0\ \mu\text{m}\ ,\ W_{neff} = 0.84\ \mu\text{m}\ ,\ L_{ndrawn} = 0.8\ \mu\text{m}\ ,\ L_{neff} = 0.8\ \mu\text{m}$$

$$W_{pdrawn} = 2.0\ \mu\text{m}\ ,\ W_{peff} = 1.34\ \mu\text{m}\ ,\ L_{pdrawn} = 0.8\ \mu\text{m}\ ,\ L_{peff} = 0.96\ \mu\text{m}$$

$$V_{Tn} = 0.84\ \text{V},\ V_{Tp} = -0.73\ \text{V, and}\ V_{DD} = 5\ \text{V}$$

The effective load is

$$C_L = C_{self} + 5\ C_{inv} + C_{wire} = 10 + 5 \cdot 8 + 5 = 55\ \text{fF}$$

Substituting into Equations (2.4) and (2.5) we get the charge and discharge times: $t_c = 1.24$ ns and $t_d = 0.61$ ns. Now, the propagation delay is

$$\tau_{pLH} = -t_c\ ln(0.5) = t_c\ 0.69 = 0.85\ \text{ns}\ ,\ \tau_{pHL} = t_d\ 0.69 = 0.42\ \text{ns}$$

Figure 2.14 shows the simulated output wave forms (SPICE) for the CMOS inverter. The propagation delays obtained from SPICE, with an effective load of 55 fF, are $\tau_{pLH} \approx 0.47$ ns and $\tau_{pHL} \approx 0.36$ ns.

With an effective load of 105 fF we get $\tau_{pLH} \approx 0.88$ ns and $\tau_{pHL} \approx 0.68$ ns. In view of the simple model this is a reasonably good estimate. There is an asymmetry in the rise and fall times since the electron mobility is about three times larger than the hole mobility. As mentioned before, the asymmetry is reduced in practice by increasing the width of the p-channel transistor.

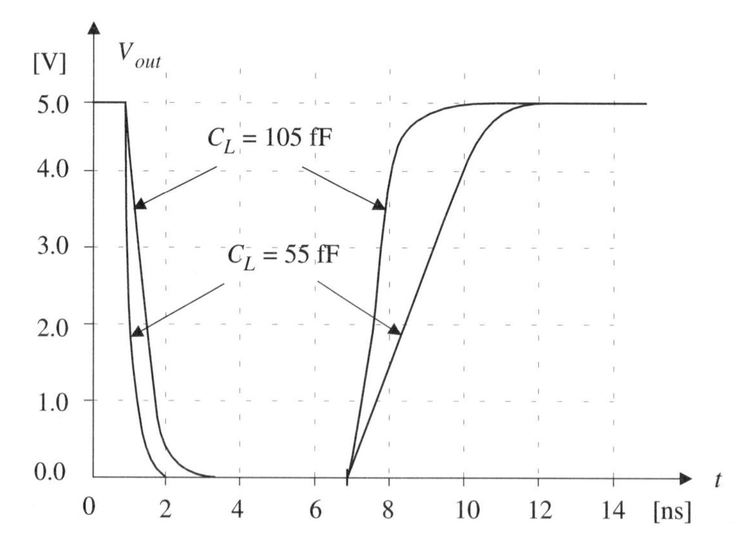

Figure 2.14 The output voltages for a CMOS inverter loaded with 55 fF and 105 fF

EXAMPLE 2.2

Use SPICE to determine the effect of finite rise and fall times of the input signal for the same CMOS inverter as in Example 2.1. Particularly, study the effect on propagation time and power consumption.

Figure 2.15 shows the simulated output wave forms of the CMOS inverter for input signals with different slopes, i.e., rise time τ_r and fall time τ_f. The load is 100 fF. Obviously, the rise and fall times increase with increasing slope of the input signal. The simulated propagation delays for inputs with different slopes are shown in Table 2.1. Note from the table that the slope of the input signal has a significant contribution to the propagation delay [20]. Signals with large slopes will not only make the circuit slower, it will also require the use of a lower clock frequency, since the transitions of the input signals take longer.

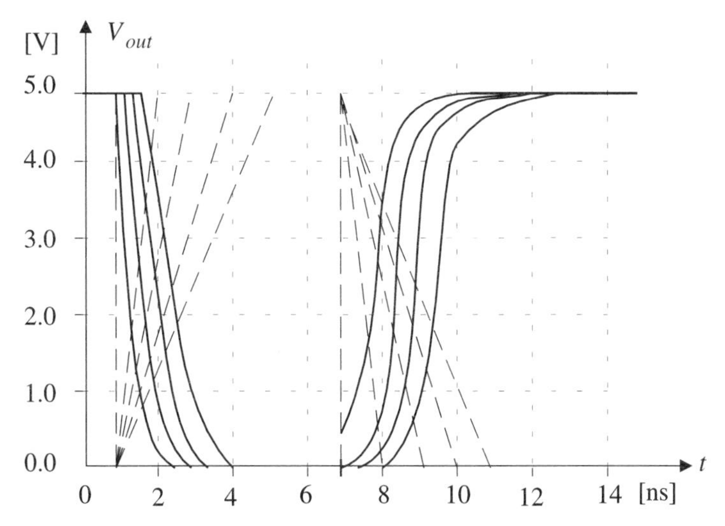

Figure 2.15 Output voltages for inputs with varying slopes for a CMOS inverter

τ_r or τ_f (ns)	τ_{pLH} (ns)	τ_{pHL} (ns)
0	0.46	0.36
1	0.63	0.49
2	0.82	0.62
3	0.99	0.69
4	1.10	0.75

Table 2.1 Propagation delays for inputs with different slopes

Fitting a linear equation to the values in Table 2.1 yields

$$\tau_{pHL} = 0.49 \times 10^{-9} + 0.146\ \tau_r \qquad \text{[s]}$$

and

$$\tau_{pLH} = 0.386 \times 10^{-9} + 0.098\ \tau_f \qquad \text{[s]}$$

The simulated drain currents through the CMOS inverter for inputs with different slopes are shown in Figure 2.16. The current and power dissipation, which have been averaged over 15 ns, are shown in Table 2.2. Also here, the slope of the input signal causes a small increase in power dissipation. The increase in power dissipation is due to the increase in the time interval when both devices conduct and a small current flows directly through the two transistors.

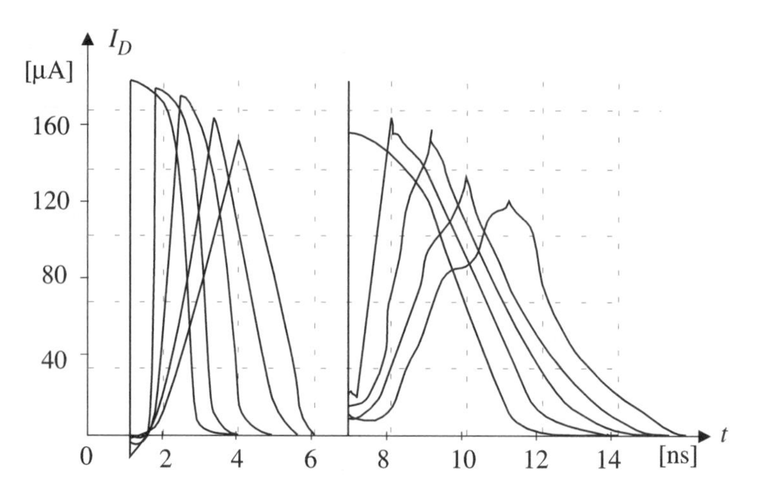

Figure 2.16 Drain currents for inputs with different slopes

τ_r or τ_f (ns)	I_{av}(μA)	P(μW)
0	32.0	160.0
1	31.9	159.5
2	33.3	166.5
3	35.1	175.5
4	37.2	186.0

Table 2.2 Average current and power dissipation for inputs with different slopes

Fitting a linear equation to the values in Table 2.2 yields

$$P = 155.9 \times 10^{-6} + 6.8 \times 103\ \tau_r \qquad \text{[W]}$$

Note, however, that the power consumption can not be reduced by simply reducing τ_r, since this would require a faster driving source which consumes large amounts of power.

2.3.4 Power Dissipation in CMOS Circuits

Accurate estimates of power dissipation are necessary for several reasons: avoiding metal migration in power (V_{DD}) and ground wires, and understanding the design trade-off between power dissipation, chip area, and switching speed. Today power consumption is one of the most important constraints in integrated circuit design.

The current in a wire must not be too large because of electromigration [2, 7]. Typically, a metal wire may not carry a current of more than ≈ 1 mA/(μm)^2. A minimum-width metal 2 wire[5] (width = 1.6 μm and thickness of 1 μm) that is used for power and ground routing can therefore only support up to 1.6 mA/35 μA ≈ 45 minimum-size CMOS inverters of the type used in Example 2.2. The lower layer (metal 1) has typically a thickness of only 0.6 μm.

As mentioned before, CMOS circuits have negligible static power dissipation. The leakage current is in the nanoampere range. Typically the power consumption due to leakage is less than 1% of the total power consumption. If the slope of the input signal is small or the transistor sizes are very large, as is the case in large buffers, a power supply current flows when both the p- and n-transistors conduct. However, for normal logic gates and input signals, the power consumption due to this current is less than 10% of the total power consumption. Significant power is dissipated when the output switches from one state to the other. The stray and load capacitances are, during a complete switch cycle, charged and discharged through the p-transistor and the n-transistor, respectively. Thus, the average power supply current is

$$I_{av} = f\,Q = f\,C_L\,\Delta V \tag{2.7}$$

if the inverter is switched with the frequency f, and ΔV is the output voltage swing. The swing for CMOS circuits is typically V_{DD}. The power dissipation associated with the CMOS inverter is

$$P = I_{av}\,V_{DD} = f\,C_L\,V_{DD}^2 \tag{2.8}$$

With C_L = 105 fF, V_{DD} = 5 V, and f = 1/16 ns = 62.5 MHz, we get an average power dissipation of 156.3 μW, which agrees well with the values in Table 2.2.

2.3.5 Precharge-Evaluation Logic

The 1990s witnessed the proposals of many variations of dynamic logic [2, 5, 7, 23–25]. They differ in various aspects, such as maximum clocking speed, transistor count, area, power consumption, and charge leakage in storage nodes. In precharge-evaluation logic, only one switch network is used together with clocking transistors.

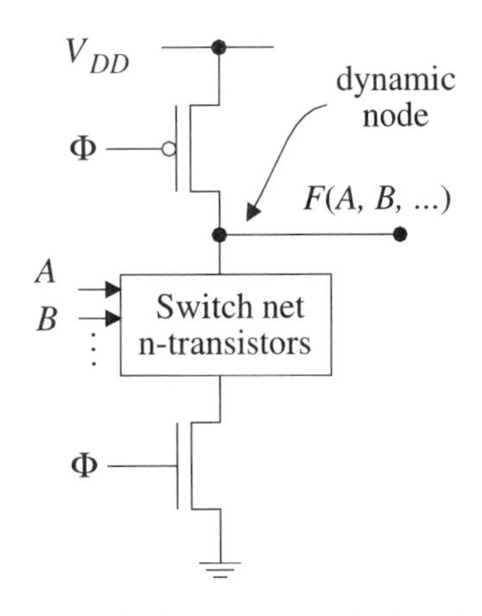

Figure 2.17 Precharge-evaluation logic

Figure 2.17 shows a precharge-evaluation circuit with a switch network with only n-transistors. Two clocking transistors are used in this circuit. The upper transistor is a precharge pMOS transistor. The lower transistor is an nMOS transistor in

5. The second metal layer, on top of the first metal layer (metal 1), is called metal 2.

series with the switch network close to the ground node. Both transistors are controlled by the clock signal, Φ. The function of the circuit is illustrated in Figure 2.18.

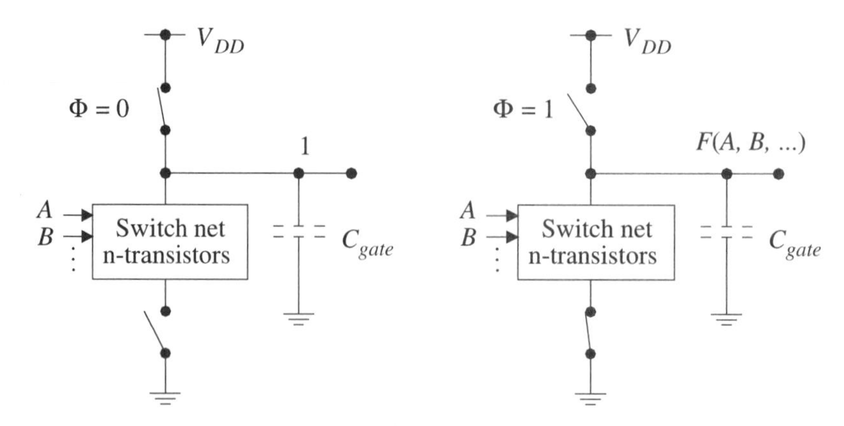

Figure 2.18 Precharge-Evaluation Logic: (a) Precharge phase, Φ = 0, (b) Evaluate phase, Φ = 1

In the precharge phase, Φ = 0, the precharge transistor is conducting. Thus, the output node is charged to V_{DD}. In the subsequent evaluation phase, the output signal value will be determined. If the expression F is evaluated to 0, the switch network will conduct and discharge the output node. On the other hand, if the expression is evaluated to 1, the switch network will not conduct, and the output is a "floating" node. Hence, the output signal stays high as long as the node is charged. The input signal must be valid and stable during the evaluation phase.

The lowest operating rate for precharge-evaluation logic is determined by the charge leakage from node F.

2.3.6 Process Variations

Integrated circuits are processed in batches of 50 to 200 wafers. Since all wafers in a batch undergo the process steps at the same time, there will be less variation in electrical properties of the dice within the batch compared to dice from different batches. Parameter variations viewed over many batches tend to follow a normal distribution. The variation in electrical parameters due to processing variations is called the process spread. Typical variation for the gate propagation due to process spread is in the range of 30 to 50%. It is common practice to sort off-the-shelf circuits into several categories based on their speed. However, in most DSP applications, there is no premium for faster circuits than necessary.

2.3.7 Temperature and Voltage Effects

Propagation delay in integrated circuits depends on junction temperature and supply voltage. Nominal values of supply voltage and ambient temperature for integrated circuits are 3.3 V and 25 °C, respectively.

To estimate junction temperature for an air-cooled integrated circuit, it is necessary to know the ambient air temperature, T_A, the air velocity, Θ, and the thermal resistance of the package, Θ_j. Table 2.3 shows the thermal resistance for some common package types.

Package	Pins	Thermal Resistance Junction-to-Case [°C/W]	Thermal Resistance Junction-to-Ambient [°C/W]
PDIP plastic	28–48	15–17	37–60
PLCC, PLCC plastic	20–84	12–32	37–90
PQFP plastic	44–160	7–24	30–85
PBGA plastic	225	–	26
CBGA ceramic	95–180	1.3–1.9	17–24

Table 2.3 Thermal resistance of some typical packages

The air temperature and velocity depend on the environment and system cabinet. The package manufacturer provides charts for different packages versus air velocity. The junction temperature is

$$T_j = T_A + \Theta_{jA} P \tag{2.9}$$

For commercial silicon circuits the junction temperature should not exceed 150 °C. If a device is operated at elevated temperature, the MTBF (mean time between failures) will be reduced. Typically it is found that the lifetime halves for every 10 °C increase in ambient temperature.

Propagation delay for CMOS circuits increases linearly (approximately) with junction temperature:

$$\frac{1}{\tau_{pd}} \frac{\partial \tau_{pd}}{\partial T_j} \approx 3.75 \times 10^{-3} \qquad [1/°C] \tag{2.10}$$

The effect of supply voltage variation on the propagation delay is

$$\frac{1}{\tau_{pd}} \frac{\partial \tau_{pd}}{\partial V_{DD}} \approx -3 \qquad [1/V] \tag{2.11}$$

EXAMPLE 2.3

Assume that the required clock frequency for an FFT processor is 192 MHz and the chip is estimated to dissipate about 195 mW. A commercial version is to be housed in a plastic chip carrier PLCC that has an effective thermal resistance of 85 °C/W. The process spread is only ±30% and the variation in power supply is ±5%. Estimate the required design margin for the clock frequency.

The excess junction temperature (above 25 °C) is

$$\Delta T_j = 85 \cdot 0.195 \approx 16.6\,°C$$

The contribution from excess temperature, T_A, process spread, and reduced voltage is

$$\tau_{CL\,req} = \tau_{CL\,design}(1 + 3.75\ 10^{-3} \cdot 16.6)(1 + 0.3)[1 + (-0.3)(-0.165)]$$

$$= \tau_{CL\,design}\ 1.0623 \cdot 1.3 \cdot 1.0495 = 1.45\ \tau_{CL\,design}$$

$$f_{CL\,design} = 1.45\ f_{CL\,req} = 1.45 \cdot 192 \approx 279 \text{ MHz}$$

Thus, the circuit should be designed with nominal parameter values to run with a clock frequency that exceeds the required frequency by 50%. Typical worst-case factors are 1.7 for commercial circuits (70 °C, 3.135 V) and 2.2 for military circuits (125 °C, 3.0 V) where the two temperatures indicate the worst-case ambient temperature.

2.4 VLSI PROCESS TECHNOLOGIES

There are several types of CMOS processes, for example, bulk CMOS and CMOS–SOI (*silicon-on-insulator*) [12, 13]. In bulk CMOS, the transistor function takes place at the surface and within the substrate while in CMOS–SOI it takes place in material placed on top of an insulator. In the early bulk CMOS technologies, the gate was formed of metal—metal gate CMOS—while in modern CMOS technologies the gate is formed of polysilicon—silicon gate CMOS. Thus, the term *MOS* itself is no longer appropriate.

2.4.1 Bulk CMOS Technology

Figure 2.19 illustrates the cross section of a silicon gate CMOS inverter [5, 12]. The p-channel transistor is formed in an n-well while the n-channel transistor is

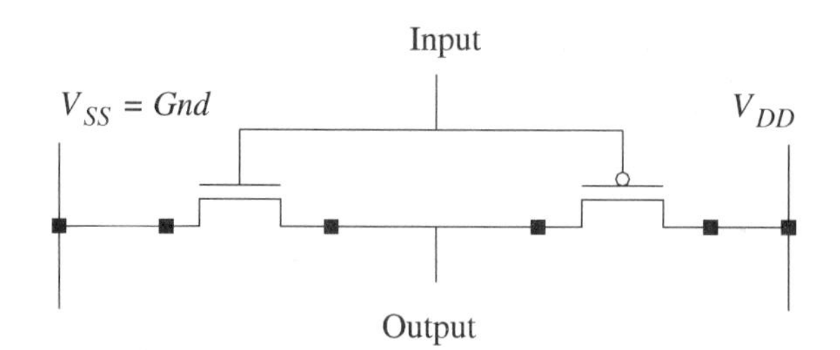

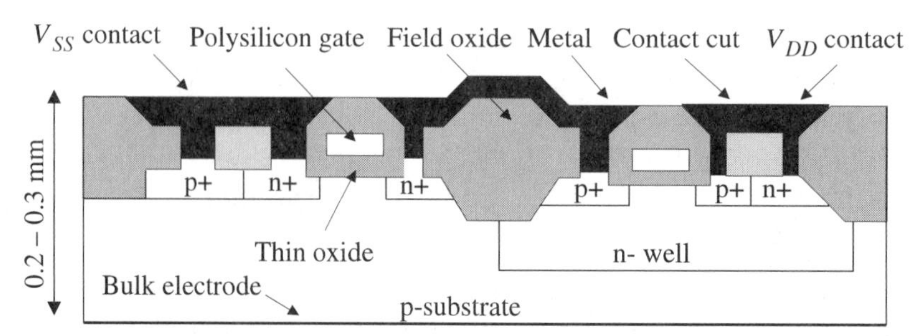

Figure 2.19 Cross section of a silicon gate CMOS inverter in a bulk n-well process

formed directly in the substrate, which is of p-type. Since hole mobility is lower than for electrons, the p-channel device is usually made wider than the n-channel device to obtain a symmetric switching behavior. The input capacitance to an inverter is about three times the gate capacitance for the n-channel device.

Latch-up due to parasitic tyristors is a problem that may appear in CMOS circuits. Latch-up causes excessive power supply current to flow and may even destroy the circuit. There is no way to turn the tyristor off except for disconnecting the power supply. An early technique to avoid latch-up was to use so-called guard bands of heavy doped material between n- and p-devices while modern processes, as the one illustrated in Figure 2.19, use field oxide to isolate the different types of devices.

2.4.2 Silicon-on-Insulation (SOI) Technology

Silicon-on-insulation technology is a serious contender for the fabrication of future integrated circuits. Compatibility with ordinary CMOS circuit styles, excellent device scalability, better device and circuit performance, low power consumption, high working temperatures (200 to 300°C), radiation-hardness, and potentially lower cost are the main reasons. An early type of SOI was SOS (silicon-on-sapphire) [12]. This material was costly and had poor quality. It was used primarily in military and aerospace applications because of the excellent radiation-hardness. SIMOX (separation by implantation of oxygen) is today the most widely used SOI technology. It is produced by an implantation of a high dose of oxygen ions into a silicon wafer, followed by an annealing at very high temperature to finally form a well-defined buried silicon dioxide layer beneath a single-crystal silicon top layer into which the devices are created. Figure 2.20 shows the cross section of an SOI inverter with sapphire substrate.

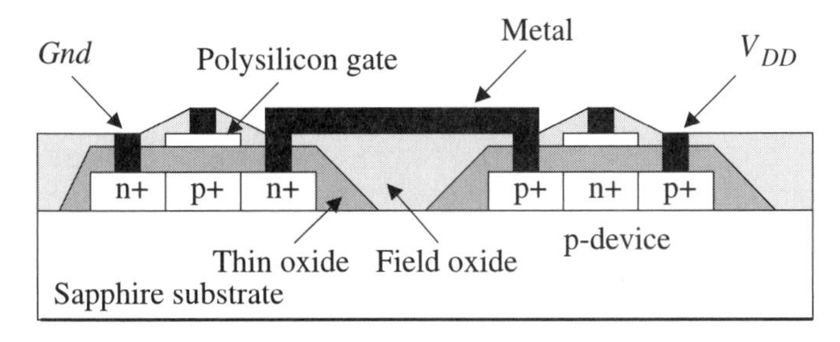

Figure 2.20 Cross section of an SOI inverter

Characteristic of SOI technologies is 30 to 40% higher switching speed as a result of small parasitic capacitances. In bulk CMOS, the main parasitic capacitances are formed between circuit elements and the conducting substrate. Since the substrate is nonconducting in SOI, the later parasitic capacitances are negligible. However, the parasitic capacitances between the circuit elements and the substrate play an important part in the suppression of switching noise. SOI circuits have therefore higher switching noise compared to bulk CMOS circuits. The nonconducting substrate also alleviates latch-up problems and allows low-voltage and high-voltage circuits on the same chip.

2.4.3 Bipolar Technologies—TTL

Figure 2.21 shows a typical three-input TTL NAND gate. Transistor–transistor logic (TTL) is now an outdated bipolar technology in which one or more transistors are driven into a highly saturated state so that excess charge is injected into the base region of the transistors. This is one of the major limiting factors of the switching speed of TTL circuits since it takes time to remove the excess charge.

Various strategies, including the use of Schottky diodes and Schottky transistors, have been introduced to reduce the amount of excess charge in the base. Another drawback of bipolar circuits is that they do not support bidirectional transmission gates and dynamic storage of signal values.

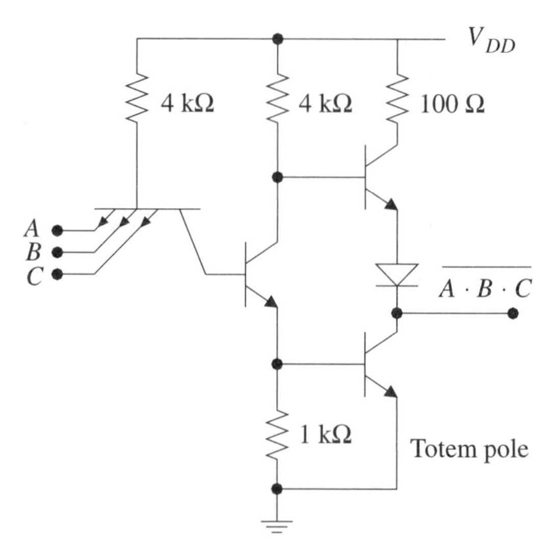

Figure 2.21 TTL NAND gate

The power consumption of bipolar circuits due to switching is

$$P \approx N f_{max} C \Delta V \tag{2.12}$$

where N is the number of gates, f_{max} is the maximum clock frequency for which the circuit has been designed, C is the equivalent load, and ΔV is the corresponding voltage swing at the load. Note that power consumption is independent of the actual clock frequency. TTL is unsuitable for VLSI due to its large power consumption.

2.4.4 Bipolar Technologies—ECL

Nonsaturating logic circuits are usually based on combinations of emitter-followers (common-collector) and common-base amplifiers. The most common type of logic is ECL (emitter-coupled logic) which is an OR-NOR gate technology. A typical ECL gate is shown in Figure 2.22.

The input stage is a differential amplifier. The OR-NOR functions are obtained by using a multi-emitter transistor in one of the branches. The crossover voltage is set by the voltage reference stage to –1.2 V. The output stage is an emitter-follower with low output impedance. The output swing is only 0.8 V which, according to Equation(2.12), results in reduced power consumption.

An advantage of ECL circuits compared to CMOS circuits is that they generate less noise on the power supply lines so that requirements on the power supply are less stringent. Power dissipation versus frequency for ECL and CMOS circuits is sketched in Figure 2.23. CMOS has lower power consumption except for circuits with very high switching frequencies. In fact, ECL is unsuitable for VLSI because

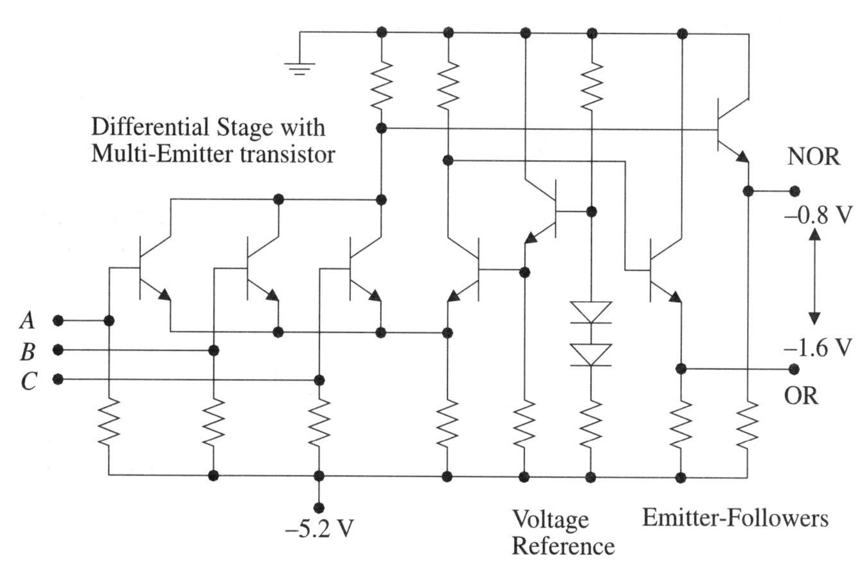

Figure 2.22 Basic ECL gate

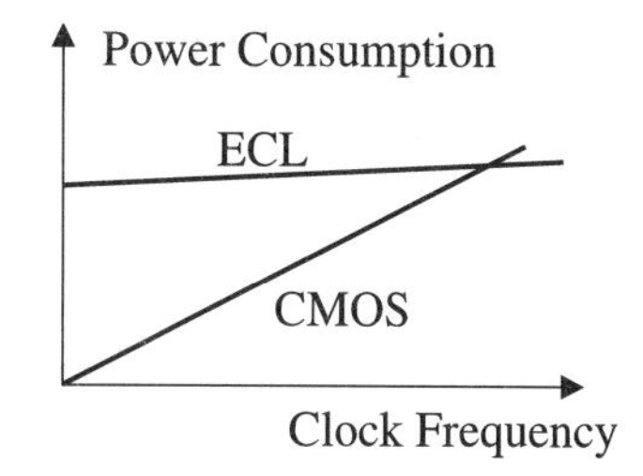

Figure 2.23 Power dissipation versus frequency for ECL and CMOS circuits

of the large power dissipation. Typical power consumption and gate delays are 20 mW/gate and < 0.1 ns, respectively. A few ECL gates are often used for time-critical parts and for driving large capacitive loads—for example, drivers for clocks and I/O. We will discuss this hybrid technology further in section 2.5.5.

2.4.5 Bipolar–CMOS Technologies—BiCMOS

Digital systems with high clock frequencies are often built with both bipolar ECL and CMOS technologies. Each technology has its advantages and disadvantages. The bipolar technology yields the fastest devices but suffers from high power consumption and low device density. For instance, the footprint of a minimum-size bipolar transistor is 20 to 40% larger than the corresponding MOS device.

In a BiCMOS technology both bipolar and MOS devices are manufactured on the same chip. This technology allows the circuit designer to exploit the advantages of both technologies. That is, the speed of the bipolar technology coupled with the complex (large) circuits and low power dissipation that is characteristic of CMOS technology. However, only a small fraction of the devices on a VLSI chip can be bipolar, because of the high power dissipation [1, 8, 17]. The bipolar devices (ECL gates) should therefore be used only in the most time-critical paths of the circuit and for driving large capacitive loads. In practice, BiCMOS circuits have a speed advantage of a factor of about 2 over pure CMOS circuits. BiCMOS technologies are also suitable for mixed analog–digital circuits, since bipolar transistors have higher transconductance and generate less noise.

The use of BiCMOS will prolong the lifetime of CMOS technologies and allow them to compete with pure bipolar and GaAs technologies in high-speed application. The fabrication process is more complex for BiCMOS compared to CMOS.

2.4.6 GaAs-Based Technologies

Gallium arsenide VLSI circuits are competing with silicon-based technologies as a viable VLSI technology [4, 6, 9, 11, 18, 26]. The potential switching speed of this technology is higher than for state-of-the-art ECL (emitter-coupled logic) while the power consumption is lower. GaAs circuits are also better suited in environments with ionizing radiation and they can operate at a wider temperature range than silicon devices. The high switching speed is due to the high electron mobility and the small parasitic capacitances that are due to the insulating substrate. The small parasitic capacitances contribute to low power consumption.

There are three major types of GaAs devices: (1) MESFET (metal-semiconductor field-effect transistor), (2) HEMT (high electron mobility transistor), also called MODFET (modulation doped FET), and (3) HBT (Heterojunction Bipolar Transistor) [3]. GaAs is basically a NOR gate technology. Buffered Direct-Coupled FET logic; (BDCFL) is the dominant logic family.

Figure 2.24 shows a typical three-input NOR gate using depletion mode transistors. The first three-input stage performs the NOR function. The second stage is an emitter-follower that provides both drive capability and a shifting of the voltage level through the Schottky diode and V_{SS}. The output voltage must be sufficiently low (negative) to turn off the input devices in the following gate. Note that two power supplies are required.

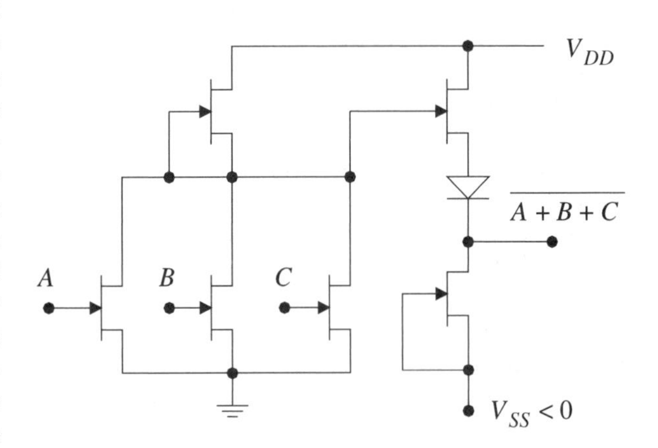

Figure 2.24 Buffered GaAs FET logic

Typical power consumption and gate delay are 0.5 mW/gate and < 0.05 ns, respectively. Thus, GaAs circuits are not much faster than ECL, but they have significantly lower power consumption. An important advantage is that light-emitting diodes and lasers can be made in GaAs. Drawbacks of GaAs circuits are lower thermal conductivity and low hole mobility compared to silicon devices. This means that GaAs is not suitable for complementary circuit styles.

2.5 TRENDS IN CMOS TECHNOLOGIES

The future isn't what it used to be.
—Arthur C. Clarke

The current trend in process technology is toward dense and high-speed CMOS circuits while at the same time the economical chip area increases. The reduction of the dimensions of MOS devices improves both the circuit performance and the packing density. As of 1998, the minimum feature sizes in state-of-the-art CMOS processes are in the range of 0.25 to 0.35 μm and are expected to be scaled down to 0.18 μm by the end of the century. Table 2.4 shows the road map for shrinking the geometries that has been agreed to by the process industry and process equipment manufacturers.

	1992	1995	1998	2001	2004	2007
Feature size (μm)	0.5	0.35	0.25	0.18	0.13	0.1
Transistors/chip ($\times 10^6$)	5	10	21	46	110	260
Chip area (mm^2)	210	250	300	360	430	520
Wire resist. (Ω/μm)	0.12	0.15	0.19	0.29	0.82	1.34
Wire cap. (fF/μm)	0.15	0.17	0.19	0.21	0.24	0.27
Frequency (MHz)	175	300	450	600	800	1000

Table 2.4 Road map for CMOS circuits

The following simple model can be used to estimate the performance of future MOS processes. The electric fields in the devices will remain essentially the same when the MOS devices are scaled with a factor S in five dimensions: three geometrical dimensions, voltages, and doping concentration. We can expect a linear increase in speed and the number of devices will increase by a factor S^2. The potential processing capacity increases as S^3. However, there are a number of second-order effects—for example, velocity saturation of the free charges and parasitic source and drain resistances that degrade the performance [3].

Another effect that becomes worse with scaling is the distributed RC delays of the wires. Ultimately, the intrinsic gate delays will no longer dictate the switching speed. Instead, interconnect and off-chip communication delays will represent far more stringent limitations. Technologies featuring shorter gate delays than CMOS (for example, gallium arsenide circuits) will therefore not necessarily prevail in the long term.

The constant field scaling model shows that smaller feature sizes will require lowering of the power supply voltage. Although it is still possible to use the standard 5-V power supply for micrometer technologies, submicron processes will require lower voltages. Typically, CMOS processes with minimum feature sizes of 0.5 μm use a supply voltage of V_{DD} = 3.3 V. However, this supply voltage can only be used down to feature sizes of about 0.4 μm. For smaller geometries, the supply voltage must be reduced even further. An advantage of low power supply voltage for battery-powered equipment is that the number of cells in the battery can be reduced, possibly to a single cell. This will significantly reduce the weight of battery-powered equipment as well as extend the usable time between battery chargings.

Lower supply voltage and smaller parasitic capacitances make dynamic logic circuits less reliable and noise problems may require careful design of the interconnection networks. Finally, it is noteworthy that the performance of integrated circuits was earlier dominated by circuit density and chip area. Now, the dominating factors are communication distances and power dissipation.

References

[1] Alvarez A.R.: *BiCMOS Technology and Applications,* Kluwer Academic Pub., Boston, 1989.

[2] Annaratone M.: *Digital CMOS Circuit Design,* Kluwer Academic Pub., Boston, 1986.

[3] Bakoglu H.B.: *Circuits, Interconnections, and Packaging for VLSI,* Addison-Wesley, Reading, MA, 1990.

[4] Di Giacomo J.: *VLSI Handbook, Silicon, Gallium Arsenide, and Super-Conductor Circuits,* McGraw-Hill, New York, 1989.

[5] Dillinger T.E.: *VLSI Engineering,* Prentice Hall, Englewood Cliffs, NJ, 1988.

[6] Glasford G.M.: *Digital Electronic Circuits,* Prentice Hall, Englewood Cliffs, NJ, 1988.

[7] Glasser L.A. and Dobberpuhl D.W.: *The Design and Analysis of VLSI Circuits,* Addison-Wesley, Reading, MA, 1985.

[8] Heimsch W., Hoffmann B., Krebs R., Müller E.G., Pfäffel B., and Ziemann K.: Merged CMOS/Bipolar Current Switch Logic (MCSL), *IEEE J. Solid-State Circuits,* Vol. SC-24, No. 5, pp. 1307–1311, Oct. 1989.

[9] Hollis E.E.: *Design of VLSI Gate Array ICs,* Prentice Hall, Englewood Cliffs, NJ, 1987.

[10] Hurst S.L.: *Custom-Specific Integrated Circuits, Design and Fabrication,* Marcel Dekker, New York, 1985.

[11] Long S. and Butner S.: *Gallium Arsenide Digital Integrated Circuit Design,* McGraw-Hill, New York, 1990.

[12] Maly W.: *Atlas of IC Technologies: An Introduction to VLSI Processes,* Benjamin/Cummings, Menlo Park, CA, 1987.

[13] McCanny J.V. and White J.C. (eds.): *VLSI Technology and Design,* Academic Press, London, 1987.

[14] Mead C. and Conway L.: *Introduction to VLSI Systems,* Addison-Wesley, Reading, MA 1980.

[15] Mukherjee A.: *Introduction to nMOS and CMOS VLSI Systems Design,* Prentice Hall, Englewood Cliffs, NJ, 1986.

[16] Ng K.K. and Brews J.R.: Measuring the Effective Channel Length of MOSFETs, *IEEE Circuits and Devices,* Vol. 6, No. 6, pp. 33–38, Nov. 1990.

[17] Pfleiderer H.J., Wieder A.W., and Hart K.: BiCMOS for High Performance Analog and Digital Circuits, *Proc. Fifth European Solid-State Circuits Conf., ESSCIRC'89,* pp. 153–167, Vienna, Sept. 1989.

[18] Preas B. and Lorenzetti M. (eds.): *Physical Design Automation of VLSI Systems,* Benjamin/Cummings, Menlo Park, CA, 1988.

[19] Pucknell D.A. and Eshraghian K.: *Basic VLSI Design—Systems and Circuits,* Prentice Hall, NJ, 1988.
[20] Shoji M.: *CMOS Digital Circuit Technology,* Prentice Hall, Englewood Cliffs, NJ1988.
[21] Sze S.M.: *VLSI Technology,* McGraw-Hill, New York, 1988.
[22] Tsividis Y.P.: *Operation and Modeling of the MOS Transistor,* McGraw-Hill, New York, 1987.
[23] Uyemura J.P.: *Fundamentals of MOS Digital Integrated Circuits,* Addison-Wesley, Reading MA, 1988.
[24] Uyemura J.P.: *Circuit Design for CMOS VLSI,* Kluwer Academic Pub., Boston, 1992.
[25] Wang N.: *Digital MOS Integrated Circuits,* Prentice Hall, Englewood Cliffs, NJ, 1989.
[26] Weste N.H.E. and Eshraghian K.: *Principles of CMOS VLSI Design—A System Perspective,* Addison-Wesley, Reading, MA, 1985.
[27] Wing O.: *Gallium Arsenide Digital Circuits,* Kluwer Academic Pub., Boston, 1990.

PROBLEMS

2.1 Estimate the relative speed and power dissipation for a CMOS circuit if the supply voltage is reduced from 5 to 3.3 V. The original circuit dissipates 1.3 W while running with a clock frequency of 100 MHz. Assume for the sake of simplicity that $V_{Tn} = |V_{Tp}| = \text{constant} = 0.75$ V, and that the capacitances in the circuit remain constant.

2.2 Determine the ID–V_{DS} regions through which the devices in an inverter pass when the input voltage goes from low to high.

2.3 NEC offers a series of cell-based ASICs based on a 0.8-μm CMOS process with 6.5 μW/gate/MHz where a typical gate is a two-input NAND gate. Supply voltages are from 3 to 5 V and up to 180,000 gates per chip are feasible.

(a) Determine the maximum number of gates that can be active simultaneously if the package that houses the chip can dissipate 2 W and the clock frequency is 80 MHz.

(b) Determine the maximum number of gates if the supply voltage is reduced to 3.3 V.

2.4 In a switch-level simulator the transistors are modeled by a voltage-controlled switch, resistors, and capacitors. Develop such a model for the minimum-size inverter used in Example 2.1.

2.5 Determine the peak current flowing through an inverter when both transistors conduct during a switching event.

2.6 Determine the relationships between propagation delay, time constant, and rise (fall) time for a first-order *RC* network.

2.7 Minimization of the power consumption is important in battery-powered applications. Assume that in a given application the throughput requirements can be met with a certain processing element (*PE*) and that the acceptable chip area used for the PEs may be increased. Show that the power consumption can be reduced by lowering the supply voltage and using two PEs in parallel.

2.8 Identify the logic style and determine the logic function that is realized by the circuit shown in Figure P2.8.

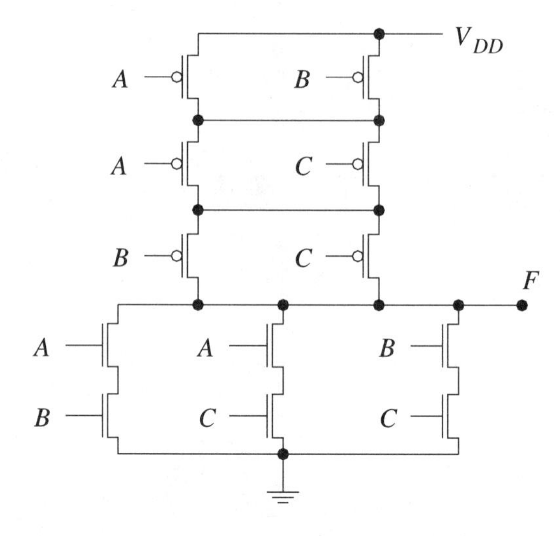

Figure P2.8

2.9 Identify the logic style and determine the logic function that is realized by the circuit shown in Figure P2.9.

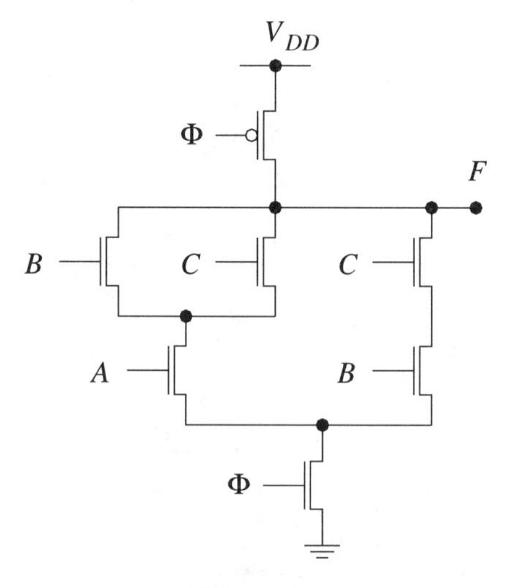

Figure P2.9

2.10 The generic circuit style illustrated in Figure P2.10 is called domino logic. Explain the operation of the circuit. This circuit style is used, for example, with advantage, in the Manchester adder discussed in Chapter 11.

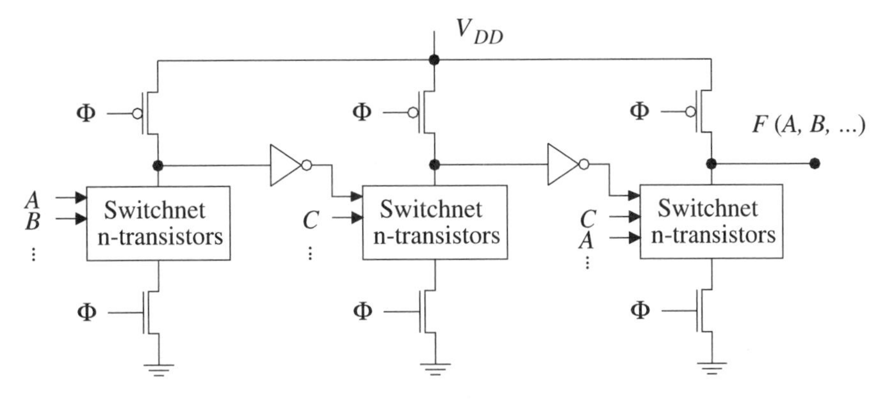

Figure P2.10

3

DIGITAL SIGNAL PROCESSING

3.1 INTRODUCTION

In Chapters 3 and 4 we will present terminology and, in brief, synthesis and analysis methods used in the design of some common digital signal processing functions. In practice, the DSP system design process comprises a mixture of empirical and ad hoc methods aimed at coping with large system complexity and minimizing the total cost [8]. Typically, the resulting DSP system is composed of blocks representing well-known "standard" functions—for example, frequency selective filters, adaptive filters, correlation, spectral estimation, discrete Fourier and cosine transforms, and sample rate converters. The design of such basic DSP functions is therefore a topic of interest.

It is useful to apply different design methodologies to some well-known DSP subsystems in order to better understand the advantages and disadvantages of a given approach. Such design studies may provide not only a better understanding of the strengths and weaknesses of the design methodology, but also a better insight into the computational properties of DSP algorithms. Further, they may provide points of reference and a basis for the design of more complex DSP systems. We will therefore carry out the first design iteration for three fixed-function subsystems that will demonstrate different implementation trade-offs.

The objective of the first design iteration is to investigate the feasibility of the selected implementation approach and estimate major system parameters, such as power consumption and chip area. Two of these functions involve discrete transforms used in many DSP applications—for example, for data compression, spectrum analysis, and filtering in the frequency domain. One of the best-known discrete transforms is the *DFT (discrete fourier transform)*. The signal processing properties of the DFT are similar to the discrete-time Fourier transform [4, 9, 17, 39, 40]. The practical usefulness of a transform is governed not only by its signal processing properties, but also by its computational complexity. The number of arithmetic operations required to directly compute the DFT is very large for long sequences. It was therefore a major breakthrough when a fast algorithm for computing the DFT was invented by Cooley and Tukey in 1965 [14]. This new algorithm was referred to as the *FFT* (*fast fourier transform*). FFTs are today widely used for digital signal processing.

Another important class of transforms is the *DCT* (*discrete cosine transforms*), mainly used in data compression schemes for speech and image signals. The DCT is also used for realization of filter banks that are used, for example, in frequency division and time division multiplexers, so-called transmultiplexers. Fast algorithms for computing the DCT have also been developed.

Both the FFT and the DCT can be implemented by using standard signal processors or ASIC processors. Many different implementation schemes, based on the direct mapping approach, have also been developed. The implementations range from fully parallel, where one PE (*processing element*) is allocated to each basic operation, to sequential computations using a single processor.

In this chapter, we will discuss two typical fixed-function subsystems that will be used as case studies. The first case study involves the design of an FFT processor for computing the DFT. The second case study concerns a two-dimensional DCT that is intended to be used in a system for transforming coding of video images in real-time. Such coding systems will be used for *HDTV* (*high-definition television*).

The third case study will be discussed in Chapter 4.

3.2 DIGITAL SIGNAL PROCESSING

Most developments in advanced signal processing for signals with bandwidths up to 50 MHz or more have since the early 1990s been based on discrete-time or digital techniques. Typical applications are in audio, video, radar, sonar, digital radio, communication, control, and measurement systems. Relatively simple signal processing functions, which traditionally have been implemented using analog continuous-time systems, have also received competition from discrete-time implementations using *SC* (*switched capacitor*) techniques which are compatible with most modern CMOS technologies.

3.2.1 Sensitivity

The major reason behind the increasing use of discrete-time and digital signal processing techniques is that problems caused by component tolerances as well as drift and aging of components are circumvented.

For analog frequency selective filters, realizations having minimum circuit element sensitivity have been developed. Thus, by using high-quality components high-performance filters can be implemented. However, there is a practical limit to the performance of analog components, e.g., the tolerances of resistors, capacitors, and amplifiers can not be arbitrarily low. Filters meeting very stringent requirements are therefore not possible to implement. At the other end of the spectrum, cheap and simple filters can be implemented using low-tolerance components. No such lower tolerance bound exists for digital signal processing techniques. In fact, the tolerances can easily be adjusted to the requirements at hand. However, it must be stressed that the ultimate performance of a composite system will be limited by the analog parts that are necessary at interfaces to the outside world.

The *flexibility* of digital signal processing, which is indirectly due to the deterministic nature of the component errors, makes a DSP system easy to change and also makes it easy to dynamically adapt the processing to changing situations. This feature can, for example, be exploited to allow hardware to be multiplexed to perform several filtering functions. Another important application is adaptive filtering. Expensive tuning procedures, which contribute significantly to the overall cost of analog circuits, are completely eliminated in a digital implementation.

Further, sensitivity to coefficient errors in an algorithm determines a lower bound on the round-off errors or *signal quality* [25, 40]. In practice, an arbitrarily good signal quality can be maintained by using sufficiently high numerical accuracy in the computations. Note that the accuracy of floating-point numbers is determined by the mantissa. Generally, a large dynamic range (large exponent) is not required in good signal processing algorithms. Therefore, DSP algorithms are normally implemented using fixed-point arithmetic.

3.2.2 Robustness

Latch-up and different types of oscillations resulting from abnormal disturbances may appear in analog systems, but most often analog systems return to normal operation when the disturbance disappears. Corresponding phenomena, so-called *parasitic oscillations*, are also present in digital signal processing algorithms, but additionally some unique phenomena occur due to finite word length effects [25]. In fact, a major design challenge is to maintain stability of the system and recover to normal operation in the presence of external disturbances. Of special interest therefore are algorithms that guarantee that the system will return to normal operation when the disturbance has subsided. Disturbances that cause abnormal behavior can originate from transients on the power supply lines, ionic radiation, initial values in the memories at the start-up, or abnormal input signals. The most important filter algorithms with guaranteed stability are wave digital filters and non-recursive FIR filters (*finite-length impulse response filters*), which will be discussed in detail in Chapter 4.

3.2.3 Integrated Circuits

The development of VLSI technology is an important prerequisite for making advanced and complex digital signal processing techniques not only viable, but also economically competitive. The use of VLSI technology also contributes to increased reliability of the system.

3.3 SIGNALS

A signal conveys information by using a *signal carrier*. Typically, information is modulated onto a physical quantity, e.g., a speech signal can be represented by a voltage variation. Voltages or currents are commonly used as signal carriers in electronic signal processing systems. In practice, several types of modulations are used to encode the information and to exploit different types of signal carriers and circuit devices. The simplest type of modulation is to let the signal carrier (voltage) vary according to the continuous time signal, e.g., a speech signal. Such a signal, which varies continuously both over a range of signal values and in time, is called an analog signal. An *analog signal* is denoted

$$t \rightarrow y: y = f(t), \quad y \in C, \quad t \in C$$

The signal may be a complex function in the complex domain, but usually both y and t are real quantities. Note that we usually refer to the independent variable(s) as "time" even if in certain applications the variable may represent, for example, the spatial points (pixels) in an image. It is not a trivial matter to extend the theory for one-dimensional systems to two- or multidimensional systems, since several basic properties of one-dimensional systems do not have a direct correspondence in higher-dimensional systems.

In many cases, the signal does not vary continuously over time. Instead, the signal is represented by a sequence of values. Often, these signals are obtained from measurements (sampling) of an analog quantity at equidistant time instances. A sampled signal with continuously varying signal values is called a *discrete-time signal*:

$$n \rightarrow y: y = f(nT), \quad y \in \mathrm{C}, \quad n \in Z, \quad T > 0$$

where T is an associated positive constant. If the sequence is obtained by sampling an analog signal, then T is called the sample period.

If signal values are restricted to a countable set of values, the corresponding signal is referred to as a *digital signal* or *sequence:*

$$n \rightarrow y: y = f(nT), \quad y \in Z, \quad n \in Z, \quad \mathrm{I} > 0$$

Unfortunately, it is common practice not to distinguish between discrete-time and digital signals. Digital signals are, in principle, a subset of discrete-time signals. Digital signals are usually obtained by measurements of some physical quantity using an A/D converter with finite resolution.

We will frequently make use of the following special sequences:

$$\text{Impulse sequence: } \delta(nT) = \begin{cases} 1 & \text{if} \quad n = 0 \\ 0 & \text{otherwise} \end{cases} \tag{3.1}$$

$$\text{Unit step sequence: } u(nT) = \begin{cases} 1 & \text{if} \quad n \geq 0 \\ 0 & \text{if} \quad n < 0 \end{cases} \tag{3.2}$$

The impulse sequence is sometimes called the unit sample sequence. Whenever convenient, we choose to drop the T and simply write $x(n)$ instead of $x(nT)$.

3.4 THE FOURIER TRANSFORM

Many signal processing systems exploit the fact that different signals and/or unwanted noise occupy different frequency bands or have in some other way different properties in the frequency domain. Examples of such systems are frequency selective filters. The Fourier transform is particularly suitable for describing (linear shift-invariant) systems, since complex sinusoidal sequences are eigenfunctions to such systems. In other words, a sinusoidal input sequence will lead to a sinusoidal output sequence with the same frequency, but with a phase-shift and possibly a different magnitude. Analysis of the frequency properties of such systems can be done using the Fourier transform.

The *Fourier transform* for a discrete-time signal is defined

$$X(e^{j\omega T}) \triangleq \sum_{n=-\infty}^{\infty} x(nT)e^{-j\omega T} \tag{3.3}$$

and the *inverse Fourier transform* is

$$x(nT) = \frac{1}{2\pi}\int_{-\pi}^{\pi} X(e^{j\omega T})e^{jn\omega T}d(\omega T) \tag{3.4}$$

if the sum in Equation(3.3) exists. Note that the Fourier transform, $X(e^{j\omega T})$, is periodic in ωT with period 2π, and that we use the angle ωT as the independent variable.

The magnitude function, corresponding to a real sequence, is shown in Figure 3.1. The magnitude and the phase function as well as the real and imaginary parts of the Fourier transform of a real sequence are even and odd functions of ωT, respectively.

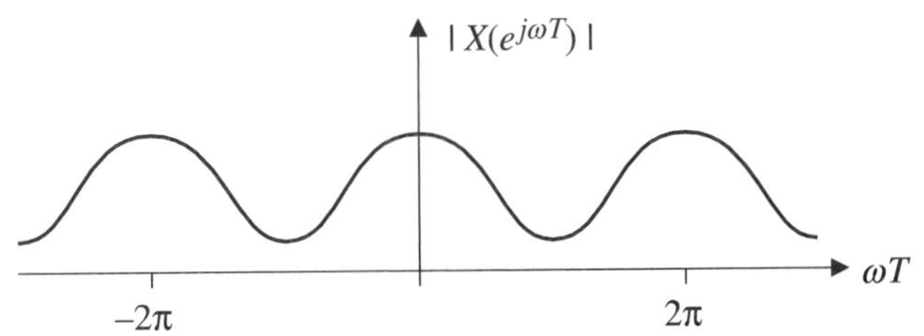

Figure 3.1 Magnitude of the Fourier transform of a real sequence

For real sequences we have

$$x_e(n) \leftrightarrow Re\{X(e^{j\omega T})\} \qquad \text{even function}$$

and

$$x_o(n) \leftrightarrow j\,Im\{X(e^{j\omega T})\} \qquad \text{odd function}$$

EXAMPLE 3.1

Determine the Fourier transform of the sequence

$$x(n) = \begin{cases} 1 & \text{if} \quad n \le 0 \le n_0 \\ 0 & \text{otherwise} \end{cases}$$

$$X(e^{j\omega T}) = \sum_{n=-\infty}^{\infty} x(nT)e^{-jn\omega T} = \sum_{n=0}^{n_0} e^{-jn\omega T} = \frac{1-e^{-j(n_0+1)\omega T}}{1-e^{-j\omega T}}$$

$$= \frac{e^{-j(n_0+1)\omega T/2}}{e^{-j\omega T/2}} \frac{\sin\left(\frac{(n_0+1)\omega T}{2}\right)}{\sin\left(\frac{\omega T}{2}\right)}$$

$$X(e^{j\omega T}) = e^{-jn_0\omega T/2} \frac{\sin\left(\frac{(n_0+1)\omega T}{2}\right)}{\sin\left(\frac{\omega T}{2}\right)} \quad \text{for } n_0 > 0$$

The magnitude function of the Fourier transform resembles the well-known function $\sin(x)/x$, but it is periodic with period 2π. For $n_0 = 0$ we get $X(e^{j\omega T}) = 1$

3.5 THE z-TRANSFORM

The Fourier transform does not exist for sequences with nonfinite energy, e.g., sinusoidal sequences. In the analog case, the Laplace transform is used for such analog signals [4, 32, 33]. An extension of the Fourier transform, used similarly to the Laplace transform for discrete-time and digital signals, is the z-transform.

The *z-transform* is defined

$$X(z) \triangleq \sum_{n=-\infty}^{\infty} x(nT)z^{-n}, \; R_+ < |z| < R_- \tag{3.5}$$

where R_+ and R_- are the *radii of convergence*. The z-transform exists, i.e. the sum is finite, within this region in the z-plane. It is necessary to explicitly denote the region of convergence in order to uniquely define the inverse z-transform.

Obviously, the z-transform and the Fourier transform coincide if $z = e^{j\omega T}$, i.e., coincide on the unit circle in the z-plane. This is similar to the analog case, where the Laplace transform and the Fourier transform coincide on the imaginary axis, i.e., $s = j\omega$.

An important property of the z-transform is that a shift of a sequence corresponds to a multiplication in the transform domain, i.e., if

$$x(n) \leftrightarrow X(z)$$

then

$$x(n-k) \leftrightarrow z^{-k} X(z)$$

EXAMPLE 3.2

Determine the z-transform for the sequence:

$$x(n) = \begin{cases} a_n & \text{if } n \geq 0 \\ 0 & \text{otherwise} \end{cases}$$

$$X(z) = \sum_{n=-\infty}^{\infty} x(n)z^{-n} = \sum_{n=0}^{\infty} a^n z^{-n} = \sum_{n=0}^{\infty} (az^{-1})^n = \frac{1}{1-az^{-1}}, \; |a\,z^{-1}| < 1$$

$$X(z) = \frac{z}{z-a}, \; |z| > |a|$$

There are several ways to find the sequence $x(n)$ from $X(z)$. Essentially $X(z)$ must be expanded into a power series in z^{-1} or z or both. The sequence values correspond to the coefficients of the power series. A formal way to obtain these values directly is to compute the following integral.

The *inverse z-transform* is

$$x(nT) = \frac{1}{2\pi j} \oint_C X(z)z^{n-1}dz \tag{3.6}$$

where C is a contour inside the region of convergence, followed counterclockwise around the origin. The inverse z-transform can be determined by using residue calculus or by partial fraction expansions in the same way as for the Laplace transform. Expansion in terms of simple geometrical series is sometimes useful [4, 6, 17, 32, 40]. In practice, the numerical inversion of a z-transform is often done by filtering an impulse sequence through a filter with the appropriate transfer function.

3.6 SAMPLING OF ANALOG SIGNALS

"Mr. Twain, people say you once had a twin brother. Is it true?"
"Oh, yes. But, when we were very young, I died, fortunately,
he has lived on, aliasing my name"

—an anecdote

Most discrete-time and digital signals are in practice obtained by uniform sampling of analog signals. The information in the analog signal, $x_a(t)$, can under certain conditions be retained and also reconstructed from the discrete-time signal. In this chapter, we will neglect possible loss of information due to finite resolution in the sample values. How to choose the necessary data word length is discussed in [6].

It is well known [6, 15, 32, 40] that the Fourier transform of the sequence $x(n)$ that is obtained by sampling, i.e., $x(n) = x_a(t)$ for $t = nT$, is related to the Fourier transform of the analog signal $x_a(t)$ by Poisson's summation formula:

$$X(e^{j\omega T}) = \frac{1}{T}\sum_{k=-\infty}^{\infty} X_a\left(\omega + \frac{2\pi k}{T}\right) \tag{3.7}$$

Figure 3.2 illustrates that the spectrum of the digital signal consists of repeated images of its analog spectrum. For the sake of simplicity we have assumed that all spectra are real. Figure 3.2 also shows that these images will overlap if the analog signal is not bandlimited or if the sample frequency is too low. In that case, the analog signal cannot be recovered.

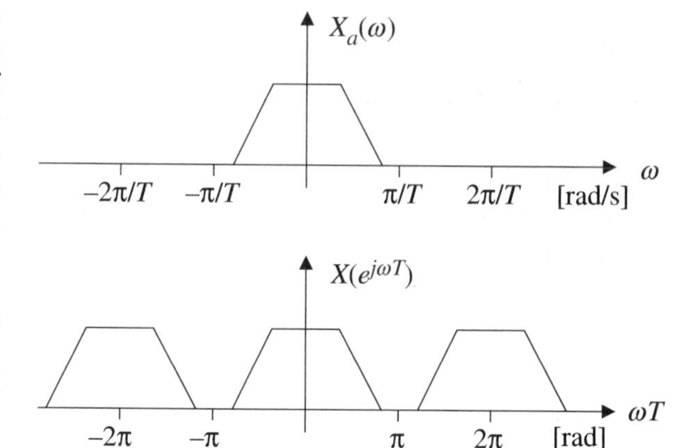

Figure 3.2 Analog and digital spectra obtained by sampling

If the images do not overlap, i.e., if $x_a(t)$ is bandlimited such that $|X_a(\omega)| = 0$ for $|\omega| > \omega_0$, then the analog signal can be reconstructed by filtering with an ideal analog lowpass filter with cutoff frequency $\omega_c = \omega_0 < \pi/T$. We get, after filtering,

$$X_a(\omega) = \begin{cases} TX(e^{j\omega T}) & \text{for } \frac{-\pi}{T} < \omega < \frac{\pi}{T} \\ 0 & \text{otherwise} \end{cases} \tag{3.8}$$

THEOREM 3.1—The Nyquist Sampling Theorem

If an analog signal, $x_a(t)$, is bandlimited so that $|X_a(\omega)| = 0$ for $|\omega| \geq \omega_0$, then all information is retained in the sequence $x(nT)$, obtained by periodic sampling $x_a(t)$ at $t = nT$ where $T < \pi/\omega_0$. Furthermore, the analog signal can, in principle, be reconstructed from the sequence $x(nT)$ by linear filtering.

Information will be lost if the requirements given by the sampling theorem are not met [15, 17]. For example, Figure 3.3 shows the spectrum for an analog sinusoidal signal with a frequency ω_0 that is larger than π/T. The spectrum for the sampled signal will appear to have come from sampling an analog signal of frequency $2\pi/T-\omega_0$. This phenomenon is called *aliasing* or *folding*.

Figure 3.4 shows another example of an analog signal that is not bandlimited. Notice that the high frequency content of the analog signal is folded and that the spectrum of the sampled signal is distorted. In this case, the original analog signal cannot be reconstructed.

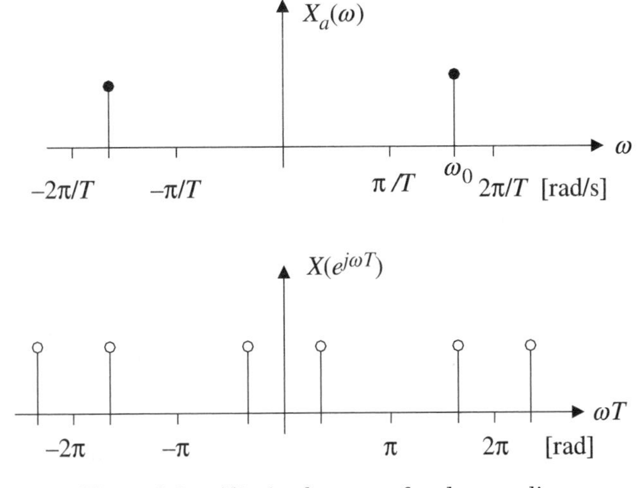

Figure 3.3 Aliasing because of undersampling

Generally, the analog signal must be bandlimited and the sample rate must be larger than twice the bandwidth. However, if the analog signal is bandpass filtered before the sampling, then the sampling frequency may be as low as twice the signal bandwidth [15, 29]. The bandlimit restriction need not be satisfied if the information contained in the digital signal is to be reduced by subsequent digital filtering.

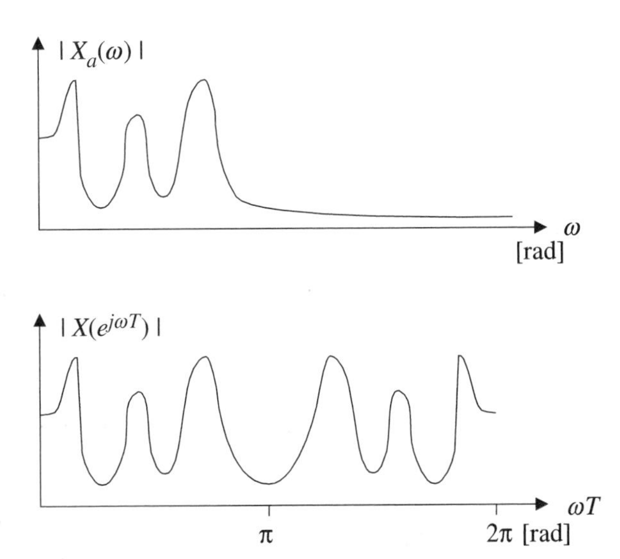

Figure 3.4 Aliasing of non-bandlimited signal

3.7 SELECTION OF SAMPLE FREQUENCY

In this section we will discuss selection of sampling frequency. To ensure that the analog signal is bandlimited, it is necessary to use an anti-aliasing filter before the *sample-and-hold* circuit (S&H) as shown in Figure 3.5. A

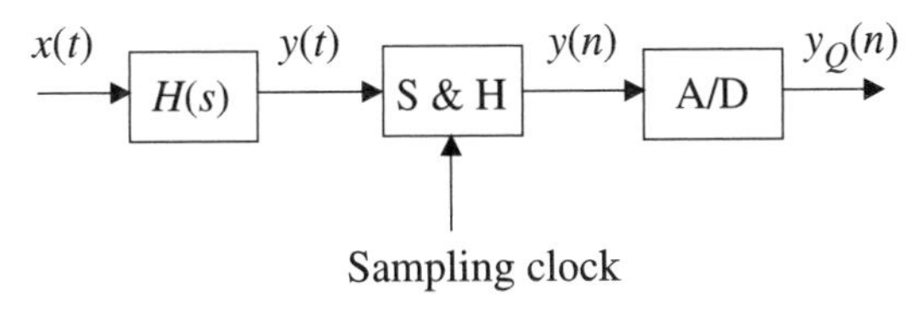

Figure 3.5 Analog-to-digital conversion

S&H circuit should always be used before the A/D converter to get samples at very precise (time) instants, since an A/D converter generally has a small, but significant, variation in its sample instants. This variation is also significant for flash converters. An error in the sampling period translates to an equivalent error in the sample value. High-resolution converters therefore require high precision in the sample instants.

Figure 3.6 shows the spectrum of an analog signal, the magnitude response of the anti-aliasing filter, and the spectrum after the analog filter. The interesting part of the input signal lies in the frequency band $0 - \omega_c$. The output signal of the anti-aliasing filter is bandlimited to the band $0 - \omega_s$. Figure 3.7 shows the spectrum of the corresponding sampled signal, the magnitude response for the digital filter, and the spectrum of the output of the digital filter. Obviously, the frequency response of the digital filter is periodic with a period of 2π. Also notice that aliasing occurs for frequencies above $2\pi - \omega_s T$, since the sample rate; is too low. However, this band will later be removed by the digital filter. In this case, the stopband edge of the anti-aliasing filter must be $\omega s \leq 2\pi/T - \omega_c$ and the passband edge ω_c. Hence, the transition band, $\omega_s - \omega_c$, can be made much larger than in the case when the sampling theorem has to be strictly fulfilled. In this case, the order of the anti-aliasing filter can be significantly reduced.

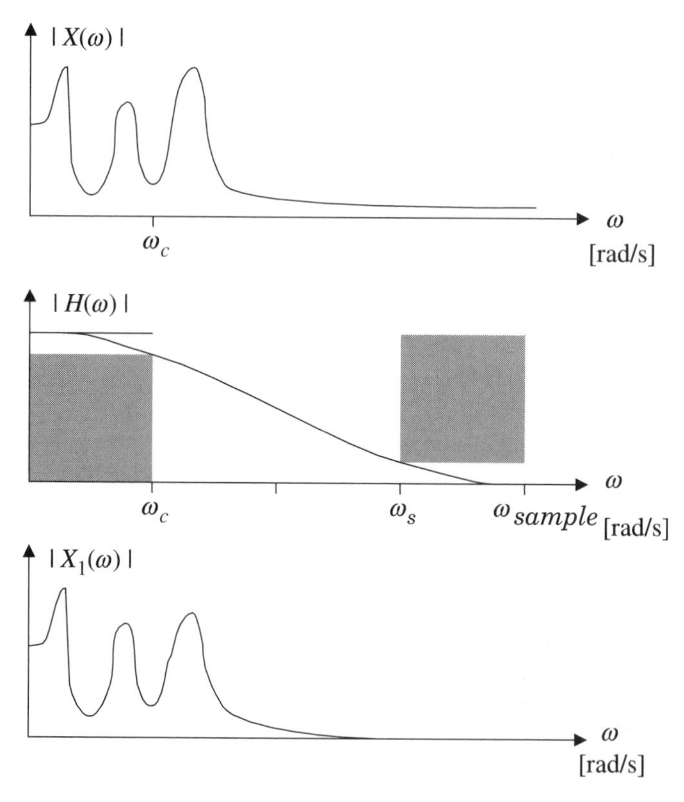

Figure 3.6 Bandlimiting due to analog filtering

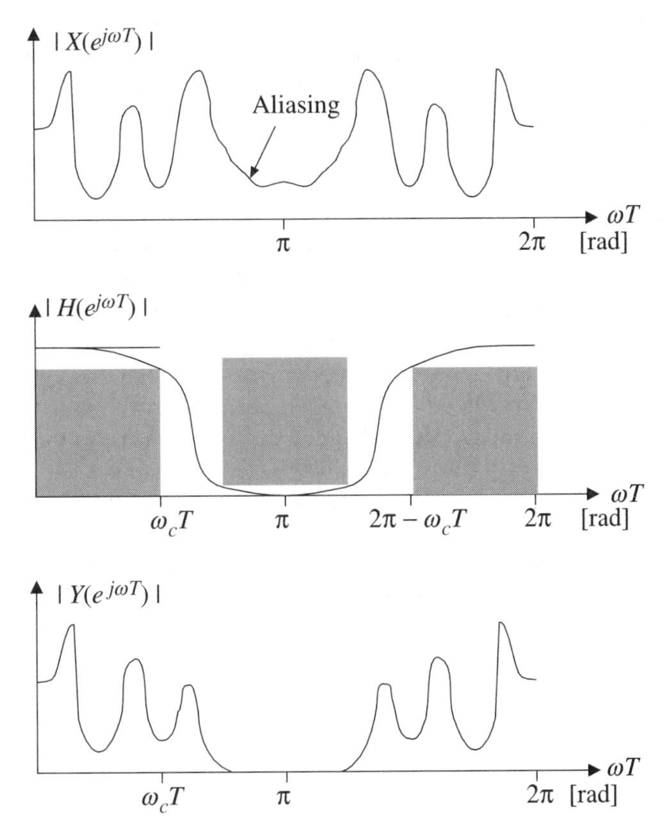

Figure 3.7 Spectrum of sampled signal, the magnitude for the digital filter, and the spectrum of the output signal

3.8 SIGNAL PROCESSING SYSTEMS

The purpose of a signal processing system is generally to reduce or retain the information in a signal. An example of the first case is a surveillance radar in which the received echo signal is contaminated by noise. The aim of the signal processing is to reduce and extract the relevant information and reject unwanted echoes and noise. Usually, the signal processing is merged with feature extraction and intelligent decision-making steps so that only "interesting" targets are called to the attention of the operator. On the other hand, in communication applications, the purpose of the signal processing is often to adapt the signal to the communication channel in order to provide a reliable and secure transmission. In the following, we will review some of the most important properties of discrete-time and digital systems [4, 17, 25, 32, 40]. Note that analog systems have similar properties.

3.8.1 Linear Systems

Most subsystems used in signal processing belong to the class of linear systems. A *linear system* is defined:

> If two proper input signals $x_1(n)$ and $x_2(n)$ yield the output signals $y_1(n)$ and $y_2(n)$, i.e., $x_1(n) \to y_1(n)$ and $x_2(n) \to y_2(n)$, then the system is linear if and only if a $x_1(n)$ + b $x_2(n) \to$ a $y_1(n)$ + b $y_2(n)$ for all *a, b* so that the combined input and output signals are within their proper domains.

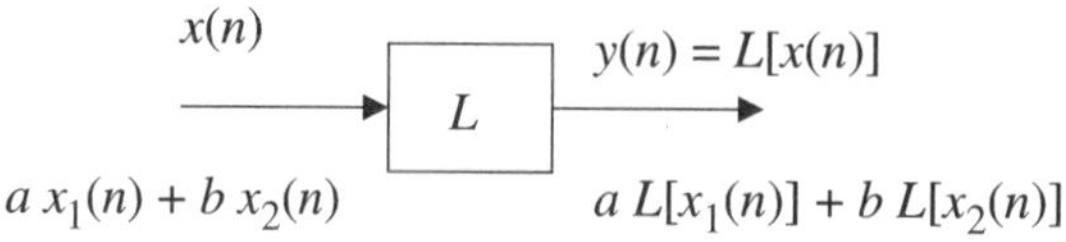

$x(n)$ → L → $y(n) = L[x(n)]$

$a\,x_1(n) + b\,x_2(n)$ → $a\,L[x_1(n)] + b\,L[x_2(n)]$

Figure 3.8 Linear system

A consequence of this property, as illustrated in Figure 3.8, is that the input signal can be expanded into an orthogonal set of basic functions, e.g., Fourier series for a periodic input signal. The effect of each Fourier component can then be determined and the output obtained by summing the contributions from these components.

3.8.2 SI (Shift-Invariant) Systems

Systems can also be classified with respect to their behavior over time. The behavior of a *shift-invariant system* (*SI*), sometimes also called a *time-invariant system* (*TI*), does not vary with time.

A system defined by $x(n) \to y(n)$

is shift-invariant if and only if $x(n - n_0) \to y(n - n_0)$ for all n_0

Thus, an SI system, as illustrated in Figure 3.9, has the same response for a given input, independent of when the input is applied to the system. The output is only shifted the same amount as the input.

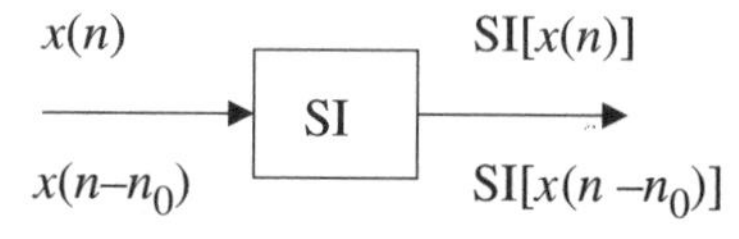

Figure 3.9 Original and shifted input and output signals for a shift-invariant system

Examples of shift-invariant operations and systems are addition, multiplication, normal digital filters, and cosine transforms. An example of a system that is not shift-invariant is a decimator that reduces the sample frequency.

3.8.3 LSI (Linear Shift-Invariant) Systems

A system that is both linear and shift-invariant is called an *LSI system*.[1] The output signal of an LSI system is determined by *convolution* of the input signal and the impulse response.

1. This abbreviation should not be confused with LSI (large scale integration).

$$y(n) = h(n) * x(n) = \sum_{k=-\infty}^{\infty} h(k)x(n-k) = \sum_{k=-\infty}^{\infty} h(n-k)x(k) \tag{3.9}$$

where the *impulse response* is denoted by $h(n)$. The impulse response of a system is obtained when the input signal is an impulse sequence, $\delta(n)$.

EXAMPLE 3.3

Determine the z-transform of $y(n) = \sum_{k=-\infty}^{\infty} h(k)x(n-k)$

We have

$$Y(z) = \sum_{n=-\infty}^{\infty} y(nT)z^{-n} = \sum_{n=-\infty}^{\infty}\sum_{k=-\infty}^{\infty} h(k)x(n-k)z^{-n}$$

$$= \sum_{n=-\infty}^{\infty}\sum_{k=-\infty}^{\infty} h(k)x(n-k)z^{-(n-k)}z^{-k}$$

$$= \sum_{k=-\infty}^{\infty} h(k)z^{-k} \sum_{n-k=-\infty}^{\infty} x(n-k)z^{-(n-k)} = H(z)X(z)$$

Hence

$$Y(z) = H(z)X(z)$$

Instead of computing the convolution, it is often simpler to first compute the product of the z-transforms of the impulse response and the input sequence, and then compute the inverse z-transform of $Y(z)$. The properties of the system, which are represented by $H(z)$, are separated in the z-domain from the properties of the input signal. Hence, we can analyze an LSI system without considering the input signal.

Another important property of LSI systems is that the order of two cascaded LSI systems can be changed, as shown in Figure. 3.10 (see Problem 3.10). This fact will allow us to rearrange the ordering of both DSP building blocks and certain arithmetic operations in order to obtain favorable computational properties.

Figure 3.10 Changing the ordering of two LSI systems

3.8.4 Causal Systems

A system is causal, if the output signal, $y(n_0)$ at the time instant n_0, depends only on the input signal for time values $n \le n_0$.

A *causal system,* as illustrated in Figure 3.11, is defined by

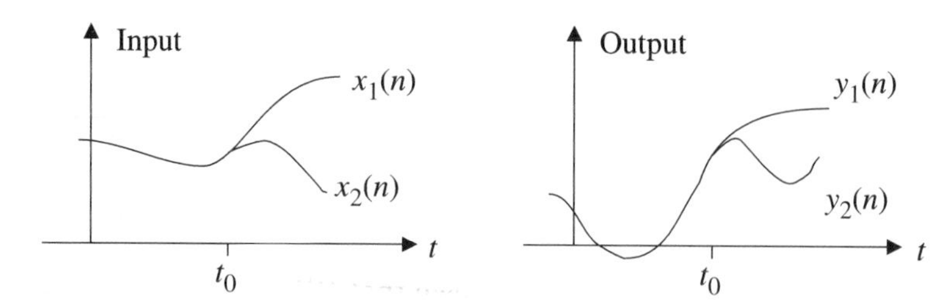

Figure 3.11 Input and output signals for a causal system

If $x_1(t) \equiv x_2(t)$ for $t \le t_0$, then for a causal system

$y_1(t) \equiv y_2(t)$ for $t \le t_0$

For a causal LSI system the impulse response $h(n) = 0$ for $n < 0$ (see Problem 3.9). Note that noncausal systems can be used in practice in non–real-time applications.

3.8.5 Stable LSI Systems

An LSI system is said to be stable in the *bounded-input bounded-output* sense (BIBO) if and only if

$$\sum_{n=-\infty}^{\infty} |h(n)| < \infty \qquad 3.10$$

This is only a weak form of stability. In Chapter 5, we will show that the issue of stability is more complicated than indicated by Equation (3.10). The poles, which will be discussed in section 3.11, of an LSI system are confined within the unit circle in the z-plane.

3.9 DIFFERENCE EQUATIONS

Many discrete-time and digital systems such as digital filters can be described by difference equations with constant coefficients. The input–output relation for an Nth-order LSI system can be described by

$$y(n) = \sum_{k=1}^{N} b_k y(n-k) + \sum_{k=0}^{M} a_k x(n-k) \qquad (3.11)$$

It is customary to distinguish between LSI filters (systems) that have an impulse response of finite or infinite duration. The first type is called FIR *(finite-length impulse response)* filters since the impulse response becomes zero after a finite number of samples. The latter type is called IIR *(infinite-length impulse response)* filters since, even though the impulse response decays toward zero, it theoretically never reaches zero.

If $b_k = 0$ for all k, the system has a finite impulse response and is therefore an FIR system. Normally, the system is an IIR system if $b_k \neq 0$ for at least one k, but there are some exceptions. If any $b_k \neq 0$, the difference equation represents a *recursive algorithm* in which some of the previously computed output values are used to compute the next output value. It can be shown that IIR filters only can be realized by using recursive algorithms while FIR filters can be realized by using recursive or nonrecursive algorithms. However, it is not generally recommended that recursive algorithms be used to realize FIR filters, because of stability problems.

3.10 FREQUENCY RESPONSE

A useful behavioral characterization of an LSI system is to describe the system response for typical input signals. Naturally, of great interest for frequency selective systems is their response to periodic inputs.

The *frequency response*, $H(e^{j\omega T})$, is obtained with a complex sinusoidal input signal, $x(n) = e^{jn\omega T}$. From Equation (3.11) we get for an LSI system

$$y(e^{jn\omega T}) = \sum_{k=-\infty}^{\infty} h(k)x(n-k) = \sum_{k=-\infty}^{\infty} h(k)e^{j\omega(n-k)T} + \sum_{k=-\infty}^{\infty} h(k)e^{-jk\omega T}e^{jn\omega T}$$

$$= H(e^{j\omega T})\, e^{jn\omega T} = H(e^{j\omega T})\, x(n)$$

The frequency response of an LSI system can also be determined from the corresponding difference equations by taking the Fourier transform of both sides of Equation (3.11). We get

$$y(e^{j\omega T}) = \sum_{k=1}^{N} b_k Y(e^{j\omega T})e^{-j\omega kT} + \sum_{k=0}^{M} a_k X(e^{j\omega T})e^{-j\omega kT} \tag{3.12}$$

and

$$H(e^{j\omega T}) = \frac{Y(e^{j\omega T})}{X(e^{j\omega T})} = \frac{\sum_{k=0}^{M} a_k e^{-j\omega kT}}{1 - \sum_{k=1}^{N} b_k e^{-j\omega kT}} \tag{3.13}$$

3.10.1 Magnitude Function

The magnitude function is related to the frequency response according to

$$H(e^{j\omega T}) = \sum_{n=-\infty}^{\infty} h(n)e^{-j\omega nT} = |H(e^{j\omega T})|e^{j\Phi(\omega T)} \tag{3.14}$$

where $|H(e^{j\omega T})|$ is the *magnitude response* and $\Phi(\omega T)$ is the phase response. The frequency response, which is the Fourier transform of the impulse response, is a rational function in $e^{j\omega T}$. The frequency response describes how the magnitude and phase of a sinusoidal signal are modified by the system.

EXAMPLE 3.4

Use MATLAB™ or any other standard filter design program to design a digital Cauer (elliptic) filter meeting the following specification:

Lowpass IIR filter with

Passband ripple: ≤ 1 dB	Stopband attenuation: > 20 dB
Passband edge: 2 kHz	Stopband edge: 2.5 kHz
Sample frequency: 10 kHz	

The following MATLAB-program

```
fcnorm = 2/5;      % The band edges in MATLAB are normalized with
fsnorm = 2.5/5;    % respect to fsample/2, e.g., fcnorm = 2fc/fsample
Amax = 1;
Amin = 20;
[N,Wn] = ellipord(fcnorm, fsnorm, Amax, Amin);
N
[Num, Den] = ellip(N, Amax, Amin, Wn);
[Z, P, K] = tf2zp(Num, Den)
```

yields a filter that has

Filter degree: N = 3
Passband ripple: A_{max} = 1.0 dB
Stopband attenuation: A_{min} = 20.0 dB

Figure 3.12 shows the magnitude response of a third-order lowpass filter of Cauer type. These types of filters will be discussed further in Chapter 4. The magnitude function can be shown in either linear or logarithmic (dB) scales, although the latter is more common. The attenuation, phase, and group delay responses are shown in Figures. 3.13, 3.14, and 3.15, respectively.

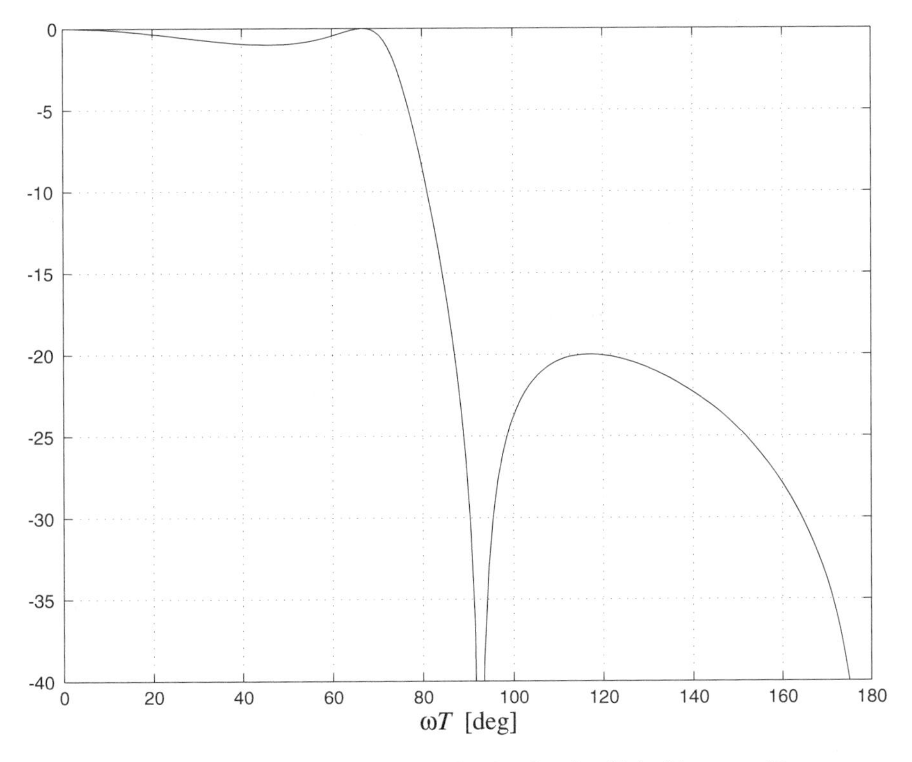

Figure 3.12 Magnitude response for the third-order digital lowpass filter

The pole-zero configuration is shown in Figure 3.16. A Cauer filter has equiripple in both the passband and the stopband. Notice the transmission zeros at $\omega T \approx 96°$ and $\omega T = 180°$, i.e., in the stopband. We will discuss digital filters in more detail in Chapter 4.

3.10.2 Attenuation Function

Instead of using the magnitude function, it is more common to use the *attenuation* function, which is defined

$$A(\omega T) \triangleq -20 \log_{10}(|H(e^{j\omega T})|), \quad \text{[dB]} \tag{3.15}$$

The magnitude function on a logarithmic scale (dB) is the same as the attenuation function, except for the sign. The maximum power that is transferred between the source and load is obtained at the frequencies of the attenuation zeros. There are two attenuation zeros in Figure 3.13: the first at $\omega T = 0$ and the second at about 65°. The transmission zeros correspond to frequencies with infinite attenuation and are therefore called *attenuation poles.*

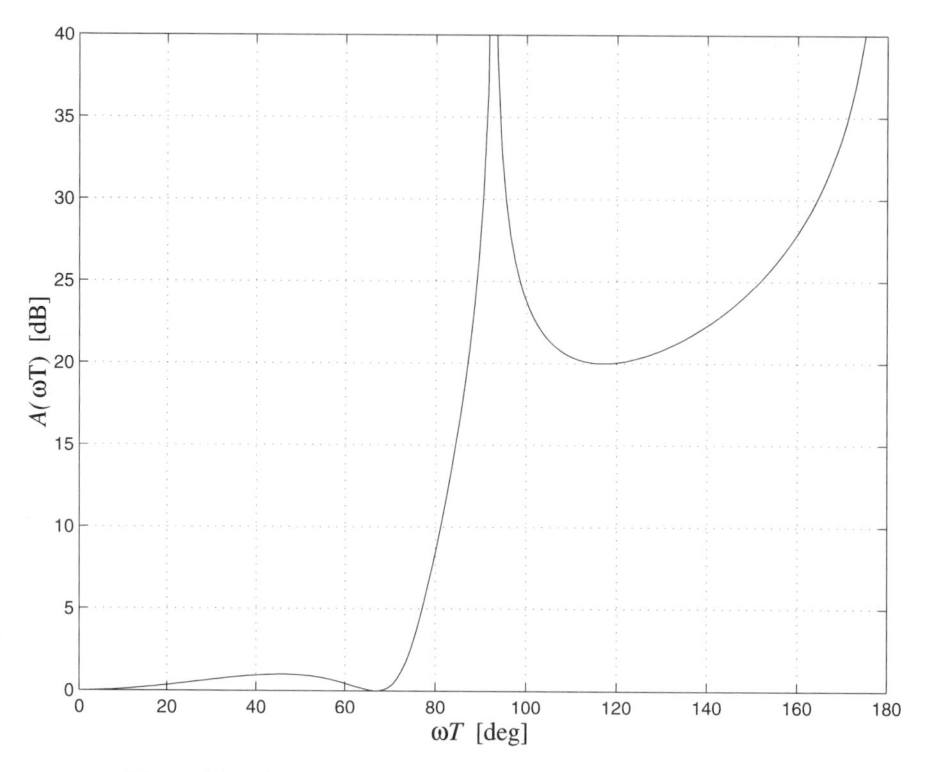

Figure 3.13 Attenuation for the third-order Cauer lowpass filter

3.10.3 Phase Function

The *phase function* of the frequency response is defined

$$\Phi(\omega T) \triangleq \arg\{\, H(e^{j\omega T}) \,\} = \arctan\left\{\frac{Im\{H(e^{j\omega T})\}}{Re\{H(e^{j\omega T})\}}\right\} \tag{3.16}$$

Figure 3.14 shows the phase response for a third-order digital lowpass filter. Notice the discontinuities in the phase response at the frequencies of the transmission zeros, the first at about 96° and the second at 180°, and note that the phase response is almost linear up to about 45°. The jump in the phase response just below 80° is an artifact due to the periodicity of the phase response.

Linear-phase response implies that the impulse response must be either symmetric or antisymmetric. Exact linear-phase response is only possible to realize using FIR filters since an IIR filter cannot have perfect linear-phase response. However, the phase response of an IIR filter can be made arbitrarily close to a linear-phase response within a finite frequency band.

Linear-phase response, which will be further discussed in the next section, is particularly important in applications where the variation in signal delay through the filter is important.

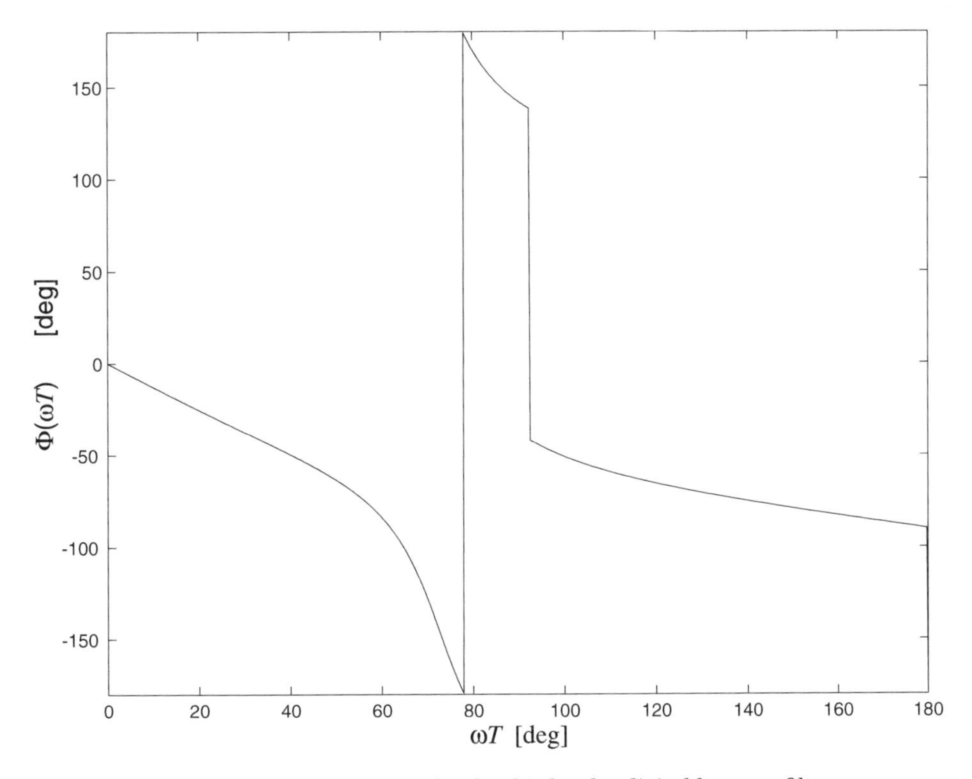

Figure 3.14 Phase response for the third-order digital lowpass filter

3.10.4 Group Delay Function

An important issue in many applications is the delay associated with the processing of a signal. A common measure of delay is the *group delay*, or envelop delay, which is defined as

$$\tau_g(\omega T) \triangleq -\frac{\partial \Phi(\omega T)}{\partial \omega} \tag{3.17}$$

The group delay should be constant in applications where the waveform of the signal is important—for example, in systems for obtaining an ECG (electrocardiogram). Images are particularly sensitive to variations in the group delay, but relatively insensitive to variations in the magnitude function. Figure 3.15 shows the group delay for the Cauer filter that was designed in Example 3.4.

The group delay can be expressed in terms of the transfer function which is defined in section 3.11 (see Problem 3.16):

$$\tau_g(\omega T) = -T\, Re\left\{z\frac{d}{dz}\ln(H(z))\right\} \qquad \text{for } z = e^{j\omega T} \tag{3.18}$$

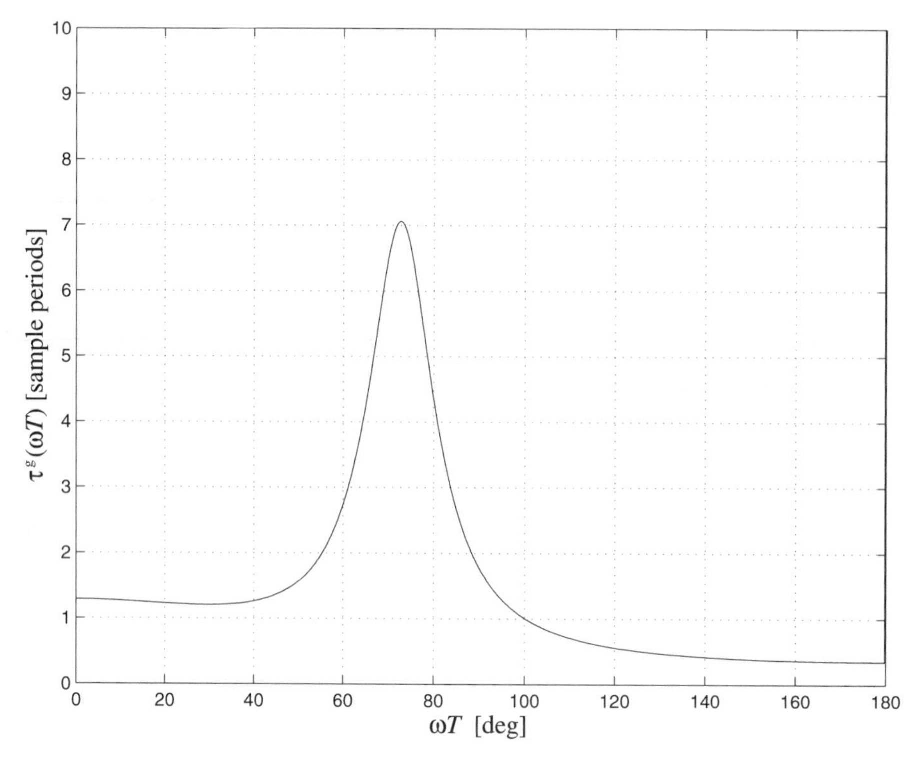

Figure 3.15 Group delay for the third-order digital lowpass filter

It should be noted that transmission zeros on the unit circle do not affect the group delay, and that the group delay is proportional to the sample period. Hence, we have

$$\tau_g(\omega T) = -\frac{\partial \Phi(\omega T)}{\partial(\omega T)} = Tf(\omega T)$$

The group delay can therefore be reduced by increasing the sampling frequency, but the system has to be adjusted accordingly.

3.11 TRANSFER FUNCTION

A behavioral description of an LSI system is the *transfer function,* which can be obtained by applying the z-transform to both sides of Equation (3.11). We get

$$H(z) = \frac{Y(z)}{X(z)} = \frac{\sum_{k=0}^{M} a_k z^{-k}}{1 - \sum_{k=1}^{N} b_k z^{-k}} \tag{3.19}$$

The transfer function for an LSI system is a rational function in z and can therefore be described by a constant gain factor and the roots of the numerator and denominator. For example, the transfer function of the digital lowpass filter designed in Example 3.4 is

$$H(z) = \frac{0.18675628(z+1)(z^2+0.09021784\ z+1)}{(z-036278303)(z^2-0.51108872z+0.73629559)},\ |z| > R_+$$

$$R_+ = \sqrt{0.73629559} = 0.85807668$$

The roots of the numerator are called *zeros*, since no signal energy is transmitted to the output of the system for those values in the z-plane.

The roots of the denominator are called *poles*. For a causal, stable system, the poles are constrained to be inside the unit circle while the zeros can be situated anywhere in the z-plane. Most zeros are in practice positioned on the unit circle, i.e., in the stopband of the filter, in order to increase the attenuation in the stopband. The poles for the digital lowpass filter designed in Example 3.4 are

$$z_{p1} = 0.36278303$$
$$z_{p2,3} = 0.25554436 \pm j\,0.81914142$$

and the zeros are:

$$z_{n1} = -1$$
$$z_{n2,3} = -0.04510892 \pm j\,0.99898207$$

Figure 3.16 shows the poles and zeros as well as the region of convergence. Another common case occurs in allpass filters where the zeros are placed outside the unit circle. Each zero has a corresponding pole mirrored in the unit circle, so that

$$z_{zero} = \frac{1}{z_{pole}}$$

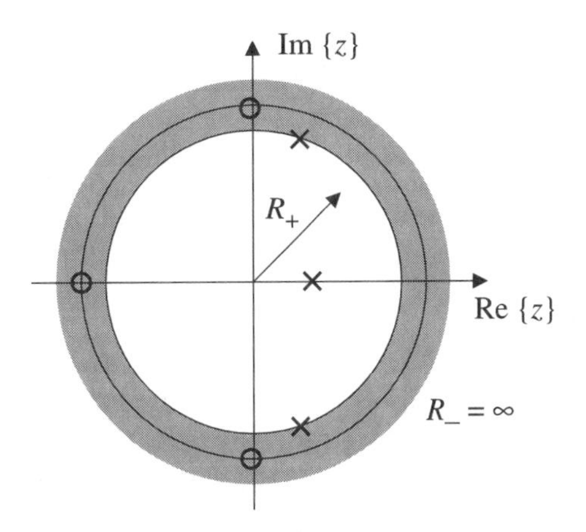

Figure 3.16 Pole-zero configuration for the third-order digital lowpass filter

3.12 SIGNAL-FLOW GRAPHS

Digital filter algorithms can be described by a system of difference equations. An alternative representation uses signal-flow graphs which more clearly illustrate the computational properties of an algorithm than difference equations. This is also the reason schematics are used to represent electrical networks rather than systems of differential-integral equations.

The variables (signals) in a *signal-flow graph* are represented by the nodes. Relations between the variables (nodes) are represented by directed branches with associate transmittances. The value of a node variable is the sum of signal values entering the node. These signal values are obtained by multiplying the node values by the corresponding transmittances. Figure 3.17 illustrates a simple signal-flow graph in both the time and frequency domains and the corresponding sets of

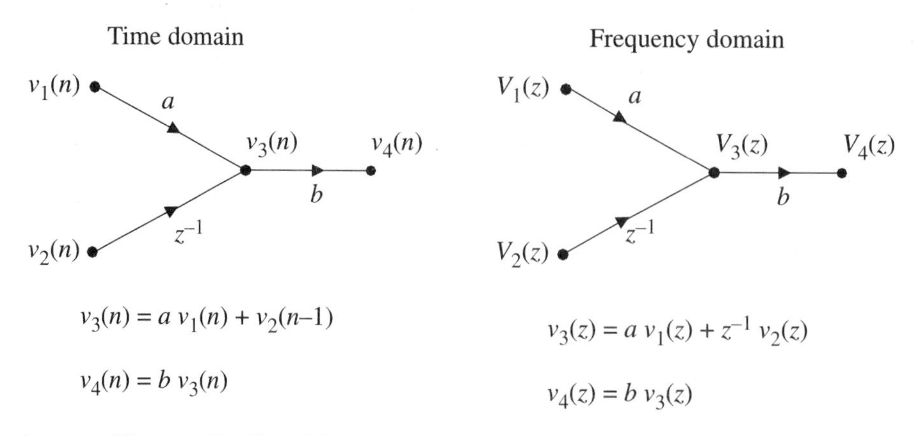

Figure 3.17 Signal-flow graphs in the time and frequency domains

equations. Note that, although not strictly correct, the delay operator z^{-1} is used in both the time and frequency domains.

The signal-flow graph is a structural description of the algorithm in terms of its operations and their ordering, whereas the transfer function is a behavioral description. Systematic procedures for deriving the transfer function are given in [32, 40]. For more general algorithms it is customary to use instead so-called *block diagrams* that resemble signal-flow graphs. The main differences are that the branches and transmittances have been replaced by blocks containing their behavioral descriptions, and that the summation nodes are denoted explicitly.

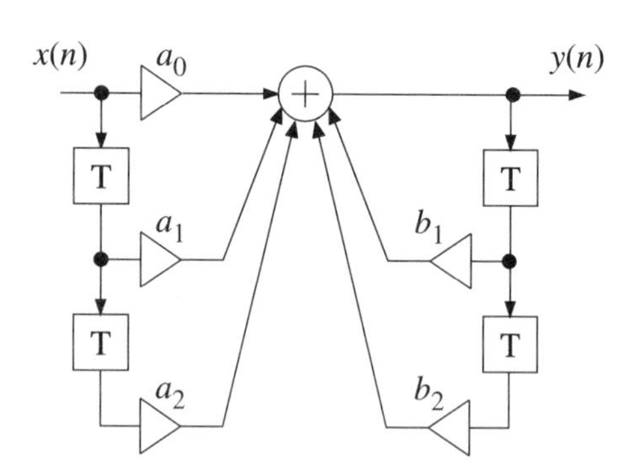

Figure 3.18 Direct form I

Figure 3.18 shows a typical block diagram for a digital filter structure called *direct form I*. For the sake of simplicity we denote the delay elements by T for block diagrams in both the time and frequency domains in order to indicate explicitly the sample period. Many textbooks use instead the shift operator z^{-1} in both domains.

3.13 FILTER STRUCTURES

A transfer function can be realized using many different algorithms. They may differ with respect to the number of arithmetic operations needed, sensitivity to deviations in the coefficient values, or influence of round-off errors in the arithmetic operations. The most important criterion for selecting a filter structure is its robustness. That is, the algorithm should always be stable even under nonlinear conditions. This issue will be further discussed in Chapter 5. Filter algorithms are usually described by signal-flow graphs instead of systems of difference equations. Such signal-flow graphs are often called filter structures.

EXAMPLE 3.5

Determine the transfer function for the filter structure direct form II. Is it an IIR or an FIR filter?

The signal-flow graph shown in Figure 3.19 represents the following set of difference equations:

$$\begin{cases} u(n) = c\,x(n) + b_1\,u(n\!-\!1) + b_2\,u(n\!-\!2) \\ y(n) = a_0\,u(n) + a_1\,u(n\!-\!1) + a_2\,u(n\!-\!2) \end{cases}$$

Taking the z-transform, we get

$$\begin{cases} U(z) = c\,X(z) + b_1\,z^{-1}\,U(z) + b_2\,z^{-1}U(z) \\ Y(z) = a_0\,U(z) + a_1\,z^{-1}\,U(z) + a_2\,z^{-1}U(z) \end{cases}$$

and eliminating $U(z)$ yields the transfer function

$$H(z) = \frac{Y(z)}{X(z)} = \frac{c(a_0 z^2 + a_1 z + a_2)}{z^2 - b_1 z - b_2}$$

A stable filter has all its poles inside the unit circle. Hence, for complex conjugate poles we must have $|b_2| < 1$. The algorithm contains recursive loops since at least one of b_1 and b_2 is nonzero. Hence, if an impulse sequence is used as input signal, it will inject a value into the recursive loop. This value will successively decay toward zero for a stable filter, but, in principle, it will never reach zero. Hence, the impulse response has infinite length and the structure is an IIR filter.

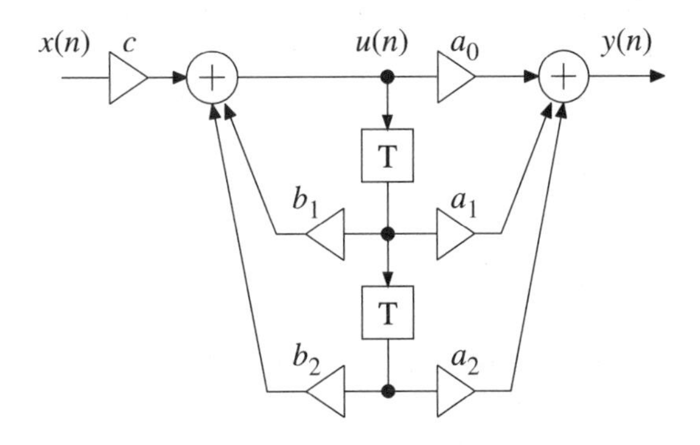

Figure 3.19 Direct form II

Some structures, such as frequency-sampling FIR structures, have finite-length impulse responses, even though they are recursive. These structures are not recommended due to severe stability problems and high coefficient sensitivities [4, 32, 40].

Alternative filter structures can be obtained by using the transposition theorem [4, 15, 32]. In essence, a new filter structure is obtained by reversing the direction of all branches in a signal-flow graph and changing inputs to outputs and vice versa. This new filter structure has the same transfer function as the original filter, but the numerical properties are generally different.

Theorem 3.2—Transposition

The transfer function of a signal-flow graph is invariant if the direction of all branches is reversed and the inputs and outputs are interchanged.

A *dual* network performs the complementary function of the original network [15]. For example, demodulators and modulators are dual operations. Given a network, its dual is found by transposing the network. Any linear time-invariant network has a dual. Interpolators and decimators, which will be further discussed in Chapter 4, perform dual operations.

Two operations are said to be *commutating* if the result is independent of the order in which the operations are performed. We will later show that, in some cases, changing the order of two operations can provide significant computational savings [15].

3.14 ADAPTIVE DSP ALGORITHMS

Adaptive DSP algorithms are used in many applications, as illustrated here. Figure 3.20a illustrates the use of an adaptive filter to cancel an unwanted (noise) component, for example, to attenuate the noise inside a moving automobile with respect to speech signals (see Problem 3.21).

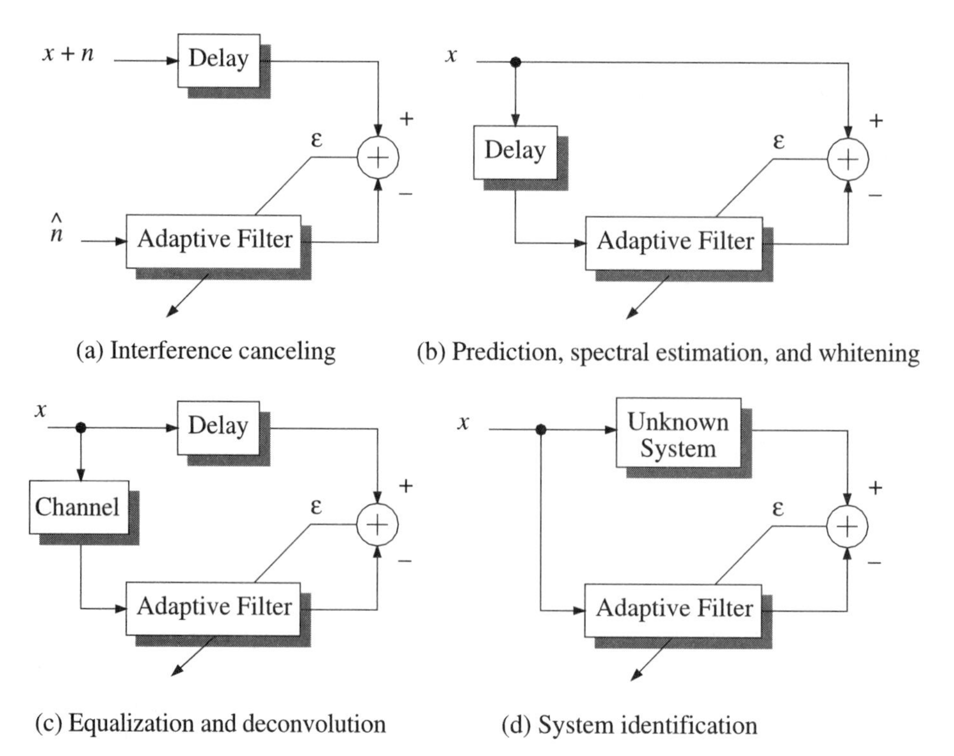

Figure 3.20 Adaptive filter applications

Figure 3.20b illustrates another group of related applications (prediction, spectral estimation, and spectral whitening), while Figure 3.20c shows the setup for equalization and deconvolution. Equalizers are required in high-speed modems to compensate for channel distortion when data are to be transmitted over a radio channel. This type of filter is therefore called a channel equalizer. Finally, Figure 3.20d depicts the system identification problem, which differs from the previous cases in that the filter coefficients are of interest, whereas in the former cases the output of the adaptive filter or error signal is the relevant result. Other applications involving adaptive DSP algorithms are echo cancellers, speech and image coders, beamforming in sensor arrays, system modeling, and control systems [12, 17, 23, 32, 36].

Both FIR and IIR filters are useful for adaptive filters, but FIR filters are more widely used because they are much simpler. The FIR filters have only adjustable zeros and they are therefore always stable. However, the stability of interest in adaptive filters is the proper adjustment of the filter coefficients. In this sense, adaptive FIR filters are not always stable. A drawback of using only FIR filters is that the required filter degree may become large if the channel characteristics are unfavorable. In such cases an adaptive IIR filter may be appropriate.

An important consideration in designing adaptive filters is the criterion used for optimizing the adjustable filter coefficients. Here we will discuss the most frequently used criteria: the least mean square error and the least square error.

3.14.1 LMS (Least Mean Square) Filters

Equalization of a transmission channel (for example, a radio channel for a mobile phone employed for digital transmission) can be achieved by placing an equalizing filter that compensates for nonideal channel characteristics in front of the receiver. Figure 3.21 shows the main block in a such a transmission system. The problem is that the time-variant channel distorts the transmitted pulses so that they interact and complicate their detection. The role of the adaptive filter is to compensate for the distortion so that the intersymbol interference is eliminated.

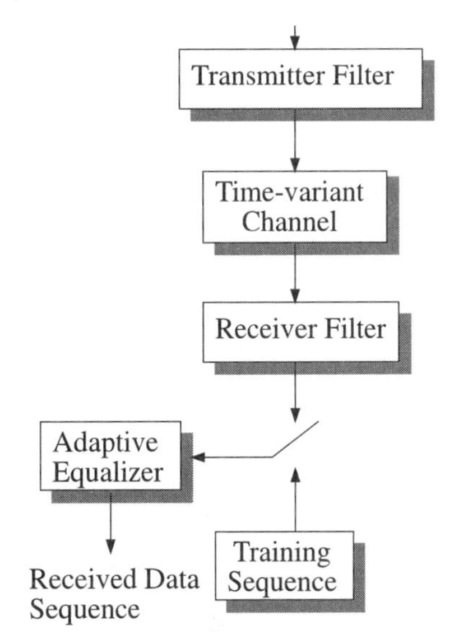

Figure 3.21 Typical transmission system

The principle of an adaptive filter algorithm is based on minimization of the difference between the filter output and a reference sequence. The filter coefficients are updated according to some algorithm so that the error is minimized. The least mean square (LMS) error criterion is often used because of its low computational complexity, but algorithms with similar complexity, using the least squares (LS) criterion, have recently become available [24, 32, 36].

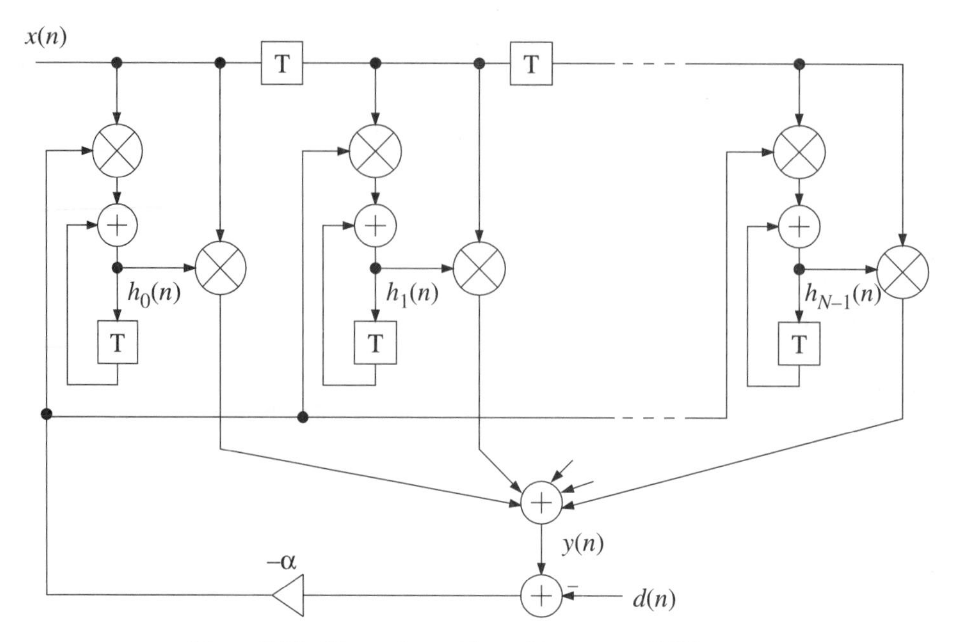

Figure 3.22 Channel equalizer of transversal FIR type

Figure 3.22 shows the structure of such an equalizer based on the LMS algorithm[2]. Generally, the signals may be complex, but here we assume, for the sake of simplicity, that they are real.

For each new input sample $x(n)$ to the FIR filter, a value $y(n)$ is computed. After subtraction of the desired, ideal response $d(n)$ of the adaptive filter, the stochastic gradient is formed by multiplying the weighted error $\alpha[y(n) - d(n)]$ by the received input samples $x(n - k)$, $k = 0, 1, \ldots, N - 1$. The values obtained are then added to the accumulators containing the impulse response values of the FIR filter.

This version of adaptive filter is called a *stochastic-gradient equalizer* [17]. Updating the filter coefficients proceeds according to

$$h_k(n+1) = h_k(n) - \alpha[y(n) - d(n)]\, x(n-k), \; k = 0, 1, \ldots, N-1$$

The operation of the equalizer begins with a training sequence. Typically a periodic sequence whose period equals the length of the FIR filter is sent to the equalizer through the channel. The same training sequence, properly delayed, is generated at the receiver and fed into the equalizer. Finally, when the equalizer has converged, the equalizer is switched into the normal mode of operation. In this mode the output values $y(n)$, which should now be close to the correct values if the channel was properly equalized, are fed back into the equalizer.

The work load involved per sample for an N-tap equalizer is $2N + 3$ multiplications and $2N + 1$ additions. Typically, N is in the range 8 to 32. Typically, 20 to 24 bits are used for the filter coefficients in order to provide the necessary dynamic

2. Also known as the Widrow–Hoff LMS algorithm (1960).

range for the adaptation process. However, only about 12 bits of the coefficients are used in multiplications within the FIR filter.

Several variations on this theme have been proposed and some have even been incorporated into commercial modems. The variations try to

- Simplify the algorithm,
- Improve the rate at which the tap values converge to their optimum values, and
- Make the steady-state tap values closer to their ideal values.

The work load can be reduced by simplifying the computation of the gradient $\alpha[y(n) - d(n)]x(n - k)$ by quantizing one or both of the two last factors to ± 1. Further, the forgetting factor α may be selected equal to a power of 2 so that it can be realized by a simple shift operation. The penalty for these simplifications is, however, a significant decrease in the rate of convergence. It can be shown that a large value of α, that depends on the channel will improve the rate of convergence, but the algorithm becomes unstable if too large a value is used. The steady-state errors in the tap values are proportional to α^2. Hence, a small value should be preferable in order to improve the accuracy of the tap values adjustment. The LMS algorithm is appropriate for slowly time-variant channels.

More recently, an equalizer that does not require an initial training period has become popular. In fact, the algorithm for adjusting the tap values does not depend on the received input sequence.

3.14.2 RLS (Recursive Least Square) Lattice Filters

Equalizers with short training times are required in many applications. A rapidly convergent algorithm is the *fast Kalman algorithm* [23, 32]. The price for this improvement is the increased complexity and sensitivity to round-off errors in the computations. The adaptation may also become unstable. The fast Kalman algorithm requires many more operations then the stochastic-gradient algorithm. The work load per input sample is about $10N$ multiplications, $9N$ additions, and $2N$ divisions. Note that division is an expensive (i.e., time-consuming) operation. Some adaptive algorithms involve square root operations which are even more expensive.

An important class of adaptive algorithms is the *RLS (recursive least square) filters*. RLS algorithms are usually realized by using so-called *lattice filters*[3]. There exist many variations of the basic RLS algorithm [33, 37]. Box 3.1 shows the so-called "direct update (error-feedback) form of an a priori RLS lattice algorithm."

The sequences $f_m(n)$ and $b_m(n)$ are the forward and backward error residuals while $Kf_m(n)$ and $Kb_m(n)$ are the corresponding reflection coefficients. The index m refers to the stage of the filter. Typically, the order of the filter, N, is in the range 8 to 12 for speech coders, but much higher filter orders may be required for channels with long delays. Besides its rapid convergence, the lattice filter is relatively insensitive to round-off errors incurred in the computations. Hence, the computations can be made using a short data word length.

---

3. This type of lattice filters should not be confused with lattice wave digital filters, which will be discussed in Chapter 4.

{ Initialization }

$\omega < 1$ (forgetting factor)

$\alpha_0(n-1) = 1, e_0(n) = d(n), f_0(n) = b_0(n) = x(n)$

$Ef_0(n) = E^b{}_0(n) = w\, E^f{}_0(n-1) + x(n)^2$

$\alpha_m(-1) = 1, K^f{}_m(-1) = K^b{}_m(-1) = 0$

$E^b{}_m(-1) = E^f{}_m(0) = \varepsilon > 0$ (to avoid division with zero)

$n = 1$

for $m = 0$ **to** $N-2$ **do**

begin

$$Kf_{m+1}(n-1) = Kf_{m+1}(n-2) - \frac{\alpha_m(n-2)f_m(n-1)b_m(n-2)}{E^b{}_m(n-2)}$$

$$K^b{}_{m+1}(n-1) = K^b{}_{m+1}(n-2) - \frac{\alpha_m(n-2)f_{m+1}(n-1)b_{m+1}(n-1)}{E^f{}_m(n-1)}$$

$f_{m+1}(n) = f_m(n) + Kf_{m+1}(n-1)\, b_m(n-1)$ (forward residue)

$b_{m+1}(n) = b_m(n-1) + K^b{}_{m+1}(n-1)\, f_m(n)$ (backward residue)

$$Ef_{m+1}(n-1) = w\, Ef_{m+1}(n-2) + \alpha_{m+1}(n-2)\, f_{m+1}^2(n-1)$$

$$\alpha_{m+1}(n-1) = \alpha_m(n-1) - \frac{\alpha_m^2(n-1)b_m^2(n-1)}{E^b{}_m(n-1)}$$

$$E^b{}_{m+1}(n-1) = w\, E^b{}_{m+1}(n-2) + \alpha_{m+1}(n-2)\, b_m^2(n-1)$$

end;

for $m = 0$ **to** $N-1$ **do**

begin

$$x_m(n-1) = x_m(n-2) - \frac{\alpha_m(n-2)b_m(n-1)e_{m+1}(n-1)}{E^b{}_m(n-1)}$$

$$e_{m+1}(n-1) = e_m(n) + x_m(n-1)\, b_m(n)$$

$$\hat{d}(n) = -\sum_{k=1}^{N-1} X_k(n-1)b_k(n)$$

end;

Box. 3.1. Pseudo-code for the direct update form of RLS lattice filter

3.15 DFT—THE DISCRETE FOURIER TRANSFORM

The discrete Fourier transform (DFT) has played an important historic role in the evolution of digital signal processing techniques. It has opened up new signal processing techniques in the frequency domain, which are not easily realizable in the analog domain [9, 10, 14].

The *discrete Fourier transform* (*DFT*) is defined as

$$X(k) \triangleq \sum_{n=0}^{N-1} x(n)W^{nk}, \quad n = 0, 1, \ldots, N-1 \tag{3.20}$$

and the inverse discrete Fourier transform (IDFT) is

$$x(n) = \frac{1}{N} \sum_{k=0}^{N-1} X(k) W^{-nk}, \quad n = 0, 1, \ldots, N-1 \tag{3.21}$$

where $W = e^{-j2\pi/N}$. Generally, both $x(n)$ and $X(k)$ are complex sequences of length N. Also note that the discrete Fourier transform is a proper transform. It is not an approximation of the discrete-time Fourier transform of a sequence that was discussed earlier. However, the two transforms are closely related.

A direct computation of the DFT or the IDFT, according to the program shown in Box 3.2, requires N^2 complex arithmetic operations.

```
Program Direct_DFT;
var
  x, Y: array[0..Nminus1] of Complex;
begin
  for k := 0 to N-1 do
    begin
      Y[k] := x[0];
      for n := 1 to N-1 do
        Y[k] := Y[k] + W^nk * x[n];
    end;
end.
```

Box 3.2. Direct computation of the DFT.

The time required for just the complex multiplications in a 1024-point DFT is

$$T_{\text{mult}} = 10242 \cdot 4 \cdot 100 \text{ ns} = 0.419 \text{ s}$$

where we have assumed that one complex multiplication corresponds to four real multiplications and the time required for one real multiplication is 100 ns. This corresponds to a sample frequency of only 2.44 kHz. Hence, the direct computation approach is limited to comparatively short sequences due to the large computation time.

3.16 FFT—THE FAST FOURIER TRANSFORM ALGORITHM

In 1965, Cooley and Tukey developed a fast algorithm which requires only $O(N\log_2(N))$ operations [14]. The difference in execution time between a direct computation of the DFT and the new algorithm is very large for large N. For example, the time required for just the complex multiplications in a 1024-point FFT is

$$\begin{aligned} T_{\text{mult}} &= 0.5\, N\log_2(N) \cdot 4 \cdot 100 \text{ ns} = \\ &= 0.5 \cdot 1024 \log_2(1024) \cdot 4 \cdot 100 \text{ ns} = 2.05 \text{ ms} \end{aligned}$$

This corresponds to a sample frequency of 500 kHz. The sequence lengths in typical applications are in the range 8 to 4096.

Several variations of the Cooley–Tukey algorithm have since been derived. These algorithms are collectively referred to as the FFT (*fast Fourier transform*). Note that the FFT is not a new transform, it simply denotes a class of algorithms for efficient computation of the discrete Fourier transform.

Originally, the aim in developing fast DFT algorithms was to reduce the number of fixed-point multiplications, since multiplication was more time consuming and expensive than addition. As a by-product, the number of additions was also reduced. This fact is important in implementations using signal processors with floating-point arithmetic, since floating-point addition requires a slightly longer time than multiplication.

Because of the high computational efficiency of the FFT algorithm, it is efficient to implement long FIR filters by partitioning the input sequence into a set of finite-length segments. The FIR filter is realized by successively computing the DFT of an input segment and multiplying it by the DFT of the impulse response. Next, an output segment is computed by taking the IDFT of the product. Finally, the output segments are combined into a proper output sequence. Note that the DFT of the impulse response needs to be computed only once. This method is competitive with ordinary time domain realizations for FIR filters of lengths exceeding 50 to 100 [6, 9, 10, 21].

3.16.1 CT-FFT—The Cooley–Tukey FFT

Box 3.3 shows a Pascal program for the CT-FFT the Cooley–Tukey FFT. This version of the FFT is called *decimation-in-time* FFT [9, 10], because the algorithm is based on the divide-and-conquer approach that is applied in the time domain. We will only discuss so-called radix-2 FFTs with a length equal to a power of 2. The radix-4 and radix-8 FFTs are slightly more efficient algorithms [7, 9, 10, 21, 33]. The two routines *Digit-Reverse* and *Unscramble* are shown in Boxes 3.4 and 3.5, respectively.

```
Program CT_FFT;
const
  N = 1024; M = 10; Nminus1 = 1023;{ N = 2^M }
type
  Complex = record
    re : Double; im : Double;
  end;
  C_array : array[0..Nminus1] of Complex;
var
  Stage, Ns, M1, k, kNs, p, q : integer;
  WCos, WSin, TwoPiN, TempRe, TempIm : Double;
  x : C_array;
begin
  { READ INPUT DATA INTO x }
  Ns := N; M1 := M;
  TwoPiN := 2 * Pi/N;
  for Stage := 1 to M do
    begin
      k := 0;
      Ns := Ns div 2;
      M1 := M1 – 1;
```

```
        while k < N do
            begin
                for q := 1 to Ns do
                    begin
                        p := Digit_Reverse( k* 2^M1);
                        WCos := cos(TwoPiN * p);          {W to the power of p}
                        WSin := -sin(TwoPiN * p);         {W = exp(-j2π/N)}
                        kNs := k + Ns;
    {Butterfly}             TempRe := x[kNs].re * WCos - x[kNs].im * WSin;
                            TempIm := x[kNs].im * WCos + x[kNs].re * WSin;
                            x[kNs].re := x[k].re - TempRe;
                            x[kNs].im := x[k].im - TempIm;
                            x[k].re := x[k].re + TempRe;
                            x[k].im := x[k].im + TempIm;
                        k := k + 1;
                    end;
                k := k + Ns;
            end;
    end;
    Unscramble; { OUTPUT DATA STORED IN x }
end.
```

Box 3.3. The Cooley–Tukey FFT. Decimation-in-time

```
function Digit_Reverse(Digit: Integer) : Integer;
var
    N, q, NewAddr, Rmbit, OldAddr : Integer;
begin
    NewAddr := 0;
    OldAddr := Digit;
    for q := 1 to M do
        begin
        Rmbit := OldAddr mod 2;
        OldAddr := OldAddr div 2;
        if Rmbit = 1 then
            NewAddr := NewAddr * 2 + 1
        else
            NewAddr := NewAddr + NewAddr;
        end;
    Digit_Reverse := NewAddr;
end;
```

Box 3.4. Digit reversal

A signal-flow graph for the Cooley–Tukey FFT for $N = 8$ is shown in Figure 3.23. The arrows represent multiplication with complex coefficients, so-called *twiddle factors*, of the form W^p, where $W = e^{-j2\pi/N}$ and p = an integer.

As just mentioned, the FFT is based on a divide-and-conquer approach. The N-point DFT is first divided into two $N/2$-point DFTs. These DFTs are then divided into four $N/4$-point DFTs, and so on. The quantity of arithmetic operations is reduced to $O(N\log_2(N))$.

```
procedure Unscramble;
var
  temp : Complex;
  k, q : integer;
begin
  for k := 0 to N – 1 do
    begin
      q := Digit_Reverse(k);
      if q > k then
        begin
          temp := x[k];
          x[k] := x[q];
          x[q] := temp;
        end;
    end;
end;
```

Box 3.5. Unscrambling

The algorithm consists of $M = \log_2(N)$ stages and a final reordering of the output sequence. The last stage, which is called *unscrambling*, is not needed in all applications [22, 35]. Each stage consists of $N/2$ basic signal-flow graphs of the type shown in Fig. 3.24.

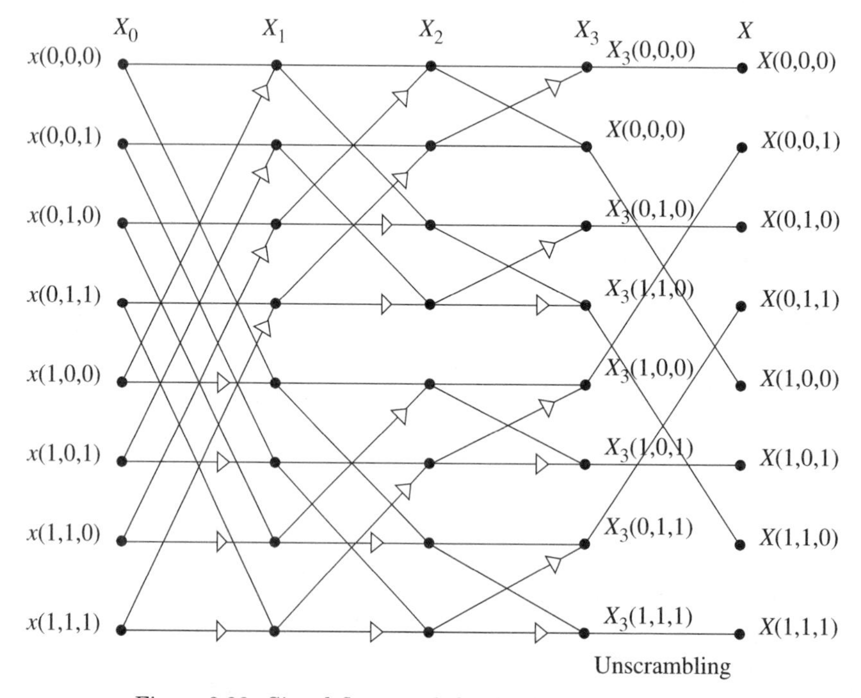

Figure 3.23 Signal-flow graph for the Cooley–Tukey's FFT

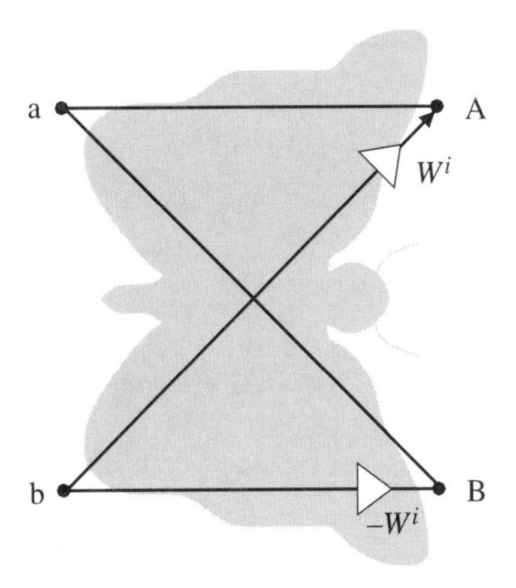

Figure 3.24 Butterfly for decimation-in-time

Note that one of the multiplications can be eliminated by a simple modification of the butterfly. Due to its appearance this flow graph is called a *butterfly.* Most hardware implementations of the FFT are based on one or several processors that implement the butterfly operations. Today, the DFT plays an important role as a building block in many digital signal processing systems. Both special purpose hardware and processors have been developed in order to facilitate real-time signal processing. Typically, a complex sequence of length $N = 1024$ can be transformed in 1 to 5 ms using standard signal processors.

Table 3.1 shows benchmarks for some standard signal processors. Note that the benchmarks do not include time for input and output of the sequences and that they may be measured under slightly different conditions and that processors with different clock frequencies are available.

Company	Model	Description	1024-point FFT [ms]
Analog Devices	ADSP-2101	16/40-bit fixed-point	2.26
	ADSP-2120	32-bit floating-point	0.77
	ADSP-21020	32-bit floating-point	0.58
AT&T	DSP32C	32-bit floating-point	2.9
Motorola	DSP56002	24-bit fixed-point	4
	DSP96002	32-bit floating-point	1.13
Texas Instruments	TMS320C25	16-bit fixed-point	15.9
	TMS320C30	32-bit floating-point	3.0
	TMS320C40	32-bit floating-point	1.0

Table 3.1 Benchmarks for standard signal processors

Table 3.2 shows benchmarks for a few commercial FFT processors. These high-performance processors generally require several support chips, (for example, external memory) to perform the FFT. Special-purpose hardware can reduce the execution time by more than an order of magnitude.

Company	Model	Description	1024-point FFT [μs]
Plessey	PDSP16510A	16-bit block floating-point	98
Array Microsystems	DaSP Chip set	Processor and Controller	130
Butterfly DSP Inc.	BDSP9124	24-bit fixed-point	54
France Telecom. CNET	FT-VD-FFT8K-4	12-bit block floating-point	52
Sharp	LH9124L LH9320LU-25	24-bit fixed-point	129

Table 3.2 Benchmarks for some FFT processors

EXAMPLE 3.6

Derive the Cooley–Tukey FFT (i.e., the decimation-in-time algorithm), for $N = 4$. Hint: Use the following binary representations:

$$k = 2k_1 + k_0$$

and

$$n = 2n_1 + n_0$$

where k_1, k_0, n_1, and n_0 are either 0 or 1.

The DFT is

$$X(k) = \sum_{n=0}^{3} x_0(n) W^{nk}, \quad k = 0, 1, 2, 3$$

Using this notation and noting that the sum over the time index, n, can be written as a double sum over n_1 and n_0

$X(k_1, k_0) = x_0(n_1, n_0)\, W^{nk}$

with

$$nk = (2n_1 + n_0)(2k_1 + k_0) = 4k_1 n_1 + 2k_0\, n_1 + (2k_1 + k_0)n_0$$

Since $W^{4k_1 n_1} = 1$, we get

$$X(k_1, k_0) = \sum_{n_0=0}^{1} \sum_{n_1=0}^{1} [x_0(n_1, n_0)\, W^{2k_0 n_1}]\, W^{(2k_1+k_0)n_0}$$

Let $X_1(k_0, n_0)$ denote the inner sum, i.e.,

$$X_1(k_0, n_0) = \sum_{n_1=0}^{1} x_0(n_1, n_0)\, W^{2k_0 n_1} = x_0(0, n_0) + x_0(1, n_0) W^{2k_0}$$

or, more explicitly,

$$X_1(0, 0) = x_0(0, 0) + x_0(1, 0)\, W^0$$
$$X_1(0, 1) = x_0(0, 1) + x_0(1, 1)\, W^0$$

$$X_1(1, 0) = x_0(0, 0) + x_0(1, 0)\, W^2$$
$$X_1(1, 1) = x_0(0, 1) + x_0(1, 1)\, W^2$$

Substituting this into the preceding expression, we get

$$X_2(k_0, k_1) = \sum_{n_0=0}^{1} X_1(k_1, n_0)\, W^{(2k_1+k_0)n_0} = X_1(k_0, 0) + X_1(k_0, 1)\, W^{2k_1+k_0}$$

or

$$X_2(0, 0) = X_1(0, 0) + X_0(0, 1)\, W^0$$
$$X_2(0, 1) = X_1(0, 0) + X_0(0, 1)\, W^2$$
$$X_2(1, 0) = X_1(1, 0) + X_0(1, 1)\, W^1$$
$$X_2(1, 1) = X_1(1, 0) + X_0(1, 1)\, W^3$$

Figure 3.25 shows the signal-flow graph for the Cooley–Tukey FFT. The last stage in Figure 3.25, called unscrambling, is needed to sort the outfput values. We get

$$X(0) = X(0, 0) = X_2(0, 0)$$
$$X(1) = X(0, 1) = X_2(1, 0)$$
$$X(2) = X(1, 0) = X_2(0, 1)$$
$$X(3) = X(1, 1) = X_2(1, 1)$$

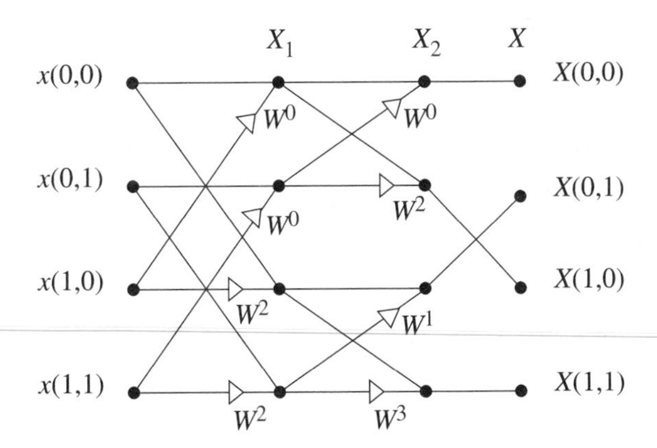

Figure 3.25 Signal-flow graph for the Cooley–Tukey FFT for $N = 4$

3.16.2 ST-FFT (The Sande–Tukey FFT)

The ST-FFT The Sande–Tukey FFT) can be derived in a similar way to the Cooley–Tukey FFT by instead performing the summations over the frequency index, k, i.e., using the divide-and-conquer approach in the frequency domain. Hence, the Sande–Tukey FFT is referred to as a *decimation-in-frequency* FFT. A Pascal program for the Sande–Tukey FFT is shown in Box 3.6.

```
Program ST_FFT;
const
   N = 1024, M = 10, Nminus1 = 1023; { N = 2^M }
type
   Complex = record
      re : Double;
      im : Double;
end;
   C_array : array[0..Nminus1] of Complex;
var
   x : C_array;
   Stage, Ns, k, kNs, n, p, q : integer;
   Wcos, Wsin, TwoPiN, TempRe, TempIm : Double;
begin
   { READ INPUT DATA INTO x }
   Ns := N;
   TwoPiN := 2 * Pi/N;
   for Stage := 1 to M do
      begin
         k := 0;
         Ns := Ns div 2; { index distance between a dual node pair }
         for q := 1 to (N div (2 * Ns)) do
            begin
               for n := 1 to Ns do
                  begin
                     p := k * 2^(Stage - 1) mod (N div 2);
                     Wcos := cos(TwoPiN * p);{W to the power of p}
                     Wsin := -sin(TwoPiN * p);{W = exp(-j2π/N) }
                     kNs := k + Ns;
   {Butterfly}          TempRe := x[k].re - x[kNs].re;
                        TempIm := x[k].im - x[kNs].im;
                        x[k].re := x[k].re + x[kNs].re;
                        x[k].im := x[k].im + x[kNs].im;
                        x[kNs].re := TempRe * Wcos - TempIm * Wsin;
                        x[kNs].im := TempIm * Wcos + TempRe * Wsin;
                     k := k + 1;
                  end;
               k := k + Ns;
            end;
      end;
      Unscramble; { OUTPUT DATA STORED IN x }
end.
```

Box 3.6. The Sande–Tukey FFT. Decimation-in-frequency.

Figure 3.26 shows the signal-flow graph for ST-FFT for N = 8. Note that two signal values are always used to compute what is called a *dual node pair*. For example, the dual node pair $x(3)$ and $x(7)$ in Figure 3.26 are used to compute the dual node pair $x_1(3)$ and $x_1(7)$. The computation of a dual node pair is independent of all other computations in the same column. It is therefore unnecessary to use any additional memory for the intermediate or the final result, except for the N (complex values)

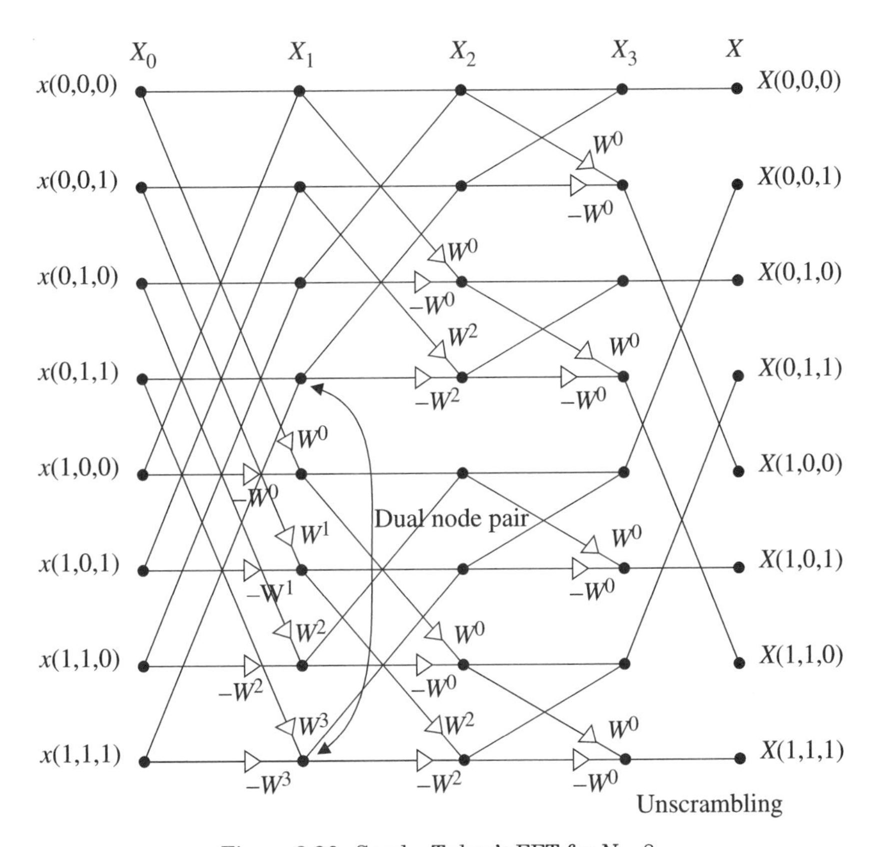

Figure 3.26 Sande–Tukey's FFT for $N = 8$.

memory cells that store the input values. The values in a dual node pair are used to compute the corresponding dual node pair in the next column. At this point, the first node pair is no longer needed so the results can be stored back into the original memory cells. This technique is called *in-place computation*. The program in Box 3.6 uses in-place computation. Note also that all of the dual node pairs in each column can be computed simultaneously. Hence, there is a relatively high degree of computational parallelism in the FFT.

We will later show that the technique of using a memory with only N complex memory cells for computing an N-point FFT results in a significant reduction in chip area. A drawback of the in-place computation is that the original input sequence is lost.

The butterfly for the Sande–Tukey FFT is shown in Figure 3.27. The Sande–Tukey butterfly can be derived from the Cooley–Tukey butterfly by using the transposition theorem.

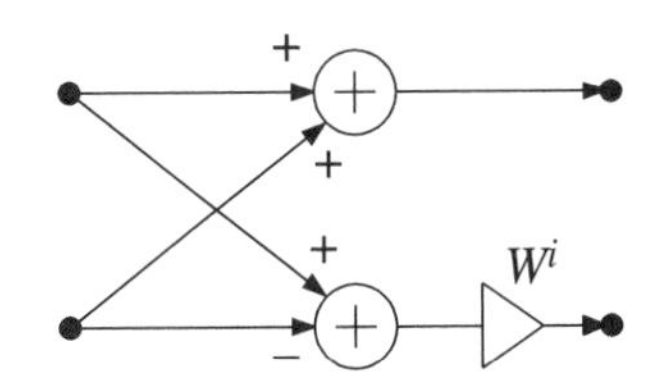

Figure 3.27 The Sande–Tukey butterfly

3.16.3 Winograd's Fast Algorithm

A more recent development is the *WFTA* (*Winograd's fast Fourier transform algorithm*) [6, 7, 10, 21], which requires only $O(N)$ multiplications. However, the number of additions is larger than for the ordinary FFTs and computations of memory addresses etc. are more irregular in the WFTA algorithm. In practice, this irregularity may more than offset the reduction in the arithmetic work load. Further, the algorithm requires that the sequence of length N can be factored into a product of the following lengths: 2, 3, 4, 5, 7, 8, 9, and 16. Each factor corresponds to a short Winograd algorithm.

3.16.4 IFFT (The Inverse FFT)

The inverse FFT can be computed by a simple modification of the input read process. The IFFT is obtained by computing the FFT of the following sequence:

$$X'(n) = \begin{cases} \dfrac{X(0)}{N} & \\ \dfrac{X(N-0)}{N} & \text{for } n = 1, 2, \ldots, N-1 \end{cases} \tag{3.22}$$

Thus, the first value is stored at address 0 while the rest are stored in reverse order. This operation can easily be implemented in hardware by changing the address lines to the memory or by using an up–down counter that is properly initiated.

Another method to compute the IFFT is to first interchange the real and imaginary parts, then perform the FFT, and, finally, interchange the real and imaginary parts. We will use this method in an implementation of an FFT processor.

3.17 FFT PROCESSOR—CASE STUDY 1

As mentioned before, FFTs are used in many DSP applications. It is therefore interesting to develop an FFT processor as a widely usable VLSI building block. In order to be flexible so that the processor can be used in a variety of applications without major redesign, the performance in terms of computational throughput, word length, and transform length should be; easily modifiable.

There are two main approaches to solving this problem. We can build either an FFT processor with characteristics that can be changed by the end-user after manufacture, by setting some parameters, or a fixed-function processor whose characteristics are defined at design time.

The first approach leads to a more flexible solution, but the FFT processor must be designed on a worst-case basis. This alternative leads to the standard or ASIC processor approaches.

In the second case, the FFT processor can be designed to meet the performance requirements exactly, and the cost in terms of chip area and power consumption can be minimized. Here we elect to design such a fixed-function high-performance FFT processor. By selecting a reasonably general design we believe it may serve as a basis for such a "parameterized" building block. The required chip area and power consumption should be minimized.

3.17.1 Specification

The FFT is a commonly used benchmark for standard signal processors. We will therefore study the implementation of an FFT processor that can compute a 1024-point complex FFT with a throughput of more than 2000 FFTs per second. We set the performance requirements slightly higher than what can be attained by one of the best standard fixed-point 16-bit signal processors. The processor shall also be able to perform both the FFT and the IFFT.

Figure 3.28 shows the environment in which the FFT processor is intended to be used. The operator interacts with the host processor to control the operation of the FFT processor—for example, in selecting- purpose 32-bit computer. The FFT processor reads the input sequence and writes the result into the main memory of the host processor. The data rate between the main memory and the FFT processor should be relatively low in order to conserve power. We assume that the data rate is a modest 16 MHz. The data word length for both input and output is selected to be 16 bits for the real and imaginary parts.

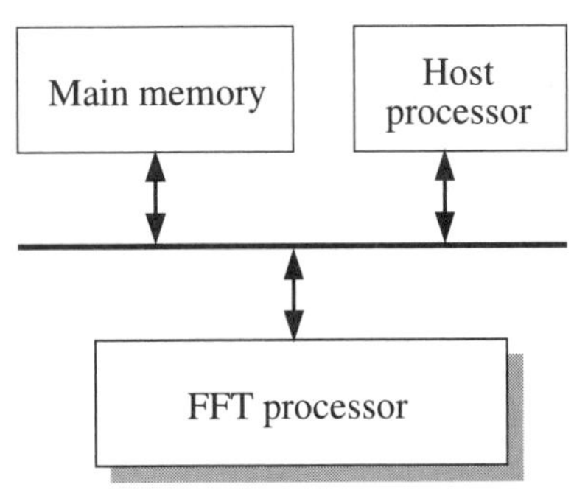

Figure 3.28 Typical environment for the FFT processor

The internal data word length is, in this early stage of the design process, estimated to 21 bits. This will allow for rounding errors incurred in the computations. Note that this will result in a much higher signal quality than that attainable using standard 16-bit signal processors. The required coefficient word length is more difficult to determine since there are no simple error criteria available in the literature. As a reasonable starting point we assume that 14 bits are sufficient. The cost of increasing the coefficient word length is relatively small. These word lengths are suitable for many FFT applications. However, if possible, the word lengths should be optimized with respect to the actual requirements. Particularly, the data word length should be minimized since the memory that stores the input sequence requires a significant amount of chip area. Furthermore, the computational throughput depends on the data word length. Finally, we arbitrarily select to implement the ST-FFT.

3.17.2 System Design Phase

The first step in the system design phase is to partition the computation of an FFT into three consecutive processes: reading the input data from the host, performing the FFT, and writing the result into the memory of the host. The requirement is that the execution time for the FFT, including I/O, should not exceed 0.5 ms. The I/O transfer frequency should be only 16 MHz. Input and output data to the FFT processor are transferred as two 16-bit real numbers. A total of 1024 complex numbers are transferred to and from the FFT processor. The time needed for the I/O is

$$t_{I/O} = \frac{2 \cdot 1024}{16 \cdot 16^6} = 0.128 \text{ ms}$$

It is possible to overlap the I/O operations, i.e., writing and reading data, and computation of the FFT, thereby effectively extending the available time for computing the FFT (see Problem 3.29). For the sake of simplicity we assume that this possibility is not exploited. The available time for the FFT is therefore 0.372 ms.

The I/O processes will handle both the rearranging of data that is required to compute the IFFT and the unscrambling of the data array in the last stage of the FFT. As mentioned in section 3.16.4, the IFFT can be obtained by interchanging the real and imaginary parts, when the data are both written into and read from the memory. We can implement this scheme by changing the addressing of the memories in the FFT processor or, alternatively, by interchanging the inputs to the butterflies in the first stage and interchanging the outputs from the butterflies in the last stage. The unscrambling can be accomplished by reversing the address lines to the memory when reading data out of the FFT processor. Hence, both of these tasks can be accomplished without any extra processing time.

3.18 IMAGE CODING

Efficient image coding techniques are required in many applications. Figure 3.29 illustrates a typical image transmission system for HDTV (high-definition TV) where the aim is to reduce the number of bits to be transmitted over the transmission channel. Modern coding techniques can compress images by a factor of 10 to 50 without visibly affecting the image quality.

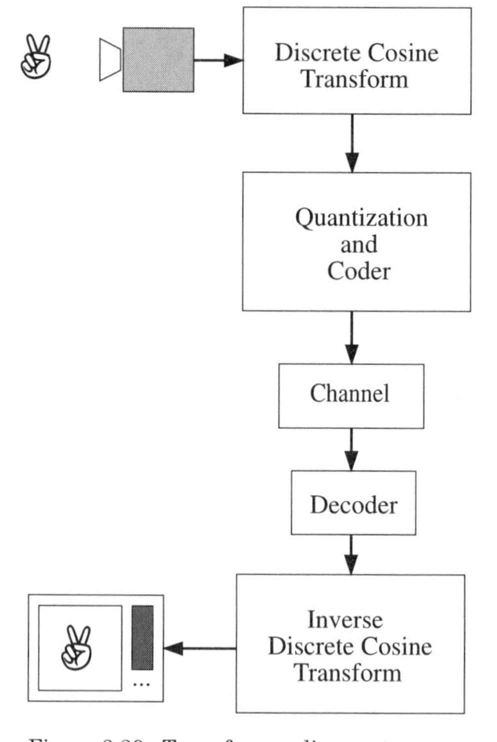

Figure 3.29 Transform coding system

Ordinary TV signals require a transmission capacity of 216 Mbit/s of which 166 Mbit/s are allocated for the video signal. The remaining bits are used for the sound signal and for synchronization purposes. The required bit rates are roughly five times larger for HDTV. Efficient data compression techniques therefore must be employed, since the amount of transmitted information is very large.

Other major applications of transform coding techniques are to be found in image storage and retrieval systems, such as digital VCRs. Multimedia, graphic arts, and desktop publishing are examples of important new applications that emerge. Efficient high-speed compression schemes are necessary in these applications to store and retrieve huge image files from the hard disk. Applications using color will particularly benefit from these compression techniques.

A transform coding; scheme based on the discrete cosine transform has been standardized and is now one of the most widely used compression techniques [5, 34]. This standard, known by the acronym *JPEG* (joint photographic expert group), establishes a standard for compression of color and grayscale still images. The JPEG scheme has been implemented both in software and by using ASIC.

In a typical transform coding scheme, an input image is divided into nonoverlapping subframes or blocks as shown in Figure 3.30. The subframes are linearly transformed by using the discrete cosine transform into the transform domain. The transform has the property that the signal is concentrated in relatively few transform coefficients compared to the number of samples in the original image. Typical subframe sizes are in the range of 8 × 8 to 16 × 16 pixels. The rationale for this data reduction is that any redundancy in the image, from a visual point of view, can more easily be removed in the frequency domain due to a masking phenomenon in the human vision system. A coarse quantization of the frequency components in a region with high power is perceived as a small distortion of the corresponding image while the human eye is more sensitive to errors in regions with low power. The energy compaction property of the discrete cosine transform is therefore important in this type of application.

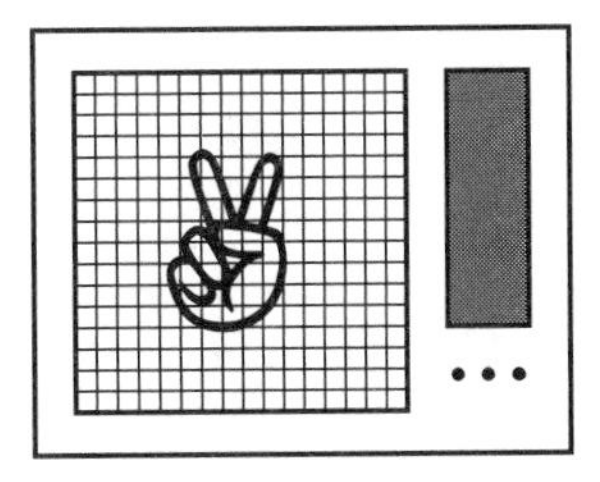

Figure 3.30 Blocking of an image

3.19 DISCRETE COSINE TRANSFORMS

The *DCT* (*discrete cosine transform*) was first proposed by Ahmed et al. [3] in 1974. The discrete cosine transform is highly suitable for transform coding of images. The main reason is that the decorrelation property of the DCT is almost as good as for the optimal transform, the Karhunen–Loéve transform (1947), but the DCT is much simpler from a computational point of view, since it is independent of the signal.

There exist several types of DCTs: even, odd, symmetric, and the modified symmetric DCT. They possess slightly different properties which are of relevance for image coding applications. In fact, only the even DCT and modified symmetric DCT are suitable for image coding [27, 30, 34, 37].

3.19.1 EDCT (Even Discrete Cosine Transform)

The EDCT (even discrete cosine transform) is defined as

$$X(k) \triangleq \sqrt{\frac{2}{N}} \sum_{n=0}^{N-1} c_k x(n) \cos\left(\frac{\pi(2n+1)k}{2N}\right), \quad k = 0, 1, \ldots, N-1 \tag{3.23}$$

Note that the denominator of the cosine term is an even number. This transform is also called DCT-II. The IEDCT (Inverse EDCT) is

$$x(n) = \sqrt{\frac{2}{N}} \sum_{k=0}^{N-1} c_k X(k) \cos\left(\frac{\pi(2n+1)k}{2N}\right), \; n = 0, 1, \ldots, N-1 \tag{3.24}$$

where

$$c_k = \begin{cases} \frac{1}{\sqrt{2}} & \text{for } k = 0 \\ 1 & \text{for } k = 1, 2, \ldots, N-1 \end{cases} \tag{3.25}$$

A careful analysis shows that the forward and inverse transforms differ. Hence two different algorithms must be implemented. Further, we are, for implementation reasons that will be evident in Chapter 11, particularly interested in the symmetry properties of the basis functions of the DCTs. Generally, the DC component of an image is very large compared with the other frequency components and must therefore not appear as an error in the higher-frequency components [30, 37, 38].

EXAMPLE 3.7

Determine if the basis functions of the EDCT are symmetric or antisymmetric and if the DC component leaks into the higher-order frequency components. For the sake of simplicity, use a DCT with $N = 4$.

For the EDCT we get

$$\begin{pmatrix} X(0) \\ X(1) \\ X(2) \\ X(3) \end{pmatrix} = \begin{pmatrix} \frac{1}{2} & \frac{1}{2} & \frac{1}{2} & \frac{1}{2} \\ \frac{1}{\sqrt{2}}\cos\left(\frac{\pi}{8}\right) & \frac{1}{\sqrt{2}}\cos\left(\frac{3\pi}{8}\right) & \frac{1}{\sqrt{2}}\cos\left(\frac{5\pi}{8}\right) & \frac{1}{\sqrt{2}}\cos\left(\frac{7\pi}{8}\right) \\ \frac{1}{\sqrt{2}}\cos\left(\frac{2\pi}{8}\right) & \frac{1}{\sqrt{2}}\cos\left(\frac{6\pi}{8}\right) & \frac{1}{\sqrt{2}}\cos\left(\frac{10\pi}{8}\right) & \frac{1}{\sqrt{2}}\cos\left(\frac{14\pi}{8}\right) \\ \frac{1}{\sqrt{2}}\cos\left(\frac{3\pi}{8}\right) & \frac{1}{\sqrt{2}}\cos\left(\frac{9\pi}{8}\right) & \frac{1}{\sqrt{2}}\cos\left(\frac{15\pi}{8}\right) & \frac{1}{\sqrt{2}}\cos\left(\frac{21\pi}{8}\right) \end{pmatrix} \begin{pmatrix} x(0) \\ x(1) \\ x(2) \\ x(3) \end{pmatrix}$$

We get after simplification

$$\begin{pmatrix} X(0) \\ X(1) \\ X(2) \\ X(3) \end{pmatrix} = \begin{pmatrix} \frac{1}{2} & \frac{1}{2} & \frac{1}{2} & \frac{1}{2} \\ \frac{1}{\sqrt{2}}\cos\left(\frac{\pi}{8}\right) & \frac{1}{\sqrt{2}}\cos\left(\frac{3\pi}{8}\right) & \frac{-1}{\sqrt{2}}\cos\left(\frac{3\pi}{8}\right) & \frac{-1}{\sqrt{2}}\cos\left(\frac{\pi}{8}\right) \\ \frac{1}{\sqrt{2}}\cos\left(\frac{2\pi}{8}\right) & \frac{-1}{\sqrt{2}}\cos\left(\frac{2\pi}{8}\right) & \frac{-1}{\sqrt{2}}\cos\left(\frac{2\pi}{8}\right) & \frac{1}{\sqrt{2}}\cos\left(\frac{2\pi}{8}\right) \\ \frac{1}{\sqrt{2}}\cos\left(\frac{3\pi}{8}\right) & \frac{-1}{\sqrt{2}}\cos\left(\frac{\pi}{8}\right) & \frac{1}{\sqrt{2}}\cos\left(\frac{\pi}{8}\right) & \frac{-1}{\sqrt{2}}\cos\left(\frac{3\pi}{8}\right) \end{pmatrix} \begin{pmatrix} x(0) \\ x(1) \\ x(2) \\ x(3) \end{pmatrix}$$

Obviously, the basis functions (rows) are either symmetric or antisymmetric. The DCT can be viewed as a set of FIR filters where the coefficients in each row represent the impulse responses. All of the filters except the first one should either be a highpass or a bandpass filter, i.e., they should have a zero at $z = 1$ in order to suppress the DC component. In this case we do have a zero at $z = 1$, since the sum of the coefficients for all rows is zero, except for the first row. We therefore conclude that the EDCT is suitable for image coding applications.

3.19.2 ODCT (Odd Discrete Cosine Transform)

The ODCT (odd discrete cosine transform) is defined as

$$X(k) \triangleq \frac{2}{\sqrt{2N-1}} \sum_{n=0}^{N-1} c_k c_n x(n) \cos\left(\frac{2\pi nk}{2N-1}\right), \quad k = 0, 1, \ldots, N-1 \tag{3.25}$$

The denominator of the cosine term is an odd number. The IODCT (inverse ODCT) is

$$x(n) = \frac{2}{\sqrt{2N-1}} \sum_{k=0}^{N-1} c_k c_n X(k) \cos\left(\frac{2\pi nk}{2N-1}\right), \quad n = 0, 1, \ldots, N-1 \tag{3.26}$$

where

$$c_k = \begin{cases} \frac{1}{\sqrt{2}} & \text{for } k = 0 \\ 1 & \text{for } k = 1, 2, \ldots, N-1 \end{cases} \tag{3.25}$$

The forward and inverse transforms are identical, but it can be shown that the basis functions are neither symmetric nor antisymmetric. Further, the DC component appears in the other components. Hence, the ODCT is unsuitable for image coding applications.

3.19.3 SDCT (Symmetric Discrete Cosine Transform)

The SDCT (symmetric discrete cosine transform) is defined as

$$X(k) \triangleq \sqrt{\frac{2}{N-1}} \sum_{n=0}^{N-1} c_k c_n x(n) \cos\left(\frac{\pi nk}{N-1}\right), \quad k = 0, 1, \ldots, N-1 \tag{3.27}$$

The ISDCT (Inverse SDCT) is

$$x(n) = \sqrt{\frac{2}{N-1}} \sum_{k=0}^{N-1} c_k c_n X(k) \cos\left(\frac{\pi nk}{N-1}\right), \quad n = 0, 1, \ldots, N-1 \tag{3.28}$$

where

$$c_k = \begin{cases} \frac{1}{\sqrt{2}} & \text{for } k = 0 \text{ or } k = N-1 \\ 1 & \text{for } k = 1, 2, \ldots, N-2 \end{cases}$$

Note that Equations(3.27) and (3.28), which describe the symmetric cosine transform, have identical forms. Hence, only one algorithm needs to be implemented. Further, the rows and columns are either symmetric or antisymmetric. This fact can be exploited to reduce the number of arithmetic operations as well as the hardware cost.

Unfortunately, the DC component will appear in some of the higher-frequency components, thus rendering the SDCT unsuitable for image coding applications.

3.19.4 MSDCT (Modified Symmetric Discrete Cosine Transform)

The MSDCT (modified discrete symmetric DCT) was developed by Sikström et al. [27, 30, 37, 38] for transform coding. Its distinguishing feature is that the basis vectors are symmetric or antisymmetric and that the DC component that may otherwise appear as an error in the other frequency components is suppressed at the expense of a slightly nonorthogonal transform. The fact that the transform is not strictly orthogonal causes an error, but it is much smaller than the error caused by quantization of the data word length by one bit. Hence, this effect is in practice negligible. Further the forward and inverse transforms are identical.

The MSDCT (modified symmetric discrete cosine transform) is defined as

$$X(k) \triangleq \sqrt{\frac{2}{2N-1}} \sum_{n=0}^{N-1} c_k x(n) \cos\left(\frac{\pi n k}{N-1}\right), \quad k = 0, 1, \ldots, N-1 \tag{3.29}$$

The IMSDCT — Inverse MSDCT is

$$x(n) = \sqrt{\frac{2}{N-1}} \sum_{k=0}^{N-1} c_k X(k) \cos\left(\frac{\pi n k}{N-1}\right), \quad n = 0, 1, \ldots, N-1 \tag{3.30}$$

where

$$c_k = \begin{cases} \frac{1}{\sqrt{2}} & \text{for } k = 0 \text{ or } k = N-1 \\ 1 & \text{for } k = 1, 2, \ldots, N-2 \end{cases} \tag{3.29}$$

Equations (3.29) and (3.30), which describe the modified symmetric cosine transform and its inverse, also have identical forms [30, 34, 37, 45]. Hence, it is necessary to implement only one algorithm.

EXAMPLE 3.8

Show that the MSDCT is not orthogonal. Use the MSDCT with $N = 8$ for the sake of simplicity.

The transform matrix can, after simplification, be written

$$\mathbb{T} = \sqrt{\frac{2}{7}} \begin{bmatrix} \frac{1}{2} & 1 & 1 & 1 & 1 & 1 & 1 & 1 \\ \frac{1}{2} & a & b & c & -c & -b & -a & -\frac{1}{2} \\ \frac{1}{2} & b & -c & -a & -a & -c & b & \frac{1}{2} \\ \frac{1}{2} & c & -a & -b & b & a & -c & -\frac{1}{2} \\ \frac{1}{2} & -c & -a & b & b & -a & -c & \frac{1}{2} \\ \frac{1}{2} & -b & -c & a & -a & c & b & -\frac{1}{2} \\ \frac{1}{2} & -a & b & -c & -c & b & -a & \frac{1}{2} \\ \frac{1}{2} & -1 & 1 & -1 & 1 & -1 & 1 & -\frac{1}{2} \end{bmatrix}$$

where $a = \cos\left(\frac{\pi}{7}\right)$, $b = \cos\left(\frac{2\pi}{7}\right)$, and $c = \cos\left(\frac{3\pi}{7}\right)$.

The odd rows (basis vectors) are symmetric and the even rows are antisymmetric. Now, the inner product

$$\mathbb{v}_1 \cdot \mathbb{v}_2 = \mathbb{v}_1 \cdot \mathbb{v}_2^{\dagger} = |\mathbb{v}_1|\ |\mathbb{v}_2| \cos(\Phi)$$

of two vectors is zero if the vectors are orthogonal. We can simultaneously compute several inner products using matrices. Hence, $\mathbb{T}\,\mathbb{T}^{\dagger}$ is a matrix with only non-zero element values on the diagonal if the row vectors are orthogonal and a unit matrix if the row vectors are orthonormal. Obviously, the basis vectors are, in our case, non-orthogonal. However, the even and odd rows are mutually orthogonal.

$$\mathbb{T}\,\mathbb{T}^{\dagger} = \frac{2}{7} \begin{bmatrix} 6.5 & 0 & -0.5 & 0 & -0.5 & 0 & -0.5 & 0 \\ 0 & 3 & 0 & -0.5 & 0 & -0.5 & 0 & -0.5 \\ -0.5 & 0 & 3 & 0 & -0.5 & 0 & -0.5 & 0 \\ 0 & -0.5 & 0 & 3 & 0 & -0.5 & 0 & -0.5 \\ -0.5 & 0 & -0.5 & 0 & 3 & 0 & -0.5 & 0 \\ 0 & -0.5 & 0 & -0.5 & 0 & 3 & 0 & -0.5 \\ -0.5 & 0 & -0.5 & 0 & -0.5 & 0 & 3 & 0 \\ 0 & -0.5 & 0 & -0.5 & 0 & -0.5 & 0 & 6.5 \end{bmatrix}$$

EXAMPLE 3.9

Show that the two-dimensional discrete cosine transform, which is defined shortly, can be computed by using only one-dimensional DCTs. What other types of operations are needed? How many operations are needed compared with a direct computation?

The two-dimensional MSDCT is defined as

$$X(p, q) \triangleq \frac{2}{N-1} \sum_{n=0}^{N-1} \sum_{k=0}^{N-1} c_k c_n x(n, k) \cos\left(\frac{\pi n p}{N-1}\right) \cos\left(\frac{\pi k q}{N-1}\right), \quad p, q = 0, 1, \ldots, N-1$$

where

$$c_k = \begin{cases} \frac{1}{\sqrt{2}} & \text{for } k = 0 \text{ or } k = N-1 \\ 1 & \text{for } k = 1, 2, \ldots, N-2 \end{cases}$$

First, an intermediate data array is computed using N steps by computing one-dimensional DCTs of the rows, i.e.,

$$X_1(n, q) = \sqrt{\frac{2}{N-1}} \sum_{k=0}^{N-1} c_k x(n, k) \cos\left(\frac{\pi k q}{N-1}\right), \quad n, q = 0, 1, \ldots, N-1$$

The final result is obtained by computing the DCTs for the columns in the intermediate array.

$$X(p, q) = \sqrt{\frac{2}{N-1}} \sum_{n=0}^{N-1} c_n X_1(n, q) \cos\left(\frac{\pi n p}{N-1}\right), \quad p, q = 0, 1, \ldots, N-1$$

Hence, as illustrated in Figure. 3.31, $2N$ one-dimensional DCTs and one matrix transposition of the intermediate data array are needed. In a direct computation, $2N^2$ multiplications are needed for each pixel. We neglect multiplications by constants. Thus, we have a total of $2N^4$ multiplications per block. A computation, based on one-dimensional DCTs, requires only $2N^3$ multiplications per block. Fortunately, similar results also hold for the number of additions.

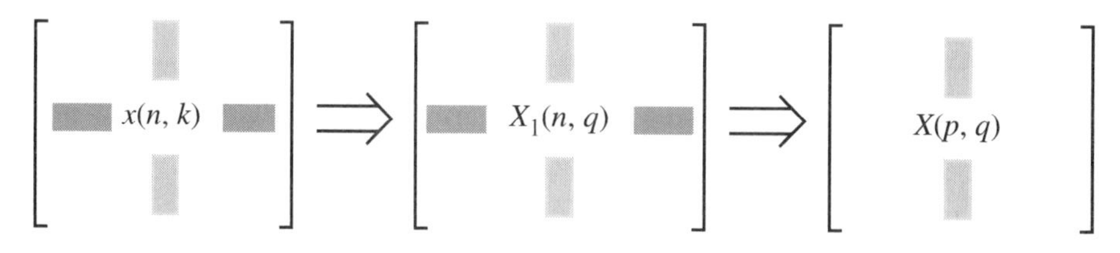

Figure 3.31 Computation of two-dimensional DCT using one-dimensional DCTs

3.19.5 Fast Discrete Cosine Transforms

Fast algorithms with a computational complexity of $O(N\log_2(N))$, have also been developed for the DCTs by using the divide-and-conquer principle [13, 20, 28, 42, 45]. However, in practice, the asymptotic complexity is of little interest for image coding applications since only short sequences with $N = 4$ to 32 are used. Fortunately, the fast algorithms that have been developed are also efficient for short

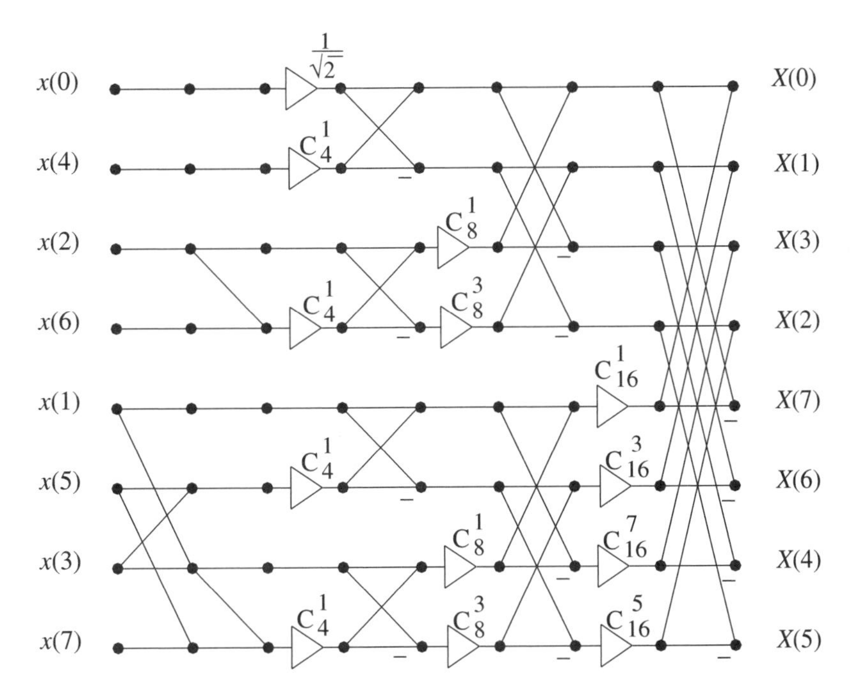

Figure 3.32 Fast even IDCT algorithm

DCTs. The number of real multiplications and additions (subtractions) for the fast DCT/IDCT are $0.5N\log_2(N)$ and $1.5N\log_2(N)-N+1$, respectively. Figure 3.32 shows the signal-flow graph for the even IDCT [28] for $N = 2^m = 8$ where

$$C_{2N}^{(2n+1)k} = \frac{1}{2\cos\left(\frac{\pi(2n+1)k}{2N}\right)}$$

Note that the input is in bit-reversed order. A fast DCT algorithm can be derived by transposing the signal-flow graph shown in Figure 3.32, i.e., reversing the directions of all branches.

In 1991, a fast algorithm for the 2-D even DCT was proposed by Cho and Lee [13]. The algorithm is regular and modular and requires only $0.5N^2\log_2(N)$ multiplications and $2.5N^2\log_2(N)-2N+2$ additions.

3.20 DCT PROCESSOR—CASE STUDY 2

As the second case study we select a two-dimensional DCT processor intended for use in a future transform coding system for real-time image processing [1, 27 30, 37, 38]. The processor is assumed to be a part of a larger integrated circuit and in this study we therefore neglect the I/O aspects of the processor. The throughput requirement for such a DCT processor varies considerably between different transform coding schemes and applications. Currently there is no standard established for HDTV.

One of the candidates is called *CIL* (common image lattice). The main parameters for the CIL system are shown in Table 3.3. The color signals (*R, G,* and *B*) are mapped linearly to a luminance (*Y*) and two chrominance signals (*U* and *V*). The discrete cosine transform is computed for these signals individually. Table 3.4 shows the main performance characteristics of some recently published DCT designs.

Field frequency [Hz]	60	50
Active lines/frame	1080	1152
Samples/active line	1920	2048
Sample frequency	74.25	74.25
Total number of lines	1125	1250
Samples/line	2200	2376
Scanning algorithm	2:1	2:1
Bits/sample	8	8
Gross bit rate [Mbit/s]	1188	1188
Active bit rate [Mbit/s]	995.3	943.72
Pixels/s [MHz]	124.4	118
16 x 16 DCTs/s	486000	461000

Table 3.3: Parameters for the CIL HDTV system

Most designs today are based on an 8 × 8 point DCT due to the stringent requirements. The required number of 8 × 8 DCTs/s for the most stringent HDTV requirement with the largest screen size is about 1.95 10^6. Obviously, none of the designs meet the requirements for HDTV, but they are usable for smaller screen sizes. However, using a modern CMOS process with smaller geometries, the throughput requirements can be met.

Company, etc.	Model	Algorithm	Description
SGS Thomson	STV 3200	8 × 8 even DCT 27 Mpixel/s	1.2-μm CMOS Serial–parallel multipliers
SGS Thomson	STV 3208	8 × 8 even DCT 27 Mpixel/s	1.2-μm CMOS Serial–parallel multipliers
Jutand et al. [26]	—	8 × 8 even DCT Fast algorithm 72 Mpixel/s	0.8-μm CMOS Active chip area ≈ 41 mm^2 16-bit data and coefficient word lengths
Jutand et al. [26]	—	8 × 8 even DCT Distributed arithmetic 72 Mpixel/s	0.8-μm CMOS Active chip area ≈ 41.4 mm^2 16-bit data and 10-bit coefficient word lengths
Defilippis [18]	—	16 × 16 MSDCT Distributed arithmetic 100 Mpixel/s	2 μm CMOS Active chip area ≈ 22.5 mm^2 16-bit data and 10-bit coefficient word lengths
Chau [11]	—	8 × 8 even DCT Distributed arithmetic 100 Mpixel/s	0.8-μm BiCMOS gate array 21.6 mm^2 < 2.2 W 16-bit data word length

Table 3.4 Benchmarks for some ASIC DCT processors

3.20.1 Specification

To meet these stringent requirements, the processor should be able to compute 486,000 two-dimensional (16 × 16) DCTs/s. Typically, input data word lengths are in the range of 8 to 9 bits and the required internal data word length is estimated to be 10 to 12 bits. The coefficient word length is estimated to be 8 to 10 bits.

3.20.2 System Design Phase

The 2-D DCT is implemented by successively performing 1-D DCTs, as discussed in Example 3.9 where the computation was split into two phases. The algorithm is illustrated in Figure 3.33.

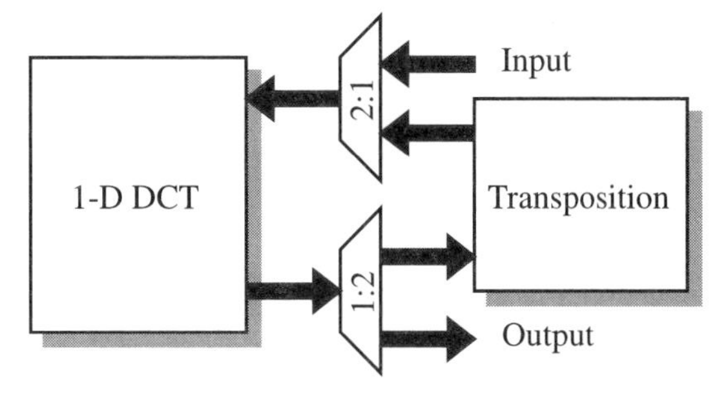

Figure 3.33 Realization of the 2-D DCT

In the first phase, the input array with 16 rows is read row-wise into the processing element, which computes a 1-D DCT, and the results are stored in the RAM. In the second phase, the columns in the intermediate array are successively computed by the processing element. This phase also involves 16 1-D DCT computations. Theoretically, a transposition of the intermediate array is required, but in practice it can be realized by appropriately writing and reading the intermediate values. See Problem 3.32. The algorithm is described by the pseudocode shown in Box 3.7.

```
Program Two_D_DCT;
var
   ...
   begin
   { Read input data, row-wise }
   for n := 1 to 16 do
      DCT_of_Row(n);
   Transpose_Intermediate_Array;
   for n := 1 to 16 do
      DCT_of_Row(n);
   { Write output, row-wise }
   end.
```

Box 3.7. Pseudo-code for the two-dimensional DCT

The required number of arithmetic operations per second in this algorithm is determined as follows: A direct computation of the 1-D DCT involves 16 multiplications and 15 additions. If the symmetry and antisymmetry in the basis vectors were exploited, the number of multiplications would be reduced to 8, but the number of additions would be increased to 23. The required number of arithmetic operations per second for the 2-D DCTs is

$$486{,}000 \cdot (16 + 16) \cdot 16 = 248.8\ 10^6 \text{ Mult/s}$$

and

$$486{,}000 \cdot (16 + 16) \cdot 15 = 233.3\ 10^6 \text{ Add/s}$$

Now, most standard signal processors are designed to perform a multiplication and an addition concurrently. Hence, an algorithm with a poor balance between the number of multiplications and number of additions will be inefficient. We therefore decide to use a direct computation of the DCTs. The transposition of the intermediate array adds a significant number of operations for address calculations in addition to overhead for I/O, etc. Hence, this work load is much too high for today's standard digital signal processors. However, the number of operations is reduced if a fast DCT algorithm is used, but even in this case the work load will be too large.

References

[1] Afghahi M., Matsumura S., Pencz J., Sikström B., Sjöström U., and Wanhammar L.: An Array Processor for 2-D Discrete Cosine Transforms, *Proc. The European Signal Processing Conf.,* EUSIPCO-86, The Hague, The Netherlands, Sept. 2–5, 1986.

[2] Agrawal J.P. and Ninan J.: Hardware Modifications in Radix-2 Cascade FFT Processors, *IEEE Trans. on Acoustic, Speech, and Signal Processing,* Vol. ASSP-26, No. 2, pp. 171–172, April 1978.

[3] Ahmed N., Natarjan T., and Rao K.R.: Discrete Cosine Transform, *IEEE Trans. on Computers,* Vol. C-23, No. 1, pp. 90–93, Jan. 1974.

[4] Antoniou A.: *Digital Filters: Analysis and Design,* McGraw-Hill, New York, 1979.

[5] Baskurt A., Prost R., and Goutte R.: Iterative Constrained Restoration of DCT-Compressed Images, *Signal Processing,* Vol. 17, No. 3, pp. 201–211, July 1989.

[6] Bellanger M.: *Digital Processing of Signals,* John Wiley & Sons, Chichester, 1984.

[7] Blahut R.E.: *Fast Algorithms for Digital Signal Processing,* Addison-Wesley, Reading, MA, 1987.

[8] Bowen B.A. and Brown W.R.: *System Design, Volume II of VLSI Systems Design for Digital Signal Processing,* Prentice Hall, Englewood Cliffs, NJ, 1985.

[9] Bracewell R.N.: *The Fourier Transform and Its Applications,* McGraw-Hill, New York, 1978.

[10] Brigham E.O.: *The Fast Fourier Transform,* Prentice Hall, Englewood Cliffs, NJ, 1988.

[11] Chau K.K., Wang I.-F., and Eldridge C.L.: VLSI Implementation of a 2-D DCT in a Compiler, *Proc. ICASSP-91,* Toronto, Canada, pp. 1233–1236, 1991.

[12] Chen C.H. (ed.): *Signal Processing Handbook,* Marcel Dekker, New York, 1988.

[13] Cho N.I. and Lee S.U.: Fast Algorithm and Implementation of 2-D Discrete Cosine Transform, *IEEE Trans. on Circuits and Systems,* Vol. CAS-38, No. 3, pp. 297–305, March 1991.
[14] Cooley J.W. and Tukey J.W.: An Algorithm for the Machine Calculation of Fourier Series, *Math. Compt.,* Vol. 19, pp. 297–301, April 1965.
[15] Crochiere R.E. and Rabiner L.R.: *Multirate Digital Signal Processing,* Prentice Hall, Englewood Cliffs, NJ, 1983.
[16] Daniels R.W.: *Approximation Methods for Electronic Filter Design,* McGraw-Hill, New York, New York, 1974.
[17] DeFatta D.J., Lucas J.G., and Hodgkiss W.S.: *Digital Signal Processing: A System Design Approach,* John Wiley & Sons, New York, 1988.
[18] Defilippis I., Sjöström U., Ansorge M., and Pellandini F.: Optimal Architecture and Time Scheduling of a Distributed Arithmetic Based Discrete Cosine Transform Chip, *Proc. The European Signal Processing Conf., EUSIPCO-90,* Barcelona, Spain, pp. 1579–1582, 1990.
[19] Duhamel P., Piron B., and Etcheto J.M.: On Computing the Inverse DFT, *IEEE Trans. on Acoustics, Speech, and Signal Processing,* Vol. ASSP-36, No. 2, pp. 285–286, Feb. 1988.
[20] Duhamel P. and Guillemot C.: Polynomial Transform Computation of 2-D DCT, *Proc. ICASSP-90,* Albuquerque, New Mexico, pp. 1515–1518, April 1990.
[21] Elliott D.F. (Ed.): *Handbook of Digital Signal Processing, Engineering Applications,* Academic Press, 1988.
[22] Evans D.M.W.: A Second Improved Digit-Reversal Permutation Algorithm for Fast Transforms, *IEEE Trans. on Acoustics, Speech, and Signal Processing,* Vol. ASSP-37, No. 8, pp. 1288–1291, Aug. 1989.
[23] Falconer D.D. and Ljung L.: Application of Fast Kalman Estimation to Adaptive Equalization, *IEEE Trans. Comm.,* Vol. COM-26, pp. 1439–1446, 1978.
[24] Haykin S.: *Adaptive Filters,* Prentice Hall, Englewood Cliffs, NJ, 1986.
[25] Jackson L.B.: *Digital Filters and Signal Processing,* Kluwer, 1986.
[26] Jutand F., Mou Z.J., and Demassieux N.: DCT Architectures for HDTV, *Proc. ISCAS-91,* Singapore, pp. 196–199, 1991.
[27] Kronander T., Matsumura S., Sikström B., Sjöström U., and Wanhammar L.: VLSI Implementation of the Discrete Cosine Transform, *Proc. Nordic Symp. VLSI in Computers and Communications,* Tampere, Finland, June 13–16, 1984.
[28] Lee B.G: A New Algorithm to Compute the Discrete Cosine Transform, *IEEE Trans. on Acoustics, Speech, and Signal Processing,* Vol. ASSP-33, No. 6, pp. 1243–1245, Dec. 1984.
[29] Lim J.E. and Oppenheim A.V. (eds.): *Advanced Topics in Signal Processing,* Prentice Hall, Englewood Cliffs, NJ, 1988.
[30] Matsumura S., Sikström B., Sjöström U., and Wanhammar L.: LSI Implementation of an 8 Point Discrete Cosine Transform, *Proc. Intern. Conf. on Computers, Systems and Signal Processing,* Bangalore, India, Dec. 10–12, 1984.
[31] Oppenheim A.V. and Schafer R.W.: *Discrete-Time Signal Processing,* Prentice Hall, Englewood Cliffs, NJ, 1989.
[32] Proakis J.G. and Manolakis D.G.: *Introduction to Digital Signal Processing,* Macmillan, New York, 1988.

[33] Ramirez R.W.: *The FFT,* Prentice Hall, Englewood Cliffs, NJ, 1985.

[34] Rao K.R. and Yip P.: *Discrete Cosine Transform, Algorithms, Advantages, Applications,* Academic Press, Boston, 1990.

[35] Rösel P.: Timing of some Bit Reversal Algorithms, *Signal Processing,* Vol. 18, No. 4, pp. 425–433, Dec. 1989.

[36] Sibul L.H. (ed.): *Adaptive Signal Processing,* IEEE Press, 1987.

[37] Sikström B.: *On the LSI Implementation of Wave Digital Filters and discrete cosine transforms,* Linköping Studies in Science and Technology, Diss. No. 143, Linköping University, Sweden, May 1986.

[38] Sikström B., Afghahi M., Wanhammar L., and Pencz J.: A High Speed 2-D Discrete Cosine Transform Chip, Integration, the VLSI Journal, Vol. 5, No. 2, pp. 159–169, June 1987.

[39] Stearns S.D. and David R.A.: *Signal Processing Algorithms,* Prentice Hall, Englewood Cliffs, NJ, 1988.

[40] Taylor F. J.: *Digital Filter Design Handbook,* Marcel Dekker, New York, 1983.

[41] Uramoto S., Inue Y., Takabatake A., Takeda J., Yamashita Y., Terane H., and Yoshimoto M.: A 100 MHz 2-D Discrete Cosine Transform Core Processor, *IEEE J. Solid-State,* Vol. 27, No. 4, pp. 492–499, April 1992.

[42] Vetterli M.: Fast 2-D Discrete Cosine Transform, IEEE Intern. Conf. Acoust., Speech, and Signal Processing, ICASSP-85, Tampa, pp. 1538–1541, 1985.

[43] Wallace G.K.: The JPEG Picture Compression Standard, *IEEE Trans. on Consumer Electronics,* Vol. 38, No. 1, Feb. 1992.

[44] Wang Z.: Fast Algorithms for the Discrete W Transform and for the Fourier Transform, *IEEE Trans. Acoust., Speech, and Signal Processing,* Vol. ASSP-32, No. 4, pp. 803–816, Aug. 1984.

[45] Weidong K. and Mark J.W.: A New Look at DCT-Type Transforms, *IEEE Trans. on Acoustics, Speech, and Signal Processing,* Vol. ASSP-37, No. 12, pp. 1899–1908, Dec. 1989.

PROBLEMS

3.1 Determine the Fourier transform of

(a) $x(n) = a^n$ for $n \geq 0$ and $= 0$ otherwise.

(b) $x(n) = -a^n$ for $n < 0$ and $= 0$ otherwise.

3.2 Determine the period and z-transform of the following sequences:

$x_1(nT)$: $0, -a, 0, a, 0, -a, \ldots$

$x_2(nT)$: $a, a, -a, -a, a, a, -a, \ldots$

$x_3(nT)$: $a, 0, -a, 0,$ a, $0, -a, \ldots$

$x_4(nT)$: $a, 0, -a, -a, 0, a, 0, \ldots$

$x_5(nT)$: $a, 2a, a, -a, -2a, -a, a, \ldots$

3.3 Show that if $X(z)$ is the z-transform of the sequence $x(n)$ then the following relationships hold:

(a) $x(n - n_0) \leftrightarrow z^{-n_0}X(z)$ (b) an $x(n) \leftrightarrow X\left(\frac{z}{-a}\right)$

(c) $x(-n) \leftrightarrow X(\frac{1}{z})$ (d) $x^*(n) \leftrightarrow X^*(z^*)$

(e) $Re\{x(n)\} \leftrightarrow 0.5[X(z) + X^*(z^*)]$ (f) $Im\{x(n)\} \leftrightarrow -0.5j\,[X(z) - X^*(z^*)]$

3.4 Determine the z-transform of

(a) The autocorrelation function: $r(k) = \sum_{n=0}^{\infty} x(n)x^*(n+k)$

(b) The convolution: $y(n) = \sum_{k=0}^{\infty} h(k)x(n-k)$

3.5 Prove Equation (1.5) assuming that the size of the subproblems is a power of c, i.e., $n = c^m$. Hint: Use the substitution $x(m) = T(c^m)/b^m$.

3.6 Determine the transfer function for a filter that has the following impulse response:

$$h(n) = \begin{cases} 0, & n < 0 \\ (0.8)^n - (0.6)^n, & n \geq 0 \end{cases}$$

3.7 Determine the step response of the filter in Problem 3.6.

3.8 The essential part of the power spectrum of an analog signal $x(t)$ lies in the range 0 to 25 kHz. The signal is corrupted by additive wide-band noise which is to be removed by a digital lowpass filter. Select a suitable sample frequency when a third-order Butterworth filter is used as the anti-aliasing filter. The attenuation of unwanted images shall be > 40 dB. The ripple in the passband of the anti-aliasing filter shall be ≤ 1 dB.

3.9 Show that $h(n) = 0$ for $n < 0$ for a causal LSI system.

3.10 Show that the ordering of two cascaded LSI systems may be interchanged.

3.11 Determine the transfer function for a system that is described by the difference equation

$$y(n) = b\,y(n-1) + a\,x(n)$$

3.12 A stable, causal digital filter has the following transfer function

$$H(z) = \frac{1.2z + 1.2}{z^2 - 1.6z + 0.63}$$

(a) Determine the impulse response.
(b) Determine the region of convergence.
(c) Plot the pole-zero configuration in the z-plane.
(d) Plot the magnitude response.
(e) Determine the step response.

3.13 Determine the transfer function of the following difference equation:

$$y(n) = -b\,y(n-2) + a[x(n) - x(n-2)]$$

Determine also the magnitude and phase responses as well as the group delay.

3.14 A digital filter has the pole-zero configuration shown in Figure P3.14a.
(a) Sketch the magnitude response between $\omega T = 0$ and $\omega T = 2\pi$ using a logarithmic scale (dB).
(b) Sketch the impulse and step responses.

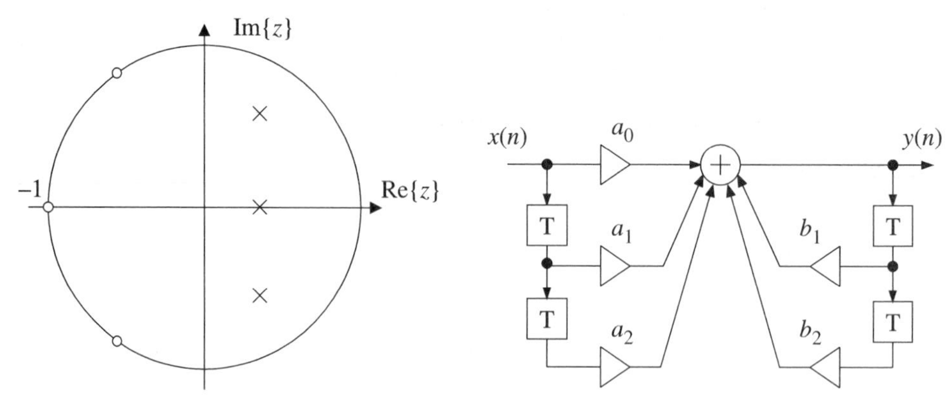

Figure P3.14a Figure P3.14b Direct form I

(c) Assume that the filter is realized in cascade form with first- and second-order direct form I sections. Further, assume that a new filter is obtained by changing the sign of the coefficients a_1 and b_1 in each section. Sketch the magnitude function of this new filter between $\omega T = 0$ and $\omega T = 2\pi$ as well as the impulse and step responses.

3.15 Prove Equation (3.18).

3.16 Give an example of an application in which the group delay properties are of major interest? What are the reasons?

3.17 Show that a filter with the transfer function

$$H(z) = \frac{1 - az}{z - a}$$

is an allpass filter. The coefficient a is real and $|a| < 1$.

3.18 Determine the transfer function for the filter shown in Figure P3.18. Sketch the pole-zero configuration in the z-plane.

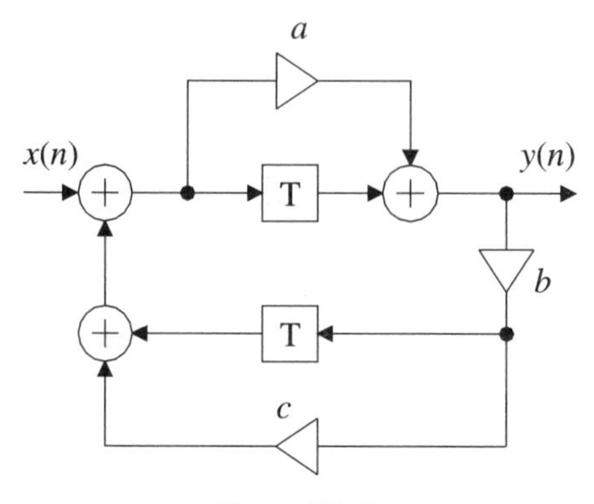

Figure P3.18

3.19 Show that the DFT of a finite-length real sequence has conjugate symmetry.

3.20 In a car phone an array of eight microphones is used to cancel noise from the speech signal. Each signal from the array is provided with an adaptive transversal FIR filter of the type shown in Figure 3.22. The length of the filter is $N = 12$ and the sample rate is 8 kHz. Determine the arithmetic work load for the adaptive filters.

3.21 Draw a block diagram for the RLS lattice filter for $N = 4$.

3.22 Show that

$$\sum_{n=0}^{N-1} W^{kn} = \begin{cases} N & k = 0 \\ 0 & \text{otherwise} \end{cases}$$

3.23 Show that the DFT for $N = 8$ can be computed by essentially using two FFTs with $N = 4$.

3.24 Identify the complex constants in the FFT in Figure 3.23.

3.25 Derive the signal-flow graph for the FFT, with $N = 4$, using the decimation-in-frequency algorithm. What is the relationship between the decimation-in-time and decimation-in-frequency algorithms?

3.26 Show that an IDFT can be computed by using a DFT algorithm, by first interchanging the real and imaginary parts of the sequence, performing the DFT, and finally interchanging the real and imaginary parts of the intermediate result.

3.27 Show that the DFT of two real (or imaginary) sequences can be computed simultaneously by only one FFT.

3.28 Determine the available time for computing the FFT in the FFT coprocessor if the I/O operations are overlapped with the FFT computation.

3.29 In a transform coding system for images, the DCT-frequency components are computed and their values are represented with different accuracy—number of bits—to reduce the amount of data. However, the zero-frequency content of most images is very large. Hence, if this component contributes an error to the nonzero frequency components, then the image will be inefficiently coded or even distorted. Determine which of the different DCTs can be used in such an application.

3.30 Determine the required channel capacity for ordinary TV signals. Determine also the bit rates for the luminance (Y) and chrominance signals (U and V).

	PAL	**NTSC**
Lines/frame	625	525
Active lines	576	480
Pixels/active line		
Luminance, Y	720	720
Chrominance: U and V	360	360
Frame rate (2:1 interlace)	50	59.94
Quantization	8-bit PCM	8-bit PCM
	220 levels	220 levels
Sample frequency Y	13.5 MHz	13.5 MHz
U and V	6.75 MHz	6.75 MHz

3.31 Show that transposition of the intermediate array can be accomplished in the DCT processor by appropriately writing and reading the intermediate values.

3.32 A 1024-point FFT is computed for a digital signal obtained from a CD player. What is the frequency resolution?

3.33 Wang [44] classified the following DCT kernels for N = power of 2:

$$\text{DCT-I:} \quad c_k\, c_n \cos\left(\frac{\pi k n}{N}\right), \qquad k, n = 0, 1, \ldots, N$$

$$\text{DCT-II:} \quad c_k \cos\left(\frac{\pi}{N} k\left(n + \frac{1}{2}\right)\right), \qquad k, n = 0, 1, \ldots, N-1$$

$$\text{DCT-III:} \quad c_n \cos\left(\frac{\pi}{N}\left(k + \frac{1}{2}\right) n\right), \qquad k, n = 0, 1, \ldots, N-1$$

$$\text{DCT-IV:} \quad c_n \cos\left(\frac{\pi}{N}\left(k + \frac{1}{2}\right)\left(n + \frac{1}{2}\right)\right), \qquad k, n = 0, 1, \ldots, N-1$$

where

$$c_m = \begin{cases} \dfrac{1}{\sqrt{2}} & \text{for } m = 0 \text{ or } k = N \\ 1 & \text{for } m \neq 0 \text{ or } N \end{cases}$$

(a) Compare the four transforms with the ones previously defined.
(b) Find the inverse transforms.

DESIGN PROJECTS

JPEG

Desktop publishing, graphic arts, color facsimile, and medical imaging are some new applications that require efficient compression of images. An international standard for compression of digital grayscale and color images has been established by JPEG. The discrete cosine transform is a key algorithm in this compression scheme.

Analyze the algorithms that are used, develop a functional description, and draw the corresponding block diagrams for the scheme used in the still picture JPEG coder–decoder. Determine also a functional specification that can be used for the design of an ASIC. Compare your findings with the performance of the *CL550* coprocessor from C-Cube Microsystems Inc. See ISO/IEC JTC11 Draft Intern. Standard 10918-1, Nov. 1991, and 10918-2, Dec. 1991.

Adaptive Filter

Develop in parallel to the FFT case study an implementation of the adaptive filter described in sections 3.14.1 and 3.14.2. Define an appropriate specification for a typical application or select for the sake of simplicity the internal data word length to 16 bits, N = 10, and f_{sample} = 32 kHz. This abbreviation should not be confused with LSI (large scale integration).

4

DIGITAL FILTERS

4.1 INTRODUCTION

In this chapter, we will review the basic properties of both FIR and IIR filters as well as the most commonly used design methods. FIR filters are straightforward to design by using CAD tools. One of the major drawbacks of FIR filters is that large amounts of memory and arithmetic processing are needed. This makes them unattractive in many applications.

IIR filters, on the other hand, require much less memory and fewer arithmetic operations, but they are difficult to design and they suffer from stability problems. Although the design is much more demanding, the use of an IIR filter may result in a lower system cost and higher performance. As a main topic we will therefore present a class of digital IIR filters, namely wave digital filters, which have highly advantageous properties and are the type of IIR filters we recommend.

A common problem is to transfer signals between two subsystems that operate with different sample frequencies—for example, between a CD player and a DAT (digital tape recorder). In other cases a system function (for example, a narrow-band filter) can be implemented more efficiently by using several sample rates. Such systems are called multirate systems. We will therefore discuss sample rate converters for changing the sample rate without affecting the information contained in the signal. Converters for increasing and decreasing the sample frequency are usually called interpolators and decimators, respectively. We elect, as the topic of the third design study, to design an interpolator based on wave digital filters.

4.2 FIR FILTERS

Digital filters can be categorized into two classes known as *FIR* (finite-length impulse response) and *IIR* (*infinite-length impulse response filters*). Advantages of FIR filters over IIR filters are that they are guaranteed to be stable (see Chapter 5) and to have a linear-phase response. Linear-phase FIR filters are widely used in digital communication systems, in speech and image processing systems, in spectral analysis, and particularly in applications where nonlinear-phase distortion cannot be tolerated. FIR filters require shorter data word length than the corresponding IIR filters. However, they require much higher orders than IIR filters for the same magnitude specification and they sometimes introduce large delays that make them unsuitable for many applications.

The transfer function of an FIR filter of order M is

$$H(z) = \sum_{n=0}^{M} h(n)z^{-n}$$

$$= \frac{h(0)z^M + h(1)z^{M-1} + \ldots + h(M-1)z + h(M)}{z^M} \tag{4.1}$$

where $h(n)$ is the impulse response. Instead of using the order of the filter to describe an FIR filter, it is customary to use the length of the impulse response, which is $N = M + 1$.

The poles of the transfer function are at the origin of the z-plane for nonrecursive algorithms. The zeros can be placed anywhere in the z-plane, but most are located on the unit circle in order to provide better attenuation. FIR filters cannot realize allpass filters except for the trivial allpass filter that is a pure delay.

Some of the poles may be located outside the origin in recursive FIR algorithms, but these poles must always be canceled by zeros. We will not discuss recursive FIR algorithms further since they generally have poor properties and are not used in practice.

4.2.1 Linear-Phase FIR Filters

The most interesting FIR filters are filters with linear phase. The impulse response of linear-phase filters exhibits symmetry or antisymmetry [3, 5, 15–18, 27]. Linear-phase response, i.e., constant group delay, implies a pure delay of the signal. Linear-phase filters are useful in applications where frequency dispersion effects must be minimized—for example, in data transmission systems. The group delay of a linear-phase FIR filter is

$$\tau_g(\omega T) = -\frac{\partial \Phi(\omega T)}{\partial \omega} = \frac{N-1}{2}T \tag{4.2}$$

FIR filters with nonlinear-phase response are rarely used in practice although the filter order required to satisfy a magnitude specification may be up to 50% lower compared with a linear-phase FIR filter [1]. The required number of arithmetic operations for the two filter types are, however, of about the same order.

To simplify the design of linear-phase FIR filters with symmetric impulse response, it is convenient to use a real function H_R which is defined by

$$H(e^{j\omega T}) = e^{j\Phi(\omega T)} H_R(e^{j\omega T}) \tag{4.3}$$

where $\Phi(\omega T) = c - \tau_g(\omega T)$ with $c = 0$ and $c = \pi/2$ for symmetric and antisymmetric impulse responses, respectively. $H_R(e^{j\omega T})$ is referred to as the *zero-phase response*. For an FIR filter with symmetric impulse response, H_R can be written

$$H_R(e^{j\omega T}) = h\left(\frac{N-1}{2}\right) + \sum_{n=1}^{(N-1)/2} 2\, h\left(\frac{N-1}{2} - n\right) \cos(\omega T n) \qquad \text{for } N \text{ odd} \tag{4.4}$$

or

$$H_R(e^{j\omega T}) = \sum_{n=1}^{N/2} 2\,h\left(\frac{N}{2} - n\right) \cos[\omega T(n - \frac{1}{2})] \qquad \text{for } N \text{ even} \qquad (4.5)$$

A corresponding function, H_{Ra}, can also be defined for an FIR filter with antisymmetric impulse response.

$$H_{Ra}(e^{j\omega T}) = \sum_{n=1}^{(N-1)/2} 2\,h\left(\frac{N-1}{2} - n\right) \sin(\omega T n) \qquad \text{for } N \text{ odd} \qquad (4.6)$$

or

$$H_{Ra}(e^{j\omega T}) = \sum_{n=1}^{N/2} 2\,h\left(\frac{N}{2} - n\right) \sin[\omega T(n - \frac{1}{2})] \qquad \text{for } N \text{ even} \qquad (4.7)$$

Even-order FIR filters with antisymmetric impulse responses are used to realize, for example, Hilbert transformers and differentiators [17, 18].

4.2.2 Design of Linear-Phase FIR Filters

Typically, an FIR specification is expressed in terms of the zero-phase function H_R as shown in Figure 4.1. The acceptable deviations are $\pm\delta_1$ and $\pm\delta_2$ in the passband and stopband, respectively. Generally, a filter may have several passbands and stopbands, but they must be separated by a transition band of nonzero width. There are no requirements on the transition band.

The most common method of designing linear-phase FIR filters is to use numeric optimization procedures to determine the coefficients in Equation (4.1) [14, 16, 18]. Alternative, but now outdated, design methods based on windowing techniques can be found in most standard textbooks [1, 3, 5, 6, 18, 27]. These techniques are not recommended since the resulting FIR filters will have a higher order and arithmetic complexity than necessary. Here we will use an optimization program that was originally developed by McClellan, Parks, and Rabiner [14]. The length of the impulse response for a linear-phase lowpass (or highpass) filter that meets the specification shown in Figure 4.1, can be estimated by

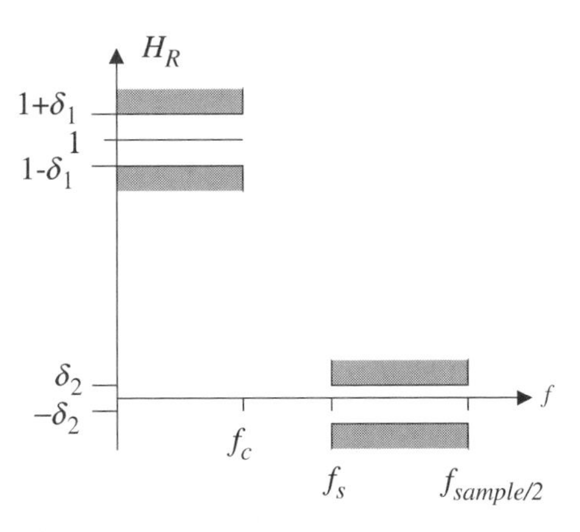

Figure 4.1 Specification for a linear-phase lowpass FIR filter

$$N \approx -\frac{2}{3}\log_{10}(10\,\delta_1\,\delta_2)\,\frac{f_{sample}}{f_s - f_c} + 1 \qquad (4.8)$$

More accurate estimates, especially for short filters, are given in [6, 18]. Obviously, the filter order will be high for filters with narrow transition bands. The filter order is independent of the width of the passband. The passband and stopband ripples have less significance than the width of the transition band. For example, with $\delta_1 = 0.01$ and $\delta_2 = 0.001$, reducing any of the ripples by a factor of 2 increases the filter order by only 6%. A decrease in the transition band by 50% will double the required filter order. The required order for a bandpass filter is essentially determined by the smallest transition band [1].

Deviations in the passbands and stopbands can also be expressed in terms of maximum allowable deviation in attenuation in the passband, A_{max}, and minimum attenuation in the stopband, A_{min}, respectively.

$$A_{max} = 20 \log_{10}\left(\frac{1+\delta_1}{1-\delta_1}\right) \approx 17.37\,\delta_1 \qquad [\text{dB}] \tag{4.9}$$

and

$$A_{min} = 20 \log_{10}\left(\frac{1+\delta_1}{\delta_2}\right) \qquad [\text{dB}] \tag{4.10}$$

Example 4.1

Determine the impulse and frequency responses for a multiple-band FIR filter of order $M = 59$ that meets the specification shown in Figure 4.2. The relationship between the acceptable deviations in the different bands is $\delta_1 = \delta_2 = 10\delta_3$.

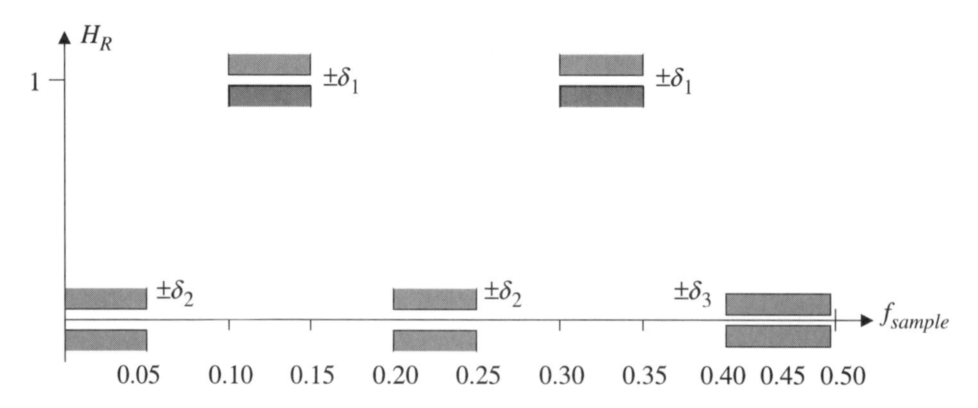

Figure 4.2 Specification for a multiple-band FIR filter

The filter is designed using the program *remez* in *MATLAB*™ which uses a version of the well-known program by McClellan, Parks, and Rabiner [14]. The designer does not have direct control over the deviation of the zero-phase function in the different bands. It is only possible to prescribe the relative deviations by selecting weighting coefficients of the deviations in the different bands. For example, the passband ripple can be decreased by increasing the corresponding weighting coefficient. The program finds a magnitude (zero-phase) response such that the weighted deviations are the same in all bands. The order of the filter must be increased if the deviations are too large.

We set all the weighting coefficients to 1, except for the last band where the weighting coefficient is set to 10, since $\delta_1 = \delta_2 = 10\delta_3$, and select the filter length $N =$ 60. The program yields the following filter parameters. Note that the attenuation in band 5 is, as expected, 20 dB larger than in the other stopbands. Note also the symmetry in the impulse response. The magnitude function is shown in Figure 4.3.

	Band 1	Band 2	Band 3	Band 4	Band 5
Lower band edge	0.00000000	0.10000000	0.20000000	0.30000000	0.40000000
Upper band edge	0.05000000	0.15000000	0.25000000	0.35000000	0.50000000
Desired value	0.00000000	1.00000000	0.00000000	1.00000000	0.00000000
Weighting	1.00000000	1.00000000	1.00000000	1.00000000	10.00000000
Deviation	0.00384571	0.00384571	0.00384571	0.00384571	0.00038457
Deviation [dB]	–48.30047209	0.06680713	–48.30047209	0.06680713	–68.30047209

* * * * Impulse Response * * * *

$h(0)$ = –0.0013347852 = $h(59)$
$h(1)$ = 0.0020944327 = $h(58)$
$h(2)$ = 0.0025711690 = $h(57)$
$h(3)$ = –0.0064205864 = $h(56)$
$h(4)$ = 0.0002917069 = $h(55)$
$h(5)$ = 0.0095955751 = $h(54)$
$h(6)$ = 0.0010398870 = $h(53)$
$h(7)$ = 0.0013729018 = $h(52)$
$h(8)$ = 0.0007270137 = $h(51)$
$h(9)$ = –0.0011216574 = $h(50)$
$h(10)$ = 0.0008927758 = $h(49)$
$h(11)$ = –0.0092173547 = $h(48)$
$h(12)$ = 0.0072828894 = $h(47)$
$h(13)$ = –0.0022192106 = $h(46)$
$h(14)$ = –0.0463359170 = $h(45)$
$h(15)$ = –0.0024767997 = $h(44)$
$h(16)$ = 0.0275492809 = $h(43)$
$h(17)$ = 0.0031292486 = $h(42)$
$h(18)$ = 0.0214422648 = $h(41)$
$h(19)$ = 0.0060822033 = $h(40)$
$h(20)$ = 0.0154276991 = $h(39)$
$h(21)$ = 0.0217612708 = $h(38)$
$h(22)$ = –0.0063723040 = $h(37)$
$h(23)$ = 0.0987030933 = $h(36)$
$h(24)$ = –0.0168138109 = $h(35)$
$h(25)$ = –0.2507191136 = $h(34)$
$h(26)$ = –0.0447299558 = $h(33)$
$h(27)$ = –0.0082915615 = $h(32)$
$h(28)$ = –0.1105607994 = $h(31)$
$h(29)$ = 0.2885732985 = $h(30)$

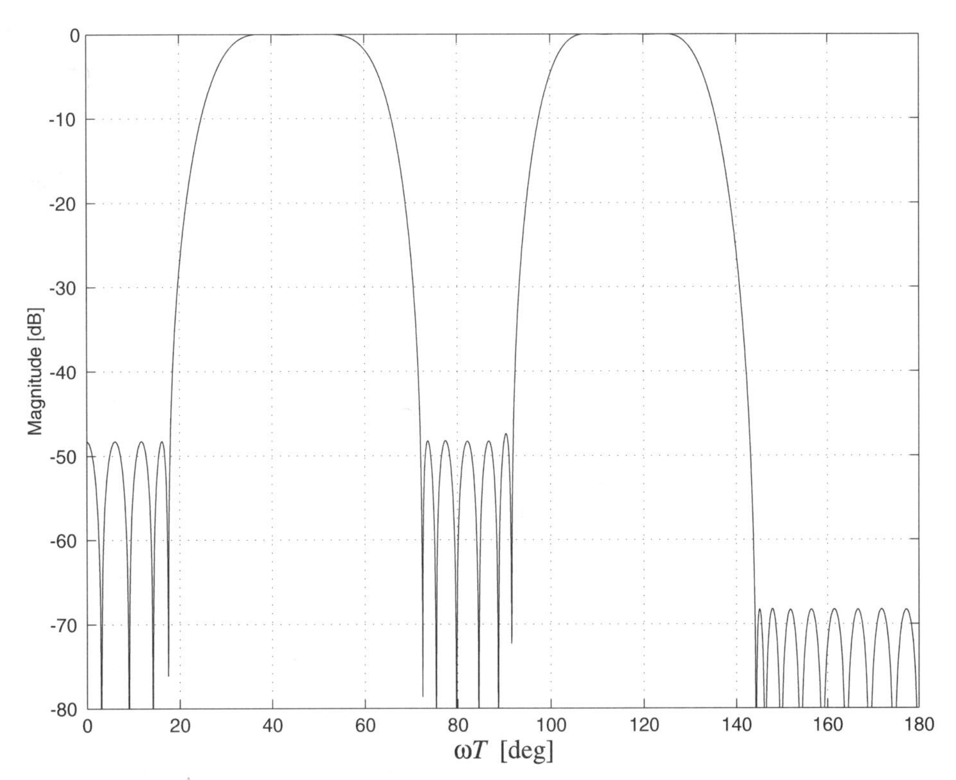

Figure 4.3 Magnitude response of the multiple-band FIR filter

4.2.3 Half-Band FIR Filters

Many DSP schemes exploit the fact that a large number of values in the impulse response of certain types of FIR filters are zero. The required number of arithmetic operations can therefore be reduced for such filters since it is unnecessary to perform multiplications by coefficients that are zero.

Theorem 4.1

If the zero-phase function of a lowpass FIR filter is antisymmetric with respect to $\pi/2$, i.e.,

$$H_R(e^{j\omega T}) = 1 - H_R(e^{j(\pi - \omega T)})$$

then every other coefficient in the impulse response is zero except for the one in the center, which is 0.5.

Such filters are called *half-band FIR filters*, since the bandwidth is about half of the whole frequency band. The symmetry implies that the relation between the cutoff angle and stopband angle is $\omega_c T + \omega_s T = \pi$, and that the passband and stopband deviations δ_1 and δ_2 are equal. Hence, if a large stopband attenuation is required, then the passband must have a small ripple and vice versa. This means that the smallest of the ripple factors will determine the filter order. The reduction in the number of arithmetic operations is significant although the required filter order will, in practice, be slightly higher than that for a corresponding linear-phase filter. The (normalized) zero-phase function is $H_R(e^{j\pi/2}) = 0.5$, i.e., the attenuation is 6 dB.

Only even-order (N = odd) half-band, lowpass FIR filters are of interest since the coefficients are nonzero in odd-order filters.

EXAMPLE 4.2

Determine the impulse response for a half-band FIR filter that meets the following specification: f_c = 400 kHz, f_s = 600 kHz, f_{sample} = 2 MHz, and A_{min} = 60 dB which corresponds to $\delta_2 \approx 0.001$. Determine also the ripple in the passband.

Due to the symmetry requirements, we must have

$$\frac{2\pi f_c}{f_{sample}} + \frac{2\pi f_s}{f_{sample}} = \pi \text{ and } \delta_1 = \delta_2$$

Using Equation (4.9) we estimate the ripple in the passband to be

$$A_{max} \approx 0.01737 \text{ dB}$$

and the required filter length, according to Equation (4.8), to be

$$N \approx -\frac{2}{3} \log_{10}(10\ \delta_1\ \delta_2)\ \frac{f_{sample}}{f_s - f_c} + 1$$

$$= -\frac{2}{3} \log_{10}(10 \cdot 0.001 \cdot 0.001) \frac{2000}{600 - 400} + 1 = 34.3$$

Hence, we select the filter length with N = 35. The program *remez* yields:

	Band 1	Band 2
Lower band edge	0.00000000	0.30000000
Upper band edge	0.20000000	0.50000000
Desired value	1.00000000	0.00000000

	Band 1	Band 2
Weighting	1.00000000	1.00000000
Deviation	0.00067458	0.00067458
Deviation in dB	0.01171863	–63.41935048

* * * * Impulse Response * * * *

$h(0) = 0.0011523147 = h(34)$ $h(9) = 0.0000205816 = h(25)$
$h(1) = 0.0000053778 = h(33)$ $h(10) = -0.0310922116 = h(24)$
$h(2) = -0.0027472085 = h(32)$ $h(11) = -0.0000213219 = h(23)$
$h(3) = -0.0000047933 = h(31)$ $h(12) = 0.0525993770 = h(22)$
$h(4) = 0.0057629677 = h(30)$ $h(13) = 0.0000279603 = h(21)$
$h(5) = 0.0000118490 = h(29)$ $h(14) = -0.0991307839 = h(20)$
$h(6) = -0.0107343126 = h(28)$ $h(15) = -0.0000262782 = h(19)$
$h(7) = -0.0000127702 = h(27)$ $h(16) = 0.3159246316 = h(18)$
$h(8) = 0.0185863941 = h(26)$ $h(17) = 0.5000310310$

Figure 4.4 shows the magnitude response of the half-band FIR filter. As expected, the actual ripple in the passband is slightly lower than required since the filter order is larger then necessary. The attenuation in the stopband is also larger than required. This so-called design margin can be used to round off the coefficient values to binary values of finite precision.

Note that all the odd-numbered coefficients are zero (rounding errors in the program cause the small, nonzero values) except for the central value: $h(17) = 0.5$. A method that improves both the accuracy and speed of the design program is described in [28].

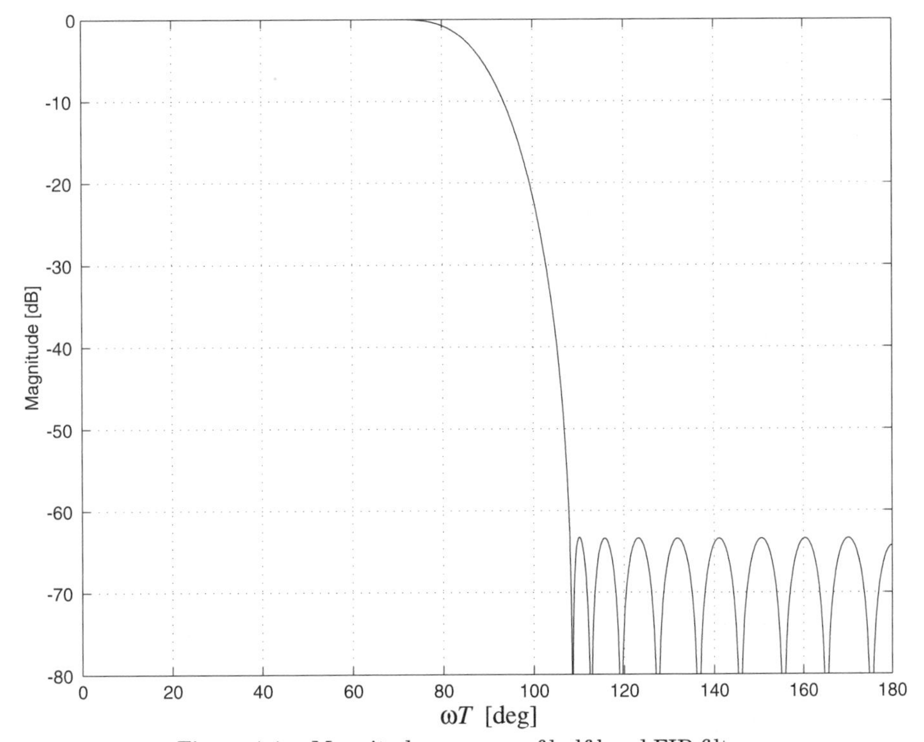

Figure 4.4 Magnitude response of half-band FIR filter

4.2.4 Complementary FIR Filters

In many applications the need arises to split the input signal into two or more frequency bands. For example, in certain transmission systems for speech, the speech signal is partitioned into several frequency bands using a filter bank. A filter bank is a set of bandpass filters with staggered center frequencies which cover the whole frequency range. The first and the last filter are a lowpass and a highpass filter, respectively. The filtered signals are then processed individually to reduce the number of bits that have to be transmitted to the receiver where the frequency components are combined into an intelligible speech signal.

A special case of band splitting filters, a lowpass and a highpass filter, is obtained by imposing the following symmetry constraints. Let $H(z)$ be an even-order (N = odd) FIR filter. The complementary transfer function H_c is defined by

$$|H(e^{j\omega T}) + H_c(e^{j\omega T})| = 1 \tag{4.11}$$

A solution to Equation (4.11) is

$$H(z) + H_c(z) = z^{-(N-1)/2} \tag{4.12}$$

The two transfer functions are complementary. Note that the attenuation is 6.02 dB at the crossover angle. We will later show that these two filters can be effectively realized by a slight modification of a single FIR filter structure.

When one of the filters has a passband, the other filter will have a stopband and vice versa. The passband ripple will be very small if requirements for the other filter's stopband is large. For example, an attenuation of 60 dB ($\delta_2 \approx 0.001$) in the stopband of the complementary filter corresponds to a ripple of only 0.00869 dB in the passband of the normal filter.

An input signal will be split by the two filters into two parts such that if the two signal components are added, they will combine into the original signal except for a delay corresponding to the group delay of the filters. A filter bank can be realized by connecting such complementary FIR filters like a binary tree. Each node in the tree will correspond to a complementary FIR filter that splits the frequency band into two parts.

4.3 FIR FILTER STRUCTURES

An FIR filter can be realized using either recursive or nonrecursive algorithms [16, 18]. The former, however, suffer from a number of drawbacks and should not be used in practice. On the other hand, nonrecursive filters are always stable and cannot sustain any type of parasitic oscillation, except when the filters are a part of a recursive loop. They generate little round-off noise. However, they require a large number of arithmetic operations and large memories.

4.3.1 Direct Form

Contrary to IIR filters, only a few structures are of interest for the realization of FIR filters. The number of structures is larger for multirate filters, i.e., filters with several sample frequencies. One of the best and yet simplest structures is the *direct form* or *transversal structure*, which is depicted in Figure 4.5.

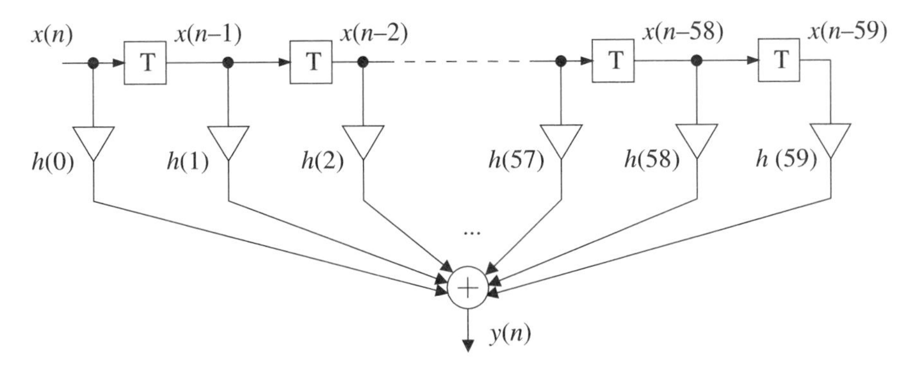

Figure 4.5 Direct form FIR structure (Transversal filter)

The direct form FIR filter structure of order $M = 59$ is described by a single difference equation:

$$y(n) = \sum_{k=0}^{59} h(k)\, x(n-k) \tag{4.13}$$

The required numbers of multiplications and additions are N and N–1, respectively. This structure is suitable for implementation on processors that are efficient in computing sum-of-products. Most standard signal processors provide special features to support sum-of-product computations, i.e., a multiplier–accumulator and hardware implementation of loops and circular memory addressing. The signal levels in this structure are inherently scaled except for the output which, for short FIR filters, is normally scaled using the "safe scaling criterion" (see Chapter 5).

4.3.2 Transposed Direct Form

The transposition theorem [3, 18] discussed in Chapter 3 can be used to generate new structures that have the same transfer function as the original filter structure. The number of arithmetic operations as well as the numerical properties are generally different from the original structure.

The transposed direct form structure, shown in Figure 4.6, is derived from the direct form structure. The amount of required hardware can be significantly reduced for this type of structure where many multiplications are performed with the same input value. This technique is discussed in Chapter 11. The filter structure is a graphic illustration of the following set of difference equations:

$$\begin{aligned}
y(n) &:= h(0)\, x(n) + v_1(n-1) \\
v_1(n) &:= h(1)\, x(n) + v_2(n-1) \\
&\vdots \\
v_{58}(n) &:= h(58)\, x(n) + v_{59}(n-1) \\
v_{59}(n) &:= h(59)\, x(n)
\end{aligned} \tag{4.14}$$

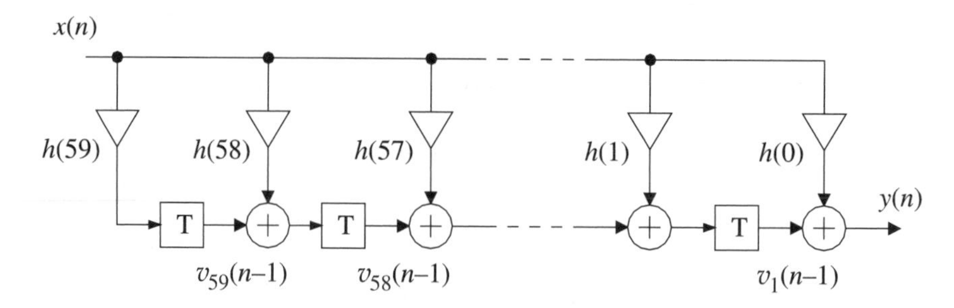

Figure 4.6 Transposed, direct form FIR structure, $N = 60$

In this case the number of arithmetic operations remains the same, but the signal level increases from left to right as more and more signal values are added. This structure is particularly favorable since the multiplier block can be significantly simplified as will be discussed in Chapter 11.

4.3.3 Linear-Phase Structure

A major reason why FIR filters are used in practice is that they can produce an exact linear-phase response. *Linear-phase* implies that the impulse response is either symmetric or antisymmetric. An advantage of linear-phase FIR filters is that the number of multiplications can be reduced by exploiting the symmetry (or antisymmetry) in the impulse response, as illustrated in Figure 4.7.

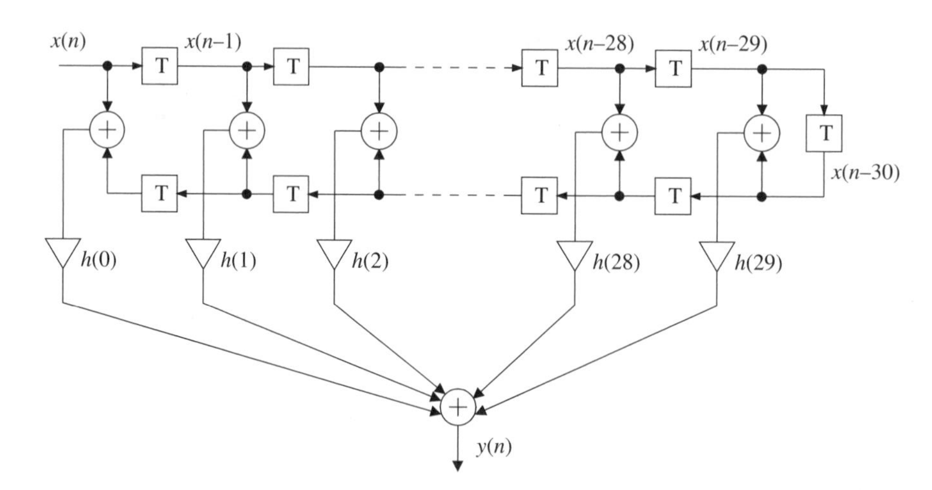

Figure 4.7 Direct form linear-phase FIR structure, $N = 60$

The structure depicts the difference equation

$$y(n) = \sum_{k=0}^{29} h(k)\,[x(n-k) \pm x(n-59+k)] \tag{4.15}$$

This structure is called a *direct form linear-phase FIR structure* since the phase response is independent of the coefficient values.

Even-order (N = odd) FIR filters are often preferred since the group delay is an integer multiple of the sample period. The number of multiplications is $(N + 1)/2$ for N = odd and $N/2$ for N = even. Thus, the number of multiplications is reduced significantly compared with the direct form, but the number of additions remains the same. Subtractors are used instead of adders if the impulse response is antisymmetric.

A major drawback is that the group delay for linear-phase FIR filters is often too large to be useful in many applications. The FIR filter previously discussed has, according to Equation (4.2), the group delay

$$\tau_g(\omega T) = \frac{N-1}{2}T = 29.5\,T$$

Using the transposition theorem a *transposed direct form linear-phase FIR structure* can be derived from the signal-flow graph in Figure 4.7. The number of multiplications and additions remains the same. The signal levels are slightly worse compared with the linear-phase direct form structure.

The number of arithmetic operations is further reduced in linear-phase, half-band filters. Each zero-valued coefficient makes one multiplication and one addition redundant. The number of multiplications and additions is reduced significantly compared with the direct form linear-phase FIR structure.

4.3.4 Complementary FIR Structures

Even more dramatic reductions of the required quantities of arithmetic operations are possible when two complementary FIR filters are needed. We again illustrate the basic ideas by an example.

The complementary filter $H_c(z)$ can be obtained from the direct form linear-phase structure by subtracting the ordinary filter output from the delayed input value, $x(nT - (N-1)T/2)$, as shown in Figure 4.8.

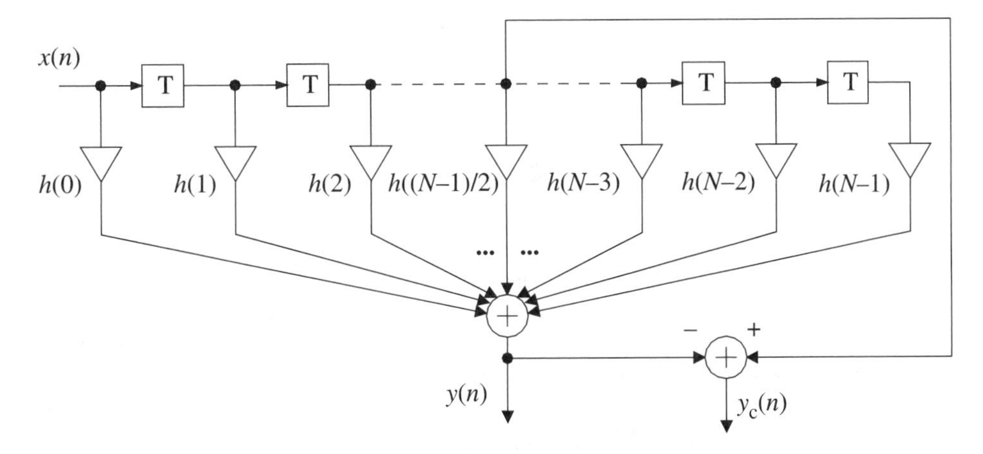

Figure 4.8 Complementary FIR filter with N = odd

The cost of realizing the complementary filter seems to be only one subtractor. However, the requirement on the normal FIR filter must usually be increased significantly in order to meet the requirements on the complementary filter. Hence, the arithmetic work load is somewhat larger than that for one single FIR filter, but is still significantly smaller than that for two separate filters.

This technique can, of course, be applied to linear-phase FIR filter structures with either symmetric or antisymmetric impulse responses and their transposes. Complementary half-band FIR filters are particularly useful, since the saving in arithmetic work load is substantial. Complementary FIR filters of odd order (N = even) are not feasible, since the value corresponding to $\tau_g(\omega T) = (N-1)T/2$ is not available.

4.3.5 Miscellaneous FIR Structures

Long FIR filters can be implemented efficiently using the fast Fourier transform (*FFT*). There are two main approaches: overlap–add and overlap–save [15–18]. These approaches are efficient if the length of the FIR filter is larger than 60 to 100.

FIR filters that are imbedded in adaptive FIR filters are often realized by a lattice structure [1, 17]. A drawback is that the number of arithmetic operations is high since there are two multiplications and two additions for each filter coefficient.

High-order IIR filters are often realized as a cascade of several low-order filters in order to reduce the sensitivity to coefficient errors. This approach can in principle also be used for FIR filters, but the benefits are offset by a decrease in dynamic signal range. Traditionally, most textbooks [15–18, 27] contain a discussion of so-called *frequency-sampling FIR structures*. These filter structures are recursive algorithms that rely on pole-zero canceling techniques. Although they may at first seem to be interesting, they should not be used due to their high coefficient sensitivity, low dynamic range, and severe stability problems.

4.4 FIR CHIPS

Table 4.1 lists some commercial FIR chips and their characteristics. The devices from Inmos and Harris are aimed at high-speed applications—for example, filtering of video and radar signals. Generally, the devices can be cascaded to obtain longer filter lengths and/or to obtain higher accuracy. For example, the A100 from Inmos has both an asynchronous parallel host interface and a high-speed synchronous interface allowing several circuits to be cascaded for high-performance applications. The circuit has two sets of coefficients that are accessible via the host interface.

The Motorola 56200 is aimed at applications using low to medium sample rates. A typical application is filtering of audio signals. Another important area where the circuits can be used is in communications systems. The Motorola DSP56200 can implement either a single FIR (SFIR) filter with up to 256 taps or dual FIR (DFIR) filter with up to 128 taps. It can also implement the adaptive LMS algorithm discussed in Chapter 3. In the adaptive FIR mode the circuit cycles through memory twice per sample—first to perform the FIR convolution, and second to update the FIR coefficients using the LMS algorithm.

Company	Model	Description
Inmos	A100	4-, 8-, 12-, or 16-bit coeff., 16-bit input data, 16 × 16-bit mult. with 36-bit accum., and 24-bit-rounded output 32-taps (fixed length) Throughput 15.00 MHz, 4-bit coeff., 3.75 MHz, 16-bit coeff 84-pin PGA–flatpack package
Harris	HSP43891	9-bit coeff., 9-bit input data, 9 × 9-bit mult. with 26-bit accum., and 26-bit output 8 taps per device Throughput 30 MHz Support for decimation by 2, 3, or 4 High-speed synchronous inputs and outputs 84-pin PGA–PLCC package
Harris	HSP43881	8-bit coeff., 8-bit input data, 8 × 8-bit mult. with 26-bit accum., and 26-bit output 8 taps per device Throughput 30 MHz Support for decimation by 2, 3, or 4 84-pin PGA–PLCC package
Harris	HSP43168	10-bit data and coeff Dual FIR with 8 taps or single FIR with 16 taps Support for decimation up to a factor 16 Throughput 40 MHz 84-pin PGA–PLCC package
Harris	HSP43220	20-bit coeff. and 16-bit input data Two-stage FIR filter for decimation. The first stage can decimate with a factor up to 1024. The second stage has up to 512 taps and can decimate with a factor 16. Throughput 30 MHz 84-pin PGA–PLCC package
Motorola	DSP56200	24-bit coeff., 16-bit input data, 24 × 16-bit mult. with 40-bit accum. 32-bit or 16-bit-rounded output. 4–256 taps (selectable) Throughput Single FIR filter 227 kHz, 32-taps, 1-device, 37 kHz, 256-taps, 1-device, 37 kHz, 1024-taps, 4-devices Two FIR filters: 123 kHz, 32-taps, 1-device, 36 kHz, 128-taps, 1-device Adaptive FIR filter: 19 kHz, 256-taps, 1-device, 19 kHz, 1024-taps, 4-devices, 115 kHz, 256-taps, 8-devices 28-pin DIP package

Table 4.1 FIR filter chips

4.5 IIR FILTERS

Digital FIR filters can only realize transfer functions with effective poles at the origin of the z-plane, while IIR filters can have poles anywhere within the unit circle. Hence, in IIR filters the poles can be used to improve the frequency selectivity. As a consequence, the required filter order is much lower for IIR as compared to

FIR filters. However, it is not possible to have exactly linear-phase IIR filters. Fortunately, neither is it usually necessary. It is only necessary to have a phase response that is sufficiently linear in the passband. In such cases it is often simpler and more efficient to cascade two IIR filters than to use a linear-phase FIR filter. One of the IIR filters is designed to meet the frequency selective requirements while the other corrects the group delay so that the two filters combined meet the linear-phase requirements. In some cases, it may be efficient to use a combination of FIR and IIR filters. The improved frequency selective properties of IIR filters are obtained at the expense of increased coefficient sensitivity and potential instability. These issues will be further discussed in Chapter 5.

4.6 SPECIFICATION OF IIR FILTERS

Frequency-selective filters are specified in the frequency domain in terms of an acceptable deviation from the desired behavior of the magnitude or attenuation function. Figure 4.9 shows a typical attenuation specification for a digital lowpass filter. The variation (ripple) in the attenuation function in the passband may not be larger than A_{max} (= 0.5 dB) and the attenuation in the stopband may not be smaller than A_{min} (= 60 dB). It is convenient during the early stages of the filter design process to use a normalized filter with unity gain, i.e., the minimum attenuation is normalized to 0 dB. The filter is provided with the proper gain in the later stages of the design process.

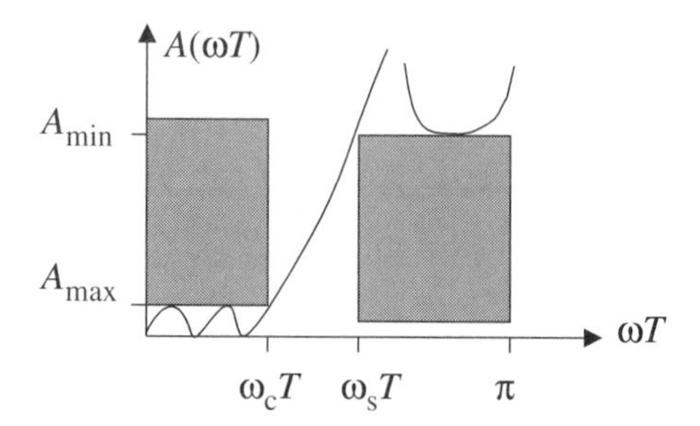

Figure 4.9 Typical specification of the attenuation for a digital lowpass filter

The passband and stopband frequencies (angles) and the acceptable tolerances in the different bands are specified. The passband for a digital lowpass filter is from 0 to $\omega_c T$ and the stopband begins at $\omega_s T$ and extends to π. The *transition band* is from $\omega_c T$ to $\omega_s T$. There are no requirements on the attenuation in the transition band.

In many applications other frequency domain characteristics such as phase and group delay requirements must also be met. In some cases additional requirements in the time domain such as step response and intersymbol interference requirements are used.

The group delay variation within the passband is typically specified to be within certain limits so that signal distortion is acceptable. The total delay is often required to be below a certain limit. For example, a speech coder must not have a delay of more than 20 ms. As mentioned before, a long delay is often not acceptable.

The synthesis of a digital filter that shall meet requirements on the magnitude or attenuation function begins from the *squared magnitude function* which can be written

$$|H(e^{j\omega T})|^2 = \frac{1}{1 + |C(e^{j\omega T})|^2} \tag{4.16}$$

where $C(e^{j\omega T})$ is the *characteristic function*. The magnitude of the characteristic function should be small in the passband and large in the stopband. We define the *ripple factors*:

$$\text{passband:} \quad |C(e^{j\omega T})| \leq \varepsilon_p \tag{4.17}$$

$$\text{stopband:} \quad |C(e^{j\omega T})| \geq \varepsilon_s \tag{4.18}$$

The attenuation requirements can be rewritten in terms of the ripple factors:

$$A_{max} = 10 \log_{10}(1 + \varepsilon_p{}^2) \tag{4.19}$$

$$A_{min} = 10 \log_{10}(1 + \varepsilon_s{}^2) \tag{4.20}$$

Synthesis of the transfer function, also referred to as the *approximation problem*, involves finding a proper characteristic function, $C(z)$, satisfying Equations (4.17) and (4.18). Various approaches to solving this problem will be discussed in the next sections.

4.6.1 Analog Filter Approximations

Many filter solutions, so-called filter approximations, have been developed to meet different requirements, particularly for analog filters [4, 24]. The main work has focused on approximations to lowpass filters, since highpass, bandpass, and stopband filters can be obtained from lowpass filters through frequency transformations [4, 15–18, 24, 27]. It is also possible to use these results to design digital filters. The classical lowpass filter approximations, which can be designed by using most standard filter design programs, are:

Butterworth The magnitude function is maximally flat at the origin and monotonically decreasing in both the passband and the stopband. The variation of the group delay in the passband is comparatively small. However, the overall group delay is larger compared to the filter approximations we will discuss shortly. This approximation requires a larger filter order than the filter approximations discussed shortly to meet a given magnitude specification.

Chebyshev I The magnitude function has equal ripple in the passband and decreases monotonically in the stopband. The variation of the group delay is somewhat worse than for the Butterworth approximation. The overall group delay is smaller than for Butterworth filters. A lower filter order is required compared to the Butterworth approximation.

Chebyshev II (Inverse Chebyshev) The magnitude function is maximally flat at the origin, decreases monotonically in the passband, and has equal ripple in the stopband. The group delay has a variation similar to the Butterworth approximation, but much smaller overall group delay. The same filter order is required as for the Chebyshev I approximation.

Cauer The magnitude function has equal ripple in both the passband and the stopband, but the variation of the group delay is larger than that for the other approximations. The overall group delay is the smallest of the four filter approximations. The Cauer filter, also called an elliptic filter, requires the smallest order to meet a given magnitude specification.

Bessel The group delay is maximally flat at the origin and monotonically decreasing in the passband. The magnitude function decreases rapidly in the passband and the stopband attenuation is poor. Analog Bessel filters are not useful prototypes for designing digital filters.

These filter approximations represent extreme cases since only one property has been optimized at the expense of other properties. In practice they are often used directly, but they can also serve as a starting point for an optimization procedure trying to find a solution that simultaneously satisfies several requirements.

It is common to use the following notation to describe standard analog lowpass filters—for example, *C*051525. The first letter denotes a Cauer filter (*P* for Butterworth, *T* for Chebyshev I, and *C* or *CC* for Cauer filters). There is no letter assigned to Chebyshev II filters. The first two digits (05) denote the filter order while the second pair denotes the reflection coefficient (15 %), and the third pair denotes the modular angle (25 degrees). The latter is related to the cutoff and stopband frequencies by

$$\Theta = \sin^{-1}\left(\frac{f_c}{f_s}\right)$$

We will show later that the reflection coefficient is related to the ripple in the passband. A Butterworth filter is uniquely described by its order, *B*05, except for the passband edge. To describe a Chebyshev I filter we also need the reflection coefficient—for example, *T*0710. The Cauer filter requires in addition the modular angle—for example, *C*071040.

4.7 DIRECT DESIGN IN THE z-PLANE

IIR filters can be designed by directly placing the poles and zeros in the z-plane such that the frequency response satisfies the requirement. In principle, approximation methods, similar to the ones that have been derived for analog filters, can also be derived for digital filters. Various numerical optimization procedures can also be used. Numerical optimization procedures are normally used in the design of FIR filters. However, it is widely recognized that it is advantageous to exploit knowledge of analog filter synthesis to synthesize digital filters. For example, widely available programs for the synthesis of analog filters can easily be extended to the design of digital IIR filters.

4.8 MAPPING OF ANALOG TRANSFER FUNCTIONS

As mentioned already, the most commonly used design methods capitalize on knowledge and experience as well as the general availability of design programs for analog filters. A summary of the classical design process is shown in Figure 4.10.

In the first design step, the magnitude specification is mapped to an equivalent specification for the analog filter. Several mappings have been proposed: bilinear, LDI, impulse-invariant, step-invariant, and the matched-z-transform [6]. However, in practice only the bilinear transformation is appropriate for mapping frequency-selective analog filters to digital filters.

The *bilinear transformation* is defined as

$$s \triangleq \frac{2}{T}\frac{z-1}{z+1} \tag{4.21}$$

The relation between the analog cutoff frequency and the cutoff angle for the digital filter is

$$\omega_{ac} = \frac{2}{T}\tan\left(\frac{\omega_c T}{2}\right) \tag{4.22}$$

Next, the approximation problem is solved for the lowpass filter, giving the analog transfer function, $H(s)$. Note that the phase response of the analog filter is distorted using the bilinear transformation [1, 6, 15–18, 27]. Finally, the poles and zeros of the analog lowpass filter are mapped to the digital domain, yielding the poles and zeros of the digital filter.

Details of this design procedure can be found in many textbooks [1, 2, 4–6, 15–18, 24, 26, 27]. This design approach only solves the problem of finding a transfer function $H(z)$ that satisfies the magnitude specification. Later we will discuss methods that will also provide good filter algorithms.

It is interesting to note that a digital lowpass filter is transformed to a bandpass filter which is symmetric around $\pi/2$ by the transformation $z \to -z^2$. The transformation can be done either by mapping the poles and zeros of the lowpass filter or by replacing the delay elements in the lowpass filter structure by two cascaded delay elements and a multiplication by –1. This is an efficient method of implementing symmetric bandpass filters since the arithmetic work load is the same as for the original lowpass filter. A symmetric stopband filter is obtained by the transformation $z \to z^2$.

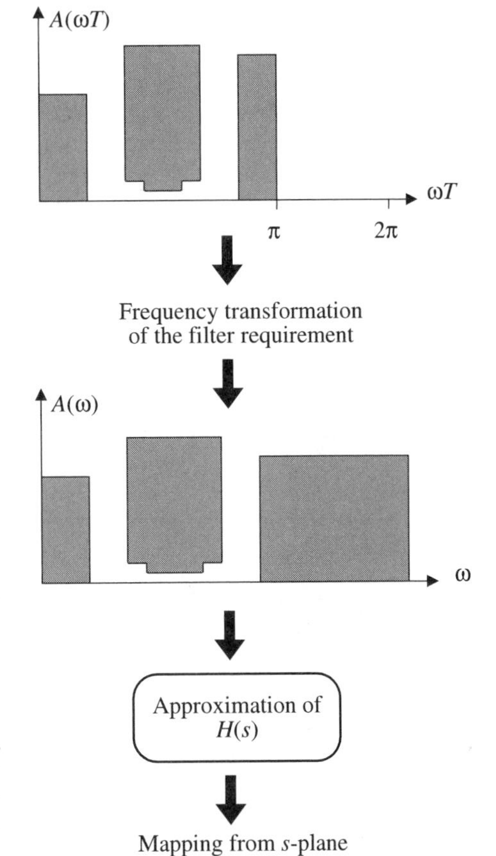

Figure 4.10 Design process based on mapping of the transfer function of an analog filter

4.8.1 Filter Order

The required filter order for an analog lowpass filter is [11]

$$N_{min} \geq \frac{c_1 \ln\left(\dfrac{c_2 \varepsilon_s}{\varepsilon_p}\right)}{\ln(c_3)} \tag{4.23}$$

where the constants are given in Table 4.2 and

$$k_0 = \sqrt{\frac{\omega_{as}}{\omega_{ac}}}$$

$$k_{i+1} = k_i^2 + \sqrt{k_i^4 - 1}, \qquad i = 0, 1, 2, 3.$$

Filter type	c_1	c_2	c_3
Butterworth	1	1	k_0^2
Chebyshev I and II	1	2	k_1
Cauer	8	4	$2\,k_4$

Table 4.2

Equation (4.23) can also be used to determine the required order for the corresponding digital IIR filter. We demonstrate the computational details with an example.

Example 4.3

Determine the required order for a digital Cauer filter that shall meet the following specification:

Passband: $A_{max} = 0.4$ dB, $f_c = 100$ kHz
Stopband: $A_{min} = 70$ dB, $f_s = 200$ kHz
$f_{sample} = 1$ MHz

We get from Equations (4.22) and (4.23)

$$k_0 = \sqrt{\frac{\omega_{as}}{\omega_{ac}}} = \sqrt{\frac{\tan\left(\frac{\pi f_s}{f_{sample}}\right)}{\tan\left(\frac{\pi f_c}{f_{sample}}\right)}} = \sqrt{\frac{\tan(0.2\pi)}{\tan(0.1\pi)}} = 1.49534878$$

$k_1 = 4.236068$ $\qquad k_2 = 35.86066$

$k_3 = 2571.973$ $\qquad k_4 = 1.323009\ 107$

$\varepsilon_p^2 = 10^{0.1A_{max}} - 1 = 0.096478$ $\qquad \varepsilon_s^2 = 10^{0.1A_{min}} - 1 = 9999999$

$$\Rightarrow \frac{\varepsilon_s}{\varepsilon_p} = 10180.88$$

Finally, we get

$$N_{min} \geq \frac{8\ln(4 \cdot 10180.88)}{\ln(2 \cdot 1.323009\ \ 10^7)} = 4.96845$$

The frequency transformations just discussed yield suboptimal solutions for bandpass and stopband filters that shall meet more general requirements—for example, different attenuation requirements in the two stopbands. Substantial improvements in such cases are obtained using iterative methods [4, 24]. In Example 4.4 we demonstrate the design of a bandpass filter with a more complicated specification.

EXAMPLE 4.4

In this example we demonstrate the synthesis of a digital IIR filter, using a so-called pole placer[1] program [4, 24], that meets the following specification:

Passband:	$A_{max} = 0.4$ dB,	$f_{c1} = 1731$ kHz,	$f_{c2} = 2395$ kHz
Lower stopband:	$A_{min1} = 70$ dB,	$f_{s1} = 1010$ kHz	
Upper stopband:	$A_{min2} = 50$ dB,	$f_{s2} = 2910$ kHz	
$f_{sample} = 18$ MHz			

The band edges are

$$\omega_{c1}T = 34.62°,\ \omega_{c2}T = 47.9°,\ \omega_{s1}T = 20.2°, \text{ and } \omega_{s2}T = 58.2°$$

The specification is illustrated in Figure 4.11. Equation (4.22) yields the corresponding band edges for the analog reference filter:

$$\omega_{ac1} = 11219644.5,\ \omega_{ac2} = 15990603.8$$
$$\omega_{as1} = 6412576.69, \text{ and } \omega_{as2} = 20037343$$

Using the program, we synthesize an eighth-order filter with different attenuation requirements in the two stopbands and with different widths of the transition bands. The program can handle stopband requirements that are piecewise constant

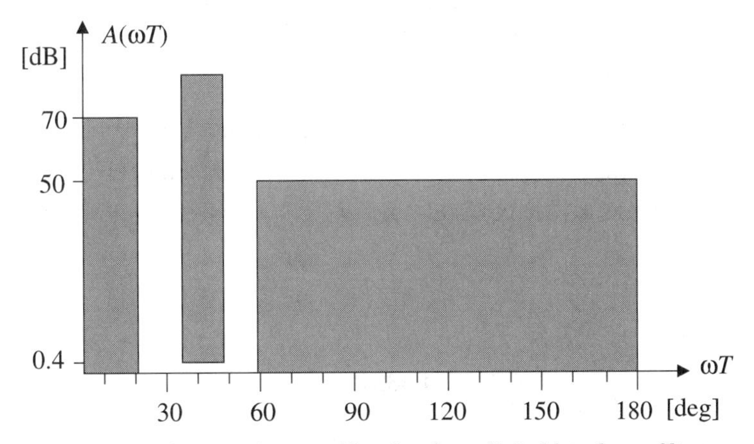

Figure 4.11 Attenuation specification for a digital bandpass filter

1. These poles are the attenuation poles, i.e., the transmission zeros.

and maximally flat or have equal ripple in the passband. In this case, we select equal ripple in the passband in order to minimize the filter order. We get

Poles	Zeros
$p_{1,2} = -384434.9838 \pm j\ 11145998.14$	$n_{1,2} = \pm j\ 6024397.4955$
$p_{3,4} = -1063192.532 \pm j\ 12441400.74$	$n_{3,4} = \pm j\ 20684703.6$
$p_{5,6} = -1154056.610 \pm j\ 14583788.02$	$n_{5,6} = \pm\ 0$ (double zero)
$p_{7,8} = -453983.571 \pm j\ 16095737.11$	$n_{7,8} = \pm\ \infty$ (double zero)
$G = 1.161876662\ 1012$ (gain constant)	

Next we can map the poles and zeros of the analog filter to the z-plane using the inverse of Equation (4.21):

$$z_i = \frac{1 + \frac{s_i T}{2}}{1 - \frac{s_i T}{2}} \tag{4.24}$$

We get the following poles:

$z_{p1,2} = 0.809095358 \pm j\ 0.55419779$

$z_{p3,4} = 0.745897626 \pm j\ 0.58606425$

$z_{p5,6} = 0.679162607 \pm j\ 0.65910842$

$z_{p7,8} = 0.652861182 \pm j\ 0.72979731$

and zeros:

$z_{n1,2} = 0.9455175 \pm j\ 0.3255714$

$z_{n3,4} = 0.5036050 \pm j\ 0.86393401$

$z_{n5,6} = 1$ (double zero)

$z_{n7,8} = -1$ (double zero)

The Pole-zero configuration is shown in Figure 4.12. The attenuation of the bandpass filter is shown in Figure 4.13 and the phase and the group delay responses for the digital bandpass filter are shown in Figures 4.14 and 4.15, respectively.

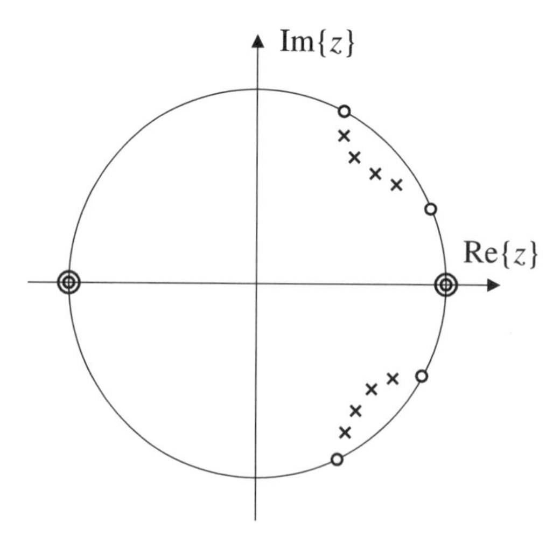

Figure 4.12 Pole-zero configuration for the eighth-order digital bandpass filter

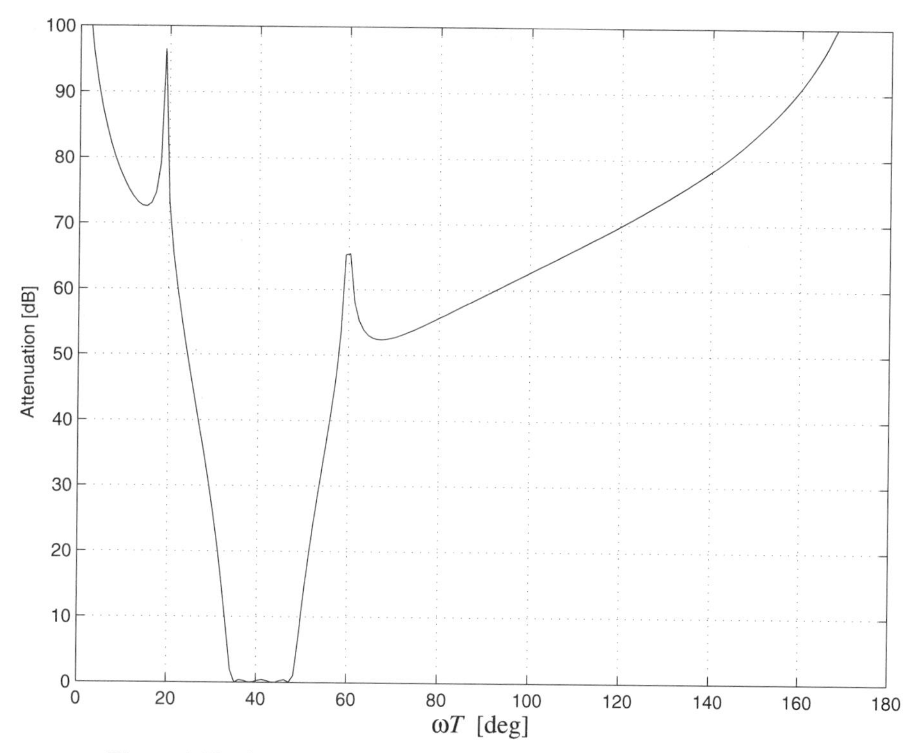

Figure 4.13 Attenuation for the eighth-order digital bandpass filter

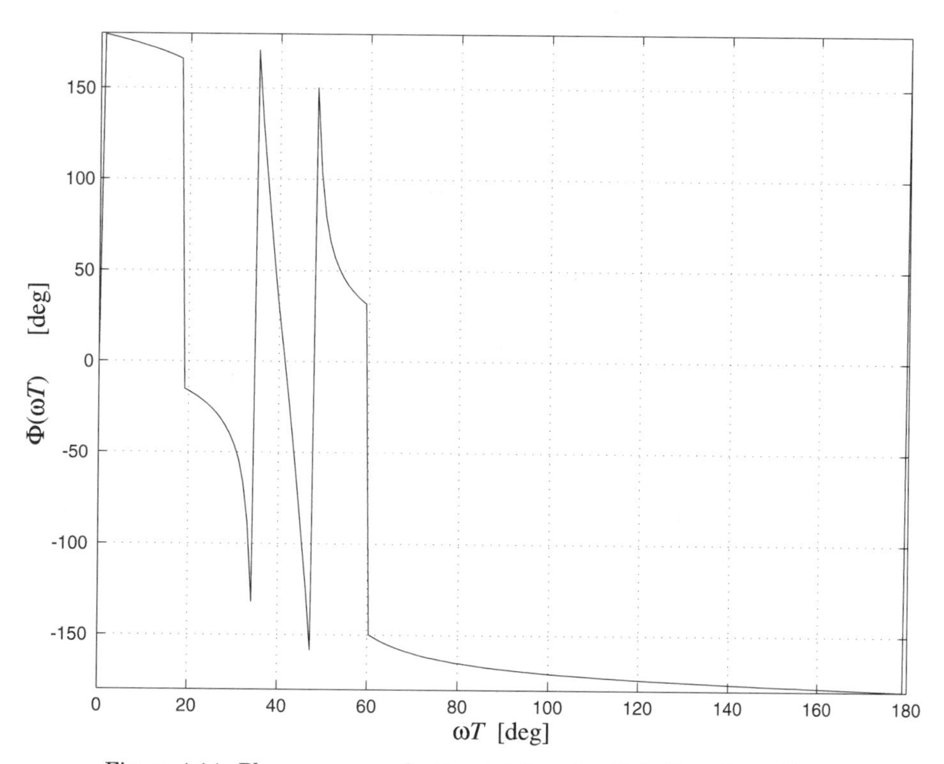

Figure 4.14 Phase response for the eighth-order digital bandpass filter

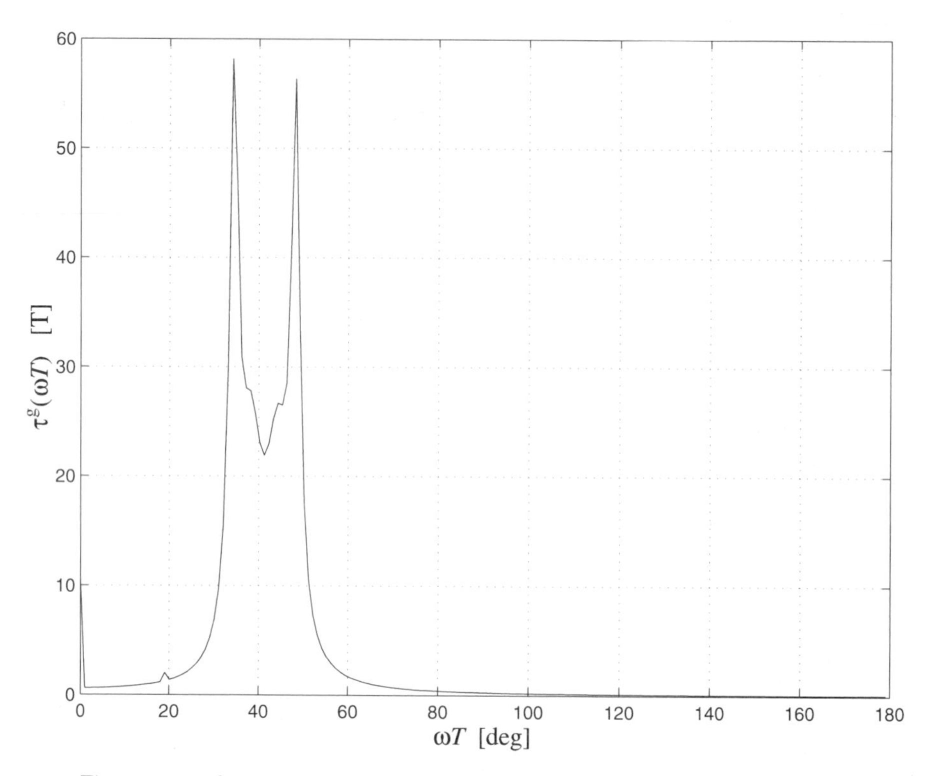

Figure 4.15 Group delay for the eighth-order digital bandpass filter

Notice the discontinuities (180°) at the frequencies of the transmission zeros, and notice that the phase varies rapidly in the passband. Also note that the group delay, which is measured in terms of the sample period, has peaks close to the passband edges. A small transition band is associated with large variations in the group delay.

Both cascade and parallel form structures are commonly used in practice. The primary reason is that they are easy to understand and design. They are also used as examples in most textbooks although their properties are not very good—e.g., lack of suppression of parasitic oscillations, high coefficient sensitivity, low dynamic range, and high computational work load.

EXAMPLE 4.5

Realize the transfer function for the filter in Example 4.4 in cascade form using second-order sections in direct form I.

The transfer function can be written

$$H(z) = \frac{(a_{01}z^2 + a_{11}z + a_{21})}{(z^2 - b_{11}z - b_{21})}\frac{(a_{02}z^2 + a_{12}z + a_{22})}{(z^2 - b_{12}z - b_{22})}\frac{(a_{03}z^2 + a_{13}z + a_{23})}{(z^2 - b_{13}z - b_{23})}\frac{(a_{04}z^2 + a_{14}z + a_{24})}{(z^2 - b_{14}z - b_{24})}$$

The filter, realized in cascade form, is shown in Figure 4.16. Note that the delay elements in two adjacent sections can be shared. The cascade form structure will be the same if, instead, direct form II sections are used, except for the first and last half-sections. The first half-section will be recursive while the last will be nonrecursive.

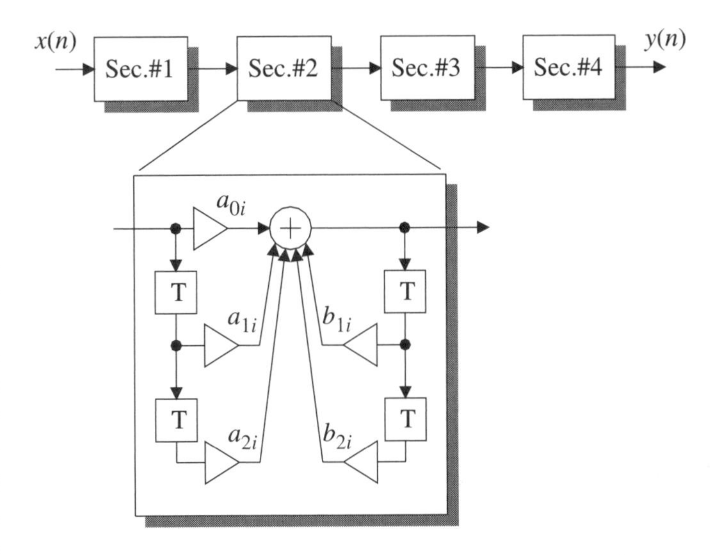

Figure 4.16 Cascade form with second-order sections in direct form I

Dynamic range is optimized by ordering the sections and pairing poles with zeros in each section. The signals at the critical nodes in the filter should be as large as possible, but not so large that overflow occurs too frequently. Hence, the gain factor G has to be distributed among the second-order sections such that the internal signal levels are optimized (see Chapter 5). A heuristic optimization procedure is given in [16]. The result of a computer-aided search over all possible combinations yields:

	Section No.			
	1	2	3	4
Pole–zero pairing	$\frac{n_3}{p_2}$	$\frac{n_2}{p_4}$	$\frac{n_1}{p_1}$	$\frac{n_4}{p_3}$
	a_{01} = 1.000000	a_{02} = 1.000000	a_{03} = 1.000000	a_{04} = 1.000000
	a_{11} = –2.000000	a_{12} = –1.006442	a_{13} = –1.890926	a_{14} = 2.000000
	a_{21} = 1.000000	a_{22} = 1.000000	a_{23} = 1.000000	a_{24} = 1.000000
	b_{11} = 1.4913900	b_{12} = 1.3056440	b_{13} = 1.6178560	b_{14} = 1.3578260
	b_{21} = –0.8997958	b_{22} = –0.9595307	b_{23} = –0.9617549	b_{24} = –0.8956455

4.9 MAPPING OF ANALOG FILTER STRUCTURES

The third method of synthesizing digital filters is based on simulating good analog filter structures. The rationale is that certain classes of lossless LC filters are optimal with respect to coefficient sensitivity and are guaranteed to be stable. Modern active RC, SC, mechanical, crystal, and ceramic filters are also based on such simulation techniques.

Figure 4.17 presents a summary of the design process based on structural mapping. The first part of the design process is the same as for the preceding methods (see Figure 4.10). First, the digital filter requirement is mapped to an analog requirement and the approximation problem is solved in the s-plane. Next, a good analog filter structure is designed. Typically, a doubly resistive terminated ladder or lattice structure is chosen. Finally, the circuit topology and circuit elements are mapped into an equivalent digital algorithm. In order to obtain a useful digital algorithm, it is necessary to use distributed circuit elements instead of lumped elements. These issues will be discussed in detail in the next sections. This design method not only solves the approximation problem, but also results in digital filter structures that may have highly favorable properties.

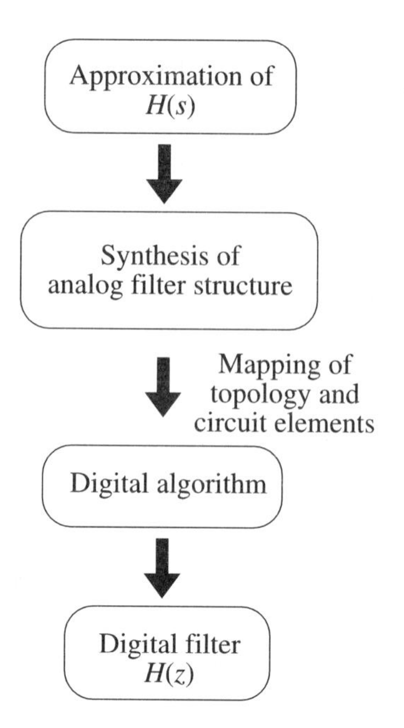

Figure 4.17 Design process based on structural mapping of an analog filter

4.10 WAVE DIGITAL FILTERS

A highly successful way to obtain a low-sensitivity digital filter structure is to simulate a low-sensitivity analog filter in such a way that the sensitivity properties are retained. Analog filters having minimum element sensitivity can be designed using the so-called *insertion loss method* [24]. Henceforth we will refer to the simulated analog filter as the *reference filter*.

An important property of wave digital filters is their guaranteed stability which is inherited from the reference filter. In practice the inductors in an *LC* ladder filter are nonlinear. Such nonlinearities may cause and sustain parasitic oscillations. However, in the passive *LC* filter such oscillations are attenuated since the filter dissipates signal power. Hence, any oscillation will eventually vanish. Wave digital filters, particularly wave digital lattice filters, are suitable for high-speed applications. They are modular and possess a high degree of parallelism which makes them easy to implement in hardware.

4.11 REFERENCE FILTERS

Figure 4.18 shows a doubly resistively terminated reactance network. Using the insertion loss method, the network is designed so that maximum power is transferred between source and load for a number of frequencies in the passband. The attenuation at these frequencies is zero with nominal element values, as shown for the filter in Figure 4.19. The filter is a fifth-order *LC* ladder structure with Chebyshev I characteristic.

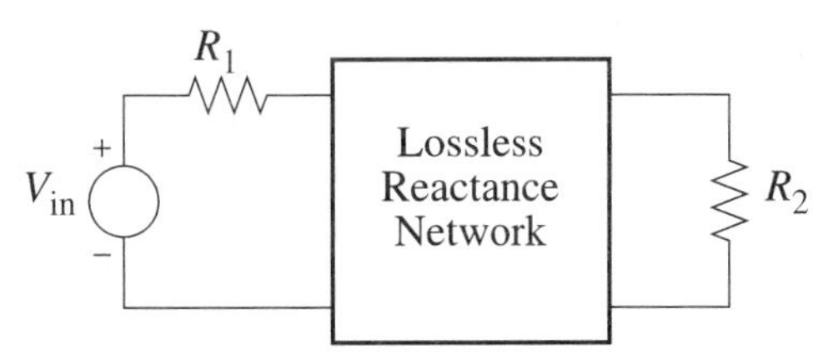

Figure 4.18 Double resistively terminated reactance network

Figure 4.18 also shows the variations in the attenuation for very large changes (±20% from the nominal value) of one of the inductances in the ladder structure. The passband ripple has been chosen to be a very large value, A_{max} = 3 dB, in order to demonstrate the low sensitivity.

A counterintuitive property of these filters is that sensitivity in the passband is reduced if the ripple is reduced. It follows from *Fettweis–Orchard's argument* that the attenuation is bounded from below (the magnitude function is bounded from above) by the maximum power transfer constraint. This lower value is usually normalized to 0 dB. Hence, if any element value in the lossless network deviates from its nominal value, the attenuation must necessarily increase and cannot decrease independently of the sign of the deviation. Thus, for these frequencies the sensitivity is zero, i.e.,

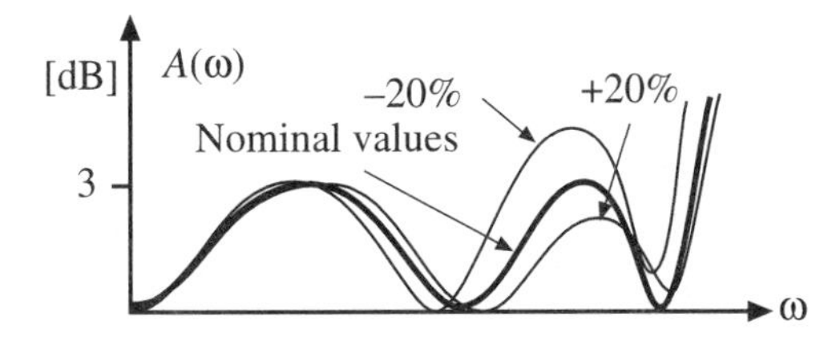

Figure 4.19 Attenuation in the passband for a fifth-order *LC* filter with nominal element values and ±20% deviation in one of the inductors

$$\frac{\partial A(\omega_{p_{max}})}{\partial e} = 0$$

where *e* represents any element in the lossless network—for example, an inductance. These filters are optimal for frequency-selective filters with "flat" passbands. The sensitivity in the stopband depends on the internal structure of the reactance network. Usually, lattice or ladder networks are used.

Many schemes for simulating doubly terminated *LC* networks have been developed for active *RC* filters [24]. Unfortunately, the problem of simulating an analog reference filter using a digital network is a nontrivial problem. Generally, a nonsequentially computable algorithm is obtained (see Chapter 6). However, Alfred Fettweis [7–10] has developed a comprehensive theory, the wave digital filter theory, which solves the problem. Indeed, it is not only a theory for design of digital filters, it is also a general theory describing the relationships between certain discrete-time networks and certain classes of lumped and distributed element networks. Furthermore, it inherently contains an energy concept[2] which can be used to guarantee stability in nonlinear digital networks.

---

[2]. There exists a Lyanuponov function for wave digital filter structures.

Historically, the rationale for developing different filter structures was the desire for low element sensitivity. Although low sensitivity is important, the most stringent requirement is that the structure is guaranteed to be stable. This issue will be further discussed in Chapter 5. Properly designed wave digital filters meet both of these requirements.

4.12 WAVE DESCRIPTIONS

Wave digital filter theory is based on a scattering parameter formalism that has been used for a long time in microwave theory for describing networks with distributed circuit elements. The one-port network shown in Figure 4.20 can be described by the incident and reflected waves instead of voltages and currents.

The steady-state *voltage waves* are defined as

$$\begin{cases} A \triangleq V + RI \\ B \triangleq V - RI \end{cases} \tag{4.25}$$

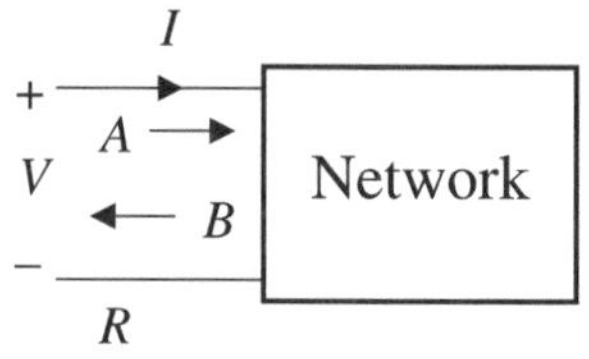

Figure 4.20 Incident and reflected waves into a port with port resistance R

where A is the *incident wave*, B is the *reflected wave*, and R is a positive real constant, called *port resistance*. Port resistance corresponds to the characteristic impedance in a lossless transmission line.

A one-port can be described by the *reflectance* function, defined as

$$S \triangleq \frac{B}{A} \tag{4.26}$$

EXAMPLE 4.6

Determine the reflectance for an impedance Z.

The voltage waves are

$$\begin{cases} A = V + RI \\ B = V - RI \end{cases}$$

and the impedance is described by $V = Z\,I$. Using Equation (4.26), we get

$$S = \frac{Z - R}{Z + R} \tag{4.27}$$

Reflectance is an allpass function for a pure reactance (see Problem 4.17).

It is not possible to directly use reference filters with lumped circuit elements, since nonsequentially computable algorithms are obtained. Instead certain classes of transmission line filters must be used. Fortunately, some of these filter structures can be mapped to classical filter structures with lumped circuit elements and we can make full use of the abundant knowledge of lumped element filters.

4.13 TRANSMISSION LINES

A special case of filter networks with distributed circuit elements is *commensurate-length* transmission line filters in which all lines have a common electrical propagation time. A *lossless transmission line* can be described as a two-port by the chain matrix

$$\begin{pmatrix} V_1 \\ I_1 \end{pmatrix} = \frac{1}{\sqrt{1-\tanh^2\left(\frac{s\tau}{2}\right)}} \begin{pmatrix} 1 & Z_0 \tanh\left(\frac{s\tau}{2}\right) \\ \frac{1}{Z_0}\tanh\left(\frac{s\tau}{2}\right) & 1 \end{pmatrix} \begin{pmatrix} V_2 \\ -I_2 \end{pmatrix} \tag{4.28}$$

where Z_0 is the characteristic impedance and $\tau/2$ is the propagation time in each direction as illustrated in Figure 4.21. Z_0 is a real positive constant cure ($Z_0 = R$) for lossless transmission lines and is therefore sometimes called the *characteristic resistance*, while lossless transmission lines are often referred to as *unit elements*. Obviously, a transmission line cannot be described by poles and zeros since the elements in the chain matrix are not rational functions in s.

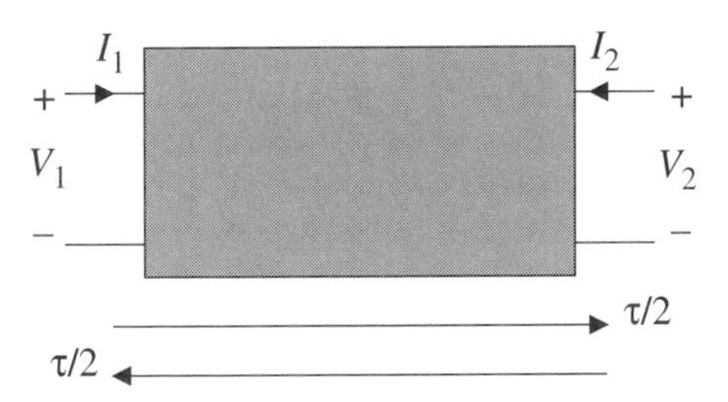

Figure 4.21 Transmission line

Wave digital filters imitate reference filters built out of resistors and lossless transmission lines by means of incident and reflected voltage waves. Computable digital filter algorithms can be obtained if the reference filter is designed using only such transmission lines. Wave digital filter design involves synthesis of such reference filters.

Commensurate-length transmission line filters constitute a special case of distributed element networks that can easily be designed by mapping them to a lumped element structure. This mapping involves *Richards' variable* which is defined as

$$\Psi \triangleq \frac{e^{s\tau}-1}{e^{s\tau}+1} = \tanh\left(\frac{s\tau}{2}\right) \tag{4.29}$$

where $\Psi = \Sigma + j\Omega$. Richards' variable is a dimensionless complex variable. The real frequencies in the s- and Ψ-domains are related by

$$\Omega = \tan\left(\frac{\omega\tau}{2}\right) \tag{4.30}$$

Notice the similarity between the bilinear transformation and Richards' variable. Substituting Richards' variable into the chain matrix yields

$$\begin{pmatrix} V_1 \\ I_1 \end{pmatrix} = \frac{1}{\sqrt{1-\Psi^2}} \begin{pmatrix} 1 & Z_0\Psi \\ \dfrac{\Psi}{Z_0} & 1 \end{pmatrix} \begin{pmatrix} V_2 \\ -I_2 \end{pmatrix} \tag{4.31}$$

The chain matrix in Equation (4.31) has element values that are rational functions in Richards' variable, except for the square-root factor. Fortunately, this factor can be handled separately during the synthesis. The synthesis procedures (programs) used for lumped element design can therefore be used with small modifications in the synthesis of commensurate-length transmission line filters.

The transmission line filters of interest are, with a few exceptions, built using only one-ports. At this stage it is therefore interesting to study the input impedance of the one-port shown in Figure 4.22. From Equation (4.31) we get the input impedance of a transmission line, with characteristic impedance Z_0, loaded with an impedance Z_2:

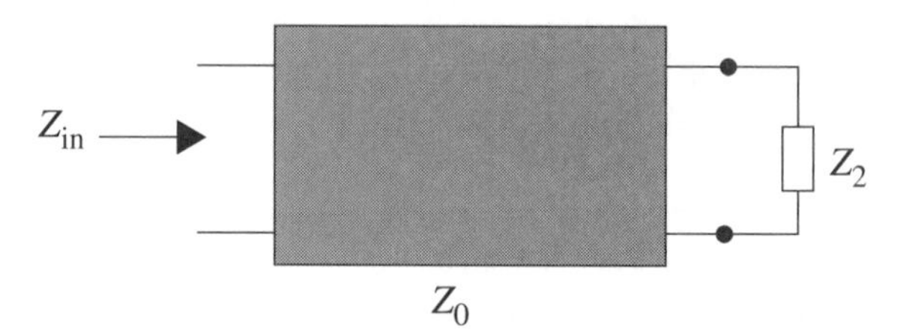

Figure 4.22 Terminated transmission line

$$Z_{in}(\Psi) = \frac{V_1}{I_1} = \frac{Z_2 + Z_0\Psi}{Z_0 + Z_2\Psi} Z_0 \tag{4.32}$$

EXAMPLE 4.7

Determine the input impedance of a lossless transmission line with characteristic impedance $Z_0 = R$ that is terminated by an impedance Z_2.

(a) $Z_2 = R$ (matched termination)

(b) $Z_2 = \infty$ (open-ended)

(c) $Z_2 = 0$ (short-circuited)

(a) $$Z_{in}(\Psi) = \frac{Z_2 + R\Psi}{R + Z_2\Psi} R = R \tag{4.33}$$

Thus, the input impedance is purely resistive and equals $Z_0 = R$.

(b) $$Z_{in}(\Psi) = \frac{Z_2 + R\Psi}{R + Z_2\Psi} R = \frac{R}{\Psi} \tag{4.34}$$

Hence, an open-ended unit element can be interpreted as a new kind of capacitor in the Ψ-domain [10].

(c) $$Z_{in}(\Psi) = \frac{Z_2 + R\Psi}{R + Z_2\Psi} R = R\Psi \tag{4.35}$$

A short-circuited unit element can be interpreted as a Ψ-domain inductor.

4.14 TRANSMISSION LINE FILTERS

Figure 4.23 shows how a commensurate-length transmission line filter is mapped to a Ψ-domain filter. Resistors are not affected since they are frequency-independent.

The synthesis of a transmission line filter starts by mapping its specification to the Ψ-domain according to Equation (4.30). In the next step, a lumped element filter is synthesized using this specification. The Ψ-domain elements are related to the normalized elements in the conventional lumped filter (s-domain) as indicated in Figure 4.24. We have

$$R_2 \leftrightarrow L_2$$
$$R_4 \leftrightarrow L_4$$
$$R_3 \leftrightarrow \frac{1}{C_3}$$

Finally, the element values for the Ψ-domain filter are obtained from the lumped filter.

In general, filters with distributed circuit elements cannot be frequency scaled, i.e., the bandwidth can not be changed by simply scaling the characteristic resistances. However, scaling commensurate-length transmission line filters can be done according to Equation (4.30) if all transmission lines are used as one-ports.

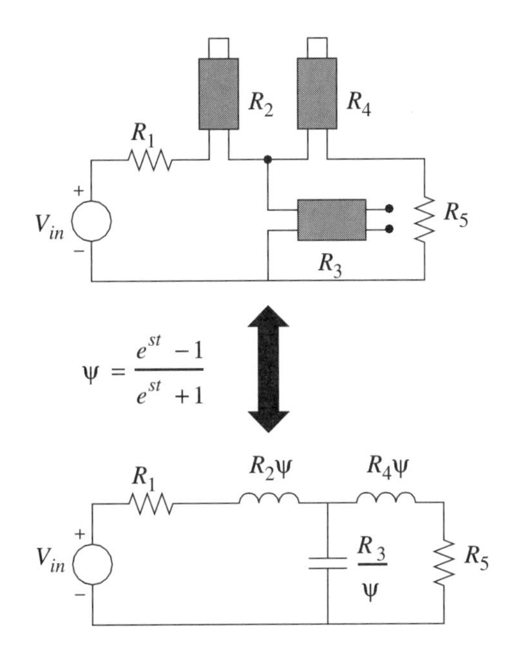

Figure 4.23 Mapping of a transmission line filter onto a Ψ-domain filter

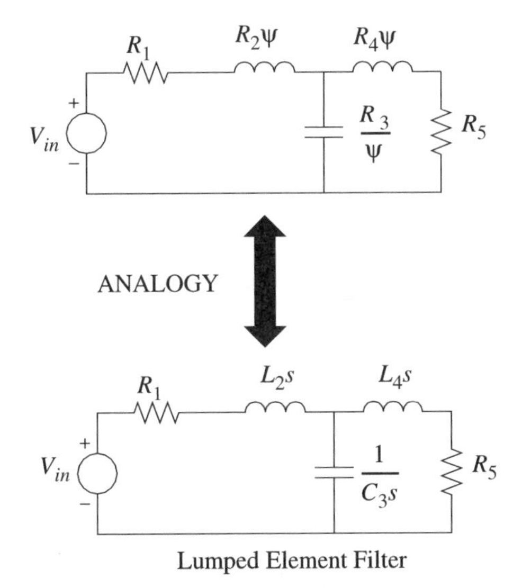

Figure 4.24 Analogy with a lumped element filter

EXAMPLE 4.8

Determine the characteristic resistances in a third-order transmission line filter of Butterworth type with a cutoff angle $\omega_c T = \pi/4$. The passband ripple is $A_{max} = 0.1$ dB.

The cutoff angle, according to Equation (4.31), is

$$\Omega_c = \tan(\pi/8) = 0.4142136$$

The corresponding lumped element filter shall have a normalized cutoff frequency:

$$\omega_c = \Omega_c$$

A ladder filter of Butterworth type with cutoff frequency ω_c has the element values

$$\begin{cases} R_1 = R_{N+2} = 1 \\ L'_{k+1} \text{ or } C'_{k+1} = 2\sin\left(\frac{(2k-1)\pi}{2N}\right), \quad k = 1, 2, \ldots, N \\ L_n = \frac{L'_n R_1}{\omega_c} \varepsilon^{1/N} \\ C_n = \frac{C'_n}{R_1 \omega_c} \varepsilon^{1/N} \end{cases}$$

$$\varepsilon = \sqrt{10^{0.1A_{max}} - 1}$$

The element values in the lumped element filter are

$R_1 = R_5 = 1$

$L'_2 = 1.0000000$	$\Rightarrow$	$L_2 = 1.2900169$
$C'_3 = 2.0000000$	$\Rightarrow$	$C_3 = 2.5803378$
$L'_4 = 1.0000000$	$\Rightarrow$	$L_4 = 1.2900169$
$\varepsilon = 0.152620419$	and	$\varepsilon^{1/3} = 0.53440545$

Using the analogy between the lumped element filter and the transmission line filter, the characteristic resistances can be determined. We get

$R_1 = R_5 = 1$	$R_2 = 1.2900169$
$R_3 = 0.3875462$	$R_4 = 1.2900169$

4.15 WAVE-FLOW BUILDING BLOCKS

The basic building blocks for the reference filter are unit elements that are either open- or short-circuited at the far end. According to Equation (4.30), the frequency response of such a unit element filter is periodic with a period of $2\pi/\tau$, i.e., the same as for a digital filter. The signals and components of the unit element filter can be mapped to a digital filter by sampling with sample period, $T = \tau$. In the following sections, we will derive the wave-flow equivalents to some common circuit elements and interconnection networks.

4.15.1 Circuit Elements

The input impedance to an open-circuited unit element (a Ψ-domain capacitor) with $Z_0 = R$ is, according to Equation (4.34),

$$Z_{in}(\Psi) = \frac{R}{\Psi}$$

Using Equation (4.27) we get the reflectance,

$$S(\Psi) = \frac{Z_{in} - R}{Z_{in} + R} = \frac{1 - \Psi}{1 + \Psi} = e^{-s\tau} \tag{4.36}$$

and

$$S(z) = z^{-1}$$

The input impedance to a short-circuited unit element (a Ψ-domain inductor) with $Z_0 = R$ is, according to Equation (4.35),

$$Z_{in}(\Psi) = R\,\Psi$$

The reflectance is

$$S(\Psi) = \frac{Z_{in} - R}{Z_{in} + R} = \frac{\Psi - 1}{1 + \Psi} = -e^{-s\tau} \tag{4.37}$$

and

$$S(z) = -z^{-1}$$

Figure 4.25 shows the two types of Ψ-domain elements and their corresponding wave-flow equivalents. An open-ended unit element corresponds to a pure delay while a short-circuited unit element corresponds to a delay and a 180-degree phase shift.

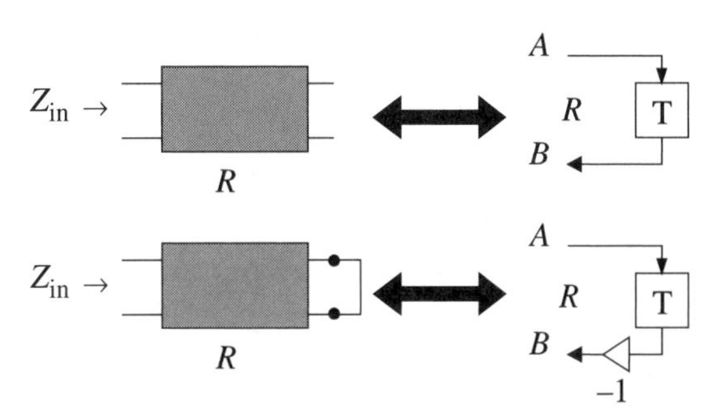

Figure 4.25 Lossless transmission lines that are open- and short-circuited at the far end and the corresponding wave-flow equivalents

The reflectance for a unit element terminated at the far end by a resistor with $Z_0 = R$ (matched) is

$$S(\Psi) = \frac{Z_{in} - R}{Z_{in} + R} = 0 \tag{4.38}$$

Hence, an input signal to such a unit element is not reflected. The corresponding wave-flow equivalent, shown to the right in Figure 4.26, is a *wave* sink.

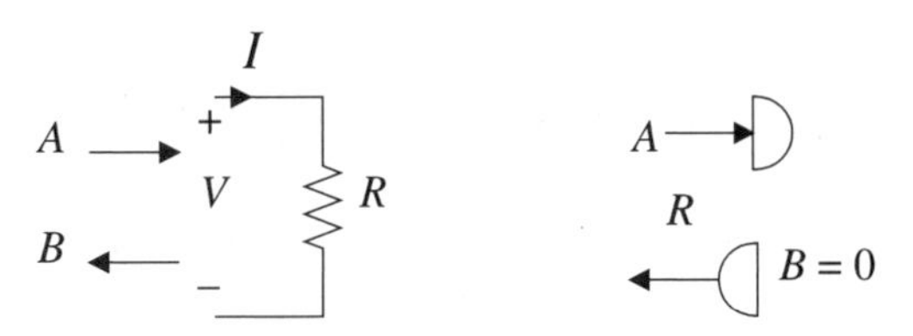

Figure 4.26 Wave-flow equivalent for a resistor

A short-circuit has the reflectance

$$S(\Psi) = \frac{Z_{in} - R}{Z_{in} + R} = -1 \tag{4.39}$$

The corresponding wave-flow graph is shown to the right in Figure 4.27. An open-circuit has the reflectance

$$S(\Psi) = \frac{Z_{in} - R}{Z_{in} + R} = 1 \quad (4.40)$$

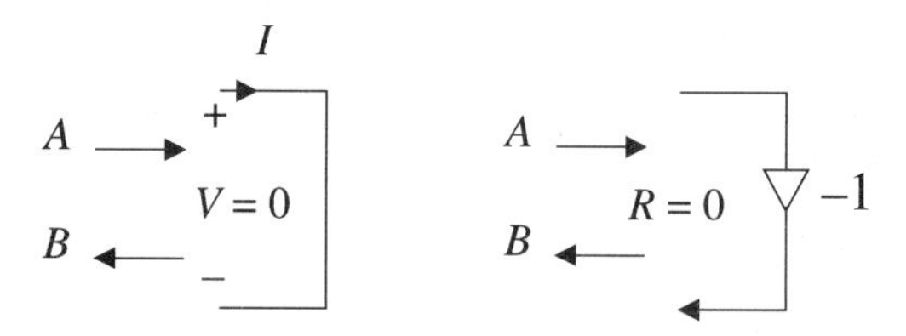

Figure 4.27 Wave-flow equivalent for a short-circuit

The wave-flow graph for an open-circuit is shown to the right in Figure 4.28.

Finally, we have for the resistive voltage source shown in Figure 4.29:

$$V_{in} = V - R\,I$$

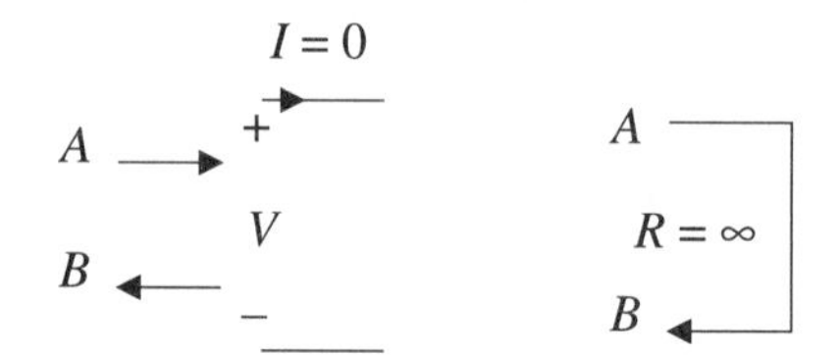

Figure 4.28 Wave-flow equivalent for an open-circuit

which yields the reflected voltage wave

$$B = V_{in} \quad (4.41)$$

The corresponding wave-flow graph is shown in Figure 4.29.

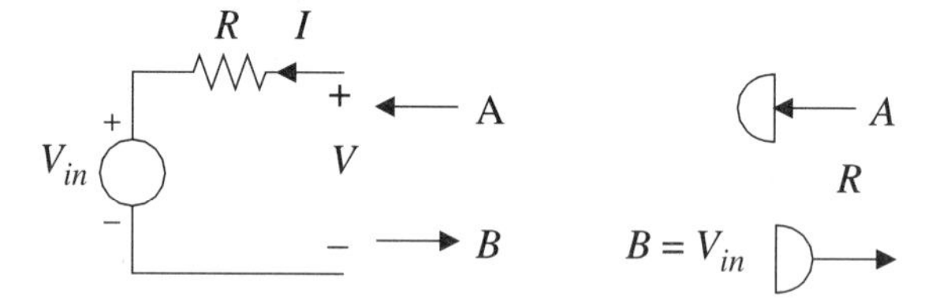

Figure 4.29 Wave-flow equivalent for a voltage source

4.15.2 Interconnection Networks

In order to interconnect different wave-flow graphs, it is necessary to obey Kirchhoff's laws at the interconnection. Generally, at a point of connection, the incident waves are partially transmitted and reflected as illustrated in Figure 4.30. Transmission and reflection at the connection point are described by a wave-flow graph called an *adaptor*.

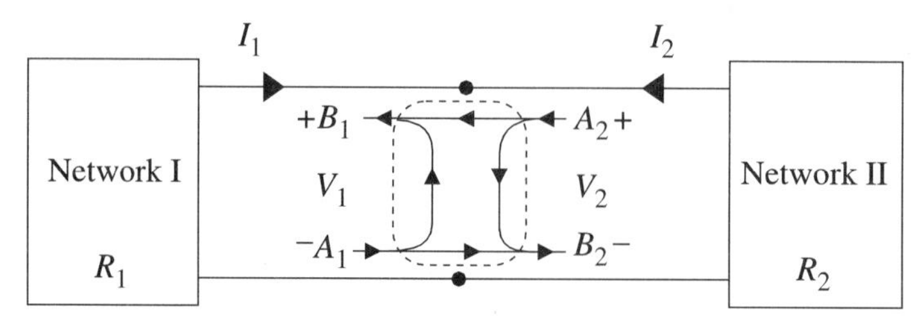

Figure 4.30 Connection of two ports

There are several types of adaptors corresponding to different types of interconnections [8]. In practice, only a few types of adaptors are used for connecting two ports. All of them represent the same voltage and current constraints, but they differ in the manner in which the reflected waves are computed. Arbitrarily complex interconnection networks can be constructed using only two-port and three-port adaptors.

Symmetric Two-Port Adaptor

Figure 4.31 shows the symbol for the symmetric two-port adaptor that corresponds to a connection of two ports;. The incident and reflected waves for the two-port are

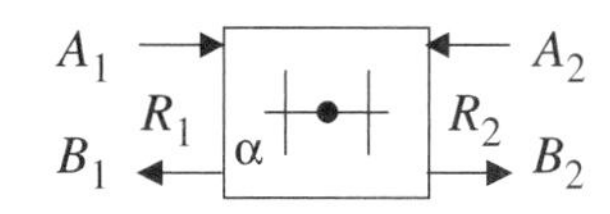

Figure 4.31 Symmetric two-port adaptor

$$\begin{cases} A_1 = V_1 + R_1 I_1 \\ B_1 = V_1 - R_1 I_1 \end{cases}$$

and

$$\begin{cases} A_2 = V_2 + R_2 I_2 \\ B_2 = V_2 - R_2 I_2 \end{cases}$$

At the interconnection we have, according to Kirchhoff's current and voltage laws,

$$\begin{cases} I_1 = -I_2 \\ V_1 = -V_2 \end{cases}$$

By eliminating voltages and currents we get the following relation between incident and reflected waves for the *symmetric two-port adaptor*

$$\begin{cases} B_1 = A_2 + \alpha(A_2 - A_1) \\ B_2 = A_1 + \alpha(A_2 - A_1) \\ \alpha = \dfrac{R_1 - R_2}{R_1 + R_2} \end{cases}$$

Equations (4.42) through (4.44) are illustrated by the wave-flow graph in Figure 4.32. The adaptor coefficient α is usually written on the side corresponding to port 1. As can be seen, the wave-flow graph is almost symmetric.

Note that $\alpha = 0$ for $R_1 = R_2$. The adaptor degenerates into a direct connection of the two ports and the incident waves are not reflected at the point of interconnection.

For $R_2 = 0$ we get $\alpha = 1$ and the incident wave at port 1 is reflected and multiplied by –1

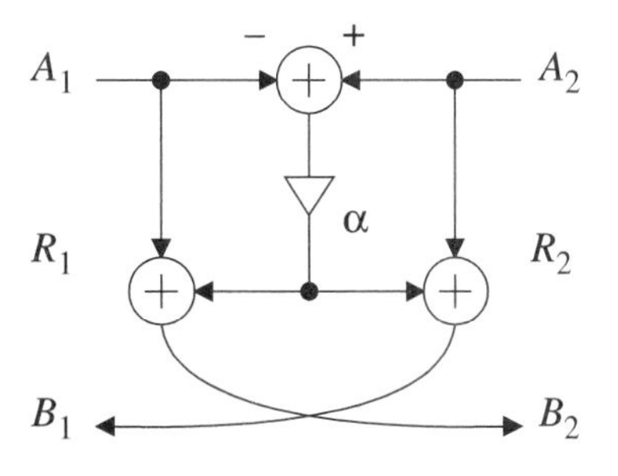

Figure 4.32 Symmetric two-port adaptor.

while for $R_2 = \infty$ we get $\alpha = -1$ and the incident wave at port 1 is reflected without a change of sign.

Series Adaptors

Ports are considered to be connected in series if the same current enters all ports while the ports have the same voltage in a parallel connection. A series connection of N ports corresponds to an adaptor described by [8, 10]

$$B_k = A_k - \alpha_k A_0 \tag{4.45}$$

where

$$A_0 = \sum_{i=1}^{N} A_i \tag{4.46}$$

and

$$\alpha_k = \frac{2R_k}{\sum_{i=1}^{N} R_i} \tag{4.47}$$

Figure 4.33 shows the symbol for the *two-port series adaptor*. In this case, each port has its own adaptor coefficient. However, the coefficients are linearly dependent since, according to Equation (4.47), we have

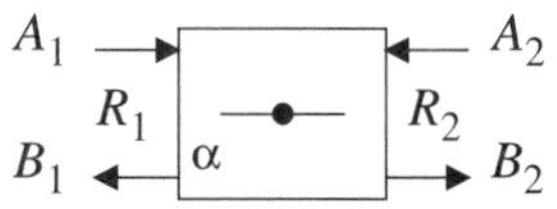

$$\alpha_1 + \alpha_2 = 2 \tag{4.48}$$

Figure 4.33 Two-port series adaptor

Hence, only one coefficient and therefore only one multiplication are required. The other coefficient can be expressed in terms of the first. The port for which the adaptor coefficient has been eliminated is called the *dependent port*. The resulting wave-flow graph, where α_2 has been eliminated, is shown in Figure 4.34. In practice, only two-, three- and four-port series and parallel adaptors are used. Arbitrarily large interconnection networks can be built using these adaptors only [8, 10]. The symbol for the *three-port series adaptor* is shown in Figure 4.35.

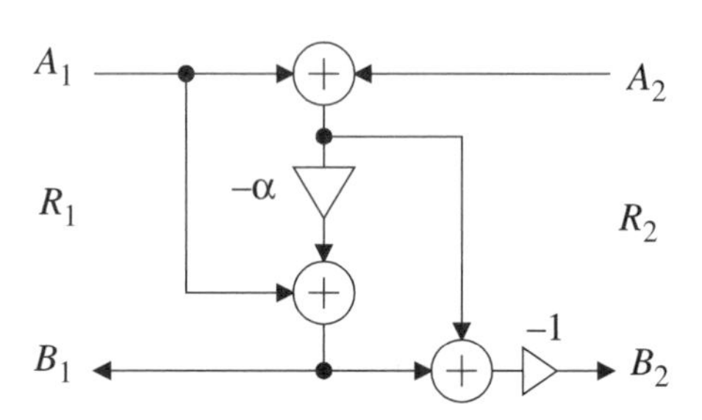

Figure 4.34 Wave-flow graph for a two-port series adaptor

A general three-port series adaptor requires two multiplications and six additions. In some cases it is possible to select one of the adaptor coefficients as 1. This means that an incident wave to this port will not be reflected. Hence, this port is called a *reflection-free port*. The number of multiplications is reduced to

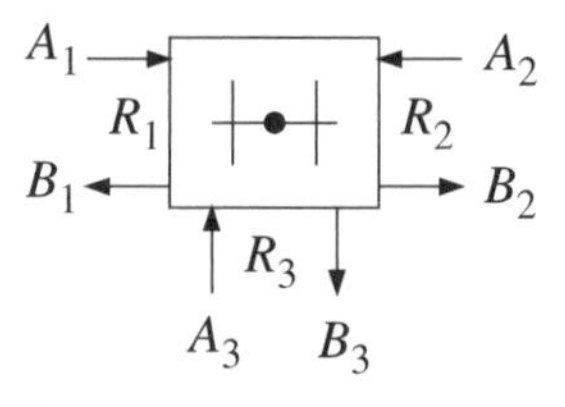

Figure 4.35 Three-port series adaptor

one and the number of additions to four in a three-port series adaptor with a reflection-free port.

Parallel Adaptors

A parallel connection of ports is characterized by the fact that all ports have the same voltage. The equations describing a parallel adaptor can be derived from the voltage wave definitions and the port voltage constraint [8, 10]. We obtain

$$B_k = A_0 - A_k \tag{4.49}$$

$$A_0 = \sum_{i=1}^{N} \alpha_i A_i \tag{4.50}$$

where

$$\alpha_k = \frac{2G_k}{\sum_{i=1}^{N} G_i} \tag{4.51}$$

and

$$G_k = \frac{1}{R_k}$$

The sum of the adaptor coefficients in an adaptor is always equal to 2 with the exception of the symmetric two-port adaptor. Hence, one of the coefficients can be expressed in terms of the others and can therefore be eliminated.

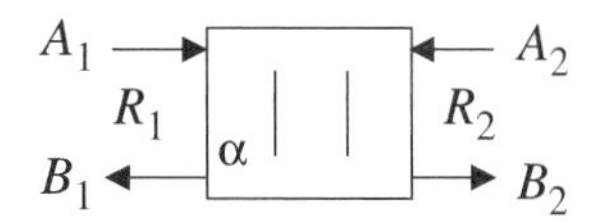

Figure 4.36 Two-port parallel adaptor

Figure 4.36 shows the symbol for a *two-port parallel adaptor* and the corresponding wave-flow graph where α_2 has been eliminated is shown in Figure 4.37. The symbol for the *three-port parallel adaptor* is shown in Figure 4.38.

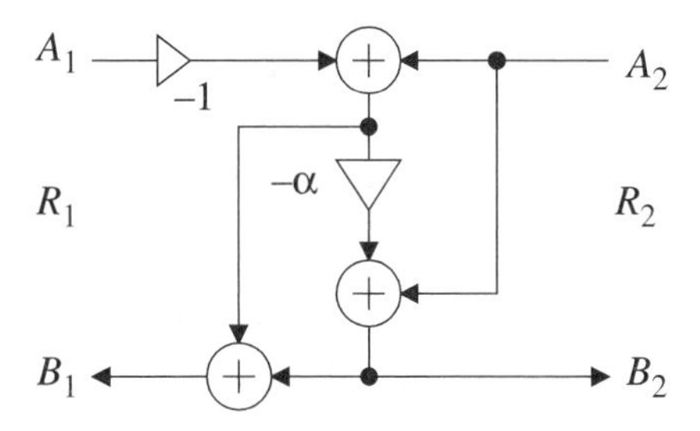

Figure 4.37 Wave-flow graph for two-port parallel adaptor

A general three-port parallel adaptor requires two multiplications and six additions. The number of multiplications is reduced to only one and the number of additions to four in a three-port parallel adaptor with a reflection-free port and a dependent port.

Adaptor networks can be modified by using various transformations [8]. The potential advantages of such transformations are improved dynamic signal range, lower element sensitivity, lower round-off noise, simpler adaptor coefficients, and improvement in the computational properties of the algorithm.

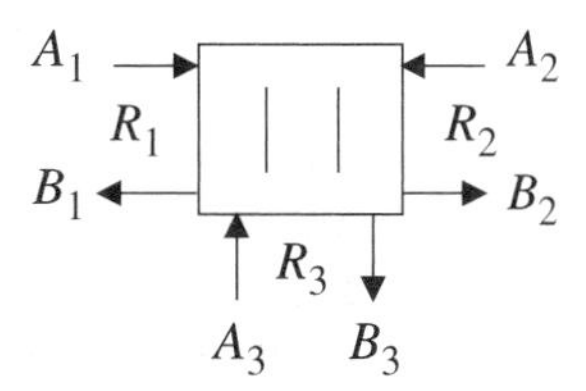

Figure 4.38 Three-port parallel adaptor

4.16 DESIGN OF WAVE DIGITAL FILTERS

The basic relationships among the wave digital filter, the corresponding reference filter, i.e., the transmission line filter, and the lumped element filter are summarized in Figure 4.39.

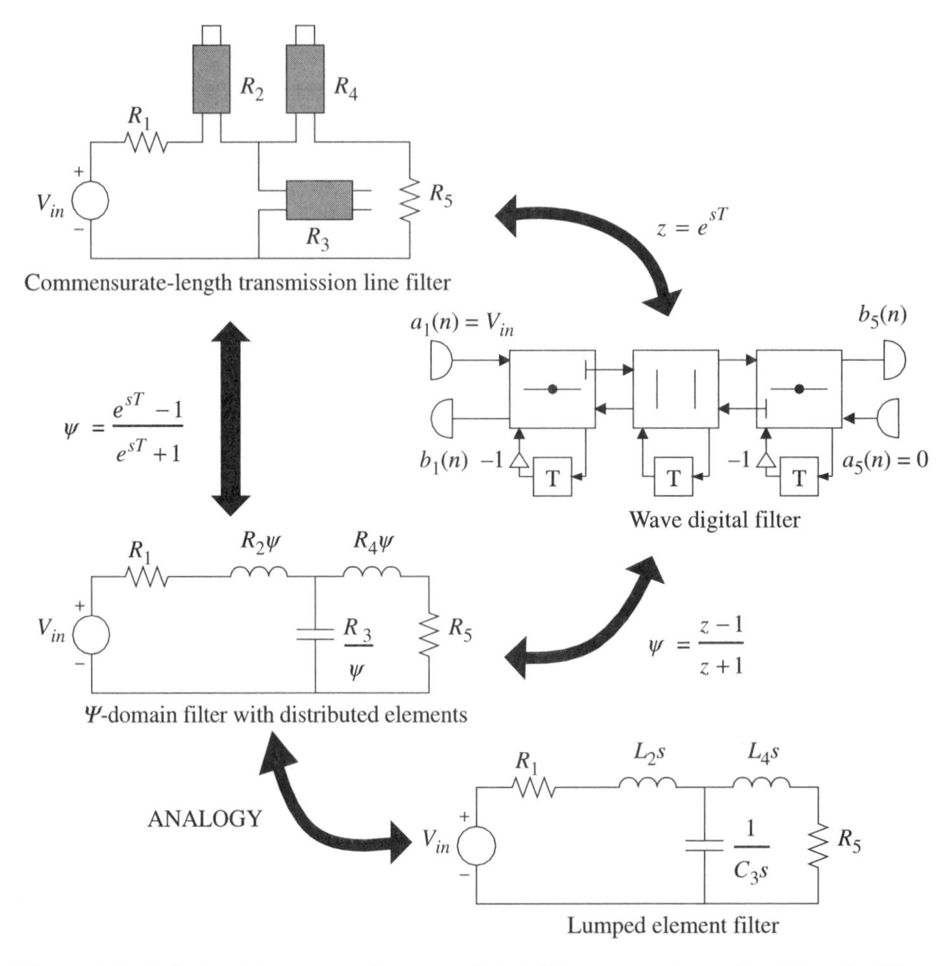

Figure 4.39 Relationships among the wave digital filter, transmission line filter, the Ψ-domain filter, and the lumped element filter

The purpose of the lumped element filter is to simplify the design process by allowing the use of conventional lumped element filter theory and design tools. In principle, all commensurate-length transmission line networks can be used as reference filters for wave digital filters. Furthermore, wave digital filters representing classical filter structures are also viable since all lumped element networks can be uniquely mapped onto commensurate-length transmission line networks using Richards' variable. However, some types of transmission line filter do not have a lumped counterpart so they must be designed directly in the Ψ-domain. Further, certain reference structures result in wave digital filter algorithms that are not sequentially computable, because the wave-flow graph contains delay-free loops (see Chapter 6).

One of the main obstacles is therefore to avoid these delay-free loops. There are three major approaches used to avoid delay-free loops in wave digital filters:

1. By using certain types of reference filters that result directly in sequentially computable algorithms. Such structures are
   a. Cascaded transmission lines, so-called Richards' structures
   b. Lattice filters with the branches realized by using circulators
   c. Certain types of circulator filters [10, 30]
2. By introducing transmission lines between cascaded two-ports [10]
3. By using so-called reflection-free ports [8, 10]

Naturally, combinations of these methods can also be used.

4.16.1 Feldtkeller's Equation

The low-sensitivity property of doubly resistively terminated *LC* filters that wave digital filters inherit can be explained by *Feldtkeller's equation*:

$$|H(e^{j\omega T})^2| + |H_c(e^{j\omega T})|^2 = 1 \tag{4.52}$$

where $H = B_5/A_1$ is the normal transfer function and $H_c = B_1/A_1$ is the *complementary transfer function*. The outputs $b_1(n)$ and $b_5(n)$ are indicated in Figure 4.39. The normal output is the voltage across the load resistor R_5 while the complementary output is the voltage across R_1 in the reference filter.

The power delivered by the source will be dissipated in the two resistors, since the reactance network is lossless. Thus, Feldtkeller's equation is a power relationship. The complementary transfer function is often called the reflection function. The maximum value of the magnitude of the reflection function in the passband is called the *reflection coefficient* and is denoted by ρ. Hence, we have

$$\rho \triangleq |H_c(e^{j\omega T})|_{max}, \ \omega T \in \text{passband} \tag{4.53}$$

Figure 4.40 shows the magnitude responses for the lowpass filter of Cauer type that was designed in Example 3.4. Note that $|H(e^{j\omega T})|$ has an attenuation peak (pole) when $|H_c(e^{j\omega T})|$ has an attenuation zero and vice versa. Also note that the attenuation at the crossover frequency is in this case 3 dB.

To achieve a reasonably large stopband attenuation for both transfer functions, the passband ripple has to be very small. Thus, the use of both transfer functions requires a higher filter order. Yet, this possibility is often highly advantageous. Note that the crossover attenuation is 3 dB in the wave digital filter and 6 dB, according to Equation (4.12), in the complementary FIR filter.

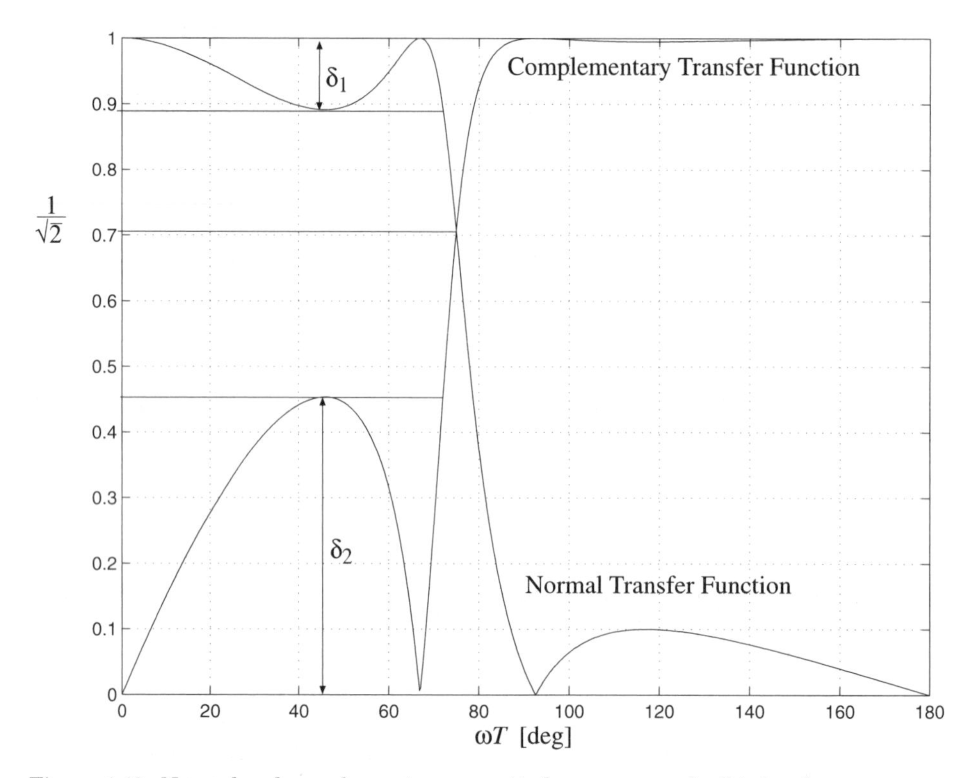

Figure 4.40 Normal and complementary magnitude responses of a third-order wave digital filter of Cauer type

Example 4.9

Determine the stopband attenuation for the complementary output in the wave digital lowpass filter just shown. The ripple in the passband of the normal output is A_{max} = 0.969 dB. Determine also the reflection coefficient.

We have

$$A_{max} = -10 \log_{10}(1 - \rho^2)$$

Let δ_1 and δ_2 denote the maximum deviation in the passband of the normal output and complementary output, respectively, as illustrated in Figure 4.40. The largest deviation in the passband is

$$A_{max} = -20 \log_{10}(1 - \delta_1) \Rightarrow (1 - \delta_1)^2 = 10^{-0.1 A_{max}}$$

From Feldtkeller's equation we have $(1 - \delta_1)^2 + \delta_2^2 = 1$

Thus, we get

$$A_{min} = -20 \log_{10}(\delta_2) = -10 \log_{10}[1 - (1 - \delta_1)^2] =$$
$$= -10 \log_{10}(1 - 10^{-0.1 A_{max}}) = 6.99 \text{ dB}$$

and

$$\delta_1 \approx 0.105556 \qquad \delta_2 \approx 0.447193$$

Hence, δ_1 must be made very small to obtain a large attenuation in the stopband of the complementary output. Further, we have

$$A_{max} = -10 \log_{10}(1 - \rho^2) = -10 \log_{10}(1 - \delta_1)^2 = -10 \log_{10}(1 - \delta_2^2)$$

Hence, the reflection coefficient is $\rho = \delta_2$ and

$$\rho^2 = 1 - (1 - \delta_1)^2 = 2\delta_1 - \delta_1^2 \approx 2\delta_1 \text{ for small values of } \delta_1$$

4.16.2 Sensitivity

It can be shown that the deviation in the passband for the doubly terminated reference filter containing only transmission lines is

$$|\Delta A(\Omega)|_{max} \leq \frac{8.686\,|\rho|_{max}\,|\varepsilon|}{|H|^2}\,\Omega\,\tau_g(\Omega) \qquad [\text{dB}] \tag{4.54}$$

where ε is the tolerance of the characteristic impedances of the unit elements. Equation (4.54) indicates that the deviation is larger close to the band edges, where the group delay is larger. Cauer and Chebyshev II filters have smaller group delays than Butterworth and Chebyshev I filters. A more important observation is that the deviation (sensitivity) becomes smaller if the filter is designed such that the complementary magnitude function, $|H_c|_{max} = \rho$, is small in the passband. This implies that the ripple in the normal magnitude function, $|H|$, is small in the passband. Hence, a filter with 3 dB ripple in the passband is more sensitive than a filter with only 0.01 dB! If the passband ripple is decreased, the filter order may have to be increased, but the filter becomes less sensitive to component errors. Thus, a trade-off between increased filter order and reduced sensitivity can be made.

4.17 LADDER WAVE DIGITAL FILTERS

Figure 4.41 shows a third-order transmission line filter of ladder type. Ladder filters can realize minimum phase transfer functions only. The corresponding ladder wave digital filter with directly interconnected three-port adaptors is shown in Figure 4.42.

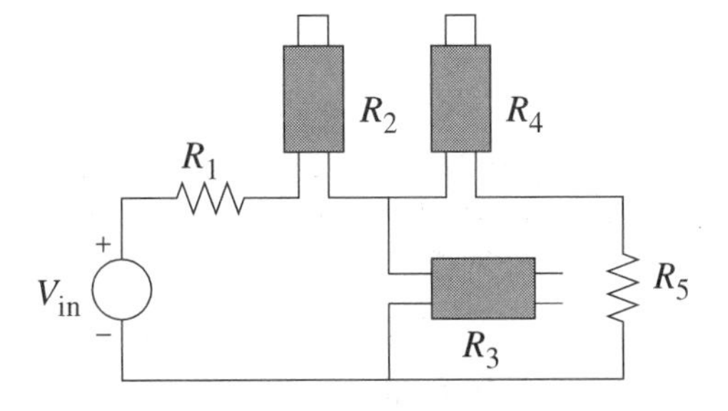

Figure 4.41 Reference filter of ladder type

Note the special symbol in the series adaptors which indicates that the corresponding port is *reflection-free* [10]. The use of reflection-free ports for ladder structures is necessary in order to obtain a sequentially computable algorithm. The input $a_1(n)$ corresponds to the voltage source V_{in}, and $b_5(n)$ is the normal output corresponding to the voltage across R_5. In some applications a second voltage source is placed in series with the load resistor R_5.

The complementary output b_1(n) corresponds to the voltage across R_1. Although ladder wave digital filter structures are very useful, we choose not to discuss the design of them in detail, since even better wave digital filter structures are available.

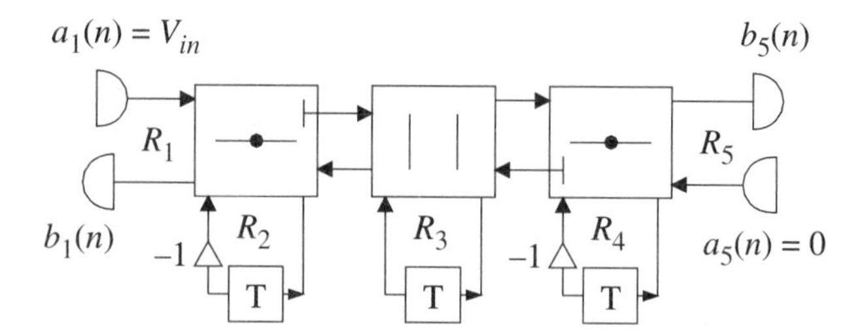

Figure 4.42 Third-order ladder wave digital filter

4.18 LATTICE WAVE DIGITAL FILTERS

Lattice wave digital filters [7, 10, 11, 30] are derived from analog lattice filters of the type shown in Figure 4.43. Impedances Z_1 and Z_2 are in practice lossless reactances. Note that the lattice structure is in fact a bridged structure. Thus, the lattice structure is extremely sensitive to element errors in the stopband. The structure is only useful if the components are extremely accurate and stable, e.g., ceramic and crystal resonators. However, this problem can be overcome when the structure is implemented using digital techniques.

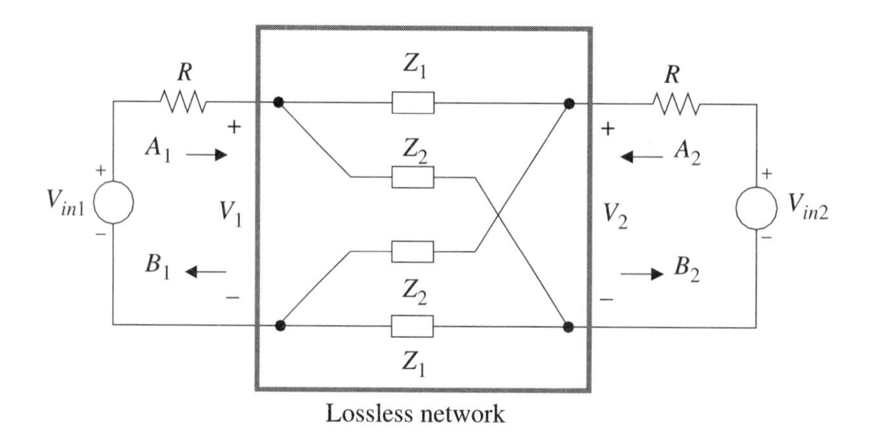

Figure 4.43 Analog lattice filter

In some applications it is useful to use two input signals simultaneously. The lattice structure can be described by the incident and reflected waves:

$$A_1 = V_1 + R\,I_1 = V_{in1} \qquad B_1 = V_1 - R\,I_1$$
$$A_2 = V_2 + R\,I_2 = V_{in2} \qquad B_2 = V_2 - R\,I_2$$

while the lossless network is described by the *scattering matrix*, $\mathbb{S}$:

$$\begin{pmatrix} B_1 \\ B_2 \end{pmatrix} = \mathbb{S} \begin{pmatrix} A_1 \\ A_2 \end{pmatrix} = \begin{pmatrix} S_{11} & S_{12} \\ S_{21} & S_{22} \end{pmatrix} \begin{pmatrix} A_1 \\ A_2 \end{pmatrix} \tag{4.55}$$

After elimination of voltages and currents we get

$$B_1 = 0.5\,[S_1(A_1 - A_2) + S_2(A_1 + A_2)] \tag{4.56}$$

$$B_2 = 0.5\,[S_1(A_2 - A_1) + S_2(A_1 + A_2)] \tag{4.57}$$

where

$$S_{11} = S_{22} = 0.5\,(S_2 + S_1)$$

$$S_{21} = S_{12} = 0.5\,(S_2 - S_1)$$

and

$$S_1 = \frac{Z_1 - R}{Z_1 + R} \tag{4.58}$$

and

$$S_2 = \frac{Z_2 - R}{Z_2 + R} \tag{4.59}$$

In practice, the impedances Z_1 and Z_2 are pure reactances. Hence, the corresponding reflectances, S_1 and S_2, are allpass functions. A lowpass lattice filter corresponds to a symmetric ladder structure and the impedances Z_1 and Z_2 can be derived from the ladder structure using Bartlett's theorem [24, 30].

Figure 4.44 illustrates the lattice wave digital filter described by Equations (4.56) through (4.59). Note that the lattice wave digital filter consists of two allpass filters in parallel. These filters have low sensitivity in the passband, but very high sensitivity in the stopband. The normal transfer function of the filter, with A_1 as input, is

$$H = S_{21} = 0.5\,(S_2 - S_1) \tag{4.60}$$

while the complementary transfer function is

$$H_c = S_{11} = 0.5\,(S_2 + S_1) \tag{4.61}$$

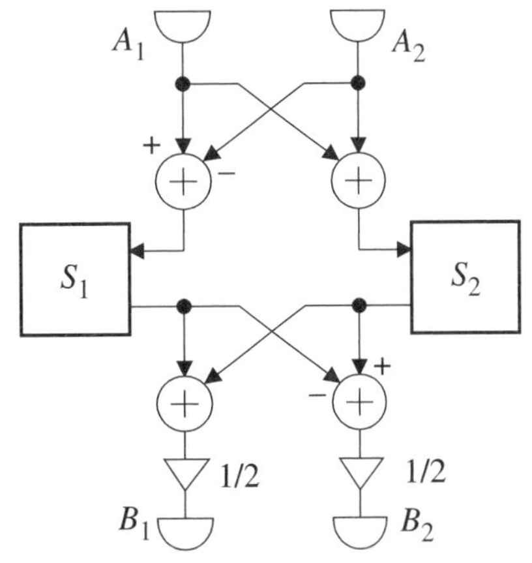

Figure 4.44 Lattice wave digital filter

It can be shown that the filter order must be odd for lowpass filters and that the transfer function must have an odd number of zeros at $z = \pm 1$ for bandpass filters.

Several methods can be used to realize canonic reactances, for example:

1. Classical *LC* structures, e.g., Foster and Cauer I and II
2. Cascaded unit elements—Richards' structures
3. Circulator structures

Figure 4.45 shows a Richards' structure, i.e., a cascade of lossless commensurate-length transmission lines, that can realize an arbitrary reactance. The far-end is either open- or short-circuited. Richards' structures are suitable for high-speed, systolic implementations of wave digital filters [12, 25].

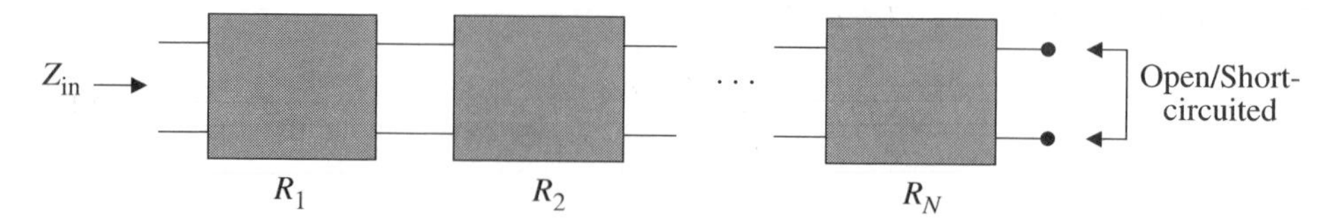

Figure 4.45 Richards' structure—cascaded unit elements

EXAMPLE 4.10

Determine the reflectance for a wave-flow graph corresponding to the Richards' structure shown in the middle of Figure 4.46. Note that the far-end is short-circuited.

It can be shown that the Richards' structure corresponds to a parallel resonance circuit in the Ψ-domain. The port resistances are

$$R_1 = \frac{R_1' R_2'}{R_1' + R_2'} \tag{4.62}$$

$$R_2 = \frac{R_1'^2}{R_1' + R_2'} \tag{4.63}$$

A direct mapping of the Richards' structure results in a wave-flow graph with three two-port adaptors. However, an adaptor with one side "short-circuited" degenerates into a multiplication by –1. The resulting wave-flow graph is shown on the right in Figure 4.46.

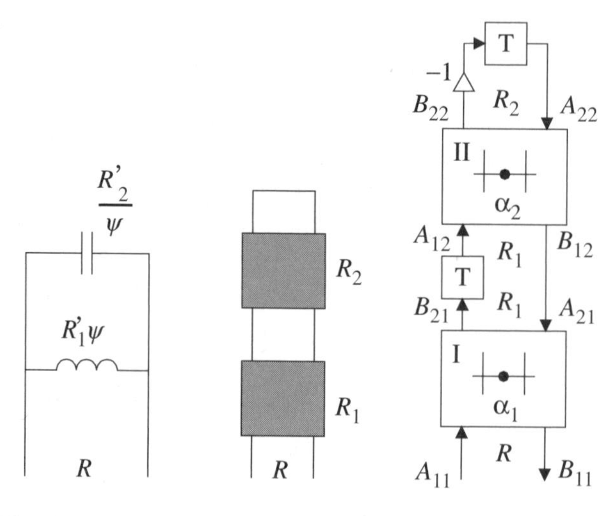

Figure 4.46 Richards' structure equivalent to second-order parallel resonance circuit with corresponding wave-flow graph

The reflectance is an allpass function:

$$S(z) = -\frac{\alpha_1 z^2 + \alpha_2(1+\alpha_1)z + 1}{z^2 + \alpha_2(1+\alpha_1)z + \alpha_1} \tag{4.64}$$

where

$$\alpha_1 = \frac{R - R_2}{R + R_2}$$

and

$$\alpha_2 = \frac{R_1 - R_2}{R_1 + R_2}$$

In some cases it may be advantageous to use a series resonance circuit since the multiplication by –1 in the wave-flow graph shown in Figure 4.46 is avoided.

Higher-order reflectances can be obtained by connecting circulators loaded with reactances as shown in Figure 4.47. The reflectance corresponding to the open-ended unit element in Figure 4.47 is

$$S(z) = \frac{-\alpha_0 z + 1}{z - \alpha_0} \tag{4.65}$$

where

$$\alpha_0 = \frac{R - R_0}{R + R_0}$$

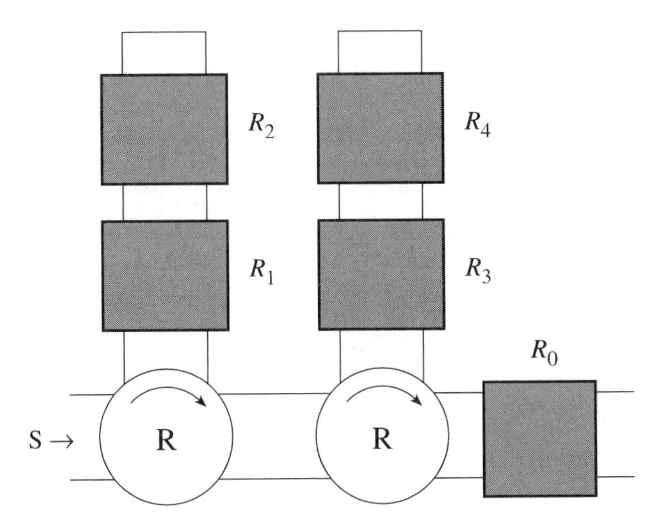

Figure 4.47 Higher-order circulator structure

Circulators can be implemented in the microwave range where they are used to, for example, direct signals from the transmitter to the antenna and from the antenna to the receiver in a radar. Ideally, no signal is directed from the transmitter to the receiver. The circulator has the property of "circulating" an incident wave to the next port.

This property of the circulator is evident from the wave-flow graph shown in Figure 4.48. Thus, the resulting wave-flow graph consists of a cascade of first- and second-order allpass (wave digital filter) sections. The combined reflectance is

$$S = (-S_1)(-S_2)(-S_3) \quad (4.66)$$

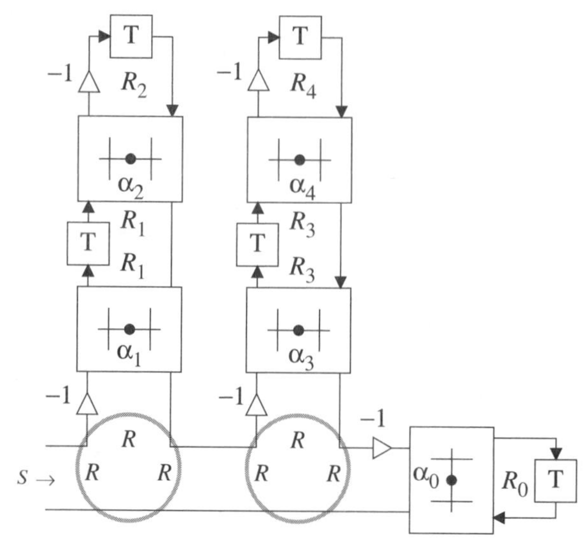

Figure 4.48 Wave-flow graph corresponding to the reference structure in Figure 4.46

Circulator structures of the type shown in Figures 4.47 and 4.48 are often preferred since they lead to computationally highly parallel and modular filter algorithms.

The multiplicatuation by –1 that appear in Figure 4.48 and Equation (4.66) are due to the definitions of port voltages and currents of the circulators and reflectances. In practice, an even number of multiplications by –1 can be removed. Hence, resulting reflectance is

$$S = -\frac{-\alpha_0 z + 1}{z - \alpha_0} \cdot \frac{\alpha_1 z^2 + \alpha_2(1 + \alpha_1)z + 1}{z^2 + \alpha_2(1 + \alpha_1)z + \alpha_1} \cdot \frac{\alpha_3 z^2 + \alpha_4(1 + \alpha_3)z + 1}{z^2 + \alpha_4(1 + \alpha_3)z + \alpha_3}$$

Characteristics of lattice wave digital filters are

- + The filter order (lowpass): $\deg\{H\} = \deg\{Z_1\} + \deg\{Z_2\}$ = odd order.
- + Number of multiplications = number of delay elements = $\deg\{H\}$.
- + Passband sensitivity is even better than for ladder structures.
- + They possess a high degree of computational parallelism and can be pipelined.
- + Simple modular building blocks are possible.
- – There is very high stopband sensitivity. However, this causes no problems in a digital implementation.

EXAMPLE 4.11

Use a program that can be derived from [11], or MATLAB, to design a digital lowpass filter of Cauer type, with A_{max} = 0.01 dB, f_c = 10 kHz, A_{min} = 65 dB, f_s = 20 kHz, and f_{sample} = 250 kHz.

A theoretical filter order of $N_{min} = 5.8707$ is required in order to meet the specification. We select $N = 7$, since only lowpass filters of odd orders are possible. The increase in filter order results in the *design margin* which can be distributed between the passband and stopband as well as between the cutoff and stopband edges. The design margin will later be used for rounding the filter coefficients to simple values. Normally, the cutoff edge is kept unchanged, but the stopband edge is reduced somewhat. The program computes the minimum stopband (frequency) edge. The stopband edge must be selected according to

$$f_{smin} = 15.434069 \leq f_s \leq f_{sreq} = 20 \text{ kHz}$$

If we select $f_s = f_{smin}$ the whole design margin is used, while if we select $f_s = f_{sreq}$ none of the margin is used, in which case the whole margin is left for the passband and stopband ripples. We select

$$f_s = 18 \text{ kHz}$$

Next, the program determines the range of the ripple factor for the passband:

$$\varepsilon_{pmin} = 0.011959065 \leq \varepsilon_p \leq 0.048012895 = \varepsilon_{preq}$$

If we select $\varepsilon_p = \varepsilon_{pmin}$ the whole design margin is used up, while if we select $\varepsilon_p = \varepsilon_{preq}$ none of the margin is used. The whole margin is left for the stopband. Since the lattice filter has very low sensitivity in the passband and very high sensitivity in the stopband, it is reasonable to allocate the major part of the design margin to the latter. We therefore select the passband ripple to

$$\varepsilon_p = 0.03$$

which corresponds to

$$A_{max} = 10 \log_{10}(1 + \varepsilon_p^2) = 0.0039069 \text{ dB}$$

Hence, $A_{max} << A_{maxreq} = 0.01$ dB.

The sensitivity of lattice wave digital filters is very large in the stopband and very small in the passband. Errors that occur when the ideal adaptor coefficients are quantized to binary values of finite precision affect the stopband much more than the passband. It may therefore be better to select a slightly larger ripple factor for the passband so that a larger part of the design margin is left for the stopband. The sensitivity in the passband will thereby increase slightly, but will still remain very low, since the ripple in the passband is extremely small.

The program determines the stopband attenuation as

$$A_{min} = 72.9885 \text{ dB} > A_{minreq} = 65 \text{ dB}$$

The lattice filter will have two parallel branches which consist of a third- and a fourth-order allpass filter. The allpass filters can be realized by using circulators, of the type shown in Figure 4.48. The resulting wave-flow graph is shown in Figure 4.49. Note the order of the sections in the upper branch has been changed compared with Figure 4.48.

The adaptor coefficients in the four allpass sections are

H_0: $\alpha_0 = 0.832865$

H_3: $\alpha_3 = 0.841873$ $\alpha_4 = -0.966577$

H_1: $\alpha_1 = 0.740330$ $\alpha_2 = -0.977482$

H_5: $\alpha_5 = 0.947965$ $\alpha_6 = -0.959780$

The poles are

$z_{p1} = 0.832865$

$z_{p2,3} = 0.850571 \pm j\,0.129845$

$z_{p4,5} = 0.890156 \pm j\,0.222475$

$z_{p6,7} = 0.934809 \pm j\,0.272208$

The zeros are

$z_{n1} = -1$

$z_{n2,3} = 0.607366 \pm j\,0.794422$

$z_{n4,5} = 0.850149 \pm j\,0.526542$

$z_{n6,7} = 0.895194 \pm j\,0.445677$

The general form of the transfer function is

$$H(z) = H_2(z)\,H_3(z) - H_0(z)\,H_1(z)$$

where $H_0(z)$ is a first-order allpass filter while the other sections are second-order allpass filters.

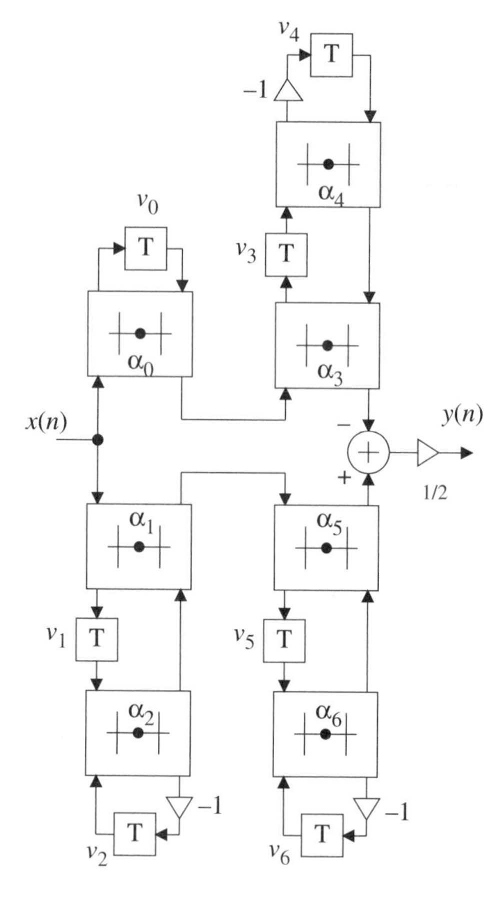

Figure 4.49 Seventh-order lowpass lattice WDF

$$H(z) = \frac{1}{2}\left[\frac{\alpha_1 z^2 + \alpha_2(1+\alpha_1)z + 1}{z^2 + \alpha_2(1+\alpha_1)z + \alpha_1} \cdot \frac{\alpha_5 z^2 + \alpha_6(1+\alpha_5)z + 1}{z^2 + \alpha_4(1+\alpha_3)z + \alpha_3} + \frac{-\alpha_0 z + 1}{z - \alpha_0} \cdot \frac{\alpha_3 z^2 + \alpha_4(1+\alpha_3)z + 1}{z^2 + \alpha_4(1+\alpha_3)z + \alpha_3}\right]$$

An important observation that simplifies the design is that the poles along the imaginary axis alternate between the upper and lower branches of the lattice filter. Thus, the poles z_{p1} and $z_{p4,5}$ are realized by the upper branch while the poles $z_{p2,3}$ and $z_{p6,7}$ are realized by the lower branch. Further, all zeros lie on the unit circle.

The attenuation is shown in Figures. 4.50 and 4.51. The passband and stopband edges are

$$\omega_c T = \frac{10 \cdot 360}{250} = 14.4° \quad \text{and} \quad \omega_s T = \frac{20 \cdot 360}{250} = 28.8°$$

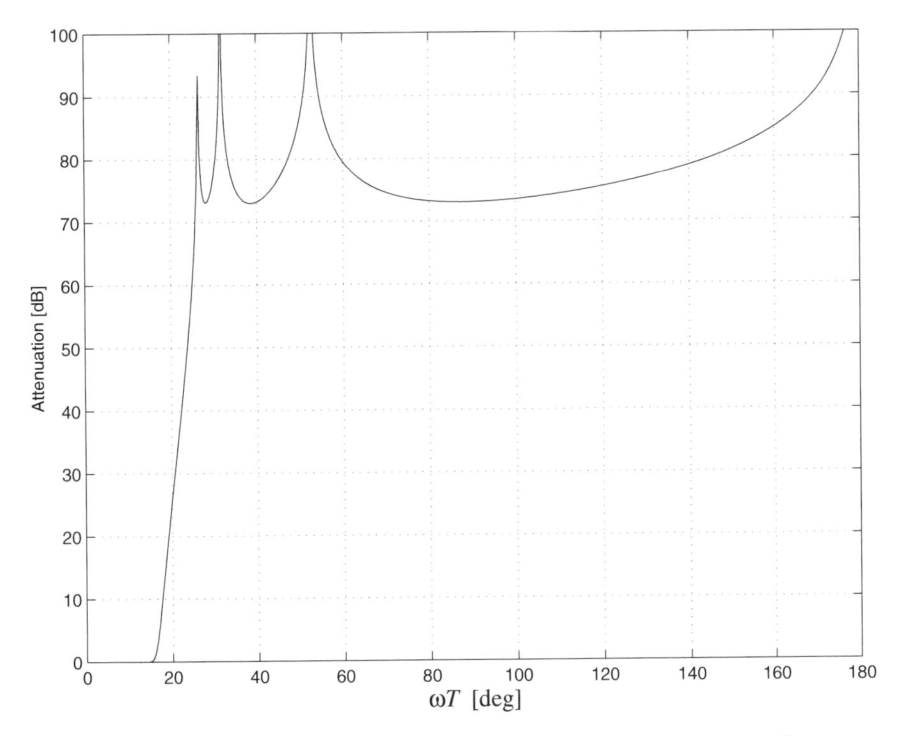

Figure 4.50 Attenuation of the seventh-order lowpass lattice WDF

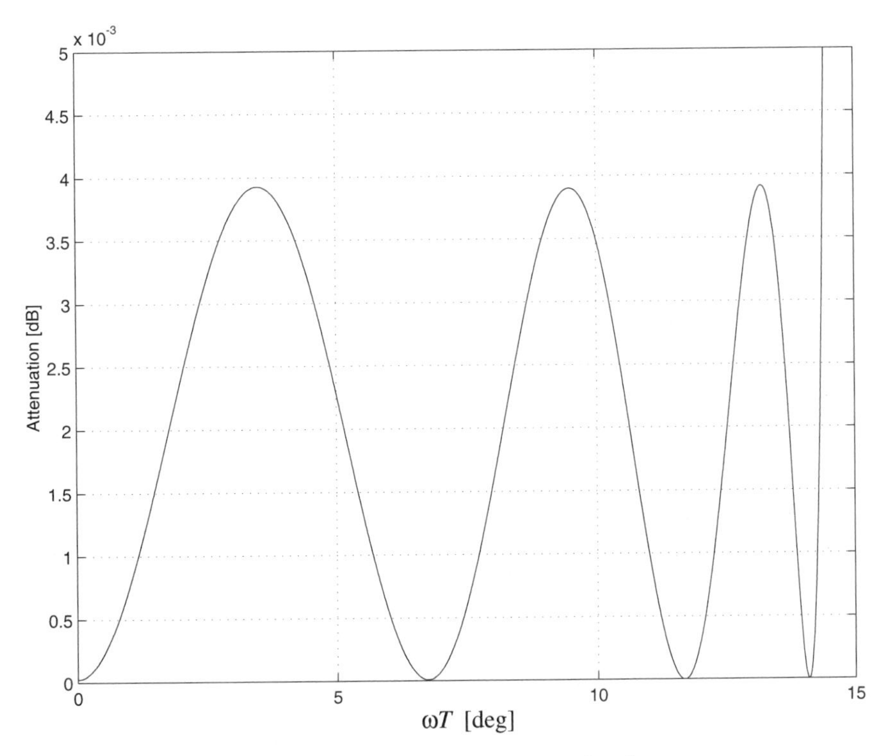

Figure 4.51 Attenuation—passband

4.19 BIRECIPROCAL LATTICE WAVE DIGITAL FILTERS

Significant simplifications of the digital filter algorithm can be made if the magnitude functions $|S_{11}|^2$ and $|S_{21}|^2$ have certain symmetric properties around $\pi/2$. Lattice wave digital filters of this type are called *bireciprocal filters*. The attenuation is always 3.01 dB at $\omega T = \pi/2$. These filters are also referred to as *half-band IIR filters*. Note that the attenuation in the passband depends on the stopband attenuation. Half-band filters are highly efficient for decimation and interpolation in multirate signal processing systems. The primary motivation for using half-band filters is the dramatically reduced computational complexity.

EXAMPLE 4.12

Use the program just mentioned [11], or MATLAB, to design a bireciprocal lowpass filter of Cauer type, with $A_{maxreq} = 0.05$ dB, $f_{creq} = 150$ kHz, $A_{minreq} = 73$ dB, $f_{sreq} = 230$ kHz, and $f_{sample} = 800$ kHz.

For bireciprocal filters we have, due to the inherent antisymmetry,

$$\varepsilon_p = 1/\varepsilon_s$$

where

$$\varepsilon_s = \sqrt{10^{0.1A_{min}} - 1} = 4466.83581$$

and

$$A_{max} = 10\log_{10}(1 + 1/\varepsilon_s^2) = 2.177\ 10^{-7}\ \text{dB}$$

The passband ripple is extremely small. Hence, the sensitivity in the passband can, according to Equation (4.54), be expected to be very low.

Now, bireciprocal filters are defined by the constraint

$$C(\Psi) = \frac{1}{C\left(\frac{1}{\Psi}\right)} \tag{4.67}$$

which implies the following symmetry constraint for the passband angle and stopband angle:

$$\omega_c T + \omega_s T = \pi$$

The corresponding relation between the stopband and passband edges in the Ψ-domain is

$$\Omega_c = \tan\left(\frac{\pi f_c}{f_{sample}}\right) = \frac{1}{\Omega_s} = \frac{1}{\tan\left(\frac{\pi f_s}{f_{sample}}\right)} = 1.268493953$$

Hence, we get

$$f_c = \frac{f_{sample}}{\pi}\arctan(1.268493953) = 170\ \text{kHz}$$

Thus, the high stopband attenuation will force the ripple in the magnitude response to be very small for frequencies up to 170 kHz. Using the program we get the required filter order

$$N_{min} \geq 10.424075$$

Lowpass lattice filters must be of odd order. We select the next higher odd integer: N = 11. The program computes that the stopband frequency must be selected in the range

$$f_{smin} = 225.693781931 \leq f_s \leq f_{sreq} = 230.000000000$$

We select f_s = 228 kHz. Thus, the passband edge will, due to the inherent symmetry, be

$$f_c = \frac{f_{sample}}{2} - f_s = \frac{800}{2} - 228 = 172 \text{ kHz}$$

The passband ripple factor can be selected in the range

$$\varepsilon_{pmin} = 0.00017056 \leq \varepsilon_p \leq 0.000223872 = \varepsilon_{preq}$$

We select ε_p = 0.000173 which corresponds to the passband ripple A_{max} = 1.255 10^{-7} dB and the stopband attenuation becomes A_{min} = 75.239 dB. It would, however, have been better to select a slightly higher value (i.e., a smaller design margin in the passband) and leave a larger part for the stopband, since the sensitivity is very large in the stopband. The wave-flow graph for the lattice filter is shown in Figure 4.52 where the order of the sections has been changed compared with Figure 4.48.

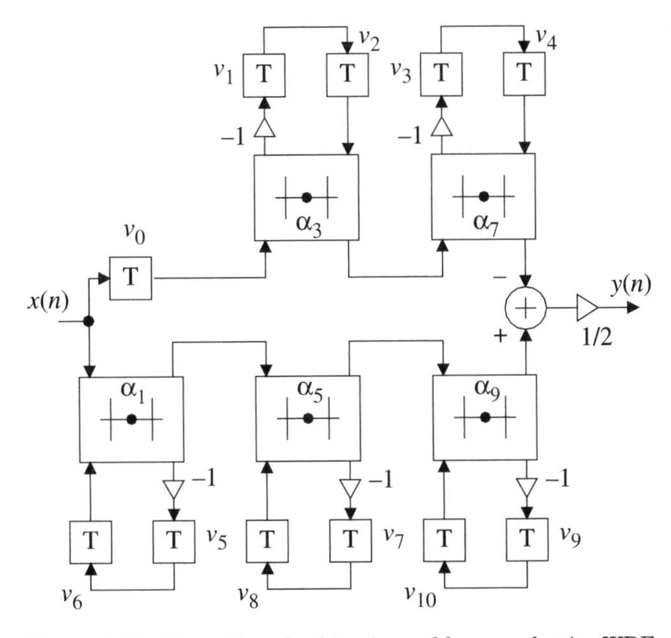

Figure 4.52 Eleventh-order bireciprocal lowpass lattice WDF

The program yields the adaptor coefficients:

$\alpha_0 = \alpha_2 = \alpha_4 = \alpha_6 = \alpha_8 = \alpha_{10} = 0$

$\alpha_1 = 0.068129$ $\alpha_3 = 0.242429$ $\alpha_5 = 0.461024$

$\alpha_7 = 0.678715$ $\alpha_9 = 0.888980$

The poles are	The zeros are
$z_{p1} = 0$	$z_{n1} = -1$
$z_{p2,3} = \pm j\,0.2610153$	$z_{n2,3} = -0.879083 \pm j\,0.476668$
$z_{p4,5} = \pm j\,0.4923708$	$z_{n4,5} = -0.633540 \pm j\,0.773710$
$z_{p6,7} = \pm j\,0.6789875$	$z_{n6,7} = -0.421103 \pm j\,0.907013$
$z_{p8,9} = \pm j\,0.8238416$	$z_{n8,9} = -0.287118 \pm j\,0.957895$
$z_{p10,11} = \pm j\,0.9428574$	$z_{n10,11} = -0.225574 \pm j\,0.974226$

The poles for bireciprocal filters are

$$z = \pm j\sqrt{\alpha_{2i-1}}$$

i.e., all poles lie on the imaginary axis in the z-plane.

The attenuations for the normal and complementary outputs are shown in Figures 4.53 and 4.54.

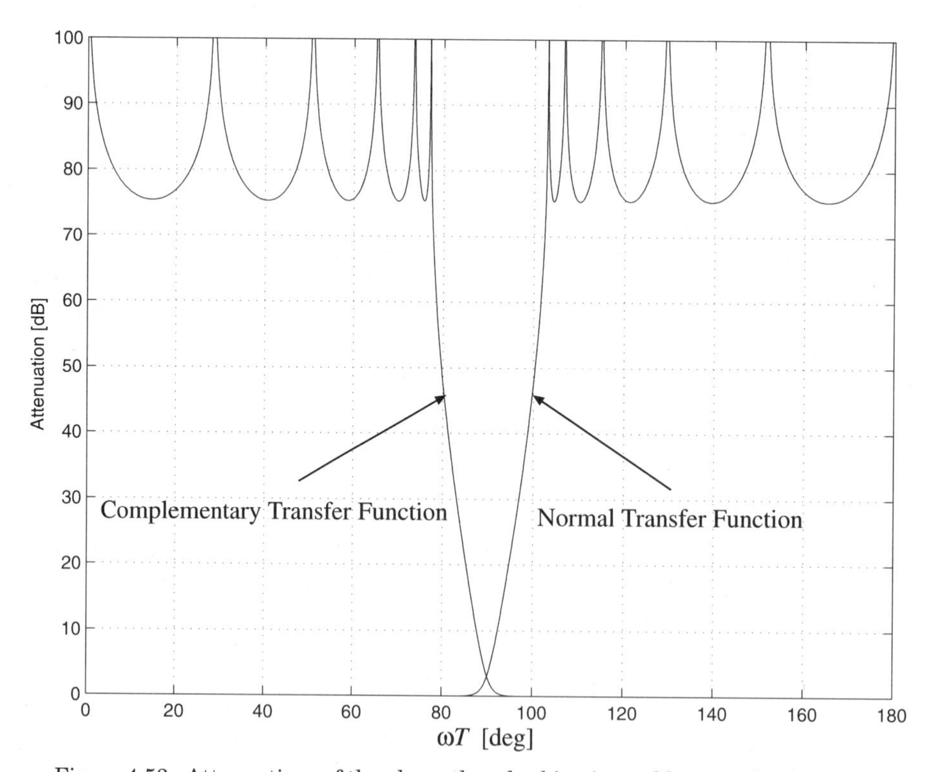

Figure 4.53 Attenuations of the eleventh-order bireciprocal lowpass lattice WDF

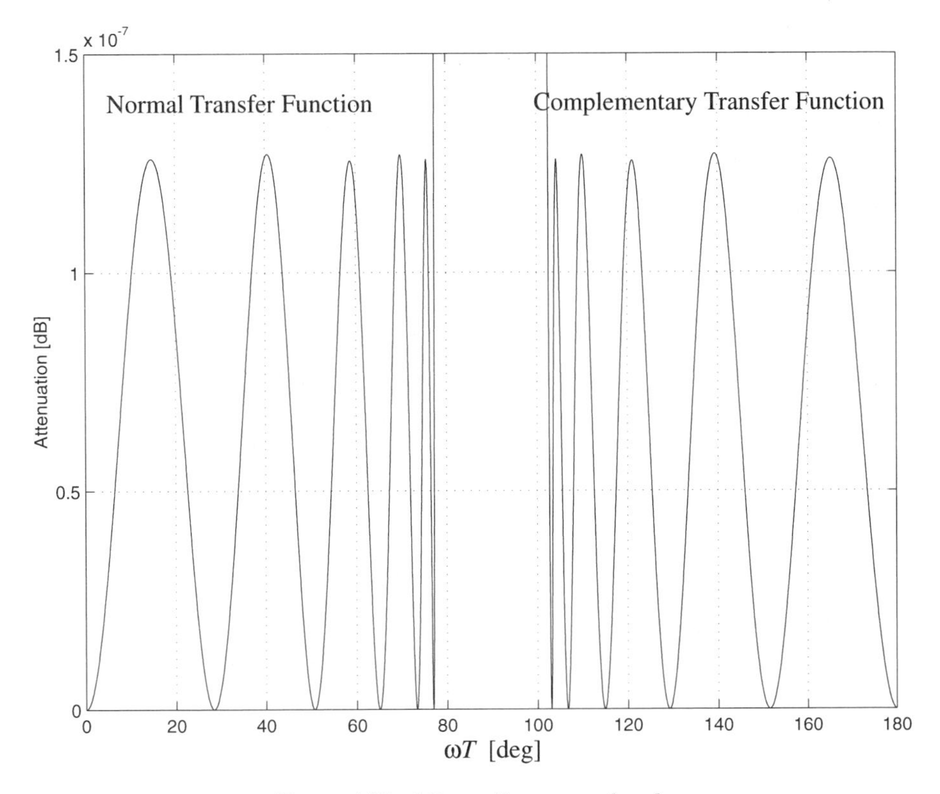

Figure 4.54 Attenuations—passbands

Note the perfect symmetry in the attenuation. The symmetry remains even after the adaptor coefficients have been quantized to binary values. The ripple in the passband will in practice be virtually unaffected by the errors in the adaptor coefficients, since the sensitivity is extremely small in the passband. In the stopband, however, the sensitivity is very large. The quantized adaptor coefficients should therefore be selected in such a way that the errors cancel each other. The adaptor coefficients can in this case be shortened to 9 bits (including the sign bit) without significant degradation in the frequency response.

The number of adaptors and multipliers is only $(N-1)/2$ and the number of adders is $3(N-1)/2 + 1$ for this type of filter (N is always odd). Bireciprocal lattice WDFs are efficient structures for decimation and interpolation of the sample frequency by a factor of two. Only magnitude responses of Butterworth and Cauer types are possible. The poles of a half-band filter lie on the imaginary axis in the z-plane. By selecting appropriate adaptor types, the overall filter can be optimally scaled in an L_∞-norm sense [11] and the wave digital filter is forced-response stable. See Chapter 5.

4.20 MULTIRATE SYSTEMS

In multirate systems the sampling frequency is changed during the signal processing. In most cases the sampling rates at the input and output differ. Such sampling rate conversions are needed when systems with different sampling frequencies are to be interconnected [1, 3, 13, 19, 20, 29]. In other cases the sampling rate is only changed internally, while the input and output rates are the same. This is done in order to improve the efficiency of the processing. These techniques are commonly used in narrow band lowpass, highpass, and bandpass filters [23], filter banks [3, 6, 13, 29], so-called transmultiplexers (converters between FDM and TDM systems), and delays of a fraction of the sample interval [3, 13]. Potential advantages of multirate signal processing are reduced computational work load, lower filter order, lower coefficient sensitivity and noise, and less stringent memory requirements. Disadvantages are more complex design, aliasing and imaging errors, and a more complex algorithm. Multirate techniques are used today in many digital signal processing systems.

4.21 INTERPOLATION WITH AN INTEGER FACTOR *L*

In many digital signal processing applications, it is necessary or computationally desirable to change the sampling frequency without changing the information in the signal. Generally, it is favorable to use as low a sampling rate as possible, since the computational work load and the required numerical accuracy will be lower.

The process of increasing the sampling rate is called interpolation. The aim is to get a new sequence corresponding to a higher sampling frequency, but with the same informational content, i.e., with the same spectrum as the underlying analog signal.

Figure 4.55 shows the original sequence $x(n)$ and an interpolated sequence y(m) that has three times as high a sampling rate. In the ideal case, both signals can be considered to be obtained by correct sampling of an analog signal, $x(t)$, but at different sampling rates. The interpolation process is essentially a two-stage process. First a new sequence $x_1(m)$ is generated from the original sequence $x(n)$ by inserting zero-valued samples between the original sequence values to obtain the desired sampling rate. Ideal lowpass filters are then used to remove the unwanted images.

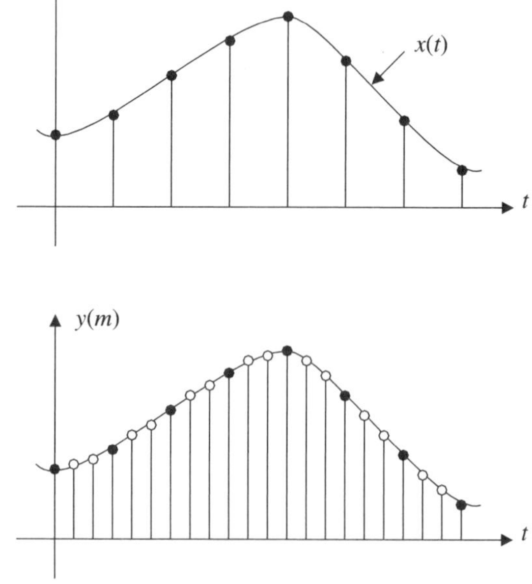

Figure 4.55 Original sequence $x(n)$ and the interpolated sequence $y(m)$

Figure 4.56 shows the symbol used for depicting the operation of interleaving L–1 zeros between each input sample. The opposite operation of removing L–1 samples is called *compression* and is depicted with a downward-pointing arrow.

Figure 4.56 Symbol for expander

The new signal, $x_1(m)$, is formed from $x(n)$ according to

$$x_1(m) = \begin{cases} x\left(\frac{m}{L}\right) & \text{if } m = 0, \pm L, \pm 2L, \ldots \\ 0 & \text{otherwise} \end{cases} \tag{4.68}$$

The sample period for the new sequence is $T_1 = T/L$. The Fourier transform of $x_1(m)$ can be expressed in terms of the Fourier transform of $x(n)$ according to

$$X_1(e^{j\omega T_1}) = \sum_{m=-\infty}^{\infty} x_1(m)e^{-j\omega m T_1} = \sum_{n=-\infty}^{\infty} x(n)e^{-j\omega n L T_1} = X(e^{j\omega T}) \tag{4.69}$$

Figure 4.57 illustrates the original sequence $x(n)$ and the corresponding interleaved sequence $x_1(m)$ that has a three times higher sampling rate. As shown in Figure 4.58, the spectrum of the sequence, $x_1(m)$, contains not only the baseband

$$\frac{-\pi}{T_1} < \omega < \frac{\pi}{T_1}$$

of the original signal, but also repeated images of the baseband.

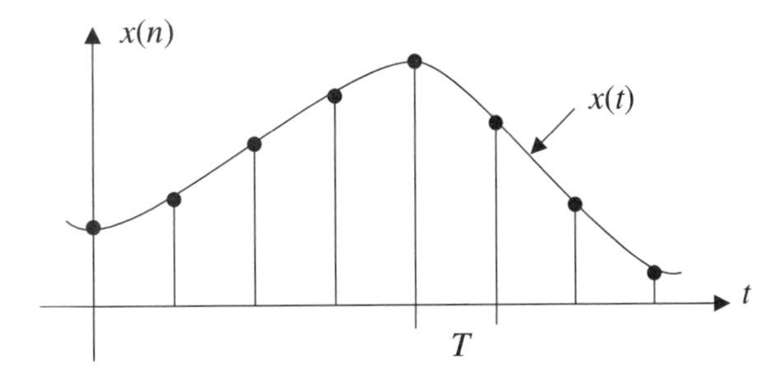

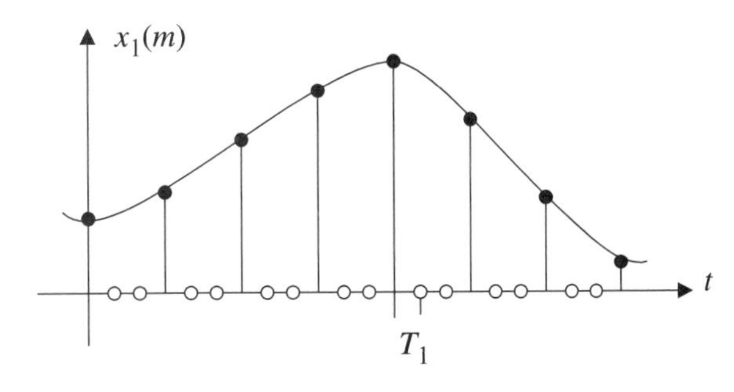

Figure 4.57 Sequences x(n) and $x_1(m)$

Obviously, the desired sequence $y(m)$ can be obtained from $x_1(m)$ by lowpass filtering. The lowpass filter is to remove the unwanted images of the baseband, as illustrated in Figure 4.59. The ideal lowpass filter shall have the stopband edge

$$\omega_s T_1 = \frac{\pi}{L} = \frac{\pi}{3}$$

In practice, the lowpass filter, $H(z)$, should have sufficient attenuation in the stopband to suppress the unwanted images of the baseband.

To summarize, an interpolator consists of an interleaving stage that generates a sequence with the correct sampling rate. This sequence is then filtered through a digital lowpass filter as illustrated in Figure 4.60.

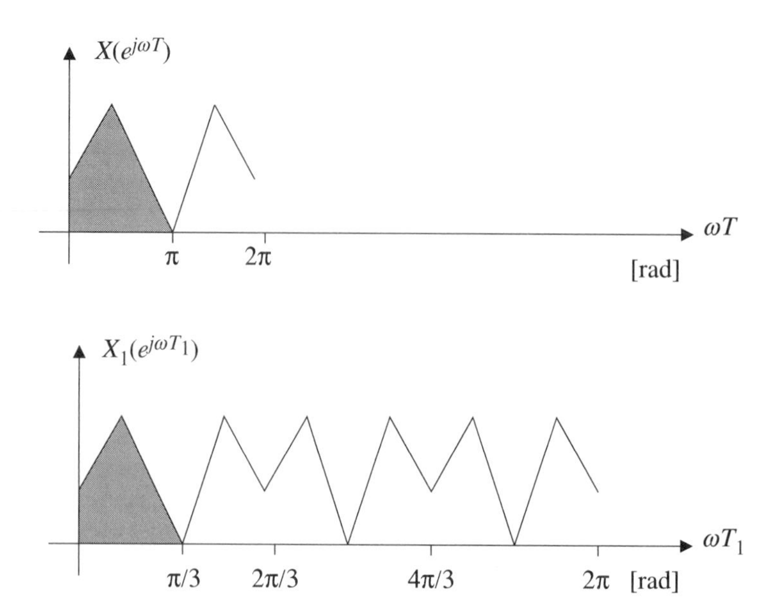

Figure 4.58 Spectrum of the original sequence and the intermediate sequence with $L = 3$

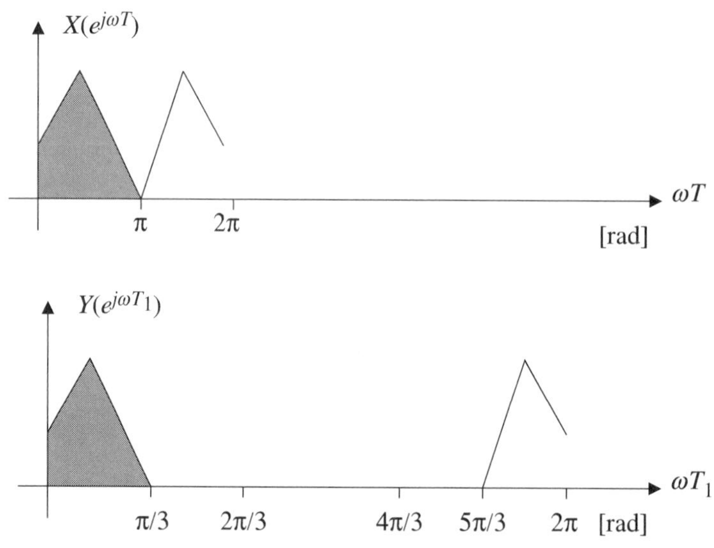

Figure 4.59 Spectrum of the sequences $x(n)$ and $y(m)$

Obviously, it is not necessary to perform arithmetic operations involving the zeros in the input sequence $x_1(m)$. Various schemes have been proposed to exploit this fact in order to reduce the computational work load. The interpolator is described by the pseudo-code shown in Box. 4.1.

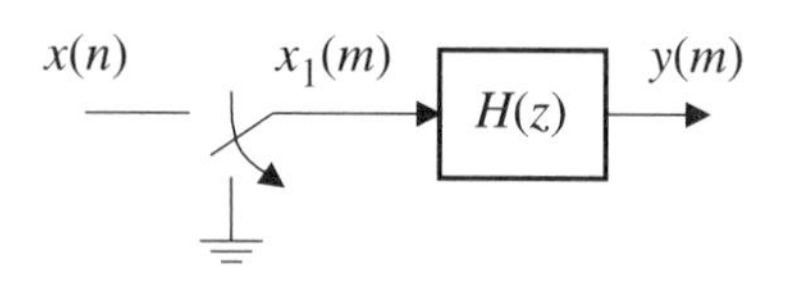

Figure 4.60 Interpolator

```
Read(x(n));
   Compute: y(n);                    { H(z) with x1(n) = x(n) }
for i := 1 to L – 1 do
   begin
x1(n + i) := 0;
      Compute: y(n + i);             { H(z) with x1(n) = 0 }
   end;
```

Box 4.1. Pseudo-code for the interpolator.

4.21.1 Interpolation Using FIR Filters

Interpolation with an integer factor can be accomplished by using FIR filters. In principle the interpolator filter is a lowpass filter that removes the unwanted images of the baseband. If the original signal is bandlimited to $\omega_0 T < \pi$, then the band $\omega_0 T$ to π will appear as bands between the images. These bands will not contain any energy from the input signal. Hence, the attenuation in these bands may be unspecified. This relaxation of the stopband requirement can be exploited to reduce the filter order.

A major drawback of higher-order FIR filters is the large group delay, approximately half the filter length. A large group delay is unacceptable in many applications. It is also associated with large memory requirement. Another drawback of FIR filters is the high computational work load, but this can be reduced by eliminating multiplications with signal values that are zero. This reduces the number of multiplications by a factor L [1]. It is also favorable to perform the interpolation in several steps of two. The lowpass filters then become half-band filters. Almost half the coefficients in a half-band filter are zero.

EXAMPLE 4.13

Determine the requirements for an FIR filter in an interpolator that interpolates the sampling frequency with a factor five. The input signal is bandlimited to 0 to $\pi/2$, and the sampling frequency is 100 kHz. Assume that image bands shall be attenuated by at least 60 dB and the passband ripple must be less than 0.5 dB.

Inserting four zeros between each input sample increases the sampling rate to 500 kHz. The spectrum for this new sequence is illustrated in Figure 4.61. The shaded images of the baseband shall be removed by the lowpass filter. From Figure 4.61, we get the requirements for the FIR filter. The passband edge is $\pi/2L$ where $L = 5$, and the stopband edge is $3\pi/2L = 3\pi/10$.

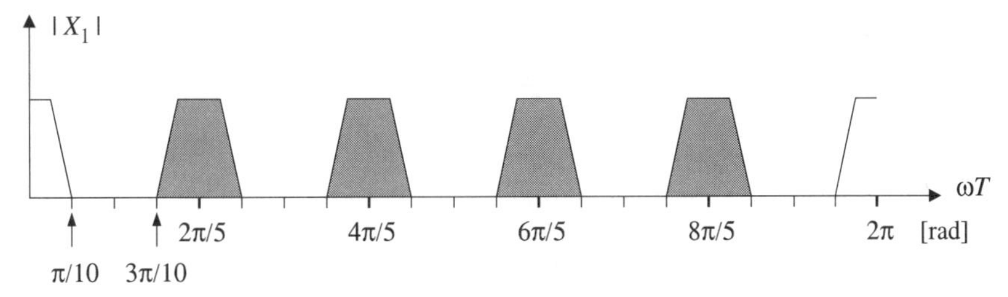

Figure 4.61 Spectrum for the interpolated sequence $x_1(m)$ with $L = 5$

Using Equations (4.8) through (4.10), with $\delta_1 \approx 0.0288$ and $\delta_2 \approx 10^{-3}$, the required filter length is estimated to $N = 24$. Note that the filter only needs to have a large attenuation in frequency bands where unwanted images appear. The frequency bands between these image bands are called don't-care bands. We get

	Band 1	**Band 2**	**Band 3**
Lower band edge	0.00000000	0.15000000	0.35000000
Upper band edge	0.05000000	0.25000000	0.45000000
Desired value	1.00000000	0.00000000	0.00000000
Weighting	1.00000000	29.00000000	29.00000000
Deviation	0.02288755	0.00078923	0.00078923
Deviation in dB	0.39766683	–62.05597565	–62.05597565

* * * * Impulse Response * * * *

$h(0) = -0.0019037865 = h(23)$ $\quad h(6) = 0.0047804355 = h(17)$

$h(1) = -0.0068186298 = h(22)$ $\quad h(7) = 0.0348714319 = h(16)$

$h(2) = -0.0119353402 = h(21)$ $\quad h(8) = 0.0743141400 = h(15)$

$h(3) = -0.0174555824 = h(20)$ $\quad h(9) = 0.1180189045 = h(14)$

$h(4) = -0.0201656641 = h(19)$ $\quad h(10) = 0.1548154097 = h(13)$

$h(5) = -0.0143036326 = h(18)$ $\quad h(11) = 0.1743385413 = h(12)$

The magnitude response is shown in Figure 4.62. Note the don't-care band between the two stopbands.

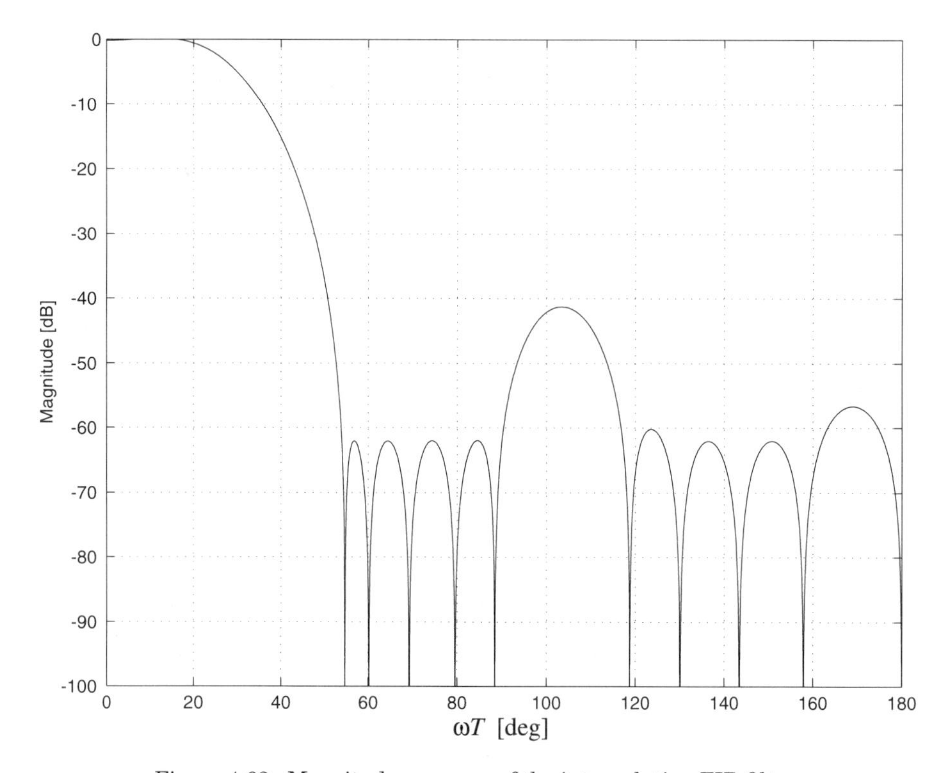

Figure 4.62 Magnitude response of the interpolating FIR filter

---

Example 4.14

Show that an FIR interpolator can be decomposed into L parallel FIR filters that operate at the input sampling rate and produce a combined output signal with an L times higher sampling rate. The resulting filter structures are often called *polyphase filters* [2, 3]. Also, determine the relation between the impulse responses of the filters.

Generally, the output of the FIR filter is described by the convolution of the input sequence $x_1(m)$ and the impulse response

$$y(m) = \sum_{k=0}^{N-1} h(k)x_1(m-k)$$

where the input sequence $x_1(m)$ is obtained from the original sequence $x(n)$ by the insertion of zeros.

Now, the original sequence is nonzero only for every Lth sample. Hence, we have for the output values at time instances, $m, m+1, \ldots, m+L-1$:

$$y(m) = \sum_{k=0}^{N-1} h(k)x_1(nL-k) = \sum_{k=0}^{M_0} h(kL)x(n-k)$$

$$y(m+1) = \sum_{k=0}^{N-1} h(k)x_1(nL-k+1) = \sum_{k=0}^{M_1} h(kL+1)x(n-k)$$

$$\ldots$$

$$y(m+L-1) = \sum_{k=0}^{N-1} h(k)x_1(nL-k+L-1) = \sum_{k=0}^{M_L-1} h(kL+L-1)x(n-k)$$

where M_i is the largest integer equal to or less than $(N-i-1)/L$. Thus, L output values are computed from the original input sequence. This corresponds to L FIR filters with the following impulse responses

FIR 1:	$h(0)$,	$h(L)$,	$h(2L)$,	$\ldots$,	$h(M_0)$
FIR 2:	$h(1)$,	$h(L+1)$,	$h(2L+1)$,	$\ldots$,	$h(M_1)$
$\ldots$					
FIR L:	$h(L-1)$,	$h(2L-1)$,	$h(3L-1)$,	$\ldots$,	$h(M_{L-1})$

Figure 4.63 shows the final signal-flow graph for the FIR interpolator. L output samples are computed for each input sample. The number of multiplications is N (the same number as in the original FIR filter) but the number of operations per second is reduced by a factor L. Only the values in the input sequence need to be stored.

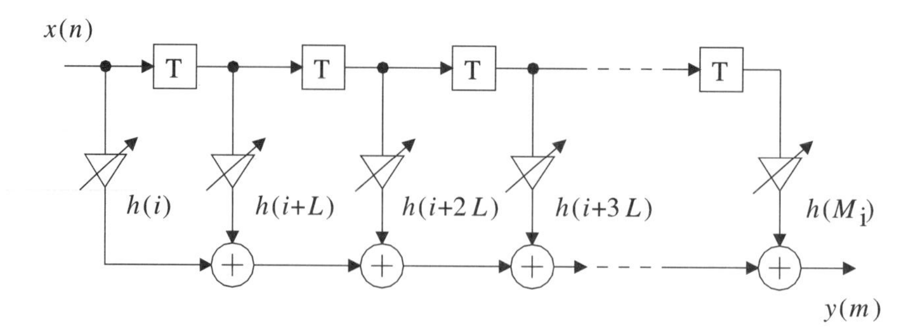

Figure 4.63 Interpolating FIR filter working at the input sample rate. $i = 0, 1, 2, ..., L - 1$ and $m = nL + i$

4.21.2 Interpolation Using Wave Digital Filters

Interpolation can also be performed efficiently using IIR filters. Lattice wave digital filters are particularly efficient for interpolation and decimation by a factor of two. An IIR filter has a much smaller group delay than its corresponding linear-phase FIR filter, but the phase response is nonlinear. Often this is of no concern, but in some cases the variation in group delay must be corrected. This can be done by placing an allpass filter in front of, or after, the lowpass filter (interpolator). In the former case, the allpass filter operates at the lower sample frequency. It is also efficient from a computational point of view to interpolate in steps of two, using half-band wave digital filters. Some improvements in stopband attenuation for wave digital filters can be obtained using special techniques [9].

An alternative to using wave digital filters is to use recursive *polyphase structures* [1, 29]. Polyphase structures consist of a number of filter branches that are periodically sampled at their outputs. The name polyphase structures comes from the fact that each branch has a different phase response.

Example 4.15

Design an interpolator that increases the sampling frequency with a factor two, from 1.6 MHz to 3.2 MHz. The energy of the input signal is contained within the band 0 to 680 kHz. No signal or noise is present above this band. The input data word length is 12 bits and the output signal should have essentially the same accuracy.

Only one interpolation stage is needed. A new signal, $x_1(m)$, is formed by interleaving the input sequence with zeros. Figure 4.64 illustrates the spectrum of the input signal and the new signal with twice the sampling rate. The lowpass filter will remove the unwanted image in the band 920–2280 kHz and have a smooth passband up to 680 kHz. The required attenuation can be estimated by noting that a signal in the unwanted frequency band should not contribute significantly to the output signal, e.g., less than $Q/2$, where Q = quantization step of the signal values. Thus,

$$A_{min} \geq -20 \log_{10}\left(\frac{Q}{2}\right) = -20 \log_{10}(2^{-12}) = 72.25 \text{ dB}$$

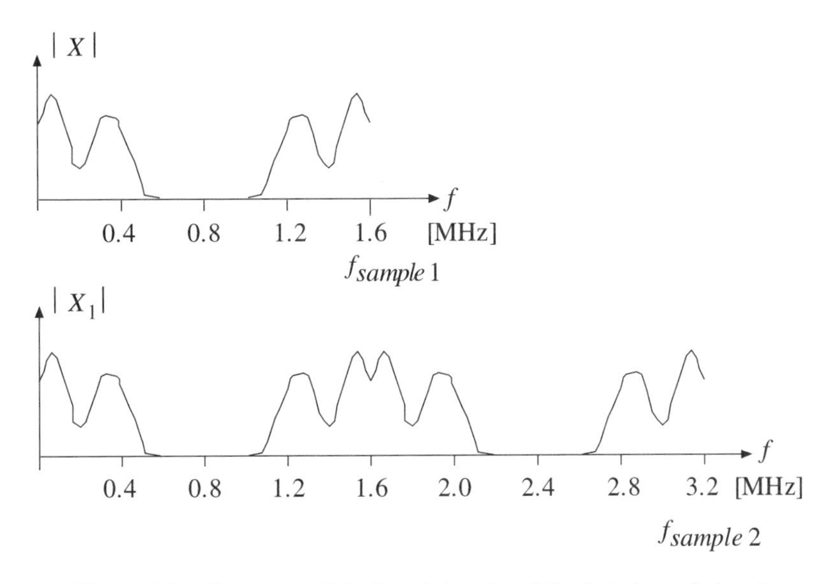

Figure 4.64 Spectrum of the input signal and the interleaved signal

It is favorable to use bireciprocal WDF lattice filters for interpolation with a factor of two. The stopband requirement can easily be satisfied and the ripple in the passband will, due to the symmetry, be very small when such filter structures are used. Note that the filter designed in Example 4.12 meets this specification.

It is, of course, unnecessary to process input values that are zero. In Example 4.16 we will show that it is possible to modify the system of difference equations so that the bireciprocal wave digital filter operates at the lower input sampling rate [3, 29].

EXAMPLE 4.16

Show that the interpolation filter in Example 4.15 can be made to operate at the lower (input) sampling rate.

The z-transform of the interleaved input sequence is

$$X_1(z) = X(z^2)$$

The transfer function of the original filter is

$$H(z) = z^{-1} H_1(z^2) + H_2(z^2) \tag{4.70}$$

Hence, the z-transform of the output is

$$\begin{aligned} Y(z) &= H(z) X_1(z) = [z^{-1} H_1(z^2) + H_2(z^2)] X(z^2) \\ &= z^{-1} H_1(z^2) X(z^2) + H_2(z^2) X(z^2) \end{aligned}$$

Note that the term $H_2(z^2) X(z^2)$ corresponds to a sequence that is zero for every other sample. Alternatively, this sequence can be obtained by filtering the

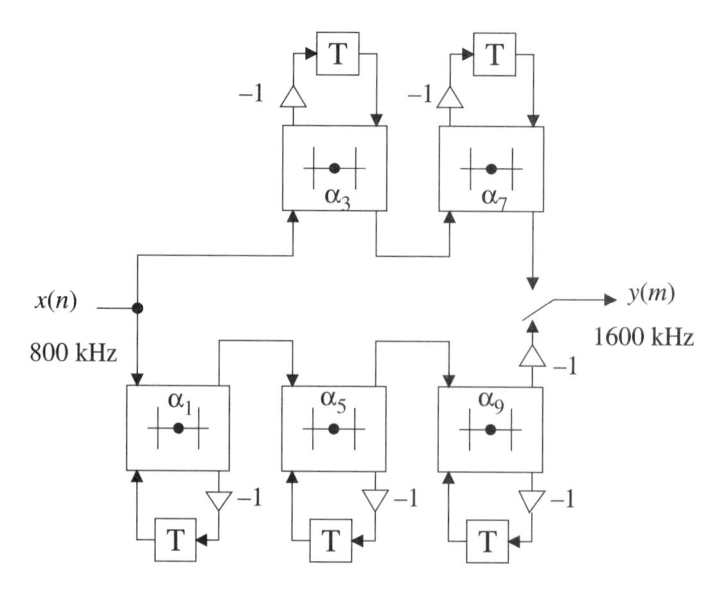

Figure 4.65 Interpolation filter working at the input sample frequency

input signal, $X(z)$, with a filter, $H_2(z)$, that operates at half the sampling rate. Finally, the output of this filter is interleaved with zeros. This scheme also applies to the first term except for a unit delay of the sequence. Thus, the output of the interpolator filter, $H(z)$, is obtained by summing the two sequences, but at each time instant only one of the two sequences has a nonzero value.

Applying these ideas to the lattice wave digital filter we get the filter shown in Figure 4.65. The interpolator is described by the pseudo-code shown in Box 4.2. The nonzero output values are taken alternately from the upper and lower branches. This technique reduces the arithmetic work load by 50 percent.

```
Read(x(n));
  Compute:        y(m) := f2(x(n));          {Lower branch, H2 }
  Compute:        y(m + 1) := f1(x(n));      {Upper branch, H1}
```

Box 4.2. Pseudo-code for the WDF interpolator

4.22 DECIMATION WITH A FACTOR *M*

The process of reducing the sampling frequency is called *decimation*[3]—even if the reduction is not by a factor of ten. Decimation is often used in A/D converters. The analog signal is sampled with a very high sampling frequency in order to reduce the requirement on the anti-aliasing filter. Usually, the sampling frequency is

---

[3]. Decimation: Military punishment used in ancient Rome to punish a whole body of troops. Every tenth man, chosen by lot, was executed.

oversampled by a factor that is a large power of two. The digital signal is then bandlimited by a digital filter with a stopband angle slightly less than $\pi/2$. The sample frequency can therefore be reduced by a factor of two by simply dropping every other sample. This process of bandlimiting and dropping every other sample is repeated until the desired sample rate is obtained.

The relation between the Fourier transforms of the decimated signal and the original signal is [3, 13, 29]

$$X_d(e^{j\omega T}) = \frac{1}{M} \sum_{k=0}^{M-1} X(e^{j(\omega T - 2\pi k)/M}) \tag{4.71}$$

where M is the decimation factor. The Fourier transform of the decimated signal consists of a sum of shifted replicas of the Fourier transform of the original signal. Generally, aliasing takes place if the original signal bandwidth is larger than π/M. Although the aliasing can be removed by a bandlimiting digital filter, the decimated signal will no longer be the same.

The decimation filter can also be realized efficiently using wave digital filters. It is efficient from a computational point of view to also do the decimation in steps of two.

4.22.1 HSP43220™

The HSP43220™ from HARRIS Corp. is a two-stage linear-phase FIR filter for decimation. The first stage, which is shown in Figure. 4.66, is a high-order decimation filter that allows decimation by factors up to 1024. The maximum input sample frequency is 30 MHz. Input data word length is 16 bits using two's-complement representation.

The second stage consists of a linear-phase direct-form FIR filter with up to 512 taps. This filter provides the proper frequency response, but can also be used to decimate by factors up to 16. The first stage can be bypassed in order to use only the linear-phase FIR filter or several chips can be cascaded to obtain higher decimation ratios or longer FIR filters.

The filtering and decimation in the first stage is accomplished by connecting up to five accumulators in cascade. The accumulators yield the transfer function

$$H_I(z) = \prod_{k=1}^{n} H_1(z) = \prod_{k=1}^{n} \frac{1}{z-1} = \frac{1}{(z-1)^n} \tag{4.72}$$

where n is the number of accumulators. The accumulators represent first-order recursive filters. Next follows the decimation. The resulting transfer function becomes

$$H(z) = \prod_{k=1}^{n} \sum_{p=0}^{M-1} \frac{1}{z^{(1-p)/M} - 1} \tag{4.73}$$

Finally, n first-order FIR filters, so-called comb filters, with the transfer function

$$H_2(z) = \frac{z-1}{z} \tag{4.74}$$

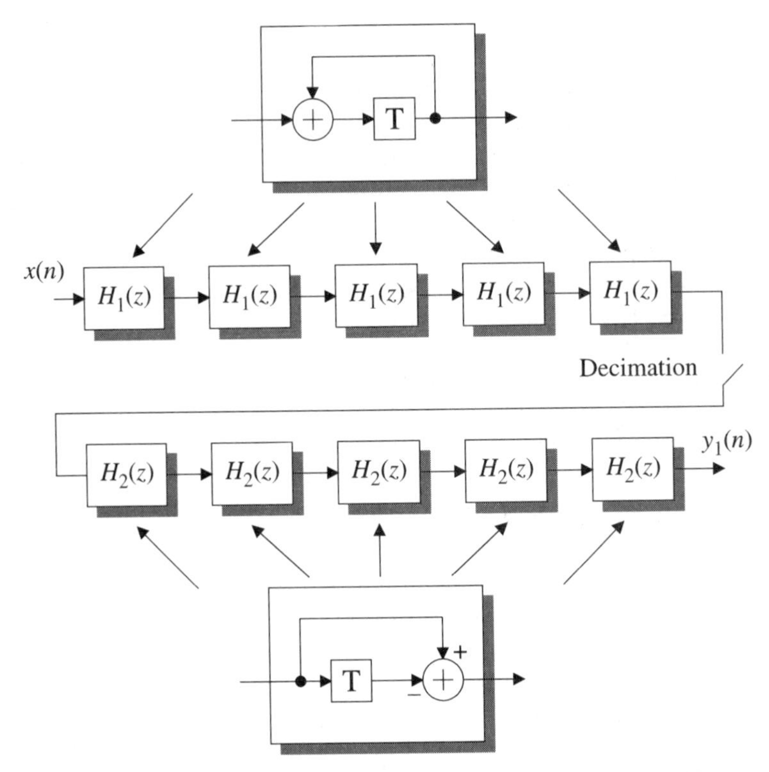

Figure 4.66 The first stage of the HSP43220 decimation FIR filter

are used to cancel the poles at $z = 1$ that are caused by accumulators. The combined effect of the first stage is a coarse lowpass filtering with an attenuation of aliasing into the passband of at least 96 dB.

A major problem with this approach is that the accumulators have poles at $z = 1$, i.e., they are marginally stable. Hence, an input signal with a nonzero mean will eventually saturate the accumulators. In order to reduce the probability of overflow in the accumulators, special scaling techniques need to be employed. In this case, word length in the first accumulator has been increased to 63 bits, the second accumulator has 53 bits, and so on. This will reduce the overflow probability, but the cost in terms of hardware is significant.

4.23 SAMPLING RATE CHANGE WITH A RATIO *L/M*

The sampling frequency can be changed by a ratio L/M, where both L and M are small integers, by combining interpolation and decimation as illustrated in Figure 4.67[1, 3, 6,

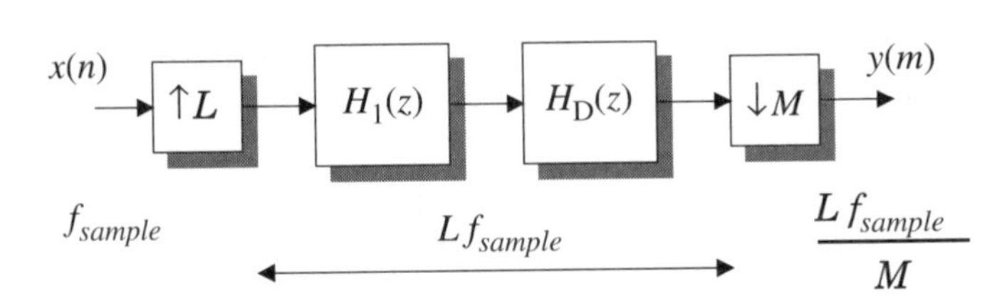

Figure 4.67 Sampling rate change by a factor L/M

13, 19]. For example, an increase in the sampling rate by a factor 3.5 can be achieved by first interpolating the sampling frequency by a factor seven and then decimating it by a factor two. The interpolating and decimating lowpass filters can be combined into a single filter. See Problem 4.32.

Note that the sampling frequencies for compact disc players and digital audio tape recorders (DAT) have been chosen such that conversion between the two systems becomes difficult. The ratio of the two sample rates is a ratio between two very large integers. Interpolation from 44.1 kHz (CD player) to 48 kHz (DAT) can be done by first interpolating by a factor 160 and then decimating by a factor 147. Hence, the sampling rate has to be interpolated to a very high rate before being decimated. This is very expensive. However, most of the sample values in the interpolator are zero and only one of the 147 output values needs to be computed, so the work load can be reduced significantly. See Problem 4.34. It is possible to interpolate and decimate with arbitrary ratios by using more advanced techniques [16, 20].

4.24 MULTIRATE FILTERS

In some cases it is efficient from a computational point of view to realize an ordinary single-rate digital filter as a multirate filter. This approach is particularly efficient to realize filters with very narrow passbands or transition bands [1, 19, 22, 23]. Such narrow band filters can be realized by a combination of decimation and interpolation stages. The stages can be of either FIR or IIR type, or a combination.

Another typical application can be found in digital filter banks that are commonly used in telecommunication systems as well as in image and speech coders.

4.25 INTERPOLATOR—CASE STUDY 3

As the third case study, we choose an application with an interpolating wave digital filter. Assume that the sampling frequency of the signal discussed in Example 4.15 shall instead be increased from 1.6 to 6.4 MHz. This can be done by interpolating the sampling rate in two steps, as shown in Figure. 4.68. The interpolator has been cascaded with an allpass filter for equalizing the group delay.

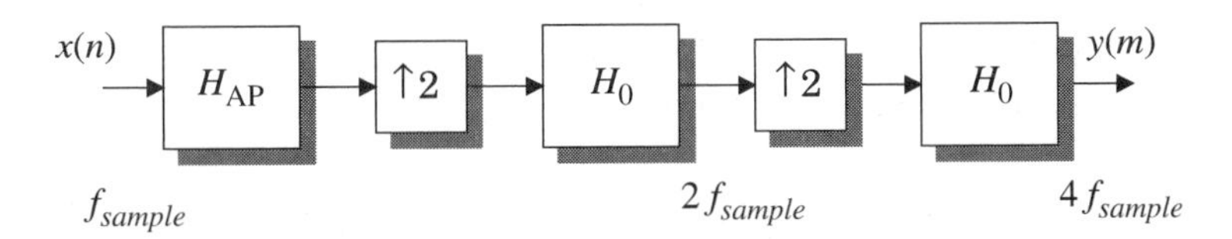

Figure 4.68 Interpolating WDF

We will for the sake of simplicity use the bireciprocal lattice wave digital filter designed in Example 4.12 for both filters, although only a ninth-order filter is required for the last stage. The transfer function for the complete interpolator is

$$\mathrm{H}_I(z) = \frac{1}{4} H_0(z)\, H_0(z^2)\, H_{AP}(z^4) \tag{4.75}$$

The transfer function is divided by $L = 4$ in order to normalize the gain. Figures 4.69 and 4.70 show the attenuation and the group delay.

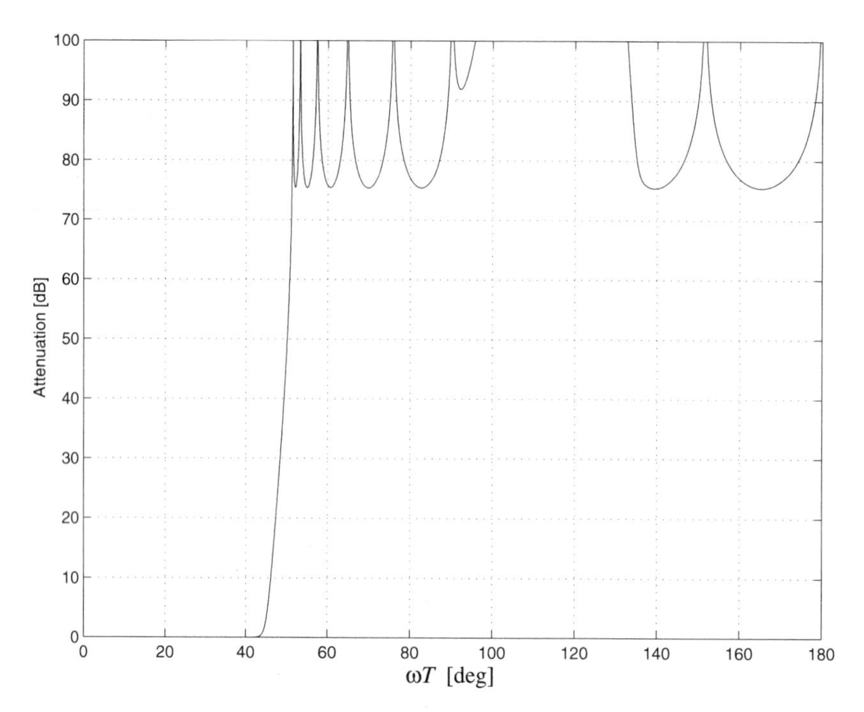

Figure 4.69 Attenuation

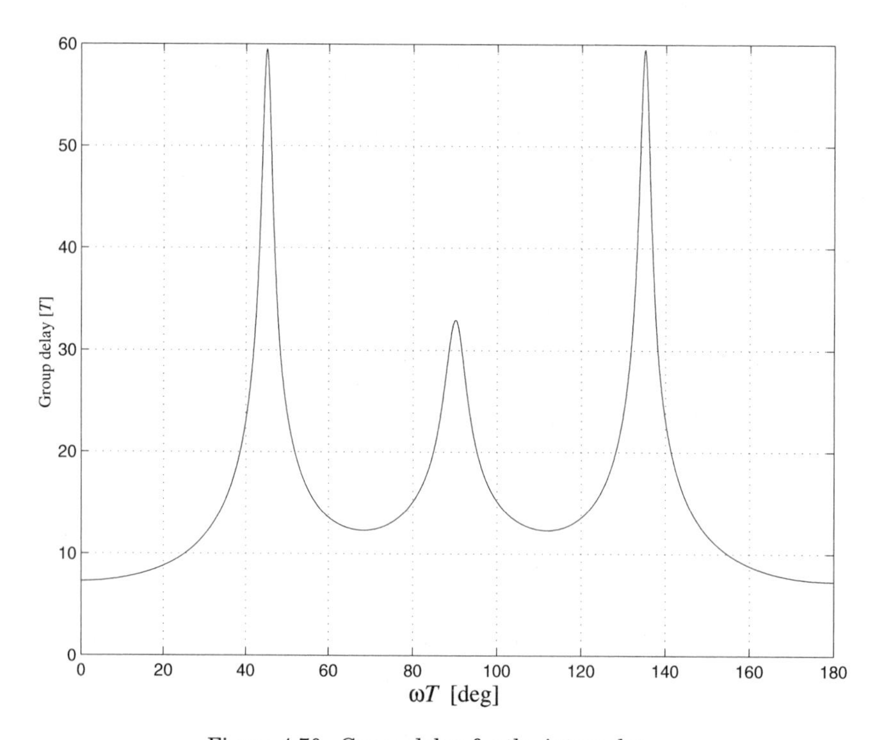

Figure 4.70 Group delay for the interpolator

The group delay varies considerably in the passband 0 to 38.25°. The group delay of the interpolator is therefore equalized by adding a seventh-order allpass filter in front of the interpolator. The resulting group delay is shown in Figure. 4.71.

The adaptor coefficients in the allpass filter are

$$\alpha_{10} = 0.4573$$

$$\alpha_{11} = 0.2098 \qquad \alpha_{12} = -0.5695$$

$$\alpha_{13} = 0.2123 \qquad \alpha_{14} = -0.0952$$

$$\alpha_{15} = 0.2258 \qquad \alpha_{16} = 0.4490$$

If instead a ninth-order filter, which represents a significantly lower work load, was used for the last stage, more of the processing capacity could be allocated to the allpass filter to further reduce the variation in the group delay.

The complete interpolator with an allpass filter for correction of group delay is shown in Figure 4.72. Alternatively, the allpass filter may be placed after the interpolator where the sampling rate is higher. This results in a slightly higher work load.

Also in this case the adaptor coefficients can be shortened to 9 bits. The arithmetic work load is 35.2 · 106 adaptors/s corresponding to 35.2 · 106 multiplications/s and 105 · 106 additions/s and a significant number of operations for

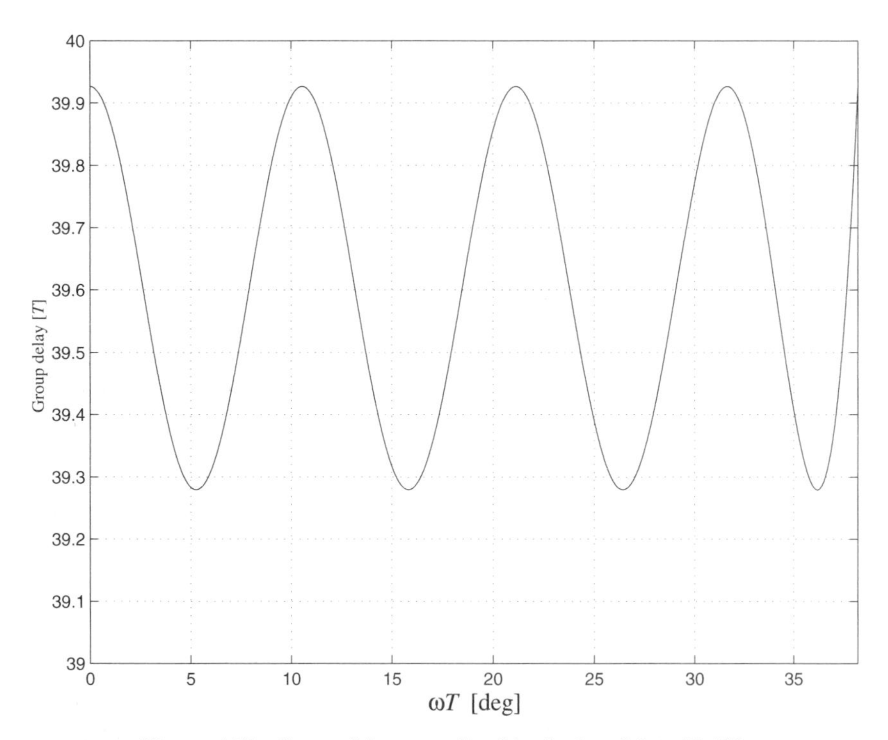

Figure 4.71 Group delay, equalized in the band 0 to 38.25°

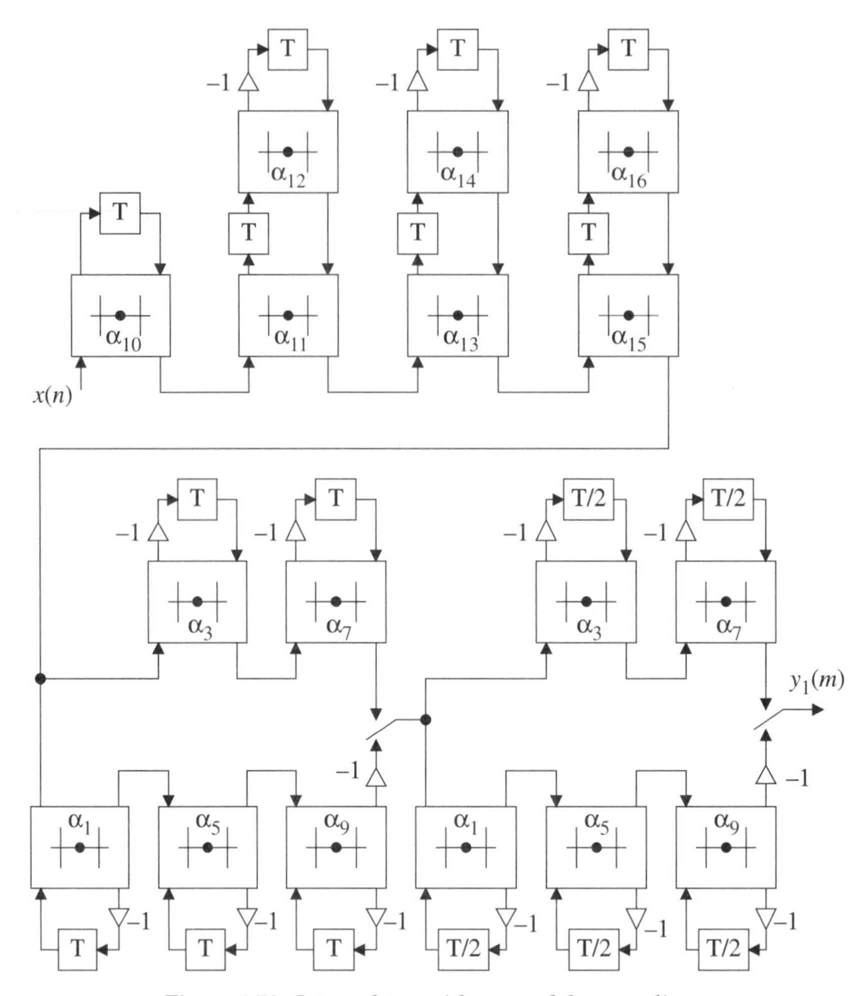

Figure 4.72 Interpolator with group delay equalizer

quantization, overflow detection, and correction. An arithmetic work load of more than 140.2 MOPS is too high for most of today's standard digital signal processors.

The interpolator is described by the pseudocode in Box 4.3.

```
Read(x(n));
Compute y0(n) := fAP(x(n));                 {Allpass filter, HAP}
Compute y1(n) := f21(y0(n));                {Lower branch in stage 1, H21}
  Compute y2(m) := f22(y1(n));              {Lower branch in stage 2, H22}
  Compute y2(m+1) := f12(y1(n));            {Upper branch in stage 2, H12}

Compute y1(n + 1) := f11(y1(n + 1));        {Upper branch in stage 1, H11}
  Compute y2(m + 2) := f22(y1(n + 1));      {Lower branch in stage 2, H22}
  Compute y2(m + 3) := f12(y1(n + 1));      {Upper branch in stage 2, H12}
```

Box 4.3. Pseudocode for the WDF interpolator

References

[1] Bellanger M.: *Digital Processing of Signals*, Theory and Practice, John Wiley & Sons, Chichester, 1984.

[2] Chen C.H. (Ed.): *Signal Processing Handbook*, Marcel Dekker Inc., New York, 1988.

[3] Crochiere R.E. and Rabiner L.R.: *Multirate Digital Signal Processing*, Prentice Hall, Englewood Cliffs, New York, 1983.

[4] Daniels R.W.: *Approximation Methods for Electronic Filter Design*, McGraw-Hill, New York, 1974.

[5] DeFatta D.J., Lucas J.G., and Hodgkiss W.S.: *Digital Signal Processing: A System Design Approach*, John Wiley & Sons, New York, 1988.

[6] Elliott D.F. (ed.): *Handbook of Digital Signal Processing, Engineering Applications*, Academic Press, San Diego, 1988.

[7] Fettweis A., Levin H., and Sedlmeyer A.: Wave Digital Lattice Filters, *Intern. J. Circuit Theory and Appl.*, Vol. 2, pp. 203–211, June 1974.

[8] Fettweis A. and Meerkötter K.: On Adaptors for Wave Digital Filters, *IEEE Trans. on Acoustics, Speech, and Signal Processing*, Vol. ASSP-23, No. 6, pp. 516–525, Dec. 1975.

[9] Fettweis A. and Nossek J.A.: Sampling Rate Increase and Decrease in Wave Digital Filters, *IEEE Trans. on Circuits and Systems*, Vol. CAS-29, No. 12, pp. 797–806, Dec. 1982.

[10] Fettweis A.: Wave Digital Filters: Theory and Practice, *Proc. IEEE*, Vol. 74, No. 2, pp. 270–327, Feb. 1986.

[11] Gazsi L.: Explicit Formulas for Lattice Wave Digital Filters, *IEEE Trans. on Circuits and Systems*, Vol. CAS-32, No. 1, pp. 68–88, Jan. 1985.

[12] Lawson S. and Mirzai A.: *Wave Digital Filters*, Ellis Horwood, England, 1990.

[13] Lim J.E. and Oppenheim A.V. (eds.): *Advanced Topics in Signal Processing*, Prentice Hall, Englewood Cliffs, New Jersey, 1988.

[14] McClellan J.H., Parks T.W., and Rabiner L.R.: A Computer Program for Designing Optimal FIR Linear Phase Digital Filters, *IEEE Trans. on Audio Electroacoustics*, Vol. AU-21, No. 6, pp. 506–526, Dec. 1973.

[15] Oppenheim A.V. and Schafer R.W.: *Discrete-Time Signal Processing*, Prentice Hall, Englewood Cliffs, New Jersey, 1989.

[16] Peled A. and Liu B.: *Digital Signal Processing: Theory, Design, and Implementation*, John Wiley & Sons, New York, 1976.

[17] Proakis J.G. and Manolakis D.G.: *Introduction to Digital Signal Processing*, Macmillan, New York, 1988.

[18] Rabiner L.R. and Gold B.: *Theory and Application of Digital Signal Processing*, Prentice Hall, Englewood Cliffs, New Jersey, 1975.

[19] Ramstad A.T.: Digital Two-Rate IIR and Hybrid IIR/FIR Filters for Sampling Rate Conversion, *IEEE Trans. on Communications*, Vol. COM-30, No. 7, pp. 1466–1476, July 1982.

[20] Ramstad A.T.: Digital Methods for Conversion between Arbitrary Sampling Frequencies, *IEEE Trans. on Acoustics, Speech, and Signal Processing*, Vol. ASSP-32, No. 3, pp. 577–591, June 1984.

[21] Roberts R.A. and Mullis C.T.: *Digital Signal Processing*, Addison-Wesley, Reading, MA, 1987.

[22] Saramäki T.: A Class of Linear-Phase FIR Filters for Decimation, Interpolation, and Narrow-Band Filtering, *IEEE Trans. on Acoustics, Speech, and Signal Processing*, Vol. ASSP-32, No. 5, pp. 1023–1036, Oct. 1984.

[23] Saramäki T., Neuvo Y., and Mitra S.K.: Design of Computationally Efficient Interpolated FIR Filters, *IEEE Trans. on Circuits and Systems*, Vol. CAS-35, No. 1, pp. 70–88, Jan. 1988.

[24] Sedra A.S. and Brackett P.O.: *Filter Theory and Design: Active and Passive*, Matrix, Champaign, IL, 1978.

[25] Sikström B.: *On the LSI Implementation of Wave Digital Filters and Discrete Cosine Transforms*, Linköping Studies in Science and Technology, Diss. No. 143, Linköping University, Sweden, May 1986.

[26] Stearns S.D. and David R.A.: *Signal Processing Algorithms*, Prentice Hall, Englewood Cliffs, New Jersey, 1988.

[27] Taylor F.J.: *Digital Filter Design Handbook*, Marcel Dekker, New York, 1983.

[28] Vaidyanathan P.P. and Nguyen T.Q.: A "Trick" for Design of FIR Half-Band Filters, *IEEE Trans. on Circuits and Systems*, Vol. CAS-34, No. 3, pp. 297–300, March 1987.

[29] Vaidyanathan P.P.: Multirate Digital Filters, Filter Banks, Polyphase Networks, and Applications: A Tutorial, *Proc. IEEE*, Vol. 78, No. 1, pp. 56–93, Jan. 1990.

[30] Wanhammar L.: *An Approach to LSI Implementation of Wave Digital Filters*, Linköping Studies in Science and Technology, Diss. No. 62, Linköping University, Sweden, April 1981.

PROBLEMS

4.1 (a) List the major features of FIR filters.
(b) List the pros and cons of recursive realizations of FIR filters.
(c) List the major features of IIR filters.
(d) List the major features of wave digital filters.
(e) What are the major advantages of lattice compared to ladder wave digital filters?

4.2 State necessary and sufficient conditions for an FIR filter to have linear phase.

4.3 Show that a lowpass FIR filter can not have an antisymmetric impulse response.

4.4 Determine both the phase response and group delay functions for an FIR filter that has the impulse response $a, b, c, b, a, 0, 0, \ldots$.

4.5 Determine the magnitude and phase responses as well as the group delay for the following FIR filter which is called a Hanning filter:

$$y(n) = 0.25[x(n) + 2x(n-1) + x(n-2)]$$

4.6 An FIR lattice filter that is part of the RLS lattice filter discussed in Chapter 3 is described by the following difference equations:

$$f_i(n) = f_{i-1}(n) + k_i\, b_{i-1}(n-1)$$
$$b_i(n) = k_i\, f_{i-1}(n) + b_{i-1}(n-1) \qquad \text{for } i = 1, 2, \ldots, N-1$$
and $f_0(n) = b_0(n) = x(n)$ and $y(n) = f_{N-1}(n)$.

(a) Draw a signal-flow graph for the FIR lattice filter.
(b) Determine the transfer function if $N = 3$.

4.7 View the DCT as an FIR filter bank and find possible reductions in the computational complexity.

4.8 Why are doubly terminated reference filters used in the design of wave digital filters? Is a Butterworth filter less or more sensitive to element errors in the passband than the corresponding Chebyshev I filter?

4.9 Determine the ripple in the passband for a doubly terminated LC filter when the reflection coefficient is 15%.

4.10 Which are the major factors that influence the passband sensitivity in a doubly terminated LC filter?

4.11 Determine the relation between the group delays of an analog filter and the corresponding digital filter that is obtained by bilinear transformation.

4.12 Determine the transfer function if each delay element in a filter structure is replaced by a structure consisting of two delay elements and a sign inversion, i.e., $-z^2$. Also, sketch the magnitude response of the filter if the original filter was a lowpass filter.

4.13 Determine the reflectance function, $S(z) = B_1/A_1$, for the structure shown in Figure P4.13. What kind of filter characteristic is realized? The detailed wave-flow graph is also shown in Figure P4.13.

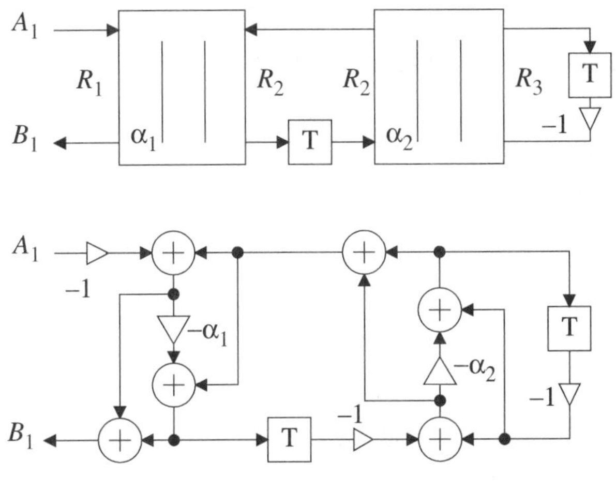

Fig. P4.13.

4.14 An analog Cauer filter with $A_{max} = 0.09883$ dB ($\rho = 15\%$), $A_{min} = 67.09$ dB $f_c = 2$ kHz, and $f_s = 4.7324$ kHz has the following poles and zeros:

$s_{p1} = -1.15102 \cdot 2\pi$ $\quad s_{n1} = \infty$

$s_{p2,3} = -0.2951 \cdot 2\pi \pm j\,2.1526 \cdot 2\pi$ $\quad s_{n2,3} = \pm j\,4.9534 \cdot 2\pi$

$s_{p4,5} = -0.86564 \cdot 2\pi \pm j\,1.4098 \cdot 2\pi$ $\quad s_{n4,5} = \pm j\,7.8014 \cdot 2\pi$ [krad/s]

The filter is uniquely described by the notation C051525. Use this filter and the bilinear transformation to design a digital lowpass filter with a passband edge of 32 kHz when the sample frequency is 256 kHz.

(a) What are A_{max}, A_{min}, $\omega_c T$, and $\omega_s T$ for the digital filter.

(b) Sketch the pole–zero configuration for both the analog and digital filters.

4.15 Determine the poles for the lattice WDF in Examples 4.11 and 4.12. Comment on the ordering of the poles in the two branches.

4.16 A third-order lowpass Butterworth filter has the transfer function $H(z)$. Sketch the pole–zero configuration as well as the magnitude response for a digital filter obtained from the lowpass filter by the transformation $z \to z^L$ for $L = 2$ and 3.

4.17 Show that the reflection function, Equation (4.26), is an allpass function for a reactance.

4.18 Derive the wave-flow graph for the two-port series and parallel adaptors from the definitions of the incident and reflected waves.

4.19 Why is it not possible to realize bireciprocal wave digital filters of Chebyshev I or II type?

4.20 A bireciprocal lattice wave digital filter of lowpass type is used to design a symmetric bandpass (BP) filter. Hint: The well-known bandpass frequency transformation $z \to -z^2$ is used.

(a) Determine the wave-flow graph for the lowpass filter.

(b) Determine the wave-flow graph for the BP filter.

(c) Determine the number of adaptors required in the BP filter.

(d) Determine the number of multiplications and additions in the BP filter.

(e) Sketch the magnitude functions for both filters.

4.21 Determine the Fourier transform for the new sequences that are derived from the sequence $x(n)$ according to:

(a) $x_1(n) = x(2n)$

(b) $$x_2(n) = \begin{cases} x\left(\dfrac{n}{2}\right) & \text{for } n = \text{even} \\ 0 & \text{for } n = \text{odd} \end{cases}$$

Also sketch the magnitude of the Fourier transform for the three sequences.

4.22 Determine and sketch the Fourier transform of a sequence $x_1(m)$ which is obtained from another sequence $x(n)$ with the Fourier transform $X(e^{j\omega T})$ according to Equation (4.67).

4.23 A digital signal can be characterized by the sum of two sinusoids:

$$x(n) \approx \sin\left(\frac{2\pi f_1 n}{f_{sample}}\right) + 0.2 \sin\left(\frac{2\pi f_2 n}{f_{sample}}\right)$$

where $f_1 = 0.8$ kHz, $f_2 = 1.6$ kHz, and $f_{sample} = 4$ kHz. The sample rate is decreased to 2 kHz by first applying a digital lowpass filter and then discarding every other output sample.

(a) Determine first the digital spectrum if the lowpass filter is not used.
(b) Determine the digital spectrum if the lowpass filter meets the following specification: $A_{max} = 0.1$ dB, $\omega_c T = \pi/4$, $A_{min} = 40$ dB, and $\omega_s T = 0.8\,\pi$.

4.24 The sample rate of the digital signal in Problem 4.23 is increased to 8 kHz by inserting zeros and removing the unwanted images using a lowpass filter.
(a) Determine first the digital spectrum after zeros have been introduced.
(b) Determine proper cutoff and stopband angles for the lowpass filter.
(c) Determine the digital spectrum if the lowpass filter meets the following specification $A_{max} = 0.1$ dB and $A_{min} = 40$ dB.
(d) Determine the required order if a Cauer filter is used as the interpolator filter.

4.25 (a) Determine the specification for an ideal interpolator that increases the sampling frequency by a factor two for a bandlimited signal contained within the band 0 to $\omega_0 T < \pi$.
(b) Determine the frequency response of the ideal interpolator filter.
(c) Determine the impulse response of the ideal interpolator filter.

4.26 Show that it is not efficient to interpolate the input sample rate by a factor two by repeating the input samples instead of introducing zeros.

4.27 Suggest a system for interpolation of the sample frequency by a factor 1.75. Determine also the cutoff and stopband edges of the filters involved.

4.28 The following scheme is used in a simple approach to double the sample frequency. The original samples are retained and new samples are created between the original samples by linear interpolation, i.e.,

$$y(n) = 0.5\,[x(n) + x(n-1)]$$

(a) Determine the difference equation for the interpolator. Hint: Create a new sequence from the original by interleaving zero samples between the original samples.
(b) Determine the transfer function for the interpolator.
(c) Determine the poles and zeros.
(d) Sketch the frequency response.
(e) Sketch the Fourier transform for the original, input, and output signals to the interpolator.
(f) Determine the impulse response of the interpolator.

4.29 Determine the work load required for the interpolator (case study) if the allpass filter is placed after the interpolator filter.

4.30 Determine the required filter order and the work load if the interpolation (case study) is done in a single step using a lattice wave digital filter.

4.31 Determine the required filter order and the work load for the interpolation (case study). Assume that the interpolation is done in
(a) A single step using an FIR filter
(b) Two steps using FIR filters

4.32 Determine the specification of the combined interpolating and decimating filter required when the sampling frequency is changed by a factor L/M. The input signal is bandlimited to the band 0 to $\omega_0 T < \pi$.

4.33 Determine the specification for the digital filter required in a sample rate converter that will increase the sampling rate by a factor 2.5. The ripple in the passband should be less than 0.5 dB and the attenuation should be larger than 40 dB. The bandwidth of the input signal is 200 kHz and the sampling frequency is 1.2 MHz.

4.34 (a) Develop a multistage scheme for a sampling rate converter between DAT (48 kHz) and CD (44.1 kHz). An FIR filter is used as a lowpass filter.
(b) Determine the reduction in work load due to multiplication by zero values.

4.35 Write a MATLAB program that can be used to design lowpass lattice WDFs[11].

DESIGN PROJECTS

1. Develop a specification and a functional description that can be used to design an interpolator for a CD player. Assume that the chip is to be used in a battery-powered application that is expected to be produced in large volumes. Compare your result with some commercial chips.
   (a) Use an FIR filter approach.
   (b) Use a lattice wave digital filter approach.
2. Redesign the interpolator using a ninth-order filter for the last stage and use the computational resources that are saved to increase the order of the allpass filter.
3. Develop a specification and a functional description that can be used to design a simple filter bank for equalization in a CD player. Assume that the chip is going to be used in a battery-powered application that is expected to be produced in large volumes. Compare your result with some commercial chips.
   (a) Use a tree of complementary FIR filters.
   (b) Use a tree of bireciprocal lattice wave digital filters.

5

FINITE WORD LENGTH EFFECTS

5.1 INTRODUCTION

In order to implement a recursive algorithm, all sequence values and coefficients must be represented with a finite word length. The effect of finite word length depends on the filter structure, pole–zero configuration, representation of negative numbers, rounding or truncation of products, overflow characteristics, and the input signal. Finite word length gives rise to a large number of phenomena caused by different types of nonlinearities.

- **Overflow of the number range:** Large errors in the output signal occur when the available number range is exceeded—*overflow*. Overflow nonlinearities can be the cause of so-called *parasitic oscillations*.
- **Round-off errors:** Rounding or truncation of products must be done in recursive loops so that the word length does not increase for each iteration. The errors that occur under normal operating conditions can be modeled as white noise, but both rounding and truncation are nonlinear operations that may cause parasitic oscillations. Floating-point addition also causes errors because of the denormalization and normalization of the numbers involved.
- **Aliasing errors:** Aliasing and imaging errors occur in A/D and D/A converters and when the sample rate is changed in multirate systems. These nonlinearities may cause nonharmonic distortion.
- **Coefficient errors:** Coefficients can only be represented with finite precision. This results in a static deviation from the ideal frequency response for a digital filter.

The effect on the filtering due to these errors is difficult to predict, except in special cases, and the acceptable distortion depends on the application. It is therefore important to determine experimentally the performance of the filter through extensive simulation. Parasitic oscillations cause various forms of nonideal filter behavior—for example, large amplitude oscillations that are independent of the input signal. In most applications, it is necessary that the filter suppress such parasitic oscillations. This means that only filter structures that guarantee the suppression of these oscillations are serious candidates for implementation. We discussed earlier two such classes of structures—namely, nonrecursive FIR filters and wave digital filters.

When implementing a filter in hardware, it is important to design the filter so that short word lengths can be used for both the internal signals and the coefficients, since short word lengths generally result in less power consumption, smaller chip area, and faster computation. We will later show that it is necessary to use filter structures with low coefficient sensitivity in order to minimize the internal data word length.

The major limitation when implementing a digital filter using standard digital signal processors with fixed-point arithmetic is their short data word length. Typically, such processors have only 16-bit accuracy except for the Motorola 56000 family, which has 24-bit. A 24-bit word length is sufficient for most high-performance applications while 16-bits is sufficient only for simple applications.

Modern standard signal processors support floating-point arithmetic in hardware. The use of floating-point arithmetic will not alleviate the parasitic oscillation problems related to nonlinearities, but will provide a large number range that can handle signals of widely varying magnitudes. The signal range requirements are, however, usually modest in most well-designed digital signal processing algorithms. Rather it is the accuracy of the calculations that is important. The accuracy is determined by the mantissa, which is 24 bits in IEEE standard 32-bit floating-point arithmetic. This is long enough for most applications, but there is not a large margin. Algorithms implemented with floating-point arithmetic must therefore be carefully designed to maintain the accuracy. To summarize, floating-point arithmetic provides a large dynamic range which is usually not required, and the cost in terms of power consumption, execution time, and chip area is much larger than that for fixed-point arithmetic. Hence, floating-point arithmetic is useful in general-purpose signal processors, but it is not efficient for application-specific implementations.

5.2 PARASITIC OSCILLATIONS

The data word length increases when a signal value is multiplied by a coefficient and would therefore become infinite in a recursive loop. The signal values must therefore be quantized, i.e., rounded or truncated to the original data word length, at least once in every loop. Quantization is therefore a nonavoidable nonlinear operation in recursive algorithms. Another type of nonlinearity results from finite number range.

Analysis of nonlinear systems is very difficult. Mainly first- and second-order sections have therefore been studied for particular classes of input signals and types of nonlinearities. Of particular interest are situations when the nonlinear and the corresponding linear system behave markedly different, for example, when the nonlinear system enters into a parasitic oscillation.

There are two kinds of *parasitic oscillation*, depending on the underlying cause of the oscillation: *overflow* and *granularity oscillations*. Parasitic oscillations are also called *limit cycles*. Granularity oscillations are caused by rounding or truncation of the signal values. Parasitic oscillations can, of course, also be caused by interaction between different types of nonlinearities. In this section, we will illustrate some of the most common phenomena. We stress that nonlinear effects are complex and by no means fully understood. For example, it has recently been shown that digital filters with sufficiently large word length exhibit near-chaotic behavior [3, 19].

5.2.1 Zero-Input Oscillations

One of the most studied cases, due to its simplicity, is when the input signal, $x(n)$, to a recursive filter suddenly becomes zero. The output signal of a stable linear filter will tend to zero, but the nonlinear filter may enter a so-called *zero-input limit cycle.*

Zero-input parasitic oscillations can be very disturbing, for example, in a speech application. At the beginning of a silent part in the speech, the input signal becomes zero. The output signal should ideally decay to zero, but a parasitic oscillation may instead occur in the nonlinear filter. The magnitude and frequency spectra of the particular oscillation depend on the values that the delay elements have at the moment the input becomes zero. Most oscillations will, however, have large frequency components in the passband of the filter and they are more harmful, from a perception point of view, than wide-band quantization noise, since the human ear is sensitive to periodic signals. Comparison of parasitic oscillations and quantization noise should therefore be treated with caution.

Zero-input limit cycles can be eliminated by using a longer data word length inside the filter and discarding the least significant bits at the output. The number of extra bits required can be determined from estimates of the maximum magnitude of the oscillations [22].

EXAMPLE 5.1

Apply a sinusoidal input to the second-order section in the direct form structure shown here in Figure 5.1 and demonstrate by simulation that zero-input limit cycles will occur. Try both rounding and truncation of the products. Use, for example, the filter coefficients

$$b_1 = \frac{489}{256} = 1.91015625$$

and

$$b_2 = \frac{15}{16} = -0.9375$$

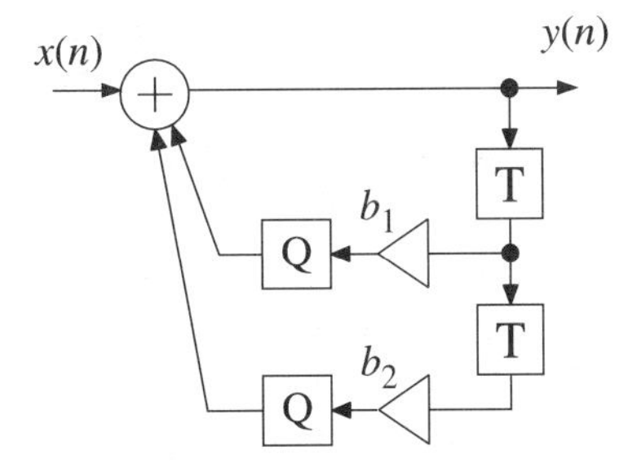

Figure 5.1 Direct form structure with two quantizations

Figure 5.2 shows a sinusoidal input signal that suddenly becomes zero and the corresponding output signal. Generally, parasitic oscillations with different magnitudes and frequency spectra occur. The zero-input parasitic oscillation shown in Figure 5.2 (rounding) is almost sinusoidal with an amplitude of $8Q$, where Q is the quantization step. Another parasitic oscillation that occurs frequently is a constant, nonzero output signal of $\pm 9Q$.

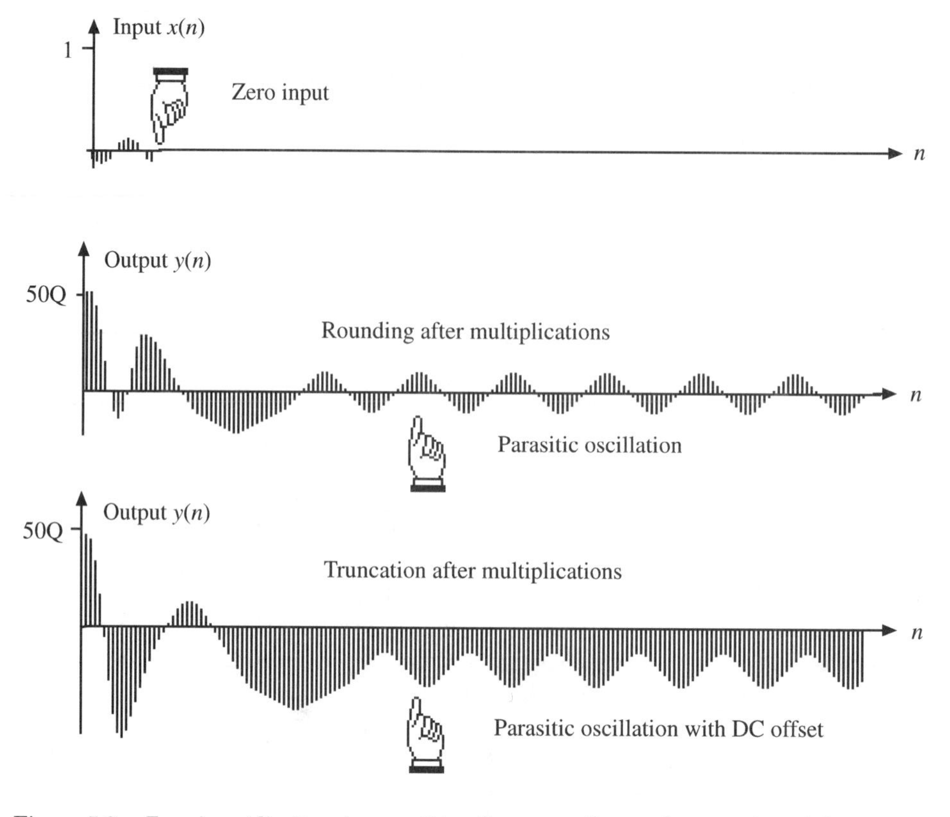

Figure 5.2 Zero-input limit cycles resulting from rounding and truncation of the products. The outputs have been multiplied by $1/Q = 512$ in order to show the parasitic oscillations more clearly.

The quantization errors resulting from truncating the two's-complement numbers have a negative average value $Q/2$ which will appear as a DC offset at the output. The offset will be large if the DC gain measured from the quantization node to the output of the filter is large. In this filter the DC gain is

$$H(1) = \left. \frac{z^2}{z^2 - b_1 z - b_2} \right|_{z=1} = \frac{1}{1 - b_1 - b_2} = 36.57$$

The average output signal is therefore $18.285Q$.

The zero-input case is in many respects the simplest case and extensive research efforts have been directed toward the problem of eliminating zero-input limit cycles in first- and second-order sections. Zero-input limit cycles can in some second-order structures be eliminated for certain pole positions [2]. Several elaborate data-dependent quantization schemes (e.g., controlled rounding [24]) have also been proposed. Another special class of oscillation, which is not so simple to analyze or suppress, is the *constant-input parasitic oscillation* which, of course, may occur when the input signal is constant [4]. Unfortunately, the work based on special input signals can not easily be extended to more general classes of signals. Except for some second-order sections, based on either state-space structures [14, 25] or wave digital filters that are free of all types of parasitic oscillations, it seems that these approaches are generally unsuccessful.

5.2.2 Overflow Oscillations

Large errors will occur if the signal overflows the finite number range. Overflow will not only cause large distortion, but may also be the cause of parasitic oscillations in recursive algorithms. A two's-complement representation of negative numbers is usually used in digital hardware. The overflow characteristic of the two's-complement representation is shown in Figure 5.3.

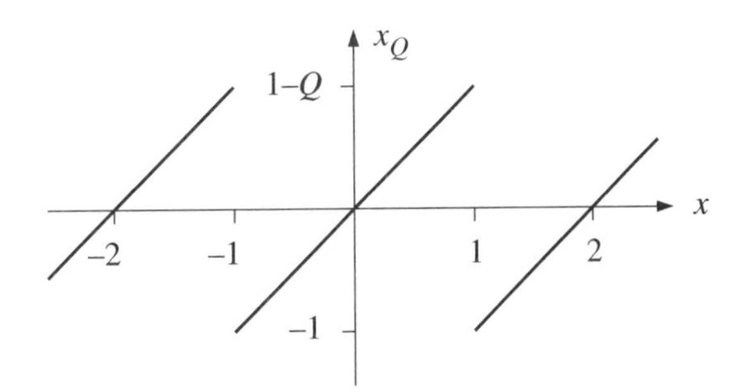

Figure 5.3 Overflow characteristic for two's-complement arithmetic

The largest and smallest numbers in two's-complement representation are $1 - Q$ and -1, respectively. A two's-complement number, x, that is larger than $1 - Q$ will be interpreted as $x - 2$, while a number, x, that is slightly smaller than -1 will be interpreted as $x + 2$. Hence, very large overflow errors are incurred.

A common scheme to reduce the size of overflow errors and their harmful influence is to detect numbers outside the normal range and limit them to either the largest or smallest representable number. This scheme is referred to as *saturation arithmetic*. The overflow characteristic of saturation arithmetic is shown in Figure 5.4. Most standard signal processors provide addition and subtraction instructions with inherent saturation. Another saturation scheme, which may be simpler to implement in hardware, is to invert all bits in the data word when overflow occurs.

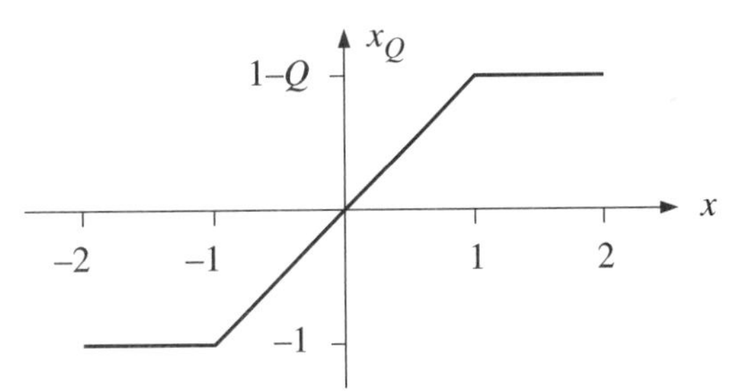

Figure 5.4 Saturation arithmetic

EXAMPLE 5.2

Figure 5.5 shows a second-order section with saturation arithmetic. Apply a periodic input signal so that the filter overflows just before the input vanishes. Compare the output with and without saturation arithmetic.

The input signal is shown in Figure 5.6. If no precautions are taken, the filter will enter into a parasitic overflow oscillation. The magnitude and spectrum of the oscillation will depend on the state of the filter at the moment the input becomes zero.

Figure 5.6 also shows the output signal when the number range has been limited using saturation arithmetic. In this case, the use of saturation arithmetic is sufficient to suppress the large parasitic overflow oscillation, but not completely.

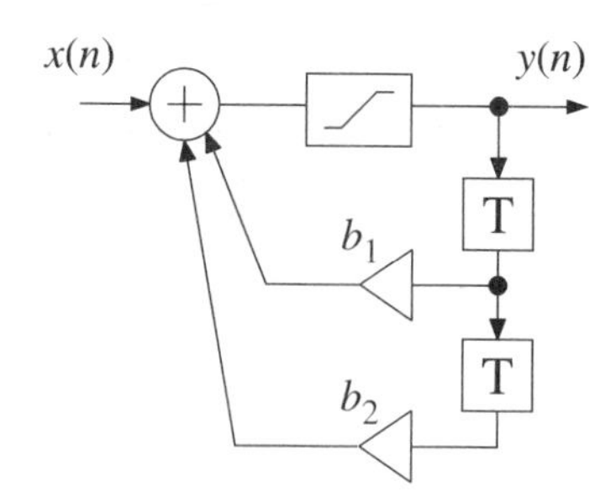

Figure 5.5 Second-order section with saturation arithmetic

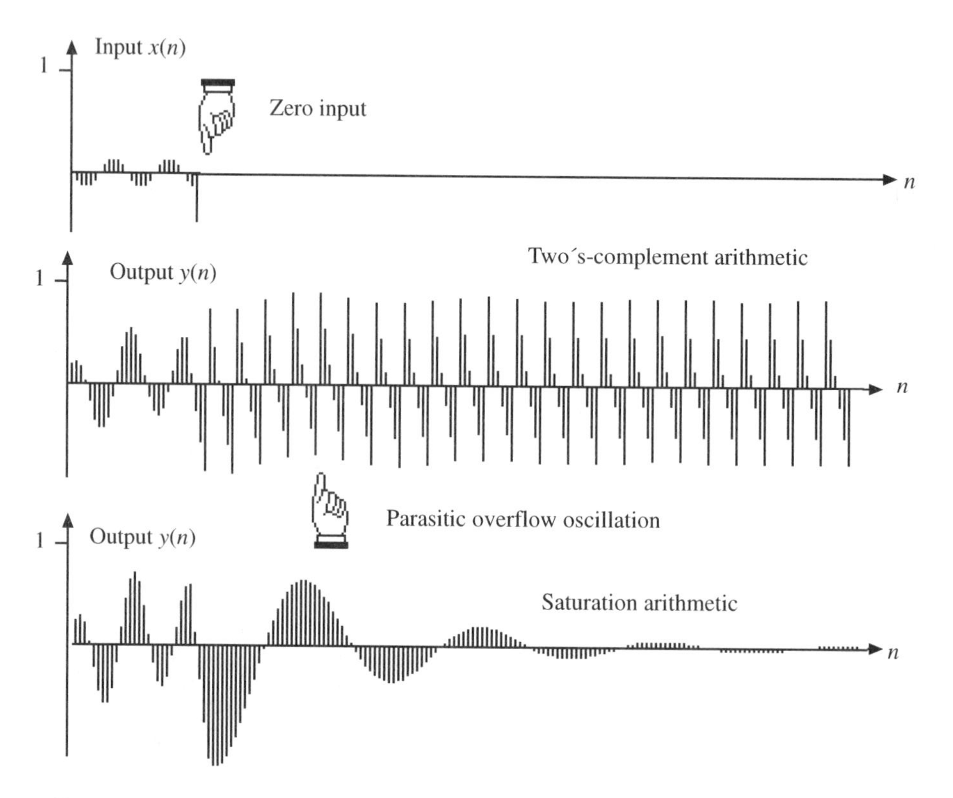

Figure 5.6 Parasitic overflow oscillation with and without saturation arithmetic

Parasitic oscillations can not, in general, be suppressed in higher-order structures by using saturation arithmetic, except for state-space structures [14, 32] and wave digital filters. The hardware expense for implementing saturation arithmetic is often non-negligible. The execution time required for standard signal processors that do not have built-in instructions with conditional saturation and quantization instructions is significant.

5.2.3 Periodic Input Oscillations

Another type of phenomenon may occur when the input signal is periodic. The cause may be either overflow or quantization nonlinearities. For example, instead of a pure output tone for a sinusoidal input, the output may contain both harmonic and subharmonic components. A subharmonic component has a period that is a submultiple of the period of the input signal. Another periodic input phenomenon is illustrated in Example 5.3.

EXAMPLE 5.3

The input to the second-order section just discussed is a sinusoid that itself does not cause overflow. However, an overflow occurs when a small disturbance is added to this input signal.

The output signal, shown in Figure 5.7, consists of two parts. The first is due to the sinusoidal input and the second is due to the disturbance. The effect of the latter would, in a linear stable system, decay to zero, but in the nonlinear system, with saturation arithmetic, the sinusoidal part and the disturbance together cause a sustained overflow oscillation. The output signal jumps between two different periodic output signals. Thus, saturation arithmetic does not suppress overflow oscillations in the direct form structure. This behavior is unacceptable in most applications.

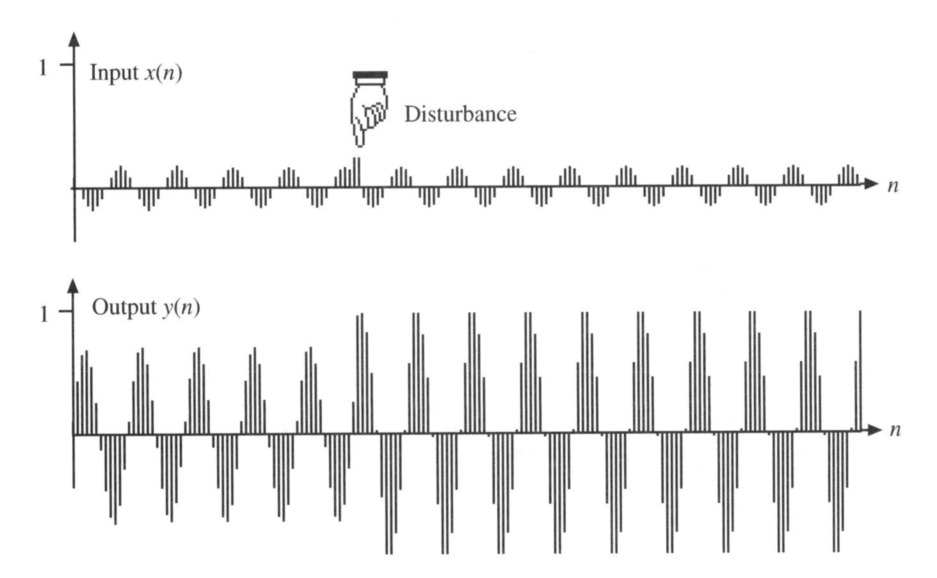

Figure 5.7 Jump phenomenon resulting from overflow in a second-order direct form section with saturation arithmetic

---

5.2.4 Nonobservable Oscillations

In some filter structures the overflow oscillation itself is not observable at the output of the filter. The presence of an internal overflow oscillation may instead be detected as an increased probability of overflow. That is, an overflow is more likely to occur if an input signal that normally does not cause overflow is applied while a nonobservable oscillation persists in the filter. We illustrate a nonobservable oscillation inside part of a ladder wave digital filter that does not use a proper quantization scheme. The design of wave digital filters that suppress parasitic oscillations will be discussed in section 5.4.

EXAMPLE 5.4

Figure 5.8 shows an example of a nonobservable, zero-input oscillation. Note that the output of the filter is zero. The quantization points are indicated with boxes denoted by Q. Quantization is performed such that the magnitude of the signal values is reduced (magnitude truncation). The signal values during one sample interval are indicated in the figure. For the sake of simplicity, they have been multiplied by $1/Q = 1024$. Verify that the signal values in the next sample interval remain the same except for a change in sign. Hence, a nonobservable zero-input oscillation persists.

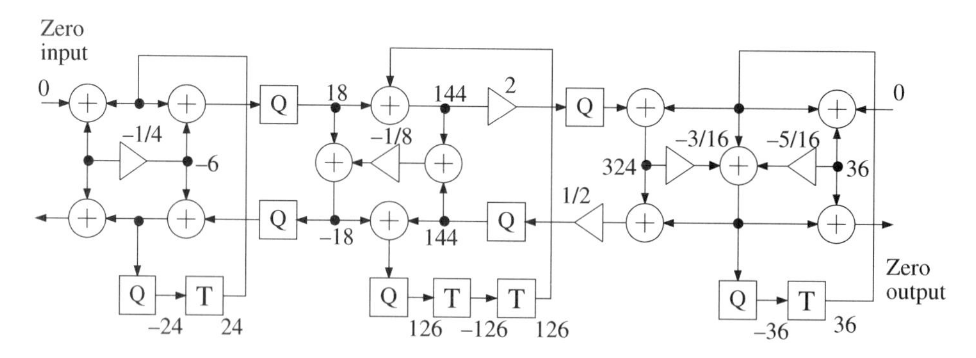

Figure 5.8 Nonobservable parasitic oscillation in a part of a ladder wave digital filter with magnitude truncation

---

5.2.5 Parasitic Oscillations In Algorithms Using Floating-Point Arithmetic

Parasitic oscillations also occur in recursive structures implemented with floating-point arithmetic. Overflow oscillations are, of course, not a problem if floating-point arithmetic is used. The parasitic oscillations are caused by nonlinearities due to quantization of products and denormalization and normalization of the mantissa that are required in addition and subtraction. The magnitude of the parasitic oscillations can be arbitrarily large in floating-point arithmetic, since the actual value of the exponent does not affect the nonlinearities. Thus, the large dynamic range of floating-point arithmetic is no remedy in avoiding undesirable phenomena caused by nonlinearities.

We will not discuss floating-point arithmetic further because it is much more expensive to implement than fixed-point arithmetic and, in general, we do not need a large number range. The number range of fixed-point arithmetic can be increased by 6 dB by providing just one extra bit in the data word length. Further, good algorithms tend to utilize the available number range effectively.

5.3 STABILITY

From the previous sections it is evident that stability is a very complicated issue [12]. Ideally, a nonlinear system should behave as closely as possible to the corresponding linear system so that the outputs from the two systems would be identical for arbitrary input signals. In other words, the difference between the outputs of a *forced-response stable* filter and the ideal linear filter should for arbitrary input signals tend to zero after a disturbance has occurred. As can be seen from Figure 5.7, a second-order section in direct form I or II with saturation arithmetic is not forced-response stable.

An even more stringent stability requirement is that independent of initial values in the delay elements, the outputs of the nonlinear filter and the linear filter should become arbitrarily close. Stability, in this sense, can be guaranteed in wave digital filters using the concept of *incremental pseudo-passivity* [23]. Large errors resulting from overflow are often more rapidly suppressed in wave digital filters as compared to the cascade and parallel forms.

A more complicated stability problem arises when a digital filter is used in a closed loop [10]. Such loops are encountered in, for example, long-distance telephone systems.

5.4 QUANTIZATION IN WDFs

Parasitic oscillations can, as already discussed, occur only in recursive structures. Nonrecursive digital filters can have parasitic oscillations only if they are used inside a closed loop. Nonrecursive FIR filters are therefore robust filter structures that do not support any kind of parasitic oscillation. Among the IIR filter structures, wave digital filters, and certain related state-space structures are of major interest as they can be designed to suppress parasitic oscillations.

Fettweis has shown that overflow oscillations can be suppressed completely in wave digital filters by placing appropriate restrictions on the overflow characteristic [8, 9]. To show that a properly designed wave digital filter suppresses any parasitic oscillation, we use the concept of *pseudo-power*[1] which corresponds to power in analog networks. Note that the pseudo-power concept is defined only for wave digital filters and not for arbitrary filter structures.

The instantaneous pseudo-power entering the adaptor network, shown in Figure 5.9, is defined as

$$p(n) \triangleq \sum_{k=1}^{N} G_k[a_k(n)^2 - b_k(n)^2] \tag{5.1}$$

where $G_k = 1/R_k$.

[1] Pseudo-power corresponds to a Lyapunov function.

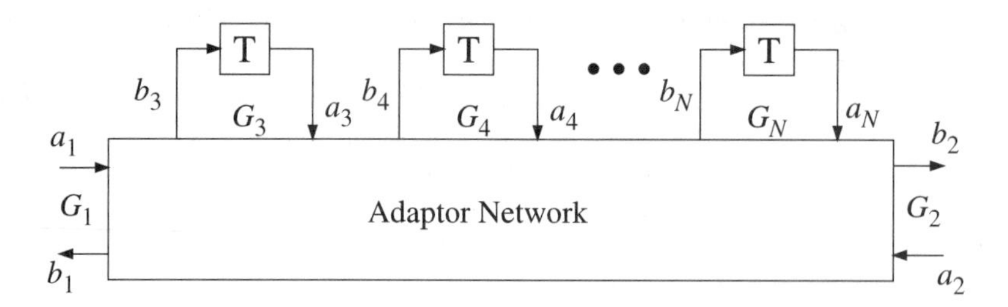

Figure 5.9 Wave digital filter

Ideally, $p(n) = 0$, since the adaptors are lossless and an arbitrary network of adaptors is also lossless.

Now, in order to suppress a disturbance caused by a nonlinearity it is sufficient to introduce losses in each port of the network such that the nonlinearities correspond to pure losses. This can be accomplished by making each term $a_k(n)^2\ b_k(n)^2 > 0$. Any parasitic oscillation will by necessity appear at one or more of the ports with delay elements. Hence, a parasitic oscillation which, of course, has finite pseudo-energy and is not supported from an external signal source will decay to zero since it will dissipate energy in the lossy ports of the adaptor network.

Lossy ports can be obtained by quantizing the reflected waves such that their magnitudes are always decreased. Hence, for each nonzero reflected wave we require that

$$\begin{cases} |\, b(n)_{\mathrm{Q}} \,| < |\, b(n)_{\mathrm{exact}} \,| \\ b(n)_{\mathrm{Q}}\, b(n)_{\mathrm{exact}} \geq 0 \end{cases} \tag{5.3}$$

where the second constraint is to assure that the signals have the same sign.

In the part of the ladder wave digital filter shown in Figure 5.8 ordinary magnitude truncation was used. Obviously this is not sufficient to suppress nonobservable oscillations since the magnitudes of the signal values, quantized to 11 bits including the sign bit, i.e., $Q = 2^{10}$, are not effected by the quantization. For example, $[72/1024]_Q = 72/1024$. To suppress all types of parasitic oscillations it is necessary to use a strict magnitude truncation, which always reduces the magnitude, e.g., $[72/1024]_Q = 72/1024$. Note that it does not help to increase the data word length. Further, it is not practical to try to detect nonobservable oscillations. These, therefore, have to be avoided by using a properly designed wave digital filter.

In order to develop a practical implementation scheme that will suppress parasitic oscillations, we first assume that there are no nonobservable oscillations of the type discussed in section 5.2.4 inside the adaptor network. For the second-order allpass section shown in Figure 5.10, the quantization constraint in Equation (5.2) can be satisfied by the following scheme:

1. Apply two's-complement truncation after the multiplications.
2. Add, to the reflected waves, a 1 in the least significant position for negative signal values.

3. Apply saturation, or invert all bits including the sign bit, if a reflected wave exceeds the signal range.
4. The internal signal range is increased by 1 bit to [2, 2], since the input to a multiplier with a noninteger coefficient may not overflow the available signal range. This word length extension is indicated by a box labeled "sign extend" in Figures 5.10 and 5.11. The word length is reduced to its normal range [1, 1[at the output of the section.

Bireciprocal lattice wave digital filters will suppress zero-input oscillations even if the second constraint is removed. The simplified scheme for a second-order allpass section with poles on the imaginary axis is shown in Figure 5.11.

In an implementation of an incremental pseudo-passive wave digital filter it is not necessary to start the filtering by first zeroing the delay elements, since any effect of initial values will fade away and can not cause a permanent oscillation. This possibility often leads to a significant simplification of the hardware, since a special start-up mode is no longer required. It also saves time and reduces code size in a software implementation. It is also possible to develop a quantization scheme for floating-point arithmetic, although this leads to long and more complicated computations.

The equations describing the two-port adaptors used in lattice wave digital filters can be modified to exploit the features of standard signal processors [11].

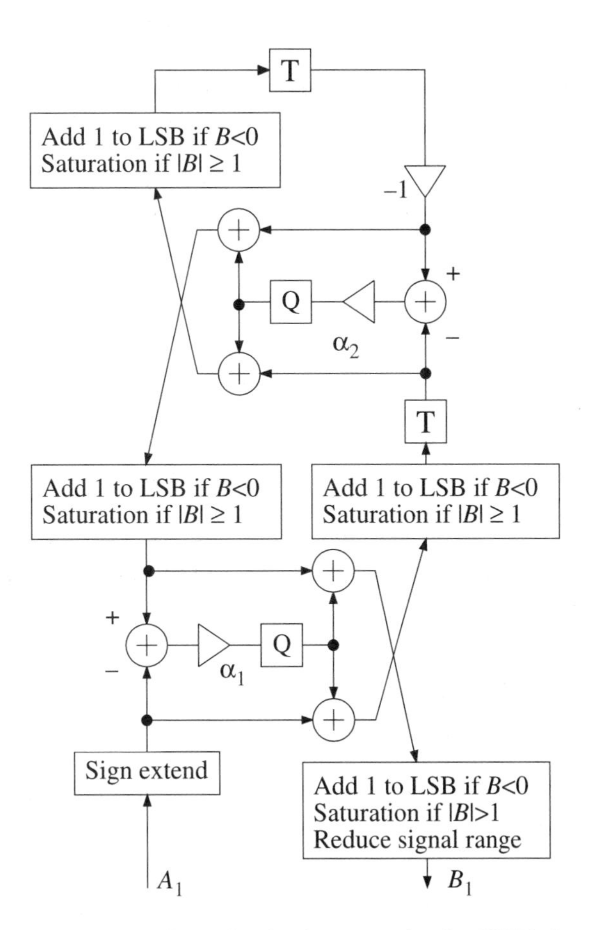

Figure 5.10 Quantization in a second-order WDF of allpass type

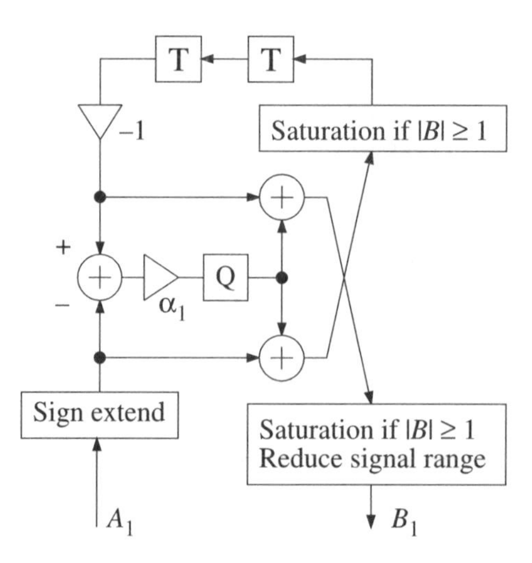

Figure 5.11 Quantization in a WDF of allpass section with all poles on the imaginary axis

5.5 SCALING OF SIGNAL LEVELS

As just discussed, parasitic oscillations can be started in fixed-point arithmetic by an overflow of the signal range. Unfortunately, for most filter structures, the parasitic oscillation persists even when the cause of the overflow vanishes. Special precautions must therefore be taken to assure that the filter does not sustain a permanent oscillation. Measures must also be taken to prevent overflow from occurring too often, since overflows cause large distortion. Typically, an overflow should not occur more frequently than once every 10^6 samples in speech applications. Of course, the duration and perceived effect of the overflow must also be taken into account. However, it is difficult to give precise guidelines for overflow probability since one overflow may increase the probability of repeated overflows.

The probability of overflow can be reduced by reducing signal levels inside the filter. This can be accomplished by inserting so-called scaling multipliers that affect signal levels inside the filter only. The scaling multiplier must not affect the transfer function. On the other hand, the signal levels should not be too low, because the SNR (signal-to-noise ratio) would then become poor, since the noise level is fixed for fixed-point arithmetic.

Scaling is not required in floating-point arithmetic since the exponent is adjusted so that the mantissa always represents the signal value with full precision.

An important advantage of using two's-complement representation for negative numbers is that temporary overflows in repeated additions can be accepted if the final sum is within the proper signal range. However, the incident signal to a multiplier with a noninteger coefficient must not overflow, since that would cause large errors. For example, the only critical overflow node in the signal-flow graph shown in Figure 5.12 is the input to the multiplier with the noninteger (1.375) coefficient. The multiplication by 2 can be regarded as an addition, and delay elements do not effect the signal values. In other types of number systems, the scaling restrictions are more severe.

Figure 5.13 shows a multiplier with a noninteger coefficient that is imbedded in a network, N_2. The input signal level to this multiplier must be properly scaled. This; is done by multiplying all signals entering the network N_2 by the scaling coefficient c, and multiplying all signals leaving the network by $1/c$.

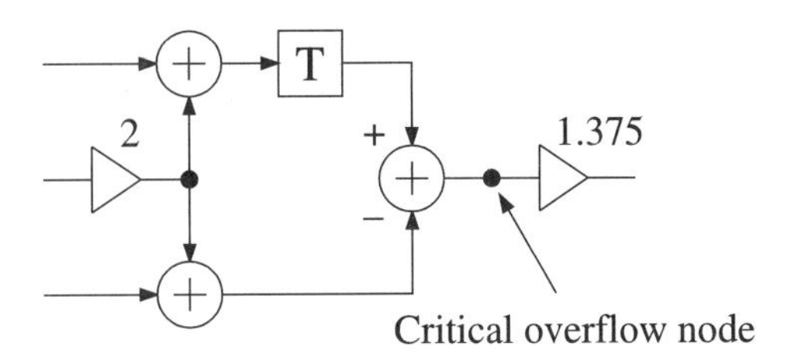

Figure 5.12 Signal-flow graph with only one critical overflow node

Scaling multiplier;s must not affect the transfer function such that the poles and zeros are changed. Only the gain from the input of the filter to the critical node may be changed. If the scaling multipliers are part of a recursive loop, it is necessary that the effect on the transfer function is eliminated by choosing c so that $c(1/c) \equiv 1$, using binary values for both coefficients. It can be shown that the only possible values are $c = 2^{\pm n}$. The scaling multipliers with associated quantizations may introduce additional round-off noise sources, but proper scaling will nevertheless improve the SNR. Additional scaling nodes may in some cases be introduced by the noninteger scaling coefficients.

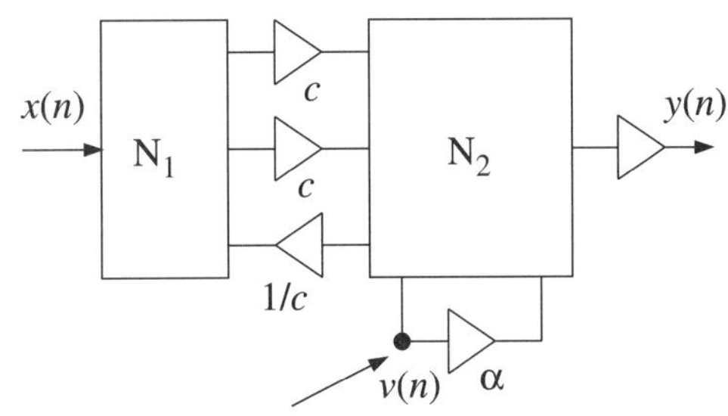

Figure 5.13 Scaling of the signal level incident to a noninteger multiplier

5.5.1 Safe Scaling

One strategy used to choose the scaling coefficient can be derived in the following way: The signal in the scaling node is given by

$$v(n) = f(n) * x(n)$$

where $f(n)$ is the impulse response from the input of the filter to the critical overflow node. The magnitude of the output signal is bounded by

$$|v(n)| = \left|\sum_{k=0}^{\infty} f(k)\, x(n-k))\right| \le \sum_{k=0}^{\infty} |f(k)||x(n-k)| \le M \sum_{k=0}^{\infty} |f(k)|$$

where

$$|x(n)| \le M$$

In this scaling approach, we insert a scaling multiplier(s), c, between the input and the critical overflow node, as shown in Figure 5.13. The resulting impulse response becomes

$$f'(k) = c f(k)$$

Now, we choose c so that

$$\sum_{k=0}^{\infty} |f'(k)| \le 1 \tag{5.3}$$

The magnitude of the scaled input signal to the multiplier will be equal to, or less than, the magnitude of the input signal of the filter. The input to the multiplier will never overflow if the input to the filter does not overflow. This scaling policy is therefore called *safe scaling*. The safe scaling method is generally too pessimistic since it uses the available signal range inefficiently. The safe scaling method is suitable for short FIR filters because the probability for overflow is high for a filter with a short impulse response.

It is sometimes argued that parasitic oscillations caused by overflow can not occur if the filter is scaled according to the safe scaling criterion. However, this is not correct since abnormal signal values can occur due to malfunctioning hardware—for example, in the memory (delay) elements due to external disturbances (e.g., ionic radiation and disturbances from the supply lines). The likelihood of temporary malfunction of the hardware increases as the size of the transistors is scaled down. In the future it will therefore be necessary to use filter structures that are guaranteed to suppress all types of parasitic oscillations and return the filter to normal operation.

EXAMPLE 5.5

Scale the signal levels in an FIR filter using the safe scaling criterion. The filter is realized using the direct form, linear-phase structure shown in Figure 4.6. The original impulse response is

$h(n)$ = 0.0030841, 0.027789, 0.066604, 0.095799, 0.066604, 0.027789, 0.0030841, 0, 0, ...

First, the input to the multipliers that have noninteger coefficients must be scaled. Hence, the input signal must be divided by 2 because the preceding additions effectively double the signal level. This operation may in practice be accomplished by either a shift operation or by simply extending the word length by one bit. Second, the output of the filter must be scaled such that the output uses the available signal range effectively. For the unscaled filter we have

$$S = \sum_{i=0}^{6} \left| \frac{h(n)}{2} \right| = 0.1453766$$

where the factor 2 comes from the scaling at the input of the filter. The new, scaled coefficients are obtained by dividing by S.

$$h'(n) = \frac{h(n)}{S}, \; n = 0, 1, ..., 6$$

We get

$h(n)$ = 0.021215, 0.19115, 0.45815, 0.65897, 0.45815, 0.19115, 0.021215, 0, 0, ...

5.5.2 FFT Scaling

In order to scale the signal levels in the FFT we first scale the inputs to the butterfly operations. We will later discuss scaling of the signals that are internal to the butterfly operations. We will use the safe scaling technique just discussed.

We have for the complex input values to the butterfly (see Figure 5.14)

$$|U| \le 0.5, \qquad |V| \le 0.5$$

and for the two output values, X and Y, we have

$$|X = |U+V| \le |U| + |V| \le 1$$

$$|Y| = |(U\ V)W^p| = |U\ V| \le |U| + |V| \le 1$$

In order to assure that the input signals to the succeeding stage of butterflies also are properly scaled, the outputs of each butterfly are divided by 2, as illustrated in Figure 5.14.

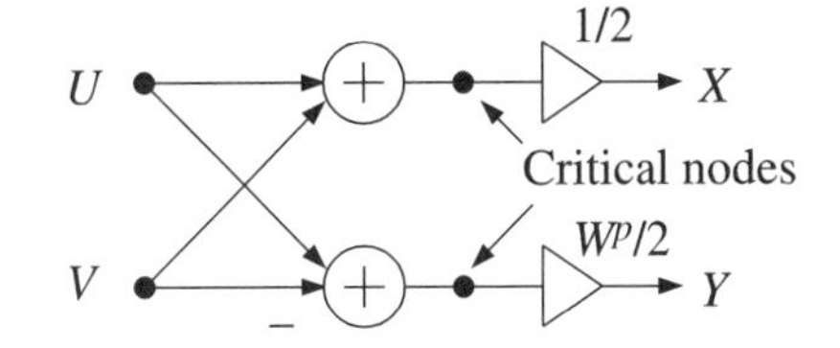

Figure 5.14 Safely scaled butterfly

We also need to scale the input signal to the FFT by dividing by 2 so that the signal level to the butterflies in the first stage becomes properly scaled. Further, the scaling of the butterflies in the last stage is not necessary. However, from an implementation point of view it may be better to retain the scaling and multiply the outputs with a factor two. The result is that the output of the scaled FFT will be divided by the factor $2^M = N$, where $M = \log_2(N)$ stages.

It is easy to show that this scaling policy is conservative and corresponds to the safe scaling just discussed. Assume that the input is a sinusoidal sequence. The DFT is

$$X(k) = \frac{1}{N}\sum_{n=0}^{N-1} x(n)W^{nk} = \frac{1}{N}\sum_{n=0}^{N-1} e^{j\omega nT}W^{nk}$$

$$= \frac{1}{N}\sum_{n=0}^{N-1} e^{j(\omega T - 2\pi k/N)n} = \begin{cases} 1 & \text{if } \omega T = 2\pi k/N \\ 0 & \text{otherwise} \end{cases}$$

Hence, a sinusoidal input sequence with amplitude A will appear in the frequency domain with amplitude A. The FFT is therefore safely scaled.

5.5.3 L_p-Norms

In the next few sections, we will discuss more efficient scaling policies based on the observation that it is the frequency properties of a signal that are of major interest in many DSP algorithms—for example, frequency selective filters. This implies that the spectral properties of the input signal are characteristic in these applications. It is therefore natural to try to exploit knowledge of how the input signal spectrum varies with frequency as a scaling criterion. We therefore need a measure of the "size" of the signals spectrum. To this purpose Jackson [15, 16] has developed a scaling procedure where L_p-norms are used as a measure of the Fourier transform.

For a sequence, $x(n)$, with Fourier transform $X(e^{j\omega T})$, the L_p-norm is defined as

$$\| X(e^{j\omega T}) \|_p \triangleq \sqrt[p]{\frac{1}{2\pi}\int_{-\pi}^{\pi} |X(e^{j\omega T})|^p d\omega T} \tag{5.4}$$

for any real $p \geq 1$ such that the integral exists. It can be shown that

$$\| X \|_p \geq \| X \|_q \qquad \text{for } p \geq q \tag{5.5}$$

Of particular interest in this context are three cases: the L_1-norm, L_2-norm, and L_∞-norm.

L_1-Norm

$$\| X(e^{j\omega T}) \|_1 = \frac{1}{2\pi}\int_{\pi}^{\pi} |X(e^{j\omega T})|^p d\omega T \tag{5.6}$$

The L_p-norm of the Fourier transform corresponding to a sinusoidal sequence

$$x(n) = A \cos(\omega n T + \Phi)$$

exists only for $p = 1$ and is

$$\| X \|_1 = A \tag{5.7}$$

The L_1-norm, which corresponds to the average absolute value, can in practice be determined either by numeric integration of the magnitude function or by computing the DFT (FFT) of the impulse response. In general, it is difficult or even impossible to obtain analytical expressions for the L_1-norm.

L_2-Norm

The L_2-norm of a continuous-time Fourier transform is related to the power contained in the signal, i.e., the rms value. The L_2-norm of a discrete-time Fourier transform has an analogous interpretation. The L_2-norm is

$$\| X(e^{j\omega T}) \|_2 = \sqrt{\frac{1}{2\pi}\int_{-\pi}^{\pi} |X(e^{j\omega T})|^2 d\omega T} \tag{5.8}$$

The L_2-norm is simple to compute by using Parsevals relation [5, 27, 33], which states that the power can be expressed either in the time domain or in the frequency domain. We get from Parsevals relation

$$\| X \|_2 = \sqrt{\sum_{n=-\infty}^{\infty} x(n)^2} \tag{5.9}$$

We frequently need to evaluate L_2-norms of frequency responses measured from the input of a filter to the critical overflow nodes. This can be done by implementing the filter on a general-purpose computer and using an impulse sequence as input. The L_2-norm can be calculated by summing the squares of the signal values in the node of interest. This method of computing the L_2-norm will be used in Example 5.6.

L_∞-Norm

The L_∞-norm corresponds to the maximum value of the magnitude function. Hence, we have

$$\|X(e^{j\omega T})\|_\infty = \lim_{p \to \infty} \|X(e^{j\omega T})\|_p = \max_{\omega T} \{|X(e^{j\omega T})|\} \tag{5.10}$$

The values of L_p-norms are illustrated in Figure 5.15 for the second-order section having the transfer function

$$H(z) = \frac{0.5z^2 - 0.4z + 0.5}{z^2 - 1.2z + 0.8}, \qquad |z| > \sqrt{0.8}$$

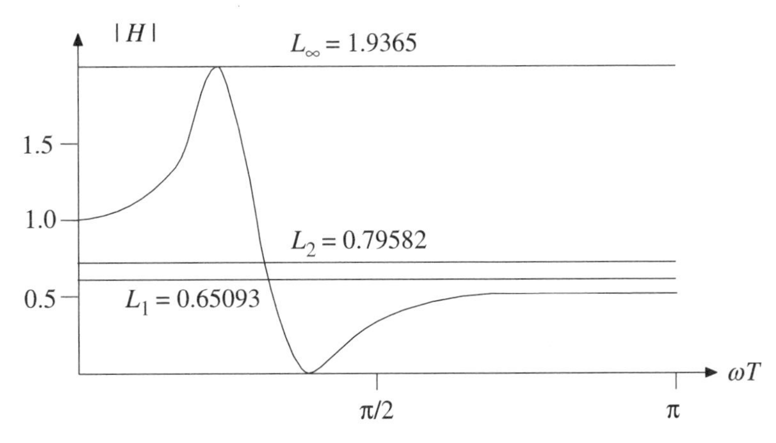

Figure 5.15 L_p-norms for a second-order section

5.5.4 Scaling of Wide-Band Signals

The first step in scaling a filter is to determine the appropriate L_p-norm that characterizes the input signal. Generally, we distinguish between wide-band and narrow-band input signals. However, it is not obvious which type of input signal should be used for scaling; the "normal" class of input signals, including noise, or some infrequently occurring disturbance. The answer to this question depends, of course, on the application.

Jackson [17] has derived the following bound on the variance of the signal in the critical node v:

$$\sigma_v^2 \le \left\| F_v(e^{j\omega T}) \right\|_{2p}^2 \left\| S_x(e^{j\omega T}) \right\|_q, \quad \frac{1}{p} + \frac{1}{q} = 1, \text{ for } p, q \ge 1 \tag{5.11}$$

where $F_v(e^{j\omega T})$ is the frequency response to the critical node v and $S_x(e^{j\omega T})$ is the power spectrum of the input signal.

A wide-band input signal is characterized by the L_∞-norm, $\| S_x \|_\infty$. See Problem 5.14. Hence, $q = \infty \Rightarrow p = 1$ and the filter should therefore be scaled such that $\| F_v \|_2 = c$, where $c \le 1$ for all critical nodes.

EXAMPLE 5.6

Scale the filter derived in Example 4.5 for wide-band input signals. Two's-complement representation is used. Only those inputs to the multipliers with noninteger coefficients, and the output, need to be scaled.

The purpose of scaling is to assure that overflow does not occur too often at multiplier inputs and at the output. Generally, we assume that the input signal level is properly scaled. Hence, the inputs to the coefficients

$$a_{01}, a_{11}, \text{ and } a_{21}$$

are assumed to have a proper signal level. However, the output of the filter and the input nodes to all other coefficients must be scaled. In order to scale the filter, the L_2-norms to the critical node have to be determined. This is easily accomplished by the program shown in Box 5.1. The program iterates the system of difference equations with an impulse sequence as input. The L_2-norms are determined by summing the squared impulse response in the critical nodes. A formal method of deriving the difference equations will be discussed in Chapter 6.

We get the L_2-norms of the outputs of the second-order sections by using the program:

first section:	$\| F_3 \|_2 = 1.892948$
second section:	$\| F_5 \|_2 = 5.560767$
third section:	$\| F_7 \|_2 = 17.289397$
fourth section:	$\| F_y \|_2 = 459.018539$

First, the coefficients a_{01}, a_{11}, and a_{21} are divided by $\| F_3 \|_2 = 1.892948$ so that the output of the first section and thereby the input to the second section become properly scaled. The new coefficients are

$$a_{01} = 0.5282766, \qquad a_{11} = 1.0565532, \qquad a_{21} = 0.5282766,$$

This will affect the signal levels in the subsequent sections and, of course, the measured L_2-norms. The L_2-norms to the outputs of the remaining sections are obtained by dividing the original L_2-norms by $\| F_3 \|_2$. We get

second section: $$\frac{\|F_5\|_2}{\|F_3\|_2} = \frac{5.560767}{1.892948} = 2.937622$$

third section: $$\frac{\|F_7\|_2}{\|F_3\|_2} = \frac{17.289397}{1.892948} = 9.133583$$

fourth section: $$\frac{\|F_y\|_2}{\|F_3\|_2} = \frac{459.018539}{1.892948} = 242.488721$$

```
program IIR_filter_scaling;
const
   a01 = 1;   a11 = 2;          a21 = 1;   b11 = 1.49139;    b21 = 0.8997958;
   a02 = 1;   a12 = 1.006442;   a22 = 1;   b12 = 1.305644;   b22 = 0.9595307;
   a03 = 1;   a13 = 1.890926;   a23 = 1;   b13 = 1.617856;   b23 = 0.9617549;
   a04 = 1;   a14 = 2;          a24 = 1;   b14 = 1.357826;   b24 = 0.8956455;
var
   x, y, F3two, F5two, F7two, Fytwo : Double;
   v, u : array[1..10] of Double;
   i : integer;
begin
   F3two := 0;    F5two := 0;   F7two := 0;   Fytwo := 0;
   for i := 1 to 10 do
      begin
         v[i] := 0;{      zero all delay elements }
         u[i] := 0;
      end;
   x := 1; (* The input is an impulse sequence*)
   for i := 0 to 500 do        {The significant part of the impulse response}
      begin                    {is assumed to be shorter than 500 samples }
         {Difference equations }
      u[3] := a01 * x + a11 * v[1] + a21 * v[2] + b11 * v[3] + b21 * v[4];
      u[5] := a02 * u[3] + a12 * v[3] + a22 * v[4] + b12 * v[5] + b22 * v[6];
      u[7] := a03 * u[5] + a13 * v[5] + a23 * v[6] + b13 * v[7] + b23 * v[8];
      u[9] := a04 * u[7] + a14 * v[7] + a24 * v[8] + b14 * v[9] + b24 * v[10];
      v[2] := v[1]; '          {Update the delay elements }
      v[1] := x;
      v[4] := v[3];
      v[3]:= u[3];
      v[6] := v[5];
      v[5]:= u[5];
      v[8] := v[7];
      v[7] := u[7];
      v[10] := v[9];
      v[9] := u[9];
      x := 0; { Reset x after the first sample to get an impulse sequence }
      F3two := F3two + u[3] * u[3]; { Calculate the squared L2-norms }
      F5two := F5two + u[5] * u[5];{      for the critical nodes }
      F7two := F7two + u[7] * u[7];
      Fytwo := Fytwo + u[9] * u[9];
   end;
   writeln(sqrt(F3two), sqrt(F5two), sqrt(F7two), sqrt(Fytwo) : 10 : 6);
end.
```

Box 5.1 Program for scaling signal levels in the filter in Example 4.5

Next, coefficients a_{02}, a_{12}, and a_{22} are scaled by dividing by $\| F_5 \|_2 / \| F_3 \|_2 = 5.560767/1.892948 = 2.937622$. The scaled coefficients are

$$a_{02} = 0.34041124, \qquad a_{12} = 0.34260417, \qquad a_{22} = 0.34041124$$

The L_2-norms to the outputs of the remaining sections are obtained by dividing the previous L_2-norms by $\| F_5 \|_2 / \| F_3 \|_2$. We get

$$\text{third section:} \qquad \frac{\|F_7\|_2}{\|F_3\|_2} = \frac{17.289397}{1.892948} = 3.109175$$

$$\text{fourth section:} \qquad \frac{\|F_y\|_2}{\|F_3\|_2} = \frac{459.018539}{5.560767} = 82.545338$$

The third set of coefficients a_{03}, a_{13}, and a_{23} are scaled by dividing the L_2-norms obtained in the previous step by $\| F_7 \|_2 / \| F_5 \|_2$. We get the scaled coefficients

$$a_{03} = 0.3216288, \qquad a_{13} = 0.608176, \qquad a_{23} = 0.3216288$$

Finally, the last set of coefficients is divided by $\| F_y \|_2 / \| F_7 \|_2$. The scaled coefficients are

$$a_{04} = 0.03766609, \qquad a_{14} = 0.07533218, \qquad a_{24} = 0.03766609$$

Finally, running the program again, but now with the new set of coefficients, shows that all the L_2-norms are unity. The resulting coefficients are slightly different from the ones given in Example 4.5 due to rounding errors. Note that rounding shall be done such that the zeros remain on the unit circle, i.e., $a_0 \equiv a_2$. The denominator coefficients are not affected by the scaling.

5.5.5 Scaling of Narrow Band Signals

The L_1-norm, $\| S_x \|_1$ characterizes, according to Equation (5.7), a sinusoidal or narrow-band signal. From Equation (5.11) we find, with $q = 1$ and $p = \infty$, that the upper bound for the variance of the signal in the critical node is determined by $\| F_v \|_\infty$. Thus, the filter should be scaled such that the maximum value of $|F_v|_{max} = \| F_v \|_\infty = c$, where $c \leq 1$ for all critical nodes. The probability of overflow can be adjusted by choosing an appropriate value for c.

EXAMPLE 5.7

Scale the signal levels for the filter used in Example 5.6, but assume now that a narrow-band input signal is used.

The appropriate scaling criterion is the L_∞-norm of the frequency responses to the critical nodes. It is difficult to determine the L_∞-norm without using a computer. In practice, the L_∞-norm is therefore determined using a program based on, for example, the DFT of the impulse responses.

We first measure the L_∞-norm from the input to the first critical node, i.e., the output of the first second-order section. We get the first L_∞-norm, $\| F_3 \|_\infty = 6.951637$. The coefficients a_{01}, a_{11}, and a_{21} are then divided by $\| F_3 \|_\infty$ so that the output of the first section and the input to the second section become properly scaled. The new coefficients are

$$a_{01} = 0.143851, \qquad a_{11} = 0.287702, \qquad a_{21} = 0.143851$$

In the second step, the next critical node, as seen from the input, is measured using the new set of coefficients in the first section. We get

$$\| F_5 \|_\infty = 4.426973$$

The coefficients a_{02}, a_{12}, and a_{22} are then divided by $\| F_5 \|_\infty$ so that the output of the second section and the input to the third section become properly scaled. The new set of coefficients are

$$a_{02} = 0.225888, \qquad a_{12} = 0.2273435, \qquad a_{22} = 0.225888,$$

By successively measuring the norms to the critical nodes using the scaled coefficients in the preceding sections we get

$$\| F_7 \|_\infty = 29.529326$$

$$a_{03} = 0.3386464, \qquad a_{13} = 0.6403554, \qquad a_{23} = 0.3386464,$$

and

$$\| F_y \|_\infty = 18.455633$$

$$a_{04} = 0.054184, \qquad a_{14} = 0.108367, \qquad a_{24} = 0.054184$$

Note that it is necessary to measure the L_∞-norms and scale the sections one at a time, from the input to the output of the filter, since the L_∞-norms depend on the scaling of the previous sections. This is not necessary if we use L_2-norms for scaling.

Comparing the coefficients obtained in Examples 5.6 and 5.7 we observe that a narrow-band input signal presents a more difficult case than a wide-band signal in the sense that the former may produce larger signal levels inside the filter. This give rise to the smaller numerator coefficients.

The two scaling methods just discussed are the ones used in practice. However, they lead only to a reasonable scaling of the filter. In practice, simulations with actual input signals should be performed to optimize the dynamic range.

5.6 ROUND-OFF NOISE

In most digital filter implementations only a few types of arithmetic operations—addition, subtraction, and multiplication by constant coefficients—are used. The advent of VLSI has fostered the use of more complex basic operations such as vector-multiplication.

In practice, only fixed-point arithmetic is used in application-specific ICs since the hardware is much simpler and faster compared with floating-point arithmetic. Also, the quantization procedure needed to suppress parasitic oscillations in wave digital filters is more complicated for floating-point arithmetic. Moreover, the analysis of the influence of round-off errors is very complicated [28, 32, 35], because quantization of products as well as addition and subtraction causes errors that depend on the signal values. Errors in fixed-point arithmetic are with few exceptions independent of the signal values and can therefore be analyzed independently. Lastly, the high dynamic range provided by floating-point arithmetic is not really needed in good filter algorithms, since filter structures with low coefficient sensitivity also utilize the available number range efficiently. We will therefore limit the discussion to fixed-point arithmetic.

Using fixed-point arithmetic with suitably adjusted signal levels, additions and multiplications generally do not cause overflow errors. However, the products must be quantized using rounding or truncation. Errors, which are often called rounding errors even if truncation is used, appear as round-off noise at the output of the filter. The character of this noise depends on many factors—for example, type of nonlinearity, filter structure, type of arithmetic, representation of negative numbers, and properties of the input signal.

A simple linear model of the quantization operation in fixed-point arithmetic can be used if the signal varies over several quantization levels, from sample to sample, in an irregular way. A simple model is shown in Figure 5.16. The quantization of a product

$$y_Q(n) = [a\, x_Q(n)]_Q$$

is modeled by an additive error

$$y_Q(n) = a\, x_Q(n) + e(n) \tag{5.12}$$

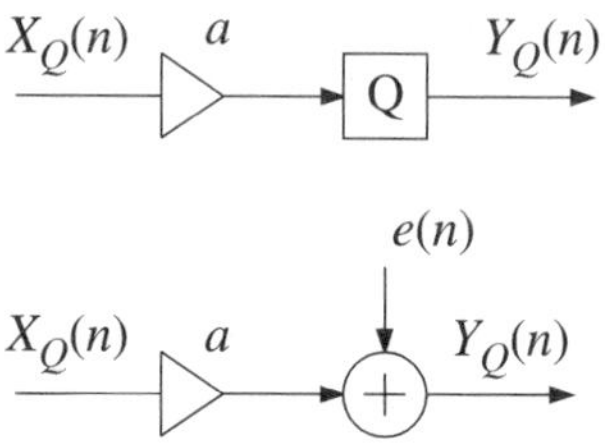

Figure 5.16 Linear noise model for quantization

where $e(n)$ is a stochastic process.

Normally, $e(n)$ can be assumed to be white noise and independent of the signal. The density function for the errors is often approximated by a rectangular function. However, the density function is a discrete function if both the signal value and the coefficient value are binary. The difference is only significant if only a few bits are discarded by the quantization. The average value and variance for the noise source are

$$m = \begin{cases} \dfrac{Q_c}{2} Q & \text{rounding} \\ \dfrac{Q_c}{2} Q & \text{truncation} \end{cases} \tag{5.13}$$

and

$$\sigma_e^2 = k_e\,(1 - Q_c^2)\sigma^2 \tag{5.14}$$

where

$$k_e = \begin{cases} 1 & \text{rounding or truncation} \\ 4 - \dfrac{6}{\pi} & \text{magnitude truncation} \end{cases} \tag{5.13}$$

$\sigma^2 = Q^2/12$, Q is the data quantization step, and Q_c is the coefficient quantization step. For long coefficient word lengths—the average value is close to zero for rounding and $Q/2$ for truncation. Correction of the average value and variance is only necessary for short coefficient word lengths, for example, for the scaling coefficients.

Generally, the output signal of a digital filter has a DC offset due to the average quantization error, as discussed already. A digital filter with M quantization points has a DC offset of

$$m_y = m \sum_{i=1}^{M} \sum_{n=0}^{\infty} g_i(n) \tag{5.15}$$

where $g_i(n)$ are the impulse responses measured from the noise sources to the output of the filter, as illustrated in Figure 5.17.

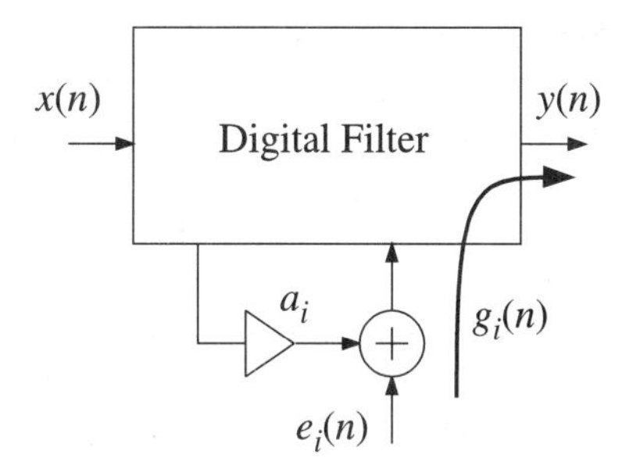

Figure 5.17 Noise model for digital filters

It is sometimes possible to select the quantization nodes in such a way that the average values tend to cancel. Note that the DC offset also can be removed by adding an appropriate offset at the output. The noise sources contribute to the noise at the output of the filter. The variance at the output, from source i, is

$$\sigma_{yi}^2 = \sigma_i^2 \sum_{n=0}^{\infty} g_i(n)^2 = \sigma_i^2 \left\| G_i(e^{j\omega T}) \right\|_2^2 \tag{5.16}$$

The variance of the round-off noise at the output is equal to the sum of the contributions from all the uncorrelated noise sources

$$\sigma_{ytot}^2 = \sum_{i=0}^{M} \sigma_{yi}^2 \tag{5.17}$$

The noise gain from a noise source to the output can easily be determined by injecting an impulse sequence into the appropriate node in the filter algorithm. The algorithm can be implemented, in a general purpose computer, using high-precision floating-point arithmetic. The squared impulse response at the output of the filter is summed over the significant part of the impulse response.

The round-off noise is generally large for narrow- or wide-band filters that have poles close to the unit circle, but filter structures vary widely in this respect.

EXAMPLE 5.8

Determine the loss in dynamic range at the output of the filters in Examples 5.6 and 5.7 due to round-off noise. Assume that rounding is performed after each multiplication.

Each section has five noise sources that effectively appear at the summation node. The noise gain, G, for these sources can be measured by successively injecting an impulse sequence at each summation node of the second-order sections and computing

$$G_i^2 = \sum_{n=0}^{\infty} g_i(n)^2$$

by iterating the difference equations. The results for the bandpass filter, with different scaling, are shown in Table 5.1.

Node	G_i^2 with L_∞-Norm Scaling	G_i^2 with L_2-Norm Scaling
u[3]	16.22323	16.05974
u[5]	25.52102	11.12439
u[7]	39.98744	19.32355
u[9]	10.38189	10.38189
Sum G_i^2	92.11358	56.88957
G_0^2	0.0749669	1.0000
ΔW	6.29	4.08

Table 5.1 L_p-norms (squared) for the bandpass filter in Example 5.6

The difference in dynamic range between the two scaling strategies is about 2.21 bits. The L_2-norm for the whole filter is G_0. The variance of the round-off noise at the output of the filter is obtained by adding the contribution from each source. The equivalent noise gain is

$$G^2 = 5\ [G_3{}^2 + G_5{}^2 + G_7{}^2 + G_y{}^2]$$

The round-off noise, corrected for the filter gain, corresponds to a loss in dynamic range of ΔW bits, where

$$\Delta W = 0.5 \log_2 \frac{G^2}{G_0^2}$$

5.6.1 FFT Round-Off Noise

Data word length will affect the cost of butterfly PEs and, more importantly, the chip area required for the memories. In fact, for large FFTs the chip area required for the PEs is small compared to the area required for memory. It is therefore important to determine accurately and, if possible, minimize the required data word length.

We assume that the computations inside the butterfly operations are performed with full accuracy using fixed-point arithmetic. The complex results are rounded at the outputs of the butterfly as shown in Figure 5.18.

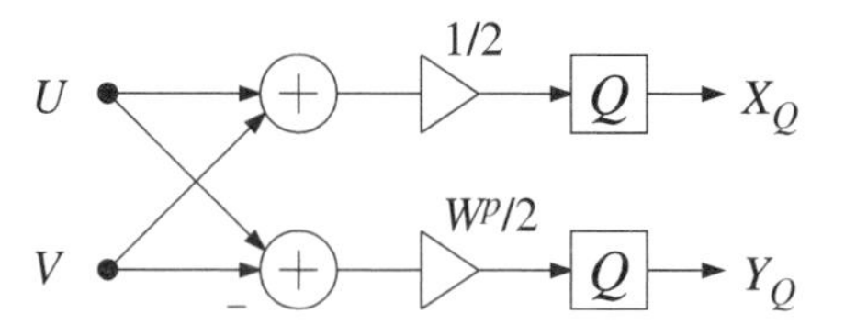

Figure 5.18 Butterfly with scaling and quantizations

For simplicity, we neglect the correction for short coefficient word lengths, Equation (5.14), and model the round-off noise at the two outputs of a butterfly with uncorrelated complex noise sources with variance

$$\sigma_B{}^2 = 2\ \frac{Q^2}{12}$$

Depending on how the arithmetic operations inside the butterfly are performed, quantization errors may be avoided for the simple coefficient values ±1 and $\pm j$. However, here we assume that we always get quantization error due to the scaling coefficient. Hence, in practice, we can expect the round-off noise to be somewhat lower than predicted by this simplified analysis.

It can be shown that the noise variance at an FFT output node is obtained by adding the contribution from no more than N_1 such noise sources. Figure 5.19 shows which butterflies contribute noise to the output node $X(0)$. Scaling the input by a factor 1/2 does not introduce any noise since the input data word length is shorter than the word length inside the FFT computations. Hence, the scaled input values need not to be quantized.

For the unscaled FFT the magnitude of the frequency response from the output of a butterfly to any output node of the FFT is unity, since $|W| = 1$. Hence, the

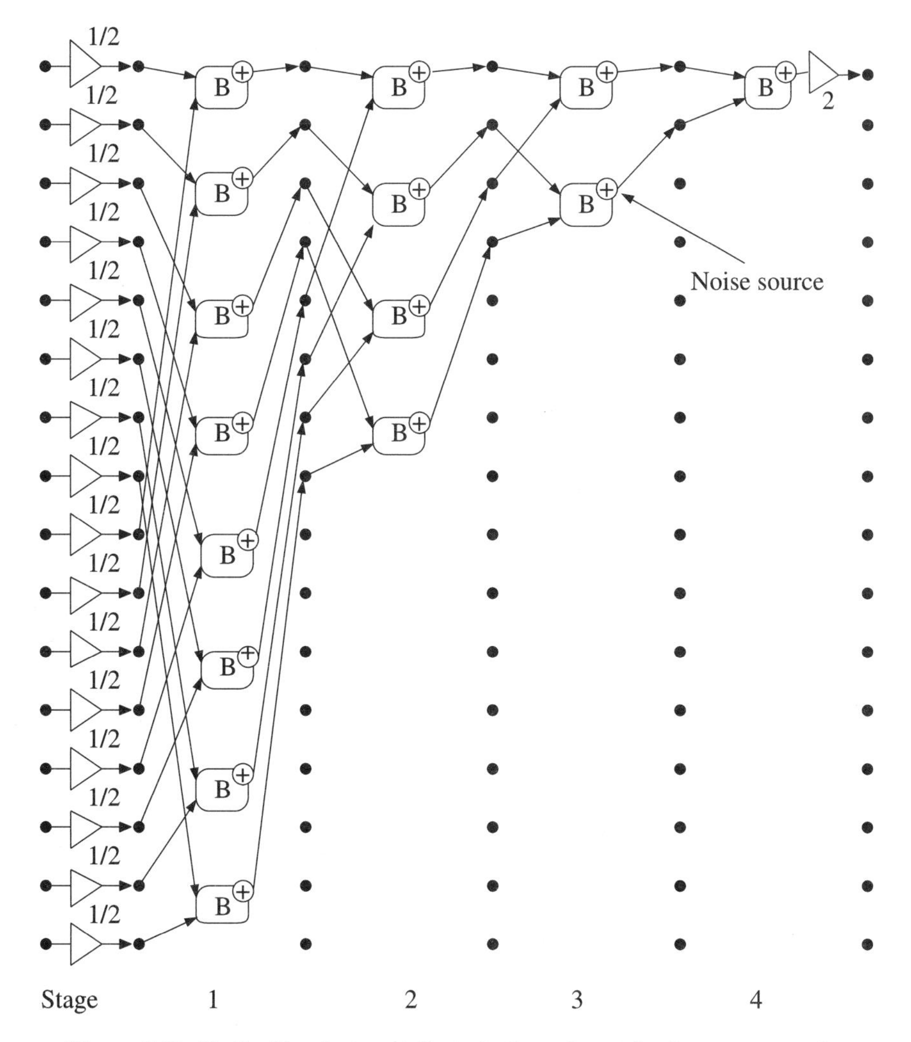

Figure 5.19 Butterflies that contribute to the noise at the first output node

noise gain is = 1. The noise variance at an output node is $(N1)\sigma_B^2$. The noise in the safely scaled FFT is determined in Example 5.9.

EXAMPLE 5.9

Determine the signal-to-noise ratio for the safely scaled, 16-point FFT.

The quantization errors are divided by two in each stage. The output of the last stage is multiplied by two. Hence the variance is divided by four. The noise gains from stages 1 to 4 to the outputs are $(1/4)^2$, $(1/4)^1$, 1, and 4, respectively. The number of noise sources are 8, 4, 2, and 1, respectively. Hence, the noise variance at the output is

$$\sigma_e^2 = [8\left(\frac{1}{4}\right)^2 + 4\left(\frac{1}{4}\right)^1 + 2 \cdot 1 + 1 \cdot 4]\ \sigma_B^2 = 7.5\sigma_B^2$$

The variance of the output signal is related to the variance of the input signal according to

$$\sigma_X^2 = \frac{1}{N}\sigma_x^2 \quad \text{where } N = 16$$

The signal-to-noise ratio is

$$\frac{\sigma_X^2}{\sigma_e^2} = \frac{1}{120}\frac{\sigma_x^2}{\sigma_B^2}$$

Hence, the SNR will be about the same at the input and output of the FFT if we use a $\log_2(\sqrt{120}) \approx 3.45$ bits longer word length inside the FFT than for the input signal. Note that is a stringent requirement and a somewhat shorter word length may be used in practice.

5.6.2 Error Spectrum Shaping

Error spectrum shaping or error feedback is a general method that can be used to reduce the errors inherent in quantization [13, 18, 29, 30]. The technique is useful for both fixed-point and floating-point arithmetic. The reduction in round-off noise is especially dramatic in narrow-band and wide-band lowpass filters, which have poles close to the unit circle. The data word length can often be reduced by several bits since the error feedback reduces the round-off noise. This is important both if the algorithm is to be implemented in ASIC, since memory is expensive, and in standard signal processors with short data word lengths.

Figure 5.20 shows a quantizer with error feedback followed by a filter. The transfer function from the quantizer to the output of the filter is

$$G(z) = \frac{\sum_{k=0}^{N} a_k z^{-k}}{1 - \sum_{k=0}^{N} b_k z^{-k}}$$

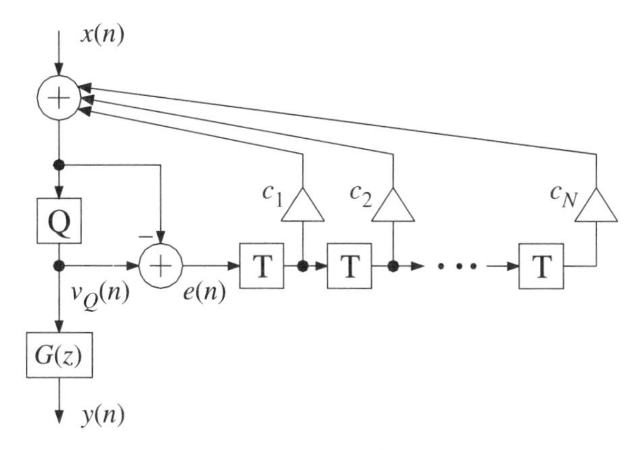

Figure 5.20 Quantizer with error feedback

The errors that occur in the quantization are saved and used to correct subsequent quantizations. The error is

$$e(n) = v_Q(n) - x(n) - \sum_{k=1}^{N} c_k \, e(n-k)$$

and

$$E(z) = V(z)\, X(z) \sum_{k=1}^{N} c_k z^{-k}\, E(z)$$

For the output of the filter we obtain

$$Y(z) = G(z)\,X(z) + G(z)\left[1 + \sum_{k=1}^{N} c_k z^{-k}\right] E(z)$$

The round-off error is affected by a different transfer function than is the input signal. There are several options to select the coefficients in the error feedback that minimize the round-off noise, simplify the implementation, etc. [18]. One option that reduces the round-off noise significantly is to select $c_k = b_k$. The error sequence is thereby filtered through an FIR filter corresponding to the denumerator of $G(z)$. Another option is to place the zeros of the error feedback polynomial in the passband of the system in order to suppress inband noise. In practice it may also be efficient to quantize the coefficients c_k to integer values, reduce the accuracy with which the errors are represented, or use a lower-order feedback path. Most standard signal processors are suited to error feedback since they have double precision accumulators.

Error feedback does not affect the sensitivity or transfer function of the filter. Neither does it affect the signal levels or overflow probability. However, the magnitude of parasitic oscillations may be reduced and even eliminated in some cases [29].

5.7 MEASURING ROUND-OFF NOISE

Round-off noise; can be measured according to the scheme shown in Figure 5.21. The systems are driven by typical input signals. Both systems have the same quantized (binary) coefficients. One of the systems, H_{Ideal}, has, in principle, infinite data word length while the other, H_{Quant}, has finite word length. In practice, the filters can be implemented in a general-purpose computer using high-precision floating-point arithmetic while the fixed-point arithmetic with quantization is simulated.

Problems arising when measuring noise are that in certain cases the quantization, e.g., magnitude truncation, will cause the two systems to have different gains and that the output noise will be correlated with the output signal. Therefore, the difference in system gain and the correlation with the input signal must be removed. Further, some types of quantization (for example, truncation) cause a DC offset at the output. In speech applications this offset should not be included in the noise, but in other applications it may be regarded as a major disturbance.

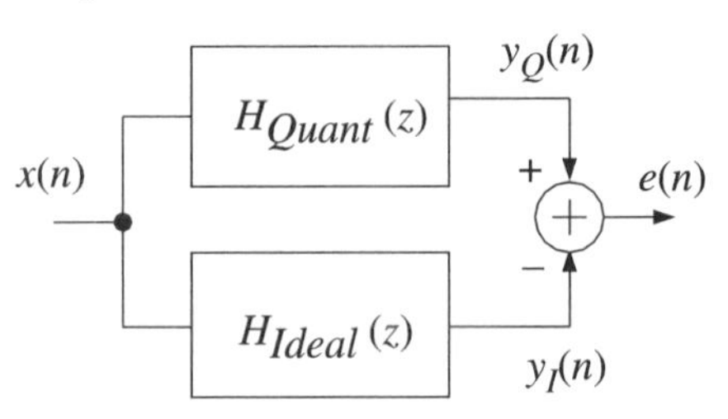

Figure 5.21 Measuring of round-off noise

Comparison and measurement of the noise in different filter structures must be done with respect to the dynamic signal range, i.e., the signal levels in the structures shall be properly scaled. The loss in dynamic range can be measured as described in Problem 5.7.

Example 5.10

Determine the required internal data word length for the bandpass filter in Example 5.6 when the dynamic signal range at the output should be at least 96 dB.

The losses in dynamic range were calculated in Example 5.8 to be 6.29 and 4.08 bits for the L_∞-norm and L_2-norm scaled filters, respectively. We define the dynamic range of a sinusoidal signal as the ratio between the peak power of the signal and the noise power. The largest output signal is

$$y(n) = 2^{N-1}Q\sin(\omega nT)$$

where the signal is represented with N bits, including the sign bit, and Q is the quantization step. The power of the signal is

$$P_s = \frac{(2^{N-1}Q)^2}{2} = 2^{2N-3}Q^2$$

The dynamic range at the output of the filter is

$$\frac{P_s}{P_n} = \frac{2^{2N-3}Q^2}{\frac{Q}{12}} = \frac{3}{2}2^{2N}$$

Thus, we get

$$SNR = 10\log_{10}\left(\frac{P_s}{P_n}\right) = 6.02\,N + 1.76 \quad \text{[dB]}$$

The required data word length inside the filter is estimated as

$$W \approx \Delta W + \frac{96 - 1.76}{6.02} = \Delta W + 15.65$$

where ΔW is the loss due the round-off noise. The required data word lengths for the two scaling strategies are 22 (21.93) and 20 (19.73) bits, respectively.

5.8 COEFFICIENT SENSITIVITY

In analog systems, sensitivity to element errors caused by unavoidable inaccuracies resulting from manufacturing tolerances, temperature variations, and aging is of major concern. These particular aspects are essentially irrelevant in digital signal processing. In fact, independence of element variation is the single most important reason for the success of digital signal processing. Sensitivity to coefficient variations nevertheless plays an important role in determining the cost of digital filters.

The frequency response is directly influenced by the accuracy of the coefficients. Although coefficient word length can in principle be chosen arbitrarily large, it is important to use a minimum word length since it has a major impact on the amount of hardware resources, maximal speed, and power consumption required.

The effect of coefficient errors on frequency response can be treated separately from the influence of round-off noise. To analyze the influence of finite coefficient word length effects on frequency response, many different measures have been used. The derivative of the transfer function $H(z)$ with respect to a coefficient in a filter structure has both practical and theoretical significance. We will therefore derive an important expression for the derivative.

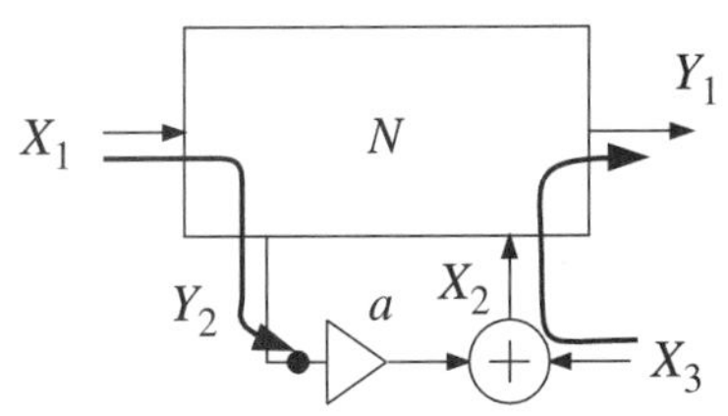

Figure 5.22 LSI network

Consider the LSI network shown in Fig. 5.22 which is described by

$$Y_1 = A\,X_1 + B\,X_2$$

$$Y_2 = C\,X_1 + D\,X_2$$

$$X_2 = a\,Y_2 + X_3$$

From these equations we get the transfer function

$$H = \left.\frac{Y_1}{X_1}\right|_{X_3 = 0} = A + \frac{aCB}{1 - aD}$$

The transfer function from the input to the multiplier is

$$F = \left.\frac{Y_2}{X_1}\right|_{X_3 = 0} = \frac{C}{1 - aD}$$

and the transfer function from the output of the multiplier to the output of the network is:

$$G = \left.\frac{Y_1}{X_3}\right|_{X_1 = 0} = \frac{B}{1 - aD}$$

Taking the derivative of the transfer function, H, with respect to the coefficient, a, leads to the main result:

$$\frac{\partial H}{\partial a} = \frac{CB}{(1 - aD)^2} = FG \tag{5.18}$$

where F is the transfer function from the input of the filter to the input of the multiplier and G is the transfer function from the output of the multiplier to the output of the filter.

Note that the Fourier transform corresponding to F is the function used in scaling the filter. Thus, the sensitivity depends on the scaling criteria used. In fact, great caution should be taken when evaluating and comparing different structures. Unfortunately, it is not uncommon that erroneous sensitivity comparisons and unjustified low-sensitivity claims are made in the literature.

5.8.1 Coefficient Word Length

It is usually advantageous to minimize the word length required to represent all the coefficients if the arithmetic operations are executed on a set of flexible processing elements. The coefficient with the longest word length therefore determines the size of the multiplier in a standard or ASIC signal processor. For a standard signal processor the coefficient word length is fixed, and it is therefore sufficient to find a set of binary coefficients that can be represented with this word length and that satisfies the frequency response requirements.

If instead each multiplication is implemented using a dedicated multiplier, it may be advantageous to minimize the number of nonzero bits in the coefficients and thereby achieve a significant reduction in cost for the multiplier. The cost is proportional to the number of nonzero bits and increases slowly with increasing coefficient word length.

In practice, the binary coefficient values are found by searching in the neighborhood of the theoretical coefficient values. Experience indicates that optimal binary values may be found far away from the theoretical values. Hence, the search for binary values may be very time consuming for filters with many coefficients. However, the time may be reduced if the search starts with a good estimate of the word length. The required coefficient word length may be estimated using statistical methods [1].

5.9 SENSITIVITY AND NOISE

Fettweis has shown that coefficient sensitivity and round-off noise are closely related [7]. Jackson [17] has derived the following lower bounds on round-off noise in terms of the sensitivity.

Let F_i be the scaled frequency response from the input of the filter to the input of the multiplier a_i, and G_i the frequency response from the output of the multiplier to the output of the filter. For a scaled filter we have

$$\| F_i \|_p = 1 \text{ for all critical nodes } i = 0, 1, \ldots, n$$

The round-off noise variance at the output of the filter is

$$\sigma_{ye}{}^2 = \sum_{i=0}^{n} \sigma_i^2 \|G_i\|_2^2 \sum_{i=0}^{n} \sigma_i^2 \|F_i\|_p \; \|G_i\|_2^2$$

We get, using Hölders inequality[2], for $p = 2$

$$\sigma_{ye}{}^2 \geq \sum_{i=0}^{n} \sigma_i^2 \|F_i G_i\|_1^2$$

and Equation (5.18)

$$\sigma_{ye}{}^2 \geq \sum_{i=0}^{n} \sigma_i^2 \left\| \frac{\partial H}{\partial a_i} \right\|_1^2 \tag{5.19}$$

This is a lower bound of the noise variance for a filter scaled for wide-band input signals. Using the fact that

$$\|F_i G_i\|_1 \leq \|F_i G_i\|_n \leq \|F_i\|_\infty \; \|G_i\|_n \quad n \geq 1 \tag{5.20}$$

and $n = 2$, we get another lower bound that is valid instead for L_∞-norm scaling:

$$\sigma_{ye}{}^2 \geq \sum_{i=0}^{n} \sigma_i^2 \left\| \frac{\partial H}{\partial a_i} \right\|_2^2 \tag{5.21}$$

This lower bound on the round-off noise corresponds to scaling for narrow-band signals.

These two bounds are important, since they show that a structure with high sensitivity will always have high round-off noise and requires a longer data word length. It is sometimes argued that high-sensitivity structures can be implemented using elaborate schemes for selecting the coefficients such that the frequency specification is met. However, in such cases a penalty is paid in terms of increased data word length.

The importance of low sensitivity to coefficient errors and low round-off noise can be appreciated by considering the cost and execution time of multipliers, which are essentially proportional to the sum of coefficients and data word lengths. Using a low-sensitivity structure, short coefficients and a short data word length can be employed. Low-sensitivity structures also tend to use the available signal range efficiently. Therefore, we stress that low-sensitivity structures should always be used for high-performance digital filters.

---

2. Hölders inequality: $\| F\, G \|_1 \leq \| F \|_p \, \| G \|_q, \; \frac{1}{p} + \frac{1}{q} = 1, \quad \text{for } p, q \geq 1$

5.10 INTERPOLATOR, CONT.

The interpolator filter will be implemented by multiplexing four processing elements (PEs). Each PE will therefore execute eight adaptor operations. Only 12 unique coefficients are needed since the two interpolator stages are identical. The coefficient word lengths can be optimized for each PE in order to minimize the cost. However, for the sake of simplicity we will here assume that the coefficient word lengths are the same and in the range of 12 to 14 bits.

Since we do not have a specific application for the interpolator in mind, we will arbitrarily select to use a data word length of 20 bits, which will be more than sufficient for most applications.

5.11 FFT PROCESSOR, CONT.

To determine the required word length in the scaled FFT we first compute the noise at the output of the FFT. Generalizing from Figure 5.19 we obtain the variance at the outputs:

$$\sigma_e{}^2 = \sigma_B{}^2 \sum_{s=1}^{M} (2^{M-s})\left(\frac{1}{2^2}\right)^{M-1-s} = 8(1-2^M)\sigma_B{}^2$$

where $M = \log_2(N)$. The first factor in the preceding sum corresponds to the number of noise sources in stage s, and the second factor corresponds to their noise gains. For large FFTs we have

$$\sigma_e{}^2 \approx 8\sigma_B{}^2$$

The variance of the output signal is related to the input signal variance according to

$$\sigma_X{}^2 = \frac{1}{N}\sigma_x{}^2 \quad \text{where } N = 1024$$

The signal-to-noise ratio is

$$\frac{\sigma_X^2}{\sigma_e^2} = \frac{1}{8N}\frac{\sigma_x^2}{\sigma_B^2}$$

Hence the SNR will be about the same at the input and output of the FFT if we use a $0.5 \log_2(8N) = 6.5$ bits longer word length inside the FFT than for the input signal. We select a data word length of 23 bits.

The coefficient word length will affect the cost of the butterfly PEs and the cost of storing the coefficients, W^p. The potential reduction in chip area and power consumption that can be achieved by optimizing the coefficient word length is relatively small. We will therefore select a coefficient word length of 14 bits [26].

5.12 DCT PROCESSOR, CONT.

The required data and coefficient word lengths required for the two-dimensional DCT have been determined experimentally to be in the ranges of 10 to 14 and 8 to

10 bits, respectively. Computations in the DCT processor will be performed by using one-dimensional vector-multipliers. This means that each output value (rows) will be computed without round-off errors and then quantized. Hence, the only round-off errors that occur are inputs to the last set of one-dimensional DCTs (columns). The round-off errors are therefore smaller than those for a conventional DCT implementation. Hence, we select a data word length of 12 bits and a coefficient word length of 11 bits.

References

[1] Bellanger M.: *Digital Processing of Signals, Theory and Practice*, John Wiley & Sons, Chichester, New York 1984.

[2] Chang T.L.: Suppression of Limit Cycles in Digital Filters Designed with One Magnitude Truncation Quantizer, *IEEE Trans. on Circuits and Systems*, Vol. CAS-28, No. 2, pp. 107–111, Feb. 1981.

[3] Chua L.O. and Lin T.: Chaos in Digital Filters, *IEEE Trans. on Circuits and Systems,* Vol. CAS-35, No. 6, pp. 648–658, June 1988.

[4] Claasen T.A.C.M., Mecklenbräuker W.F.G., and Peek J.B.H.: Effects of Quantization and Overflow in Recursive Digital Filters, *IEEE Trans. on Acoustics, Speech, and Signal Processing,* Vol. ASSP-24, No. 6, pp. 517–529, Dec. 1976.

[5] DeFatta D.J., Lucas J.G., and Hodgkiss W.S.: *Digital Signal Processing: A System Design Approach,* John Wiley & Sons, New York 1988.

[6] Elliott D.F. (ed.): *Handbook of Digital Signal Processing, Engineering Applications*, Academic Press, San Diego, 1988.

[7] Fettweis A.: On the Connection between Multiplier Word Length Limitation and Roundoff Noise in Digital Filters, *IEEE Trans. on Circuit Theory,* Vol. CT-19, No. 5, pp. 486–491, Sept. 1972.

[8] Fettweis A.: Suppression of Parasitic Oscillations in Wave Digital Filters, *IEEE Trans. on Circuits and Systems,* Vol. CAS-22, No. 3, March 1975.

[9] Fettweis A. and Meerkötter K.: Suppression of Parasitic Oscillations in Half-Synchronic Wave Digital Filters, *IEEE Trans. on Circuits and Systems,* Vol. CAS-23, No. 2, pp. 125–126, Feb. 1976.

[10] Fettweis A.: On Parasitic Oscillations in Digital Filters under Looped Conditions, *IEEE Trans. on Circuits and Systems,* Vol. CAS-24, No. 9, pp. 475–481, Sept. 1977.

[11] Fettweis A.: Modified Wave Digital Filters for Improved Implementation by Commercial Digital Signal Processors, *Signal Processing,* Vol. 16, No. 3, pp. 193–207, 1989.

[12] Fettweis A.: On Assessing Robustness of Recursive Digital Filters, *European Trans. on Telecomm. and Related Tech.,* Vol. 2, No. 1, pp. 103–109, March-April 1990.

[13] Higgins W.E. and Munson D.C.: Noise Reduction Strategies for Digital Filters: Error Spectrum Shaping versus the Optimal Linear State-Space Formulation, *IEEE Trans. on Acoustics, Speech, and Signal Processing,* Vol. ASSP-30, No. 6, pp. 963–973, Dec. 1982.

[14] Hwang S.Y.: Minimum Uncorrelated Unit Noise in State-Space Digital Filtering, *IEEE Trans. on Acoustics, Speech, and Signal Processing,* Vol. ASSP-25, No. 4, pp. 273–281, Aug. 1977.

[15] Jackson L.B.: Roundoff-Noise and Analysis for Fixed-Point Digital Filters Realized in Cascade or Parallel Form, *IEEE Trans. on Audio Electroacoust.*, Vol. AU-16, pp. 413–421, Sept. 1968.

[16] Jackson L.B.: On the Interaction of Roundoff Noise and Dynamic Range in Digital Filters, *Bell Syst. Techn. J.*, Vol. 49, pp. 159–184, Feb. 1970.

[17] Jackson L.B.: Roundoff Noise Bounds Derived from Coefficient Sensitivities for Digital Filters, *IEEE Trans. on Circuits and Systems*, Vol. CAS-23, No. 8, pp. 481–485, Aug. 1976.

[18] Laakso T.I. and Hartimo I.O.: Noise Reduction in Recursive Digital Filters Using High-Order Error Feedback, *IEEE Trans. on Signal Processing*, Vol. SP-40, No. 5, pp. 1096–1107, May 1992.

[19] Lin T. and Chua L.O: On Chaos of Digital Filters in the Real World, *IEEE Trans. on Circuits and Systems*, Vol. CAS-38, No. 5, pp. 557–558, May 1991.

[20] Liu B. and Kaneko T.: Error Analysis of Digital Filters Realized Using Floating Point Arithmetic, *Proc. IEEE*, Vol. 57, No. 10, pp. 1735–1747, Oct. 1969.

[21] Liu B.: Effect of Finite Word Length on the Accuracy of Digital Filters—A Review, *IEEE Trans. on Circuit Theory*, Vol. CT-18, No. 6, pp. 670–677, Nov. 1971.

[22] Long L.J. and Trick T.N.: An Absolute Bound on Limit Cycles due to Roundoff Errors in Digital Filters, *IEEE Trans. on Audio Electroacoust.*, Vol. AU-21, No. 1, pp. 27–30, Feb. 1973.

[23] Meerkötter K.: Incremental Pseudopassivity of Wave Digital Filters, *Proc. European Signal Processing Conf.*, EUSIPCO-80, pp. 27–32, Lausanne, Switzerland, Sept. 1980.

[24] Mitra D. and Lawrence V.B.: Controlled Rounding Arithmetics, for Second-Order Direct-Form Digital Filters That Eliminate All Self-Sustained Oscillations, *IEEE Trans. on Circuits and Systems*, Vol. CAS-28, No. 9, pp. 894–905, Sept. 1981.

[25] Mullis C.T. and Roberts R.A.: Synthesis of Minimum Roundoff Noise Fixed Point Digital Filters, *IEEE Trans. on Circuits and Systems*, Vol. CAS-23, No. 9, pp. 551–561, Sept. 1976.

[26] Oppenheim A.V. and Weinstein C.J.: Effects of Finite Register Length in Digital Filtering and the Fast Fourier Transform, *Proc. IEEE*, Vol. 60, pp. 957–976, Aug. 1972.

[27] Proakis J.G. and Manolakis D.G.: *Introduction to Digital Signal Processing*, Macmillan, New York, 1988.

[28] Rao B.D.: Floating Point Arithmetic and Digital Filters, *IEEE Trans. on Signal Processing*, Vol. SP-40, No. 1, pp. 85–95, Jan. 1992.

[29] Renfors M.: Roundoff Noise in Error-Feedback State-Space Filters, *Proc. Intern. Conf. Acoustics, Speech, and Signal Processing*, ICASSP-83, Boston, pp. 619–622, April 1983.

[30] Renfors M., Sikström B., and Wanhammar L.: LSI Implementation of Limit–Cycle-Free Digital Filters Using Error Feedback Techniques, *Proc. European Conf. Signal Processing*, EUSIPCO-83, Erlangen, F.R.G., pp. 107–110, Sept. 1983.

[31] Samueli H. and Willson Jr. A.N.: Nonperiodic Forced Overflow Oscillations in Digital Filters, *IEEE Trans. on Circuits and Systems*, Vol. CAS-30, No. 10, pp. 709–722, Oct. 1983.

[32] Smith L.M., Bomar B.W., Joseph R.D., and Yang G.C.: Floating-Point Roundoff Noise Analysis of Second-Order State-Space Digital Filter Structures, *IEEE Trans. on Circuits and Systems,* Vol. CAS-39, No. 2, pp. 90–98, Feb. 1992.
[33] Taylor F.J.: *Digital Filter Design Handbook,* Marcel Dekker, New York, 1983.
[34] Tran-Thong and Liu B.: Fixed-Point FFT Error Analysis, *IEEE Trans. on Acoustics, Speech, and Signal Processing,* Vol. ASSP-24, No. 6, pp. 563–573, 1976.
[35] Zeng B. and Neuvo Y.: Analysis of Floating Point Roundoff Errors Using Dummy Multiplier Coefficient Sensitivities, *IEEE Trans. on Circuits and Systems,* Vol. CAS-38, No. 6, pp. 590–601, June 1991.

Problems

5.1 A two's-complement number is multiplied by a factor 0.5 and then quantized to the original word length. Determine both the average value and the variance of the quantization error.

5.2 Show that the quantization scheme, discussed in section 5.4, will suppress parasitic oscillations.

5.3 Show that the simplified scheme, discussed in section 5.4, will suppress zero-input parasitic oscillations for second-order allpass sections with poles on the imaginary axis in the z-plane.

5.4 Show that a two-port adaptor is pseudo-lossless.

5.5 (a) Scale the signal levels in the FIR filter shown in Figure P5.5. Use safe scaling since the filter length is short. Two's-complement representation shall be used.
(b) Also, scale the filter using L_2-norms.

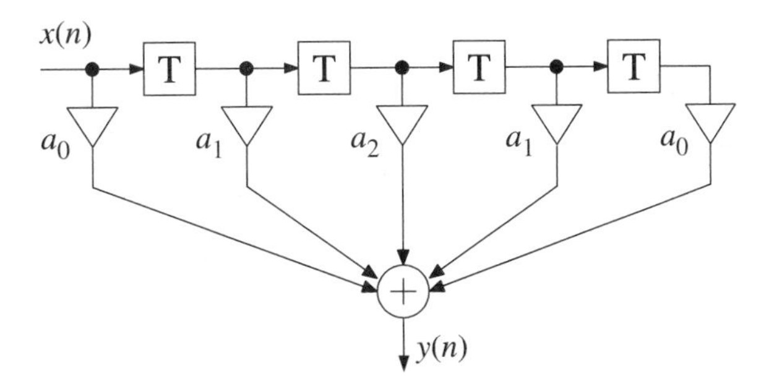

Figure P5.5 Direct form FIR

5.6 (a) Scale a direct form II second-order section using the safe scaling criterion. The filter is a half-band filter of Butterworth type.
(b) Scale the filter assuming a narrow-band input signal.
(c) Scale the filter assuming a wide-band input signal.

(d) Compare the three scaling procedures.
(e) Determine the round-off noise for the three filters in terms of SNR assuming both rounding and truncation are performed after the summation. Comment on the result!

5.7 Rank the different scaling criteria in terms of severity.

5.8 (a) Show that $\left\| |F(e^{j\omega T})|^2 \right\|_p = \left\| F(e^{j\omega T}) \right\|_{2p}^2, \ p \geq 1$

(b) Show that $\|FG\|_p \leq \|F\|_\infty \|G\|_p, \ p > 1$

5.9 Scale the filter shown in Figure P5.9 using
(a) L_2-norm scaling
(b) L_∞-norm scaling

$|b| < 1$

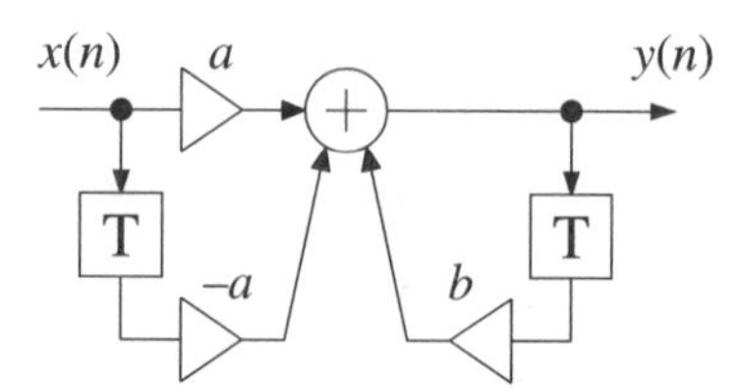

Figure P5.9 Direct form I

5.10 A properly scaled filter structure generates low round-off noise, i.e., measurements show that σ_{out}^2 is small. What conclusion can be drawn about the coefficient word length required to satisfy a stringent magnitude specification?

5.11 Scale the filter shown in Figure P5.11 using the safe scaling criterion. Also, determine the coefficients if instead L_∞-norm scaling is used.

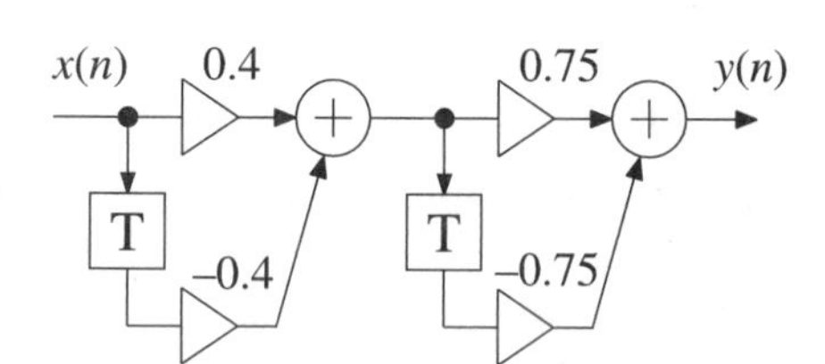

Figure P5.11 Cascade form

5.12 (a) Scale the filter shown in Figure P5.12 using L_2-norms. The data word length W_d is 8 bits, including the sign bit. Two's-complement representation is used. The coefficients are

$b = 0.98$ and $a_0 = a_1$

For which type of input signals is this scaling policy appropriate?
(b) Also, determine the variance for the round-off noise at the output of the filter when rounding is performed after the multiplications.

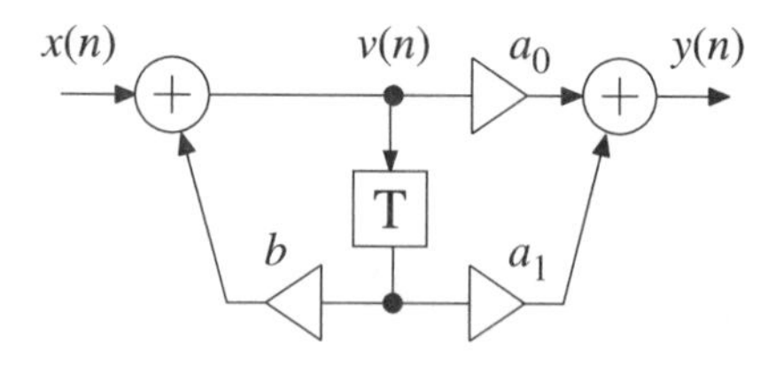

Figure P5.12 Direct form II

5.13 (a) Show that $\|S_x\|_\infty$ characterizes a wide-band signal.

(b) Show that $\|S_x\|_1$ characterizes a narrow-band signal.

5.14 Determine the required order for a wave digital ladder filter of Cauer type that meets the specification $A_{max} = 2$ dB, $A_{min} = 70$ dB, $f_c = 50$ kHz, $f_s = 60$ kHz, and $f_{sample} = 200$ kHz. Now, assume that you are unable to find a set of quantized coefficients with sufficiently short word length. Propose a remedy to the problem.

5.15 Write a program to determine the DC offset at the output of the 1024-point Sande–Tukeys FFT.

5.16 (a) Determine the coefficients for the one-dimensional MSDCT.
(b) Scale the coefficients.
(c) Determine the average value and variance of the round-off errors in the two-dimensional DCT.

5.17 A bandpass filter is cascaded with a fifth-order allpass filter in order to equalize the group delay. The allpass filter is realized by using first- and second-order sections of type direct form I.
(a) Scale the allpass sections using a suitable L_p-norm for a wide band input signal.
(b) Determine an ordering of the allpass sections that yields the best dynamic signal range.

5.18 The requirement on a system for sampling an analog audio signal is very high. The sample frequency is 44.1 kHz with an accuracy of 18 bits.

The following scheme is therefore proposed as an attempt to reduce the cost and meet the stringent requirements on accuracy and the anti-aliasing filter.
(a) Determine the required sample frequency such that the new system generates equal or less quantization noise compared to the original system.

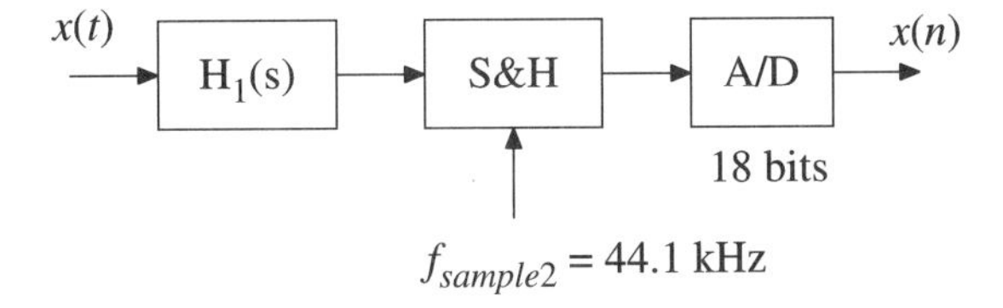

Figure P5.18a

(b) Assume that the decimation is done by factor of 1024. Determine the sample frequency and suggest a realization of the decimation filter that has a low workload, i.e., operations per sample.

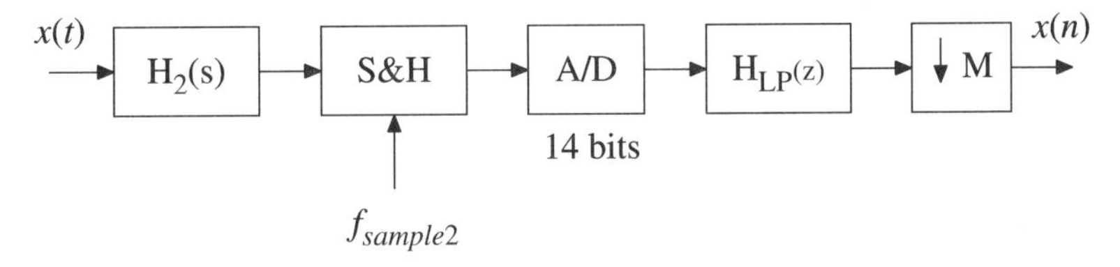

Figure P5.18b

(c) Determine the requirements for the anti-aliasing filters for the two systems.

(d) Determine the required filter orders for the two anti-aliasing filters if a Butterworth approximation is used. The audio signal has a bandwidth of 20 kHz.

5.19 An FIR filter with linear phase, i.e., symmetric impulse response, has the frequency response:

$$H(e^{j\omega T}) = e^{j\omega(K+1)T}\left[h_0 + 2\sum_{n=1}^{K} h_n \cos(\omega n T)\right]$$

where $N = 2K+1 = 201$. Determine the maximum deviation of attenuation in the passband when the coefficients are quantized with an accuracy of $Q = 2^{10}$. Assume that the filter is realized in

(a) direct form

(b) linear-phase direct form

5.20 Show that the variance of the output signal of an accumulator goes to infinity when the input is a zero mean Gaussian process.

5.21 A D/A converter in an early circuit for CD players (TDA 1540) manufactured by Phillips has a resolution of only 14 bits while the digital signal has a word length of 28 bits. Hence, 14 bits have to be discarded. In order to improve the SNR, the following features are implemented:

1. The output sampling frequency before the converter is increased by interpolation to 4×44.1 kHz. The unwanted images are attenuated by at least 50 dB using a digital filter.
2. The analog output signal of the zero-order hold A/D converter is filtered through a third-order Bessel filter with a 3 dB passband edge of 30 kHz.
3. The 14 bits that are discarded are added to the next output sample before it is rounded to 14 bit precision.

Determine the improvement that is obtained in the SNR.

6

DSP ALGORITHMS

6.1 INTRODUCTION

In this chapter, we will discuss the computational properties of digital signal processing systems. We will describe a DSP system as a hierarchy of computational structures. Each structure (algorithm) involves a set of basic operations called processes. At the higher levels of the hierarchy these processes correspond to large DSP tasks (for example FFTs, DCTs, and digital filters) while the lowest-level processes involve arithmetic operations. The computational properties of the algorithms at the different hierarchical levels will limit the achievable performance of the system. We will introduce a number of graphs that will allow us to determine the computational properties of a DSP algorithm. Properties of interest are the order of execution of the arithmetic operations (processes), the number of operations that can be executed in parallel, and the shortest possible execution time. Some of these graphs will serve as a basis for the scheduling of the operations that will be discussed in Chapter 7. The computational properties of the higher-level processes can be analyzed in a manner similar to the arithmetic operations. We will also discuss a formal method to derive the difference equations in a computable order from the signal-flow graph. In this form the set of difference equations can be translated directly into code for a standard signal processor or a multicomputer. Various methods of improving the computational properties of recursive algorithms will also be discussed.

6.2 DSP SYSTEMS

Generally, any reasonably complex DSP system can be described, as illustrated in Figure 6.1, by a hierarchy of cooperating processes. We use the concept of process to denote a structural component that performs a function at a given hierarchical level. Examples of processes at the higher levels in the system hierarchy are FFTs, DCTs, and digital filters. The inner structure of these processes can be further described by another computational structure. At the lower levels of the hierarchy processes typically describe the algorithms that realize the basic arithmetic operations. Note that storage may also be described as a process. Typically, several algorithms at various hierarchical levels can be identified within a DSP system.

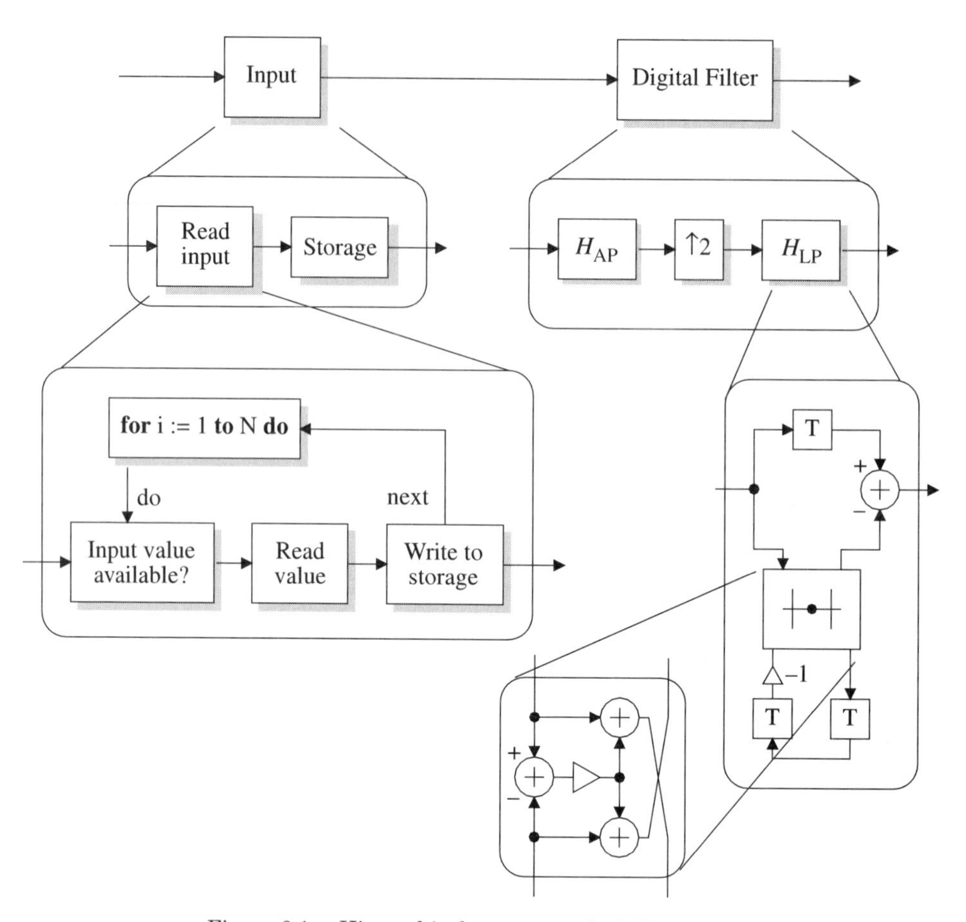

Figure 6.1 Hierarchical structure of a DSP system

6.2.1 DSP Algorithms

A *DSP algorithm*[1] is a computational rule, f, that maps an ordered input sequence, $x(nT)$, to an ordered output sequence, $y(nT)$, according to

$$x(nT) \rightarrow y(nT), \qquad y(nT) := f(x(nT))$$

Generally, in hard real-time applications the mapping is required to be causal since the input sequence is usually causal. We stress that the computational rule is an unambiguously specified sequence of operations on an ordered data set as summarized in Figure 6.2. An algorithm is based on a set of basic operations—for example, additions and multiplications. The algorithm also contains a detailed description of numerical accuracy and number ranges including the rounding and overflow nonlinearities involved.

1. The word *algorithm* is derived from the name of the ninth-century Arab mathematician, Al-Khuwarizmi.

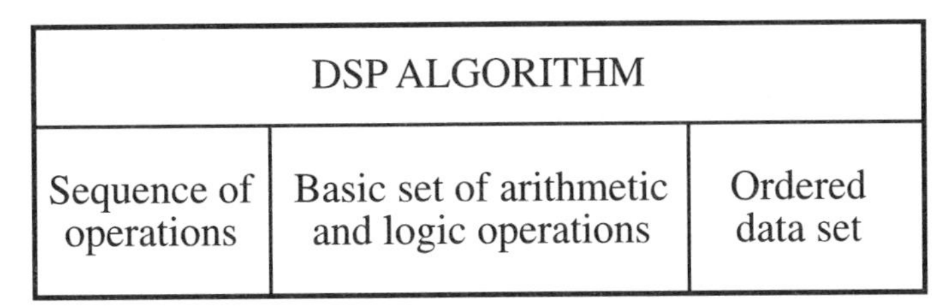

Figure 6.2 The three parts of an algorithm

Generally, a DSP algorithm can be described by a set of expressions, as shown shortly. The sign ":=" is used to indicate that the expressions are given in computational order.

$$\begin{cases} x_1(n) := f_1[-, x_1(n-1), ..., x_p(n-1), x_p(n-2), ..., a_1, b_1, ...] \\ x_2(n) := f_2[-, x_1(n-1), ..., x_p(n-1), x_p(n-2), ..., a_2, b_2, ...] \\ x_3(n) := f_3[-, x_1(n-1), ..., x_p(n-1), x_p(n-2), ..., a_3, b_3, ...] \\ x_N(n) := f_N[-, x_1(n-1), ..., x_p(n-1), x_p(n-2), ..., a_N, b_N, ...] \end{cases}$$

Note that the arguments on the right-hand side of the two first expressions contain only old (i.e., known) values. These two expressions can therefore be computed immediately. Once these have been evaluated, the remaining expressions can be computed successively.

DSP algorithms can be divided into *iterative processing* and *block processing* algorithms. An iterative algorithm performs computations on a semi-infinite stream of input data, i.e., input data arrive sequentially and the algorithm is executed once for every input sample and produces a corresponding output sample. The period between two consecutive iterations is called the iteration period or sample interval. In block processing, a block of output samples is computed for each input block, which results in a large delay between input and output. Block processing is commonly used for DSP algorithms based on matrix operations, but it can also be used to increase the maximum sample rate. Block processing algorithms can be used if the input samples arrive in batches (blocks) or in non–realtime applications. Here we will mainly discuss the former type of algorithm which is characterized by the following properties:

- The input and output data rates are high.
- The input and output values are synchronized with the sample period.
- The sequence of operations is data independent.
- The algorithm must be executed periodically.
- Hard real-time operation with a deadline equal to the sampling period.

On the surface, most DSP algorithms look very simple since they consist of a set of basic arithmetic expressions evaluated repeatedly. Complicated data structures are seldom used. These facts are often misunderstood by novice hardware designers who may be tempted to modify an algorithm to suit the idiosyncrasies of the hardware, design techniques, or tools. Such uninitiated and sometimes undocumented changes of the algorithm must be avoided, since underlying the DSP algorithm is a complex set of requirements that may be violated by even a simple change in the algorithm. For example, intricate bit manipulations and nonlinear operations are used in most speech coding systems. A simple change of the quantization scheme may drastically affect speech quality. In fact, the quality of most DSP algorithms is strongly related to various finite word length effects. An even more important aspect concerns the stability of the algorithm with respect to various types of nonlinearities.

Recursive algorithms contain nonlinearities that may cause parasitic oscillations. Temporary malfunction of the hardware may also cause parasitic oscillations, as discussed in Chapter 5. Sophisticated theories and methods are therefore invoked in algorithm design to assure that such oscillations cannot persist. Examples of filter algorithms that suppress all parasitic oscillations are wave digital filters and nonrecursive FIR filters. To avoid dangerous pitfalls, it is necessary to use a design method that guarantees the integrity at each design step involved in transforming the DSP algorithm into a working system.

6.2.2 Arithmetic Operations

At each level of system hierarchy it is convenient to define a set of basic operations that can be executed on the PEs of that level. The operations are assumed to be atomic and can not be further divided. Once an operation begins execution on a PE, it will not be interrupted by another operation. Thus, these operations represent the smallest granularity of which parallelism can be exploited at the system level.

The basic arithmetic operations used in the DSP algorithm can again be considered as lower-level algorithms acting on binary numbers, as illustrated in Figure 6.3. Algorithms for realizing the arithmetic operations will be further discussed in Chapter 11.

The properties of the arithmetic operations will determine the computational properties of the algorithm at the next higher level in the system hierarchy. It is

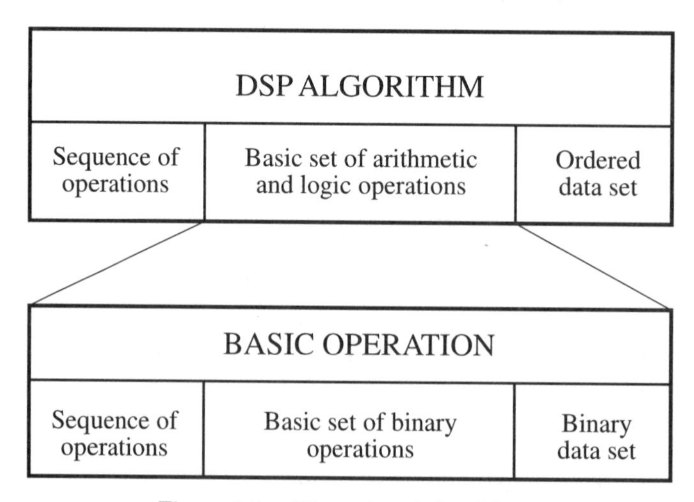

Figure 6.3 Hierarchy of algorithms

therefore important that the algorithms at the different hierarchical levels are carefully designed in order to achieve a high overall system performance.

6.3 PRECEDENCE GRAPHS

A signal-flow graph representation of a DSP algorithm is not directly suitable for analysis of its computational properties. We will therefore map the signal-flow graph to other graphs that can better serve to analyze computational properties such as parallelism and minimum execution time.

The graph shown in Figure 6.4 is a *precedence graph* which describes the order of occurrence of events: $A, B, \ldots, F$. The directed branches between the nodes denote the ordering between the events that are represented by the nodes [12]. For example, the directed branch between nodes E and B shows that event E precedes event B. E is therefore called a *precedent* or predecessor to B. Node E also precedes event A, and therefore node E is a second-order precedent to node A. In a similar manner, we define an event as *succedent*, or successor, to another event. For example, event B is a succedent to event E. An *initial node* has no precedents and a *terminal node* has no succedents, while an *internal node* has both. If two events are not connected via a branch, then their precedence order is unspecified.

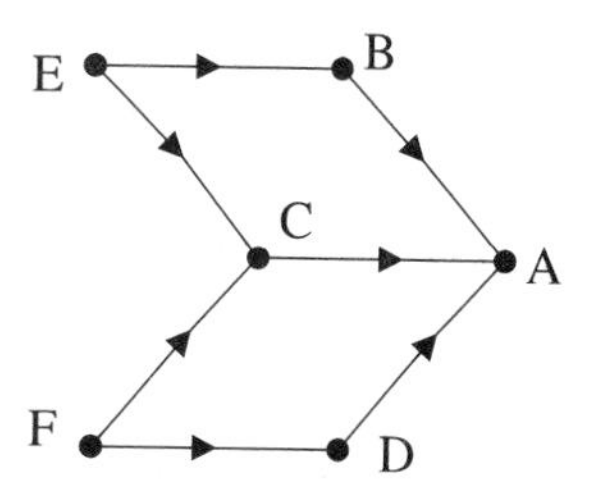

Figure 6.4 Precedence graph with activity on the nodes

Sometimes, it may be more convenient to let the branches represent the events and the nodes represent the precedence relations. Such precedence graphs are called *AOA* (*activity on arcs*) graphs, while the former type of graphs, with activity on the nodes, are called *AON* (*activity on nodes*) graphs. Notice that an activity in an AON graph may correspond to several branches in an AOA graph. For example, the event E in Figure 6.4 corresponds to two branches shown in Figure 6.5.

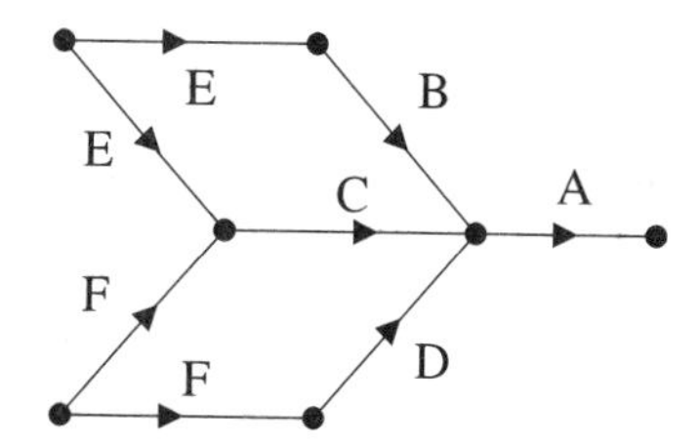

Figure 6.5 Precedence graph with activity on arcs

6.3.1 Parallelism in Algorithms

Parallelism can be used to significantly reduce the power consumption of a system by reducing the clock frequency and power supply voltage at the expense of silicon area. Precedence relations between operations are unspecified in a *parallel algorithm*. Figure 6.6 illustrates two examples of parallel algorithms. In the first case, the additions have a common precedent, while in the second case they are independent. Two operations (algorithms) are *concurrent* if their execution times overlap. We will later show that in some cases, another more parallel algorithm can be derived from an algorithm with little parallelism by allowing the algorithmic delay between the input and output to increase.

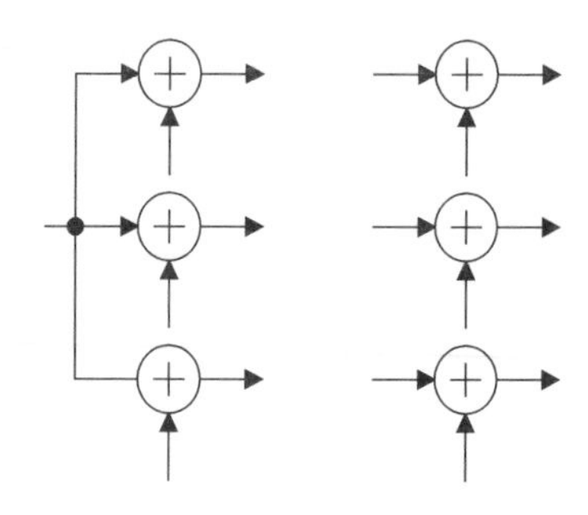

Figure 6.6 Parallel algorithms

Additional computational parallelism may also appear at both the lower and higher system levels illustrated in Figure 6.1 [8]. At the lower level, several bits of an arithmetic operation may be computed in parallel. At a higher level (for example, in a DCT-based image processing system), several rows in a 2-D DCT may be computed concurrently. At a still higher level, several DCTs may be computed concurrently. Hence, large amounts of parallelism and concurrency may be present in a particular DSP system, but are often difficult to detect and therefore difficult to exploit.

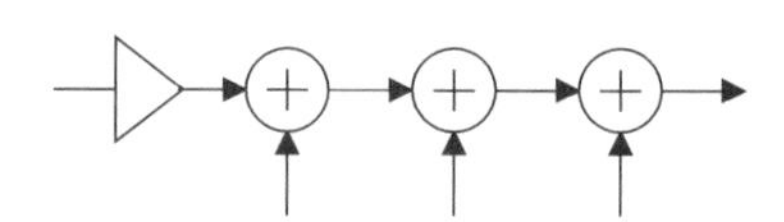

Figure 6.7 Sequential algorithm

An algorithm having a precedence graph for its operations according to Figure 6.7 is completely sequential. In a *sequential algorithm* every operation, except for the first and the last, has only one precedent and one succedent operation. Thus, the precedence relations are uniquely specified for a sequential algorithm. Obviously, a sequential algorithm is not suitable for high sample rate applications since it can utilize only a single processing element. A sequential algorithm, however, is suitable for low sample rate applications. The chip area can be reduced since only one processing element is required.

6.3.2 Latency

We define *latency* of an algorithm as the time it takes to generate an output value from the corresponding input value. Note that latency refers to the time between input and output, while the algorithmic delay refers to the difference between input and output sample indices.

The throughput (samples/s) of a system is defined as the reciprocal of the time between successive outputs. It is often possible to increase the throughput of an algorithm—for example, by using pipelining. Figures 6.8 and 6.9 illustrate the latency and throughput for a multiplication using bit-parallel and bit-serial arithmetic, respectively. For the sake of simplicity we assume that the input signals arrive simultaneously, i.e., in the same clock cycle, and that there is only one output signal.

Bit-serial multiplication can be done either by processing the least significant or the most significant bit first. The former is the most common since the latter is more complicated and requires the use of so-called *redundant arithmetic*. The latency, if the LSB is processed first, is in principle equal to the number of fractional bits in the coefficient. For example, a multiplication with a coefficient $W_c = (1.0011)_2$ will have a latency corresponding to four clock cycles. A bit-serial addition or subtraction has in principle zero latency while a multiplication by an integer may have zero or negative latency. However, the latency in a recursive loop is always positive, since the operations must be performed by causal PEs. In practice the latency may be somewhat longer, depending on the type of logic that is used to realize the arithmetic operations, as will be discussed shortly.

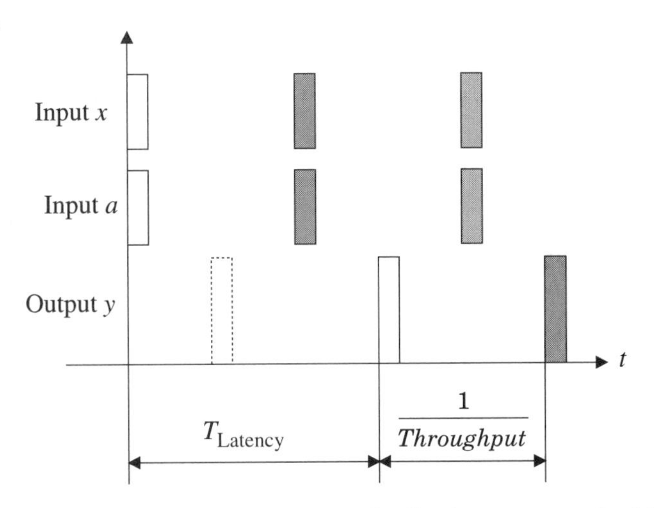

Figure 6.8 Latency and throughput for a multiplication $y = a \cdot x$ using bit-parallel arithmetic

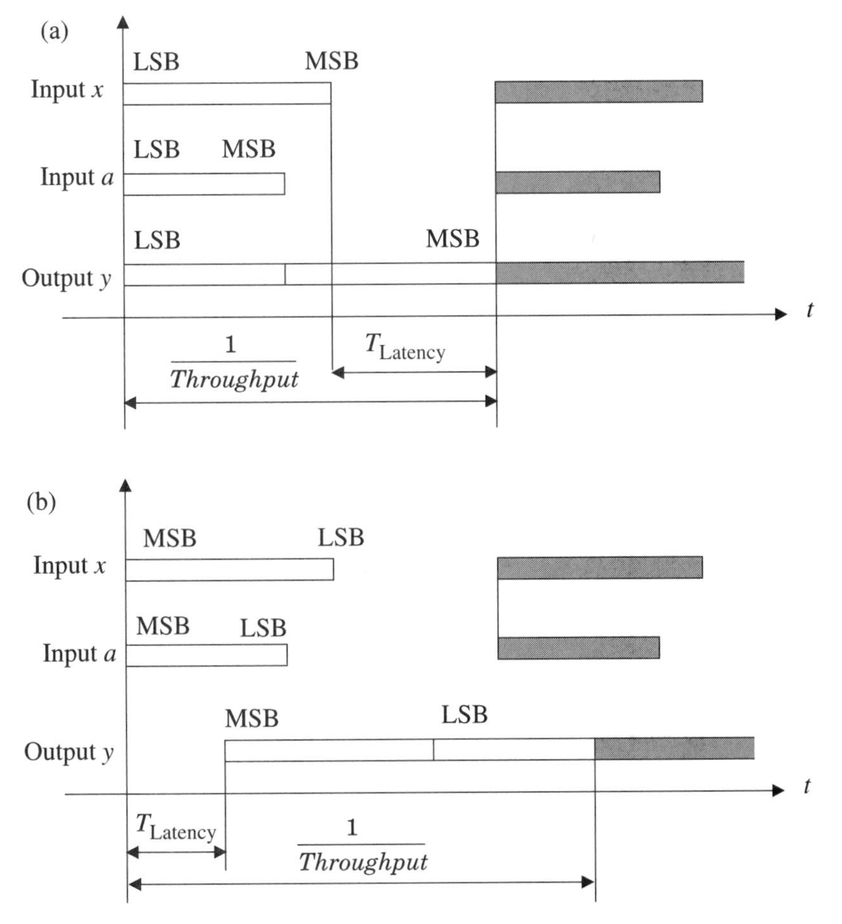

Figure 6.9 Latency and throughput for a multiplication $y = a \cdot x$ using bit-serial arithmetic with: (a) least significant bit (LSB) first, and (b) most significant bit (MSB) first

In redundant arithmetic, which processes the MSB first, the carry will propagate for a few bits only. The latency is determined by the length (number of bits) of the carry propagation. We will later show that throughput in a recursive algorithm will be determined by the total latency of the operations in the recursive loops and not by the throughput of the basic operations (multiplications and additions) within the loop.

Here we define two latency models for bit-serial arithmetic. In Chapter 11, we will discuss bit-serial arithmetic in more detail. Two latency models for a bit-serial adder are shown in Figure 6.10. In model 0, which can be used to model a static CMOS logic style without pipelining of the gates, the latency is equal to the gate delay of a full adder. In model 1, which can be used to model a dynamic CMOS logic style, or a static CMOS logic style with pipelining on the gate level, the full adder, followed by a D flip-flop, causes the latency to become one clock cycle. Model 1 generally results in faster bit-serial implementations, due to the shorter logic paths between the flip-flops in successive operations. Note that the longest logic paths can in some cases occur between the arithmetic operators. In these cases the maximal clock frequency will be determined by these paths and not by the arithmetic units. In some cases it can be useful to have two D flip-flops at the output, i.e., use a model 2.

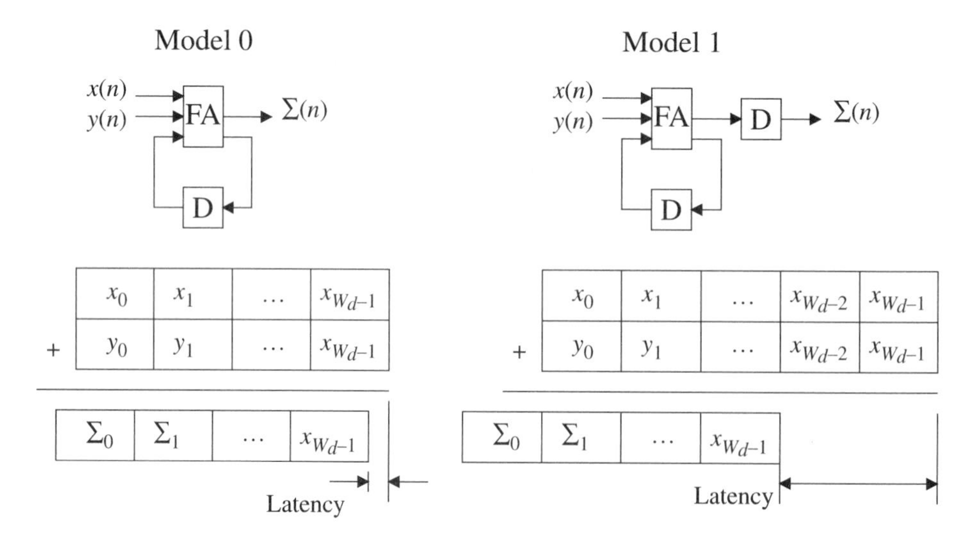

Figure 6.10 Latency models for a bit-serial adder

For multiplication, a simplified serial/parallel multiplier that uses bit-serial adders may be used. The corresponding latency models for a serial/parallel multiplier are shown in Figure 6.11. Denoting the number of fractional bits of the coefficient $W_{\alpha f}$, the latencies become $W_{\alpha f}$ for latency model 0, and $W_{\alpha f} + 1$ for latency model 1. We will later show that it is important to minimize the coefficient word lengths, since they affect both the minimum sample period and the power consumption.

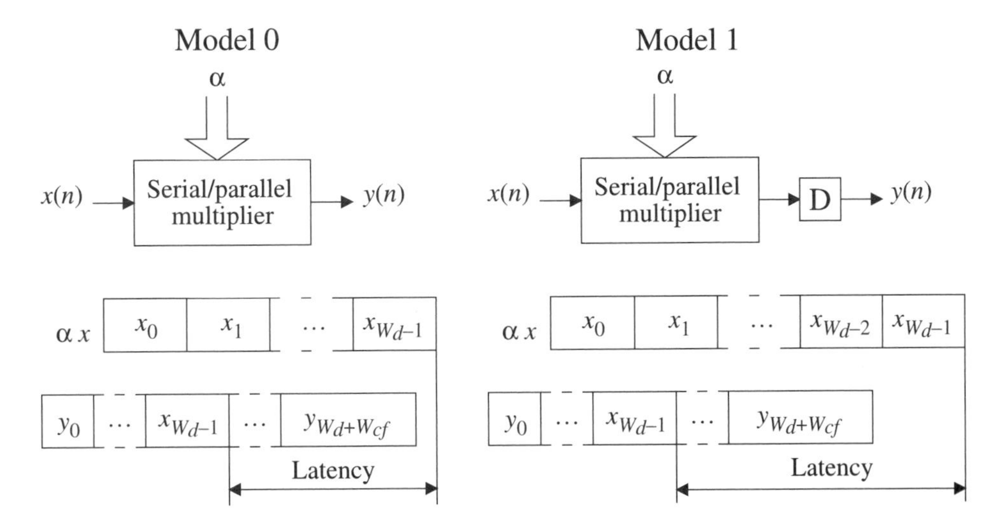

Figure 6.11 Latency models for a serial/parallel multiplier

6.3.3 Sequentially Computable Algorithms

A precedence graph may be contradictory in the sense that it describes an impossible ordering of events [6 – 7, 9]. An example of such a precedence graph is shown in Figure 6.12. Event A can occur only after event C has occurred. However, event C can only occur when event B has occurred, but event B can not occur until event A has occurred. Hence, this sequence of events is impossible since the sequence cannot begin.

In a digital signal processing algorithm, events correspond to arithmetic operations. In a proper algorithm, at least one of the operations in each recursive loop in the signal-flow graph must have all its input values available so that it can be executed. This is satisfied only if the loops contain at least one delay element, since the delay element contains a value from the preceding sample interval that can be used as a starting point for the computations in the current sample interval. Thus, there must not be any delay-free loops in the signal-flow graph. For a sequentially computable algorithm the corresponding precedence graph is called a *directed acyclic graph* (*DAG*).

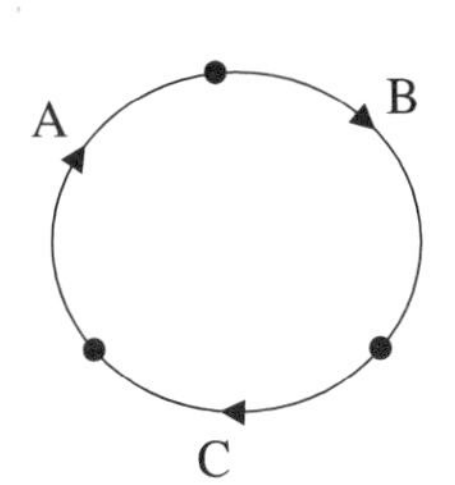

Figure 6.12 Directed, cyclic precedence graph

Theorem 6.1

A necessary and sufficient condition for a recursive algorithm to be *sequentially computable* is that every directed loop in the signal-flow graph contains at least one delay element.

An algorithm that has a delay-free loop is *non-sequentially computable* [9]. Such an algorithm can be implemented neither as a computer program nor in digital hardware. These algorithms are therefore of no practical use. Algorithms must therefore be checked for delay-free loops since such loops may occur in certain synthesis procedures.

6.3.4 Fully Specified Signal-Flow Graphs

Usually, a signal-flow graph is not fully specified from a computational point of view. For example, the order in which three or more values are added is usually not specified in signal-flow graphs. The ordering of the additions is, under certain conditions, not important if two's-complement representation is used. These conditions will be further discussed in Chapter 11. However, the ordering of the additions may affect the computational properties of the algorithm. For example, the maximum sample rate may change. In a *fully specified signal-flow graph*, the ordering of all operations is uniquely specified as illustrated in Figure 6.13. Generally the signal-flow graph should be described in terms of the operations that can be executed by the processing elements that are going to be used in the implementation.

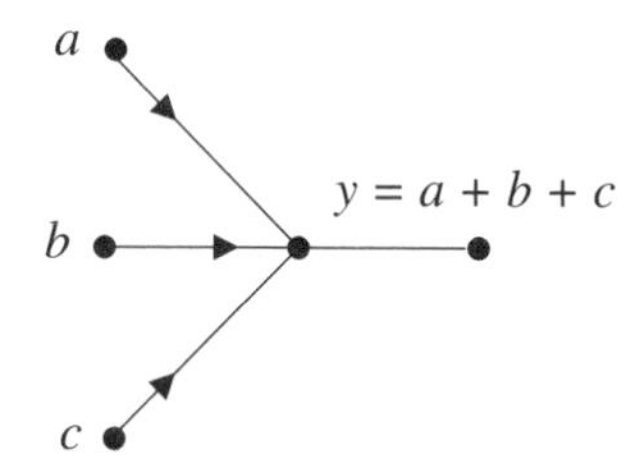

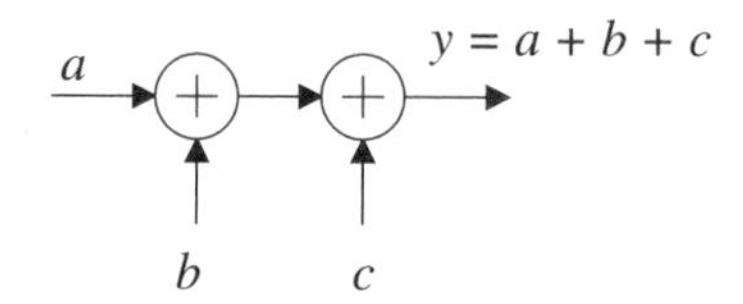

Figure 6.13 Signal-flow graph and the corresponding fully specified signal-flow graph

6.4 SFGs IN PRECEDENCE FORM

In this section we will present a method of redrawing the signal-flow graph into a precedence form that yields the order in which the node values must be computed [6, 7]. A procedure for deriving the expressions for the node values in a sequentially computable order is shown in Box 6.1.

In the next section we will use the precedence form to derive the corresponding set of difference equations in a form that can easily be mapped to code for a general-purpose signal processor. Note, however, that the precedence form represents a simplified view of an algorithm and does not provide full insight into the computational properties of the signal-flow graph. We will develop a better insight into the true properties later in this chapter and in the next chapter.

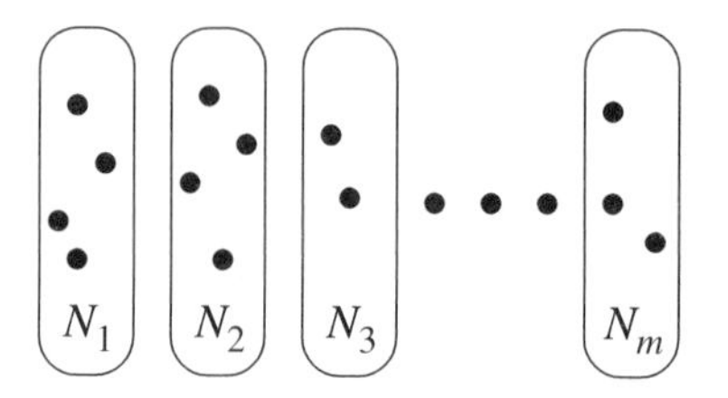

Figure 6.14 Sets of nodes that are sequentially computable

If the procedure shown in Box 6.1 is terminated prematurely due to lack of initial nodes, then the remaining part of the precedence graph contains a cycle that corresponds to a delay-free loop in the signal-

1. **1.** Collapse unnecessary nodes in the fully specified signal-flow graph by removing all branches with transmittance = 1. Transmittances of 1 can often be propagated into adjacent branches. This will give an equivalent, and potentially simpler, signal-flow graph with fewer nodes and branches.
2. **2.** Assign node variables to all nodes in the fully specified signal-flow graph according to
   - Input and output nodes with variables, $x_i(n)$ and $y_i(n)$, respectively.
   - Contents of the delay elements (outputs) with variables, $v_i(n)$.
   - Outputs from the basic operations, i.e., all the remaining nodes, with variables, u_i

   The computational order for the branches, corresponding to arithmetic operations, is determined in steps 3 to 7.
3. **3.** Remove all branches with delay elements in the signal-flow graph. Let $j = 1$.
4. **4.** Choose all initial nodes in the (remaining) signal-flow graph and denote this set of nodes by N_j, as shown in Figure 6.14.
5. **5.** Delete the branches that correspond to basic operations that are executable (that is, operations for which all inputs are initial nodes). Remove all initial nodes that no longer have any outgoing branches.
6. **6.** If there are any nodes left, let $j \leftarrow j + 1$ and repeat from step 4. The algorithm is not sequentially computable if there are some operations but no initial nodes left. Hence, the precedence form does not exist.
7. **7.** Connect nodes with branches (basic operations) according to the signal-flow graph. Note that the node variables $v_i(n)$ always belong to the first set of nodes, while the node variables u_i belong to the other sets.

Box 6.1 Procedure for deriving the precedence form of a signal-flow graph.

flow graph. Hence, the algorithm is not sequentially computable. Such algorithms have no practical use.

EXAMPLE 6.1

Derive the signal-flow graph in precedence form for the second-order section in direct form II shown in Figure 6.15.

The necessary quantization in the recursive loops is shown but not the ordering of the additions. Hence, the signal-flow graph is not fully specified. We get a fully specified signal-flow graph by ordering the additions according to Figure 6.16. Note that there are several ways to order the additions.

The first step, according to the procedure in Box 6.1, does not apply since, in this case, there are no multiplications by –1. The second step is to first assign node variables to the input, output, and delay elements. Next we identify and assign node variables to the basic operations and the nodes that correspond to the outputs (results) of the basic operations. We assume that multiplication, two-input addition, and quantization are the basic operations. The resulting variable assignments are shown in Figure 6.16.

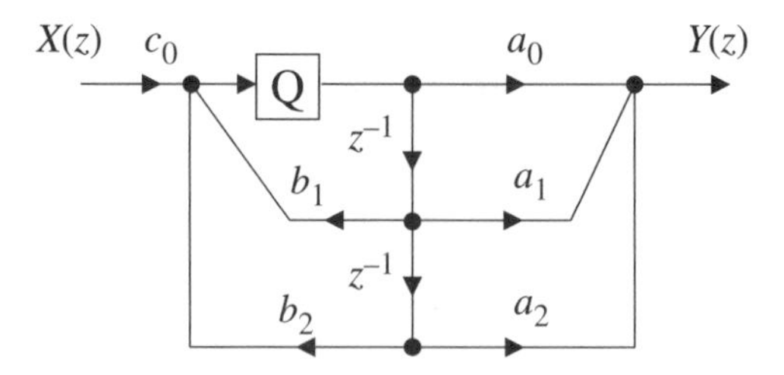

Figure 6.15 Second-order section in direct form II

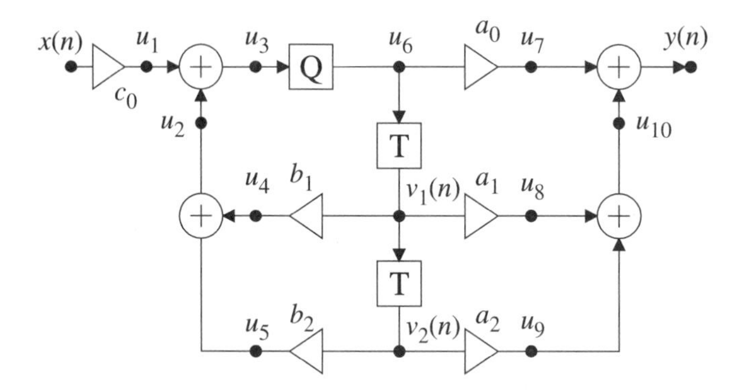

Figure 6.16 Fully specified signal-flow graph with node variables

An operation can be executed when all input values are available. The input, $x(n)$, and the values stored in the delay elements, $v_1(n)$ and $v_2(n)$, are available at the beginning of the sample interval. Thus, these nodes are the initial nodes for the computations. To find the order of the computations we begin (step 3) by first removing all delay branches in the signal-flow graph. The resulting signal-flow graph is shown in Figure 6.17.

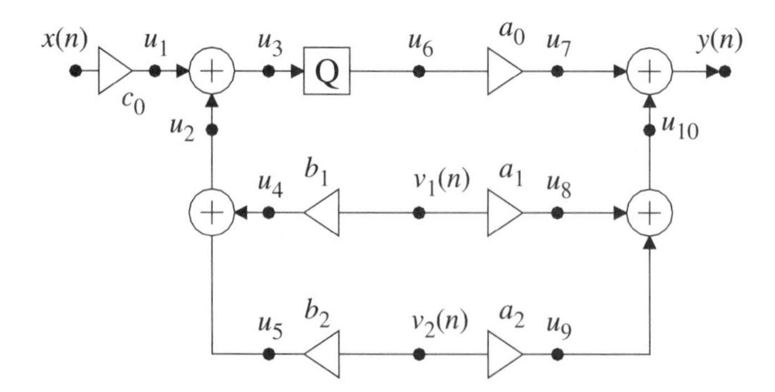

Figure 6.17 Signal-flow graph with removed delay branches

In step four, all initial nodes are identified, i.e., $x(n)$, $v_1(n)$, and $v_2(n)$. These nodes are assigned to node set N_1, as shown in Figure 6.24.

In step five we remove all executable operations, i.e., all operations that have only initial nodes as inputs. In this case we remove the five multiplications by the coefficients: c_0, a_1, a_2, b_1, and b_2. The resulting graph is shown in Figure 6.18.

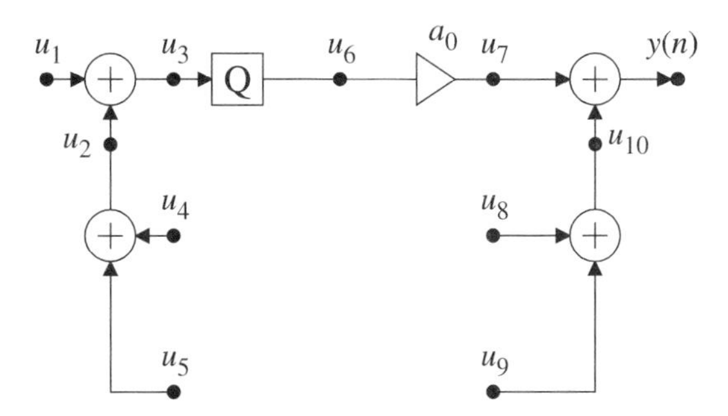

Figure 6.18 The operations c_0, b_1, b_2, a_1, and a_2 have been removed

Repeating steps 4 and 5 we successively get the graphs shown in Figures 6.19 through 6.23. The sets of nodes shown in Figure 6.24 illustrate the order in which the nodes must be computed.

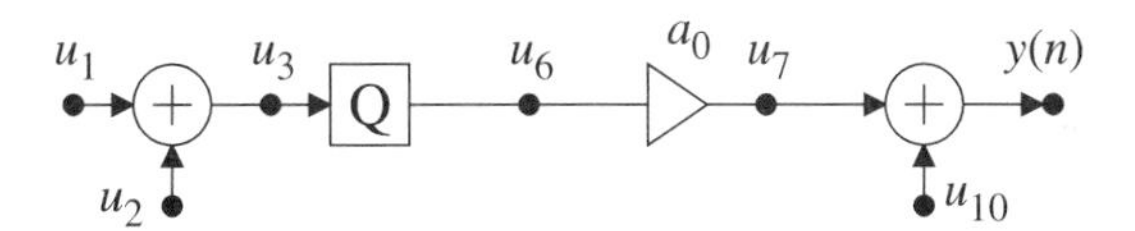

Figure 6.19 Two additions have been removed

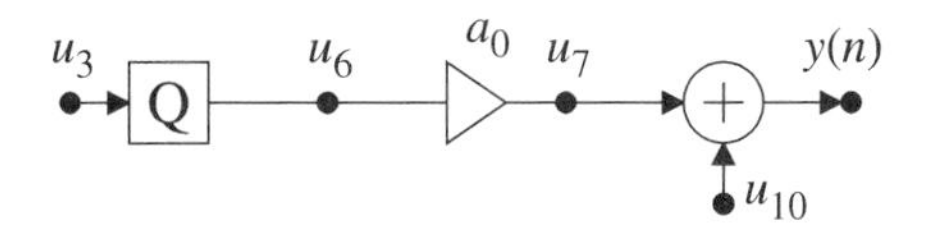

Figure 6.20 One addition has been removed

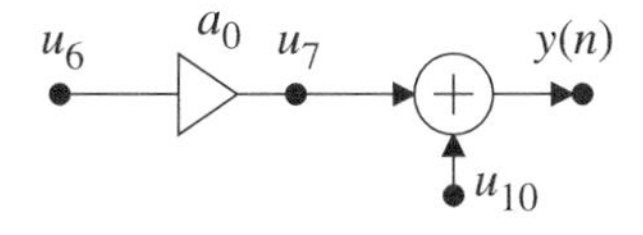

Figure 6.21 The quantization has been removed

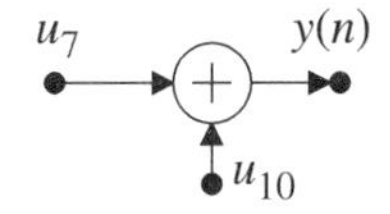

Figure 6.22 One multiplication has been removed

Finally, we obtain the signal-flow graph in precedence form by connecting the nodes in Figure 6.24 with the delay branches and the arithmetic branches according to the original signal-flow graph. Notice the two registers that are needed for the delay elements.

Figure 6.23 The last addition has been removed

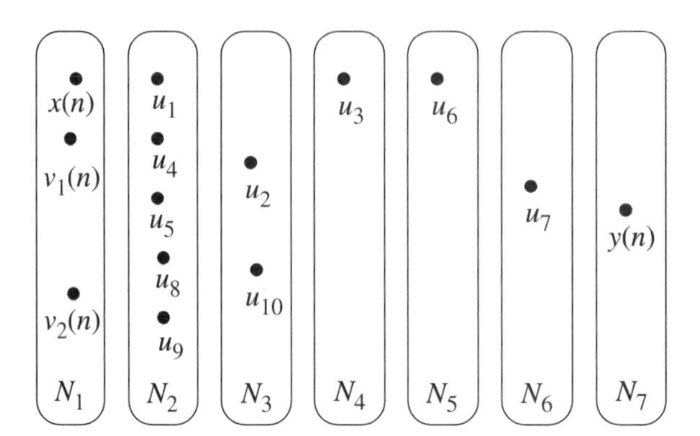

Figure 6.24 Precedence of node sets

The precedence form is shown in Figure 6.25. Note that the computations take place from left to right. A delay element is here represented by a branch running from left to right and a gray branch running from right to left. The latter indicates that the delayed value is used as input for the computations belonging to the subsequent sample interval.

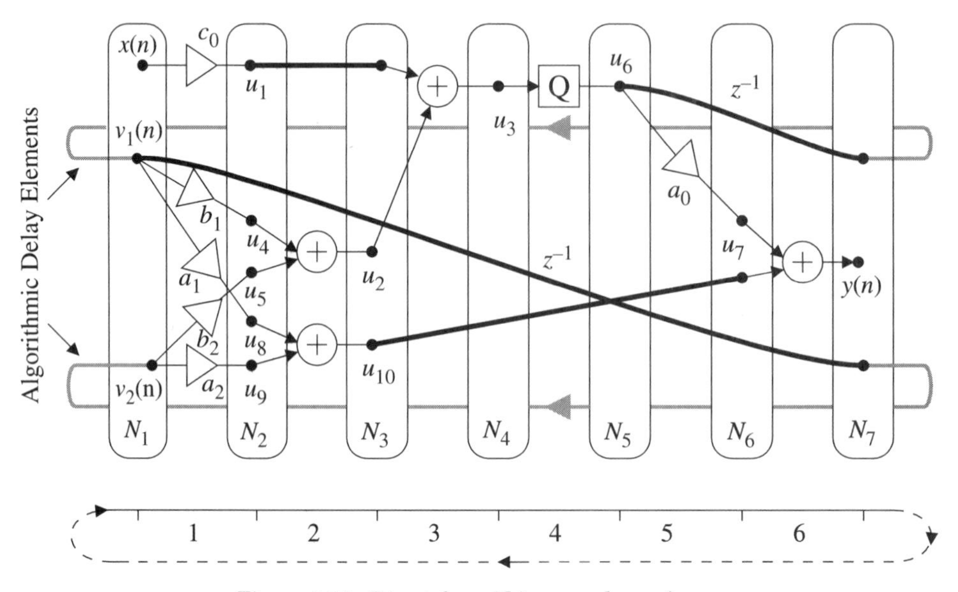

Figure 6.25 Direct form II in precedence form

If a node value (for example, u_1) is computed earlier than needed, an auxiliary node must be introduced. The branch connecting the nodes represents storing the value in memory. The sample interval is assumed here to begin with execution of the leftmost operations and end with the computation of the nodes belonging to set N_7. Once the latter have been computed, the computations belonging to the next sample interval can begin. In fact, the nodes in sets N_1 and N_7 can be regarded as belonging to the same node set. Six time steps are required to complete the operations within one sample interval.

It is illustrative to draw the precedence form on a cylinder, as shown in Figure 6.26, to demonstrate the cyclic nature of the computations. The computations are imagined to be performed repeatedly around the cylinder. The circumference of the cylinder corresponds to a multiple of the length of the sample interval.

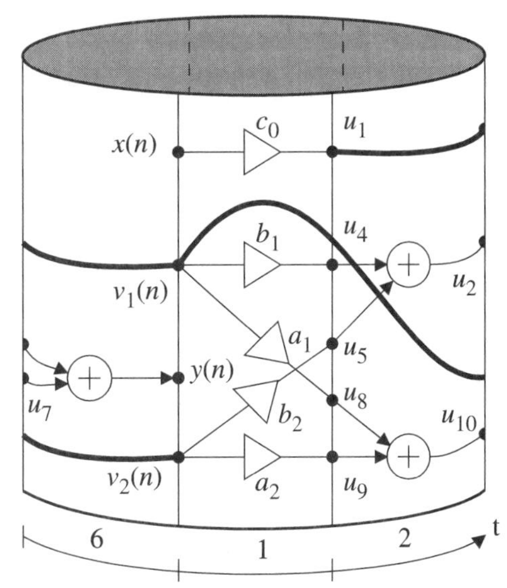

Figure 6.26 Illustration of the cyclic nature of the computations

As just mentioned, a gray delay branch running from right to left in Figure 6.25 transfers values between different sample intervals. The gray branches are artifacts and do not appear in Figure 6.26. Instead a delay element will correspond to a branch around the whole cylinder, or a part of it. Storage is indicated by heavy lines in Figure 6.26. In this example, the first delay element represents storage of the node value u_6 from time step 5 to time step 1, i.e., during only two time steps. The second delay element stores the value $v_1(n)$ during a complete sample interval. The branches in Figure 6.26 represent either arithmetic operations or temporary storage of values.

6.5 DIFFERENCE EQUATIONS

A digital filter algorithm consists of a set of difference equations to be evaluated for each input sample value. In this section we will present a method to determine the order in which these equations must be evaluated. Usually, some of the equations can be evaluated simultaneously while other equations must be evaluated sequentially. The computational ordering of equations describing those other types of DSP algorithms not described by difference equations (for example, FFTs) can also be derived by the method presented shortly. In both cases, this ordered set of equations is a useful starting point for implementing the algorithm in software using a standard signal processor or a multicomputer.

The difference equations for a digital filter can be obtained directly from its signal-flow graph in precedence form. The signal values corresponding to node set N_1 are known at the beginning of the sample interval. Hence, operations having output nodes belonging to node set N_2 have the necessary inputs and can therefore be executed immediately. Once this is done, those operations having output nodes belonging to node set N_3 have all their inputs available and can be executed. This process can be repeated until the last set of nodes has been computed. Finally, the signal values corresponding to node set N_1 can be updated via the delay elements. This completes the operations during one sample interval.

A method for ordering the updating of the node values that correspond to delay elements is shown in Box 6.2. Delay elements connected in series are updated sequentially starting with the last delay element, so that auxiliary memory cells are not required for intermediate values.

Operations with outputs belonging to the different node sets must be executed sequentially while operations with outputs belonging to the same node set can be executed in parallel. Similarly, updating of node values that correspond to delay elements must be done sequentially if they belong to different subsets of N_{1k}. The updating is done here at the end of the sample interval, but it can be done as soon as a particular node value is no longer needed. We will illustrate the method with an example.

1. Extract the delay branches in the original signal-flow graph. Note that all delay branches terminate on node set N_1. Let $k = 1$.
2. Find the set of terminal nodes in the delay element graph and denote this subset N_{1k}. Update these node values and delete all incident delay branches.
3. If there are any branches left, let $k \leftarrow k + 1$ and repeat from step 2.

Box 6.2 Procedure for deriving the order for updating the delay elements

EXAMPLE 6.2

Determine the system of difference equations in computable order for the second-order digital filter in Example 6.1.

The system of difference equations can be obtained directly from Figure 6.25. All nodes belonging to node set N_1 are initial values. Hence, all nodes in set N_2 can be computed immediately. In the next step, the nodes belonging to set N_3 can be computed using the values from node sets N_1 and N_2 as inputs to those operations with output values that belong to node set N_3. In the same way, the remaining node sets can be computed successively.

Lastly, we update the two cascaded delay elements. According to the procedure shown in Box 6.2, the updating must begin with node v_2, i.e., copy the value $v_1(n)$ into the storage cell for v_2, in order to release the storage cell used for v_1 so that it can be overwritten. Auxiliary memory cells are therefore not needed. The result is summarized in Table 6.1.

The operations are here assumed to be executed as soon as their input values are available. However, the execution of an operation can be postponed until the result is needed as input to a subsequent operation. For example, u_{10} need only be computed prior to $y(n)$. In Chapter 7 we will discuss scheduling techniques that

Node Set	Equations
N_2	$u_1 := c_0\, x(n)$ $u_4 := b_1\, v_1(n)$ $u_5 := b_2\, v_2(n)$ $u_8 := a_1\, v_1(n)$ $u_9 := a_2\, v_2(n)$
N_3	$u_2 := u_4 + u_5$ $u_{10} := u_8 + u_9$
N_4	$u_3 := u_1 + u_2$
N_5	$u_6 := [u_3]_Q$
N_6	$u_7 := a_0\, u_6$
N_7	$y(n) := u_7 + u_{10}$
N_{11}	$v_2(n+1) := v_1(n)$
N_{12}	$v_1(n+1) := u_6$

Table 6.1 Difference equations in computational order for direct form II

explore this flexibility to minimize the amount of hardware resources required to execute the algorithm.

In the method just discussed, a large number of simple expressions are computed and assigned to intermediate variables. Hence, a large number of intermediate values are computed that must be stored in temporary memory cells and require unnecessary store and load operations. If the algorithm is to be implemented using, for example, a standard signal processor, it may be more efficient to eliminate some of the intermediate results and explicitly compute only those node values required. We demonstrate how this can be done by the means of an example.

EXAMPLE 6.3

Show that some of the intermediate values can be eliminated in the system of difference equations derived in Example 6.2. Use the simplified system of difference equations to write a Pascal program that implements the second-order section in direct form II. The program shall emulate a data word length of 16 bits and use saturation arithmetic.

Obviously, the only values that need to be computed explicitly are

- ❑ Node values that have more than one outgoing branch
- ❑ Inputs to some types of noncommutating operations, and
- ❑ The output value.

The only node with two outgoing branches in Figure 6.25 is node u_6. The remaining node values represent intermediate values used as inputs to one subsequent operation only. Hence, their computation can be delayed until they are needed. Such node values appear in only one of the subsequent expressions. We get

$$\begin{cases} u_3 := c_0\, x(n) + b_1\, v_1(n) + b_2 v_2(n) \\ u_6 := [u_3]_Q \\ y(n) := a_0\, u_6 + a_1\, v_1(n) + a_2 v_2(n) \\ v_1(n+1) := u_6 \end{cases}$$

In this algorithm, the only operation that is noncommutating with its adjacent operations is the quantization operation. Generally, the inputs to such operations must be computed explicitly, but in this case u_3 appears in only one subsequent expression. Hence, u_3 can be eliminated. We get the following expressions that can be translated directly into a program:

$$\begin{cases} u_3 := [c_0\, x(n) + b_1\, v_1(n) + b_2 v_2(n)]_Q \\ y(n) := a_0\, u_6 + a_1\, v_1(n) + a_2 v_2(n) \\ v_2(n+1) := v_1(n) \\ v_1(n+1) := u_6 \end{cases}$$

A pseudo-Pascal program that implements the second-order section in direct form II is shown in Box 6.3.

```
Program Direct_form_II;
  const
    c0 = ...; a0 = ...; a1 = ...; a2 = ...; b1 = ...; b2 = ...; NSamples = ...;
  var
    xin, u6, v1, v2, y : Real;
    i: Longint;

  function Q(x: Real): Real;
    begin
      x := Trunc(x*32768); { Wd = 16 bits => Q = 2^ – 15 }
      if x > 32767 then x := 32767;
      if x < –32768 then x := –32768;
      x := x/32768;
    end;
  begin
    for i := 1 to NSamples do
      begin
        Readln(xin);                          {Read a new input value, xin }
        u6 := Q(c0*xin + b1*v1 + b2*v2);      {Direct form II }
        y := a0*u6 + a1*v1 + a2*v2;
        v2 := v1;
        v1 := u6;
        Writeln(y);   {Write the output value }
      end;
  end.
```

Box 6.3 Pseudo-Pascal program for emulation of a second-order section in direct form II with two's-complement truncation and saturation arithmetic

6.6 COMPUTATION GRAPHS

It is convenient to describe the computational properties of an algorithm with a *computation graph* that combines the information contained in the signal-flow graph with the corresponding precedence graph for the operations. The computation graph will be used in Chapter 7 as the basis for scheduling the operations. Further, the storage and communication requirements can be derived from the computation graph. In this graph, the signal-flow graph in precedence form remains essentially intact, but the branches representing the arithmetic operations are also assigned appropriate execution times. Branches with delay elements are mapped to branches with delay. As will be discussed in the following sections, additional branches with different types of delay must often be inserted to obtain consistent timing properties in the computation graph. These delays will determine the amount of physical storage required to implement the algorithm.

6.6.1 Critical Path

By assigning proper execution times to the operations, represented by the branches in the precedence graph, we obtain the computation graph. The longest (time) directed path in the computation graph is called the *critical path* (*CP*) and its execution time is denoted T_{CP}. Several equally long critical paths may exist. For example, there are two critical paths in the computation graph shown in Figuure. 6.25. The first CP starts at node $v_1(n)$ and goes through nodes u_4, u_2, u_3, u_6, u_7, and $y(n)$ while the second CP begins at node $v_2(n)$ and goes through nodes u_5, u_2, u_3, u_6, u_7, and $y(n)$.

Note that the critical path will not necessarily determine the shortest time in which the algorithm can be computed since it represents only the precedence between operations within a sample interval and neglects intersample interval precedences. The minimum sample period is determined by the recursive loops in the algorithm (see section 6.6.4). The critical path may in some cases, therefore, be longer than the minimum sample period. Later, in section 6.8.2, we will discuss methods of breaking the critical path into smaller pieces so that the maximum sample rate can be achieved.

6.6.2 Equalizing Delay

Assume that an algorithm shall be executed with a sample period T and the time taken for the arithmetic operations in the critical path is T_{CP}, where $T_{CP} < T$. Then additional delay, which is here called *equalizing delay*, has to be introduced into all branches that cross an arbitrary vertical cut in the computation graph, as illustrated in Example 6.4. This delay accounts for the time difference between a path and the length of the sample interval. The required duration of equalizing delay in each such branch is

$$T_e = T - T_{CP}$$

The amount of equalizing delay, which usually corresponds to physical storage, can be minimized by proper scheduling of the operations, which will be further discussed in Chapter 7.

6.6.3 Shimming Delay

If two paths in a computation graph have a common origin and a common end node, then the data streams in these two paths have to be synchronized at the summation node by introducing a delay in the fastest path. For example, execution of the two multiplications, shown in Figure 6.27, may be started at time instances t_{0a} and t_{0b}, and the times required for multiplication are t_a and t_b, respectively. This means that the products will arrive at the inputs of the subsequent adder with a time difference

$$\Delta t = (t_{0b} + t_b) - (t_{0a} + t_a)$$

A delay must therefore be inserted in the upper branch of the computation graph so that the products arrive simultaneously. These delays are called *shimming delays* or *slack*. Shimming delays usually correspond to physical storage.

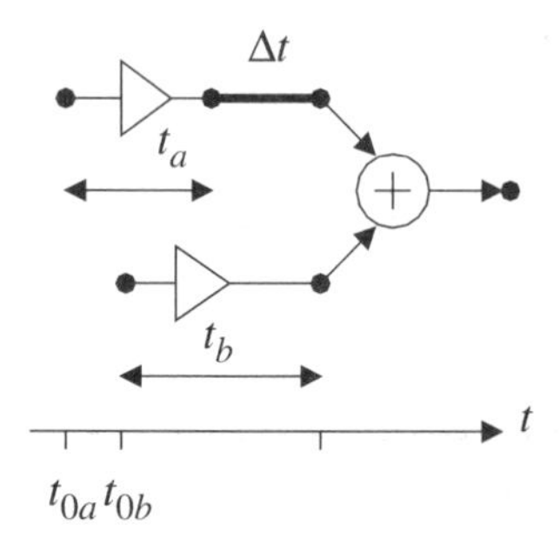

Figure 6.27 Shimming delay

The amount of shimming delay required to implement an algorithm can be minimized by using equivalence transformations, which will be discussed in section 6.9. This is particularly important if the algorithm is implemented with bit-serial PEs using an architecture that is isomorphic to the signal-flow graph. Since the delay branches correspond to moving data (i.e., shift registers) this will consume large amounts of power.

Example 6.4

Determine the computation graph for the second-order section in direct form II. Assume that the operations have the same execution time (i.e., 1 time unit) and indicate equalizing and shimming delays.

The computation graph shown in Figure 6.28 is found directly from the precedence form shown in Figure 6.25. Note that two critical paths exist: one path starting with the multiplication by coefficient b_1 and the other path starting with the multiplication by b_2. Both paths go through two additions, quantization, multiplication by a_0, and, finally, an addition.

Two branches with shimming delays have to be inserted. Three branches with equalizing delay have been inserted to prolong the sample interval. Here we have assumed that the input and output should be separated by precisely one sample period. Note that the branches representing the delay elements in the signal-flow

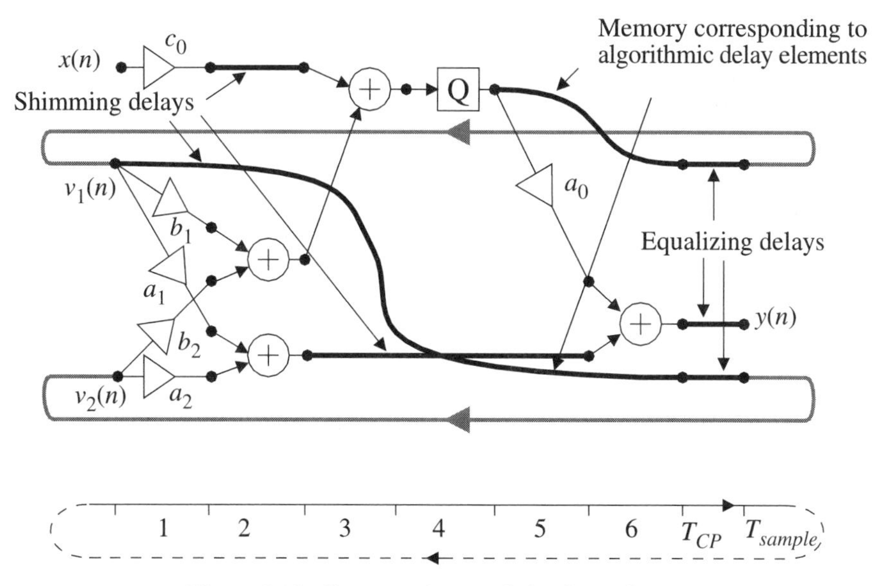

Figure 6.28 Computation graph for direct form II

graph correspond to execution time for operations and shimming and equalizing delays in the computation graph.

6.6.4 Maximum Sample Rate

The maximum sample rate of an algorithm is determined only by its recursive parts according to Theorem 6.2 [9, 23, 24]. Nonrecursive parts of the signal-flow graph (e.g. input and output branches) generally do not limit the sample rate, but to achieve this limit additional delay elements may have to be introduced into the nonrecursive branches. This problem will be discussed in more detail in section 6.8.

Theorem 6.2

The *minimum sample period* for a recursive algorithm that is described by a fully specified signal-flow graph is

$$T_{min} = \max_{i} \left\{ \frac{T_{opi}}{N_i} \right\} \tag{6.1}$$

where T_{opi} is the total latency of the arithmetic operations etc. and N_i is the number of delay elements in the directed loop i.

The minimum sample period is also referred to as the *iteration period bound*. Loops that yield T_{min} are called *critical loops*. This bound can directly be found from the signal-flow graph by inspection. The maximum sample rate can also be found from the intrinsic coefficient word length [28].

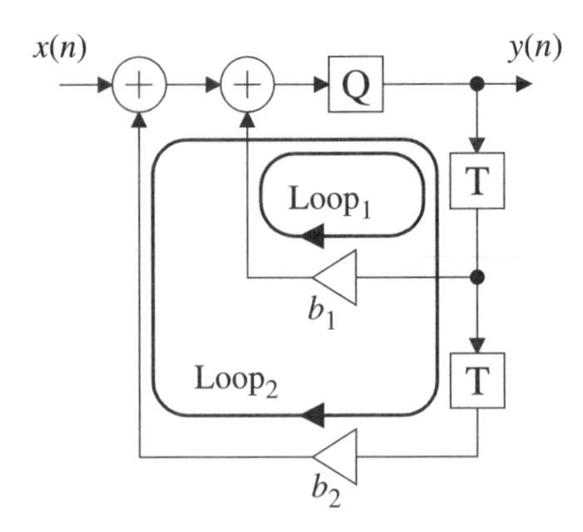

Figure 6.29 Signal-flow graph with two directed loops

Figure 6.29 shows a fully specified signal-flow graph that may be a part of a larger graph. The signal-flow graph has two loops. Loop 1 is the critical loop if

$$\frac{T_{b2} + 2T_{add} + T_Q}{2} < \frac{T_{b1} + T_{add} + T_Q}{1}$$

otherwise loop 2 is the critical loop. This ordering of the additions is often the most favorable one.

EXAMPLE 6.5

The filter shown in Figure 6.30 has been implemented using so-called redundant, bit-parallel arithmetic in which long carry propagations are avoided [11]. The results obtained using a 2-μm, double metal CMOS process were

Input data word length, 8 bit
Output data word length, 11 bit
Maximum sample rate, ≈ 35 MHz
Number of devices, ≈ 9000
Power consumption, 150 mW
Chip area, 14.7 mm^2 (including pads)

We will later discuss alternative implementations of this filter aimed at video applications. The filter is a lowpass filter, but it can also be used as a highpass filter as well as for decimation or interpolation of the sample frequency by a factor two, with minor changes in the hardware. Determine the maximum sample frequency for the filter. The adaptor coefficient is $\alpha = 0.375 = (0.011)_2$.

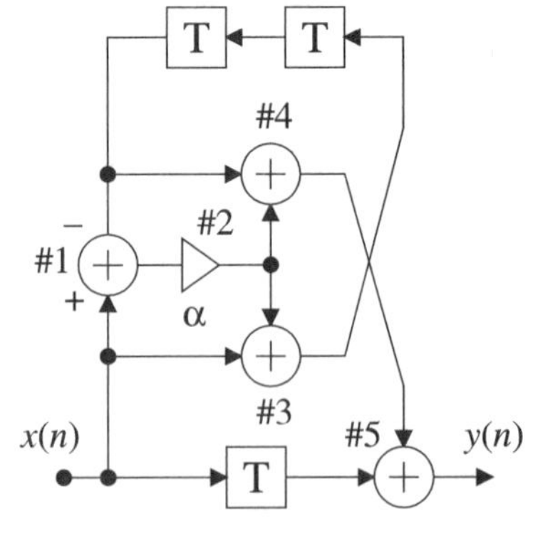

Figure 6.30 Third-order lattice wave digital filter.

Assume here that the data word length is 12 bits and that the filter is implemented with bit-serial arithmetic using so-called *true single-phase clock* (*TSPC*). This type of logic is modeled by model 1, i.e., the latencies for addition, subtraction, and multiplication in this logic style are one clock cycle longer that the theoretical minimum. Thus, we have in this case T_{add} = 1 clock cycle and T_{mult} = 4 clock cycles.

The delay (latency) due to the arithmetic operations in the recursive loop is

$$T_{op} = 2T_{add} + T_{mult} = 6 \text{ clock cycles}$$

and the number of delay elements in the loop is N_i = 2. Hence, we get the iteration period bound

$$T_{min} = \frac{6}{2} = 3 \text{ clock cycles}$$

and

$$f_{sample\ max} = \frac{1}{T_{min}} = \frac{f_{CL}}{3}$$

where f_{CL} is the clock frequency. The bit-serial implementation must therefore have a clock frequency of at least 105 MHz to be as fast as the implementation using bit-parallel, redundant arithmetic. This represents a relatively modest clock frequency.

The iteration period bound for this algorithm using a static logic style (i.e., model 0, having minimum latencies T_{add} = 0 clock cycle and T_{mult} = 3 clock cycles) is

$$T_{min} = \frac{2T_{add} + T_{mult}}{2} = \frac{2 \cdot 0 + 3}{2} = 1.5 \text{ clock cycles/sample}$$

It is important to note that the length of the clock cycles are different between the two models. In practice, the implementation corresponding to model 1 may be the fastest, because of shorter propagation paths between the memory elements.

6.7 EQUIVALENCE TRANSFORMATIONS

Computation graphs describing DSP algorithms usually contain only shift-invariant operations or processes. Hence, the ordering of an operation in series with a memory element may be changed. This simple property is the basis for scheduling operations that aim at minimizing the amount of resources required to implement the algorithm. Scheduling will be discussed in more detail in Chapter 7. Another related transformation, also based on interchanging shift-invariant operators, is pipelining. We will discuss various forms of pipelining in section 6.8.

In some cases operations in DSP algorithms represent associative, distributive, or commutative operators [5, 20–22, 25]. In this section we will show how these properties can be exploited to improve the computational properties of an algorithm. Of course, the maximum sample rate of a fully specified recursive algorithm is fixed and cannot be improved. However, an algorithm can in some cases be used as a starting point to derive another algorithm that is faster.

6.7.1 Essentially Equivalent Networks

The condition under which a shift-invariant network in series with a set of delay elements may be changed is given next [9]. This modification is also called *retiming* [15].

Theorem 6.3

If an arbitrary, nonlinear, time-varying discrete time network, N_n, has delay elements in series with all inputs (outputs), then all the delays can be moved to the outputs (inputs) and the properties of N_n shifted, according to Figure 6.31, without changing the input–output behavior of the composite system. N_n and N_{n+1} denote the properties of the network with reference to samples n and $n+1$, respectively.

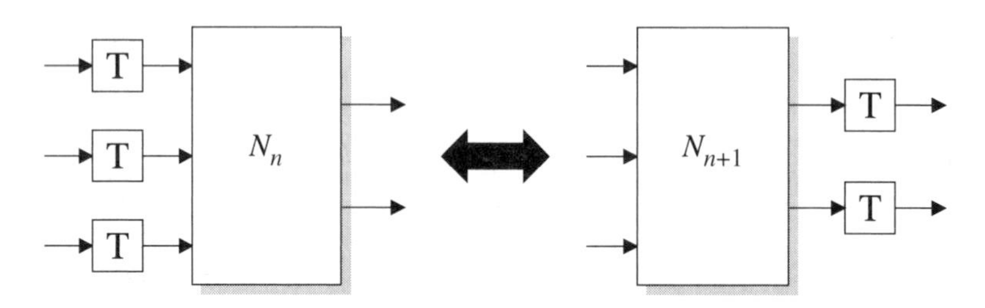

Figure 6.31 Networks with equivalent input-output behavior

For a shift-invariant system we have $N_n = N_{n-n_0}$ for all n_0. Two networks that can be transformed into one another by the equivalence transformations just described, except for different (positive or negative) delays appearing in their input and output branches, are called *essentially equivalent* networks [9]. A delay element cannot be propagated into a recursive loop. However, the positions of the delay elements in a recursive loop can be changed as shown in Example 6.6. The latency of the algorithm may be affected by such a change, but the maximum sample rate is unaffected.

EXAMPLE 6.6

A common substructure that appears in the design of wave digital filters based on Richards' structures is shown in Figure 6.32. The basic theory results in noninteger delay elements.

By introducing a *T*/2-delay element in series with the output of the rightmost adaptor and then applying Theorem 6.3, the *T*/2-delay elements at the outputs to the rightmost adaptor can be moved to the inputs, as shown in Figure 6.33.

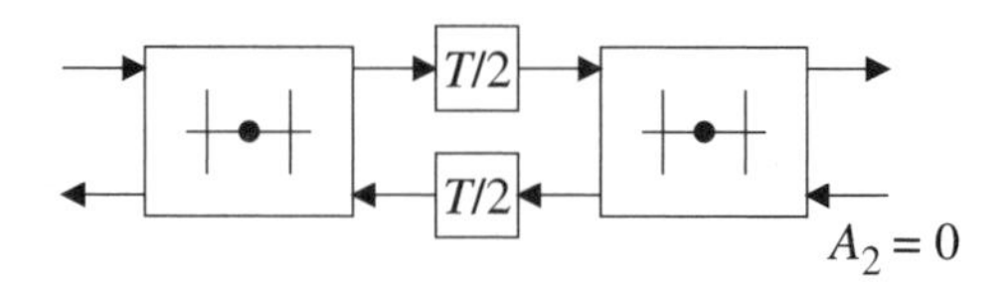

Figure 6.32 Structure with non-integer delay elements

One integer delay element is obtained between the adaptors and one $T/2$-delay element in series with the A_2-input. The latter serves no purpose and can therefore be removed, since the input signal $A_2 = 0$. Note that the total delay in the recursive loop is unchanged. The outputs have been delayed by $T/2$ with respect to the initial network. In fact, this modification is a form of pipelining, which will be discussed in section 6.8.

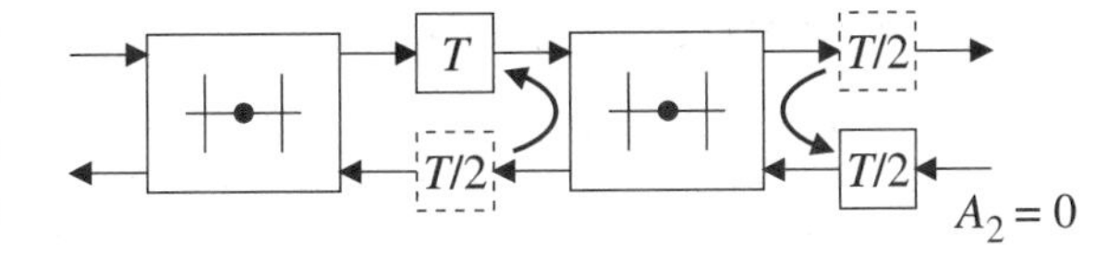

Figure 6.33 Pipelined structure with integer delay elements

6.7.2 Timing of Signal-flow Graphs

Equivalence transformations can be used to determine the proper timing of the operations in signal-flow graphs as well as proper control signals. Such signal-flow graphs can easily be transformed into proper computation graphs. Of particular interest is to determine the required amount of shimming delay as well as the positions of the D flip-flops. The technique presented next leads to the same results as methods based on *cut set transformations* [15]. We will demonstrate the technique by means of an example.

EXAMPLE 6.7

Determine the minimum number of shimming delays inside the adaptors in Example 4.12. Assume that the data word length is $W_d = 21$ bits, the coefficient word length is $W_c = 10$ bits, and that bit-serial arithmetic, implemented in single-phase logic, is used. Further, determine the number of stages of D flip-flops required in the delay elements.

First, we assume that the arithmetic operations have zero execution time. These operations are indicated in gray in Figure 6.34. We place an auxiliary delay element in series with the input of the two-port adaptor. This is not necessary, but it is often advisable to include a delay of 21 clock cycles from input to output of the adaptor.

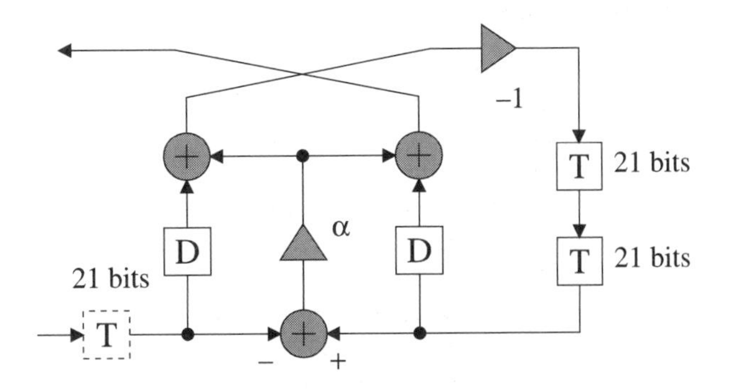

Figure 6.34 Initial configuration with auxiliary delay element at the input

In the first step, we propagate the multiplication by –1 through the delay elements. The upper rightmost adder becomes a subtractor. The two sign inversions at the inputs to the lower adder are removed by changing the sign of the adaptor coefficient, as shown in Figure 6.35. In the next step one D flip-flop, taken from the delay elements, is propagated forward in the signal-flow graph. At a take-off node a D flip-flop is propagated into both the outgoing branches, as shown in Figure 6.35. In this case, the adder absorbs one D flip-flop from each input, since the addition takes 1 clock cycle.

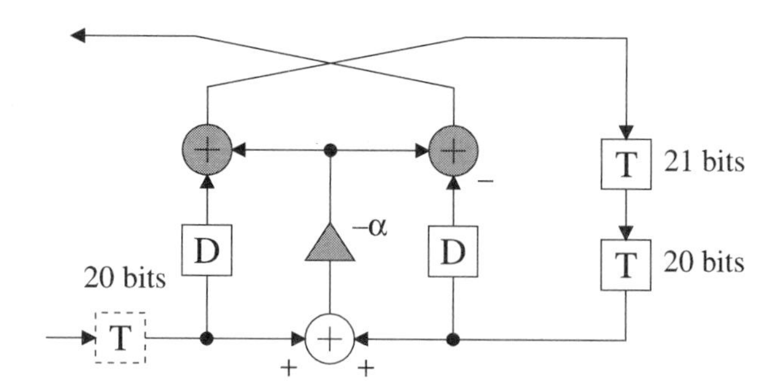

Figure 6.35 First step. One D flip-flop, taken from each of the delays, is propagated through the first adder

The magnitude of the adaptor coefficient is always less than 1. Hence, the latency for the multiplication is $W_c - 1$ clock cycles. In the same way, we propagate nine D flip-flops past the adder and the multiplier, as shown Figure 6.36. The multiplier absorbs nine D flip-flops.

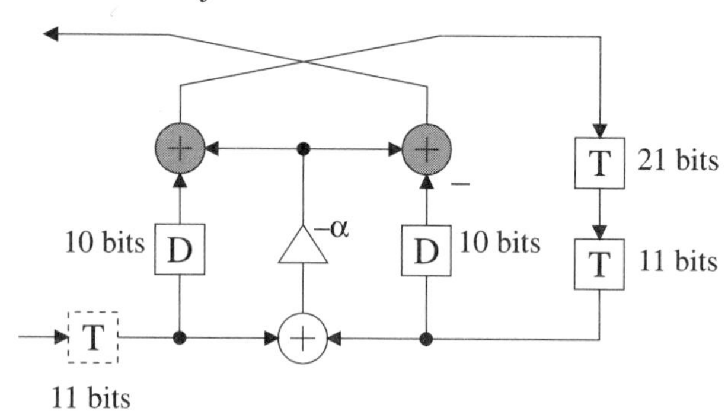

Figure 6.36 Second step. Propagate $W_c - 1 = 9$ D flip-flops, taken from each of the delays, through the adder and multiplier

The final wave-flow graph, shown in Figure 6.37, is obtained in the last step by propagating one pair of D flip-flops past the adder and multiplier, to finally be absorbed in the last adder and subtractor. This is the last step, since all arithmetic operations have absorbed the proper number of D flip-flops. Here, 20

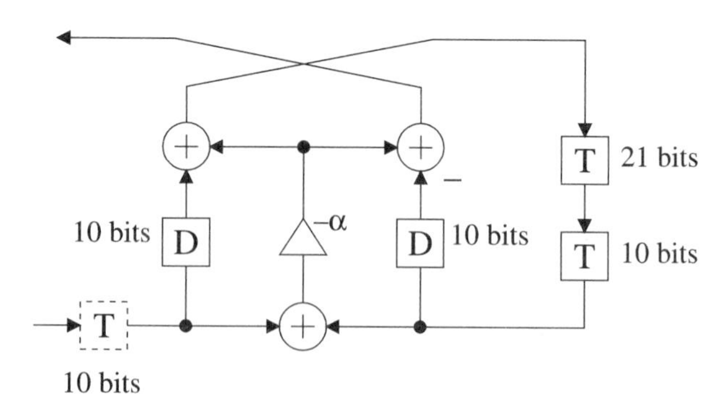

Figure 6.37 Final step. Propagate one D flip-flop, taken from each of the delays, through the adder, multiplier, and the last adder and subtractor

shimming delays (D flip-flops) are needed, but 11 of these are taken from the delay elements. The minimum delay between input and output is

11 clock cycles. Altogether, 51 D flip-flops are required. The final wave-flow graph, shown in Figure 6.37, with proper timing can be readily transformed into a computation graph.

6.7.3 Minimizing the Amount of Shimming Delay

Minimizing the required shimming delay in the computation graph is important in the isomorphic implementation approach based on bit-serial arithmetic, which will be discussed in Chapter 9. The main reason is that the shimming delays in bit-serial arithmetic represent moving signal values (shift registers) that consume significant amounts of power. The amount of shimming delay is of less importance in bit-parallel implementations since data can be stored for arbitrarily long times in static registers, i.e., the power consumption does not depend on how long the values are stored.

The amount of shimming delay can be minimized by using Theorem 6.4, which is based on the fact that the ordering of additions (subtractions) is irrelevant if two's-complement representation is used and if the final sum is properly scaled, as was discussed in Chapter 5.

Theorem 6.4

For a multi-input adder (subtractor) that has inputs that arrive at times $t_1 < t_2 < \ldots < t_N$ the minimum latency and the minimum amount of shimming delay are attained if, recursively, the two earliest inputs or a previously computed partial sum are added (subtracted) using two's-complement representation, as illustrated in Figure 6.38.

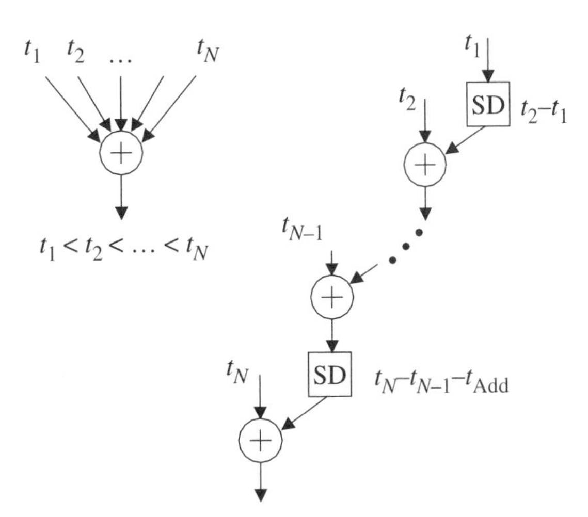

Figure 6.38 Adder tree with minimum latency and amount of shimming delay

The required amount of shimming delay can often be reduced by simple changes to the computation graph that do not affect the numerical properties of the algorithm [5]—for example, reordering the additions.

6.7.4 Maximally Fast Critical Loops

In many cases, the order of arithmetic operations can be changed without affecting the numerical properties of the algorithm. Generally, the associative, commutative, and distributive properties of the operations can be exploited to reduce the iteration period bound.

As discussed in Chapter 5, the order of fixed-point two's-complement additions can be changed as long as the final sum is within the proper range. This property can be used to reorder the additions, as illustrated in Figure 6.39, so that the true minimum iteration period bound is obtained. Note that this bound can also be found using the *intrinsic coefficient word length* concept [28].

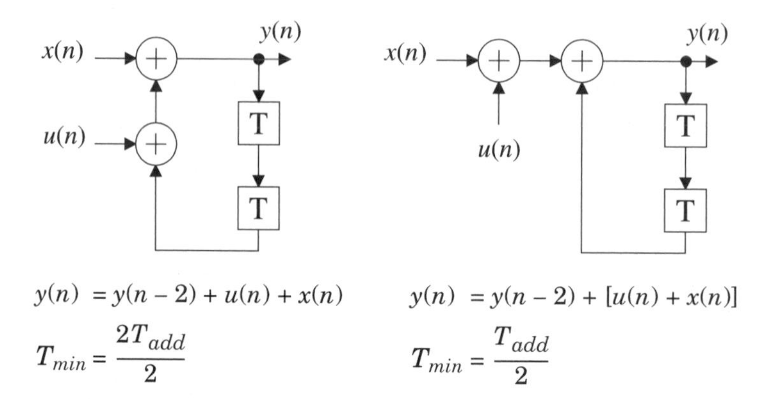

Figure 6.39 Numerical equivalence transformation (associatively) used to reduce the critical loop. a) Original signal-flow graph, b) Transformed graph with lower iteration period bound

Distributivity does not itself reduce the iteration period bound, but it may make other transformations feasible. An example of the use of distributivity is shown in Figure 6.40. In some cases, it can be advantageous to duplicate certain nodes.

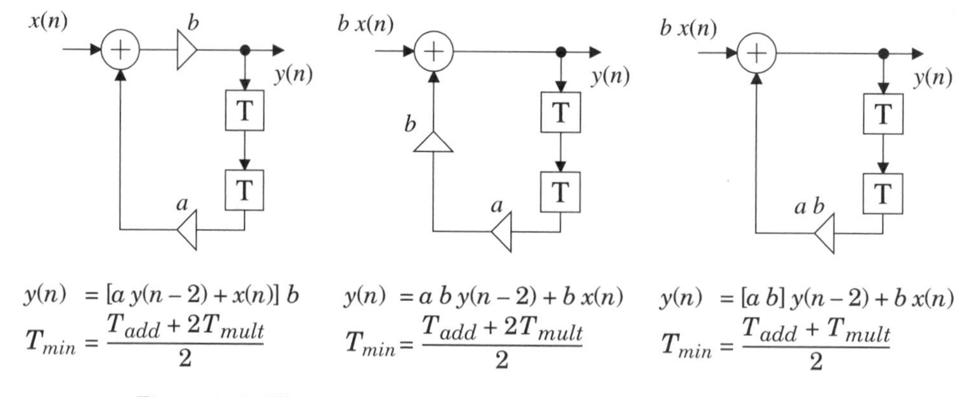

Figure 6.40 Illustration of the use of distributivity and associativity

Generally, the critical loops will contain only one addition and one multiplication, but some structures may have loops that interlock in such a way that rearranging the operations is hindered. We conclude that the iteration bound is uniquely defined only for a fully specified signal-flow graph.

EXAMPLE 6.8

Show that it is possible to reduce the length of the critical loop in the filter shown in Figure 6.30, and thereby reduce the required number of clock cycles per sample, by performing a numerical equivalence transformation (reordering of the additions) on the wave-flow graph.

In this case, the critical loop contains two such additions. To demonstrate the transformation we first move addition #3 across the delay elements as shown in Figure 6.41. The input to the multiplier becomes $x(n) - x(n-2)u(n-2)$. Now, by first computing $x(n) - x(n-2)$, as shown in Figure 6.42, only one addition is required in the critical loop. However, operation #3 shall add $u(n-2)$ and $x(n-2)$. We therefore subtract this sum from $x(n)$ and use it as an input to operation #4. The transformed wave-flow graph is shown in Figure 6.42.

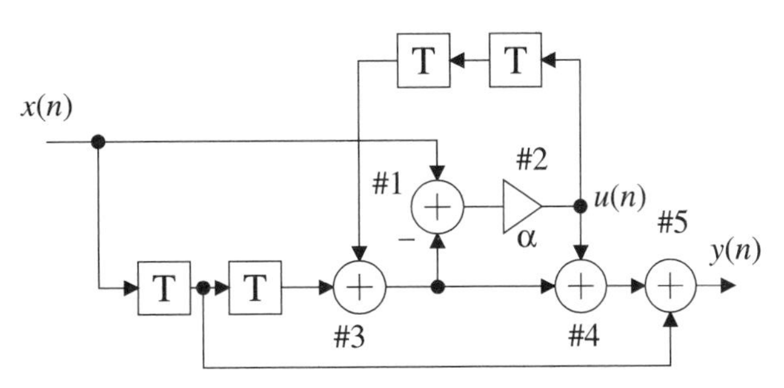

Figure 6.41 First transformation step

The minimum sample period is by this transformation reduced to 2.5 clock cycles. It should be stressed that the values computed in the modified wave-flow graph are the same as in the original. We gain an increase in maximum sample frequency of 17% at the expense of one addition. Generally, the required amount of physical memory is affected by this transformation.

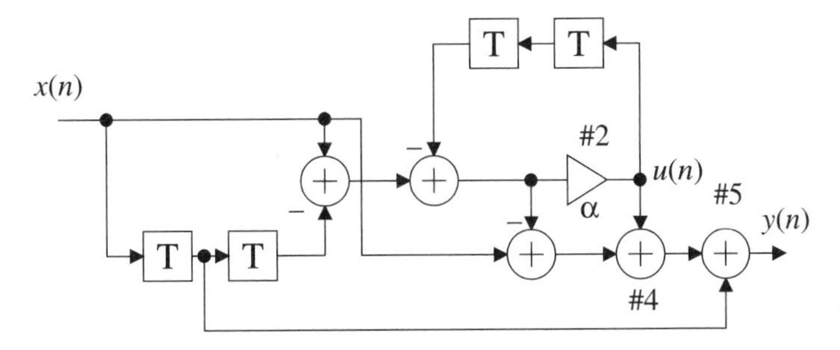

Figure 6.42 Transformed wave-flow graph with higher maximum speed

6.8 INTERLEAVING AND PIPELINING

The throughput of a recursive algorithm is always bounded by the critical loop(s), as discussed in section 6.6.4. However, input, output, and other nonrecursive branches and parts of loops in the signal-flow graph may have a critical path with a computation time longer than T_{min}. In that case, the critical path(s) will limit the sample rate. However, the minimum iteration period can be achieved by interleaving and/or pipelining.

Interleaving and pipelining are commonly used techniques to modify a partly sequential algorithm so that the data throughput of the new algorithm becomes higher. Both techniques can be simultaneously applied at different levels within a system—for example, at the algorithm level and at the hardware level.

6.8.1 Interleaving

Interleaving is a technique to increase the throughput of sequential algorithms. Consider the sequential algorithm shown in Figure 6.43. The execution time for the whole algorithm (i.e. the latency) is determined by the critical path, and the throughput is the inverse of the length of the critical path, T_{CP}. The throughput can be increased if the three independent processes P_1, P_2, and P_3 are mapped onto three processors as illustrated in Figure 6.44.

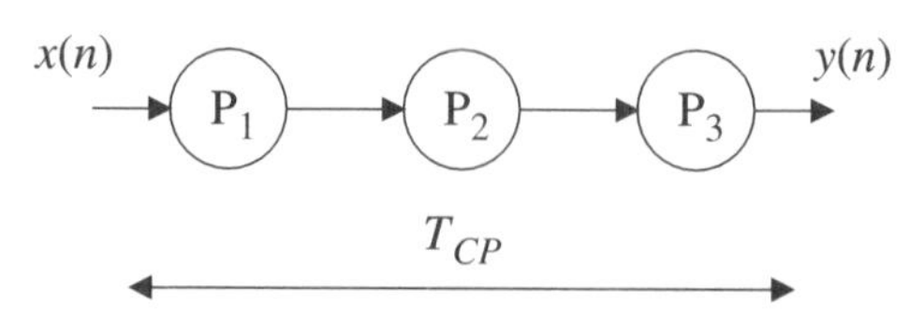

Figure 6.43 Sequential algorithm

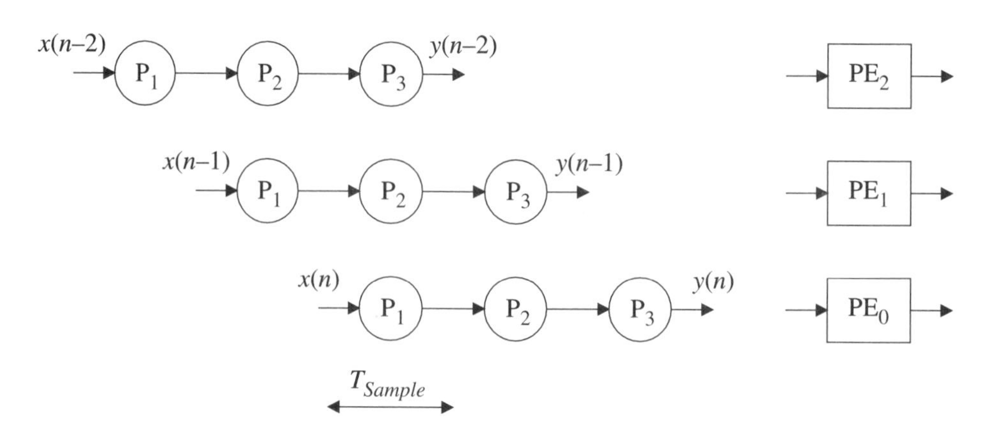

Figure 6.44 Interleaving of processes

Now, since three PEs are used to compute three consecutive outputs, the throughput is increased by a factor three, but the latency remains the same. Interleaving of the computations can be used to increase the throughput of a sequential algorithm by an arbitrarily large amount. However, it requires a corresponding increase in resources since operations corresponding to several different time indices must be computed concurrently.

Interleaving is an inefficient technique in terms of resource utilization if the three processes for some reason cannot be mapped onto a general-purpose processor that is capable of executing all the processes. One such reason is that the processes may require different types of operations so that a general-purpose processor becomes slow and inefficient. In such cases it may be better to use specialized PEs for the different types of operations.

6.8.2 Pipelining

Pipelining, which is another method of increasing the throughput of a sequential algorithm, can be used if the application permits the algorithmic delay to be increased. This is usually the case when the system (algorithm) is not inside a recursive loop, but there are many cases when the algorithmic delay must be kept within certain limits.

Pipelining of the three processes shown in Figure 6.43 is accomplished by propagating algorithmic delay elements into the original critical path so that a set of new and shorter critical path(s) is obtained as illustrated in Figure 6.45.

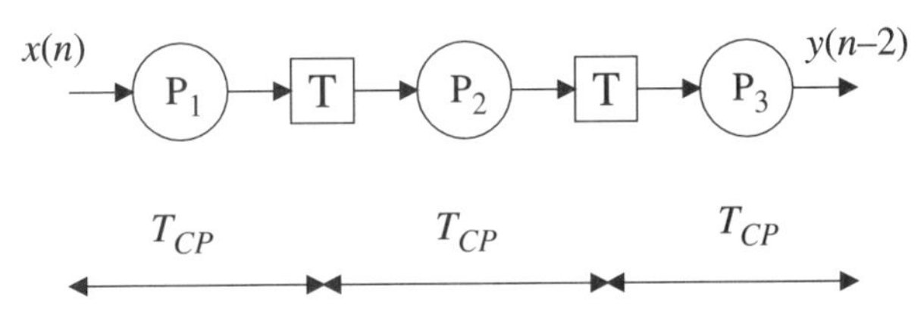

Figure 6.45 Pipelining of three processes

Ideally the critical path is broken into paths of equal length. Obviously, for each of the three processes a new computation can begin as soon as the results of the current computations are stored in the corresponding (output) delay elements. At steady state, three successive output value computations are performed concurrently. A pipeline with n stages allows n computations to be processed concurrently and attains a speedup of n over sequential processing. Throughput is the inverse of the longest critical path, i.e., the same throughput as with interleaving. Note that the maximum sample rate is, in principle, "infinite" for a nonrecursive structure (for example, an FIR filter), but this requires that the critical path be divided into infinitesimally small pieces. Thus, the group delay of the filter will be infinite.

The latency is changed only if the critical paths are of unequal length. In this case, equalizing delays must be inserted into the shorter paths so that the effective lengths become equal. The latency is thereby increased by the same amount as the sum of all inserted equalizing delays.

The main differences between interleaving and pipelining are that in the latter case the output is delayed by two sample periods and that the amount of resources is somewhat reduced. The number of PEs is still three, but each PE is now required to execute only one process per sample period. Hence a specialized, and thereby less expensive, PE may be used for each process. In pipelining, each PE operates on different parts of each process while in interleaving a PE operates on every nth sample.

Pipelining can be applied to sequential algorithms at any level of abstraction. Typically, both basic DSP algorithms and arithmetic operations are pipelined. At the digital circuit level pipelining corresponds to inserting latches between different circuit levels so that each level has the same transistor depth. Note that parallelism in a structure is a fundamental property that cannot be changed. By inserting delay elements into the critical path (for example, by pipelining), a new structure with a higher degree of parallelism is obtained.

EXAMPLE 6.9

Determine the precedence relations for the operations in the structure shown in Figure 6.46 [6, 7]. Introduce pipelining and compare the parallelism between the two structures.

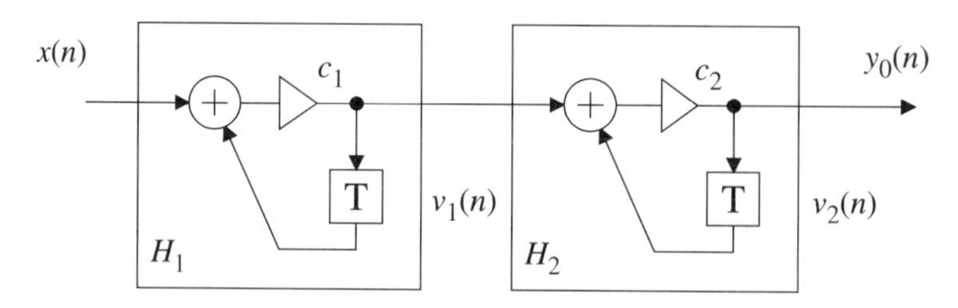

Figure 6.46 Initial signal-flow graph

The structure is completely sequential with respect to the operations according to the precedence graph in Figure 6.47. The length of the critical path, which is equal to the latency, is

$$T_{CP} = 2\,(T_{add} + T_{mult})$$

Note that all operations lie on the critical path. The transfer function is

$$H_0(z) = H_2(z)\,H_1(z), \quad |\,z\,| > R_+$$

where

$$H_1(z) = \frac{c_1}{1 - c_1 z^{-1}} \quad \text{and} \quad H_2(z) = \frac{c_2}{1 - c_2 z^{-1}}$$

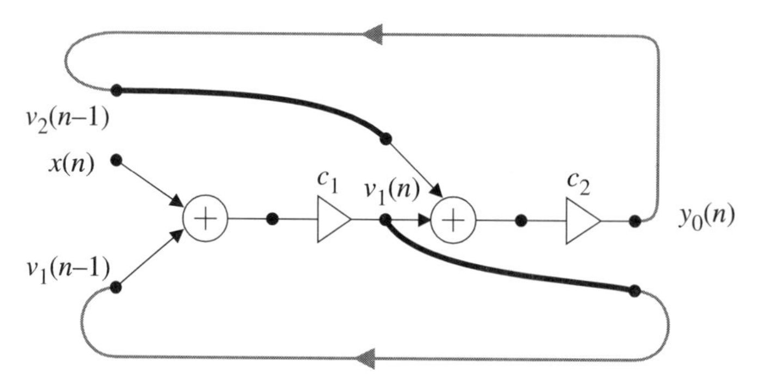

Figure 6.47 Precedence graph for the operations in the structure in Figure 6.46

The original filter, $H_0(z)$, is now cascaded with a single delay element. A cascaded delay element will increase the group delay by a constant factor, T. This extra delay is acceptable in many, but not all, applications. We get

$$H_P(z) = H_0(z)\,z^{-1} = z^{-1}\,H_2(z)\,H_1(z), \quad |\,z\,| > R_+$$

The order of the linear, shift-invariant filters and the delay element can be interchanged. Hence, we get

$$H_P(z) = H_2(z)\, z^{-1} H_1(z) = \frac{c_2}{1 - c_2 z^{-1}} = z^{-1}$$

The signal-flow graph for the new, pipelined structure is shown in Figure 6.48. This structure can be simplified as shown in Figure 6.49. The maximum sample rate is determined by the recursive loops. The minimum iteration period bound for both loops is

$$T_{min} = T_{add} + T_{mult}$$

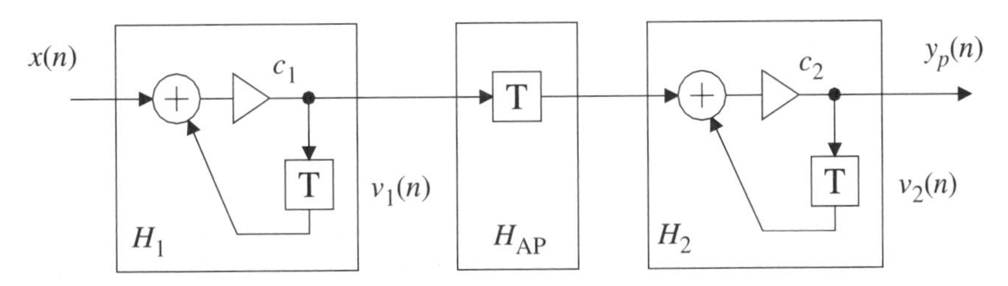

Figure 6.48 Pipelined structure

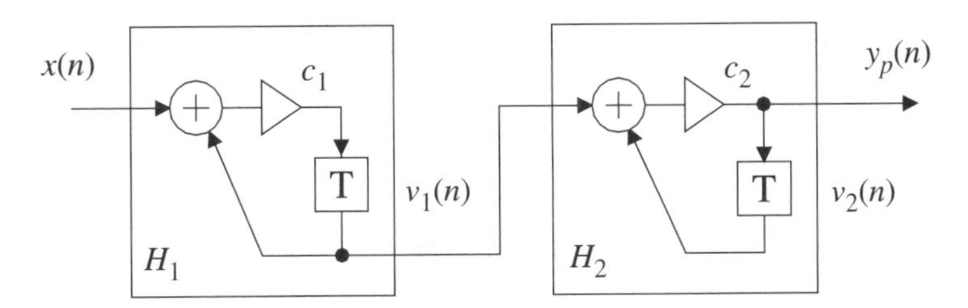

Figure 6.49 Simplified, pipelined structure

Figure 6.50 shows the precedence relationship for the operations in the pipelined structure. The precedence graph has two equally long critical paths of length T_{min}. The new algorithm has a higher degree of parallelism compared to the original structure, which is completely sequential. The throughput is twice that of the original algorithm. The algorithmic delay has increased, but the latency has remained the same. Notice that the delay elements have completely been absorbed by the arithmetic operations. However, still two registers are needed to store the variables v_1 and v_2.

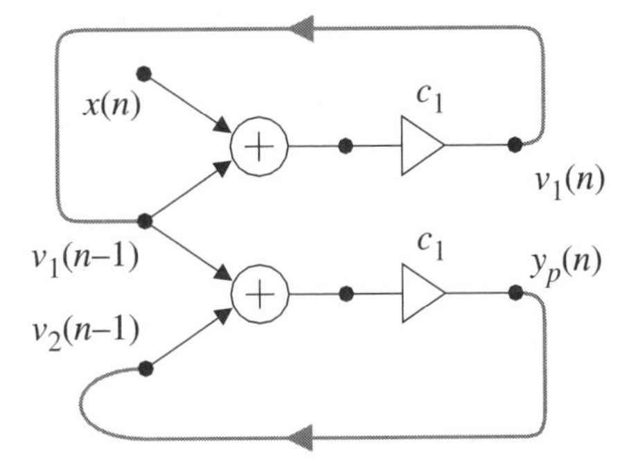

Figure 6.50 Precedence relationship for the operations in the pipelined structure

EXAMPLE 6.10

Determine how many effective pipelining stages can be introduced into the lattice wave digital filter shown in Figures 6.51. Assume that all adaptor coefficients have the same word length and that latches can be inserted inside the multipliers, i.e., they can be pipelined.

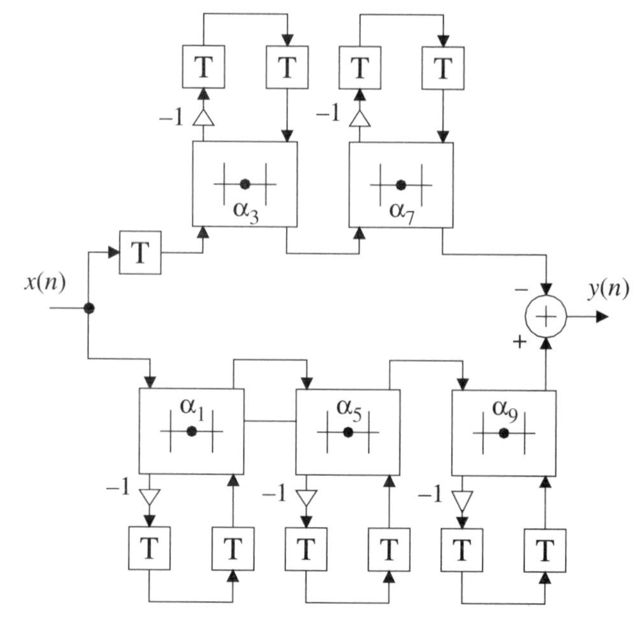

Figure 6.51 Bireciprocal wave digital filter

First, we note that the critical loop(s) determines the maximum sample rate and that it is not possible to further increase the sample rate by pipelining. We have the same recursive computation paths through the symmetric adaptors

$$T_{min} = \frac{T_{add} + 2T_{mult} + T_{add}}{2} = T_{add} + \frac{T_{mult}}{2}$$

Hence, the sequential paths from the input to the output shall be partitioned into pieces of length T_{min}. The lengths of the upper and lower paths, except for the last adder, are

$$T_{min} + 2\,(T_{add} + T_{mult} + T_{add}) = 5\,T_{min}$$

and

$$3\,(T_{add} + T_{mult} + T_{add}) = 6\,T_{min}$$

Six delay elements can be propagated into each of the filter branches. Thus, only seven stages of pipelining can be used, as illustrated in Figure 6.52, where only those operations belonging to the critical paths are shown.

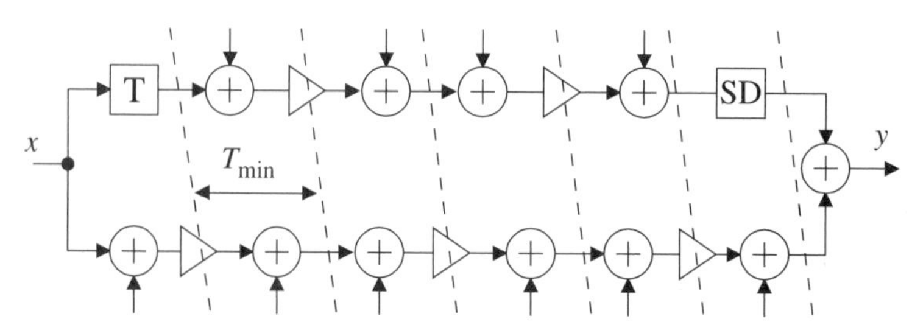

Figure 6.52 Pipelining levels obtained by propagating delays from the input side. Note that only the critical paths in the two branches are shown

To summarize, the two critical paths are broken into pieces of equal size. The last stage consists of only one addition. The wave-flow graph is partitioned into 10 paths, each performing an addition and a "half"-multiplication. Hence, the multipliers should have two pipeline stages. The last stage in the upper branch corresponds to a shimming delay of length T_{min}. The group delay of the filter has, because of the pipelining, increased by a constant factor $\Delta\tau_g = 6\ T_{min}$.

6.8.3 Functional and Structural Pipelines

Both algorithms and hardware structures can be pipelined. In classical pipelines, only one operation is executed by each pipeline stage (PE) separated by registers (delay elements). A PE is allocated to each stage of the pipeline. The pipeline stages do not share resources. This leads to a fixed and simple communication pattern between the processors. This case is often referred to as a *structural pipeline*. Figure 6.53 illustrates a pipelined algorithm with three stages and the corresponding structural pipeline where each process has been assigned to its own PE.

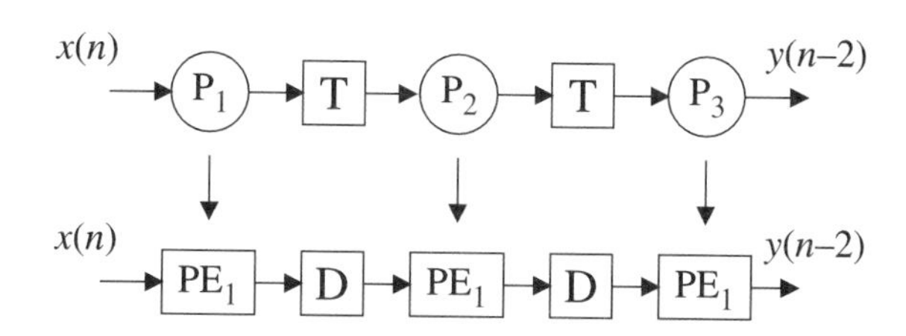

Figure 6.53 Structural pipeline

A more general case is to partition the pipeline stages into several smaller subtasks (operations) that are subsequently mapped onto a hardware structure. The operations can of course be executed by a structural pipeline, but a more efficient way to organize the hardware structure in terms of utilization of the resources is to allow the pipeline stages to share resources. That is, operations within a pipeline stage are not bounded to a particular PE, and operations that are not concurrent may share resources. This is referred to as a *functional pipeline*.

Figure 6.54 shows three processes that require different amounts of operations within the different stages. For simplicity we neglect precedence relations between operations inside the processes. Obviously, nine PEs (six adders and three multipliers) are required if a structural pipeline is used.

Only three PEs (two adders and one multiplier) are required if, instead, the operations can be scheduled as shown in Figure 6.54. Pipelining is used here to increase the parallelism in the algorithm so that the operations may be scheduled more freely.

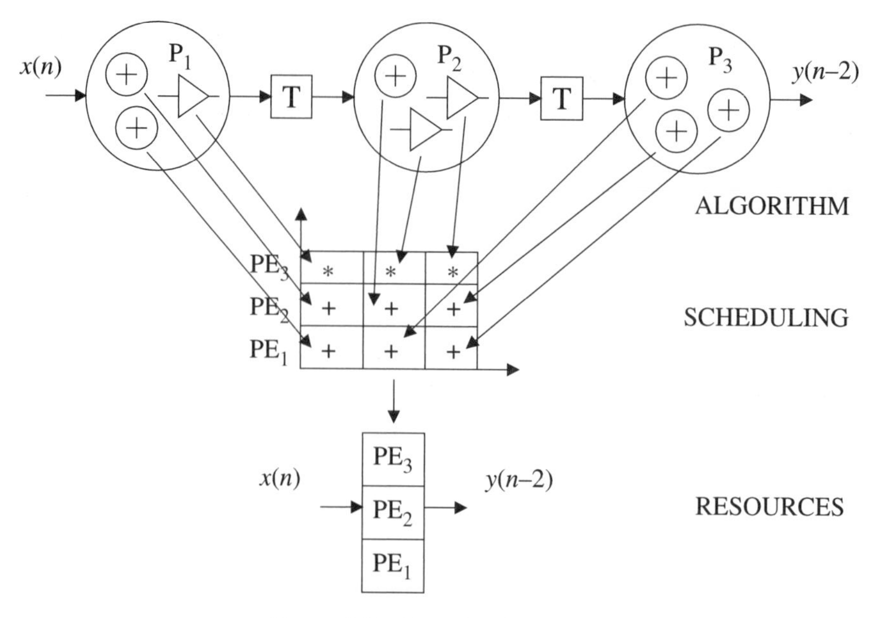

Figure 6.54 Functional pipeline

6.8.4 Pipeline Interleaving

Pipeline interleaving can be used, for example, when several signals are to be processed by the same algorithm [13, 14, 18, 19]. We demonstrate pipeline interleaving by means of an example. Consider a second-order section in direct form with the transfer function

$$H(z) = \frac{z^2}{z^2 - b_1 z - b_2}$$

Now, if each delay element is replaced by two delay elements, we obtain a structure with the transfer function

$$H_2(z) = \frac{z^4}{z^4 - b_1 z^2 - b_2}$$

Hence, the transfer function is modified, but now we also interleave the input sequence with zeros. The new input sequence becomes

$$X_2(z) = X(z^2)$$

The output of the pipelined filter becomes

$$Y_2(z) = H_2(z)\, X_2(z)$$

Hence, the proper output sequence is obtained by decimating the sequence $Y_2(z)$ by a factor two. The maximum sample rate has been increased by a factor two, but only every other output value is of interest. In reality we have not gained anything. However, two independent sequences can be filtered by the same filter as illustrated in Figure 6.55. This technique can be used to map algorithms onto systolic and wave front arrays as well as for programming heavily pipelined stan-

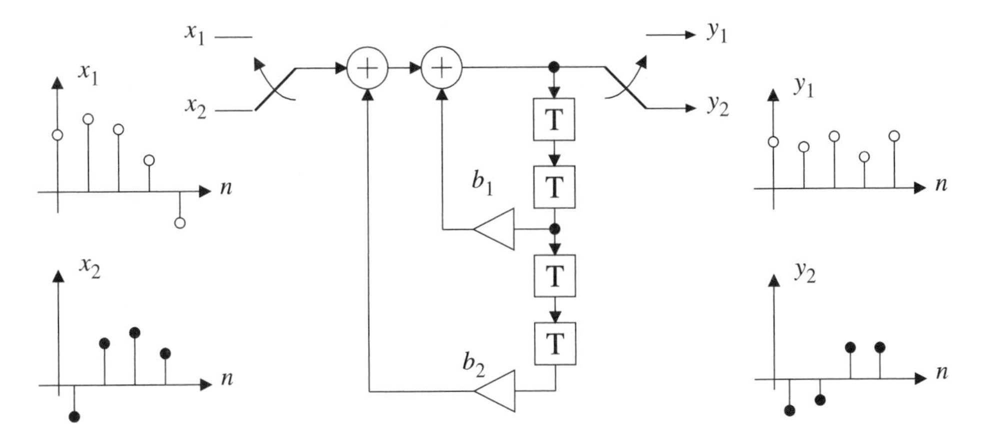

Figure 6.55 Pipeline interleaving of a second-order section

dard signal processors. Pipeline interleaving is useful in some applications, but, of course, the maximum sample rate bound is not improved.

6.9 ALGORITHM TRANSFORMATIONS

In this section we will discuss various techniques used to derive new faster filtering algorithms from a given algorithm.

6.9.1 Block Processing

Parallelism in the computations can be increased beyond that of the basic algorithm by partitioning the input and output sequences into blocks of length L [25, 27]. Burrus has presented a block processing approach for recursive digital filters based on the FFT [3]. However, this approach is efficient only for filter orders of 10 or more and for block lengths of 100 or more.

Another approach is the *block processing* or *state-decimation* technique [27, 29, 30]. We assume for the sake of simplicity that a digital filter is described by the state-space equations

$$\begin{cases} \mathbb{x}(n+1) = \mathbb{A}\,\mathbb{x}(n) + \mathbb{B}\,\mathbb{u}(n) \\ \mathbb{y}(n+1) = \mathbb{C}\,\mathbb{x}(n) + \mathbb{D}\,\mathbb{u}(n) \end{cases}$$

where $\mathbb{x}(n)$ is the state vector and $\mathbb{A}$, $\mathbb{B}$, $\mathbb{C}$, and $\mathbb{D}$ are matrices and vectors of appropriate dimensions. Iterating Equation (6.2) one step yields

$$\begin{cases} \mathbb{x}(n+2) = \mathbb{A}\,\mathbb{x}(n+1) + \mathbb{B}\,\mathbb{u}(n+1) = \mathbb{A}^2\,\mathbb{x}(n) + \mathbb{A}\,\mathbb{B}\,\mathbb{u}(n+1) + \mathbb{B}\,\mathbb{u}(n+1) \\ \mathbb{y}(n+1) = \mathbb{C}\,\mathbb{x}(n+1) + \mathbb{D}\,\mathbb{u}(n+1) = \mathbb{C}\,\mathbb{A}\,\mathbb{x}(n) + \mathbb{C}\,\mathbb{B}\,\mathbb{u}(n) + \mathbb{D}\,\mathbb{u}(n+1) \end{cases}$$

and by iterating L steps ahead we get

$$\begin{cases} \mathbb{x}(n+L+1) = \mathbb{A}'\mathbb{x}(n) + \mathbb{B}'\mathbb{u}'(n) \\ \mathbb{y}(n+L) = \mathbb{C}'\mathbb{x}(n) + \mathbb{D}'\mathbb{u}'(n) \end{cases}$$

where

$$\mathbb{A}' = \mathbb{A}^L$$

$$\mathbb{B}' = \sum_{m=0}^{L-1} \mathbb{A}^{L-m-1}\mathbb{B} = [\mathbb{A}^{L-1}\mathbb{B}, \mathbb{A}^{L-2}\mathbb{B}, \ldots, \mathbb{A}\mathbb{B}, \mathbb{B}]$$

$$\mathbb{C}' = [\mathbb{C}, \mathbb{C}\mathbb{A}, \mathbb{C}\mathbb{A}^2, \ldots, \mathbb{C}\mathbb{A}^{L-1}]^T$$

$$\mathbb{D} = \begin{bmatrix} \mathbb{D}, 0, 0, 0, \ldots, 0 \\ \mathbb{C}\mathbb{B}, \mathbb{D}, 0, 0, \ldots, 0 \\ \mathbb{C}\mathbb{A}\mathbb{B}, \mathbb{C}\mathbb{B}, \mathbb{D}, 0, \ldots, 0 \\ \mathbb{C}\mathbb{A}^{L-2}\mathbb{B}, \mathbb{C}\mathbb{A}^{L-3}\mathbb{B}, \mathbb{C}\mathbb{A}^{L-4}\mathbb{B}, \mathbb{C}\mathbb{A}^{L-5}\mathbb{B}, \ldots, \mathbb{C}\mathbb{B}, \mathbb{D} \end{bmatrix}$$

$$\mathbb{u}'(n) = [\mathbb{u}(n), \mathbb{u}(n+1), \mathbb{u}(n+2), \mathbb{u}(n+3), \ldots, \mathbb{u}(n+L)]^T$$

and

$$\mathbb{y}'(n) = [\mathbb{y}(n), \mathbb{y}(n+1), \mathbb{y}(n+2), \mathbb{y}(n+3), \ldots, \mathbb{y}(n+L)]^T$$

We assume that these matrices can be precomputed and neglect the fact that the word length of the matrix coefficients increases. Hence, Equation (6.3) can be iterated with an iteration period that is the same as the original set of state-space equations, but L output values are computed at each iteration. The effective iteration bound has therefore been decreased. We define a high-speed digital filter structure as a structure whose throughput is not bounded by the multiplication cycle time.

Theorem 6.5

The *minimum sample period* for block processing of an Nth-order state-space algorithm is

$$T_{min} = \frac{T_{mult} + \lceil \log_2(N+1) \rceil T_{add}}{L} \tag{6.4}$$

where $T_{mult} + \lceil \log_2(N+1) \rceil T_{add}$ is the total delay (latency) due to the arithmetic operations and L is the block size.

The number of operations per sample is

$$N_s = \frac{N^2 + \frac{L^2}{2} + 2NL + \frac{L}{2}}{L} + \frac{N(N-1) + \frac{L^2}{2} + 2NL - \frac{L}{2}}{L} \tag{6.5}$$

where the first term is the number multiplications, the second term is the number of additions, and N is the filter order.

The operation rate is minimized if $L = \sqrt{N(2N-1)} \approx \sqrt{2}\ N$, which corresponds to $N_s \approx 2(2+\sqrt{2})N$. Hence, by using a large number of PEs, many times larger than the filter order, block structures can produce several output values per multiply–add cycle.

An advantage of block processing is that the round-off noise is reduced and that overflow oscillations can be avoided in some cases [2]. It is well known that conventional structures with low round-off noise have low coefficient sensitivity; see Equations (5.20) and (5.21). However, block structures implicitly depend on pole-zero cancellations, and the cancellations may not occur in the presence of coefficient errors. Thus, block structures may have exceedingly high coefficient sensitivity [1]. Further, coefficient errors may change the nominal time-invariant filter into a time-variant one. Notice also that the numerical properties of the original algorithm are not retained in block processing and that most algorithms do not have a state-space representation numerically equivalent to the original algorithm.

Block processing is particularly suitable for implementation on processors that support vector multiplication. Block processing of FIR and certain IIR filters (for example, lattice wave digital filters with branches realized using Richards' structures) can be implemented efficiently. The use of a state-space representation is particularly suited to digital filters having a lower triangular or quasi-triangular state matrix. Such state matrices can be obtained either by an orthogonal similarity transformation of any state matrix or, directly, by using a numerically equivalent state-space representation of a wave digital circulator structure. Block processing is also suitable for decimation and interpolation of the sample rate.

6.9.2 Clustered Look-Ahead Pipelining

In this section we will discuss the clustered look-ahead pipelining technique [16, 18, 26]. For simplicity we will consider only the recursive part of an algorithm, since the nonrecursive parts do not limit the sample rate. Further, we assume that the algorithm has only one such part described by

$$y(n) = \sum_{k=1}^{N} b_k y(n-k) + x(n)$$

The sample rate is limited by the time for one multiplication and one addition. The corresponding transfer function is

$$H(z) = \frac{1}{1 - \sum_{k=1}^{N} b_k z^{-k}}$$

We modify the transfer function by multiplying both the denominator and numerator by the same polynomial:

$$H(z) = \frac{\sum_{k=0}^{N} c_k z^{-k}}{\left(1 - \sum_{k=1}^{N} b_k z^{-k}\right)\left(\sum_{k=0}^{N} c_k z^{-k}\right)} = \frac{\sum_{k=0}^{N} c_k z^{-k}}{1 - \sum_{k=1}^{N+M} d_k z^{-k}} \tag{6.6}$$

Coefficients c_k are selected such that coefficients d_k are zero for $k = 1, \ldots, M$. The maximum sample rate has thereby been increased by a factor M. The corresponding difference equation is

$$y(n) = \sum_{k=M+1}^{N+M} d_k y(n-k) + \sum_{k=0}^{M} c_k x(n-k) \tag{6.7}$$

The output is computed from the N past values that are clustered in time, i.e., $y(n-M-1)$, $y(n-M-2)$, $y(n-M-3), \ldots, y(n-M-N)$. We illustrate the clustered look-ahead pipelining with two examples.

EXAMPLE 6.11

Use clustered look-ahead pipelining to increase the maximum sample rate of a first-order filter by a factor two.

The original transfer function is

$$H(z) = \frac{a_0 z + a_1}{z - b_1}$$

We form the new transfer function

$$H(z) = \frac{(a_0 + a_1 z^{-1})(1 + c_1 z^{-1})}{(1 - b_1 z^{-1})(1 + c_1 z^{-1})} = \frac{(a_0 + a_1 z^{-1})(1 + c_1 z^{-1})}{1 - (b_1 z^{-1}) z^{-1} - b_1 c_1 z^{-2}}$$

and select $c_1 = b_1$, which yields

$$H(z) = \frac{(a_0 + a_1 z^{-1})(1 + b_1 z^{-1})}{1 - b_1^2 z^{-1}} = \frac{(a_0 z + a_1)(z + b_1)}{z^2 - b_1^2} = \frac{(a_0 z + a_1)(z + b_1)}{(z + b_1)(z - b_1)}$$

as expected. The pole-zero configuration is shown in Figure 6.56 where $a_0 = a_1 = 1$ and $b_1 = 0.75$. The corresponding difference equation is

$$y(n) = b_1^2\, y(n-2) + a_0\, x(n) + (a_1 + b_1)\, x(n-1) + a_1\, b_1\, x(n-2)$$

The new coefficient b_1^2 can be computed off-line. If the multiplication by b_1^2 can be performed in the same time as the original multiplication by b_1, the maximum sample rate has been increased by a factor two. Note that the word length of coefficient b_1^2 is twice as long as that of b_1. The round-off properties and coefficient sensitivities of the original algorithm are not retained.

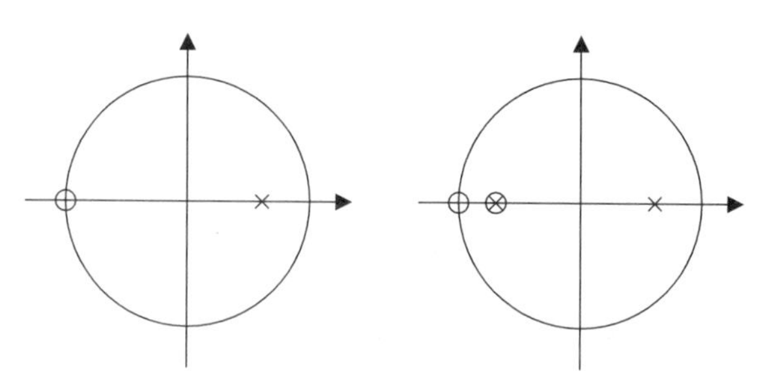

Figure 6.56 Pole-zero configuration obtained by clustered look-ahead pipelining

EXAMPLE 6.12

Use clustered look-ahead pipelining to increase the maximum sample rate of the second-order filter by a factor two.

The original transfer function is

$$H(z) = \frac{z^2}{z^2 - b_1 z - b_2}$$

We form the new transfer function

$$H(z) = \frac{1}{(1 - b_1 z^{-1} - b_2 z^{-2})} \frac{(1 + c_1 z^{-1})}{(1 + c_1 z^{-1})}$$

$$= \frac{1 + c_1 z^{-1}}{1 - (b_1 - c_1) z^{-1} - (b_1 c_1 + b_2) z^{-2} - b_2 c_1 z^{-3}}$$

An increase in the maximum sample rate by a factor of two is obtained for $c_1 = b_1$. Figure 6.57 illustrates the pole–zero configuration for $b_1 = 1.5$ and $b_2 = 0.625$. Unfortunately, the resulting filter will only be stable if $|b_1| < 1$. Hence, this approach is only useful for a limited class of transfer functions. In order to arrive at a stable filter we form the transfer function

$$H(z) = \frac{1}{(1 - b_1 z^{-1} - b_2 z^{-2})} \frac{(1 + c_1 z^{-1})(1 + c_2 z^{-1})}{(1 + c_1 z^{-1})(1 + c_2 z^{-1})}$$

$$= \frac{(1 + c_1 z^{-1})(1 + c_2 z^{-1})}{1 - e_1 z^{-1} - e_2 z^{-2} - e_3 z^{-3} - e_4 z^{-4}}$$

where

$$e_1 = b_1 \ c_1 \ c_2 \qquad e_2 = b_1(c_1 + c_2) + b_2 + c_1 c_2$$

$$e_3 = b_2(c_1 + c_2) + b_1 c_1 c_2 \qquad e_4 = b_2 c_1 c_2$$

The desired increase in the maximum sample rate is obtained for $e_1 = 0$, i.e., $b_1 = c_1 + c_2$. We get a stable filter only if $|c_1| < 1$ and $|c_2| < 1$. If we select $c_1 = c_2$, the poles are placed as far as possible from the unit circle. In some cases it is possible to force c_1 and c_2 simultaneously to zero, but the resulting filter may not be stable. An alternative strategy that often leads to stable filters is to successively remove the coefficients e_i one at a time.

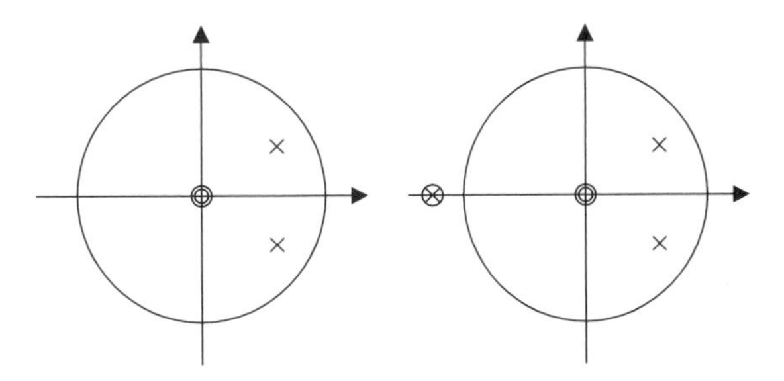

Figure 6.57 Clustered look-ahead pipelining leading to an unstable filter

6.9.3 Scattered Look-Ahead Pipelining

In the scattered look-ahead pipelining approach, the transfer function is modified such that the pipelined filter has poles with equal spacing around the origin [16, 18, 19, 26]. The effect of the added poles is canceled by the corresponding zeros. The denominator will contain only powers of z^M where M is the number of pipeline stages. The filter output is therefore computed from past values that are scattered in time, i.e., $y(n-M)$, $y(n-2M)$, $y(n-3M)$,... , $y(n-NM)$ where N is the order of the original filter. In fact, this technique is the same as the state-decimation technique discussed in section 6.9.1. This approach always leads to stable filters. We illustrate the scattered look-ahead pipelining technique by a few examples.

EXAMPLE 6.13

Apply scattered look-ahead pipelining to a first-order filter. The pipeline shall have four stages.

The original filter is described by the transfer function

$$H(z) = \frac{z}{z-b}$$

which corresponds to the difference equation

$$y(n) = b\,y(n-1) + x(n)$$

We add $M-1$ poles and zeros at $z = b\,e^{j2\pi k/M}$ for $k = 1, 2,..., M-1$. The transfer function becomes

$$H(z) = \left(\frac{z}{z-b}\right)\frac{\prod_{k=1}^{M-1}(z-be^{j2\pi k/M})}{\prod_{k=1}^{M-1}(z-be^{j2\pi k/M})} = \frac{z\prod_{k=1}^{M-1}(z-be^{j2\pi k/M})}{\prod_{k=0}^{M-1}(z-be^{j2\pi k/M})}$$

$$= \frac{z\prod_{k=1}^{M-1}(z-be^{j2\pi k/M})}{z^M-b^M} = \frac{z^4+bz^3+b^2z^2+b^3z}{z^4-b^4}$$

where $M = 4$. The resulting pole–zero configuration is shown on the right in Figure 6.58. The corresponding difference equation is

$$y(n) = b^4\,y(n-4) + x(n) + b\;x(n-1) + b^2\,x(n-2) + b^3\,x(n-3)$$

The numerator can always be factored into a product of polynomials representing symmetrically placed zeros. We get

$$H(z) = \frac{z^4+bz^3+b^2z^2+b^3z}{z^4-b^4} = \frac{z(z+b)(z^2+b^2)}{z^4-b^4}$$

This transfer function can be realized by a recursive structure followed by a number of cascaded FIR structures as shown in Figure 6.59. Note that the FIR filters do not limit the maximum sampling frequency. The pipelined structures require

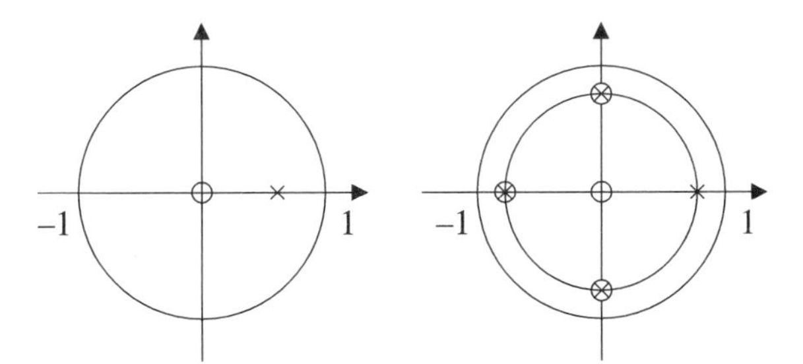

Figure 6.58 Pole–zero configurations for the original and the pipelined filter obtained by scattered look-ahead pipelining

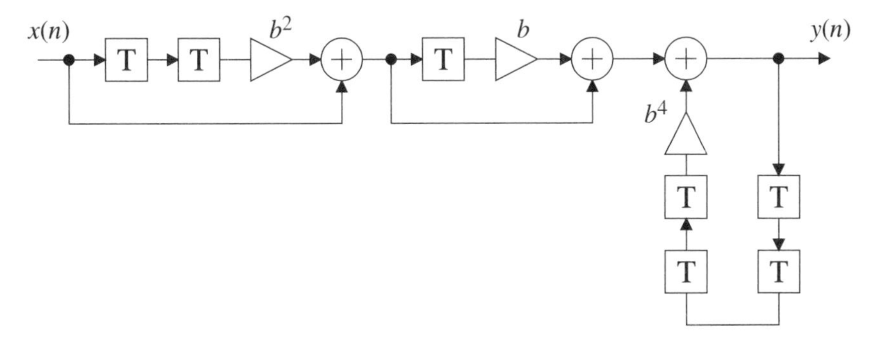

Figure 6.59 First-order scattered look-ahead pipelined structure

data and coefficient word lengths a few bits longer than the original filter to maintain the same round-off noise and appropriate pole–zero cancellation. Look-ahead pipelining can also be applied to a restricted set of nonlinear recursions [17, 20-22].

6.9.4 Synthesis of Fast Filter Structures

An alternative, and possibly more efficient, approach is to use a computer to directly synthesize a transfer function of the desired form. The poles are placed symmetrically around the origin in the z-plane such that the recursive loops have the desired number of delay elements. In the method just discussed, the effect on the frequency response from some of the poles was canceled by placing zeros at the same locations. Now the zeros of the transfer function can be used more efficiently if they are placed on the unit circle instead. Their positions can be found by using a computer-based optimization procedure. Digital filters with this type of transfer function will have both high speed and low arithmetic work load.

6.10 INTERPOLATOR, CONT.

In order to schedule the operations and determine the necessary amount of computational resources for the interpolator, we first need to determine the precedence

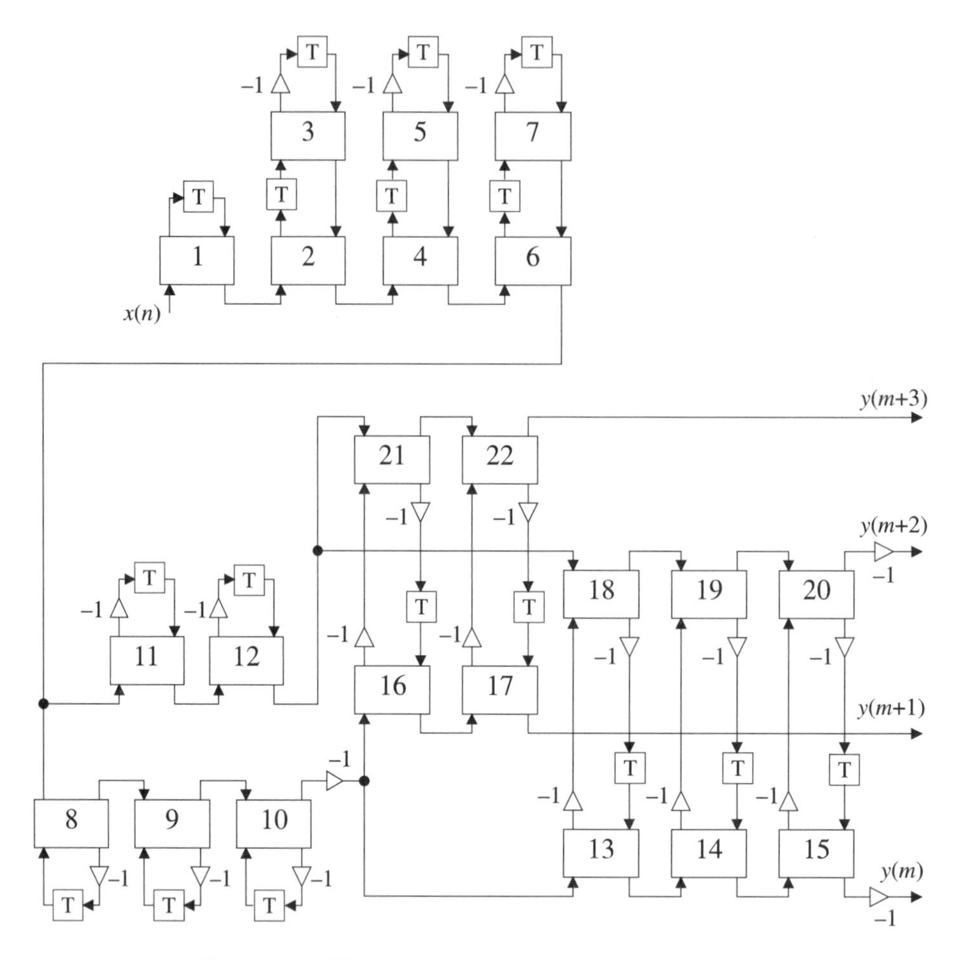

Figure 6.60 Wave-flow graph for the interpolator filter

relations for the operations. We choose to use the adaptors as atomic operations, i.e., an operation with three inputs (two data and one coefficient) and two output values. First, we draw the wave-flow graph without switches, as shown in Figure 6.60, to demonstrate the algorithm more clearly.

Adaptors that have valid inputs can be executed immediately. In the first time slot, only adaptors 1, 3, 5, and 7 have valid inputs. Only adaptor 2 can be executed in the second time slot, and so on. Altogether, 11 time slots are required to complete the whole filter. The number of concurrent operations ranges between 1 and 4. The average number is only two adaptors per time slot.

The precedence relations for the adaptors are shown in Figure 6.61. The recursive loops are indicated explicitly to show more clearly the inherent speed limitations in the algorithm. In order to determine the maximum sample rate for the interpolator, we assume that a multiplication takes 21 clock cycles and an addition takes one clock cycle. The computation paths through an adaptor involve two additions and one multiplication, i.e., (21+2) clock cycles. The usable clock fre-

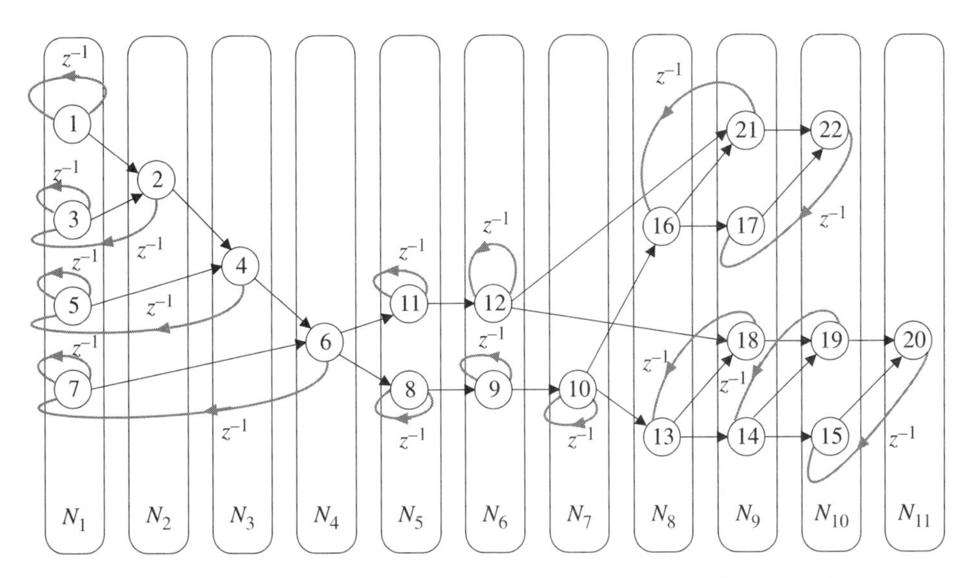

Figure 6.61 Precedence graph for the adaptor operations in the interpolator

quency for the target technology is estimated to be at least 220 MHz. The critical path for the interpolator is through 11 adaptors, i.e.,

$$T_{CP} = \frac{11(21+2)}{220 \cdot 10^6} = 1.15 \text{ µs} \quad \text{or} \quad 869.6 \text{ kHz}$$

The throughput will be too low, since the required input sampling frequency is 1.6 MHz. Four output values are generated for each input sample. Hence, the output sampling frequency is only 3.478 MHz. We will therefore use pipelining of the interpolator to reach the required output sample frequency of 6.4 MHz.

From Figure 6.60 it is evident that the minimum input sampling period is determined by a loop through two adaptors. We get

$$T_{min} = \frac{2(21+2)}{220 \cdot 10^6} = 0.209 \text{ µs} \quad \text{or} \quad 4.785 \text{ MHz}$$

It is therefore possible to achieve the required sample frequency by breaking the critical path into smaller pieces using pipelining.

References

[1] Arun K.S. and Wagner D.R.: High-Speed Digital Filtering: Structures and Finite Wordlength Effects, *J. of VLSI Signal Processing,* Vol. 4, No. 4, pp. 355–370, Nov. 1992.

[2] Barnes C.W. and Shinnaka S.: Finite Word Effects in Block-State Realizations of Fixed-Point Digital Filters, *IEEE Trans. on Circuits and Systems,* Vol. CAS-27, No. 5, pp. 345–350, May 1980.

[3] Burrus C.S.: Block Realization of Digital Filters, *IEEE Trans. on Audio Electroacoust.,* Vol. AU-20, pp. 230–235, Oct. 1972.

[4] Chung J. and Parhi K.: Pipelined Lattice and Wave Digital Recursive Filters, Kluwer Academic Pub., Boston 1996.

[5] Claesen L., De Man H.J., and Vandewalle J.: Delay Management Algorithms for Digital Filter Implementations, *Proc. Sixth European Conf. on Circuit Theory and Design,* ECCTD-83, pp. 479–482, Stuttgart, FRG., Sept. 1983.

[6] Crochiere R.E.: *Digital Network Theory and Its Application to the Analysis and Design of Digital Filters,* Ph.D. Diss., MIT, Cambridge, Mass., May 1974.

[7] Crochiere R.E. and Oppenheim A.V.: Analysis of Linear Digital Networks, *Proc. IEEE,* Vol. 63, No. 4, pp. 581–595, April 1975.

[8] Danielsson P.-E. and Levialdi S.: Computer Architectures for Pictorial Information Systems, *Computer Magazine,* pp. 53–67, Nov. 1981.

[9] Fettweis A.: Realizability of Digital Filter Networks, *Archiv. Elektr. Übertragungstechnik,* Vol. 30, No. 2, pp. 90–96, Feb. 1976.

[10] Hatamian M. and Parhi K.K.: An 85-MHz Fourth-Order Programmable IIR Digital Filter Chip, *J. Solid-State Circuits,* Vol. SC-27, No. 2, pp. 175–183, Feb. 1992.

[11] Kleine U. and Böhner M.: A High-Speed Wave Digital Filter Using Carry-Save Arithmetic, *Proc. ESSCIRC87,* Bad-Soden, pp. 43–46, 1987.

[12] Langefors B.: *Theoretical Analysis of Information Systems,* Auerbach Pub., Philadelphia, PA, 1973.

[13] Lee E.A. and Messerschmitt D.G.: Pipeline Interleaved Programmable DSPs: Architecture, *IEEE Trans. Acoust., Speech, and Signal Processing,* Vol. ASSP-35, No. 9, pp. 1320–1333, Sept. 1987.

[14] Lee E.A. and Messerschmitt D.G.: Pipeline Interleaved Programmable DSPs: Synchronous Data Flow Programming, *IEEE Trans. Acoust., Speech, and Signal Processing,* Vol. ASSP-35, No. 9, pp. 1334–1345, Sept. 1987.

[15] Leiserson C.E., Rose F.M., and Saxe J.B.: Optimizing Synchronous Circuitry by Retiming, *3rd Caltech Conf. on VLSI,* pp. 87–116, Pasadena, March 1983.

[16] Loomis H.H. and Sinha B.: High Speed Recursive Digital Filter Realizations, *Circuits, Systems, and Signal Processing,* Vol. CSSP-3, No. 3, pp. 267–294, 1984.

[17] Parhi K.K.: Look-Ahead in Dynamic Programming and Quantizer Loops, *ISCAS-89,* pp. 1382–1387, Portland, OR, May 1989.

[18] Parhi K.K. and Messerschmitt D.G.: Pipeline Interleaving and Parallelism in Recursive Digital FiltersPart I: Pipelining Using Scattered Look-Ahead and Decomposition, *IEEE Trans. Acoust., Speech, and Signal Processing,* Vol. ASSP-37, No. 7, pp. 1099–1117, July 1989.

[19] Parhi K.K. and Messerschmitt D.G.: Pipeline Interleaving and Parallelism in Recursive Digital FiltersPart II: Pipelining Incremental Block Filtering, *IEEE Trans. Acoust., Speech, and Signal Processing,* Vol. ASSP-37, No. 7, pp. 1118–1134, July 1989.

[20] Parhi K.K.: Algorithm Transformation Techniques for Concurrent Processors, *Proc. IEEE,* Vol. 77, No. 12, pp. 1879–1895, Dec. 1989.

[21] Parhi K.K.: Pipelining in Dynamic Programming Architectures, *IEEE Trans. Acoust., Speech, and Signal Processing,* Vol. ASSP-39, No. 4, pp. 1442–1450, June 1991.

[22] Parhi K.K.: Pipelining in Algorithms with Quantizer Loops, *IEEE Trans. on Circuits and Systems,* CAS-38, No. 7, pp. 745–754, July 1991.

[23] Reiter R.: Scheduling Parallel Computations, *J. Ass. Comp. Machn.*, Vol. 15, No. 4, pp. 590–599, Oct. 1968.

[24] Renfors M. and Neuvo Y.: The Maximal Sampling Rate of Digital Filters Under Hardware Speed Constraints, *IEEE Trans. on Circuits and Systems,* CAS-28, No. 3, pp. 196–202, March 1981.

[25] Roberts R.A. and Mullis C.T.: *Digital Signal Processing,* Addison-Wesley., Reading, MA, 1987.

[26] Soderstrand M.A., Loomis H.H., and Gnanasekaran R.: Pipelining Techniques for IIR Digital Filters, *ISCAS-90,* pp. 121–124, 1990.

[27] Voelcker H.B. and Hartquist E.E.: Digital Filtering via Block Recursion, IEEE *Trans. on Audio Electroacoust.*, Vol. AU-18, pp. 169–176, 1970.

[28] Wanhammar L.: Implementation of Wave Digital Filters Using Vector-Multipliers, *Proc. First European Signal Processing Conf., EUSIPCO-80,* pp. 21–26, Lausanne, Switzerland, Sept. 1980.

[29] Wanhammar L.: *An Approach to LSI Implementation of Wave Digital Filters,* Linköping Studies in Science and Technology, Diss. No. 62, Sweden, April 1981.

[30] Zeman J. and Lindgren A.G.: Fast Digital Filters with Low Roundoff Noise, *IEEE Trans. on Circuits and Systems,* Vol. CAS-28, No. 7, pp. 716–723, July 1981.

PROBLEMS

6.1 (a) Determine the precedence form of the structure shown in Figure P6.1.

(b) Write the system of difference equation in computable order.

(c) Write a program, using a high-level language of your choice, to realize the filter structure.

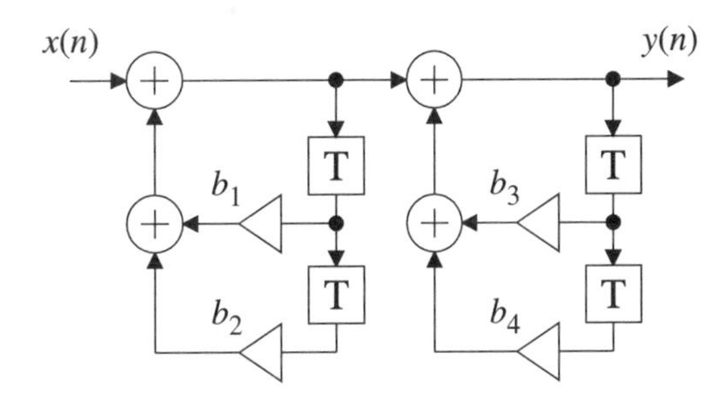

Figure P6.1 Cascade Form

6.2 Find the system of difference equations in computable order for the filter in Example 4.11.

6.3 Find the system of difference equation in computable order for the filter in Example 4.12.

6.4 A running-sum FIR filter has the impulse response

$$h(n) = \begin{cases} 1 & \text{for } 0 \le n \le M \\ 0 & \text{otherwise} \end{cases}$$

Derive the corresponding difference equation and write a program, using a high-level language of your choice, to realize the filter.

6.5 Derive the system of difference equations in computational order for the filter in Example 4.16
   (a) From the filter in Example 4.11, assuming that only every other output sample is retained.
   (b) Directly from Figure 4.64

6.6 (a) Determine the precedence form for the parallel form structure shown in Figure P6.6.
   (b) Write the system of difference equations in computable order.
   (c) Write a program, using a high-level language, to realize the filter structure.

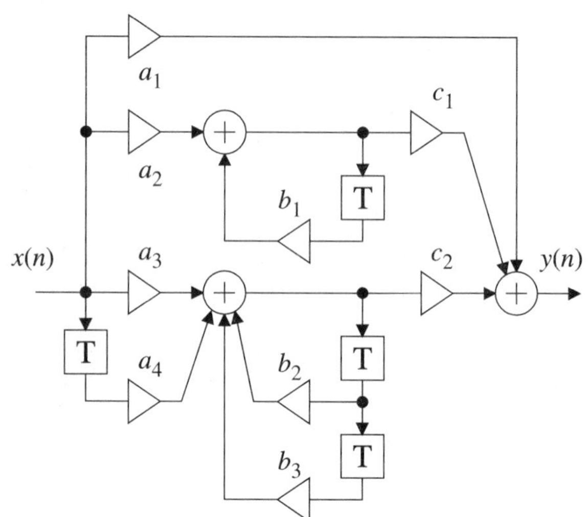

Figure P6.6 Parallel form

6.7 Determine the system of difference equations in computable order for the lattice wave digital filter shown in Figure P6.7.

6.8 Describe the following concepts, by means of an example:
   (a) AON
   (b) AOA
   (c) Precedence graph
   (d) Critical path
   (e) Computational graph
   (f) Sequentially computable
   (g) Pipelining

6.9 Derive the last two stages of the wave-flow graph shown in Figure 6.53 from the wave-flow graph shown in Figure 4.64.

6.10 Determine the set of difference equations, in computational order, for the filter in Problem 3.21.

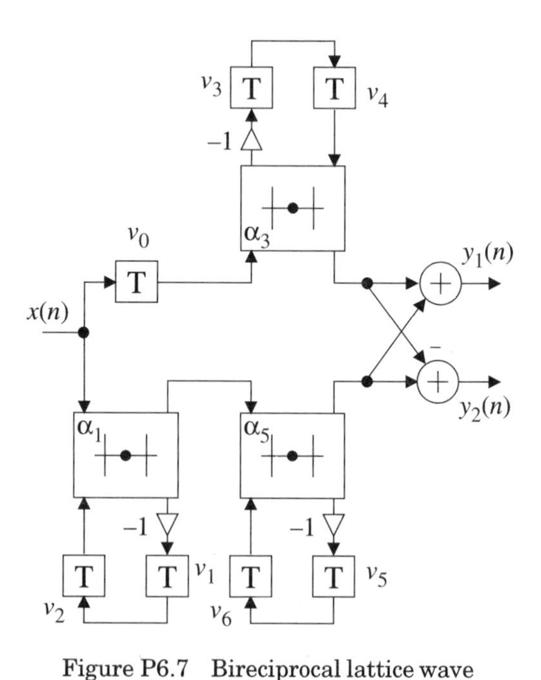

Figure P6.7 Bireciprocal lattice wave digital filter

6.11 (a) Introduce pipelining into the signal-flow graph used in Problem 6.1 and determine its precedence form.

(b) Write the system of difference equation in computable order.
(c) Write a program, using a high-level language, to implement the filter structure.

6.12 Insert shimming delays into the Cooley–Tukey butterfly. Assume that a real addition takes 1 time unit and a real multiplication takes 24 time units. Further, a complex multiplication is realized with four real multiplications. Note that more efficient ways exist to realize a complex multiplication using fewer real multiplications.

6.13 Determine the number of shimming delays and D flip-flops needed to implement the second-order allpass sections shown in Figure 5.10, using bit-serial arithmetic. Assume that the coefficient word length W_c = 13 bits, data word length W_d = 19 bits, and that twos-complement truncation is performed following the multiplications. Further, a correction is made in front of the delay elements by adding a 1 to the least significant bit if the value is negative; otherwise a 0 is added.

6.14 (a) Determine the wave-flow graph for the three-port adaptor, shown in Figure P6.14, in precedence form when the basic operations are multiplication and two-input addition.
(b) Bit-serial adders and multipliers are used. A multiplication takes 12 clock cycles, which correspond to a coefficient word length of 12 bits, while an addition takes one clock cycle. Determine the maximum sample frequency when the maximum clock frequency is 60 MHz.
(c) The data word length W_d = 21 bits. Determine the required amount of shimming delays.

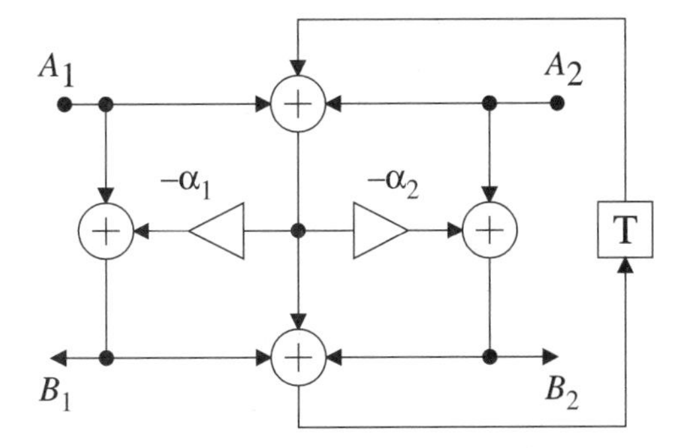

Figure P6.14 Tree-port series adaptor. Port 3 is the dependent port

6.15 Compare the maximum sample rate for a digital filter realized with
(a) Cascade form with first- and second-order sections in direct form I or direct form II.
(b) Parallel form with first- and second-order sections in direct form I or direct form II.
(c) A lattice wave digital filter with the branches realized using circulator structures (cascaded allpass sections).
d) A lattice wave digital filter with the branches realized using Richards structures (cascaded unit elements).

6.16 Determine the maximum sampling frequency for the filter structure shown in Figure P6.16 if a bit-serial addition and a multiplication have latencies of one and eight clock cycles, respectively. The maximum clock frequency is assumed to be at least 100 MHz.

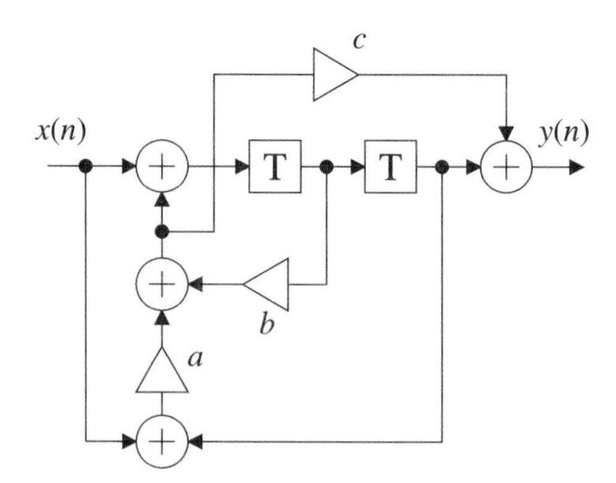

Figure P6.16

6.17 Determine the maximum sample rate for
(a) The LMS algorithm,
(b) The RLS algorithm that was discussed in Chapter 3.

6.18 How is the frequency response of a digital filter affected by pipelining?

6.19 Discuss the meaning of pipelining at the algorithmic and digital circuit levels. Discuss especially pipelining of bit-serial arithmetic.

6.20 Show how the filter in Example 4.16 can be pipelined. Assume that addition and multiplication are atomic operations. Also, find the system of difference equation in computable order for the pipelined filter.

6.21 How many useful pipeline levels can be introduced in the filters in Examples 4.11, 4.12, and 4.16, respectively? Assume that the two-port adaptors are atomic operations.

6.22 Determine the shimming delays required in a third-order FIR filter realized in
(a) Direct form,
(b) Transposed direct form.

The filter is implemented using bit-serial arithmetic. The data word length is 16 bits and the coefficient word length is 12 bits. Multiplication takes 12 and addition 1 clock cycle.

6.23 Show that for block processing of an Nth order filter
(a) The number of operations per sample is given by Equation (6.5).
(b) The minimum operation rate is $N_s \approx 2(2 + \sqrt{2}\,)N$ which is obtained for $L = \sqrt{N(2N-1)} \approx \sqrt{2}\,N$.
(c) The number of multipliers required to achieve T_{min} is $(N + L)^2$.

6.24 Determine the transfer function and the corresponding difference equation for a pipelined first-order all-pole section using:
(a) Clustered look-ahead pipelining.
(b) Scattered look-ahead pipelining.

The pipeline shall have three stages. By how much is the maximum sample frequency increased?

6.25 Determine the transfer function and the corresponding difference equation for a pipelined second-order all-pole section using

(a) Clustered look-ahead pipelining,

(b) Scattered look-ahead pipelining.

The pipeline shall have three stages. By how much is the maximum sample frequency increased?

6.26 Derive the structures in Problem 6.25 in transposed form. Compare the advantages of the original and the transposed structures.

6.27 Discuss the relationships between the following concepts: pipelining, latency, sampling period, throughput, number of resources, and their utilization.

6.28 There are three distinct and possibly complementary ways to increase the maximum throughput of a recursive algorithm. Discuss the various versions of these ways.

6.29 The signal-flow graph shown in Figure P6.29 is a part of a larger graph.

(a) Determine the maximum sample frequency when a multiplication takes two time units and an addition takes one time unit.

(b) Modify the graph so that the true maximum sample frequency is attained.

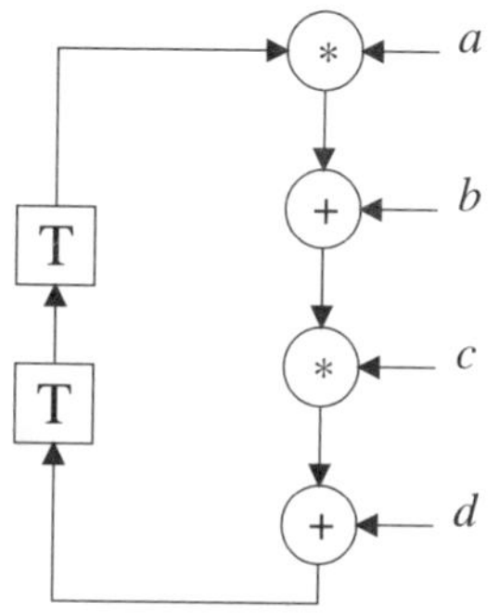

T T * + * + a b c d

Figure P6.29.

6.30 Determine the maximum (output) sample frequency that can be achieved by the interpolator discussed in section 6.10.

6.31 (a) Derive the system of difference equations for the filter in Figure 6.28 when it is being used for decimation with a factor of two.

(b) Derive the corresponding state-space representation.

(c) Derive a scheme based on block processing to process two samples at a time.

7

DSP SYSTEM DESIGN

7.1 INTRODUCTION

In this chapter we will discuss some of the most important steps in the system design phase. As illustrated in Figure 7.1, the system design phase follows the specification phase and involves a number of design steps that ultimately lead to a logical description of the architectural components. After the system design phase follows the integrated circuit design phase, which involves circuit design, electrical validation, floor planning, layout, etc. In this chapter we assume that the specification and algorithm design phases have been completed. The first step in the system design phase involves hierarchically partitioning the DSP system into a set of cooperating processes. These processes are then scheduled in such a way that they can be mapped onto suitable software–hardware structures. The design of the hardware structures will be discussed in subsequent chapters.

The purpose of the scheduling is to distribute the processes in time so that the amount of hardware required is minimized. The underlying assumption is that minimizing the amount of hardware will also minimize the power consumption. Because of the complexity involved, in practice often only the number of PEs is minimized. We will present a design approach that also makes it possible to minimize the memory and communication requirements.

In Chapter 9, we will present methods of synthesizing optimal architectures based on schedules derived in this chapter. Good scheduling techniques are essential for the synthesis of efficient architectures that match the algorithm well. Most digital signal processing algorithms allow a static scheduling of operations which avoids the significant work load that dynamic scheduling generally represents in real-time systems.

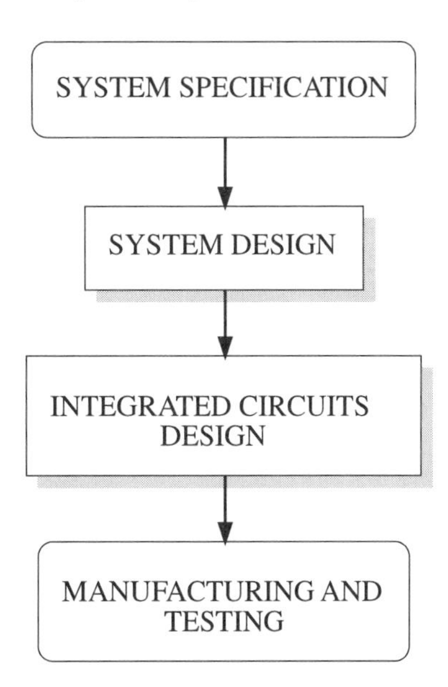

Figure 7.1 The main phases in the design of a DSP system

In this chapter, we will first discuss a highly structured design approach based on a direct mapping technique. We stress that the scheduling, resource allocation, and assignment steps are crucial to a successful design. We will discuss different formulations of the scheduling problem as well as some common optimization techniques. We will also demonstrate the shortcomings of conventional scheduling techniques and present a formulation of the scheduling problem that facilitates optimal scheduling.

7.2 A DIRECT MAPPING TECHNIQUE

Generally, an algorithm can be mapped to a software–hardware architecture either by direct mapping or by first mapping the algorithm to an intermediate, *virtual machine* followed by mapping to a programmable hardware structure. The first approach typically aims at high-performance, fixed-function systems whereas the second method sacrifices performance in an attempt to reduce both the design effort and cost by using standard digital signal processors.

Next, we outline the direct mapping technique that is suitable for design of fixed-function systems. The main steps in the direct mapping technique are shown in Figure 7.2. We will follow this approach in the case studies.

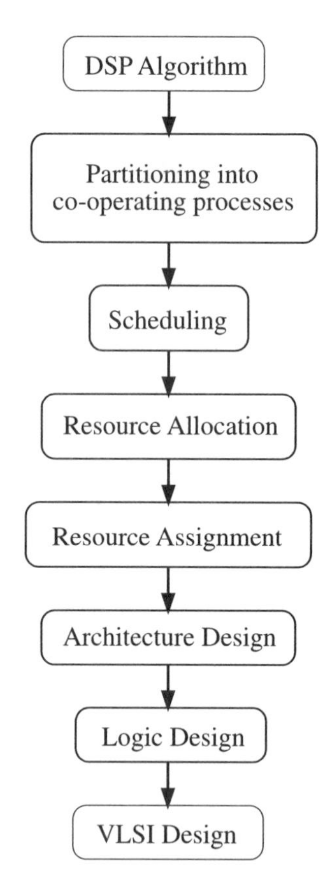

Figure 7.2 The main steps in the DSP system design phase—idealistic view

1. Ideally, the specification of the system contain a complete specification of the system and its components. In reality, however, the specifications of the components from which the system is built must be derived from the system specification. In a top-down design approach component specification at a given system level will depend on the results of the synthesis done at the higher levels. Hence, the component specifications will evolve as the design progresses.
2. A good DSP algorithm is selected and tuned to the application. Usually, a high-level language is used to develop and validate the correctness of the DSP algorithm. Note that a software implementation of the DSP algorithm can serve as specification for the next design phase.
3. In the next phase, the algorithm is successively partitioned into a hierarchy of processes. We refer to the tasks in the algorithm as processes, for example, storage of data is considered as a

process. The complexity of the processes (*granularity*) can vary from atomic (indivisible) operations (*fine granularity*) up to complex processes (*coarse granularity*). Typically the higher-level processes will be mapped onto virtual machines, while the lower-level processes will be mapped onto hardware components such as PEs and memories. The partitioning of the system into a hierarchy of processes is a crucial design step that will have a major influence on both the system cost and performance.

The partitioning should be done so that the processes can be mapped onto an appropriate software–hardware structure. An important issue is interprocess communication. Generally, it is favorable to partition the algorithm in such a way that communication between processes is minimized, since a data transfer between two processes requires the processes to be synchronized at that point in time. Hence, one of the processes must wait for the other, and the hardware structure associated with the first process may become idle.

Some DSP algorithms (for example, digital filters) are usually described using signal-flow graphs, which depict the data flow in the algorithm. The partitioning of such systems is relatively simple, since it mainly involves arithmetic operations, and the number of levels in the hierarchy is low. However, the signal-flow graph is a less suitable representation for algorithms that have more complicated control structures—for example, the FFT. Many algorithms like the FFT that are described using a high-level programming language (for example, Pascal or *C*) contain "for"-loops. The corresponding implementation will therefore have a control structure that implements a "for"-loop by repeatedly executing the hardware structure that performs the operations within the loop. Hence, a loop in the algorithm will generally be identified as a control process. In some cases, however, it may be advantageous or necessary to increase the computational parallelism by unwinding the loop.

4. In practice, most complex DSP algorithms are derived in sequential form by using an imperative high-level programming language. The sequential algorithm must therefore be transformed into a more parallel description. The processes in the parallel description must be scheduled in time so that the specified performance is met and the amount of resources required is minimized. Often, only the number of PEs is minimized, but it is also possible to minimize the memory and communication resources. In fact, it is possible to minimize the total amount of hardware as well as the power consumption by proper scheduling.
5. Synthesis of an optimal architecture starts from the schedule that determines the number and type of PEs, memories, and communication channels for a class of optimal architectures. It turns out that the generic architecture is a shared-memory architecture.
6. Adequate resources are allocated according to the given schedule. The number of PEs required is determined by the number of concurrent operations while the number of memories, or memory ports, is determined by the largest number of simultaneous memory transactions. This design step is called *resource allocation*.

7. In the next step, the *resource assignment* step, the operations are assigned to specific PEs and the results of the operations are assigned to specific memories and memory cells. The transactions between PEs and memories are assigned to appropriate communication channels. The methods we will use are based either on clique partitioning or on the left-edge algorithm used for optimizing memory utilization. The left-edge algorithm is a well-known algorithm used for wire routing.

   In this step, we minimize the cost of the PEs, memories, and communication circuitry. In Chapter 9 we will show that it is advantageous to use specialized bit-serial PEs and bit-parallel RAMs for data storage. The bit-serial PEs must therefore communicate with the memories through serial–parallel converters that act as cache memories.
8. The next step involves logic design of the modules in the architecture—i.e., the PEs, memories, control units, etc. Control signals, which are also derived from the schedule, are defined in this step.
9. In the circuit design step, the modules are designed at the transistor level. Transistor sizes are optimized with respect to performance in terms of speed, power consumption, and chip area.
10. The last step, the VLSI design phase, involves layout—i.e., floor planning, placement, and wire routing. A key concept is that neither system nor circuit design is done in this phase.

The scheduling, resource allocation, and assignment problems are, in general, NP-complete problems. They are therefore, in practice, solved in sequence and the whole design sequence, steps 3 through 9, must be iterated until a satisfactory solution is found.

We stress that design is an iterative process. We can not expect to go through the preceding steps only once. Instead, we must be prepared to redesign the system at a higher level if the lower levels reveal nonsolvable problems or lead to an inefficient solution. The estimate of system parameters (e.g., chip area or power consumption) will become more and more accurate as design progresses. In the first pass through a design level we may do a preliminary, crude design and successively refine it in subsequent design iterations. Generally, we are also interested in exploring the design space by investigating a large variety of design alternatives. Simulation and other means of validation are required to validate and evaluate the design decisions at all levels.

However, these design steps are highly interdependent. For example, the cost function that is minimized by the scheduling assumes certain costs for communication, PEs, and memories. Further, the selection of a particular algorithm restricts the class of usable architectures. Therefore, the synthesis process must, in practice, be followed by an evaluation phase and a subsequent design iteration. The design process may need to be reiterated several times in order to achieve a satisfactory solution. It may also require extension to a lower level of abstraction in order to improve the performance of the hardware components.

7.3 FFT PROCESSOR, CONT.

In this chapter we will perform the main steps in the system design phase of the FFT processor. This means that we will synthesize the architecture with its hardware modules step by step, starting from the algorithm. We begin by partitioning the algorithm, followed by scheduling, resource allocation, resource assignment, and finally, synthesis of a matching architecture. We will first perform a crude design, then subsequently improve it until a satisfactory solution is reached.

7.3.1 First Design Iteration

In a top-down design approach we start by identifying the major tasks (processes) to be executed. The aim is to successively describe the whole DSP system as a hierarchy of communicating processes. In practice, it is important that the system description can be executed in order to validate its correctness. High-level languages such as VHDL can be used for this purpose. Unfortunately, many conventional languages such as Pascal and *C* are unsuitable for this purpose due to a lack of static variables between procedure calls.

At the top level we partition the FFT processor into four communicating processes: input, FFT, output, and memory process as shown in Figure 7.3. Generally, it is useful to introduce a memory process at this level explicitly. The input process is responsible for reading data into the FFT processor from the outside world.

The input sequence and intermediate results are stored by the memory process. Conversely, the output process writes data from the memory to the outside world. The output process also handles the unscrambling of the data array in the last stage of the FFT. The FFT process only computes the FFT since the IFFT is computed by interchanging the real and imaginary parts of the data in the input and output processes.

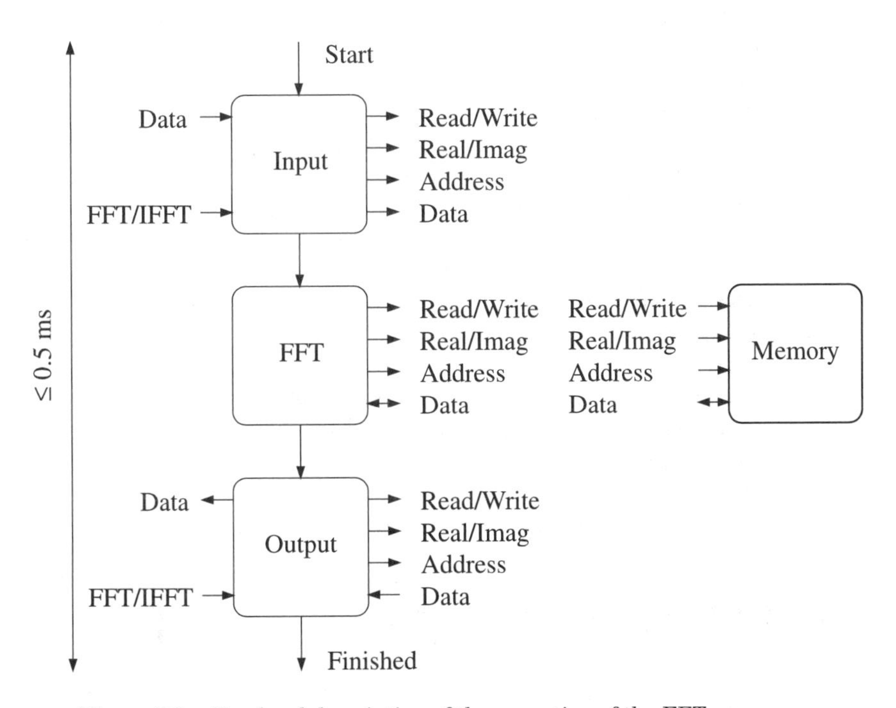

Figure 7.3 Top-level description of the operation of the FFT processor

These processes are then partitioned into a hierarchy of simpler processes. Partitioning should be performed such that the processes at the lowest level in the hierarchy can be easily mapped onto hardware resources. Both the *input* and *output* processes are simple processes needing little further refinement. The partitioning should be done in steps where each step is sufficiently small as to make the transformation obvious, thereby avoiding design errors.

Resource Allocation

Next we determine the resources required. The partitioning assumes that the input, FFT, and output processes are executed sequentially. It is possible to overlap their execution and speed up the FFT processor slightly, but the complexity of the design would increase significantly. The number of butterfly operations for each FFT is

$$\frac{N}{2}\log_2(N) = 5120$$

In Chapter 11 we will show that a bit-serial butterfly PE can be implemented using $W_d = 24$ clock cycles and we assume that the usable clock frequency, $f_{CLPEmax}$, is at least 220 MHz. The time available for the FFT is $t_{FFT} = 0.372$ ms. The number of butterfly PEs is

$$N_{PEb} = \frac{W_d \cdot \frac{N}{2}\log_2(N)}{t_{FFT} \cdot f_{CLPEmax}} = \frac{24 \cdot 5120}{0.372 \cdot 10^{-3} \cdot 220 \cdot 10^6} \approx 1.5$$

Thus, we need only two butterfly PEs to meet the required throughput. The minimum required clock frequency for the PEs is

$$f_{CLPE} = \frac{W_d \cdot \frac{N}{2}\log_2(N)}{N_{PE} \cdot t_{FFT}} = \frac{24 \cdot 5120}{2 \cdot 0.372 \cdot 10^{-3}} = 165.2 \text{ MHz}$$

We can also estimate the required data rate for the memory process. For each butterfly operation we must read and write two complex values. Hence, the data rate will be

$$f_{memory} = \frac{(2+2) \cdot \frac{N}{2}\log_2(N)}{t_{FFT}} = 55 \cdot 10^6 \text{ complex words/s}$$

In principle, it is possible to use only one logical memory. Small memories with this data rate can easily be implemented. However, we choose to use two logical memories since this will simplify the implementation of the memories and increase the memory bandwidth, which is a critical factor. Also, it is desirable that the memory clock frequency is a multiple of the I/O frequency. This will make it possible to use the same clock frequency for the memories throughout the input, FFT, and output processes. We therefore decide to use a memory clock frequency of 32 MHz, i.e. twice the I/O data rate. The time required for the FFT then becomes

$$T_{FFT} = \frac{(2+2) \cdot \frac{N}{2}\log_2(N)}{2 \cdot 32 \cdot 10^6} = 0.32 \text{ ms}$$

and the minimum required clock frequency for the PEs is

$$f_{CLPE} = \frac{24 \cdot 5120}{2 \cdot 0.372 \cdot 10^{-3}} = 192 \text{ MHz}$$

The throughput becomes

$$\frac{1}{t_{I/O} + t_{FFT}} = \frac{1}{0.128 \cdot 10^{-3} + 0.32 \cdot 10^{-3}} = \frac{1}{0.448 \cdot 10^{-3}} = 2232 \text{ FFTs/s}$$

7.3.2 Second Design Iteration

Note that both the original FFT program and the description using communicating processes are sequential descriptions. Execution of the statements (operations) in the Pascal program is determined by the ordering of the statements, while execution of the processes is controlled by the availability of valid inputs. The following design iteration therefore aims to modify the original sequential algorithm into a parallel description that can be mapped efficiently onto the hardware resources. We must modify the original program so that two butterfly operations can be executed concurrently. We must also compute two coefficients W^p concurrently. There are many possible ways to do this. Let us therefore first consider a smaller problem.

Figure 7.4 shows the signal-flow graph for a 16-point Sande–Tukey's FFT where the butterfly operation has been chosen as the basic operation. There are $M = \log_2(N)$ stages, excluding the unscrambling stage, with $N/2$ butterfly operations each. The coefficients (i.e., the twiddle factors) in the two parts are the same.

The relation between the coefficients for the butterflies in rows p and $p + N/4$ in the first stage of the FFT is

$$W^{p + N/4} = W^{N/4}\, W^p = -j\, W^p$$

Thus, the twiddle factor for one of the butterfly PEs can be obtained from the twiddle factor of the corresponding butterfly PE by simply changing the sign and interchanging the real and imaginary parts. In Chapter 11 we will show that these modifications can be implemented with an insignificant amount of hardware. We will therefore partition the FFT to have two concurrent butterfly processes while exploiting the fact that one W^p-process is sufficient. In general, only one processing element is required to generate the twiddle factors.

We begin by rewriting the original Pascal code for the ST-FFT using VHDL. The new representation is shown in the Box 7.1.

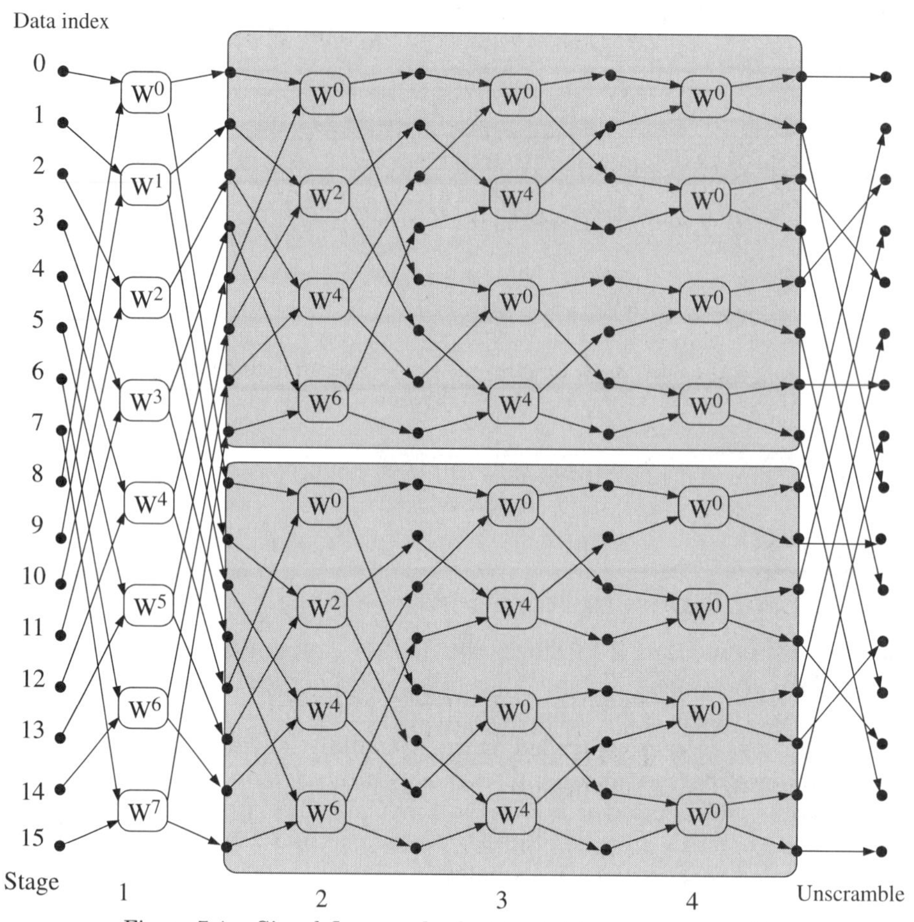

Figure 7.4 Signal-flow graph of a 16-point Sande–Tukey's FFT

```
--Package containing functions, types and constants to FFT-proc
library ieee;
use ieee.math_real.all;

package FFT_pck is
   type Complex is record
      re : real;
      im : real;
end record;
constant M : integer := 10;
constant N : integer := 1024;
constant Nminus1 : integer := 1023;
constant TwoPiN : real := 2.0*MATH_PI/real(N);
type complexArr is array(0 to Nminus1) of complex;
function Digit_Reverse(digit : integer) return integer;
procedure Unscramble(variable xx : inout complexArr);
```

```
procedure Butterfly(variable xtmp : inout complexArr;
                    variable k,kNs : integer; variable Wcos,Wsin : real);
end FFT_pck;

--package body for function bodies to FFT-proc.
package body FFT_pck is
function Digit_Reverse(digit : integer) return integer is
  variable NewAddr, Rmbit, OldAddr : integer;
  begin
    NewAddr := 0;
    OldAddr := Digit;
  for i in 1 to M loop
    Rmbit := OldAddr mod(2); --requires both operands to be integer
    OldAddr := OldAddr/2;   --integer division
    if (Rmbit = 1) then
      NewAddr := NewAddr + NewAddr + 1;
    else
      NewAddr := NewAddr + NewAddr;
    end if;
  end loop;
  return NewAddr;
end Digit_Reverse;

procedure Butterfly(variable xtmp : inout complexArr;
                    variable k, kNs : integer;
                    variable Wcos, Wsin : real) is
  variable TempRe, TempIm : real;
  begin
      TempRe := xtmp(k).re – xtmp(kNs).re;
      TempIm := xtmp(k).im – xtmp(kNs).im;
      xtmp(k).re := xtmp(k).re + xtmp(kNs).re;
      xtmp(k).im := xtmp(k).im + xtmp(kNs).im;
      xtmp(kNs).re := TempRe * Wcos – TempIm * Wsin;
      xtmp(kNs).im := TempIm * Wcos + TempRe * Wsin;
  end Butterfly;

procedure Unscramble(variable xx : inout complexArr) is
  subtype int_type is integer range 0 to Nminus1;
  variable temp : complex;
  variable it : int_type;
  begin
    for k in 0 to Nminus1 loop
      it := Digit_Reverse(k);
      if it > k then
        temp := xx(k);
        xx(k) := xx(it);
        xx(it) := temp;
      end if;
    end loop;
  end Unscramble;

end FFT_pck;
```

```
--Fast Fourier Transform–Sande-Tukey
library ieee;
use ieee.math_real.all;
use work.fft_pck.all;

entity ent_FFT is
   port(x_in : in complexArr; x_ut : out complexArr);
end ent_FFT;

architecture beh_ST of ent_FFT is
   begin
      Main:process(x_in)
      variable Ns, Stage, k, kNs, p : integer;
      variable Wcos, Wsin : real;
      variable x : complexArr;
         begin
         x := x_in;                   --copy insignal
         Ns := N;
         for Stage in 1 to M loop
            k := 0;
            Ns := Ns/2;               --index distance between a dual node pair
            for q in 1 to (N /(2*Ns)) loop
               for n in 1 to Ns loop
                  p := k*2**(Stage–1) mod (N/2);
                  Wcos := cos(TwoPiN*real(p)); --W to the power of p
                  Wsin := –sin(TwoPiN*real(p));--W = exp(–j2π/N)
                  kNs := k + Ns;
                  Butterfly(x,k,kNs,Wcos,Wsin);
                  k := k + 1;
               end loop;
               k := k + Ns;
            end loop;
         end loop;
         Unscramble(x);  --output data
         x_ut <= x;-     -output
      end process Main;
   end beh_ST;
```

Box 7.1 Sande–Tukey's FFT

The VHDL description in Box 7.1 is sequential and must be modified into a description with two parallel butterfly processes in order to meet the performance constraints. The original code contains three loops: the *Stage* loop, the q loop, and the n loop. The *Stage* loop makes sure the FFT is computed stage by stage. The q and n loops are used to compute indices to the variables $x(n)$ in each stage. Since we want two parallel butterfly processes in each stage, we must modify the q and n loops. First we note that the index k can be expressed in terms of q and n:

$$k = 2N_S\,(q-1) + (n-1) = 2N_S\,q' + n'$$

and introduce two new integers q' and n'. The integers N_S, n', and q' can be expressed in terms of *Stage* and N:

$$N_S = N\,2^{-Stage}$$
$$0 \le q' < 2^{Stage-1}$$
$$0 \le n' < N_S = N\,2^{-Stage}$$

Note that the binary representations of q' and n' require *Stage* – 1 and $\log_2(N)$ – *Stage* bits, respectively. Hence, the total number of bits is $\log_2(N) - 1 = 9$ bits. We introduce another integer variable, m, with the range $0 \le m < N/2 = 512$. The integers q', n', and k can now be written

$$q' := \lfloor m/N_S \rfloor \qquad \text{(Integer division)}$$
$$n' := m \ \mathbf{mod}(Ns)$$
$$k := 2Ns \lfloor m/N_S \rfloor + m \ \mathbf{mod}(Ns)$$

Thus, the q and n loops can be replaced with a single loop only controlled by the integer m, as shown in Box 7.2. It is now easy to modify the inner loop so that it will contain two butterfly operations.

```
--Fast Fourier Transform–Sande-Tukey–modified
--Filename: FFT_ST0.VHD

architecture beh_ST0 of ent_FFT is
  begin
    Main: process(x_in)
    variable Ns, Stage, m, k, kNs, p : integer;
    variable Wcos, Wsin : real;
    variable x_tmp : complexArr;
    begin
      x_tmp := x_in;          --copy input signal
      Ns := N;
      for Stage in 1 to M loop
        Ns := Ns/2;               --index distance between a dual node pair
        for m in 0 to ((N/2) – 1) loop
          k := 2 * Ns * (m/Ns) + m mod(Ns);      -- Integer division
          p := k*2**(Stage – 1) mod(N/2);
          Wcos := cos(TwoPiN*real(p));            --W to the power of p
          Wsin := –sin(TwoPiN*real(p));           --W = exp(–j2pi/N)
          kNs := k + Ns;
          Butterfly(x_tmp, k, kNs, Wcos, Wsin);
        end loop;         --for loop
      end loop;--for loop
      Unscramble(x_tmp);
      x_ut <= x_tmp;      --output signal
    end process Main;
  end beh_ST0;
```

Box 7.2 Modified Sande–Tukey's FFT

First Alternative

In this and the next section we will investigate two different alternative choices of order for the computations of the butterflies in the inner loop. Having two concurrent butterflies means that we must generate two indices k in parallel: k_1 and k_2. In the first alternative, we choose to execute butterflies p and $p + N/4$ concurrently.

Since two butterfly operations will be performed for each pass through the inner loop, the range of m is reduced to $0 \le m < N/4$ and the integers k_1 and k_2 become

$$k_1 := 2N_s \lfloor m/N_s \rfloor + m \ \mathbf{mod}(N_s)$$

$$k_2 := \begin{cases} k_1 + N/4 & Stage = 1 \\ k_1 + N/4 & Stage \ge 2 \end{cases}$$

The modified program is shown in Box 7.3.

```
--Fast Fourier Transform – Sande-Tukey, ALTERNATIVE #1
--Filename: FFT_ST1.VHD

architecture beh_ST1 of ent_FFT is
  begin
    Main: process(x_in)
    variable Ns, Stage, m, k1, k2, k1Ns, k2Ns, p : integer;
    variable Wcos, Wsin : real;
    variable x_tmp : complexArr;
    begin
      x_tmp := x_in;          --copy input signal
      Ns := N;
      for Stage in 1 to M loop
        Ns := Ns/2;           --index distance between a dual node pair
        for m in 0 to ((N/4) – 1) loop
          k1 := 2 * Ns * (m/Ns) + m mod(Ns);
          if (stage = 1) then
            k2 := k1 + N/4;
          else
            k2 := k1 + N/2;
          end if;
          p := k1*2**(Stage – 1) mod(N/2);
          Wcos := cos(TwoPiN*real(p));--W to the power of p
          Wsin := –sin(TwoPiN*real(p));--W = exp(–j2pi/N)
          k1Ns := k1 + Ns;
          k2Ns := k2 + Ns;
          Butterfly(x_tmp, k1, k1Ns, Wcos, Wsin, Stage); --Concurrent
          Butterfly(x_tmp, k2, k2Ns, Wcos, Wsin, Stage); --Butterflies
        end loop;--for loop
      end loop;--for loop
      Unscramble(x_tmp);
      x_ut <= x_tmp;          --output signal
    end process Main;
  end beh_ST1;
```

Box 7.3 Alternative #1

Second Alternative

In the second alternative, we execute p and $p + N_S/2$ concurrently in the first stage while p and $p + N_S$ are executed concurrently in the following stages, as shown in Box 7.4. The integers k_1 and k_2 become

$$k_1 := 4N_S \lfloor m/N_S \rfloor + m \ \mathbf{mod}(N_S)$$

$$k_2 := \begin{cases} k_1 + N_S/4 & Stage = 1 \\ k_1 + 2N_S & Stage \geq 2 \end{cases}$$

```
--Fast Fourier Transform – Sande-Tukey, ALTERNATIVE #2
   --Filename: FFT_ST2.VHD

architecture beh_ST2 of ent_FFT is
   begin
      Main: process(x_in)
      variable Ns, Stage, m, k1, k2, k1Ns, k2Ns, p : integer;
      variable Wcos, Wsin : real;
      variable x_tmp : complexArr;
      begin
         Ns := N;
         for Stage in 1 to M loop
            Ns := Ns/2;       --index distance between a dual node pair
            for m in 0 to ((N/4) – 1) loop
               k1 := 4 * Ns * (m/Ns) + m mod(Ns);
               if (stage = 1) then
                  k2 := k1 + Ns/2;
               else
                  k2 := k1 + Ns * 2;
               end if;
               p := k1*2**(Stage – 1) mod(N/2);
               Wcos := cos(TwoPiN*real(p));--W to the power of p
               Wsin := –sin(TwoPiN*real(p));--W = exp(–j2pi/N)
               k1Ns := k1 + Ns;
               k2Ns := k2 + Ns;
               Butterfly(x_tmp, k1, k1Ns, Wcos, Wsin, Stage); --Concurrent
               Butterfly(x_tmp, k2, k2Ns, Wcos, Wsin, Stage); --Butterflies
            end loop;-          -for loop
         end loop;              --for loop
         Unscramble(x_tmp);
         x_ut <= x_tmp;         --output signal
      end process Main;
   end beh_ST2;
```

Box 7.4 Alternative 2

In both alternatives the two concurrent butterflies will use coefficients that either are identical or can easily be computed from each other, as just discussed. One of the butterflies must be able to switch the real and imaginary parts of the twiddle factor in the first stage. The two alternatives will lead to similar solutions. The only difference is the index generation, as just indicated.

An interesting observation is that for both alternatives the additions in the index computation can be implemented by setting appropriate bits in the binary representation of the index. For example, the index computations

$$k_2 := k_1 + \lfloor N_s/2 \rfloor$$
$$k_2 := k_1 + 2N_s$$
$$k_{1N_s} := k_1 + N_s$$
$$k_{2N_s} := k_2 + N_s$$

in the second alternative need not be implemented as additions [15]. In the first stage, k_1 is in the range 0 to 255, but $N_s/2$ is 256. Thus, it is sufficient to set bit 7 to 1.

We just assumed that integer m was incremented by 1 starting from 0 and ending at $N/4 - 1$. However, this scheme is unnecessarily restrictive. It is sufficient that m take on all values in this range, the order is not important. One alternative we will investigate later is to let the order of m be that of a Gray counter.

7.3.3 Third Design Iteration

In this step we collect index computations into a process, *Addresses*. The behavior of this process depends on the chosen index generation scheme. We also specify the behavior of the input and output processes in more detail by partitioning them into the processes *Memory*, *Memory_Read*, and *Memory_Write*. The program corresponds to a set of cooperating processes, as illustrated in Figure 7.5.

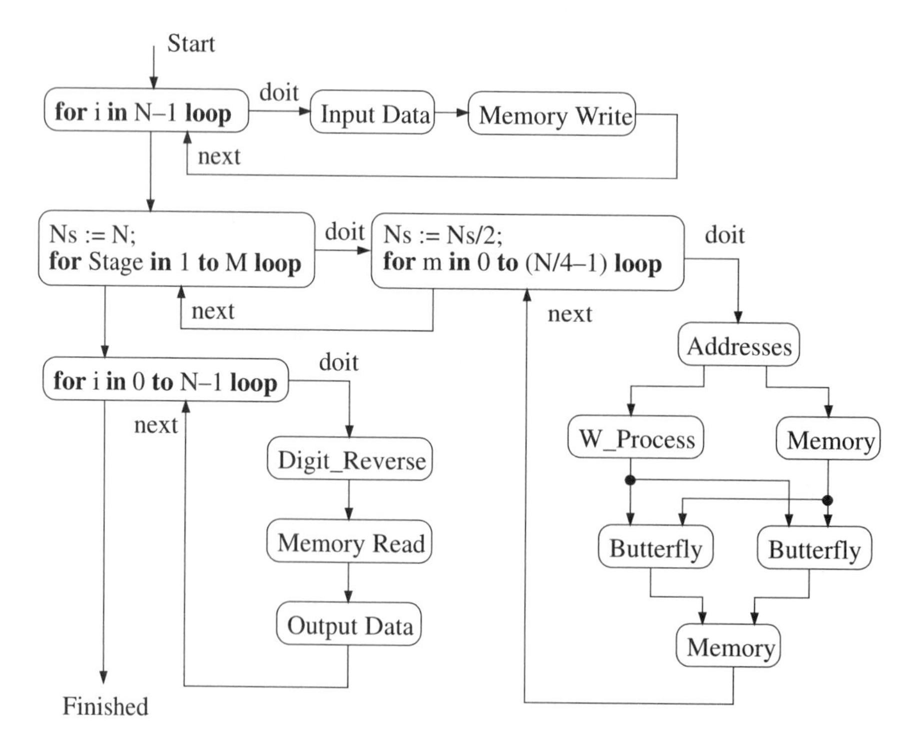

Figure 7.5 Processes derived in third design iteration

Each of these processes will later be mapped to a hardware structure. We notice that each of the input and output phases is controlled by a single process that can be realized by a counter (0 to *N*–1). In fact, the counter can be of arbitrary type, i.e., up or down counter or in any order as long as all values are counted. This fact will later be used to simplify the implementation.

The FFT phase is controlled by two nested control processes that can be realized by using two counters and some simple logic. Hence, using this approach the control structure becomes obvious and simple to map to an efficient hardware structure.

The program for the third and final iteration is shown in Box. 7.5.

```
--Sande-Tukey FFT. Third design iteration and a new entity
-- entity ent_third is
-- port(data : in complex; data_ut : out complex);
-- end ent_third;
--Filename: FFT_ST3.VHD

architecture beh_ST3 of ent_FFT is
 begin
   Main: process(data)
   variable Ns, Stage, ind, m, k1, k2, k1Ns, k2Ns, p : integer;
   variable Wcos, Wsin, TempRe, TempIm : real;
   variable x_tmp : complexArr;
   begin
     for i in 0 to N–1 loop
       Input_Data(data);
       Memory_Write(i, data);
     end loop;
     Ns := N;
     for Stage in 1 to M loop
       Ns := Ns/2;        --index distance between dual node pairs
       for m in 0 to ((N/4) – 1) loop
         Addresses(p, k1, k1Ns, k2, k2Ns, m, Stage);
           Wcos := cos(TwoPiN*real(p));--W to the power of p
           Wsin := –sin(TwoPiN*real(p));--W = exp(–j2pi/N)
         Butterfly(xtmp_1, Wcos, Wsin, Stage);--Concurrent
         Butterfly(xtmp_2, Wcos, Wsin, Stage); --Butterflies
         Memory(Wflag, xtmp_1, xtmp_2, k1, k1Ns, k2, k2Ns);
         Memory(Rflag, xtmp_1, xtmp_2, k1, k1Ns, k2, k2Ns);
       end loop;-         -for loop
     end loop;           --for loop
     for i in 0 to N – 1 loop
       ind := Digit_Reverse(i);
       Memory_Read(i, data);
       data_ut <= Output_Data(data);
     end loop;
   end process Main;
 end beh_ST3;
```

Box 7.5 Third design iteration

7.4 SCHEDULING

So much to schedule, so little time.

Scheduling operations is a common problem that appears in a variety of settings—for example, operating systems, real-time systems, and distributed–parallel computer systems. A process is an arbitrarily large, or small, computational task, and is therefore a more general concept than a simple operation. A process that is allowed to be halted while it is executed and continued in another PE, or in the same PE at a later stage, is called a *preemptive process*, in contrast to the opposite, a *nonpreemptive process*. A preemptive process can be split into smaller processes. Processes to be scheduled are characterized by their start or release time, duration, or deadline, and they may arrive at random time instances or periodically. *Hard real-time processes* are processes for which the deadline must be met.

If all processes are known in advance, it is possible to perform scheduling before run time, yielding a *static schedule* that will not change. Static schedules need only be done once. Conversely, a *dynamic schedule* must be used whenever the process characteristics are not completely known. Scheduling must therefore be done dynamically at run time by a scheduler that runs concurrently with the program. The use of dynamic schedulers may cause a significant overhead since most scheduling problems are NP-complete. Fortunately, most DSP algorithms are amenable to static scheduling. In fact, for such algorithms it is possible to synthesize optimal architectures.

The processes shall be mapped according to the schedule onto a hardware structure. Due to the complexity of the problem, the scheduling, resource allocation, and assignment are usually separated into distinct problems. Resource allocation involves the allocation of physical resources such as PEs, memories, and communication channels. Assignment involves the selection of a particular resource to perform a given function.

The following five criteria of optimality are commonly used for static scheduling.

- Sample period optimal
- Delay optimal
- Resource optimal
- Processor optimal
- Memory optimal

A schedule is *sample period optimal* or *rate optimal* if it has a sample period equal to the iteration period bound.

A schedule is *delay optimal* (latency) if it has a delay from input to output equal to the delay bound. The delay bound is the shortest possible delay from input to output of the algorithm.

A *resource optimal* schedule yields a minimum amount of resources for given performance constraint, e.g., sample rate and/or latency. It is interesting to note that general-purpose computing corresponds to a *resource limited scheduling* problem while real-time processing, including most DSP applications, corresponds to a *resource adequate scheduling* problem.

A schedule is *processor optimal* if it uses as minimum number of PEs of each type. The minimum number of PEs of type i, called the *processor bound*, is

$$P_i = \left\lceil \frac{D_{op\ i}}{T_{min}} \right\rceil \tag{7.1}$$

where $D_{op\ i}$ is the total execution time for all operations of type i and T_{min} is the minimal sample period.

7.5 SCHEDULING FORMULATIONS

The maximum amount of usable computational resources is limited by the parallelism in the algorithm, while the minimum corresponds to the single-processor case, as indicated in Figure 7.6. Generally, there are a large number of possible solutions between these two extremes.

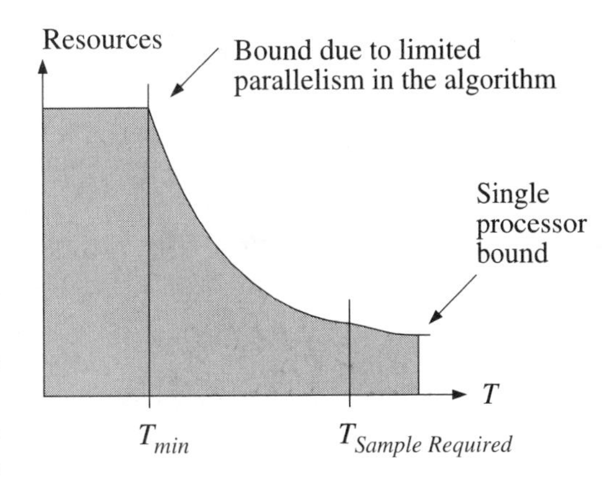

Figure 7.6 Resources versus sample rate for different schedules

The design problem of interest here is to find an operation schedule that allows the operations to be mapped to a minimum-cost hardware structure that meets the performance constraints. Other factors to consider include design effort, simple and regular processing elements, and availability of building blocks, and CAD tools etc.

In this section we will discuss the following formulations of the scheduling problems:

- Single interval formulation
- Block formulation
- Loop-folding
- Periodic formulation

The aim of the scheduling is to minimize a *cost function*. The most commonly used cost function is the number of processors. However, it is also important to include other hardware resources that consume significant amounts of chip area and power. For example, memory is comparatively expensive in terms of chip area and may therefore be a limitation in many applications. Generally, power consumption is the major limiting factor, but we will choose to minimize the amount of hardware resources (i.e., the number of PEs, memories, and communication and control circuitry) with the assumption that this translates to small chip area and low power consumption.

7.5.1 Single Interval Scheduling Formulation

The starting point for scheduling is the computation graph derived from the fully specified signal-flow graph. The first step is to extract all delay elements, as shown in Figure 7.7. The remaining part, which contains only the arithmetic operations and delays, is denoted by N. The computation graph N is a *DAG (directed acyclic graph)*.

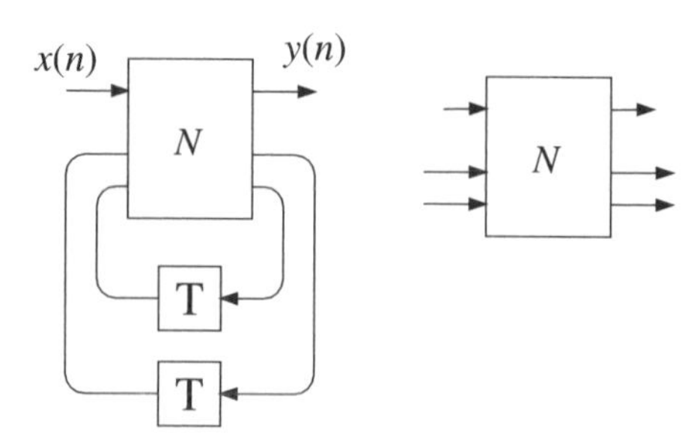

Figure 7.7 SFG and the corresponding computation graph

It appears that an analysis of this graph, which represents the arithmetic operations during one sample interval, would reveal the computational properties of the algorithms, such as the parallelism, and the maximum sample rate. However, this is not the case. In fact, it is not sufficient to consider the operations during one sample interval only. We will demonstrate the shortcomings of considering only one sample interval with some examples.

As a special case, consider the computation graph shown in Figure 7.8. The input signals to the computation graph are to be applied at the beginning of the sample interval and all output values are to have been computed by the end of the sample interval. Thus, the boundaries of the schedule over one sample interval are uniform as shown in Figure 7.8. Since the algorithm is recursive (i.e., the input and output pairs, (1, 1') and (2, 2'), are connected via the delay elements) the two results are used as inputs in the next sample interval. For simplicity, we assume that all operations have a unit execution time. The critical path is indicated with thick lines. The computational paths in the computation graph will form constraints on the sample period. With this formulation we can not use a sample period shorter than the critical path, which in this case is three time units. Note that, according to Theorem 6.2, the minimum sample period is two time units.

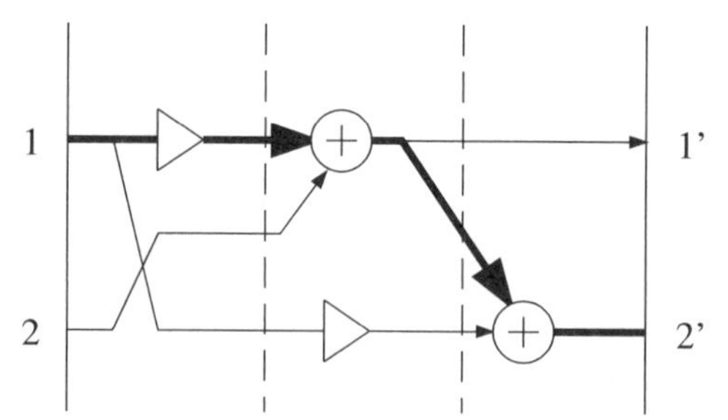

Figure 7.8 Schedule with uniform boundaries

Another schedule is obtained if nonuniform boundaries are introduced. Here, the only restriction is that the paths from an input to the corresponding output, 1 – 1' or 2 – 2', must be of equal length. This is because the shape of the boundaries at the start and the end must be compatible in order to allow the schedule to be periodically repeated. The longest path of this type is two time units long. This fact can be used to achieve the schedule in Figure 7.9. In this case, the computational paths have been skewed in time (retimed) in such a way that the sample period becomes two time units, which is equal to the minimum sample period.

Another problem with this simple view is that the paths from the inputs to outputs correspond to loops with only one delay element. This means that if the critical loop of the SFG contains more than one delay element, the critical path restricting the sample period can not be found. Also, non-uniform boundaries make it more difficult to evaluate the cost function since operations belonging to different sample intervals are performed concurrently.

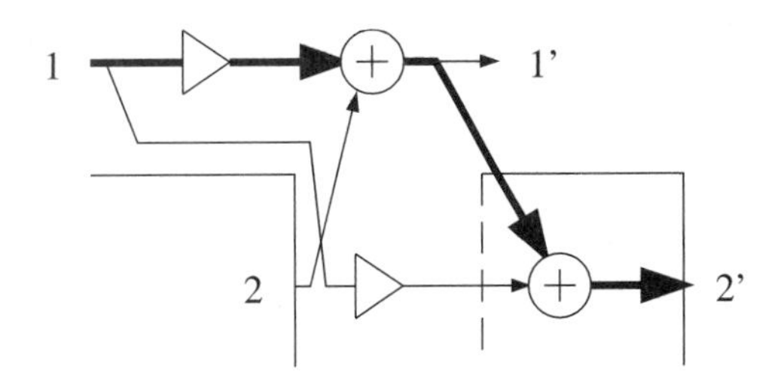

Figure 7.9 Schedule with nonuniform boundaries

EXAMPLE 7.1

Figure 7.10 shows a second-order section in direct form II. The basic operations are multiplication and addition. A multiplication is assumed to take four time units and an addition takes one time unit. The required sample period is T_{sample} = 10 time units.

The critical loop and the critical path of this algorithm are $T_{mult} + 2T_{add} = 6$ and $2T_{mult} + 3T_{add} = 11$ time units, respectively. Hence, to achieve the desired sample rate, we must either duplicate some computational resources or introduce delays to divide the critical path into smaller pieces (pipelining). We choose to introduce pipelining. The pipelined structure is shown in Figure 7.11 and the computation graph is shown in Figure 7.12. The dots indicate the starting and finishing times of the operations.

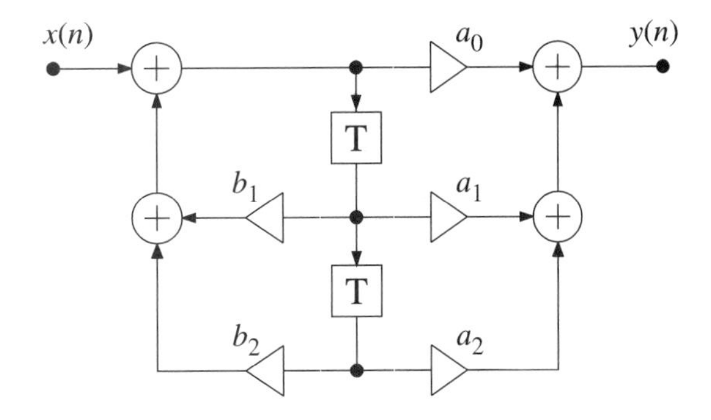

Figure 7.10 Second-order section in direct form II

Equalizing and shimming delays have been inserted in order to meet the specified sample period by extending all computational paths to 10 time units. An equalizing delay of four time units has been placed after the last adder. A shimming delay, with unit time, has been placed after the multiplication by a_0. The output in the recursive loop has been delayed by an equalizing delay of four time units, so that the time difference between the input and output is 10 time units. The

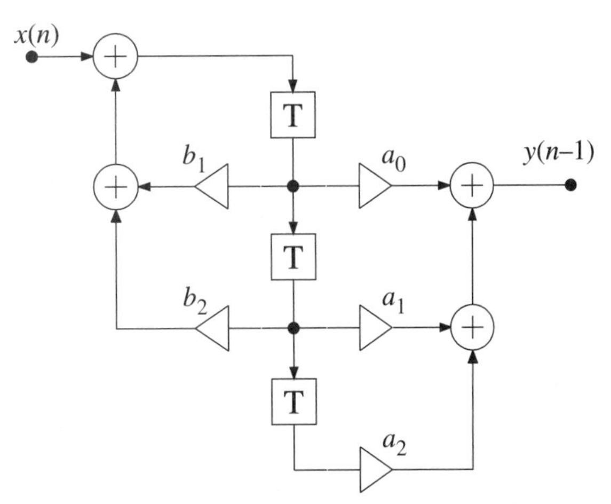

Figure 7.11 Second-order section with pipelining

gray lines indicate that the delay elements carry signal values from one sample interval to the next.

The computation graph shown in Figure 7.12 can be directly interpreted as a schedule for the arithmetic operations. Five multipliers and two adders are needed. The number of time units of storage is 34. Clearly, a better schedule can be found. The operations can be scheduled to minimize the number of concurrent operations of the same type. The scheduling is done by interchanging the order of operations and delays in the same branch. Note that the order of two shift-invariant operations can be changed, as was discussed in Chapter 6.

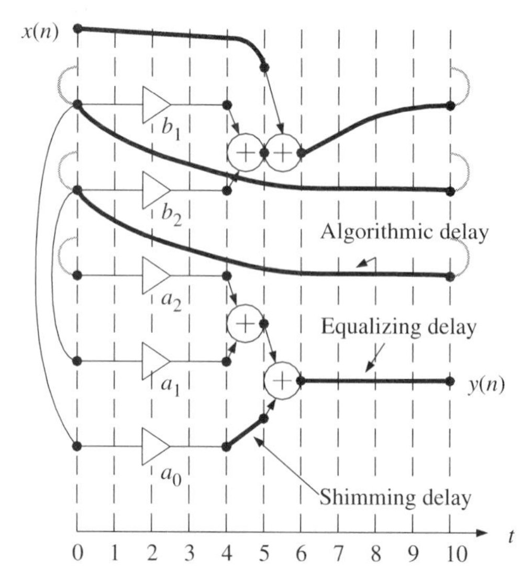

Figure 7.12 Computation graph for direct form II with pipelining

An improved schedule is shown in Figure 7.13. The amount of storage has increased to 38 time units while the numbers of multipliers and adders have been reduced to three and one, respectively. Thus, fewer PEs, but more storage, are needed for this schedule. It is obvious from Figure 7.13 that there are many possible schedules that require the same number of processing elements, but have different requirements on storage and communication resources, i.e., communication channels between processing elements and memories.

Note that the average number of multipliers required is only $(5 \cdot 4)/10 = 2$. This suggests that it may be possible to find a more efficient schedule that uses only two multipliers. However, it is not possible to find such a schedule while considering the operations during a single sample interval only. In the following sections we will show that it is necessary to perform the scheduling over several sample intervals.

Generally, a trade-off between computational resources and the amount of memory can be made by proper scheduling. Often, the communication cost is significant and should therefore also be taken into account.

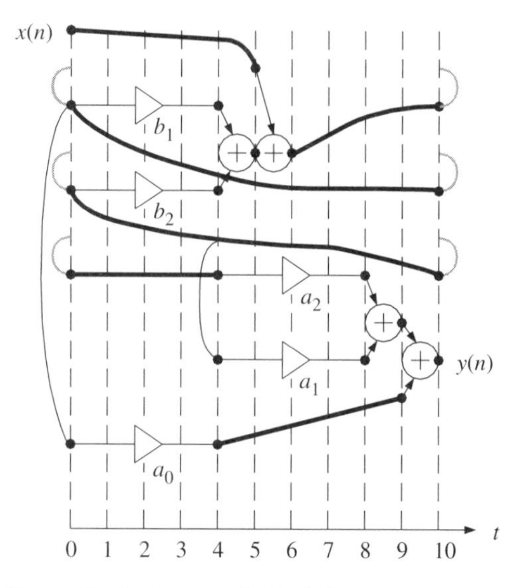

Figure 7.13 Improved schedule requiring fewer PEs

7.5.2 Block Scheduling Formulation

The signal-flow graph and computation graph shown in Figure 7.7 represent the operations that must be executed during a single sample interval. However, the DSP algorithm will, in principle, be executed repeatedly. From a computational point of view, the repeated execution of the algorithm can be represented by an infinite sequence of computation graphs, as illustrated in Figure 7.14, each representing the operations within a single sample interval [4, 8, 24]. The computation graphs are connected via the inputs and outputs of the delay elements since the values that are stored in the delay elements are used as inputs to the next sample interval.

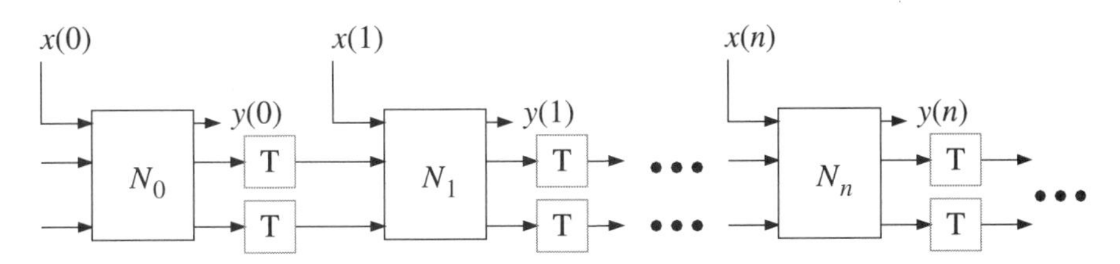

Figure 7.14 Blocking of computation graphs

The computation graph in Figure 7.14 contains a number of delay-free paths of infinite length, since the delay elements just represent a renaming of the input and output values. The longest average path is the *critical path (CP)*. Note that the input $x(n)$ and the output $y(n)$ do not belong to the critical path. The average computation time of the CP is equal to the iteration period bound.

The block scheduling formulation allows the operations to be freely scheduled across the sample interval boundaries. Hence, inefficiency in resource utilization due to the artificial requirement of uniform scheduling boundaries can be reduced. The price is longer schedules that require longer control sequences.

7.5.3 Loop-Folding

Operations inside a (possibly infinite) "for"-loop correspond to the operations within one sample interval in a DSP algorithm. *Loop-folding* techniques have been used for a long time in software compilers to increase the throughput of multiprocessor systems. Here we will show that loop-folding can be used in real-time systems to increase the throughput or reduce the cost of the implementation. We illustrate the basic principle by the means of an example.

Figure 7.15 shows the kernel of a loop containing three multiplications and two additions. The loop has a critical path of length three and the operations are scheduled in three different control steps, i.e., the sample

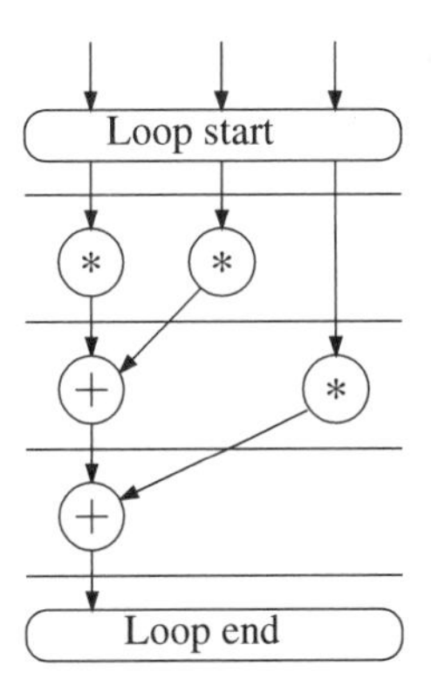

Figure 7.15 Original loop

period is three time units. We need two multipliers and one adder. It is not possible to reschedule operations within the loop to further reduce the number of resources. The operations are divided into two sections so that the critical paths are broken into pieces of lengths one and two. The loop is folded as shown in Figures 7.16 and 7.17.

There are two main options: We can either *minimize loop time* (sample period) or *minimize the resources*. Figure 7.16 illustrates the first case—the loop time is decreased to two.

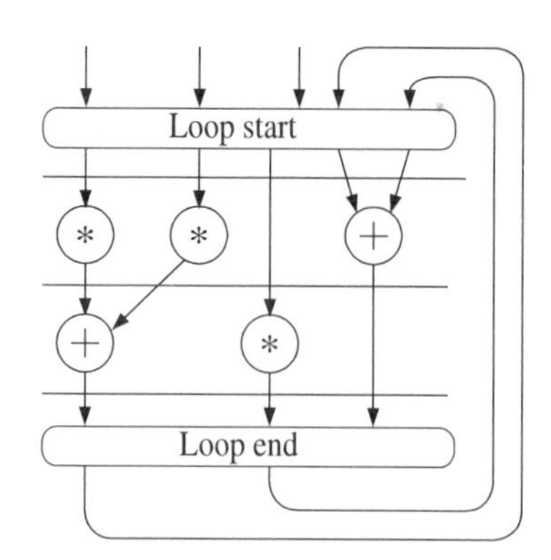

Figure 7.16 Loop-folding that minimizes the loop time

In the second case, illustrated in Figure 7.17, the number of PEs is minimized. Only one multiplier and one adder are required. The drawback is that latency from input to output has been increased from the original three to six. Unfortunately, excessive delay is a major problem in many signal processing applications. Loop-folding is equivalent to recognizing that some operations belonging to an interval or iteration can be started earlier than others and be executed at the same time as operations from previous intervals. We do not have to execute all operations in an interval before we start to execute the operations belonging the next interval. This resembles scheduling a cyclic graph instead of a DAG. Loop-unfolding is a simplified version of the more general cyclic scheduling formulation, which is described in the next section.

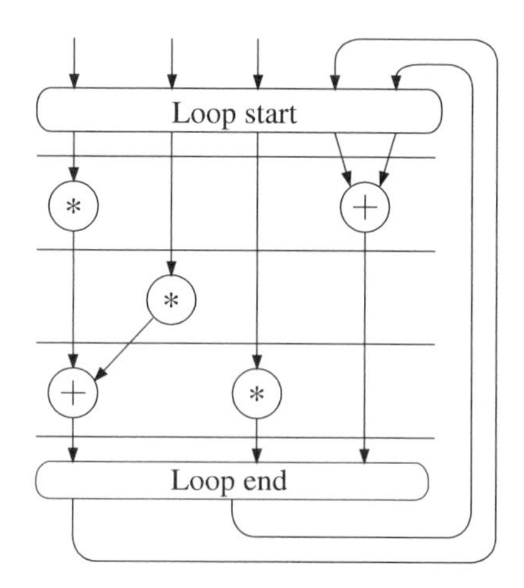

Figure 7.17 Loop-folding that minimizes the number of PEs

7.5.4 Cyclic Scheduling Formulation

In the single interval and blocking formulations we did not take into account the fact that the schedule is inherently periodic. It is therefore convenient to connect K computation graphs in a closed loop, as shown in Figure 7.18 [24]. This will result in a periodic scheduling formulation that allows a schedule period that is K times longer then the sample interval. This cyclic formulation will under certain conditions result in a maximally fast schedule. This is the most general scheduling formulation and it can be used to find resource-optimal schedules using arbitrary constraints on, for example, sample rate and latency.

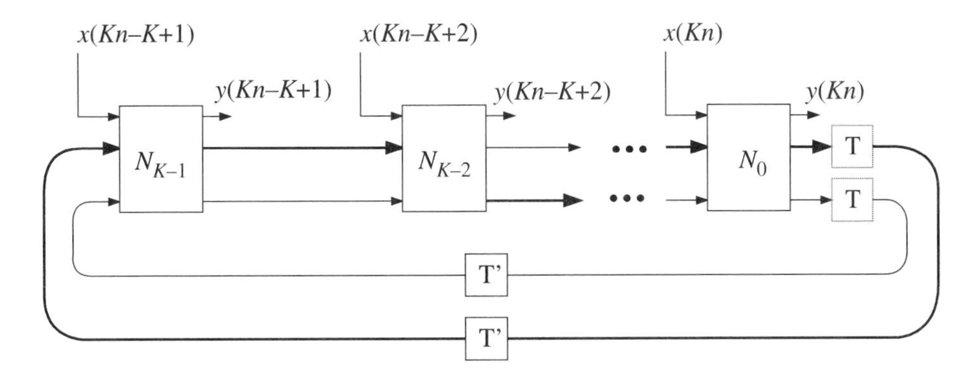

Figure 7.18 K circularly concatenated computation graphs

Each computation graph represents the operations and the shimming and equalizing delays within one sample interval. The delay elements in the critical loop are replaced by new delay elements, representing a delay of $T' = K\,T$. If the required sample rate is lower than the iteration period bound, an appropriated amount of equalizing delay must be inserted into each loop so that their lengths equal a multiple of the desired sample period. It may help to imagine the computation graphs drawn on a cylinder of circumference equal to the scheduling period (which is a multiple of the sample period, as was discussed in Chapter 6).

This problem formulation forces the schedule to be periodic—i.e., the length of the schedule is proportional to K. Unfortunately, it is not possible to determine the best choice of K in the general case. In order to attain the minimum sample period, it is necessary to perform cyclic scheduling of the operations belonging to several sample successive intervals if

- ❑ The computation time for a PE is longer than T_{min} or
- ❑ The critical loop(s) contains more than one delay element.

Generally, the critical loop should be at least as long as the longest execution time for any of the PEs in the loop. For bit-serial arithmetic, the execution times of the processing elements are normally longer than the minimum sample period.

The minimum sample period, using an unlimited amount of resources, is achieved for $K = N_{iCP}$. A more interesting case is to obtain T_{min} when using a minimum number of resources. This is achieved if only one CP exists, and a proper scheduling is done, with

$$K = m\,N_{iCP} \tag{7.2}$$

where m = integer. In order to find the actual resource minimum the best schedules, with minimum resources, must be found for all reasonable values of m. Typical values for m are in the range 1 to 10.

If several CPs exist, the scheduling shall instead be done for

$$K = m\,\{N_{1CP}, N_{2CP}, \ldots\} \tag{7.3}$$

where {. } denotes the least common multiple. Generally, a search over all reasonable values of K must be done in order to find the minimum resource schedule.

EXAMPLE 7.2

Find a periodic schedule, with period two sample intervals, for the second-order section used in Example 7.1 that minimizes the number of PEs.

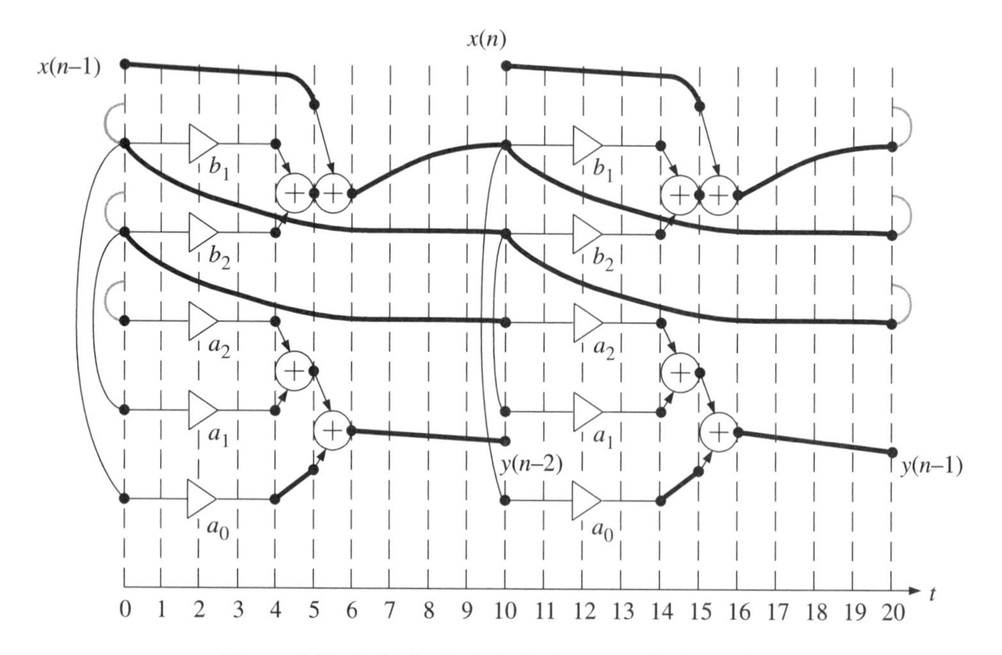

Figure 7.19 Initial schedule for two sample intervals

Figure 7.19 shows an initial schedule obtained by cascading two computation graphs. The storage requirements and the number of multipliers and adders remain the same as in Example 7.1. A better schedule is shown in Figure 7.20. It is obtained by propagating the various types of delays throughout the graph so that the number of concurrent multiplications is reduced. In this case, we obtain a schedule that requires only two multipliers and one adder. The number of delay elements is 92/2 = 46 units per sample. However, some delays can be shared between the branches so that only 71/2 = 35.5 units of delay are required per sample. Generally, many alternative schedules requiring the same amount of PEs exist, but with different memory and communication requirements.

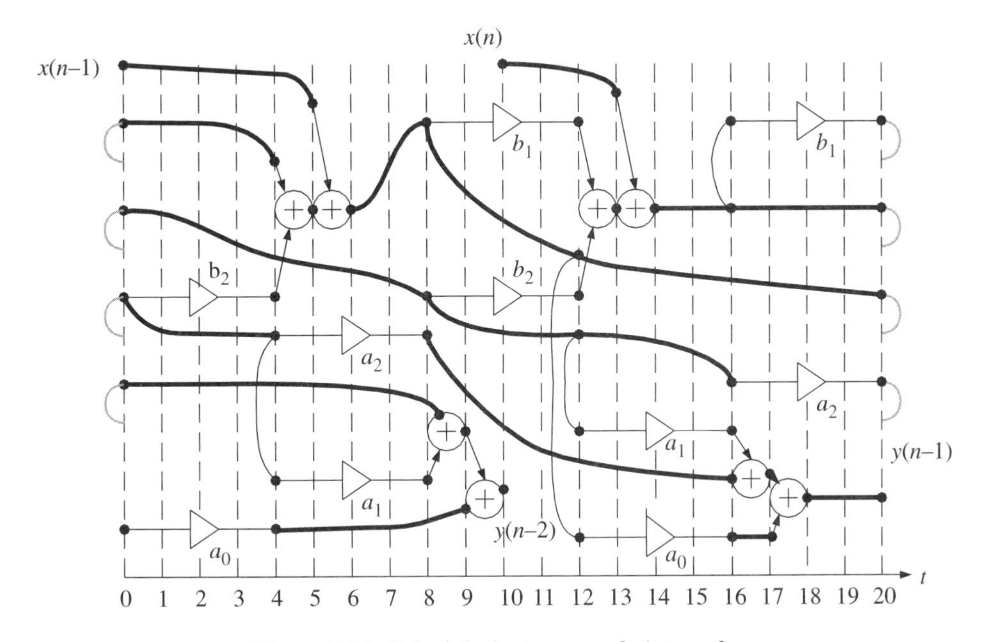

Figure 7.20 Schedule for two sample intervals

EXAMPLE 7.3

Find a periodic schedule for the third-order lattice wave digital filter used in Example 6.5.

There is only one critical loop in this case. It contains two additions, one multiplication, and two delay elements. This implies that the minimum sample period is $(2T_{add} + T_{mult})/2$. In bit-serial arithmetic, the operations are pipelined at the bit-level. Hence, T_{add} and T_{mult} are the time (latency) it takes for the least significant bit to emerge on the output of the corresponding PE. Thus, T_{mult} depends only on the coefficient word length and is independent of the data word length. An addition will delay the least significant bit by 1 clock cycle while the multiplier will cause a delay of $W_c = 4$ clock cycles. The minimum sample period is therefore 3 clock cycles.

Generally, the critical loop in Figure. 6.30 should be at least as long as the longest execution time for any of the PEs. This is equivalent to scheduling over at least five sample periods, since the multiplier requires $W_c + W_d - 1 = 15$ clock cycles and $T_{min} = 3$ clock cycles. A resource minimal schedule with a sample period of 3 clock cycles is shown in Figure 7.21.

Implementation of this schedule requires five multipliers, 20 adders, and 70 D flip-flops. In this case, the multiplier corresponds to a sign-extension circuit, one bit-serial adder with an implicit D flip-flop, and one D flip-flop. The input and output to the filter consist of five bit-serial streams skewed in time. An

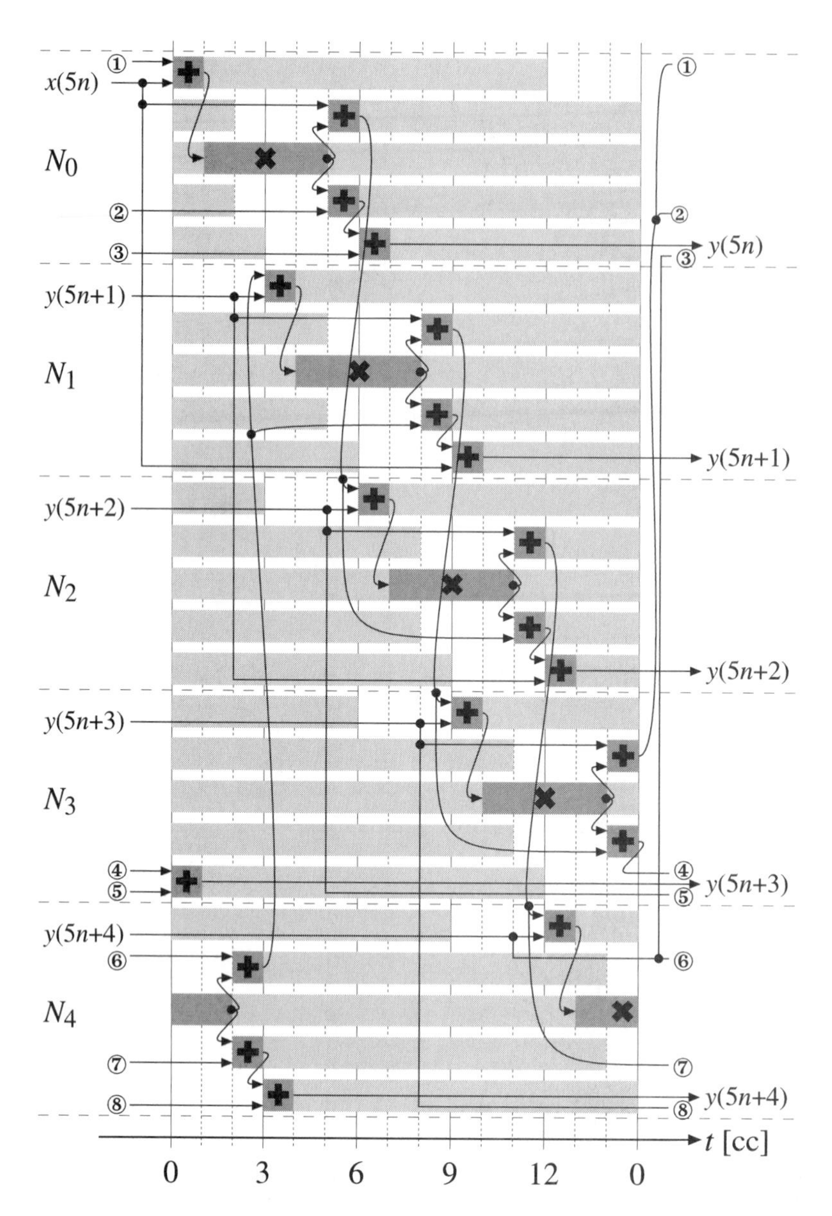

Figure 7.21 Resource minimal schedule

implementation with a bit-parallel input and output format can be implemented using shift registers as serial-parallel converters.

Figure 7.22 shows more explicitly the logic realization based on single-phase logic [26]. The estimated device count is 2400 for the filter with an additional 400 transistors required for the control circuitry. The control unit is implemented using a 15-bit shift register. About 2800 transistors are required for the complete filter. The power consumption is estimated to be 0.8 nJ/sample. A sample frequency of 35 MHz corresponds to a clock frequency of 105 MHz and 30 mW. The

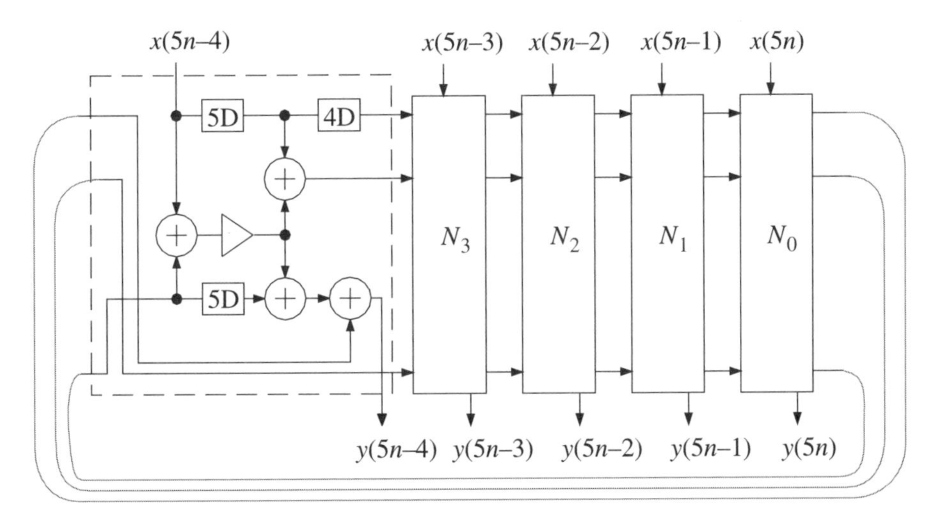

Figure 7.22 Logic implementation

required chip area using a 0.8-μm, double metal CMOS process (AMS—Austria Micro Systeme Intern. Gmb) is less than 1.0 mm^2.

Simulation as well as measurement shows that these circuits can be clocked at very high speeds. Clock frequencies of well above 400 MHz are feasible with careful circuit design. Hence, a sample frequency of more than 130 MHz is feasible with a power consumption of about 325 mW.

Notice that there appears to be a delay-free loop, which would make the implementation impossible, in Figure 7.22. Fortunately, this is not the case. We will discuss how this issue can be resolved in section 7.5.5.

EXAMPLE 7.4

Find a periodic, maximally fast schedule for the third-order lattice wave digital filter used in Example 6.8. In this case we must schedule over six sample intervals. The resource minimal schedule for the transformed algorithm is shown in Figure 7.23. This schedule has an average sample period of only 2.5 clock cycles which is the true minimum. The input and output to the filter consist of six bit-serial streams that are skewed in time.

This implementation, which is shown in Figure 7.24, requires six multipliers, 30 adders, and 78 D flip-flops. The estimated device count is 3000 for the filter with an additional 400 transistors for control circuitry. Thus, about 3400 transistors are required for the complete filter. The required chip area is less than 1.2 mm^2. The increase in maximum speed results in a corresponding increase in area. The expected power consumption is the same as for the filter in Example 7.3 with equivalent sample frequencies.

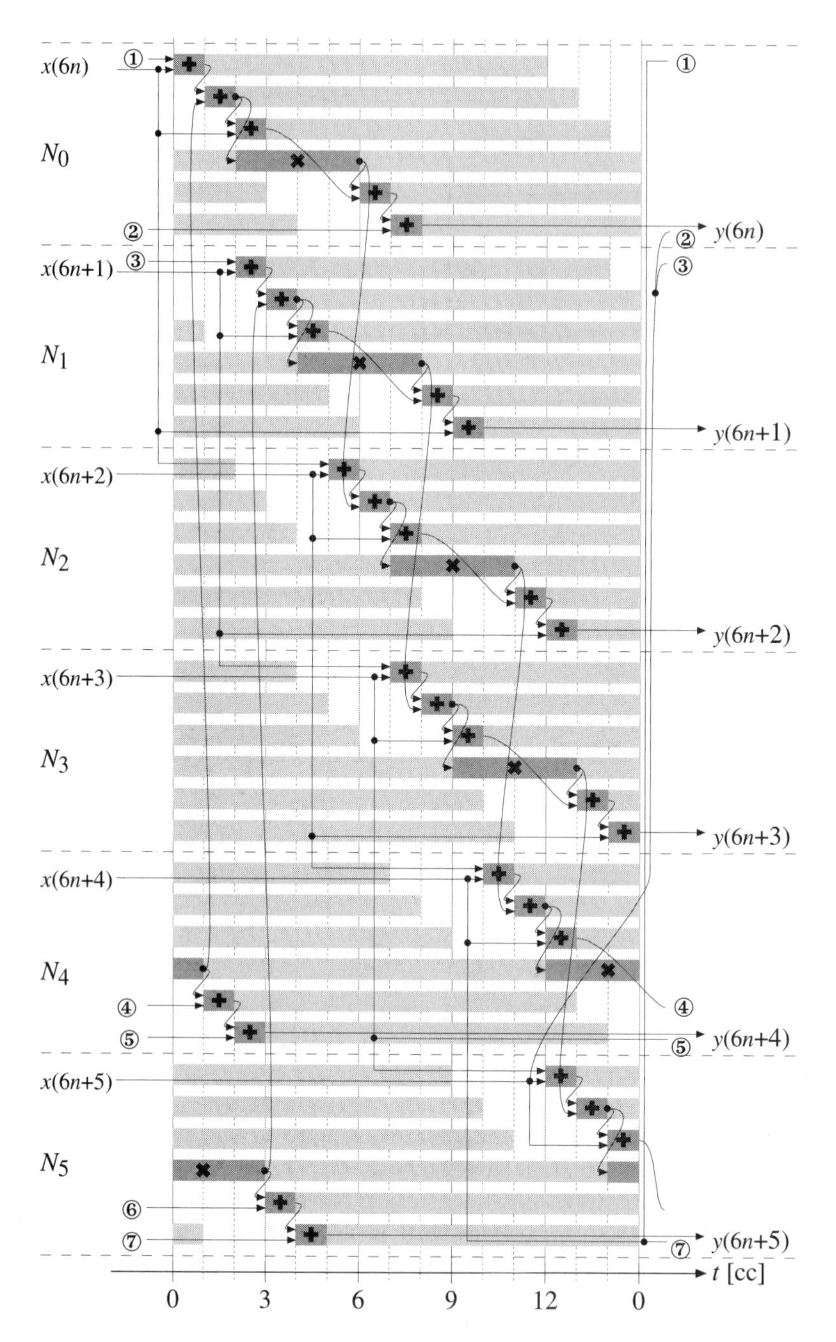

Figure 7.23 Maximally fast, resource minimal schedule

It is interesting to compare this implementation with the bit-parallel implementation that used one clock cycle (35 MHz two-phase clock) per sample and had seven bit-parallel carry-save adders. The bit-parallel implementation requires 77 full-adders while the bit-serial requires only 36 full-adders and uses only 2.5

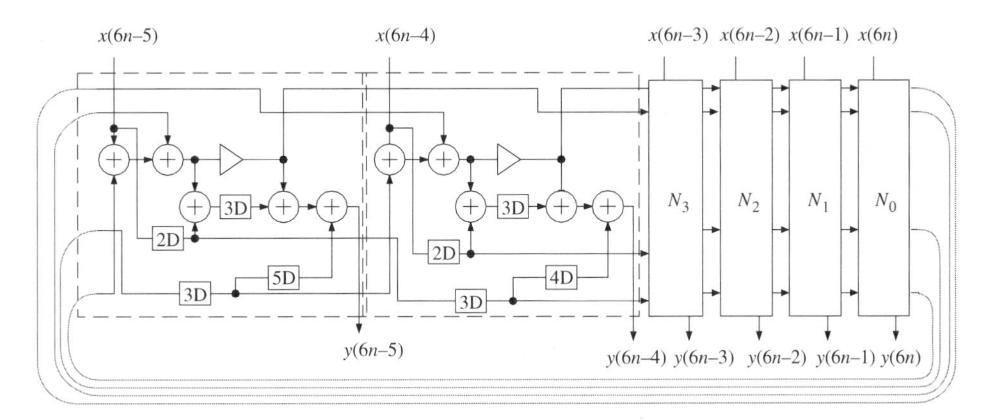

Figure 7.24 Logic implementation

clock cycles (400 MHz) per sample. This leads to significantly higher speed as well as less chip area and power consumption compared to the bit-parallel implementation.

The preceding examples show that the hardware resources can be minimized by proper scheduling of the operations. However, there are still several difficult questions unanswered. For example, what are suitable cost measurements in terms of PEs, memory, and communication and over how many sample intervals the planning shall be performed. Further, it is important to note that cyclic planning problems are much more difficult than acyclic problems.

7.5.5 Overflow And Quantization

Due to the finite word lengths, the processing elements must handle overflow and quantization of data. Generally, overflow is detected by introduction of guard bits by extending the number of sign-bits (two's-complement). An overflow can be detected by comparing the sign-bit to the guard bits. If the bit values differ, an overflow has occurred. Minimizing the impact of an overflow could, for example, be performed by setting the data to the largest number that can be represented (or the largest negative number for negative overflow), i.e., using saturation arithmetic. In bit-serial processing, the detection of a possible overflow and correction must be delayed until the result is available. This becomes a problem in the recursive parts of an algorithm, where it causes a decrease in throughput. A solution to the overflow problem is to use an increased word length in the loop so that overflow can not occur.

We will describe an approach to handle overflow and quantization by the means of the recursive algorithm in Figure 7.25. Here, the output node of the adder may assume values twice as large as the input magnitude. Assuming an input word length of W_d bits, overflow can be prevented by increasing the word length to $W_d + 1$ bits.

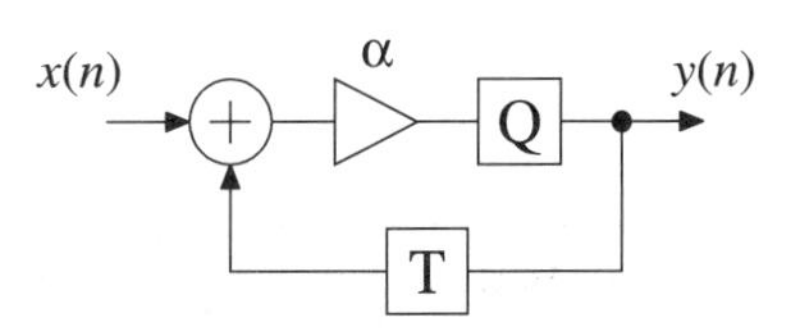

Figure 7.25 First-order recursive filter

Quantization of data must also be performed in recursive parts of an algorithm. The solution is to truncate the data somewhere in a loop. Rounding can be performed by adding a one to the bit-position to the right of the least-significant bit before the truncation takes place.

The filter in Figure 7.25 has a coefficient of $\alpha = (0.5)_{10} = (0.1)_2$, i.e., a word length W_c of 2 bits. Generally, an input to a serial/parallel multiplier needs to be sign-extended to $W_d + W_c - 1$ bits, i.e., one additional bit is required in the example. It is advantageous to move the sign-extension across the addition as shown in Figure 7.26.

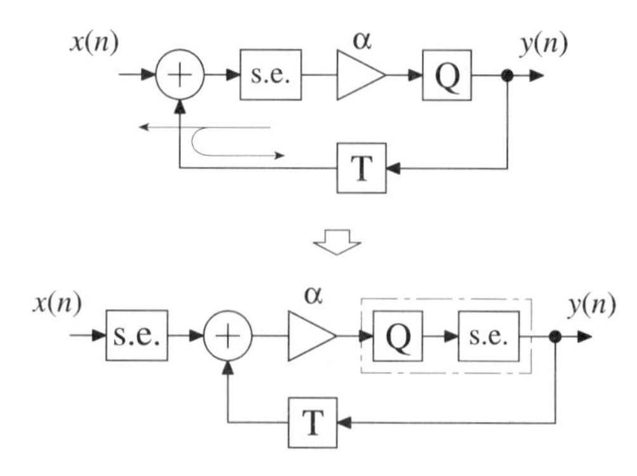

Figure 7.26 Moving the sign-extension (s.e.)

First, the result of the addition is represented with a word length of $W_d + 1$ bits, thus preventing overflow at the multipliers input. Second, it is sufficient to represent the output word with W_d bits, allowing the sign-extension in the loop to be moved to the output of the multiplier. Then, the truncation may be performed at the same time as the sign-extension, without additional delay in the loop. Timing of the bits in the loop for a three-bit input word is shown in Figure 7.27. Observe that the truncation is performed by replacing the least-significant bit with the sign-bit from the previous operation. In fact, the multiplication by 0.5 does not require any hardware. A direct realization of the example would include a latch at the location of the sign-extension, three delay flip-flops in the loop, and a carry–save adder for the addition as shown in Figure 7.28. Here, activation of the control signal c_0 clears the carry D flip-flop and sign-extends the previous multiplier output whenever a new sample is input.

sign-extended input	$x_t(n)$	$x_1(n)$	$x_0(n)$	s.e. $x_0(n)$	$x_2(n+1)$
output of the multiplier	$y_3(n)$	$y_2(n)$	$y_1(n)$	$y_0(n)$	$y_3(n+1)$
output after sign-extension	$y_0(n-1)$	$y_2(n)$	$y_1(n)$	$y_0(n)$	s.e. Q $y_0(n)$

Figure 7.27 Sign-extension in the filter

Since the output of the filter is sign-extended, cascades of such filter sections do not require sign-extension at the inputs, if the word lengths are selected to be equal in all sections. The computation graph is shown in Figure 7.29. Here, model 0 has been used in the derivation of latencies. It is assumed that the input word length is 3 bits. The shaded areas indicate execution time for the operations, with darker shaded areas indicating latency.

If there is a time difference between the start and the end of a branch, an implementation would require memory to be inserted in order to save the output value until it is needed. The time difference is referred to as *shimming delay*. A branch corresponding to a short difference in time may be implemented as a cascade of D flip-flops. For long cascades of D flip-flops, the power consumption becomes high, since all D flip-flops perform a data transaction in each clock cycle. Then a cyclic memory may be more power efficient, since only one input and one output transaction are required in each clock cycle.

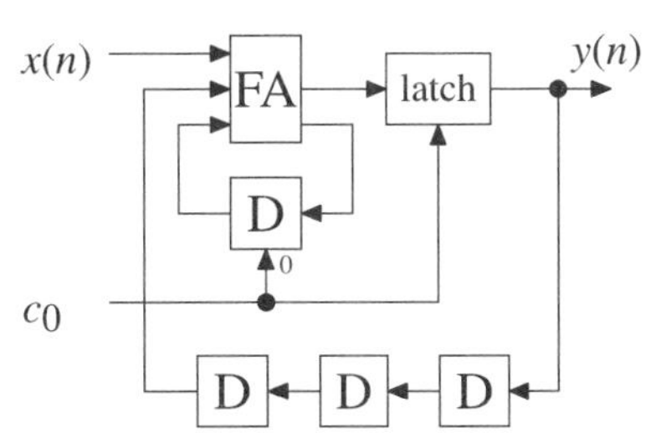

Figure 7.28 Direct realization of the example filter

Returning to the computation graph in Figure 7.29, the word length in the critical loop is four bits. Since the minimum sample period is one clock cycle for this example, the schedule must at least contain 4 concatenated computation graphs. An initial ASAP schedule for this case is shown in Figure 7.30.

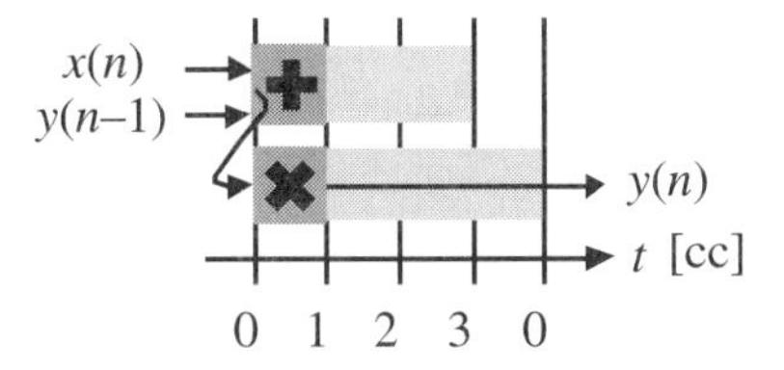

Figure 7.29 Computation graph

Since it is graphically difficult to visualize the example drawn on a cylinder, the cylinder has been cut at the time instant 0 clock cycles. The branch passing through the cut has been labeled ①, which represents a direct connection. Due to the simplicity of this example, the start times of the operations do not have to be changed in order to reach the minimum sample period. It is sufficient to extend the execution times over the cut, as shown in Figure 7.31. Here, the sample period is one clock cycle, which is equal to the minimum sample period. Thus, the schedule is maximally fast.

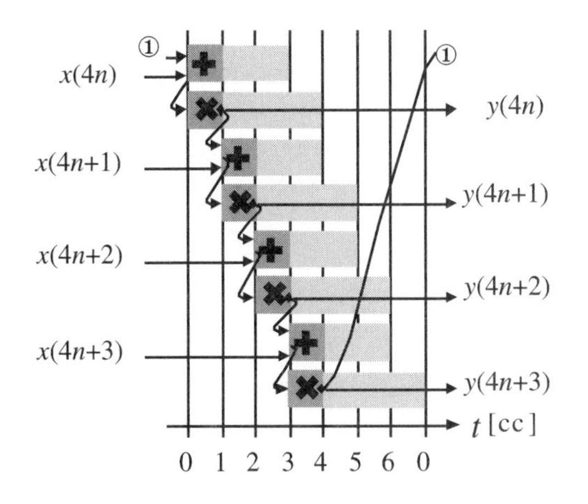

Figure 7.30 Initial ASAP schedule

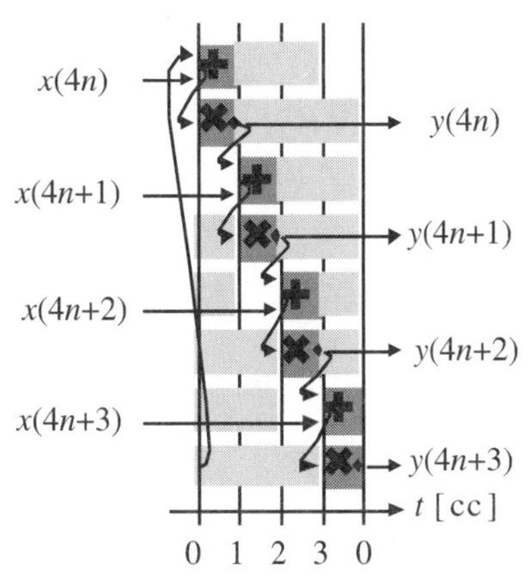

Figure 7.31 Final maximally fast schedule

Logic Realization Of Loops

Continuing with the example schedule shown in Figure 7.32, an isomorphic mapping results in the arithmetic realization shown in Figure 7.33. Here, four computation graphs are connected in a loop. This result may seem strange because of the apparently delay-free loop in the realization, which would cause it to be nonsequentially computable. Fortunately, this is not the case. In fact, quantization of the products have to be included as well. Further, sign-extended signals have to be provided at the multiplier inputs. By moving the sign-extension circuits to the outputs of the multipliers as just described, the algorithm can be realized in the same way. Then the filter would contain a carry-save adder for each addition, no hardware for the multiplications, and a latch for each sign-extension after the multiplications. This logic realization is shown in Figure 7.33.

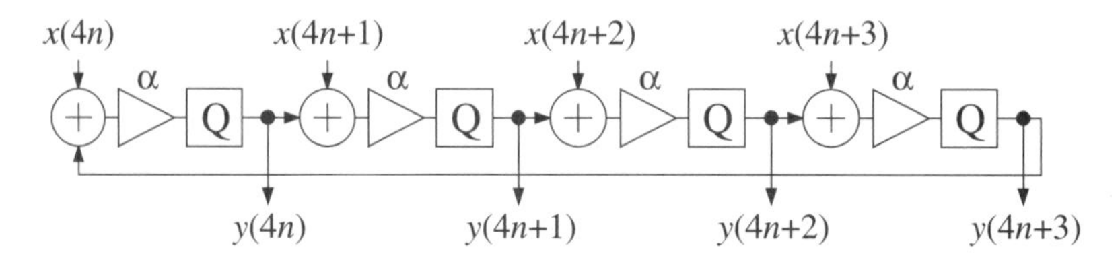

Figure 7.32 Algorithmic realization of a maximally fast filter

Here, control signals c_i have been added. The operation is illustrated in Figure 7.34. In the figure, the function of the latches and the D flip-flops have been pre-evaluated from the control signals to illustrate the function. It is assumed that the control signals activate the latches and the D flip-flops instantly. In a real implementation this could easily be changed to a synchronous operation, where the clock signal is evaluated on a clock edge.

In the first step (clock cycle), the least-significant bit of an input word arrives to the leftmost stage, for which the control signal is activated. The active control signal clears the D flip-flop in the carry path, and activates the sign-extension circuit. The result from this operation is truncated by sign-extension of the previously output bit. In the second step, the next input bit arrives at the leftmost stage. The control signal is deactivated, which causes the carry-save adder to

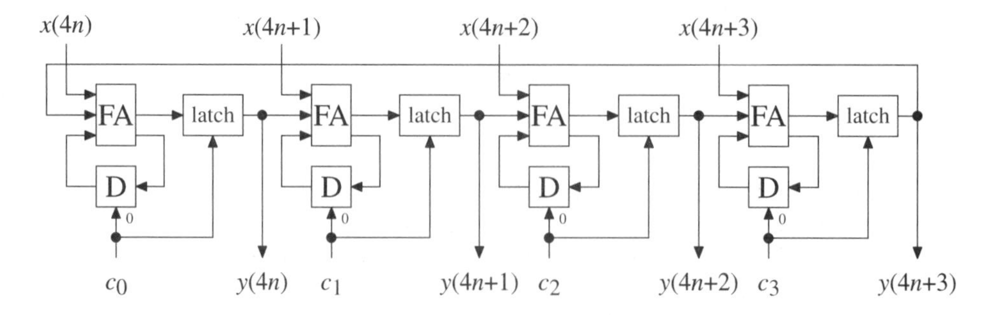

Figure 7.33 Detailed logic realization of the filter

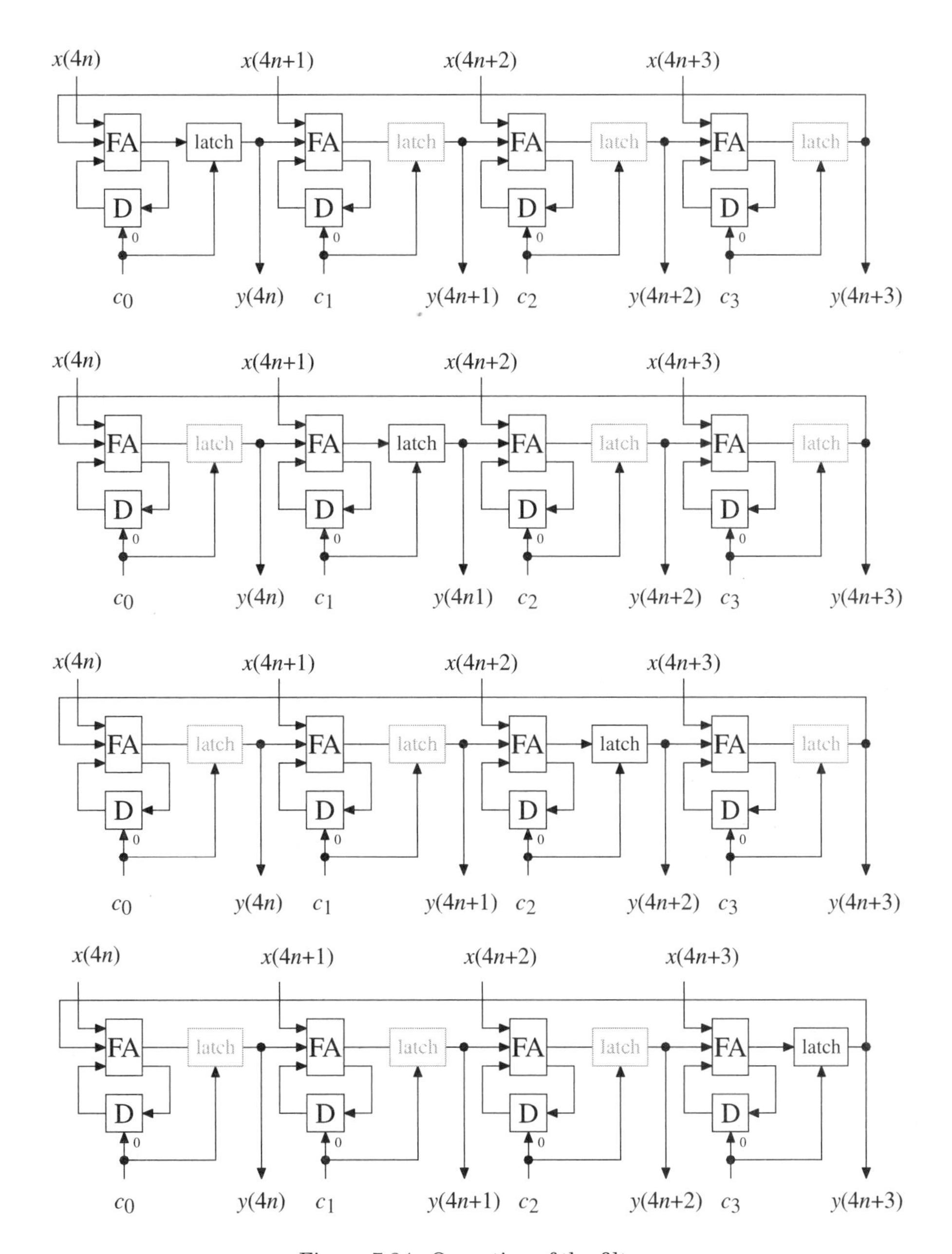

Figure 7.34 Operation of the filter

include the carry bit in the addition. The sign-extension is released, which ends the truncation of the output. In the third step, the original sign-bit arrives to the input of the leftmost stage, where the control signal still is deactivated. Finally, in the last step, the sign-extended input bit arrives to the leftmost stage, where the control signal is still deactivated. The input bit from the rightmost stage is currently a sign-extended result from that stage.

Each stage in this scheme works identically to the leftmost stage, except for the difference in the timing of one clock cycle between two stages. Thus, the input to the filter is four bit-serial streams skewed in time. Figure 7.34 illustrates that there is always an active latch that copies a sign-bit in the loop.

7.5.6 Scheduling of Lattice Wave Digital Filters

In this section we will discuss scheduling of bit-serial, lattice wave digital filters so that the maximal sample frequency is obtained. An Nth order lattice wave digital filter consists of a number of cascaded first- and second-order allpass sections. Since the allpass sections are recursive, the throughput of the filter is bounded by the critical loops.

The loop in a first-order section is indicated with the thick line in Figure 7.35. This loop contains two additions, one multiplication with α_0, and one delay element, yielding the minimum sample period bound

$$T_{0min} = 2T_{add} + T_{\alpha 0}$$

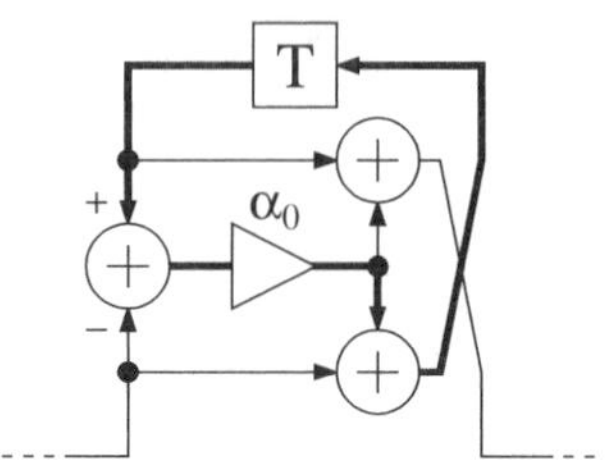

Figure 7.35 The critical loop in a first-order section

In Figure 7.36, the critical loop in a second-order section is indicated with a thick line. For this loop, four additions, two multiplications, and one delay element yield the bound

$$T_{i/2min} = T_{\alpha i} + T_{\alpha i-1} + 4T_{add}$$

The minimum sample period of the filter is then bounded by the section with the lowest throughput, i.e., the loop i that yield the higest T_{min}, since each section has to operate with the same throughput.

Since a processing element only processes 1 bit in each clock cycle, while a minimal sample period of T_{min} clock cycles requires a sample of length W_d bits to be processed, operations belonging to m sample periods need to be included in the schedule. Therefore, m must be selected as

$$m \geq \left\lceil \frac{W_d}{T_{min}} \right\rceil \tag{7.4}$$

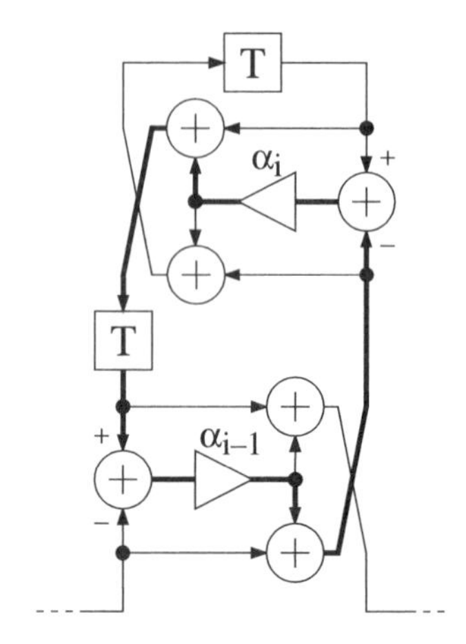

Figure 7.36 The critical loop in a second-order section

in order to match the bandwidths between the processing elements and the sample period. The scheduling period for a maximally fast schedule becomes $m\,T_{min}$. The resulting scheduling formulation for the complete lattice wave digital filter is illustrated in Figure 7.37. Here, each allpass section is scheduled with a scheduling period of $m\,T_{min}$. The input and output to the filter consist of m parallel bit-serial streams that are distributed in time over the scheduling period. Note that the scheduling of the inputs and outputs of two consequtive sections must match.

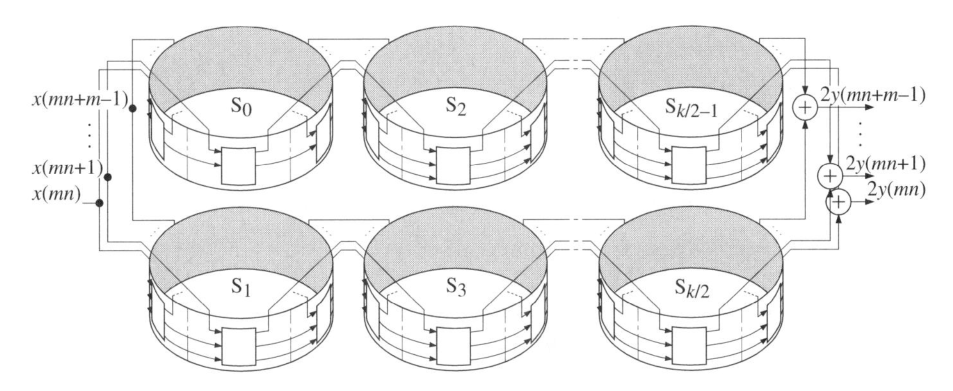

Figure 7.37 Maximally fast filter scheduling formulation

Figure 7.38 shows a maximally fast scheduling of the operations belonging to m sample periods in the first-order allpass section. The shaded areas indicate execution time for the operations, with darker shaded areas indicating latency. The corresponding, maximally fast schedule for the second-order section is shown in Figure 7.39.

For the allpass sections with critical loops with lower T_{min} than T_{min} for the complete filter, the throughput has to be lowered to equalize the throughput in all sections. This can be achieved in two ways. First, m can be selected equal in all

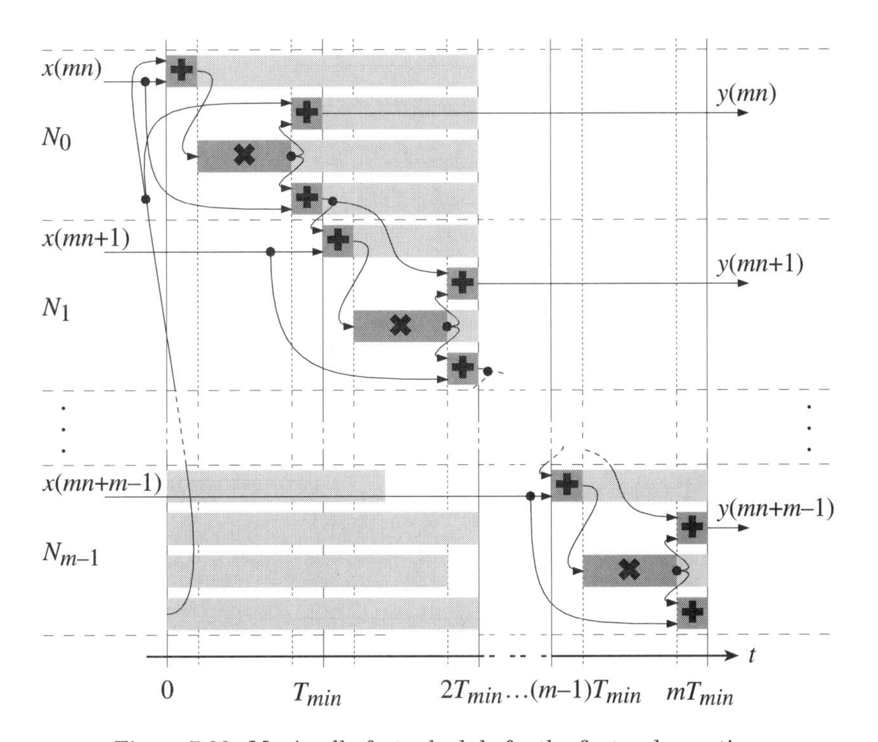

Figure 7.38 Maximally fast schedule for the first-order section

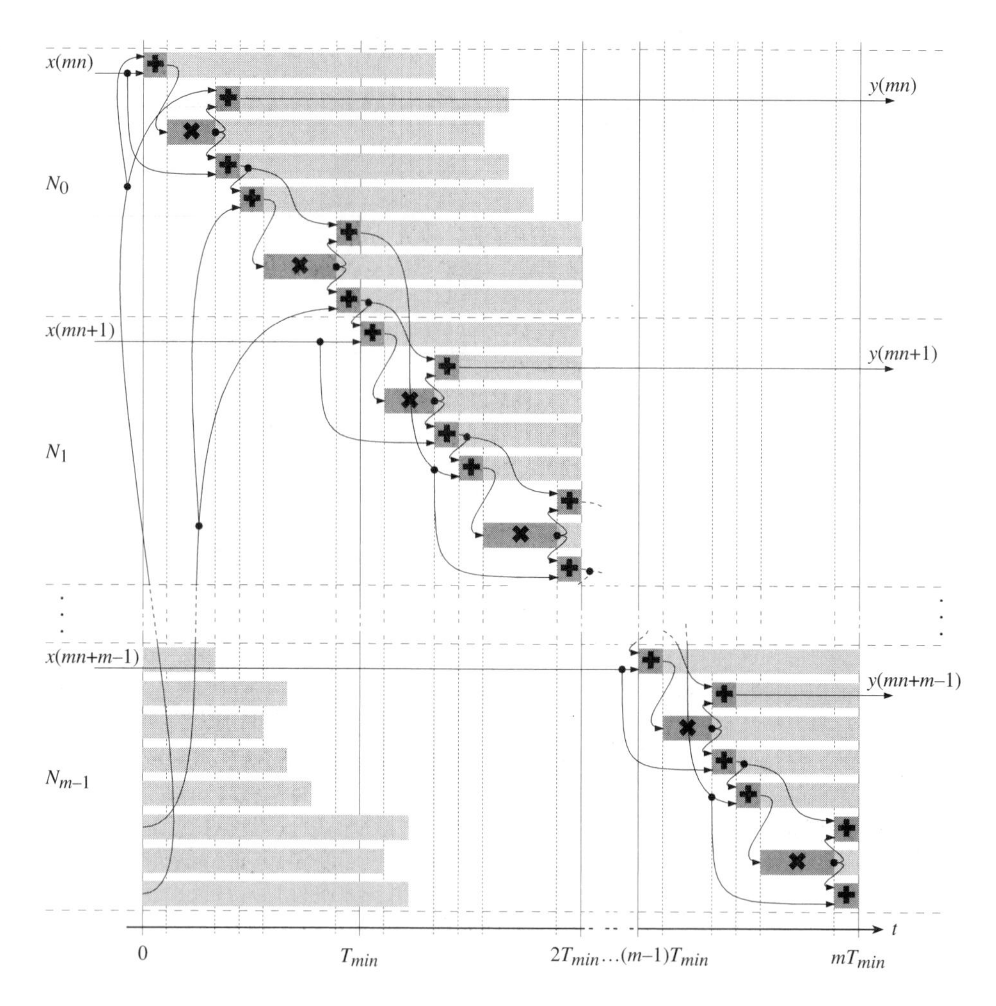

Figure 7.39 Maximally fast schedule for the second-order section

sections, which will require additional equalizing delays in the schedule to lengthen the scheduling period. Second, each section can be scheduled to be maximally fast, and then the clock period can be increased to match the bandwidths. The obvious choice is the first alternative, since the second alternative requires excessive hardware in terms of multiple clocks, bit-buffers, and multiplexers–demultiplexers between sections with different m in order to adjust the number of sample inputs.

Additional equalizing delays should preferably be inserted in cuts with few branches to minimize the amount of extra storage. Suitable locations for additional delay in the preceding schedules are between the processor sets N_i, since these cuts only contain the branches corresponding to the loops in the algorithm.

Since the regular structure of lattice wave digital filters also is found in the maximally fast schedule, it should be a reasonably simple task to develop software tools for automatic synthesis of such filters directly from the transfer function.

7.6 SCHEDULING ALGORITHMS

Scheduling algorithms are basically combinatorial optimization techniques. They search the solution space for a solution with favorable properties. Combinatorial optimization methods can be characterized as

- ❑ Heuristic methods
  - Constructive methods
  - Iterative methods
- ❑ Nonheuristic methods

Heuristic methods use a set of rules that limit the search in the solution space in order to shorten the required search time. This is often necessary since most scheduling problems have been shown to be NP-complete. A drawback of heuristic methods is that they may get stuck in local optima. Conversely, *nonheuristic methods* will find the optimal solution, but they will, in general, require excessive run times since they traverse the whole solution space. Exhaustive searches can be made effectively by depth-first or breadth-first approaches [3].

Constructive methods construct a solution whereas *iterative methods* produce better solutions iteratively from older ones. Constructive methods can be used to find an initial solution for an iterative method. Both these types of methods traverse only a part of the solution space and yield suboptimal solutions.

The following optimization techniques are commonly used for scheduling:

- ❑ ASAP (as soon as possible) and ALAP (as late as possible) scheduling
- ❑ Earliest deadline and slack time scheduling
- ❑ Linear programming methods
- ❑ Critical path list scheduling
- ❑ Force-directed scheduling
- ❑ Cyclo-static scheduling
- ❑ Simulated annealing
- ❑ Genetic algorithms

7.6.1 ASAP and ALAP Scheduling

Many real-life projects start with an ASAP schedule
but are completed according to an ALAP schedule.

Ragnar Arvidsson, Ericsson Radar Electronics

ASAP (as soon as possible) scheduling [8, 14, 17], simply schedules operations as soon as possible. A computation can be performed as soon as all of its inputs are available—i.e., all predecessors have been performed. The aim is to obtain the shortest possible execution time without considering resource requirements. Figure 7.40 shows an example of an ASAP schedule.

ALAP (as late as possible) is a method similar to ASAP. Figure 7.41 shows an example of an ALAP schedule. In ALAP the operations are scheduled as late as possible. ASAP and ALAP scheduling are often used to determine the time range in which the operations can be scheduled. This range is often referred to as *life-span*.

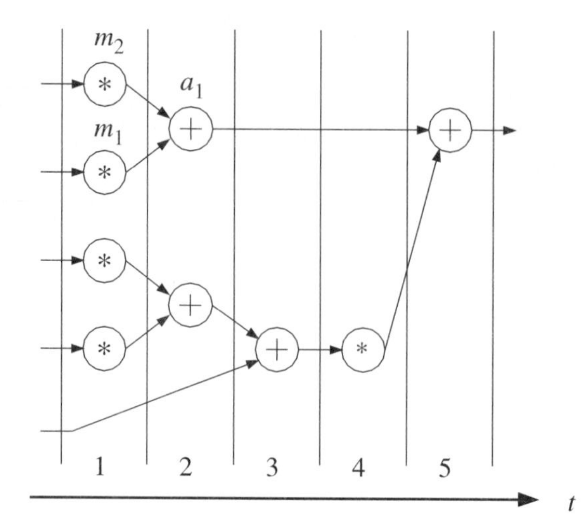

Figure 7.40 Example of ASAP schedule

For example, multiplications m_1 and m_2 and addition a_1 are the only operations in Figures 7.40 and 7.41 that can be rescheduled. The reduction in lifespans of the operation results in a reduction in execution time of the scheduling algorithm.

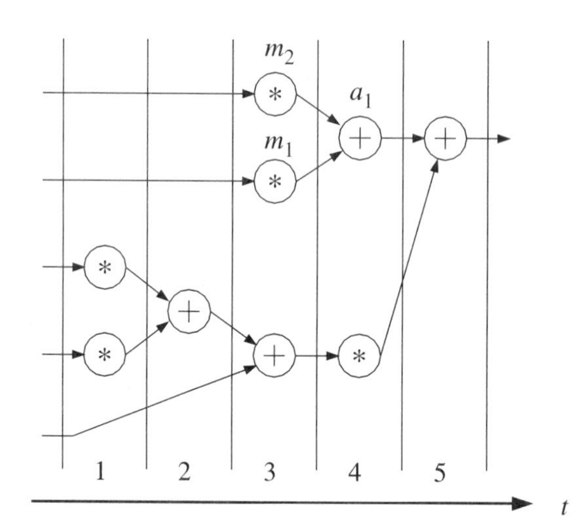

Figure 7.41 Example of ALAP schedule

7.6.2 Earliest Deadline and Slack Time Scheduling

These methods can be used to schedule periodically or randomly occurring processes. The *earliest deadline* scheduling algorithm [2] schedules, in each time step, the processes whose deadline is closest. In general, this scheduling technique requires that the execution of the process may be preempted. This method has proven to be execution-time–optimal for a single processor.

The *slack time* algorithm [11] resembles the earliest deadline algorithm. In each time slot, it schedules the process whose slack time is least. Slack time is the time from the present to the deadline minus the remaining processing time of a process. It is equivalent to shimming delay. This method has been proven to give solutions that are better than or equally as good as solutions obtained using the earliest deadline algorithm in cases where more than one processor is used.

7.6.3 Linear Programming

The scheduling problem is a combinatorial problem that can be solved by integer linear programming (*LP*) methods [1, 13]. These methods (for example, the *simplex method* and the *interior point methods*) find the optimal value of a linear cost function while satisfying a large set of constraints. The precedence relations for the operations to be scheduled represent a set of inequalities in the form of nonfeasible time intervals. Using linear programming methods, the optimal solution can be obtained. Problems with a few thousand constraints and several thousand variables can be tackled on a small PC while workstations can often handle problems with variables in the tens of thousands or even greater. However, very large problems with many operations might be intractable due to NP-completeness of the scheduling problem and the cost function being limited to a linear function. Linear programming methods are nonheuristic.

Integer LP models are ones where the answers may not take fractional values. This is a very much harder problem than ordinary LP.

7.6.4 Critical Path List Scheduling

Critical path list scheduling [6–7, 14], is related to ASAP scheduling. It is one method in the class of list scheduling methods. The first step in list scheduling methods is to form an ordered list of operations to be performed. Secondly, the operations are picked one by one from this list and assigned to a free resource (PE). In critical path list scheduling, the list is formed by finding the critical paths in the DAG of the algorithm. Operations are ordered according to path length from the start nodes in the DAG. The method takes into account the fact that operations on the time critical path must be given a higher priority than others. In order to achieve as short an execution time as possible, not necessarily equal to the minimum sample period, operations on the critical path must be scheduled first. Critical path list scheduling is a constructive and heuristic method that leads to optimal schedules if the precedence graph is a tree and all processing times are equal. Other forms of list scheduling methods are obtained by using different heuristics for scheduling the operations do not belong to the critical paths.

7.6.5 Force-Directed Scheduling

The force-directed scheduling method attempts to distribute the operations in time so that the resources required are minimized under a fixed execution time constraint [18, 19].

The first step in force-directed scheduling is to perform an ASAP and an ALAP schedule on a fully specified signal-flow graph. This will give the earliest and latest times for all operations and their execution time frames. Each operation is given a probability to be assigned eventually to a time slot in its execution time frame as shown in Figure 7.42. The probabilities for each time slot are added into a distribution graph, (*DG*), one for each type of operation, as shown in Figure 7.43.

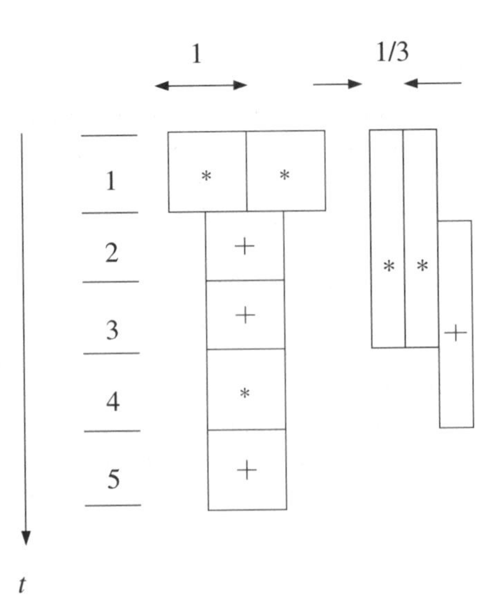

Figure 7.42 DG of multiplication and addition

Now, forces are associated with the restriction of the original time frame for operations $[t_0, t_1]$ and $[t_a, t_b]$ where $t_a \geq t_0$ and $t_b \leq t_1$ are calculated using the formula

$$Force(t_a, t_b) = -\sum_{i-t_a}^{t_b}\left[\frac{DG(i)}{t_b-(t_a+1)}\right]-\sum_{i-t_0}^{t_1}\left[\frac{DG(i)}{t_1-(t_0+1)}\right] \tag{7.5}$$

where $DG(i)$ is the value of the distribution graph for that type of operation in time slot i. For an assignment to a single time slot, k, Equation (7.5) reduces to

$$Force(k) = Force(k, k) = DG(k) - \sum_{i-t_0}^{t_1}\left[\frac{DG(i)}{(t_1-t_0+1)}\right] \tag{7.6}$$

This force is associated with the cost of assigning an operation to time slot k. However, assigning an operation will have an influence on the time frames of adjacent operations. Therefore indirect forces representing the cost of the assignment will restrict the time frames of other operations. This calculation is done according to Equation (7.6). The total force (self force plus indirect forces) is used to guide the scheduling. Scheduling is done by assigning the operations one at a time while always choosing the operation with the least total force.

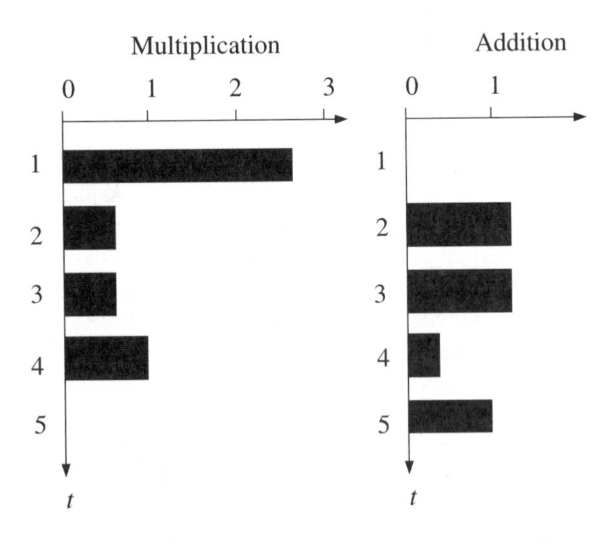

Figure 7.43 Execution time frames and associated probabilities

EXAMPLE 7.5

Calculate the forces in the preceding example.

The operations to the left in Figure 7.42 are all fixed. Assigning one of these will not alter any time frames. This means that these operations all have a total force equal to zero. The remaining operations, m_1, m_2, and a_1, will have forces according to Figure 7.44. The first of the three columns for an operation is the self force, the second is the indirect force, and the third is the total force.

The least total force is associated with assigning m_1 or m_2 to time slot 3. We choose to assign m_1. This will lock a_1 to time slot 4 and the only remaining moveable operation will be m_2. When an operation has been assigned, the time frames of the adjacent operations are changed—i.e., we have to recalculate the distribution graph and the forces for adjacent operations before proceeding with the next operation. In our case we recalculate the time frame of m_2 and the DG for the multiplication.

	$\mathbf{m_1}$			$\mathbf{m_2}$			$\mathbf{a_1}$		
	Self	**Indirect**	**Total**	**Self**	**Indirect**	**Total**	**Self**	**Indirect**	**Total**
1	$\frac{4}{3}$	0	$\frac{4}{3}$	$\frac{4}{3}$	0	$\frac{4}{3}$	—	—	—
2	$-\frac{2}{3}$	$-\frac{1}{6}$	$-\frac{5}{6}$	$-\frac{2}{3}$	$-\frac{1}{6}$	$-\frac{5}{6}$	$\frac{1}{3}$	$\frac{8}{3}$	3
3	$-\frac{2}{3}$	$-\frac{2}{3}$	$-\frac{4}{3}$	$-\frac{2}{3}$	$-\frac{2}{3}$	$-\frac{4}{3}$	$\frac{4}{3}$	$\frac{2}{3}$	1
4	—	—	—	—	—	—	$-\frac{4}{3}$	0	$-\frac{2}{3}$

Figure 7.44 Forces associated with assigning operations m_1, m_2, and a_1 to different control steps. Self, indirect, and total force for each combination

7.6.6 Cyclo-Static Scheduling

The theory of cyclo-static processor schedules was developed by Schwartz and Barnwell [22, 23]. An operation schedule is a specification of which operations should be performed at all time instances, while a processor schedule is a specification of which operations are performed on each processor at all time instances. Thus, a processor schedule is an operation schedule including processor assignment.

The processor schedule for one sample interval is considered as a pattern in the processor–time space, $\mathbb{P} \times \mathbb{T}$. The processors are indexed cyclically, i.e., if we have N processors, the processors are indexed modulo(N). In general, the processor space is multidimensional. Further, a displacement vector is defined in processor space between two iterations. Together, the time and processor displacements form the principal lattice vector, $\mathbb{L}$, in the processor–time space $\mathbb{P} \times \mathbb{T}$.

A cyclo-static schedule is characterized by a pattern in the processor–time space for one iteration and its principal lattice vector. The pattern is repeated and displaced in both the time and processor domains by the vector $\mathbb{L}$. The first component of the principal lattice vector gives the spatial displacement. When the processor component is zero, the schedule is called static schedule. The second component vector denotes the time shift from one iteration to the other. Without blocking of the operations belonging to several sample intervals, this component is equal to the iteration period.

EXAMPLE 7.6

Consider the second-order section in direct form II shown in Figure 7.10. Assume that two types of processors are available: multipliers with delay 2 and adders with delay 1.

The iteration period bound is

$$T_{min} = \max\left\{\frac{4}{1}, \frac{4}{2}\right\} = 4$$

The processor bounds for adders and multipliers are $\lceil 1 \cdot 4/4 \rceil = 1$ and $\lceil 2 \cdot 5/4 \rceil = 3$, respectively.

A cyclo-static schedule is shown in Figure 7.45. The processor space is two-dimensional, one dimension for multipliers and one for adder, so the processor-time space will in fact be three-dimensional. The principal lattice vector is $\mathbb{L} = (0, 1, 4)$, i.e., no displacement for adders, displacement equal to 1 for multipliers, and sample period equal to 4. Figure 7.46 shows the processor schedule for three consecutive intervals. The three shadings denote operations belonging to successive sample intervals.

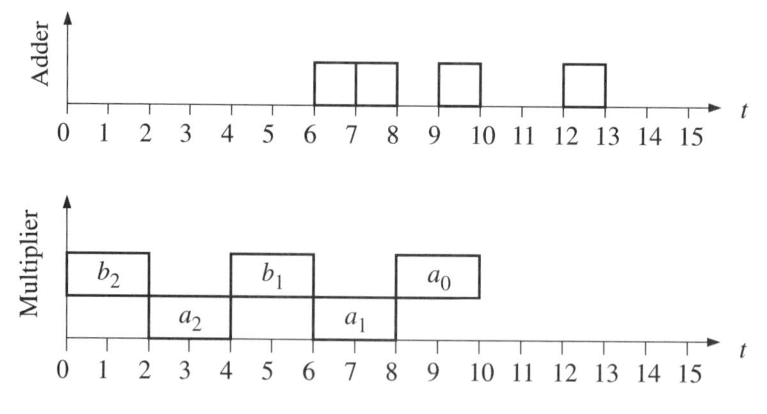

Figure 7.45 Cyclo-static processor schedule with $\mathbb{L} = (0, 1, 4)$

Obviously, the scheduling of two operations in the processor schedule shown in Figure 7.45 are not allowed to be a multiple of the principal lattice vector, because then the corresponding operations belonging to different intervals would

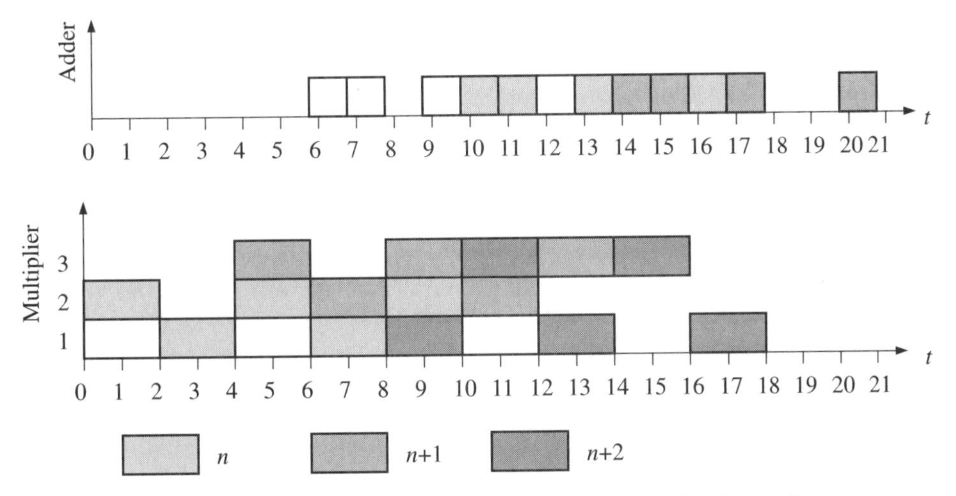

Figure 7.46 Processor schedule for three consecutive intervals

overlap in Figure 7.46. Note also that operations are folded due to the cyclic indexing of the multiplier processor space.

The schedule is both rate optimal and processor optimal since three multipliers and one adder are used. The multipliers have a utilization of $10/12 \approx 83\%$.

The processor schedule can be found by a depth-first search for valid cyclo-static schedules in the processor-time space with an iteration period equal to the iteration period bound and number of processors equal to the processor bound. Only solutions that are processor optimal are considered.

This approach consists of the following steps:

1. Determine the iteration period bound.
2. Depth-first search in the discrete space of
   (a) Valid processor optimal and iteration period optimal schedules.
   (b) Valid cyclo-static processor assignments.

The same method can be used to find non-rate optimal solutions. In such cases either the number of processors or the iteration period should be chosen. Cyclo-static scheduling is nonheuristic since it includes a search of the whole solution space. The search has a worst-case exponential complexity due to the NP-completeness of the scheduling problem. However, the size of real problems makes the solution space reasonably small.

Cyclo-static processor and rate optimal solutions exist for all recursive shift-invariant signal-flow graphs. The existence of solutions meeting other combinations of optimality criteria may not exist, and are problem dependent. Further, cyclo-static solutions can be found for a subset of the optimality criteria: processor optimality, rate optimality, and delay optimality. However, the approach requires that the PEs are homogeneous.

7.6.7 Maximum Spanning Tree Method

The maximum spanning tree method, which was developed by Renfors and Neuvo [20], can be used to achieve rate optimal schedules. The method is based on graph-theoretical concepts. The starting point is the fully specified SFG. The SFG corresponds to a computation graph N, which is formed by inserting the operation delays into the SFG. Further, a new graph N' is formed from N by replacing the delay elements by negative delay elements ($-T$). In this graph we find the *maximum distance spanning tree*—i.e., a spanning tree where the distance from the input to each node is maximal. Next, shimming delays are inserted in the link branches (i.e., the remaining branches) so that the total delay in the loops becomes zero. Finally, remove the negative delays. The remaining delays, apart from the operation delays, are the shimming delays. These concepts are described by an example.

EXAMPLE 7.7

Consider the second-order section as in Example 7.6 and derive a rate optimal schedule using the maximum spanning tree method.

Figure 7.47 shows the computation graph N'—i.e., the computation graph with the operation delays inserted and T replaced by $-T$. Note that the graphs are of the type activity-on-arcs and that the delay of adders is assigned to the edge immediately after the addition. The maximal spanning tree is shown in Figure 7.48 where edges belonging to the tree are drawn with thick lines and the link branches with thin lines. Next, insert link branches one by one and add shimming delays so that the total delay in the fundamental loops that are formed becomes zero. Finally, we remove the negative delays and arrive at the scheduled computation graph shown in Figure 7.49.

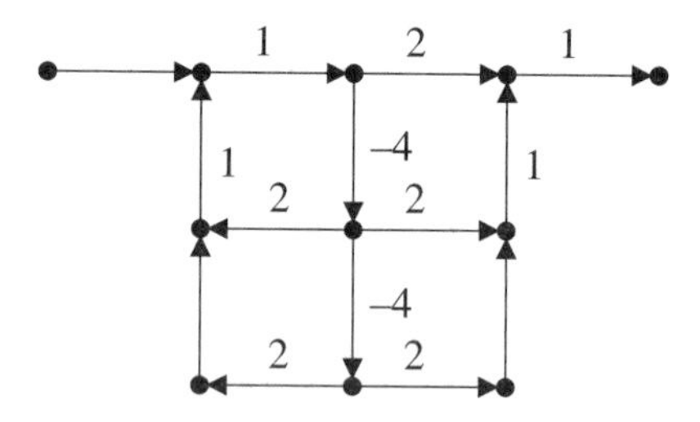

Figure 7.47 Modified computation graph N'

Note that there are no shimming delays in the critical loops. All operations are scheduled as early as possible, which in general will be suboptimal from a resource

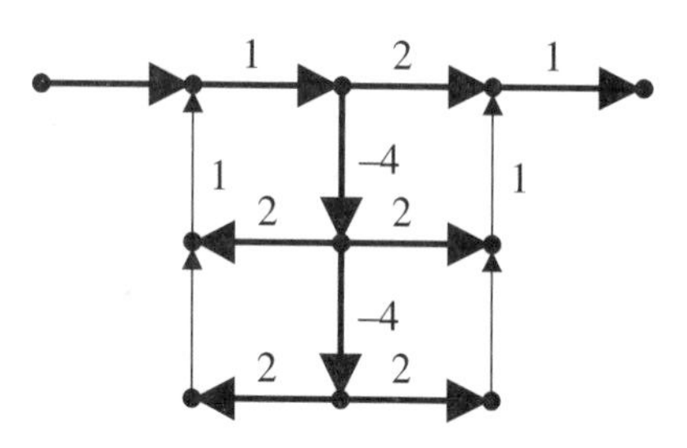

Figure 7.48 Scheduled computation graph

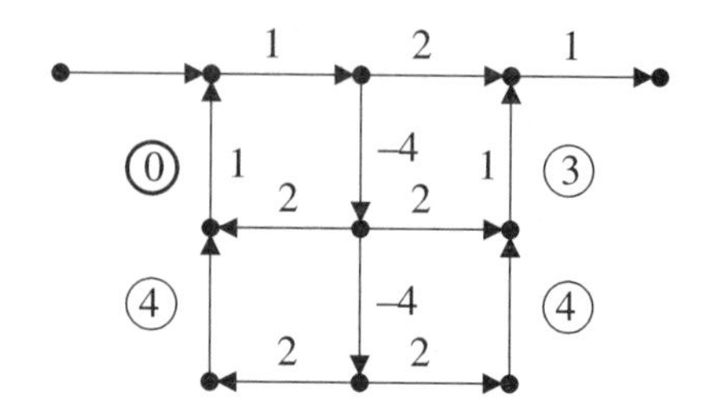

Figure 7.49 The maximum spanning tree of N'

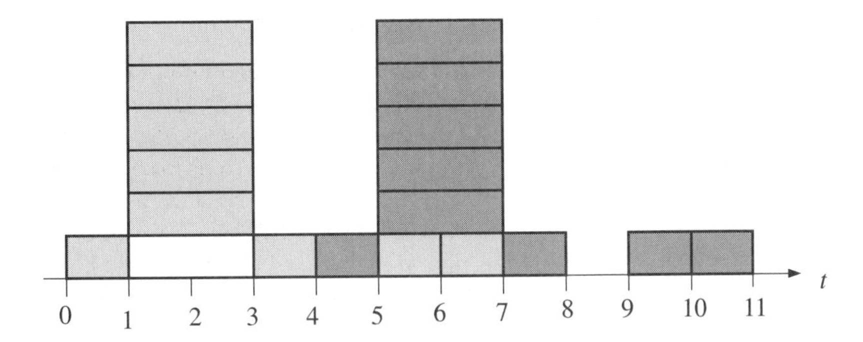

Figure 7.50 Rate optimal operation schedule

point of view. The resulting schedule is shown in Figure 7.50. Thus, this approach does not contain a minimization step—for example, minimization of the amount of resources. It only finds a feasible rate optimal schedule. However, it may be possible in a subsequent design step to move operations in time so that the required number of processors is minimized.

7.6.8 Simulated Annealing

In 1953, Metropolis *et al.* proposed an algorithm for simulation of how a solid reaches thermal equilibrium. Almost 30 years later, Kirkpatrick *et al.* and Černy realized that there is a simple analogy between certain combinatorial optimization problems and the slow cooling of a solid until it reaches its thermal equilibrium.

For example, an almost perfect monolithic *Si* crystal is grown from a seed crystal dipped into the silicon melt and a large crystal rod is formed if the seed is pulled out of the melt sufficiently slowly. At high temperatures, the molecules in the crystal will move about randomly. When the temperature is slowly reduced, the molecules tend to move to positions corresponding to a lower energy state (perfect crystal) and it becomes less likely that they move away from these positions as the temperature decreases. However, there is a small possibility that a molecule may temporarily move to a position with higher energy. Such moves to unfavorable positions are important for the annealing process in order to allow the perfect crystal to be formed. This means that the optimization algorithm does not get stuck in local minima. This property distinguishes simulated annealing from gradient methods which seek a solution in the direction of the largest derivative—i.e., the latter methods try to decrease the cost function as fast as possible.

Finally, at zero temperature, the molecules occupy states corresponding to the global energy minimum. The process of cooling a physical system until it reaches its low-energy ground state is called annealing. By simulating such a cooling process, near global-optimum solutions can be found. Simulated annealing is useful for solving large combinatorial problems with many degrees of freedom [10, 16, 21]. The simulated annealing algorithm is shown in Box 7.6.

```
Program Simulated_Annealing;
    ...
    begin
        Initialize;
        M := 0;
        repeat
            repeat
                Perturb(state(i) → state(k), ΔCost(i, k));
                if ΔCost(i, k) ≤ o then accept
                else
                    if exp(–ΔCost(i, k)/T_M) > Random(0, 1) then accept;
                if accept then Update(state(k));
            until Equilibrium; {Sufficiently close }
            T_M+1 := f(T_M);
            M := M + 1;
        until Stop_Criterion = True; {System is frozen }
    end.
```

Box 7.6 Simulated annealing algorithm

The algorithm starts from a valid solution (state) and randomly generates new states, *state*(i), for the problem and calculates the associated cost function, *Cost*(i). Simulation of the annealing process starts at a high fictitious temperature, T_M. A new state, k, is randomly chosen and the difference in cost is calculated, $\Delta Cost(i, k)$. If $\Delta Cost(i, k) \leq 0$, i.e., the cost is lower, then this new state is accepted. This forces the system toward a state corresponding to a local or possibly a global minimum. However, most large optimization problems have many local minima and the optimization algorithm is therefore often trapped in a local minimum.

To get out of a local minimum, an increase of the cost function is accepted with a certain probability—i.e., if

$$\exp\left(\frac{-\Delta Cost(i, k)}{T_M}\right) > \text{Random}(0, 1) \tag{7.7}$$

then the new state is accepted. The simulation starts with a high temperature, T_M. This makes the left-hand side of Equation (7.7) close to 1. Hence, a new state with a larger cost has a high probability of being accepted. For example, starting in state i, as illustrated in Figure 7.51, the new state k_1 is accepted, but the new state k_2 is only accepted with a certain probability. The probability of accepting a worse state is high at the beginning and decreases as the temperature decreases. This hill climbing capability is the key to escaping local minima.

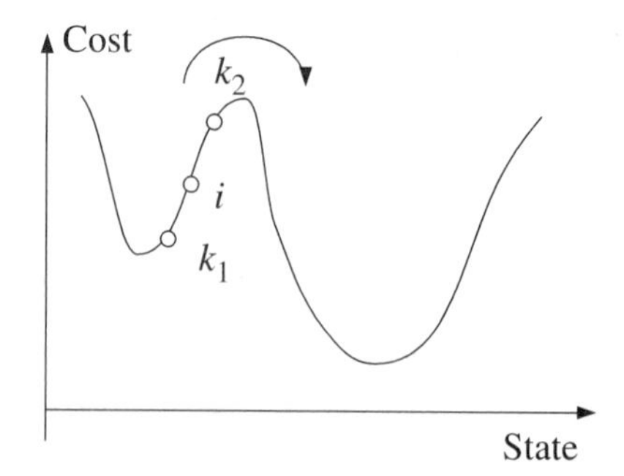

Figure 7.51 Selection of new states in simulated annealing

For each temperature, the system must reach an equilibrium—i.e., a number of new states must be tried before the temperature is reduced typically by 10%. It can be shown that the algorithm will, under certain conditions, find the global minimum and not get stuck in local minima [10, 16, 21].

Simulated annealing can be applied to optimize a cyclic schedule. We can find an initial schedule with almost any of the methods presented earlier. The only restriction is that the schedule must be feasible. Further, we can reduce the search space by performing both an ASAP and an ALAP schedule to determine the life-span of the operations. The simulated annealing algorithm can be directly applied to the cyclic scheduling problems just discussed. The operations are randomly moved within their feasible time frame so that the cost function is minimized. Generally, for any optimization method it is difficult to find a good cost function that has both local and global properties such that the global minimum is reached.

7.7 FFT PROCESSOR, Cont.

In this section we will schedule the processes of the FFT to satisfy speed constraints while respecting constraints on the available resources. We have already decided to use two butterfly PEs and two logical memories.

Figure 7.52 shows the global schedule for the processes in the FFT processor. The time estimates included in the figure will be explained later. The FFT processor

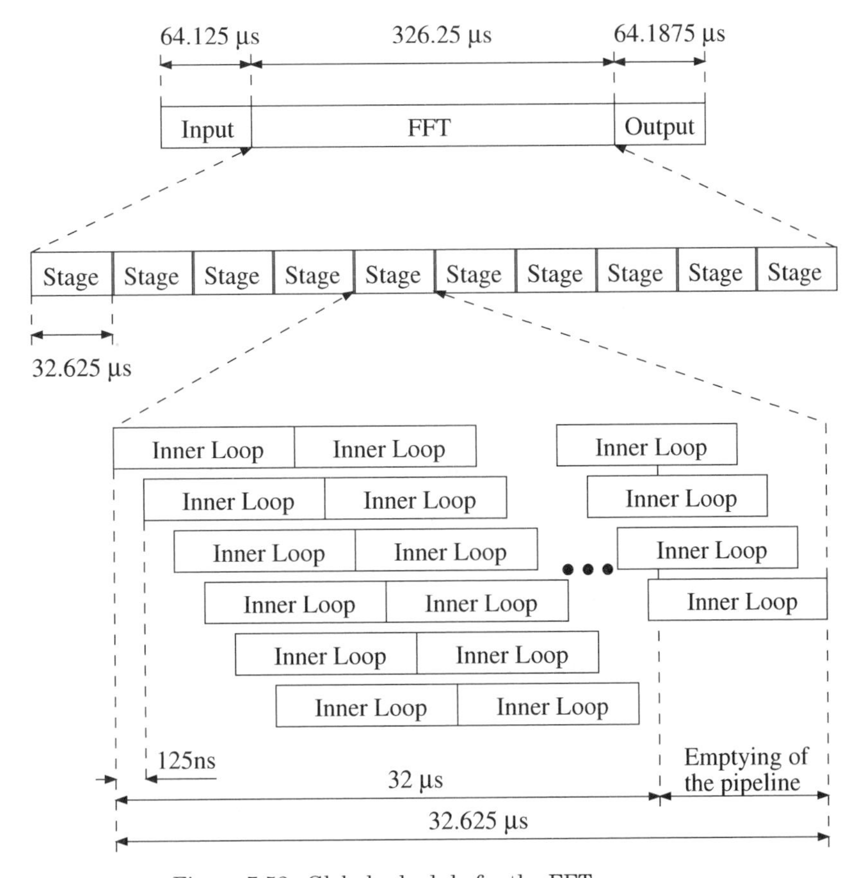

Figure 7.52 Global schedule for the FFT processor

has three sequential processes on the top level: input, FFT computation, and output. The FFT consists of ten stages that are computed sequentially. Each stage contains 256 inner loops. We will later schedule the inner loops so that a new computation can start every fourth memory clock cycle. The time interval between two successive inner loops is 125 ns. The inner loops will be mapped onto a pipelined hardware structure with six stages. If the executions of two stages are not allowed to overlap, we will need five extra loop cycles in each stage to empty the pipeline. This means that the execution time of a stage will be $(256 + 5) \cdot 125 \cdot 10^{-9} = 32.625$ µs. The FFT computation consists of 10 stages. Hence, the execution time of the FFT computation alone will be 326.25 µs. Scheduling the input and output processes that contain the input and output loops (denoted by *Input L* and *Output L*) is shown in Figure 7.53.

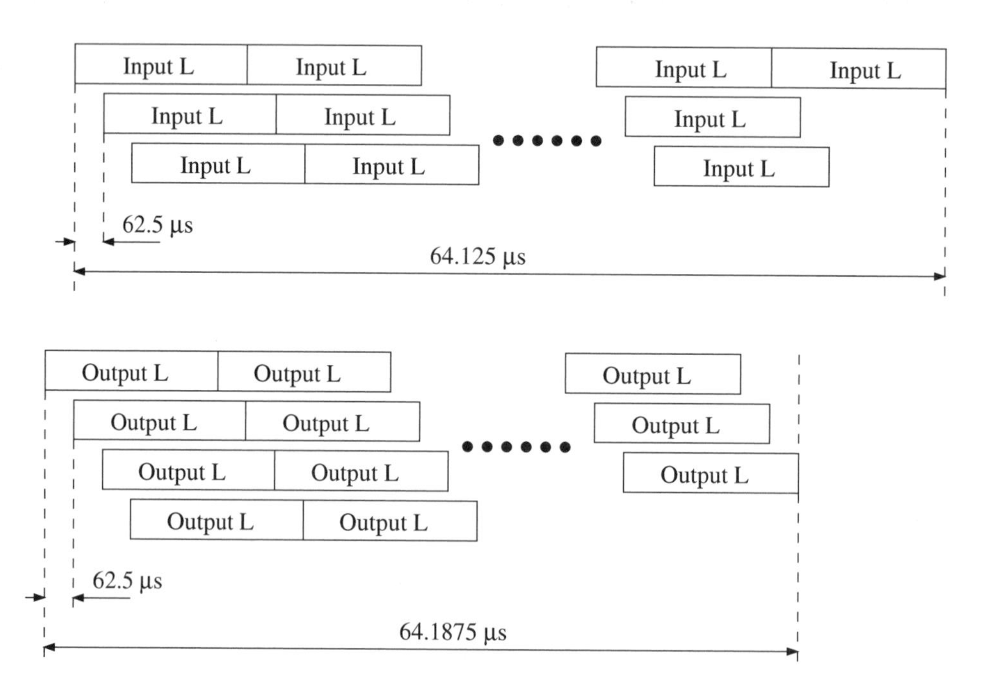

Figure 7.53 Schedule of input and output loops in the input and output processes

The input and output loops will be scheduled for one loop computation every second memory clock cycle. This means that the period of the loops is 62.5 ns, which corresponds to the 16 MHz I/O frequency. These loops will be mapped onto pipelined hardware structures with three and four stages, respectively. The output loop is similar to the input loop except for the digit reversal operation. The computation time for each stage in the pipelines will correspond to two memory clock cycles ($T_M = 31.25$ ns). The throughput for the FFT processor is

$$\frac{1}{64.125 \cdot 10^{-6} + 326.25 \cdot 10^{-6} + 64.1875 \cdot 10^{-6}} = \frac{1}{454.5625 \cdot 10^{-6}} \approx 2200 \text{ FFTs/s}$$

7.7.1 Scheduling of the Inner Loops

The processes in the inner loop are shown in Figure 7.54, where the memory processes have been explicitly shown as read and write processes. The precedence relationships between the loops has been denoted with dashed arrows. For scheduling purposes we have included the estimated lifetimes (latency) of the processes. The lifetimes are measured in units of memory clock cycles, T_M = 31.25 ns. We assume that the memory and W^p processes are pipelined internally so that the throughput is one data item per unit of time. Thus, a read or write cycle can be performed every memory cycle, but the internal pipeline causes the data item to be ready two memory cycles after the corresponding address has been applied. More specifically, the memory pipeline latches will be placed within the address decoders of the RAMs to allow higher throughput rates. The butterfly processes are internally pipelined to support one computation per four memory cycles.

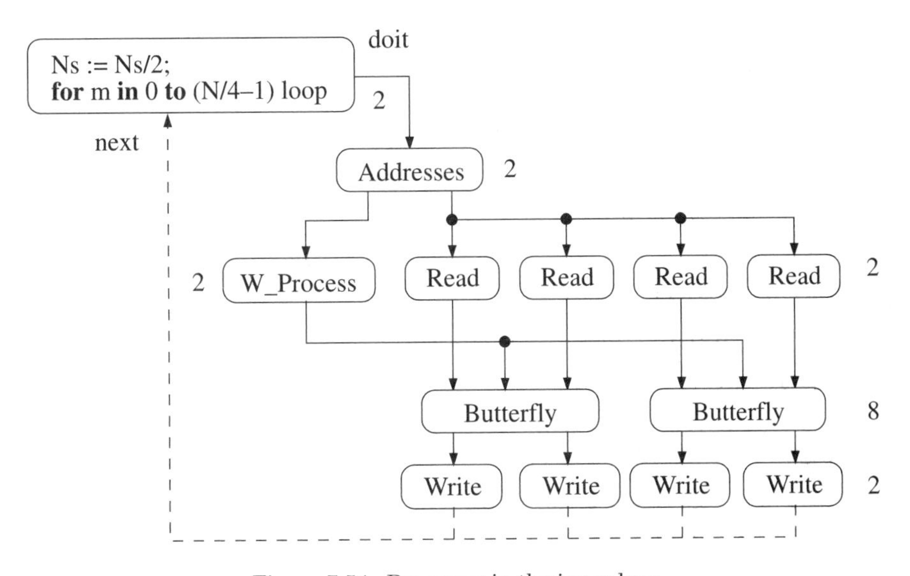

Figure 7.54 Processes in the inner loop

The inner loop in Figure 7.54 is too time consuming since the loop time is 16 time units. The FFT computation would take $16 \cdot 5120T_M/2$ = 1.28 ms. This is too long since only 0.32 ms is available. Loop folding has therefore been applied by overlapping successive loops, as illustrated in Figure 7.55, so that the loop time becomes equal to four memory cycles. The reads and writes for the ASAP schedule shown in Figure 7.55 have been denoted by R_1 to R_4 and W_1 to W_4 while the *W_Process* is denoted with W^p.

At the most two concurrent read and write operations are possible, since we have selected to use only two memories. In order to show more clearly how many concurrent processes the initial schedule contains, we fold the schedule modulo(4)—i.e., modulo the desired loop time. The folded schedule is shown in Figure 7.56.

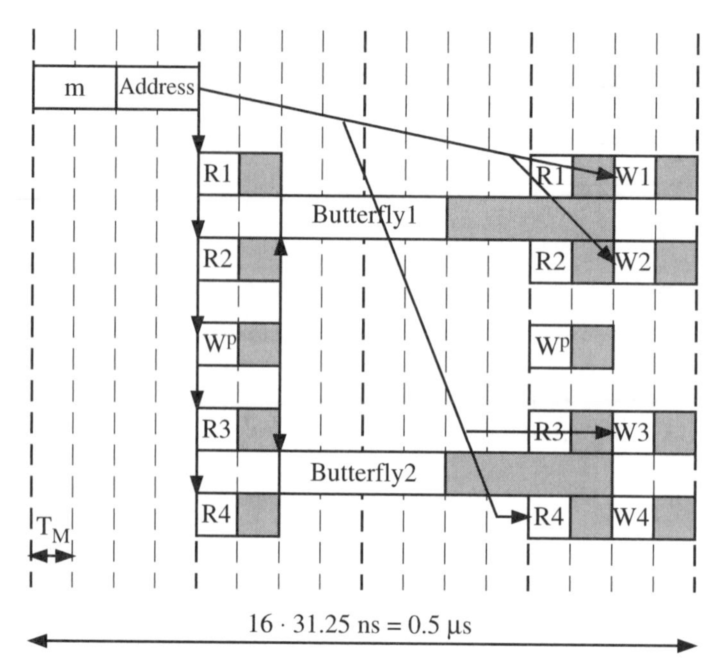

Figure 7.55 ASAP schedule of the inner loop

Obviously, this memory access schedule is unacceptable since there are four concurrent read or write operations. This schedule therefore requires four memories. Fortunately, the memory accesses are not evenly distributed within the folded schedule. We can therefore reschedule the processes so that the memory accesses are distributed more evenly in time and still meet the memory constraint. This can be done by introducing extra slack time (shimming delay), as shown in Figure 7.57. The corresponding folded schedule is shown in Figure 7.58.

The increased latency will increase the storage requirement. The *Address* must be stored for either 15 memory cycles or only four cycles if the address is recomputed. However, this requires extra hardware. The *W^p processes* (W^p) must be stored for two memory cycles or be scheduled to be computed two cycles later.

The inputs and outputs of the butterfly operations must also be stored as shown in Figure 7.59. Thus, only four registers, one for each complex data word to and from a butterfly, are required in the corresponding architecture.

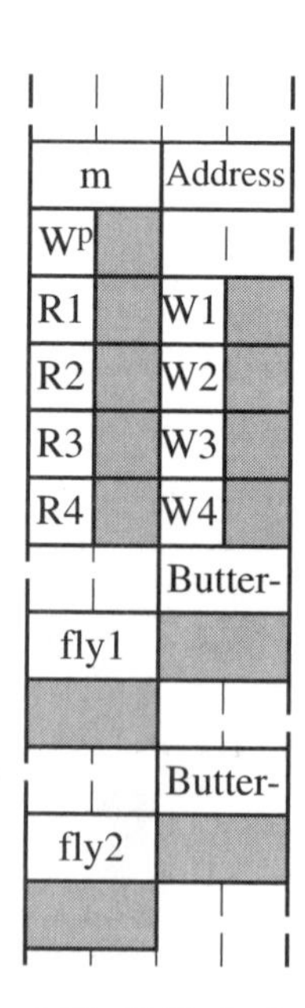

Figure 7.56 Folded schedule

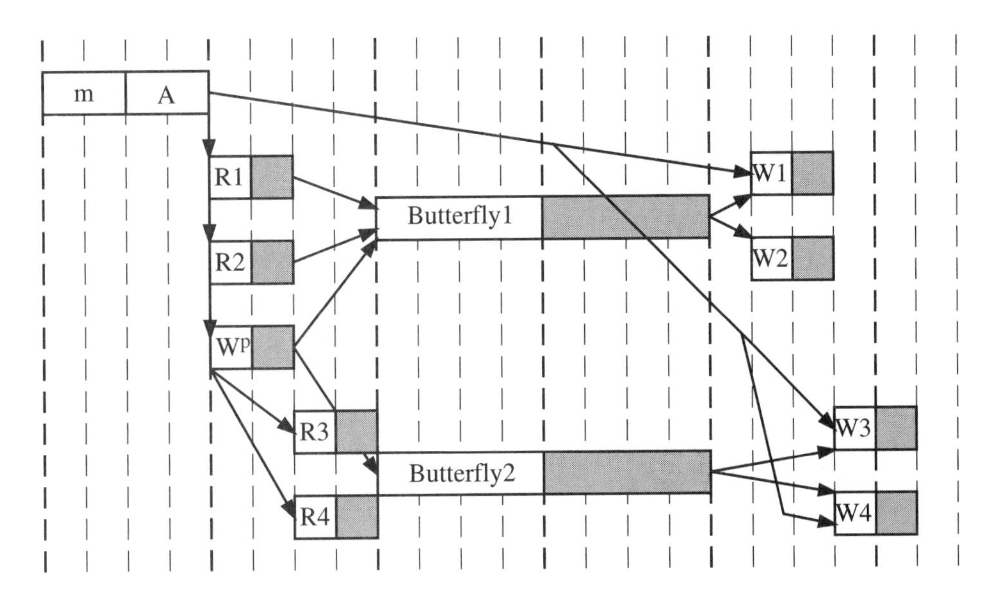

Figure 7.57 Improved schedule for inner loop processes

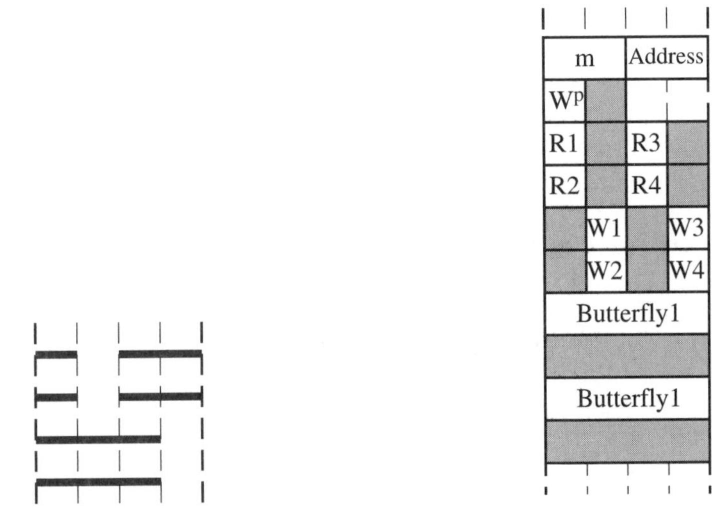

Figure 7.58 Improved folded schedule

Figure 7.59 Folded schedule for storage of the input and output variables of the butterflies

7.7.2 Input and Output Processes

The processes in the input and output loops must also be scheduled. Figure 7.60 shows the processes in the input loop with estimates of the execution time measured in memory clock cycles. The I/O frequency of the FFT processor has been chosen to be 16 MHz and the memory clock frequency to be 32 MHz. Hence, a write operation must be performed every other memory clock cycle. The execution

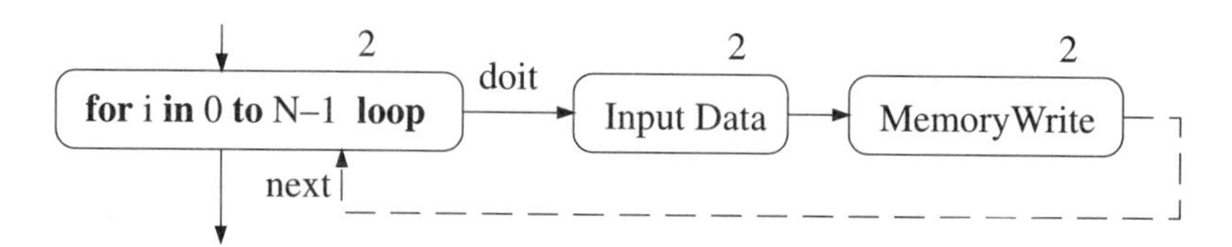

Figure 7.60 Processes in the input loop

time for the input loop is six memory clock cycles. The operations in the loop are therefore folded (pipelined), as shown in Figure 7.61, so that one loop can be executed every other memory clock cycle. The pipeline will have three stages. The output loop with execution time estimates, as well as the folded, pipelined schedule, is shown in Figure 7.62. The pipeline will have four stages.

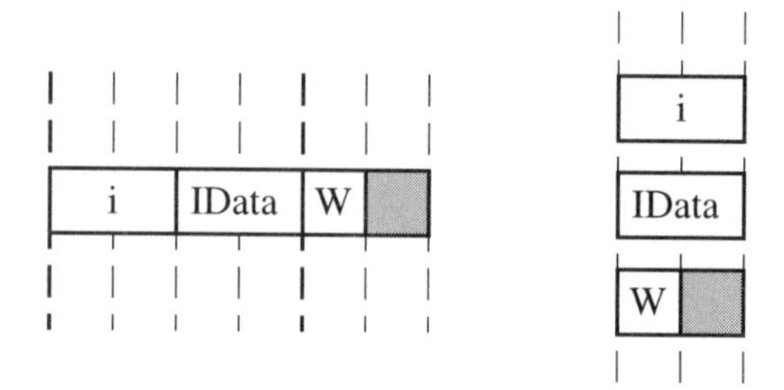

Figure 7.61 Schedule and folded schedule for the input loop

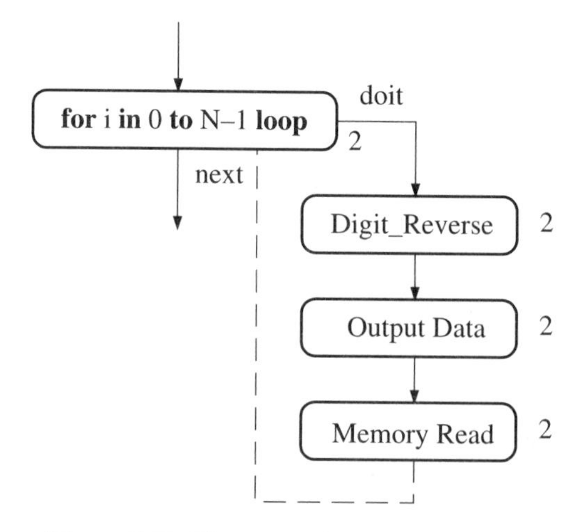

Figure 7.62 The output loop with schedule

7.8 RESOURCE ALLOCATION

In the resource allocation step the amount of resources required to execute the different types of processes is determined. We will refer to the time interval during which a process is executed as the *lifetime* of the process. Note that storing a value in memory is a (storage) process.

The amount of resources can be minimized by letting processes that do not overlap in time share resources. The number and type of resources can be determined from the operation schedule and corresponding lifetime table for the signal values. Usually, the largest number of concurrent processes determines the number of resources required. In fact, both the resource allocation and assignment steps are usually simple one-to-one mappings where each process can be bounded to a free resource. However, as will be shown in Example 7.8, more resources than there are concurrent processes are required in some cases. This fact makes it difficult to evaluate the cost function to be minimized by the scheduling.

EXAMPLE 7.8

Consider the periodic algorithm whose schedule is shown in Figure 7.63. The iteration period is assumed to be 3. Obviously, there are at most two concurrent operations. Show that three PEs are required.

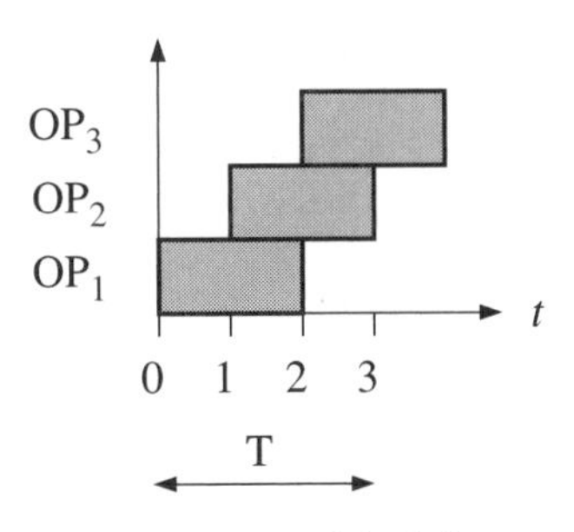

Figure 7.63 Schedule

Figure 7.64 shows the same schedule folded modulo(3). Still, there are at the most two concurrent operations. Now, let each row represent a PE. Obviously, it is not enough to use only two PEs, since a process that is folded modulo(3) must be continued on the same row in the diagram, i.e., continued in the same PE. It is illustrative here to draw the schedule on the circumference of a cylinder. In this case it is not possible to find a schedule that uses fewer than three PEs assuming that the schedule is nonpreemptive. A possible PE assignment with three PEs is shown in Figure 7.65.

The total amount of storage required can be determined from the lifetime table. Obviously, the largest number of concurrent storage processes is a lower bound of the storage requirement. In practice, it is important early in the design process to determine the number of logical memories required. This number equals the largest number of simultaneous memory transactions. The logical memories may in a low-speed application be implemented by a single physical memory, while in high-speed applications several physical memories must be used.

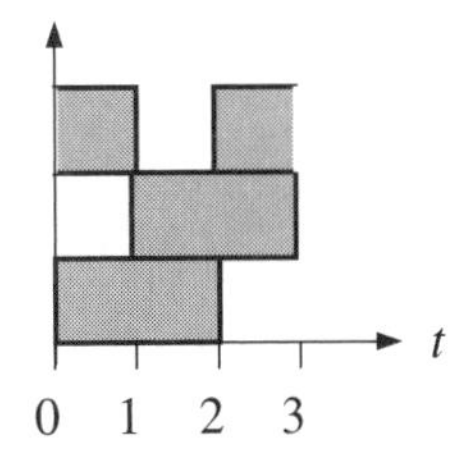

Figure 7.64 Folded schedule

We stress that scheduling the processes is a crucial design step, since it determines the number of PEs, memories, and communication channels required.

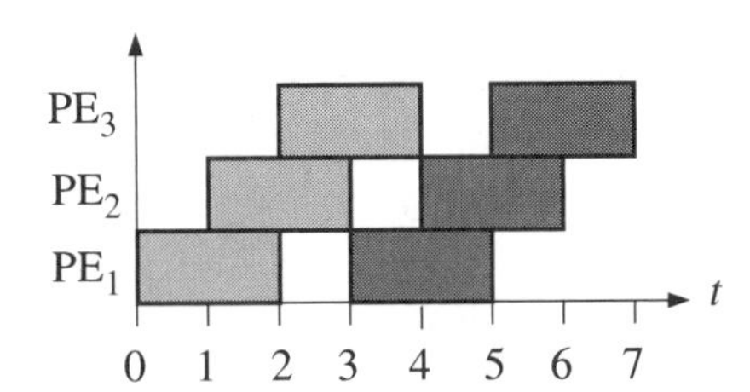

Figure 7.65 PE schedule for two consecutive sample periods

7.8.1 Clique Partitioning

The resource allocation problem can be solved by using different types of graphs to determine whether two processes of the same type may, or may not, share a resource. For example, two multiplications (processes) may be executed on a single multiplier (resource). The *connectivity graph* (which is also called a *compatibility graph)* is obtained by connecting two vertices with an edge if the lifetimes of the corresponding processes do not overlap. This implies that the processes may share a resource. A separate connectivity graph may be used for each type of resource if there are different types of processes that require different types of resources.

In order to determine which processes can share resources, we partition the connectivity graph into a number of cliques where a *clique* is defined as a fully connected subgraph that has an edge between all pairs of vertices. Thus, the processes corresponding to the vertices in a clique may share the same resource. Now, we choose the cliques so that they completely cover the connectivity graph—i.e., so that every vertex belongs to one, and only one, clique. Hence, each clique will correspond to a set of processes that may share a resource and the number of cliques used to cover the connectivity graph determines the number of resources.

EXAMPLE 7.9

Determine if the schedule shown in Figure 7.66 can be implemented by using two and three PEs. Also determine which processes should be allocated to the same PE in the two cases.

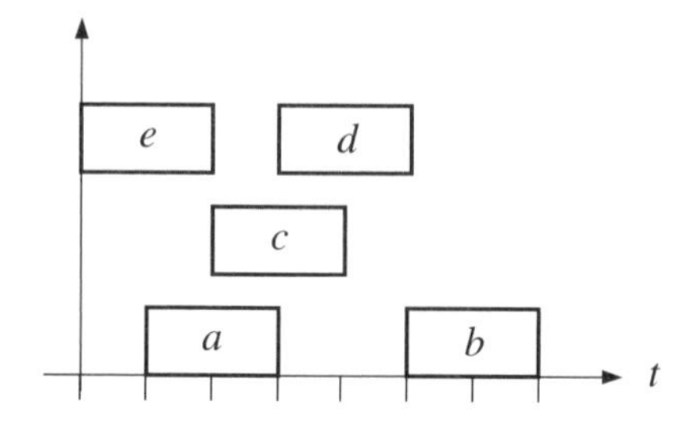

Figure 7.66 Schedule with five operations (processes)

The schedule has only five processes. Let each process correspond to a vertex and connect the vertices with a branch if the corresponding processes do not overlap. For example, processes a and c overlap, but processes a and b do not. Hence, we get a branch between vertex a and b, but not between a and c. We get the connectivity graph shown in Figure 7.67. We assume that the operations can be mapped to a single type of PE. Figure 7.68 shows two of many possible clique selections. The leftmost selection has only two cliques while the other has three cliques. The first choice requires only two PEs while the second requires three. Resource utilization is poor in the latter case.

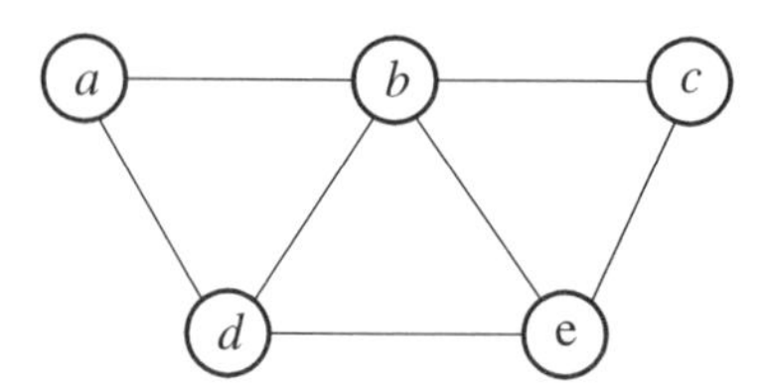

Figure 7.67 Connectivity graph

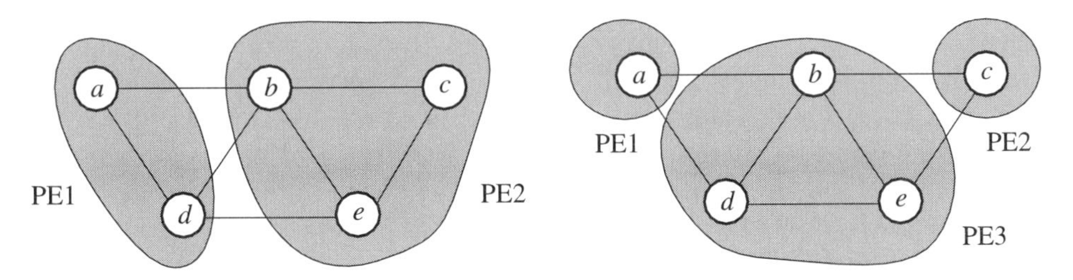

Figure 7.68 Two different clique selections leading to two and three PE requirements

There are several ways to select the vertices that should belong to a clique. This freedom can be used to minimize the number of cliques in order to minimize the number of resources. The problem is to find as large cliques as possible even though the amount of work represented by each clique becomes larger. Generally, the problem of finding an optimal set of cliques in a graph is NP-complete.

Sometimes it is more convenient to use another type of graph, which is called an *exclusion graph*. In the exclusion graph we connect two vertices with an edge if the corresponding processes can not share a resource. To perform resource allocation we try to find sets of vertices that are not connected. All processes corresponding to unconnected vertices may share a resource.

7.9 RESOURCE ASSIGNMENT

The resource assignment step involves selection of a particular resource from a pool of resources to execute a certain process. Ideally this assignment should be made at the same time as the scheduling, but this is not done in practice. Because of the complexity of the problem, it is instead divided into separate scheduling, allocation, and assignment steps.

The problem of assigning a class of processes to a corresponding set of resources can be formulated as a scheduling problem, where processes with compatible lifetimes are assigned to the same resource. Graph coloring techniques may be used to find optimal assignments. The aim is here to assign processes so that several processes can share the same resource—i.e., to maximize the resource utilization. Generally, the assignments should be made such that the total cost is reduced. For example, communication costs should be included.

7.9.1 The Left-Edge Algorithm

We will use a heuristic list-scheduling method related to the well-known left-edge algorithm commonly used in the VLSI design phase for routing wires. There the goal is to pack the wires in a routing channel such that the width of the channel is minimized. The width will here correspond to the number of resources. The algorithm assigns processes to resources in such a way that the resources become fully utilized. However, the left-edge algorithm does not always lead to an optimal solution.

The list of sorted processes is searched repeatedly for a process that can be assigned to the first resource. Such a process must have a lifetime that does not overlap the lifetimes of processes already assigned to this resource. The search for such processes is continued until we reach the end of the list. Then all processes that can be assigned to this resource have been assigned. This procedure for assigning processes is repeated for a new resource until all processes in the list have been assigned to a resource. The left-edge algorithm is based on the heuristic rule that processes with longer lifetimes should be assigned first. The left-edge algorithm is described in Box 7.7.

1. Sort the processes into a list according to their starting times. Begin the list with the processes having the longest lifetimes if the schedule is cyclic. Processes with the same starting time are sorted with the longest lifetime first. Let $i = 1$.
2. Assign the first process in the list to a free resource i, determine its finishing time, and remove it from the list.
3. Search the list for the first process that has
  (a) a starting time equal to, or later than, the finishing time for the previously assigned process; and
  (b) a finishing time that is no later than the starting time for the first process selected in step 2.
4. **If** the search in step 3 fails **then**
    if there are processes left in the list **then**
        let $i \leftarrow i + 1$ and repeat from step 2
    else
        Stop
    else
        assign the process to resource i, determine its finishing time, remove it from the list, and repeat from step 3.
    end.

Box 7.7 The left-edge algorithm

EXAMPLE 7.10

Figure 7.69 shows a schedule for a set of similar processes that shall be assigned to a set of homogeneous resources.

The first step is to sort the processes according to their starting time. The list of sorted processes is illustrated in Figure 7.70. To the first resource we begin by first assigning process A. The next process that can be assigned to this resource must have a starting time later than 5 and a finishing time earlier than 0 (15). Process B is the first such process in the list. No more processes can be assigned to the first resource.

To the second resource we assign process E. The next process that can be assigned to this resource must have a starting time later than 9 and a finishing

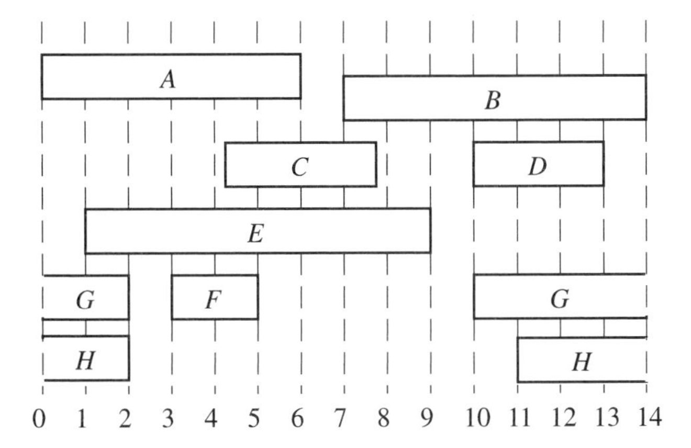

Figure 7.69 Processes to be assigned

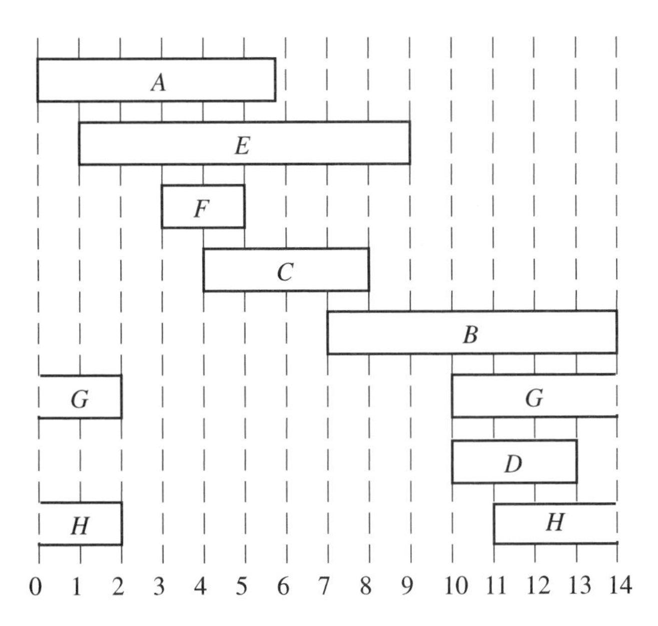

Figure 7.70 Processes sorted according to their left edge (starting time)

time earlier than 1. Process *D* in the only acceptable process in the list. Process *G* has a finishing time that is too late. No more processes can be assigned to the second resource. To the third resource we first assign process *F* and then *G*.

Finally, the two remaining processes are assigned to the fourth resource. The resulting assignments are shown in Figure 7.71. Note that there are often several assignments that yield the same number of resources. Additional criteria may therefore be used to select the best among these alternatives.

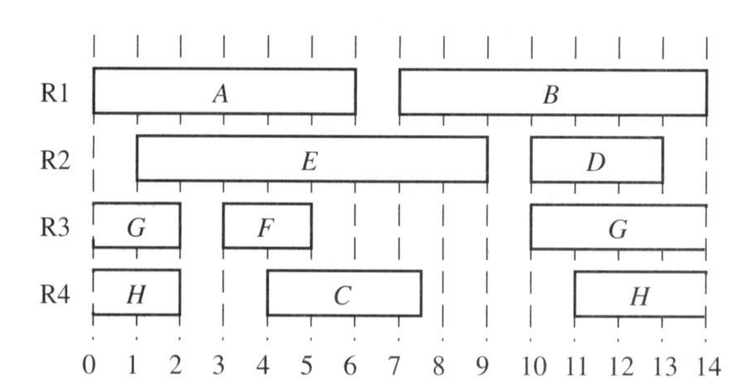

Figure 7.71 Resource assignment

7.10 INTERPOLATOR, CONT.

In this section we will determine a suitable operation schedule, and perform resource allocation and assignment, for the interpolator. We start by first assuming that the PEs have two pipeline stages. This means that the outputs of an adaptor can not be used directly as inputs to an adjacent adaptor, since the signal values are inside the pipeline. An operation can be executed in each time slot, but the result will be first available after two time slots, as indicated in Figure 7.72. Using the precedence relations shown in Figure 6.61, we get the ASAP schedule shown in Figure 7.72. In this case it is enough to perform the scheduling over only one sample interval, since the critical loops contain only one delay element.

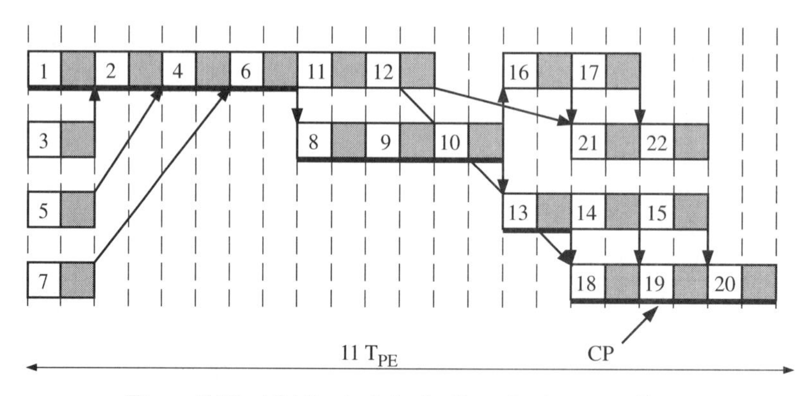

Figure 7.72 ASAP schedule for the adaptor operations

The PEs that execute the adaptor operations are implemented by using bit-serial arithmetic. The execution time for such a PE is $W_d + 3$ clock cycles, where W_d is the data word length. The latency is $2W_d = 2$ time units, because of the pipelining. We estimate that typically $W_d = 21$ bits are required and that the minimum clock frequency for the PEs is at least 220 MHz. The design of such PEs will be discussed in Chapter 11.

The minimum number of PEs is

$$N_a = \frac{(W_d + 3)N_{\text{op}} f_{sample}}{f_{CL}} = \frac{24 \cdot 22 \cdot 1.6 \cdot 10^6}{220 \cdot 10^6} \approx 3.84$$

We choose to use four PEs to perform the 22 adaptor operations. Hence, $T_{sample} = 6\ T_{PE}$.

Figure 7.73 shows the ASAP schedule folded modulo(T) = $6T_{PE}$. The number of concurrent operations is: 8, 0, 7, 0, 7, and 0. Obviously, a better schedule can be found. We therefore reschedule the operations using the loop folding technique in order to reduce the number of concurrent operations to only four. We assume that these operations can be assigned to the four PEs.

Note that the critical path is through 11 adaptors—i.e., the length is 22 time units. If four PEs are to be used, we must introduce four pipeline stages into the algorithm. The new schedule, which is shown in Figure 7.74, is $13T_{PE}$ long. For the sake of simplicity, data transfers between adaptors have not been shown explicitly in Figure 7.74.

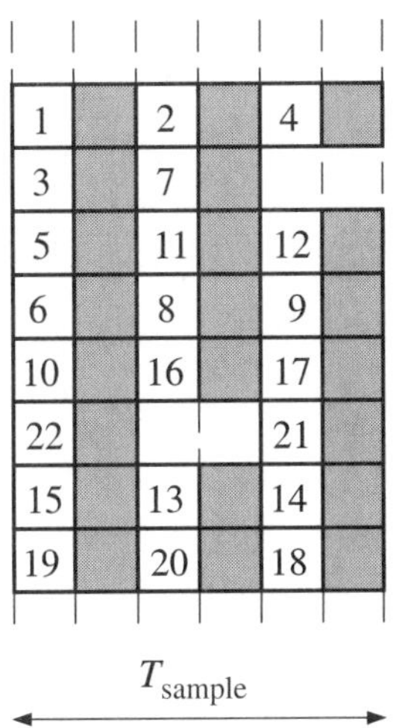

Figure 7.73 Folded ASAP schedule

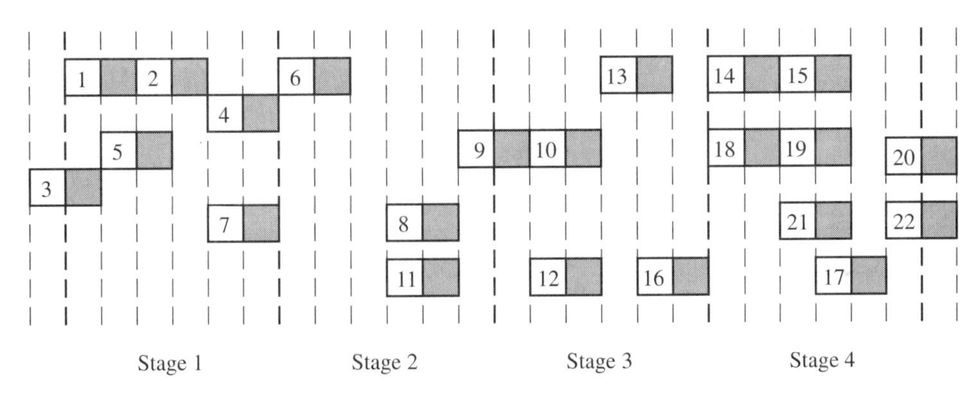

Figure 7.74 Improved schedule for the adaptor operations

We redo the scheduling modulo(T_{sample}), as shown in Figure 7.75. As wanted, there are only four concurrent operations. This indicates that four PEs may be enough. However, the nonpreemption constraint on the operations might mean that we will need extra PEs. Here we have unit delay operations only, and this will make four PEs sufficient. This is due to the fact that unit delay operations can not overlap operations starting at another time instance. Only concurrent operations can overlap.

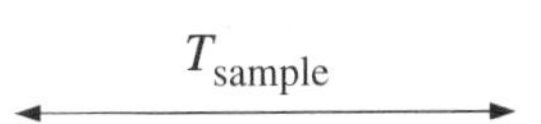

Figure 7.75 Folded schedule

7.10.1 Processor Assignment

The next design step is processor assignment—i.e., mapping the processes onto processors. We assume that the PEs will share a set of common memories. Hence, the PEs can only access the memories one at a time. The executions of the PEs are therefore skewed in time, as shown in Figure 7.76. The sample interval is divided into 24 time slots. In each time slot a PE writes its result to the memory and receives a new set of inputs from the memory.

There are many feasible PE assignments. This freedom in assignment may be used to minimize the amount of memory required and to simplify the control. In this case, the clique assignment scheme becomes too involved to be presented in detail. We therefore revert to a simple ad-hoc scheme. Two adjacent adaptor operations, where the output of the first is used as input to the next, are assigned to PE_i and $PE_{(i+1)\mathbf{mod}(4)}$.

We choose the PE assignment shown in Figure 7.76. Two of the PEs are idle one-sixth of the time. Hence, it is possible to increase the order of the allpass filter to 9 and get a smoother group delay with almost no penalty in terms of the number of PEs, power consumption, and execution time.

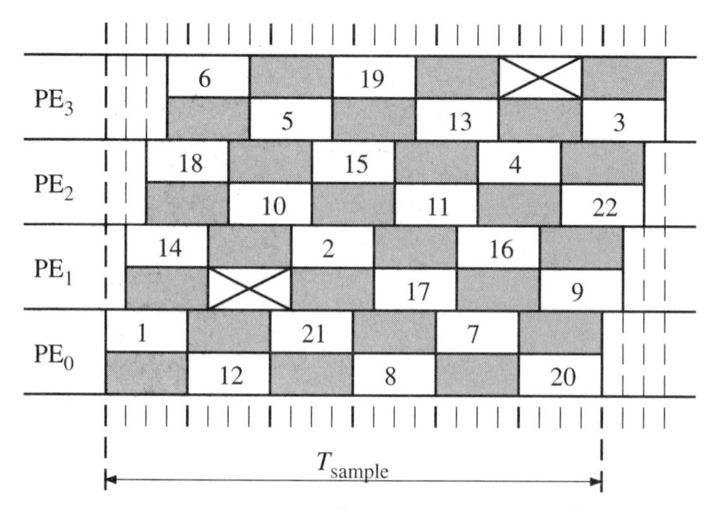

Figure 7.76 Skewed execution of the PEs

7.10.2 Memory Assignment

The constraints on memory allocation and assignment can be derived from the PE schedule. The memory–PE transactions can be extracted from the PE schedule shown in Figure 7.76 and the precedence form shown in Figure. 6.61. The arrows in Figure. 6.61 correspond to a transaction—i.e., one write and one read operation. There are four transactions (two inputs and outputs, each requiring one read and one write operation) taking place in each T_{PE} time slot. The memory schedule will therefore be over $2 \cdot 4 \cdot 6 = 48T_M$ time slots, where T_M is the time required for a memory read or a write operation. The lifetimes for the memory variables are shown in Figure 7.77. This chart is called a *lifetime table* for the memory variables.

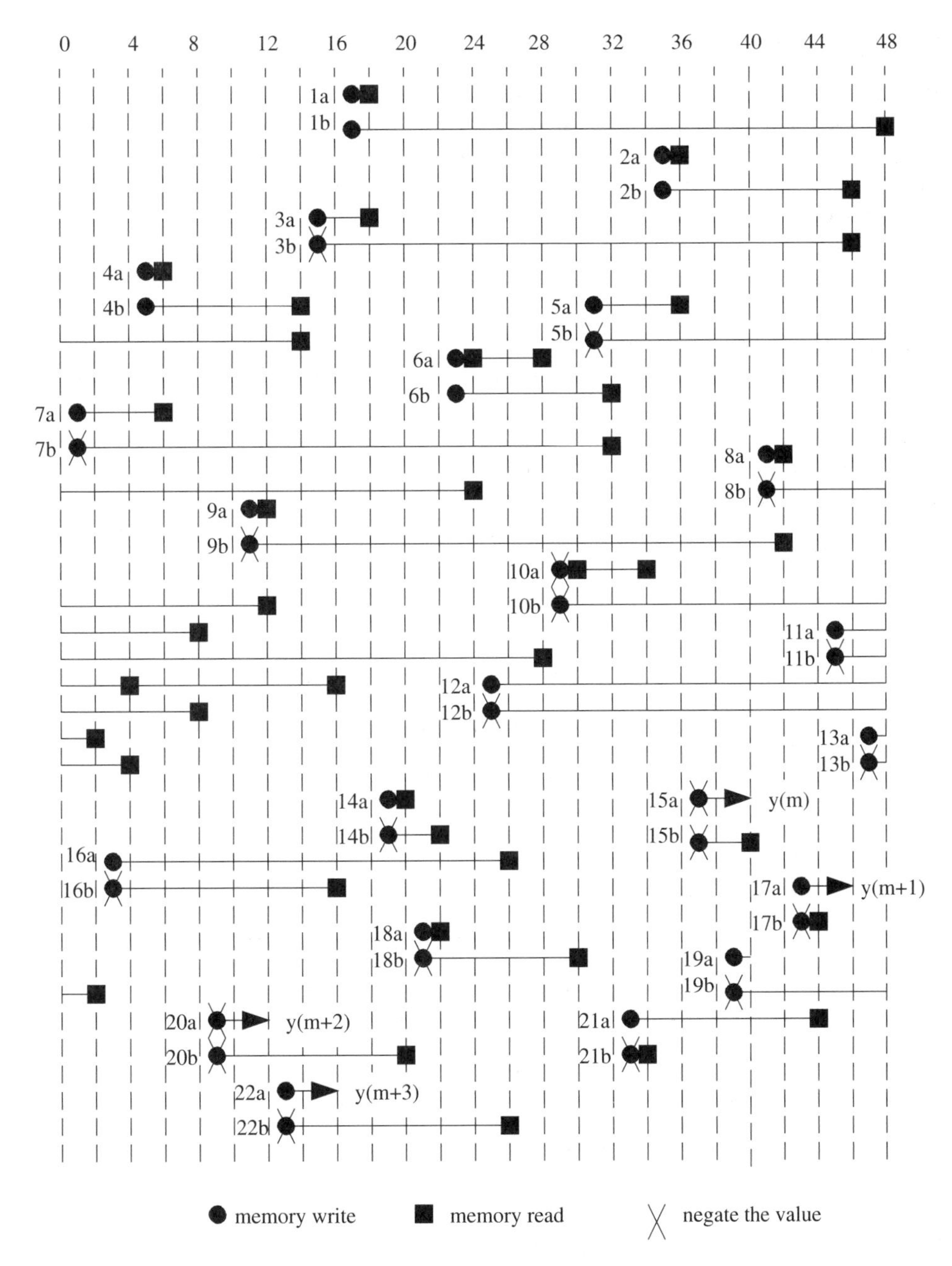

Figure 7.77 Lifetime table for memory variables

The lifetime table has been folded modulo($48T_M$). The horizontal lines represent storage of variables and their numbers correspond to the adaptor numbers. The wave-flow graph contains a number of multiplications by –1. These multiplications can easily be performed when the corresponding values are either written to or read from the memories.

The operation of the memory is as follows: First the result from a PE is written into the memory before the operands to the subsequent operation are read from the memory. Hence, two memory variables that share a memory cell according to the memory schedule must be separated by $2T_M$ time slots.

Some of the PE-memory transactions require only one memory cycle. These resources (i.e., a memory cell and a memory cycle) are not required if direct interprocessor communication is allowed in the target architecture. However, here we assume that this is not the case.

We have chosen to use two memories, since the adaptors have two inputs and two outputs. We will use a modified version of the clique partitioning technique just described to assign the variables to the memories. Instead of drawing a line between processes that may share resources, we will draw a line between vertices (memory variables) that can not be stored in the same memory or are accessed at the same time instances. The reason for this is that the number of branches in the graph otherwise becomes unmanageably large. The resulting *exclusion graph* is shown in Figure 7.78.

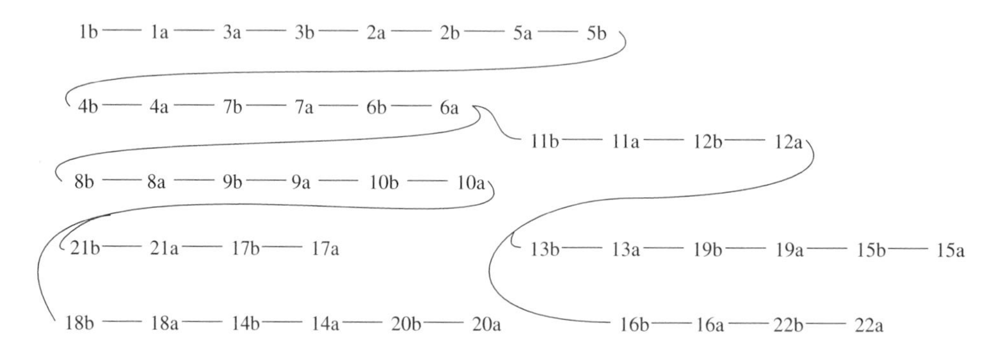

Figure 7.78 Exclusion graph for the memory variables

This graph must now be partitioned into two cliques (two memories) by drawing a line that separates all connected vertices. This procedure separates variables that must be stored in different memories. This is simple to do in this case. We pick every other variable, starting with variable 1a, to belong to the first clique and the rest to the other clique. There is only one way to choose the cliques in this case. However, we can change the PE assignment and get a new exclusion graph.

Figures 7.79 and 7.80 show the two sets of memory variables obtained by this choice. The variables have been sorted according to their starting times. Note that the amount of storage differs significantly between the two memories.

7.10.3 Memory Cell Assignment

The two memories can be minimized—i.e., we can do a memory cell assignment—by using an algorithm that resembles the left-edge algorithm used for wire routing on the chip.

First, we sort the variables in the lifetime table in a list according to their left edge—i.e., they are sorted according to their start times. Next we pick the first variable—i.e. the variable with the earliest starting time and the longest lifetime.

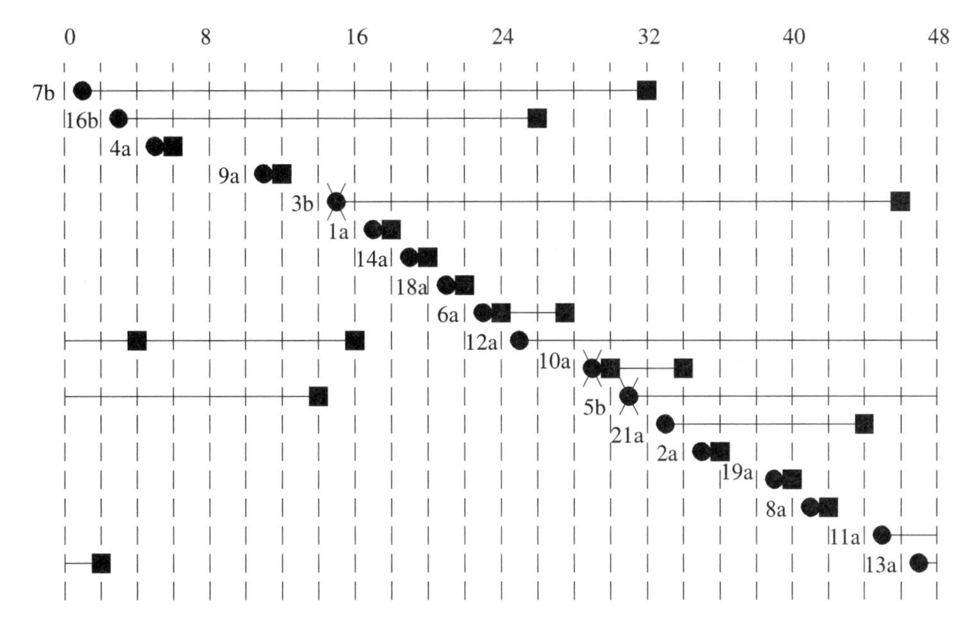

Figure 7.79 Lifetime table for memory 1 folded modulo($48T_M$)

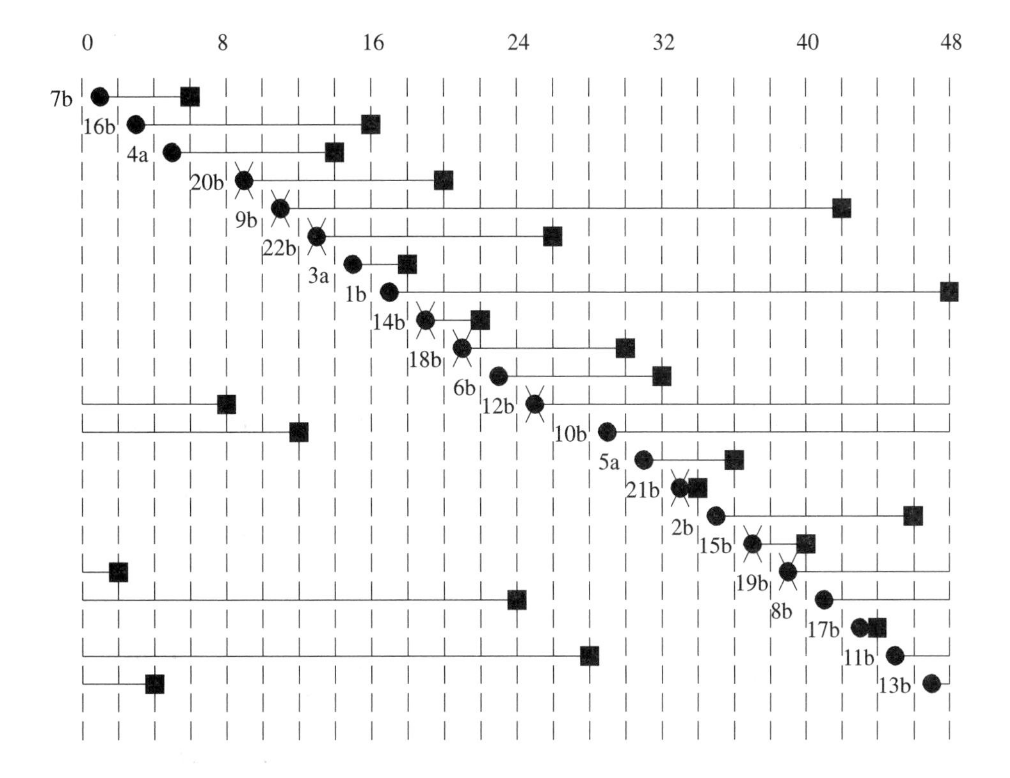

Figure 7.80 Lifetime table for memory 2 folded modulo($48T_M$)

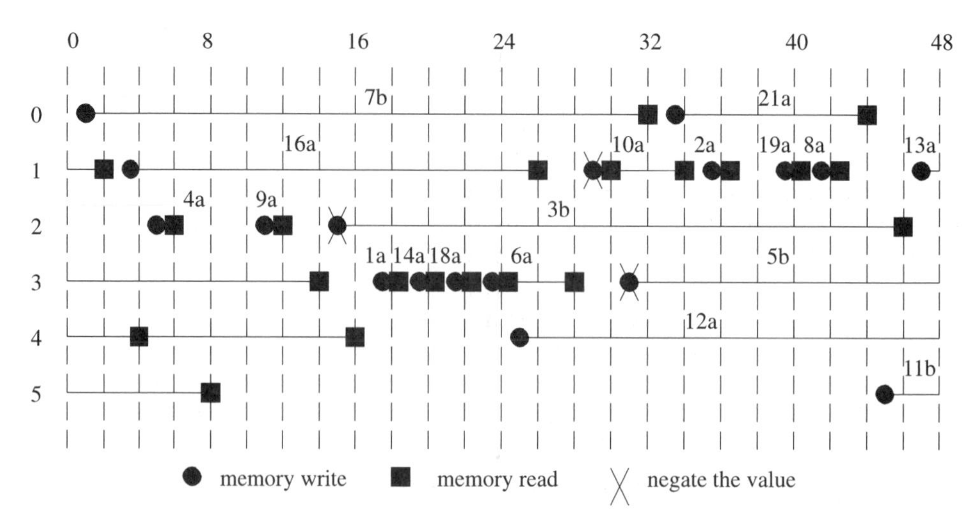

Figure 7.81 Cell assignment for memory 1

According to the left-edge algorithm, we combine this variable with the next variable in the list that has a starting time later than the finishing time of the first variable. However, each time step consists of a write followed by a read operation. Two variables in Figures 7.79 and 7.80 that shall be stored in the same cell must therefore be separated by one time slot. Otherwise the variables will overlap.

The results, after the left-edge algorithm has been applied, are shown in Figures 7.81 and 7.82. The two memories must according to this scheme have six

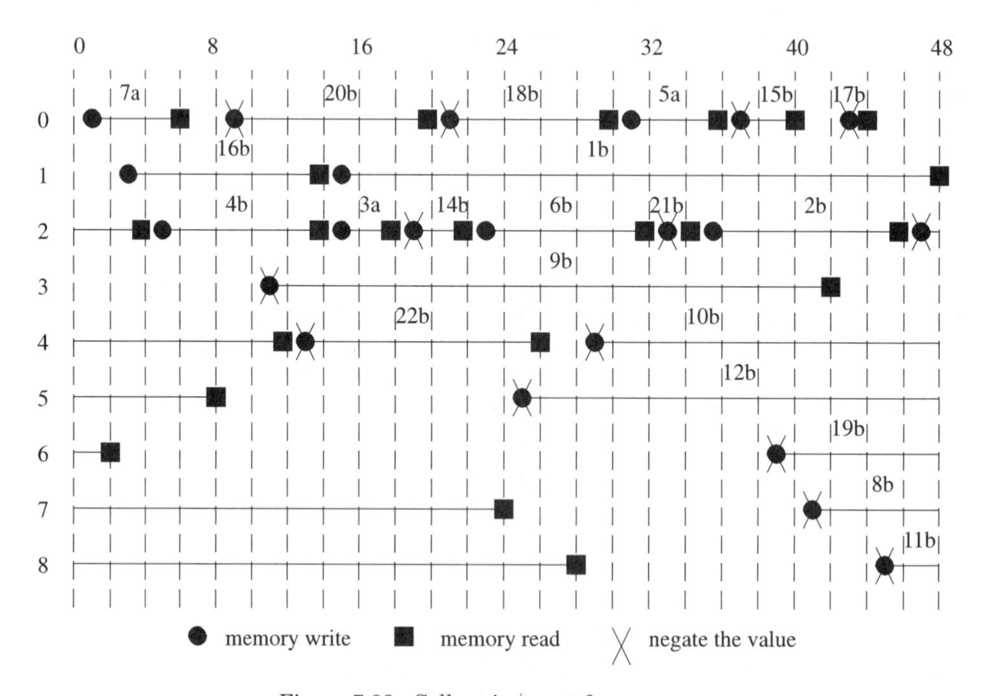

Figure 7.82 Cell assignment for memory

and nine memory cells, respectively. Note that the left-edge algorithm does not in general lead to minimum memory size. The required access rate for the memories is

$$f_{RAM} = \frac{1}{T_M} = \frac{48}{T_{sample}} = 48 \cdot 1.6\ 10^6 = 76.8 \text{ MHz}$$

The lifetime will overlap itself in the folded lifetime table if it is longer than the scheduling period. In principle, this problem can be solved in either of two ways: either by allowing preemption or by duplicating the memory resources used to store these values. Preemption of storage requires that values are transferred between memory cells. Of course, this is not an efficient scheme.

The second solution is similar to duplicating resources in an operation schedule when operations are longer than the scheduling period. The longest storage time in this case is only $34T_M$ long. Hence, this is not a problem in this case.

7.11 FFT PROCESSOR, CONT.

In this section, we discuss various alternatives in assigning processes to hardware resources. In particular we will assign butterfly operations to PEs and data to memories. The assignments will differ with respect to complexity of the control structures in the architecture and the interconnection network that connects the PEs and the memories. In fact, the scheduling, resource allocation, and assignment steps specify the main features of the corresponding class of optimal architectures. We will therefore defer evaluation of the alternatives to Chapter 9.

7.11.1 Memory Assignment

The FFT algorithm allows in-place computation—i.e., the results of a butterfly operation can always be written into the memory cells that stored the inputs to the butterfly. A value in the memory is used as input to one butterfly operation only. It is therefore necessary to store only N complex values in the memories.

Because of the speed requirement we have already decided to use two logical memories (RAMs). In practice, large RAMs are implemented as a set of smaller memories.

Memory Assignment 1

There are many feasible memory assignments for the ST-FFT. It is not practically possible to evaluate all assignments. We will here investigate two alternative memory assignments for the ST-FFT. The first assigns the first $N/2$ variables ($0 \le i < N/2$) to RAM_0 and the remaining values ($N/2 \le i < N$) to RAM_1. The assignment for a 16-point FFT is illustrated in Figure 7.83. However, this assignment requires a trivial modification of the schedule shown in Figure 7.57.

The FFT is a divide-and-conquer algorithm—i.e., the FFT is recursively divided into smaller and smaller FFTs. However, the first stage of the FFT differs from the other stages. This memory assignment will not cause any difficulties in the second to fourth stages of the FFT, since both input values to a butterfly operation will be read from the same memory. However, in the first stage, a butterfly operation must receive a value from each of the memories. This implies a switching function in the interconnection network. The switching will also require some control circuitry. These factors will influence the cost of the implementation.

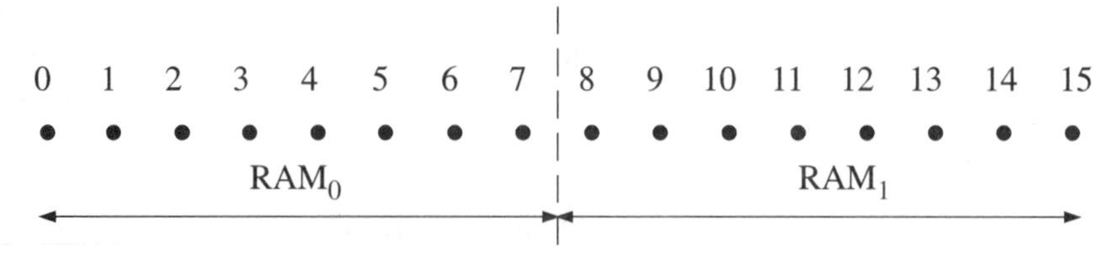

Figure 7.83 First RAM assignment alternative

Memory Assignment 2

The second alternative is to always store the two input and output values to a butterfly in different memories. To develop such an assignment scheme we will first study some small FFTs. Note that the unscrambling stage need not to be considered since this operation will be performed when data are written out of the FFT processor.

There are two input values to a butterfly operation, hence the input values must be stored in different RAMs. The corresponding exclusion graph will therefore have a branch between vertices 0 and 1, as shown to the right in Figure 7.84. For simplicity we select to store the values with indices 0 and 1 in RAM_0 and RAM_1, respectively. Figure 7.84 shows the exclusion graph for the values in an in-place computation of a two-point FFT.

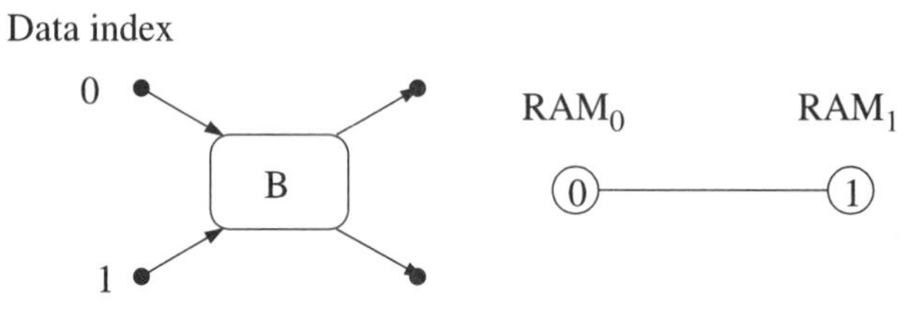

Figure 7.84 A two-point FFT and exclusion graph for memories

A four-point FFT can be built out of two two-point FFTs by adding a first stage, as shown in Figure 7.85. The constraints in the exclusion graph will come from the first stage—i.e., there are exclusion branches between vertices 0–2 and 1–3. The constraints from the second stage are similar to the constraints obtained for the two-point FFT.

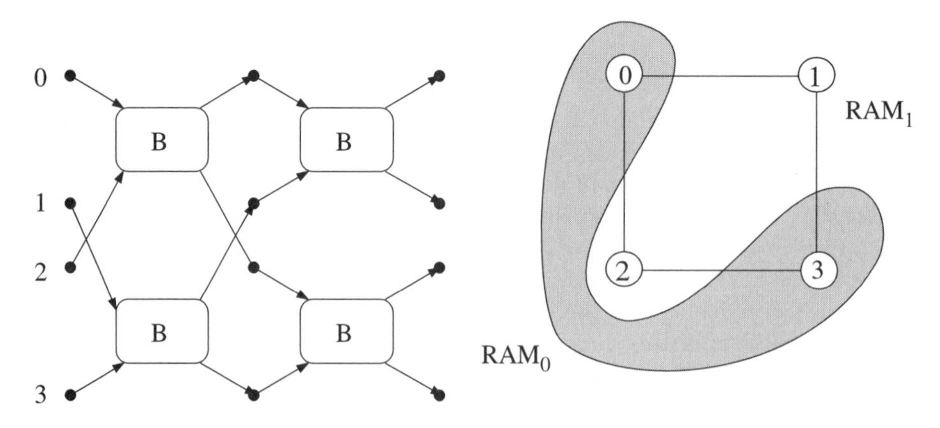

Figure 7.85 A four-point ST-FFT and exclusion graph for memories

To determine possible N-memory assignments, we partition the graph into N parts by assigning the vertices (variables) with exclusion branches to different memories. There is only one way to partition this graph into two such parts. Hence, there is only one possible memory assignment, except for a trivial interchange of the memories.

Note that the new exclusion branches in the four-point exclusion graph compared to the two-point exclusion graph are due to the first stage. This becomes even more clear in the exclusion graph of the eight-point FFT illustrated in Figure 7.86. The effect of adding a first stage is that an exclusion branch will occur between the corresponding indices in the two smaller FFTs.

Generally, there will be an exclusion branch between vertices corresponding to the values with index i and $i + N/2$ for $0 \leq i < N/2$.

For a two-point FFT, the variables must be assigned to different memories, as shown in Figures 7.84 and 7.87. This must also hold for each of the two-point FFTs in a four-point FFT, but the corresponding variables in the two-point FFTs must also be stored in different memories. For example, $x(0)$ in the first two-point FFT in Figure 7.87 is stored in RAM_0. Hence, $x(2)$ must be stored in RAM_1. This pattern repeats for larger FFTs as illustrated in Figure 7.87. The two-RAM assignment [15] can be expressed in terms of the binary representation of $i = i_L\, i_{L-1} \cdots i_3\, i_2\, i_1\, i_0$ where $L = \log_2(N) - 1$.

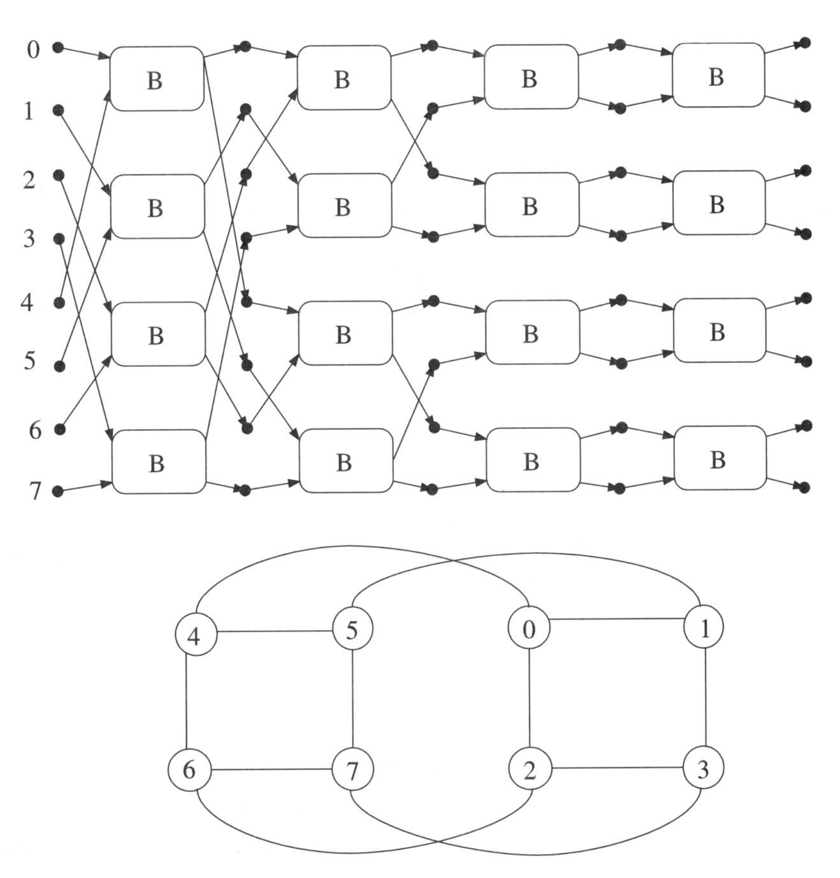

Figure 7.86 An eight-point ST-FFT and exclusion graph for memories

A variable with index i is assigned to the $RAM_{P(i)}$ where

$$P(i) = i_0 \oplus i_1 \oplus i_2 \oplus \cdots \oplus i_L \tag{7.8}$$

The function P is the parity of the index i.

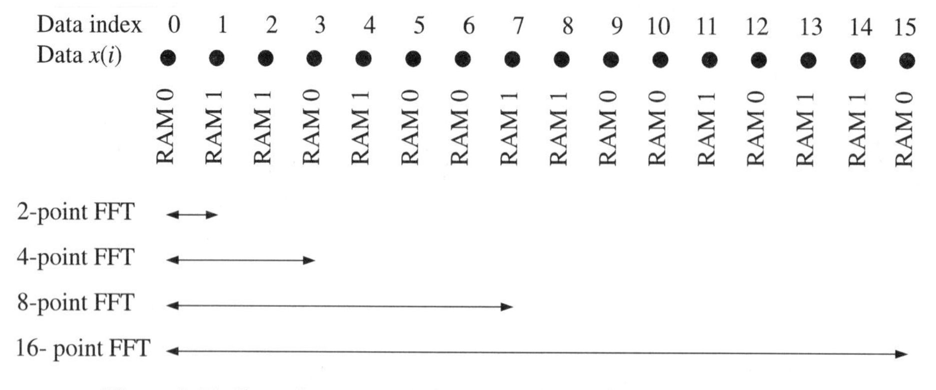

Figure 7.87 Second memory assignment alternative for various FFTs

7.11.2 Butterfly Processor Assignment

In section 7.3, the ST-FFT was partitioned in such a way that two butterfly operations could be executed concurrently. In this section we will discuss two alternative ways to assign the butterfly operations to the PEs. Generally, there are many possible PE assignments that yield different implementation costs. The PE assignment will mainly affect the complexity of the interconnection network and the control structure.

PE Assignment 1

The seemingly simplest and most obvious assignment of butterfly processes to PEs is shown in Figure 7.88 for the case $N = 8$. It possible with this assignment to use only one W^p-PE and to use both memory assignments discussed in section 7.3.2. The main disadvantage with this scheme is that switches in the interconnection network can not be avoided.

PE Assignment 2

The second PE assignment scheme, which aims to simplify the switch network, can be derived as described next. First, we notice that both memory assignment schemes use in-place computation so that the input and output variables in each row of butterfly operations are assigned to the same memory cells. Hence, the outputs of a butterfly operation are written into the same RAM that the input came from.

Now, we add the following constraint to the butterfly assignment: Assign the butterflies to PEs in such a way that a memory (RAM) is never connected to both inputs of a PE (butterfly operation). This constraint will reduce the need for switches in the interconnection network. Further, we assume that all butterflies in a row are assigned to the same PE.

The set of butterflies that can not be executed in the same PE can be found in a way similar to the second memory assignment alternative. Figure 7.89 illus-

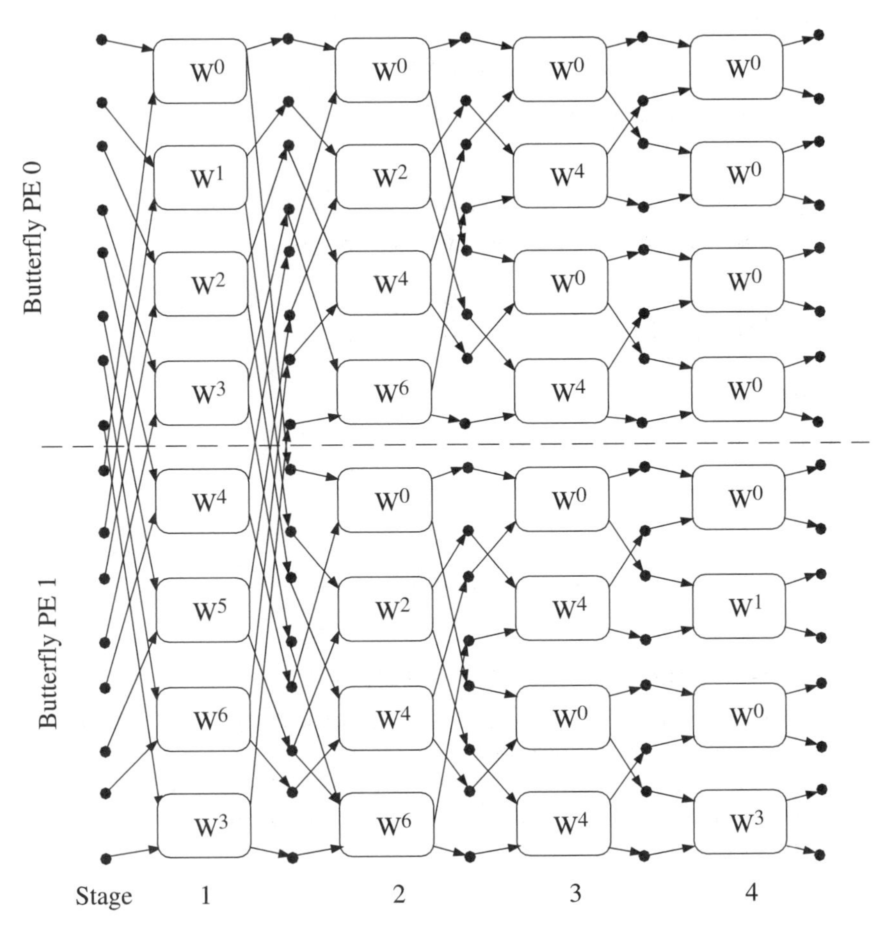

Figure 7.88 First butterfly assignment

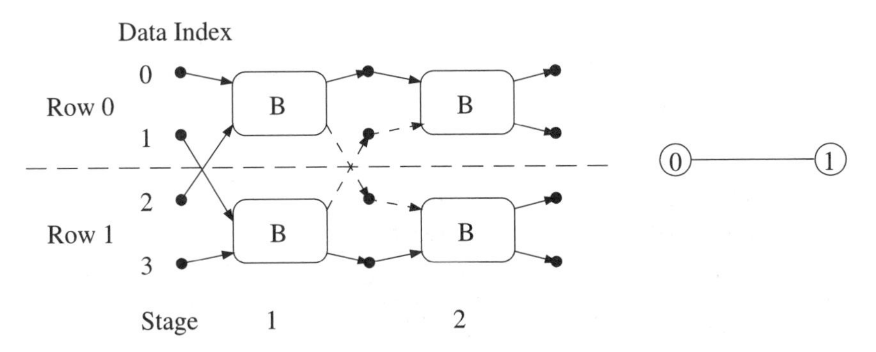

Figure 7.89 Exclusion graph for PEs for a four-point

trates the derivation of the exclusion graph for a four-point FFT. The data with index 1 after stage 1 is assigned to one of the memories. If the butterflies in rows 0 and 1 were assigned to the same PE, this memory would have to communicate with both the top and the bottom inputs of the PE. This is denoted with the dashed

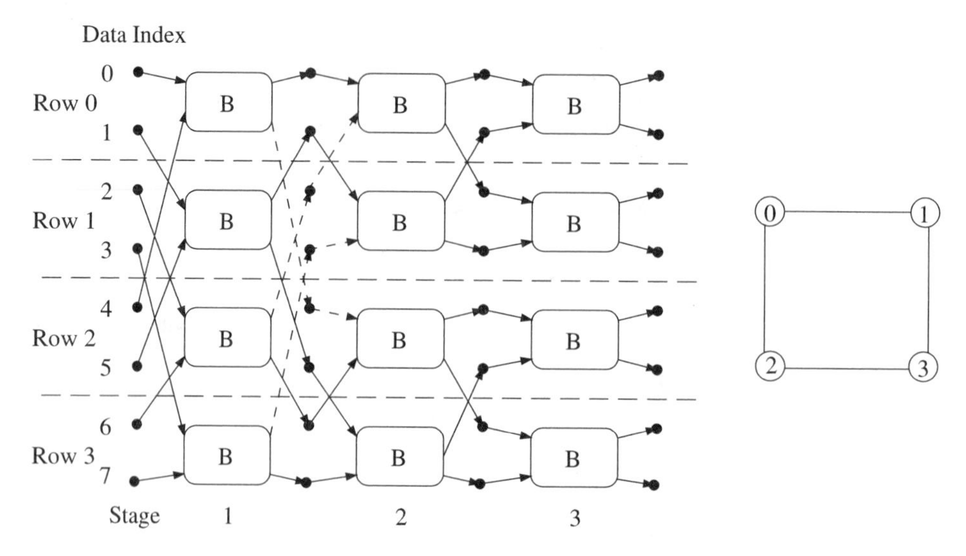

Figure 7.90 Exclusion graph for PEs for an 8-point FFT

arrows. The butterflies in rows 0 and 1 must therefore exclude each other according to the constraint just discussed. This is represented by a branch in the exclusion graph in Figure 7.89. The two vertices that denote rows 0 and 1 are assigned to PE_0 and PE_1, respectively.

Figure 7.90 shows the exclusion graph for an eight-point FFT. The dashed arrows represent the new exclusion constraints. The new exclusion branches are due to the first stage. Hence, the effect of doubling the size of the FFT is similar to the effect when we examined the first memory assignment alternative. The exclusion graph for a 16-point FFT is shown in Figure 7.91 and the PE assignment is show 2 in Figure 7.92. The graphs show those rows of butterflies that are mutually exclusive. The graphs are identical to the memory exclusion graphs for the second memory assignment alternative. Hence, the problems have the same solution.

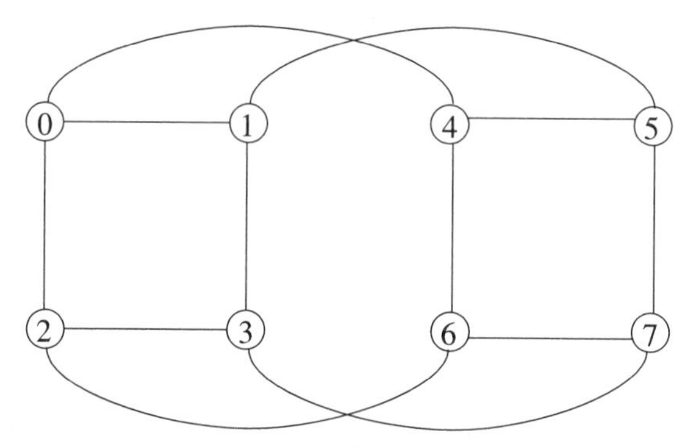

Figure 7.91 Exclusion graph for PEs for a 16-point FFT

This PE assignment can use the same concurrent butterfly processes as the first PE assignment. Either butterflies in rows p and $p + N/4$ can be executed in parallel or butterflies in rows p and $p + N_S/2$ an be ececuted in the first stage and thos in rows p and $p + N_S$ in the following stages. This means that we can use both the alternatives of index generation described in section 7.3.2.

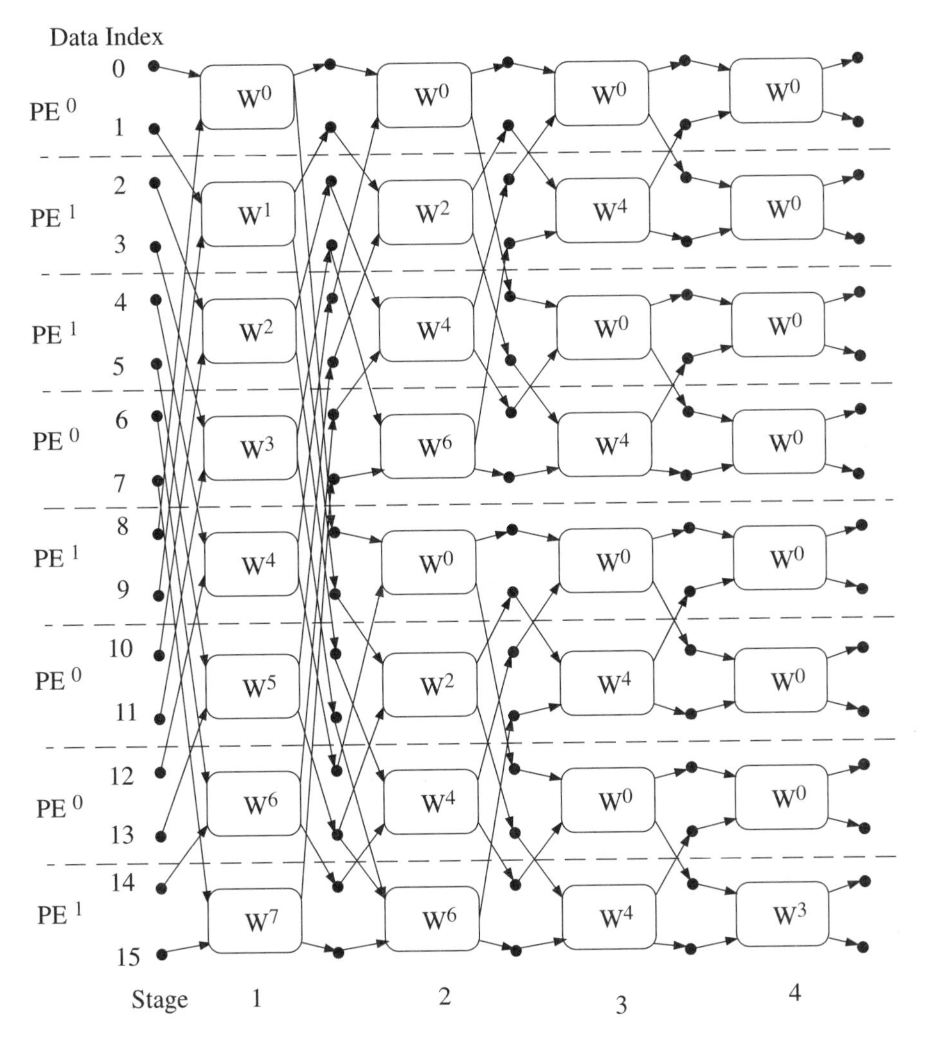

Figure 7.92 Second PE assignment

The assignment of butterfly operations can be determined from the binary representation of the row number. A butterfly operation in row r is assigned to the $PE_{P(r)}$ where

$$P(r) = r_0 \oplus r_1 \oplus r_2 \oplus \cdots \oplus r_L \tag{7.9}$$

where $L = \log_2(N/2)-1$.

7.11.3 Input and Output Process Assignment

The allocation and assignment of the input and output stages is trivial. One PE is required for each process in the loops. Also, only one memory will be active at a time.

7.12 DCT PROCESSOR, CONT.

A 2-D DCT system should be able to process an HDTV image of 1920 × 1080 pixels in real time in order to satisfy the most stringent requirement in Table 3.3. The image frequency is 60 Hz. The image is partitioned into subframes of 16 × 16 pixels which are processed in the 2-D DCT system. Hence, the requirement is

$$\frac{\text{image size}}{N^2} \cdot f_{image} = \frac{1920 \cdot 1080}{16 \cdot 16} \cdot 60 = 486 \cdot 10^3 \qquad \text{[2-D DCT/s]}$$

The line memory stores 16 complete lines of image and supports the DCT chip with data, as illustrated in Figure 7.93. The processing element in the 2-D DCT system is a 1-D DCT bit-serial PE. The PE can compute one 16-point DCT in $W_d = 12$ clock cycles = 1 time unit. The feasible clock frequency is at least 220 MHz. As discussed in Chapter 3, the computation of a 2-D DCT requires 32 time units. Two additional time units are lost since the pipelined PEs must be emptied between the row and column computations. Thus, one such PE can compute

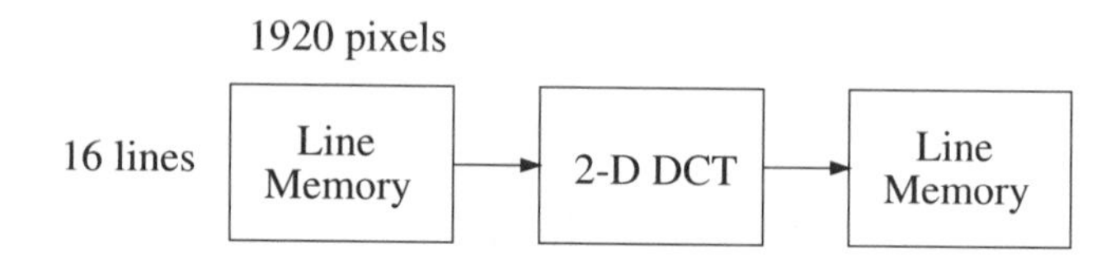

Figure 7.93 Environment for the 2-D DCT

$$\frac{f_{CL}}{(32+2)W_d} = \frac{220\ 10^6}{34 \cdot 12} = 539 \cdot 10^3 \qquad \text{[2-D DCT/s]}$$

Obviously, this is enough, but we still select to use two 2-D DCT PEs. The reason is the reduction in speed requirement on both the PEs and the RAMs. Further, the use of two PEs facilitates a more regular architecture. The pixels have a resolution of 8 to 9 bits while internal computations in the 2-D DCT use 12 bits, including the guard bit.

For testing purposes we will use standard FIFOs, which are available in organizations of $2^n \times 9$-bits words (Am7204-25 with a cycle time of 35 ns), as line memories. Allowing 15 ns for I/O, the FIFOs can be clocked at 20 MHz. The FIFOs will directly feed data into the PEs. The required number of FIFOs is

$$N_{FIFO} = \frac{34}{32} \cdot \frac{16 \cdot 16 \cdot 486 \cdot 10^3}{20 \cdot 10^6} = 6.61$$

We choose to use eight FIFOs, each of 4096 words. The total amount of memory is

$$8 \cdot 4096 \cdot 9 = 294,912 \text{ bits}$$

of which we use only

$$16 \cdot 1920 \cdot 9 = 276,480 \text{ bits}$$

The use of FIFOs simplifies communication with the surrounding circuitry. However, it is not necessary to use FIFOs. RAMs can also be used. The DCT PEs receive data from the FIFOs and produce intermediate data during the first 16 time units. Two time units are lost in the pipeline, then in the last 16 time units the output is produced and directed to the output port. Consequently, by interleav-

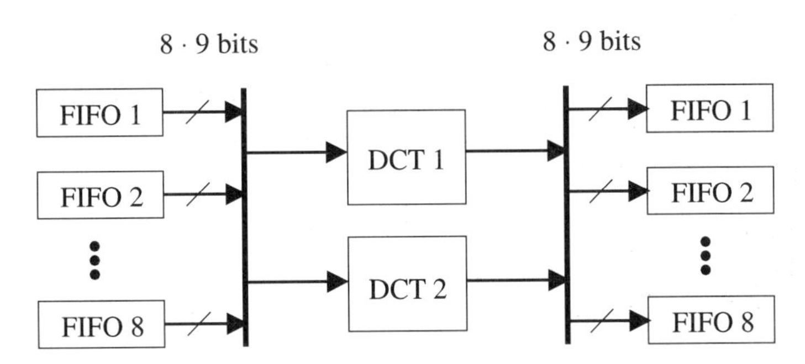

Figure 7.94 FIFO–DCT–FIFO communication

ing the operations of the two PEs, only one set of FIFOs is needed as illustrated in Figure 7.94.

Figure 7.95 shows the resulting schedule for the PEs and I/O. Two images are processed simultaneously. Hence, the two internal memories must each hold 16 × 16 12-bit words (3072 bits) each.

The required access rate is

$$f_{RAM} = (16 \cdot 16 + 2 \cdot 16 + 16 \cdot 16)\ 243\ 10^3 = 132.2 \text{ MHz}$$

This access rate is too large for one RAM. Therefore we choose to interleave the operation of the two RAMs. Interleaving of memories will be discussed in Chapter 8. Each RAM must have an access rate of at least 66.1 MHz. This is also a relatively high access rate. The size of the four RAMs is 128 12-bit words = 1536 bits each. Alternatively we may interleave four RAMs in order to reduce the access rate to 33.05 MHz.

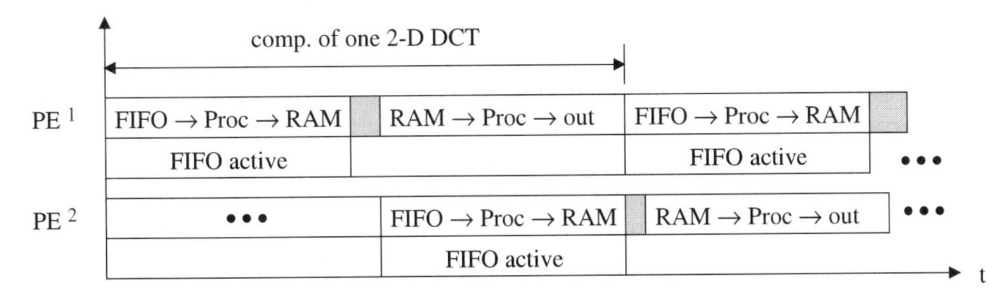

Figure 7.95 Schedule for the DCT PEs and I/O

REFERENCES

[1] Chaudhuri S., Walker R.A., and Mitchell J.E.: Analyzing and Exploring the Structure of the Constraints in the IPL Approach to the Scheduling Problem, *IEEE Trans. on VLSI Systems,* Vol. 2, No. 4, pp. 456–471, Dec. 1994.

[2] Chetto H. and Chetto M.: Some Results of the Earliest Deadline Scheduling Algorithm, *IEEE Trans. on Software Engineering,* Vol. 15, No. 10, Oct. 1989.

[3] Cormen T.H., Leiserson C.E., and Rivest R.L.: *Introduction to Algorithms,* MIT Press, 1993.

[4] Crochiere R.E.: *Digital Network Theory and Its Application to the Analysis and Design of Digital Filters*, PhD Diss., Dept. Elec. Eng., MIT, Cambridge, MA., May 1974.

[5] Crochiere R.E. and Oppenheim A.V.: Analysis of Linear Digital Networks, *Proc. IEEE,* Vol. 63, No. 4, pp. 581–595, April 1975.

[6] Davidson S., Landskov D., Shriver B.D., and Mallett P.W.: Some Experiments in Local Microcode Compaction for Horizontal Machines, *IEEE Trans. on Computers,* Vol. C-30, No. 7, July 1981.

[7] Granski M., Koren I., and Silberman G.M.: The Effect of Operation Scheduling on the Performance of a Data Flow Computer, *IEEE Trans. on Computers,* Vol. C-36, No. 9, Sept. 1987.

[8] Heemstra de Groot S.M. and Herrmann O.E.: Evaluation of Some Multiprocessor Scheduling Techniques of Atomic Operations for Recursive DSP Filters, *Proc. European Conf. on Circuit Theory and Design,* pp. 400–404, Brighton, UK, Sept. 5–8, 1989.

[9] Heemstra de Groot S.M. and Herrmann O.E.: Rate-Optimal Scheduling of Recursive DSP Algorithms Based on the Scheduling-Range Chart, *Proc. IEEE Intern. Symp. on Circuits and Systems,* pp. 1805–1808, New Orleans, LA, May 1–3, 1990.

[10] Laarhoven van P.J.M. and Aarts E.H.L.: *Simulated Annealing: Theory and Applications,* D. Reidel., 1987.

[11] Leung J.Y.-T.: A New Algorithm for Scheduling Periodic, Real-Time Tasks, *Algorithmica,* No. 4, pp. 209–219, 1989.

[12] Lee E.A. and Messerschmitt D.G.: Synchronous Data Flow, *Proc. IEEE,* Vol. 75, No. 9, pp. 1235–1245, Sept. 1987.

[13] Lee J.-H., Hsu Y.-C., and Lin Y.-L.: A New Integer Linear Programming Formulation for the Scheduling Problem in Data Path Synthesis, *Proc. IEEE Intern. Conf. on Computer-Aided Design,* pp. 20–23, Santa Clara, CA, Nov. 5–9, 1989.

[14] McFarland M.C., Parker A.C., and Camposano R.: Tutorial on High-Level Synthesis, *25th ACM/IEEE Design Automation Conf.,* pp. 330–336, 1988.

[15] Nordhamn E.: *Design of an Application-Specific FFT Processor,* Linköping Studies in Science and Technology, Thesis No. 324, Linköping University, Sweden, 1992.

[16] Otten R.H.J. and van Ginneken L.P.P.P.: *The Annealing Algorithm,* Kluwer Academic Pub., 1989.

[17] Paulin P.G. and Knight J.P.: Force-Directed Scheduling for the Behavioral Synthesis of ASICs, *IEEE Trans. on Computer-Aided Design,* Vol. 8, No. 6, June 1989.

[18] Paulin P.G. and Knight J.P.: Scheduling and Binding Algorithms for High-Level Synthesis, *Proc. 26th ACM/IEEE Design Automation Conf.,* pp. 1–6, Las Vegas, NV, June 25-29, 1989.

[19] Paulin P.G. and Knight J.P.: Algorithms for High-Level Synthesis, *IEEE Design and Test of Computers,* pp. 18–31, Dec. 1989.

[20] Renfors M. and Neuvo Y.: The Maximum Sample Rate of Digital Filters under Hardware Speed Constraints, *IEEE Trans. on Circuits and Systems,* CAS-28, No. 3, March 1981, pp. 196–202.

[21] Rutenbar R.A.: Simulated Annealing Algorithms: An Overview, *IEEE Circuits and Devices Magazine,* pp. 19–26, Jan. 1989.

[22] Schwartz D.A. and Barnwell III T.P.: Cyclo-Static Multiprocessor Scheduling for the Optimal Realization of Shift-Invariant Flow Graphs, *Proc. Inter. Conf. on Acoustics, Speech and Signal Processing,* Tampa, FL, Vol. 3, pp. 1384–1387, March 26-29, 1985.

[23] Schwartz D.A. and Barnwell III T.P.: Cyclo-Static Solutions: Optimal Multiprocessor Realizations of Recursive Algorithms, *VLSI Signal Processing, II*, IEEE Press, 1986.

[24] Wanhammar L., Afghahi M., and Sikström B.: On Mapping of DSP Algorithms onto Hardware, *IEEE Intern. Symp. on Circuits and Systems,* ISCAS-88, pp. 1967–1970, Espoo, Finland, June 1988.

[25] Wanhammar L., Sikström B., Afghahi M., and Pencz J.: A Systematic Bit-Serial Approach to Implement Digital Signal Processing Algorithms, *Proc. 2nd Nordic Symp. on VLSI in Computers and Communications,* Linköping, June 2–4, 1986.

[26] Yuan J. and Svensson C.: High-Speed CMOS Circuit Technique, *IEEE J. on Solid-State Circuits,* Vol. SC-24, No. 1, pp. 62–70, Feb. 1989.

PROBLEMS

7.1 Consider a 16-point ST-FFT.

(a) Determine which pairs of butterflies are computed in the two alternatives discussed in section 7.3.2.

(b) Determine the range of the following indices: m, k_1, k_{1Ns}, k_2, k_{2Ns}, and p.

(c) Show how the computations of these indices can be simplified.

(d) What happens if the index m is incremented in Gray-order?

7.2 Modify the program in Box 7.2 so that four butterflies can be computed concurrently.

7.3 Schedule the operations in the second-order filter shown in Figure P7.3 so that T_{min} is reached.

(a) Assume $T_{add} = T_{mult} = 1$ t.u.

(b) Assume $T_{mult} = 3\ T_{add}$ and $T_{add} = 1$ t.u.

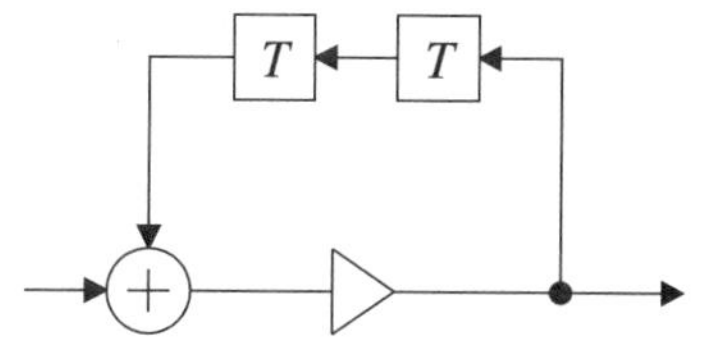

Figure P7.3. Second-order filter

7.4 Assume that the adders have unit delay. The times taken for the multiplications are indicated in Figure P7.4. What is the minimum sample period? Sketch a possible scheduling of the PEs that yields a minimum iteration period assuming that the multipliers are nonpreemptive. How many PEs are needed and what is their degree of utilization?

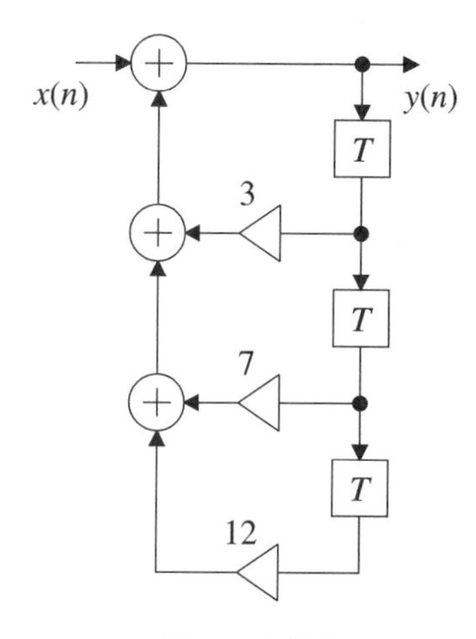

Figure P7.4

7.5 (a) Derive the computation graph for the maximally fast second-order section in direct form II.

(b) Schedule the operations over one sample interval so that the number of concurrent operations is minimized.

(c) Schedule the operations over two sample intervals so that the number of concurrent operations is minimized.

(d) Compare your result with the results obtained in Example 7.2.

7.6 In Figure P7.6's filter section, the time taken for a multiplication is 10 t.u. and the time for an addition is 1 t.u. What is the minimum sample period for the filter? How long is the time critical path? Find a suitable schedule for the PEs both without and with pipelining.

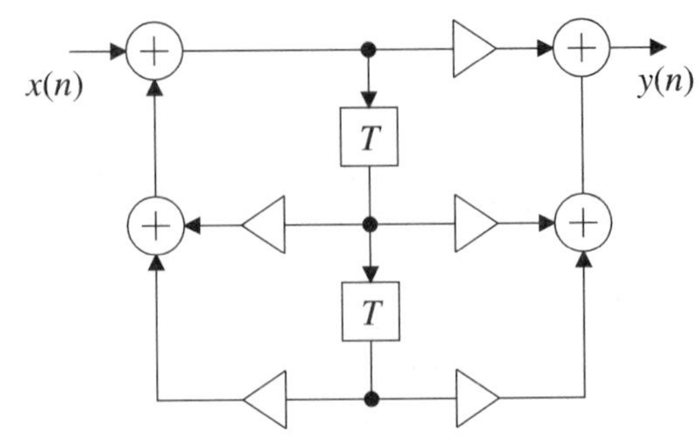

Figure P7.6 Direct form II

7.7 The execution times for the process in Figure P7.7 are

$$T_{P1} = T_{P2} = T_{P3} = T_{P4} = T_{P8} = 4 \text{ and } T_{P5} = T_{P6} = T_{P7} = T_{P9} = 1$$

(a) Determine the iteration period bound, T_{min}.

(b) Determine the number of PEs in a PE optimal schedule when T_{sample} = 8. The processors are homogeneous.

(c) Let T = 8 and determine a schedule such that the largest number of concurrent PEs is a minimum. The latency may not increase more than four time units.

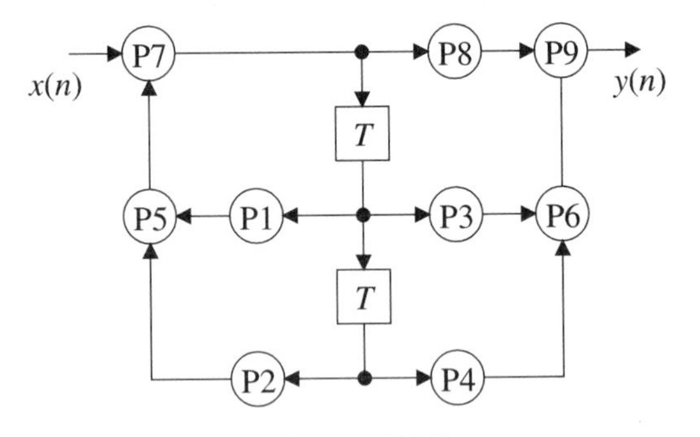

Figure P7.7

(d) Perform PE allocation and assignment by using clique partitioning.

(e) Draw the lifetime table for the variables.

(f) Perform memory assignment and determine the required number of cells.

7.8 Schedule the signal-flow graph shown in Figure P7.8 for the three cases when the addition takes 1 t.u. and the multiplier has two pipeline stages. The latency for the multiplier is 4 t.u.

(a) Minimum iteration period.

(b) An iteration period of 6 t.u.

(c) The same as in (b) but only one adder and one multiplier are allowed.

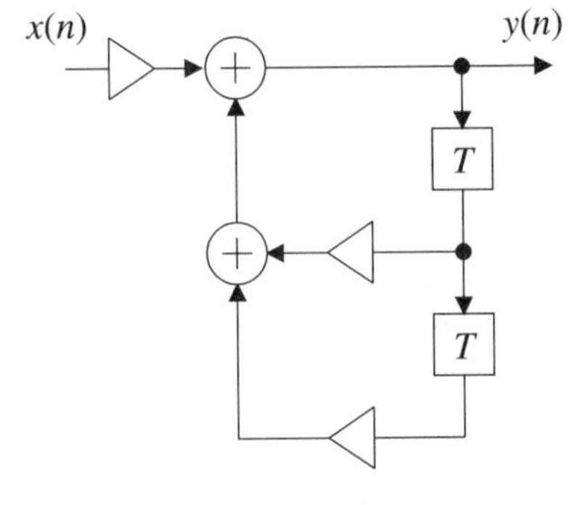

Figure P7.8

7.9 How many different schedules with maximum speed are possible for the algorithm in Figure 7.40? Assume that the algorithm is recursive—i.e., the output is fed back to one of the inputs of one of the multiplications.

7.10 Introduce shimming delays into the structures shown in Figures 7.22 and 7.24.

7.11 (a) Show that the realizations shown in Figures 7.22 and 7.24 yield sample frequencies corresponding to 3 and 2.5 clock cycles, respectively.
(b) Determine the required number of full-adders and D flip-flops in the realization shown in Figure 7.21.

7.12 The filter in Figure P7.12 shall be implemented using only one PE that can execute a multiplication and an addition as illustrated here.
(a) Define a fully specified signal-flow graph using only this type of operation and that has the shortest possible critical path.
(b) Assume that the PE can execute one operation every 20 clock cycles and that it can be clocked at 100 MHz. The required sample frequency is 1.25 MHz. Assume that a result from a PE can not immediately be used by a PE due to pipelining. Hence, the result must be written into the memory before it can be used again. Schedule the operations so that a minimum number of PEs is required. Only one level of pipelining may be introduced.

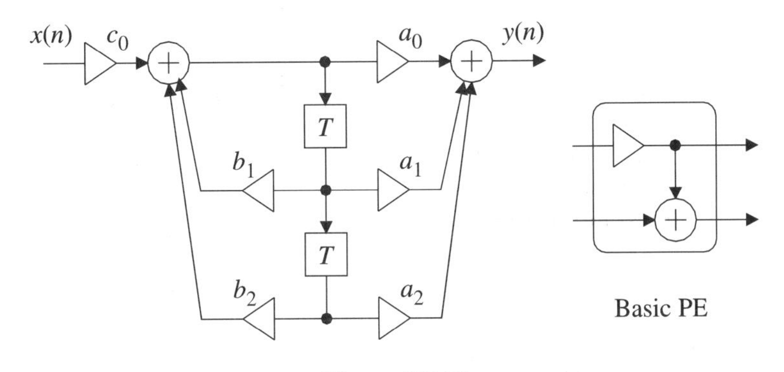

Figure P7.12

7.13 Show by using clique partitioning that it is not possible to use only two PEs to implement the algorithm in Example 7.8.

7.14 Determine by using clique partitioning the number of resources required for the processes described by the exclusion graph shown in Figure. P7.14.

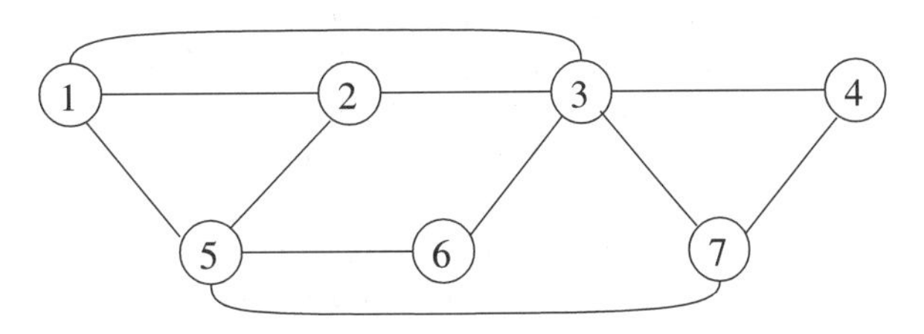

Figure P7.14 Exclusion graph

7.15 Perform the memory allocation and assignments design steps for the filter in Example 7.2.

7.16 (a) Perform a memory assignment for a 16-point ST-FFT with four memories.
(b) Perform a PE assignment for a 16-point ST-FFT with four PEs.

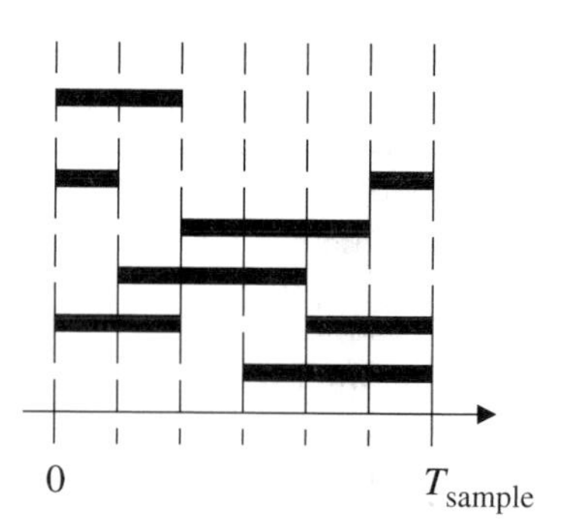

Figure P7.17 Lifetime diagram

7.17 (a) Determine a minimum memory assignment for the six variables shown in the lifetime diagram in Figure P7.17.
(b) Determine the minimum iteration period when only one logical one-port memory is used. The access time is 40 ns.

7.18 The variables shown in the lifetime diagram in Figure P7.18 are to be stored in a single memory. Determine the minimum number of cells required.

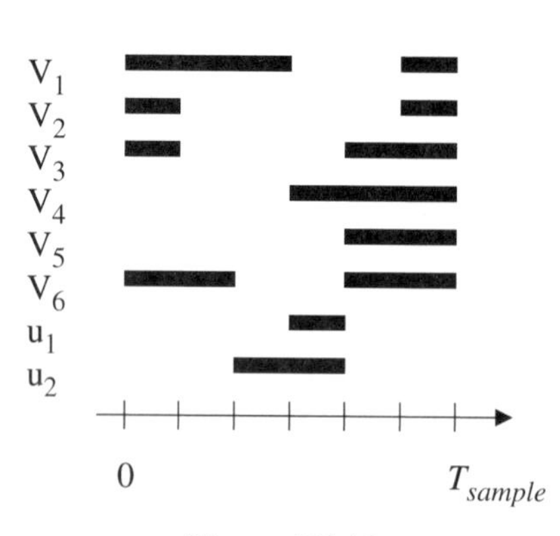

Figure P7.18

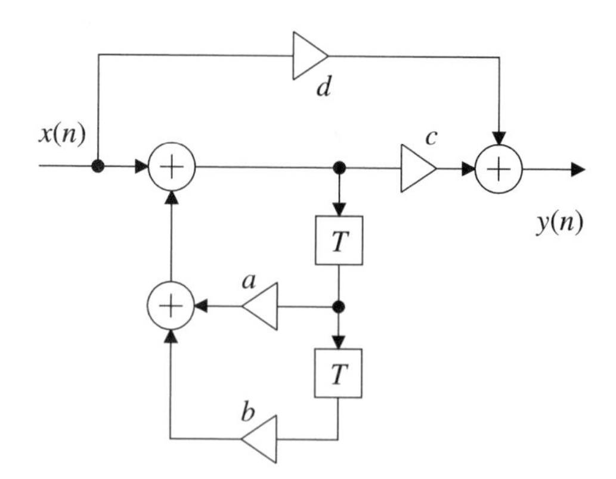

Figure P7.19

7.19 (a) Determine the precedence graph for the filter in Figure 7.19.
(b) Determine the system of difference equations in computable order. Eliminate unnecessary variables.
(c) Perform an ALAP scheduling of the operations. All operations take equal time.

7.20 Find an operation schedule using the maximum spanning tree method for the algorithm shown in Figure P7.7 when

$$T_{P1} = 4\,,\ T_{P2} = 10\,,\ T_{P3} = T_{P4} = T_{P8} = 5$$

$$T_{P5} = T_{P6} = T_{P7} = T_{P9} = 1$$

and $T_{sample} = 8$.

7.21 The multiplication in Figure P7.21 has a latency of three clock cycles and the additions one clock cycle. The required sample period should be three clock cycles. Schedule the operations so that the maximal sample frequency is reached.

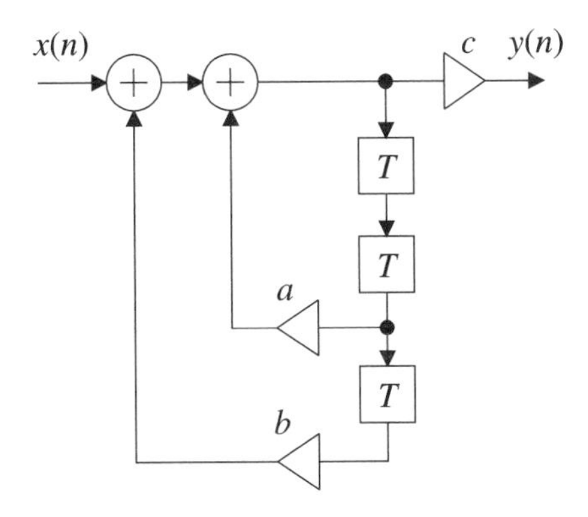

Figure P7.21

7.22 The variables in the lifetime diagram shown in Figure P7.22 shall be stored in a shared memory. Assume that two variables can be stored adjacent in time in the memory cells and that the sample frequency is 15 MHz. The memory can either read or write in one cycle.

(a) Determine an upper and lower bound on the required number of memory cells.

(b) Allocate and assign the variable to memory cells using the left-edge algorithm.

(c) Determine the required access time for the memory.

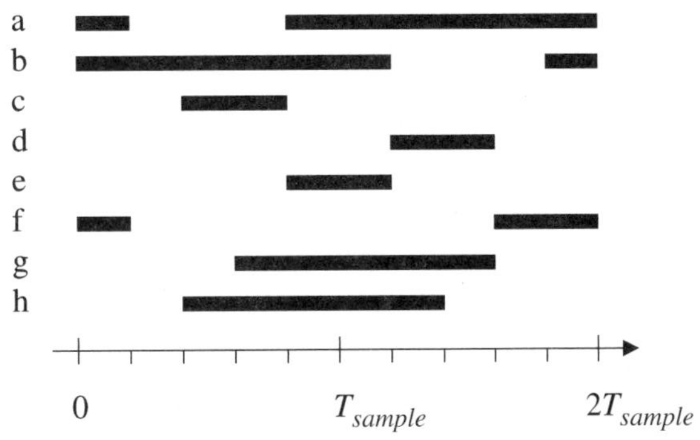

a
b
c
d
e
f
g
h

0 T_{sample} $2T_{sample}$

Figure P7.22 Lifetime diagram

7.23 The filter in Figure P7.23 is to be implemented using nonpreemptive, homogeneous processing elements. The multiplications a, d, and e have a latency of three time units and the multiplications b and c have a latency of seven time units, and each addition has a latency of one time unit. The filter is to be implemented with $T_{sample} = 5$ time units.

(a) Determine the minimal sample period T_{min}.
(b) Determine the precedence graph of the filter.
(c) What is the minimal number of required processing elements?
(d) Schedule the operations.

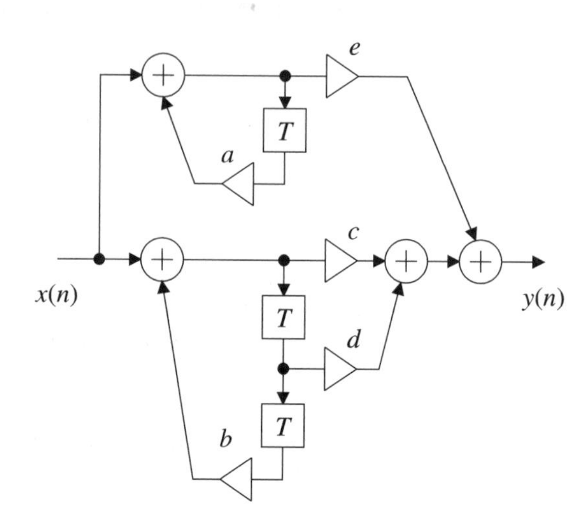

Figure P7.23

8

DSP ARCHITECTURES

8.1 INTRODUCTION

Many digital signal processing applications, particularly video applications, represent very high computational work loads. Fortunately, most DSP algorithms exhibit a high degree of parallelism and are therefore suitable for implementation using parallel and/or distributed architectures. Such parallel architectures may exploit future improvements in integrated circuit technologies—e.g., the fact that it will be possible to put more circuitry onto a single chip. The expected increase in speed of integrated circuits resulting from scaling down the size of the switching devices will also contribute to increased performance. However, the expected performance increase resulting from faster circuitry will be significantly smaller than the increase due to the increase in the number of gates.

In this chapter we will discuss the properties of multiprocessor and multicomputer architectures suitable for DSP systems. Of particular interest are architectures for hard real-time DSP systems—i.e., systems in which the correctness of the system depends not only on the result of the computation, but also on the time at which the results are produced. The architectures must therefore have resources adequate to perform their functions.

8.2 DSP SYSTEM ARCHITECTURES

The work load must usually be shared by several processors in order to achieve the throughput required in many real-time applications. Most DSP systems are, at the top levels, inherently composed of several connected subsystems. These subsystems can be mapped, using an isomorphic mapping, onto a similar set of hardware structures or virtual machines, which communicate through an interconnection network, as illustrated in Figure 8.1.

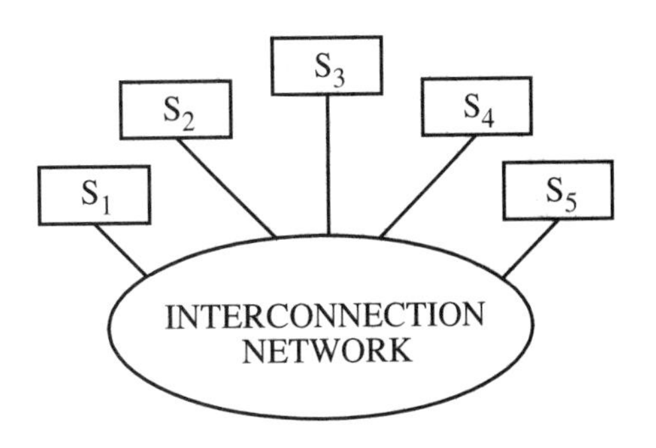

Figure 8.1 Top view of a DSP system

We will discuss different types of interconnection networks in more detail in section 8.6.1.

The difference between system and subsystem architectures is essentially due to the difference in granularity of the processes executed. Process granularity tends to be larger at the higher levels and lower at the subsystem levels. For example, a process at the system level may be an FFT and at the lowest level an arithmetic operation. In practice, the most common architectures at the higher system levels are the cascade form and the parallel form, shown in Figures 8.2 and 8.3, respectively.

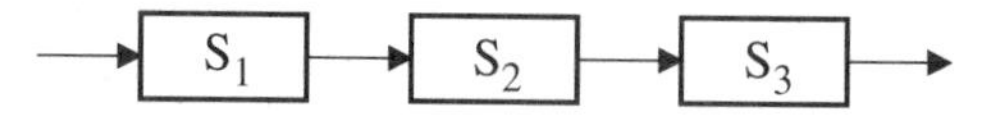

Figure 8.2 Cascade form

Other common system architectures have feedback loops and may have several levels of hierarchies. Large DSP systems are usually organized as hybrids of these forms. From a computational point of view, the cascade form is well suited for pipelining while the subsystems can be executed concurrently in parallel form.

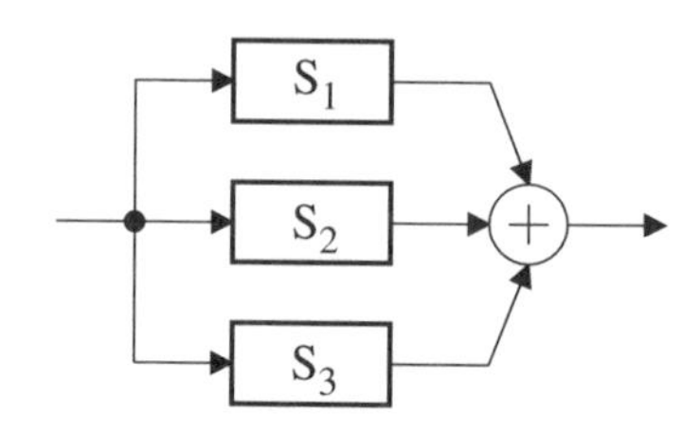

Figure 8.3 Parallel form

Most decomposition approaches at the higher system levels exploit the fact that many DSP algorithms are naturally partitioned into parts with small intercommunication requirements. The first design step is therefore to identify these parts. Internally these parts may have high communication demands, for example, a recursive digital filter. Such tightly coupled parts of an algorithm must be implemented using shared-memory architectures. These architectures will be discussed in section 8.9.

A message-based system architecture, with the subsystems as components, can be chosen if the communication requirements between the subsystems are small. Otherwise, the system must be implemented with a shared-memory architecture. Message-based architectures will be discussed in section 8.6.

The subsystems can be implemented using, for example, a standard digital signal processor (uniprocessor architecture) or an ASIC processor if the work load is small, whereas a specialized multiprocessor architecture is required for larger work loads. A complex DSP system can be composed of several architectural alternatives as illustrated in Figure 8.4.

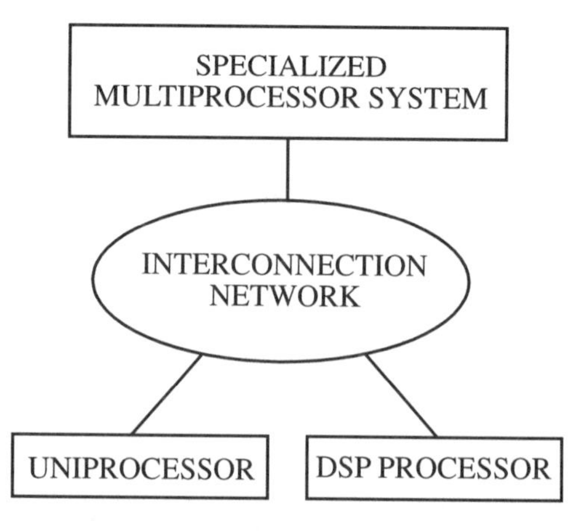

Figure 8.4 DSP system implemented using several types of architectures

8.3 STANDARD DSP ARCHITECTURES

Current computer architectures are commonly classified as RISCs (reduced instruction-set computers) and CISCs (complex instruction-set computers). The latter has a large number of powerful instructions while a RISC processor has fewer instructions and is defined as performing a register-to-register move in one machine cycle. RISC computers are today beginning to replace CISCs, because they achieve higher throughput by using highly pipelined execution of simple instructions and efficient compilers. The time and cost for developing RISCs are lower than for CISCs. Hence, time between two RISC generations can be made shorter and the latest VLSI technology can be used to achieve high performance.

Standard DSP processors have many RISC-like features, but they are special-purpose processors whose architecture is designed to operate in a computationally demanding, real-time environment. RISCs, on the other hand, are general-purpose processors, even though they are beginning to find their way into some embedded applications. A standard DSP processor executes several operations in parallel while a RISC uses heavily pipelined functional units that can initiate and complete a simple instruction in one or two clock cycles. Note, however, that the latency of a RISC instruction (for example, a floating-point addition) may be much longer and that the maximal iteration period is bounded by the latency. In practice it may be difficult to make efficient use of long pipelines.

Standard DSP processors are generally characterized by the following architectural features:

1. A fast on-chip multiplier that can perform multiply–and–accumulate type operations in one instruction cycle. An instruction cycle is generally one or two clock cycles long. Both fixed-point and floating-point arithmetic DSPs are available.
2. Several functional units that perform several parallel operations, including memory accesses and address calculations. The functional units typically include the main (*ALU*) together with two or more address generation units. The functional units have usually their own set of registers and most instructions execute in a single instruction cycle.
3. Several large on-chip memory units (typically two to three) used to store instructions, data, or look-up tables. Each memory unit can be accessed once every instruction cycle. Large numbers of temporary registers are used to store values used for long periods of time. Many modern DSPs have some form of instruction cache, but they do not have data caches and do not support virtual memory.
4. Several on-chip system buses to increase memory transfer rate and avoid addressing conflicts.
5. Support for special addressing modes, especially modulo and bit-reversed addressing needed in the FFT. Modulo addressing allows for fast circular buffers that reduce the overhead in recursive algorithms. Dedicated hardware for address calculations.
6. Support for low-overhead looping and fast interrupt handling, especially that arising from arithmetic or I/O operations. On-chip serial ports.
7. Standby power-saving modes during which only the peripherals and the memory are active.

Most DSP routines are fairly small and can reside either in the on-chip program memory or in a small external EPROM. Most DSP routines are hand-coded to fully exploit the DSPs architectural features [6]—for example, to reduce the inefficiencies associated with flushing long pipelines, unwinding short loops and interleaving several independent loops to keep the pipelines full [22].

Branch instructions waste processing time in heavily pipelined architectures. The typical solution is to use dedicated hardware that allows zero-overhead loop. However, hardware support is often provided for only one loop due to the cost.

8.3.1 Harvard Architecture

In the classical von Neumann architecture the ALU and the control unit are connected to a single memory that stores both the data values and the program instructions. During execution, an instruction is read from the memory and decoded, appropriate operands are fetched from the memory, and, finally, the instruction is executed. The main disadvantage is that memory bandwidth becomes the bottleneck in such an architecture.

The most common operation a standard DSP processor must be able to perform efficiently is multiply-and-accumulate. This operation should ideally be performed in a single instruction cycle. This means that two values must be read from memory and (depending on organization) one value must be written, or two or more address registers must be updated, in that cycle. Hence, a high memory bandwidth is just as important as a fast multiply-and-accumulate operation.

Several memory buses and on-chip memories are therefore used so that reads and writes to different memory units can take place concurrently. Furthermore, pipelining is used extensively to increase the throughput. Two separate memories are used in the classical Harvard architecture as shown in Figure 8.5. One of the memories is used exclusively for data while the other is used for instructions. The Harvard architecture therefore achieves a high degree of concurrency. Current DSP architectures use multiple buses and execution units to achieve even higher degrees of concurrency. Chips with multiple DSP processors and a RISC processor are also available.

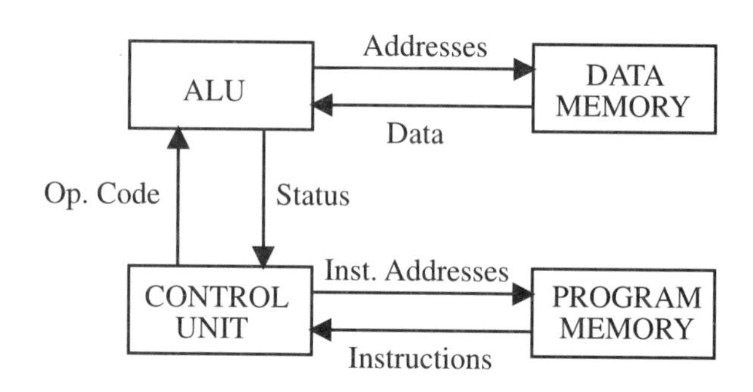

Figure 8.5 Harvard architecture

8.3.2 TMS32010™

The Texas Instruments TMS32010 family [14] of low-cost fixed-point DSPs is based on a single-bus architecture as shown in Figure 8.6. The family, introduced in 1983, has various members with differing RAM and ROM configurations. The architecture has 32-bit registers, one 16–bit wide bus, and serial I/O ports. This architecture

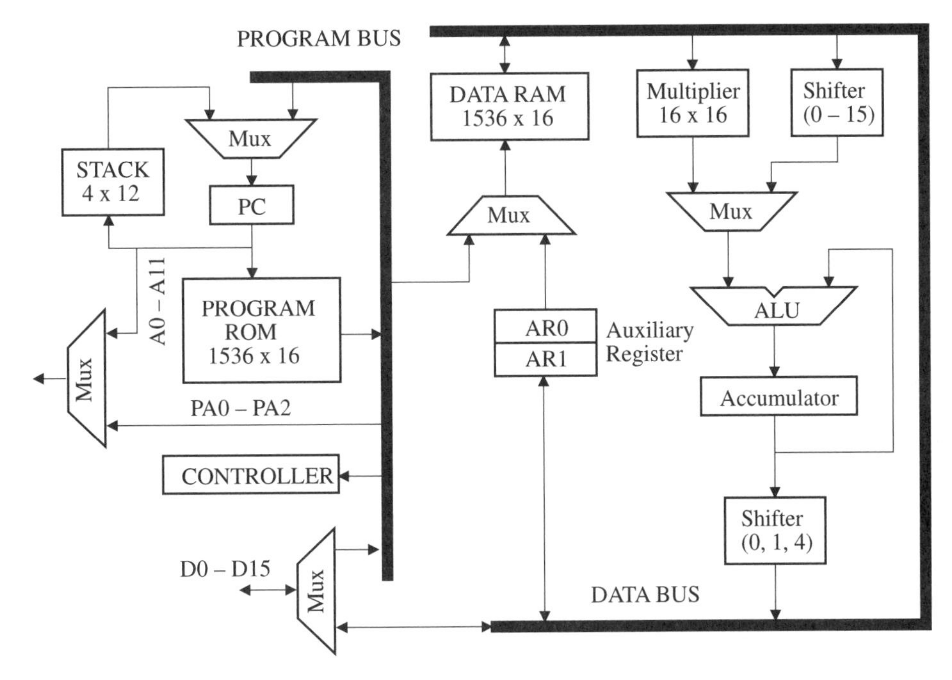

Figure 8.6 Single-bus architecture of TMS32010

requires a small chip area at the expense of performance. In particular the short data word length may be a severe limitation in many applications. The throughput is also relatively low (5 MOPS) since each multiply-and-accumulation operation requires at least two cycles since the two multiplier operands must be fetched sequentially. In the first step, the first operand is transferred from the memory to the temporary register. In the next step, the second operand is transferred to the multipliers input. Clearly, the performance of this architecture is limited by the single bus.

The on-chip memory is only 144 × 16-bits words and the external memory may be expanded up to 4k × 16-bits words at full speed. The TMS320C10 is manufactured in a 2-μm CMOS process and has an instruction cycle time of 200 ns. Typical power consumption is 100 mW @ 20 MHz. The TMS320C10 is available in a 40-pin PLCC package. The TMS320C10-25 is a faster version that has an instruction cycle time of 160 ns.

8.3.3 TMS320C25™ and TMS320C50™

The TMS32020 family [14] are second-generation DSPs with similar architecture to the TMS32010 family, but the on-chip data and program memory have been increased. The throughput has been increased by using a single-cycle multiplier–accumulator with data move option, auxiliary registers, separate arithmetic unit, and faster I/O. The on-chip memory is increased to 544 × 16-bits words and the external memory may be expanded up to 128k × 16-bits words (64k words program and 64k words data). The overall improvement is two to three times that of the TMS32010. The chip, which is manufactured in a 4-μm nMOS process, has a typical power consumption of 1.2 W @ 20.5 MHz. The instruction cycle time is 200 ns.

The TMS320C25, which is pin-compatible with the TMS32020, is a 16-bit fixed-point DSP with 32-bit registers. The chip uses a Harvard architecture and is manufactured in a 1.8-μm CMOS process. The current version can be clocked at 50 MHz. The instruction cycle time is 80 ns—i.e., 12.5 MIPS @ 50 MHz. The TMS320C25 is available in a 68-pin PLCC package. Typical performance of the TMS320C25 is as follows: An N-tap FIR filter requires N+8 clock cycles while N second-order sections require 15N+4 clock cycles. A 91-tap FIR filter can be executed in 7.92 μs and a sixth-order filter in 3.72 μs. A 256-point and a 1024-point FFT, without bit-reversal, take 1.8 and 15.6 ms, respectively.

The TMS320C50 is an enhanced version of the TMS320C25 with twice as large a throughput. Distinguishing features are low loop overhead and 10 kwords of on-chip RAM.

8.3.4 TMS320C30™

The TMS320C30 is a multibus 32-bit floating-point DSP that also supports 24-bit fixed-point arithmetic. The clock frequency is 40 MHz and the instruction cycle time is 50 ns. Distinguishing features are 40-bit registers, DMA controller, dual serial ports, and support for processor arrays. The architecture of the TMS320C30 is illustrated in Figure 8.7.

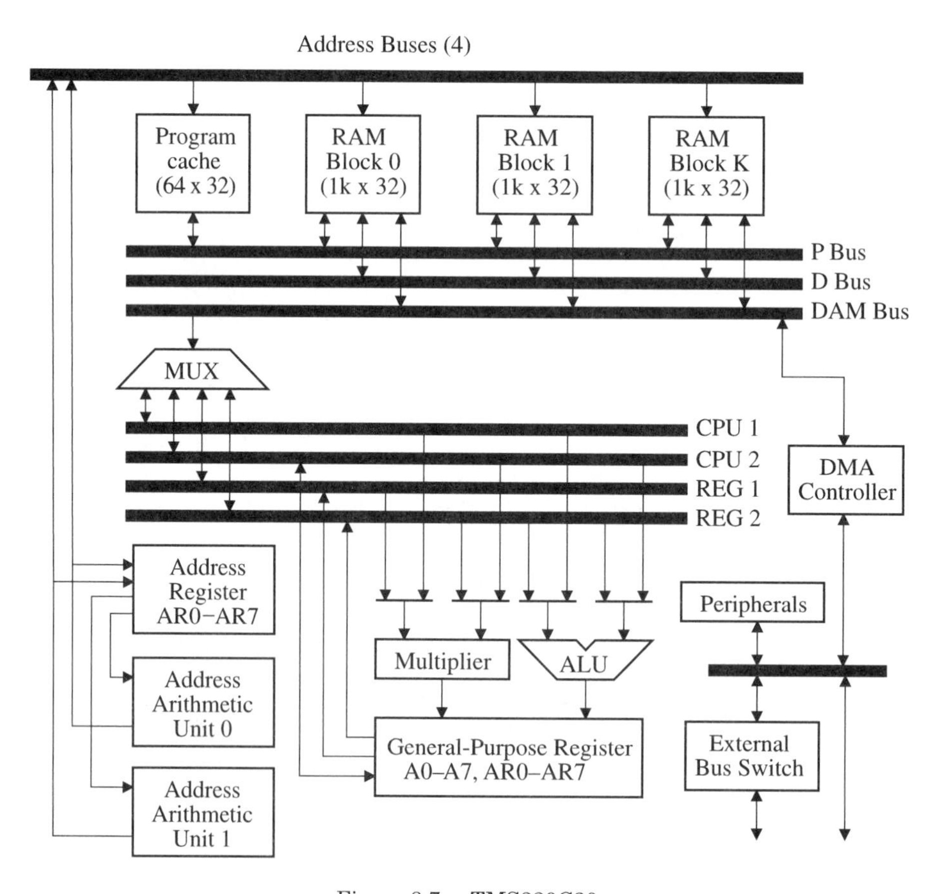

Figure 8.7 TMS320C30

An N-tap FIR filter requires N+5 clock cycles while N second-order sections require 6N+24 clock cycles. A 91-tap FIR filter can be executed in 4.0 μs and a sixth-order filter in 1.8 μs. A 256-point and a 1024-point FFT, without bit-reversal, take 0.68 and 1.97 ms, respectively.

8.3.5 TMS320C40™

The TMS320C40 floating-point DSP provides extensive parallel processing support through six buffered byte-wide 20 Mb/s links and a six-channel DMA coprocessor. It runs at 50 MHz and can perform up to eight operations per instruction cycle, which corresponds to 200 MFLOPS. The DMA coprocessor can perform up to 75 MFLOPS.

The TMS320C40 presents a more conventional architecture, where parallelism is achieved by pipelining and the ability to execute certain instructions in parallel.

8.3.6 Motorola DSP56001™ and DSP56002™

The Motorola DSP56001 and DSP56002 use triple-bus Harvard architecture, which has a high degree of parallelism. The architecture shown in Figure 8.8 has two buses (XDB and YDB) between the ALU and the data memories [17]. The DSP56002, which uses 24-bit fixed-point arithmetic, has a clock frequency of 40 MHz. Two clock cycles are required per instruction—i.e., the instruction cycle time is 50 ns. The 24-bit memory word length is sufficient for most applications. The Data ALU has a single-cycle 24 × 24-bit multiplier–accumulator and two 56-bit accumulators. A special bit manipulation unit is also provided. The on-chip memory consists of two independent data memories, each containing a 256 × 24-bit data RAM, a 256 × 24-bit data ROM, and a 512 × 24-bit program RAM. The two ROMs are used for a four-quadrant sine wave table and positive Mu- and A-law to linear tables, respectively. The largest external memories are 128k × 24-bit data

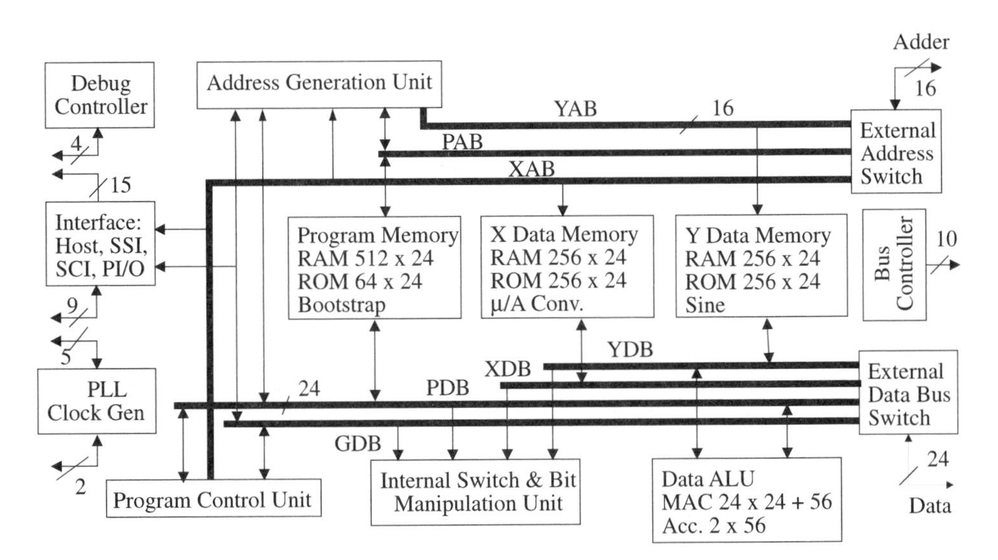

Figure 8.8 Motorola DSP56002

RAM and 64k × 24-bit program memory. The DSP56002 is available in a 132-pin pin–grid array package or plastic-quad flat-pack (PQFP).

The Motorola 56002 has three functional units: (a) multiply–accumulate unit, (b) address generation unit, and (c) program control unit. This architecture makes it possible to perform two memory operations concurrently: a multiplication and an addition. The address generation unit is responsible for calculating the different addresses and contains its own set of registers (the address registers of the DSP). The ALUs used for address calculations can only perform the simple addition operations, but can also support modulo and bit-reversed address calculations. The program control unit is responsible for fetching and decoding instructions, while at the same time containing different control-registers such as the program counter, the stack pointer, and the status register. The program control unit also contains the hardware to support low-overhead looping and interrupt handling.

Some DSP chips use a VLIW[1] with separate fields in the instruction control word to control the different units. For example, the Motorola DSP56001 and DSP56002 perform an instruction prefetch using a three-stage instruction pipeline, a 24 × 24-bit multiply, a 56-bit addition, two data moves, and two address pointer updates in a single instruction cycle. The DSP56002 can perform up to six operations simultaneously, which corresponds to a peak performance of 120 MOPS @ 40 MHz. Typical power consumption is less than 0.5 W.

Typical performance of the DSP56002 is that an N-tap FIR filter requires N+7 clock cycles while N second-order sections in direct form II require 5N+1 clock cycles. A 91-tap FIR filter can be executed in 4.9 μs and a sixth-order filter in 0.8 μs @ 40 MHz. A 256-point and a 1024-point FFT, without bit-reversal, take 0.78 and 3.89 ms, respectively.

The DSP56002 is an enhanced, software-compatible, version of the DSP56001. The current versions have clock frequencies of 40 to 80 MHz. A low-power version, DSP56L002, that operates with a power supply voltage of 3.3 V is also available. The main new features, compared to the DSP56001, are an on-chip emulator, a phase-locked loop (*PLL*) clock oscillator, improved external memory interface, and some new instructions. The new instructions include INC (increment) and DEC (decrement), double-precision multiply, and debug instructions. The latter makes it possible to examine registers, memory, and on-chip peripherals during program development.

8.3.7 Motorola DSP96001™ and DSP96002™

The DSP96001 is a floating-point version of the DSP56001 aimed at multimedia applications. It can be used as a stand-alone device or in multiprocessor configurations. The clock frequency of the current version is 40 MHz and the instruction cycle time 50 ns with a 40-MFLOP peak performance. The Motorola DSP96001 has three functional units operating in parallel: (a) IEEE 32-bit floating-point multiply-and-accumulate unit, (b) address generation unit, and (c) program control unit. In addition, two channels of DMA, six on-chip memories, and various external inter-

1. VLIW (Very Long Instruction Word): A parallel architecture that uses multiple, independent functional units and packages multiple operations into one very long instruction. The instruction has a field for each PE.

faces are provided. The on-chip memory consists of two independent memories, each containing a 512 × 32-bit data RAM, a 512 × 32-bit data ROM, and a 512 × 32-bit program RAM. The two ROMs are used for two-quadrant sine and cosine wave tables. The DSP96001 is available in a 163-pin pin–grid array package.

The DSP96002 is a dual-port version of the DSP96001. The two expansion ports effectively double the off-chip bus bandwidth [18]. The size of the on-chip program RAM has been increased from 512 to 1024 words. The DSP96002 can perform up to 10 operations simultaneously. The peak performance is 200 MOPS and 50 MFLOP @ a 40-MHz clock frequency. Typical power consumption is less than 1 W. The DSP96002 is available in a 223-pin pin–grid array package. Typical performance of the DSP96001 is as follows: An N-tap FIR filter requires N+7 clock cycles while N second-order sections in direct form II require 5N+5 clock cycles. A 91-tap FIR filter can be executed in 4.9 μs and a sixth-order filter in 1.0 μs. A 1024-point FFT, without bit-reversal, takes 1.04 ms, while the DSP96002 requires less than 0.8 ms.

8.4 IDEAL DSP ARCHITECTURES

At the architectural level, the main interest is the overall organization of the system which is described in terms of the high-level primitives: processors, memories, communication channels, and control [13]. These primitives are characterized by parameters such as memory capacity, access time, word length, transmission rates, and processor execution time. At this level, the designer is also interested in the global data flow between the high-level primitives in the system. The objective is to estimate system performance and identify potential bottlenecks in data transmission within the system. At the end of the architectural design phase, timing relations between the hardware components are of more interest. Also test pattern generation and fault simulation are a concern at this stage of the system design. Figure 8.9 illustrates the main architectural components.

The fact that most DSP algorithms lack data-dependent branching operations facilitates perfect scheduling of the operations. The scheduling step is followed by the resource allocation and assignment steps, as discussed in Chapter 7. For simplicity we assume that the schedule for the subsystem is fixed. In practice, the schedule may have to be iterated for different sets of basic operations, processors, and architectures. The design decisions taken in these steps determine the major parameters of the architecture. For example, the operation schedule determines the number of processors, their type, and performance. The variable assignment determines the required amount of memory and its organization. Both

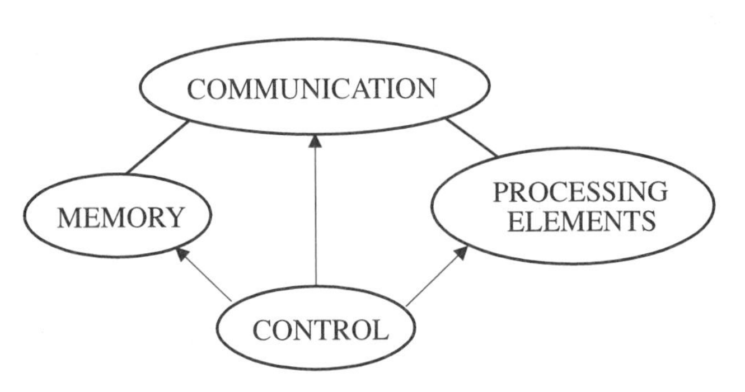

Figure 8.9 Architectural components

of these assignments determine the communication requirement—i.e., communication channels and their bandwidth, etc. Hence, the minimum requirements are specified in these design steps. Therefore, to each static schedule corresponds a class of ideal multiprocessor architectures.

An *ideal DSP architecture* belongs to a class of architectures that implements the static schedule. An ideal architecture has processing elements that can execute the operations according to the schedule and is supported with appropriate communication channels and memories.

Note that there may be several architectures that implement a given schedule, and that a new class of architectures is obtained if the schedule is changed. Algorithms that require dynamic scheduling lead to architectures that either must handle worst-case situations or are optimized in a statistical sense. However, the execution time must be predictable since the sample period constraint must be met in hard real-time applications [13]. The latter type of architectures are therefore difficult to use.

8.4.1 Processing Elements

Processing elements (PEs) usually perform simple, memoryless mappings of the input values to a single output value. The arithmetic operations commonly used in DSP algorithms are

Add/sub, add/sub-and-shift

Multiply, multiply-and-accumulate

Vector product

Two-port adaptor

Butterfly

We will reserve the more general term *processor* to denote a PE with its internal memory and control circuitry. Hence, a processor is able to perform a task independently of other processors.

If several processing elements always operate on the same inputs, it may be advantageous to merge these into one PE with multiple inputs and outputs—for example, two-port adaptors and butterflies. Experience indicates that it is advantageous to use the largest operations possible (i.e., large PE granularity) since this tends to reduce the communication. However, flexibility in scheduling the operations is reduced and resource utilization may become poor if the operations chosen are too large. As always, a good trade-off is the best.

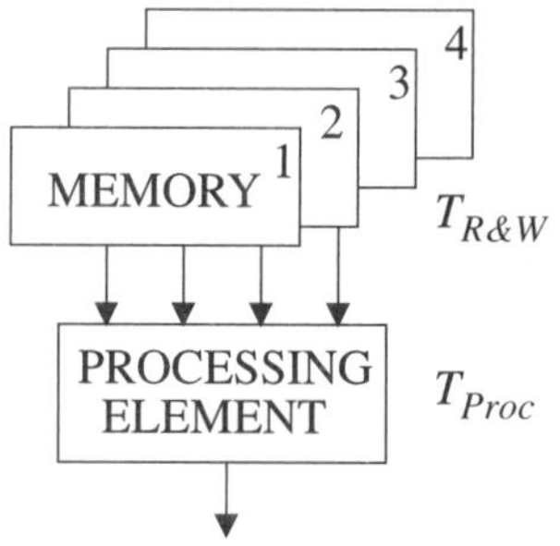

Figure 8.10 Processing element with multiple inputs

At this point it is interesting to note that the execution time for processing elements and the cycle time (read and write) for memories manufactured in the same technology are of the same order. Hence, to fully utilize a multiple-input processing element, as shown in Figure 8.10, one memory or memory port must be provided for each input and output value.

8.4.2 Storage Elements

The *storage elements* (*SE*) or *memory elements* (*M*) shall store data so that the processing element, in addition to implementing the algorithm, can access appropriate data without loss of any computational time slots. Since the processing elements normally require several simultaneous inputs and outputs, we assume that the memories are partitioned into several independent memories, or have several ports, which can be accessed in parallel. The important issues are the access mechanism for the memories, the bandwidth, and the number of ports. Furthermore, the storage shall be efficient, since memories are expensive in terms of chip area. The memories must also have reasonable form factors—i.e., the ratio of the number of words (height) and the word length (width).

8.4.3 Interconnection Networks

The *interconnection network* (*ICN*) shall provide the communication channels needed to supply the PEs with proper data and parameters, and store the results in the proper memories. The data movement should be kept simple, regular and uniform. Major design issues involve the topology of the communication network and its bandwidth.

8.4.4 Control

It is harder to command than to obey.
F. Nietzsche

There are two types of *control signals*: one type for setting up control and communication paths and another type for loading information into the memories [2]. The first type needs to be valid only in time intervals during which the second type has significant transitions. The second type is used to capture values and store them into the memories. Since they define the basic units of time, they are called *clocks*.

An important issue is the *control strategy* used to coordinate activities between the architectural components. Control strategy is mainly concerned with the manner in which control signals direct the data flow in the system. In a *centralized control* scheme, all the control signals come from a single source. Obviously, the central controller is a critical part in a system and may become a bottleneck that affects the performance and reliability of the entire system. The central controller must therefore be carefully designed to achieve good system performance. These drawbacks can be somewhat alleviated by using a *distributed control* strategy in which a small controller is associated with each operational unit in the system. In multiprocessor systems, which are further discussed in sections 8.5 through 8.9, control of crossbar interconnection networks is usually decentralized while multiple-bus interconnection networks can use either centralized or decentralized control.

At present, there is no general architecture suitable for the design of all types of control units. In [2], different implementation techniques commonly used in microprocessors are discussed. However, the simple and static computational structure that characterizes most DSP algorithms allows the use of far less complex control schemes than the one used in general-purpose computers.

Many DSP algorithms have an inherent hierarchy that can be mirrored directly by the control structure. For example, the nested loops in the FFT have such a hierarchical structure. It is often efficient to map a hierarchy of signals to a hierarchical control structure.

8.4.5 Synchronous and Asynchronous Systems

> There was no "One, two, three, and away," but they began running when they liked and left off when they liked, so that it was not easy to know when the race was over.
>
> Lewis Carroll

Timing philosophy; is one of the most important attributes characterizing a computing system. There are two types of possible timing schemes—*synchronous* and *asynchronous*.

Synchronous timing techniques, which are the most commonly used techniques, are characterized by the existence of a global clock that defines the basic unit of time. This time reference is used to order the activities into a proper sequence. The whole system is an isosynchronous domain. This is in practice accomplished by a central, global clock that broadcasts clock signals to all units in the system so that the entire system operates in a lock-step fashion.

Asynchronous timing techniques operate without a global clock. Communication among operational units is performed by means of interlock handshaking schemes. Asynchronous systems are difficult to design, and communication between synchronous and asynchronous systems is difficult and may cause malfunction.

8.4.6 Self-Timed Systems

> A man with a watch knows the time.
> A man with two watches is not sure.

Asynchronous, or self-timed, circuitry is often used in CMOS circuits because of its performance and area efficiency [8]. Synchronization between self-timed systems or modules is obtained by causal handshaking operations instead of by using a global clock. Communication between two modules can be established using two signals and an SR latch, as shown in Figure 8.11. The *Enable* signal sets the latch,

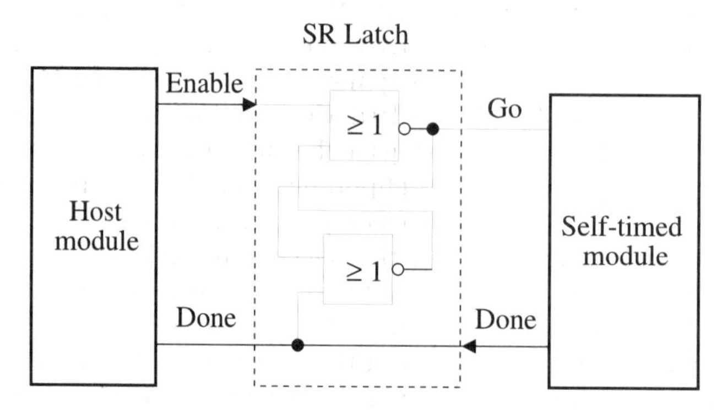

Figure 8.11 Principle of self-timed systems

which in turn releases the self-timed module. When the module has finished its task, it issues the *Done* signal. The *Done* signal can then act as an *Enable* signal so that the host can initiate the next transaction [8].

A problem with asynchronous and self-timed systems is that the time difference between the *Enable* and *Done* signals may be so small, or even negative, that the handshaking malfunctions. Also noise in the signals may cause errors or put the latches into a metastable state where neither of the outputs are within their low or high voltage ranges. In theory, the latch may stay in this undefined state indefinitely.

8.4.7 Autonomous Bit-Serial PEs

A problem associated with bit-serial PEs is their high clock frequency—e.g., 200 to 700 MHz. It is not possible to feed these clocks into a chip through the external pins, so they must be generated internally.

One approach is to let each PE generate its own clock by using a private ring counter as illustrated in Figure 8.12. The execution of an operation begins by loading an appropriate number of input data into the cache memory of the PE—i.e., into a set of shift registers. When the PE has received all input data from the memories, the clock is started, and the PE released to execute its operation. Because of pipelining the result of the operation is stored into cache memory one operation later.

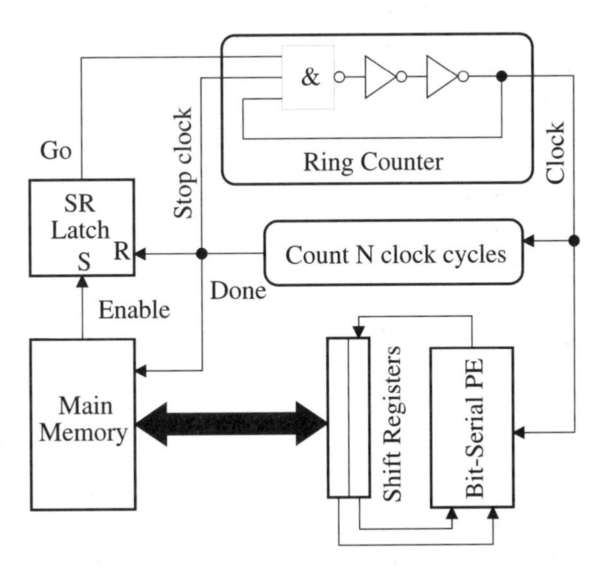

Figure 8.12 Self-timed PE with cache memory

A bit-serial PE requires a fixed number of clock cycles to perform its operation. Hence, a counter is used to stop the clock after the appropriate number of clock cycles [15]. The requirement on the PE is that it completes its operation within the specified time limit. Hence, the PE need only meet a minimum speed requirement. The PE operates asynchronously using its local clock, and when it has put the result of the operation into cache memory it stops and waits for the next set of input data. A faster PE will not improve the throughput, but more importantly it will not consume more power, since the power consumption is proportional to the number of clock cycles per operation. Further, the design problem becomes much easier. Instead of designing a high-speed circuit with a well-defined speed, it is sufficient to design a circuit that is sufficiently fast. The ring counter is designed so that the clock frequency is guaranteed to be lower than the maximal operating frequency of the PE. Speed variations due to process variations can be minimized if the ring counter and PE are placed close to each other on the chip.

Timing and synchronization problems are alleviated in this approach due to the inherent decomposition of the system into independent isochronous parts. The clock skew problem is changed into a communication problem between isochronous zones. We will use this technique of handshaking in the three case studies to interface the fast bit-serial processing elements with the main memories.

8.5 MULTIPROCESSORS AND MULTICOMPUTERS

General-purpose parallel or distributed computer systems can be divided into two categories: *multiprocessors* and *multicomputers* [9]. The main difference between these two categories lies in the way in which communication between the processors is organized.

All processors share the same memory space in a multiprocessor system while in a multicomputer system each processor has its own private memory space. Figures 8.13 and 8.14 illustrate the difference between the two generic architectures. There must be explicit communication interfaces between processors in the multicomputer system. This implies that there is a higher latency of communication between processors than would be the case if direct memory access was possible.

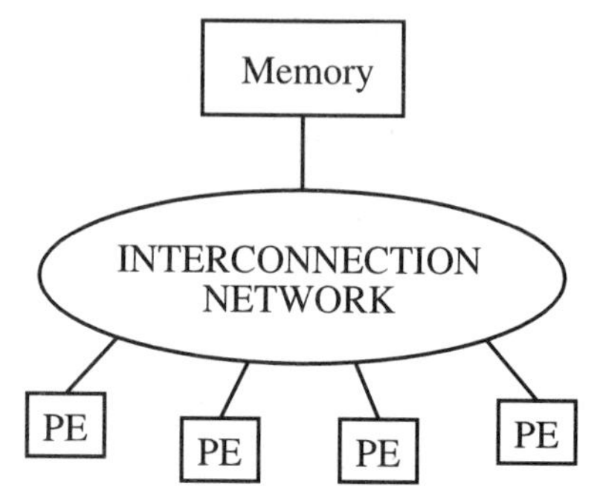

Figure 8.13 Multiprocessor architecture

Multiprocessor systems can be subdivided into tightly coupled and loosely coupled systems. A tightly coupled system has a common main memory so that the access time from any processor to memory is the same. In a loosely coupled system the main memory is distributed among the processors, although the processors share the same memory space. A processor can address another part of the main memory directly, but the access time is higher compared to accessing the local part of the main memory. As a result, partitioning and allocation of program segments and data play a crucial role in the design.

There are several ways to map the basic operations onto the processors. The simplest choice is a generalized processor type that can execute all types of operations. Architectures of this type are called *homogeneous architectures*. This choice simplifies processor assignment, since each operation can be mapped onto any of the free processors.

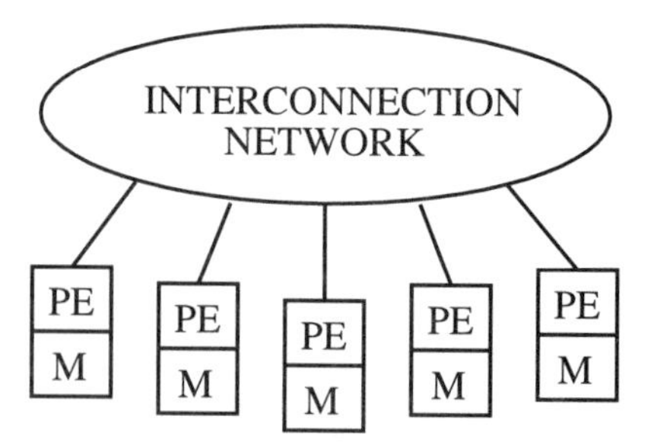

Figure 8.14 Multicomputer architecture

Another choice is to use dedicated processors, which can execute only one or a few types of operations. The performance of these processors can be optimized with respect to their function, but they may be poorly utilized if they are too specialized, with the work load for each type becoming low. The design problem is to find a number of feasible implementations—i.e., points in the resource-time domain—by proper scheduling and by using alternate processor types.

8.6 MESSAGE-BASED ARCHITECTURES

As mentioned before, each processing element (PE) in a multicomputer has its private memory. The PE–memory pairs are connected to certain neighboring processors via links. Communication between the processors occurs by passing messages between the processors (nodes). Multicomputer architectures are therefore often called *message-based architectures.* A message can be communicated only over a link that connects two nodes, and a given message may have to pass through several such nodes before it raches its destination. Hence, the distance between processors is not constant. Only those algorithms that can be partitioned into parts having low interprocessor communication needs are well suited to multicomputer architectures, since each interaction involves some communication overhead.

Shared-memory architecture is more versatile than direct link architecture, but it may be more expensive for architectures with many PE–memory pairs. A direct link architecture is shown in Figure 8.15.

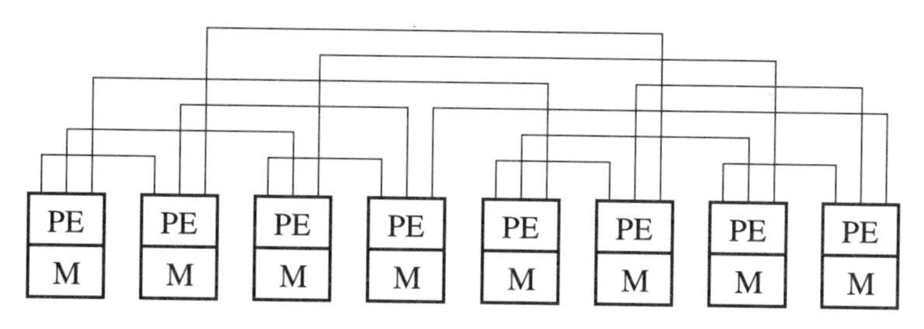

Figure 8.15 Direct link architecture

Usually, it is argued that the best choice is

Tens of PEs $\rightarrow$ shared memory

Hundreds of PEs $\rightarrow$ hybrid

More than hundreds of PEs $\rightarrow$ direct link

The term *processor array* is used when the processing elements are connected in a regular manner [7]. A processor array is a block of memory with processing elements distributed among it. The processors usually execute different types of operations. This should not be confused with the term array processor. An *array processor* is an array of processing elements doing the same operation but on different data items. The array is usually controlled by a common master processor.

A common way to classify architectures is single or multiple instruction stream, or single or multiple data stream. There are four combinations. SISD (single-instruction stream and single-data) architectures represent the classical serial computers. In MISD (multiple-instruction stream, single data) architectures, multiple instructions operate on a single datum. This case is generally deemed impractical. A more practical case is SIMD (single-instruction stream, multiple-data) architectures. SIMD architectures are particularly suitable for applications with high data parallelism, such as image processing. In MIMD (multiple-instruction stream, multiple-data) architectures, multiple processors operate on different data.

8.6.1 Interconnection Topologies

To exploit parallelism efficiently, a parallel or distributed system must be designed to minimize communication overhead between processors. A given communication strategy might support one application well but be inefficient for others. A complete interconnection between all processors (for example, using the crossbar shown in Figure 8.16) might be cost-prohibitive while a shared single-bus interconnection, as shown in Figure 8.17, might be inefficient. Hence, most applications call for a solution whose cost and performance lie somewhere between the two extremes. The interconnection network must be efficient, reliable, and cost-effective. Topologies that can accommodate an arbitrary number of nodes are called scalable architectures.

Many different interconnection networks have been proposed [1, 3, 9]. Some common examples are shown in Figures 8.18 through 8.21. Mesh-connected processor arrays have simple and regular interconnections and are therefore preferred architectures for special-purpose VLSI designs [7].

Special, single-chip, 32-bit processors called *Transputers*™ have been introduced by Inmos for use in such networks [11]. The transputer, which is aimed at multicomputer applications, has four high-speed communication ports. Hence, it is well suited for square mesh and Boolean 4-cube topologies. However, the Transputer is a general RISC processor and is therefore not optimized for DSP applications. Some of the more recent standard DSPs are provided with I/O ports to support multicomputer applications.

A variation of a Boolean cube is the cube connected cycles (CCC) where each node in the Boolean cube consists of a number of PE-memory pairs where each PE-memory pair needs only three ports.

Multistage interconnection networks (*MICN*) and multiple-bus topologies are also commonly used [2, 7]. The omega network, shown in Figure 8.20, has multiple shuffle exchange stages of switches. The source processor generates a tag that is the binary representation of the destination. The connection switches in the *i*th stage

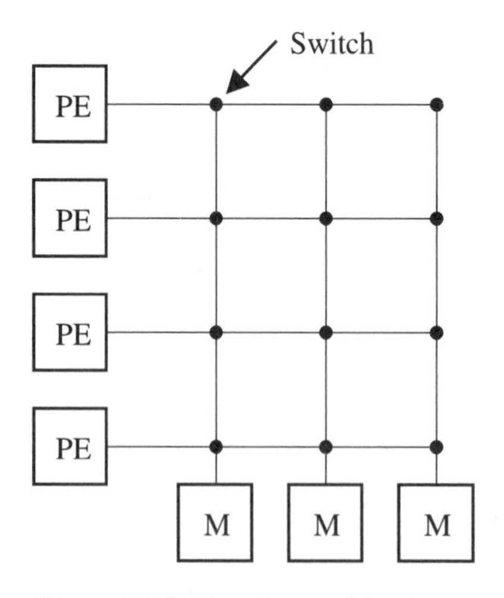

Figure 8.16 Crossbar architecture

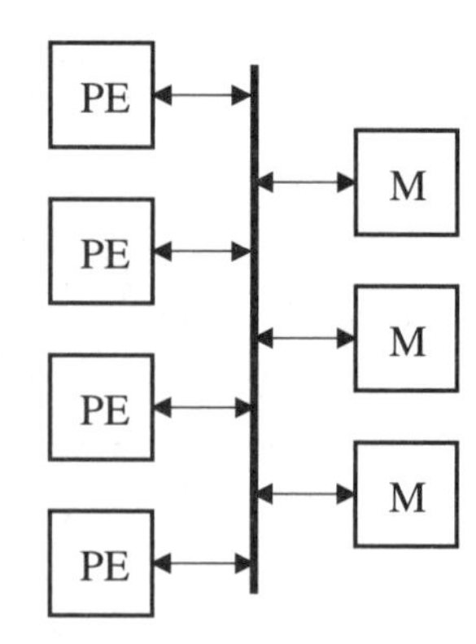

Figure 8.17 A single-bus structure

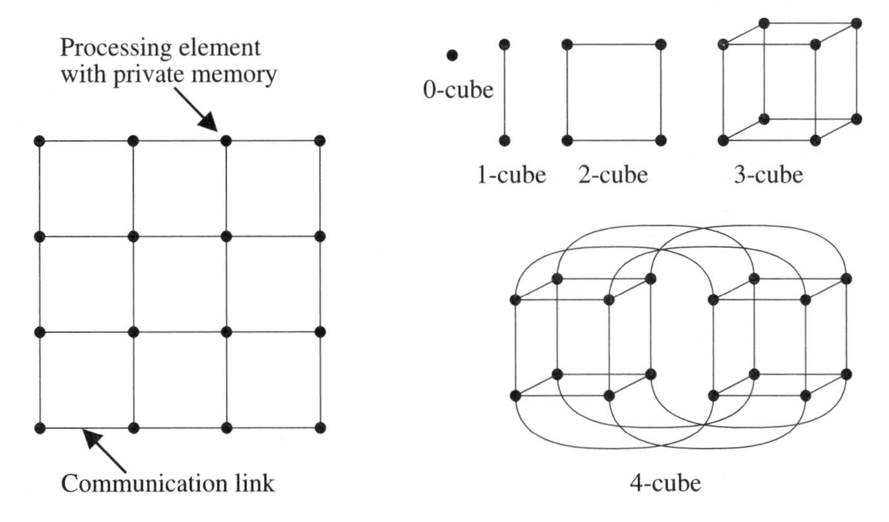

Figure 8.18 Square mesh

Figure 8.19 Boolean cubes

are controlled by the ith bit of the binary tag. The self-routing property of an MICN avoids the need for a central controller, making it suitable for multiprocessors.

Figure 8.21 shows two interconnection networks of the tree type. In the binary trees with N processors, a processor can be accessed in at most $\log_2(N)$ time compared to a one-dimensional array, $O(N)$, and $O(\sqrt{N})$ for a rectangular array.

The number of processors available in a tree increases exponentially at each level. If a problem being solved has this growth pattern, then tree geometry may be suitable. FFTs are suitable for implementation using a massively parallel approach with 2^N PEs arranged in a hypercube where N is the length of the sequence.

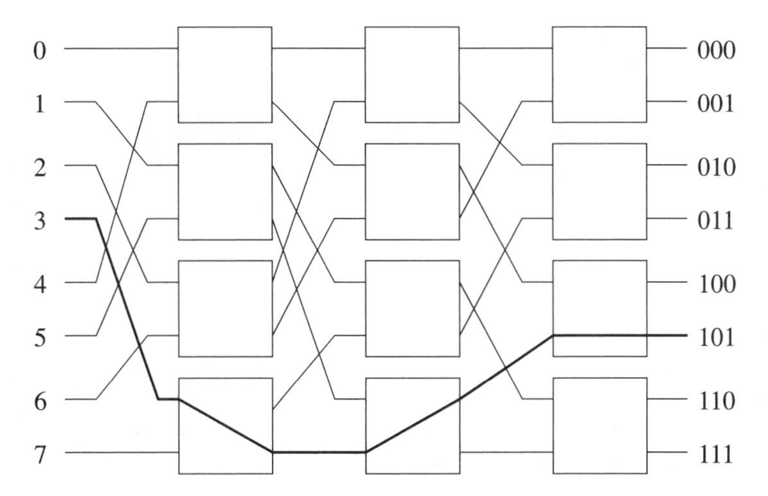

Figure 8.20 An eight-port omega network. Input 3 is connected to output 5

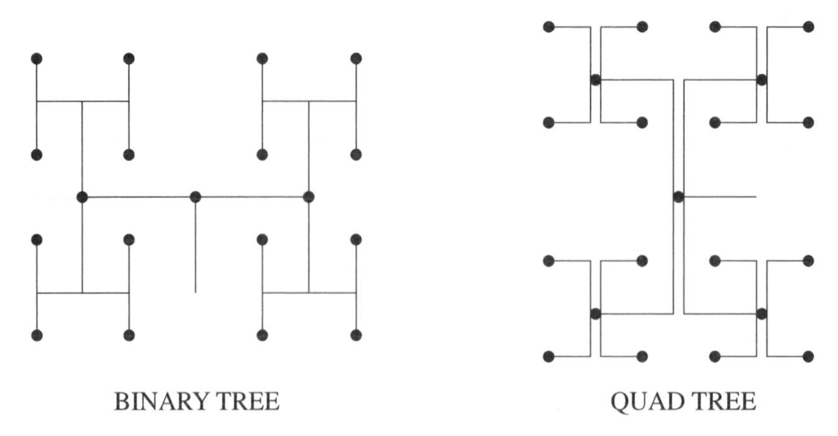

Figure 8.21 Binary and quad trees

Interconnection networks are also characterized by communication, control, and switching strategies. The communication strategy used in the interconnection network can be either synchronous or asynchronous while control, of the data flow can either be centralized or distributed.

There are two major switching methodologies: *packet switching* and *circuit switching*. In packet switching, a message is partitioned into small packets which are transmitted through the network. A packet experiences a random delay at each switching point, depending on the traffic in the network along its path to the destination. Conversely, circuit switching sets up a physical path between source and destination. A time delay is incurred in setting up the path. Once the path is established, it is held for the entire data transmission. Generally, circuit switching is more suited to long messages, and packet switching to short messages. Flexibility in multiple PE architectures is often low—e.g. it is difficult to use and exploit varying data word lengths in an algorithm.

8.7 SYSTOLIC ARRAYS

Systolic and wave front arrays are suitable for applications with very high throughput requirements. Typical applications can be found in sonar, radar, seismic, video, and image processing where the amount of data is very large. Further, real-time video applications require high sample rates. Matrix and vector operations are suitable for implementation using systolic or wave front array architectures [1, 7, 9, 12, 16, 19].

A *systolic array* is defined as a lattice of synchronous and locally connected PEs that can perform iterative algorithms with regular data dependencies. A systolic algorithm is basically described by a set of indexed calculations performed in a lattice space. Indexing of the dependent variables is local and regular. Ordering or scheduling of calculations is established by preserving the dependencies. By selecting appropriate projections of the index set along a direction, lattice points in the index space can be mapped onto one- or two-dimensional structures. These time-space mappings result in a scheduled, regular, connected PE structure.

Usually, the PEs are identical and perform identical operations, but in some cases several types of PEs or operations are required. Unlike other parallel architectures employing a lattice of PEs, a systolic array is characterized by a regular data flow. Typically, two or more data streams flow through the cells of the systolic arrays with various speeds and directions. Data items from different streams interact with each other and trigger computations in the PEs where they meet. A systolic array is an n-dimensional structural pipeline with synchronous communication between the PEs. Thus, a systolic array simultaneously exploits both pipelining and parallelism in the algorithm.

We illustrate the basic principle of systolic arrays by a simple example.

Example 8.1

Figure 8.22 illustrates the basic principle of systolic arrays.

The array implements a matrix multiplication

$$\mathbb{C} = \mathbb{A}\,\mathbb{B}$$

where the sizes of $\mathbb{A}$ and $\mathbb{B}$ are 4 × 4. The $\mathbb{A}$ and $\mathbb{B}$ matrices are applied from the left side and from the top of the array. The PEs accept data from the left and top

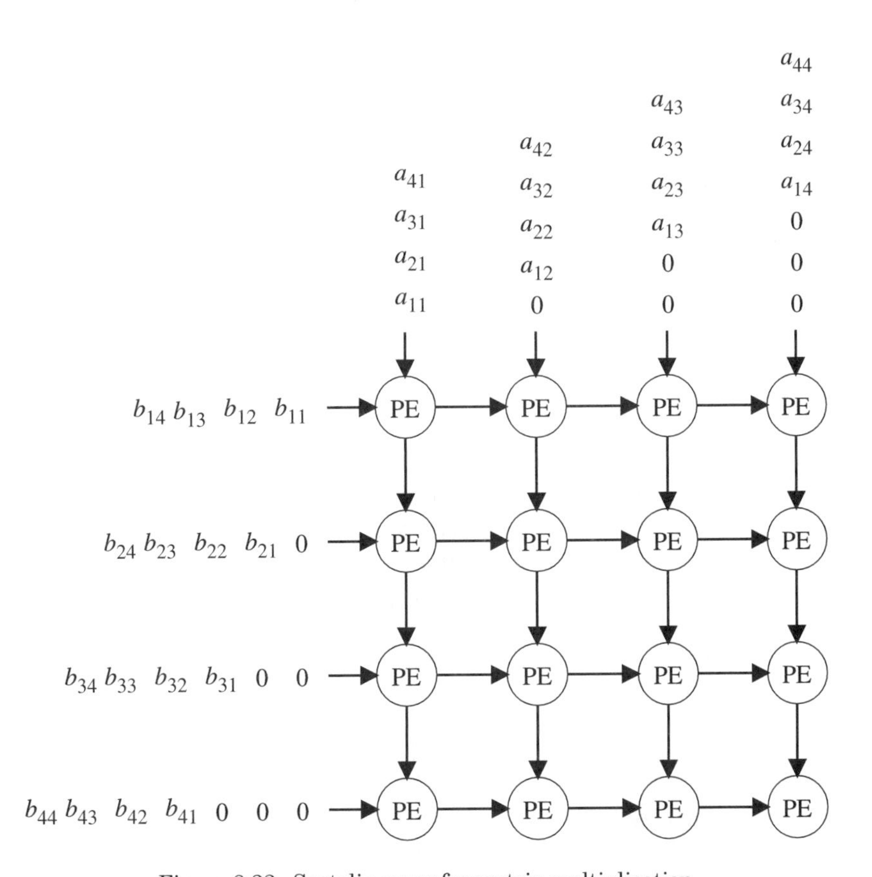

Figure 8.22 Systolic array for matrix multiplication

and pass them along to PEs on the right and bottom, respectively. In each time step the PEs compute the sum of products:

$$c(n) = c(n-1) + a(n)\, b(n)$$

The result of the computation is stored in each PE. After the whole matrix operation is complete, the result can be shifted out, for example, through the right-hand side of the array.

Locality in communication is an important property in large systems. Pipeline structures exploit locality by providing direct communication paths between communicating functional units. The crux of the systolic array approach is to ensure that once a data item is brought out from the system memory, it can be used effectively at each PE it passes. A high computational throughput can therefore be obtained by using many memories, each with a modest memory bandwidth. The ability to use each input data item a number of times is just one of the many advantages of a systolic architecture. Other advantages, including modular expandability, are simple and regular data and control flow and use of simple and almost homogeneous PEs. Elimination of global broadcasting and fan-in are also characteristic. However, global clock distribution in n-dimensional arrays is difficult.

Large systolic arrays usually need a large internal data word length. Arrays exploit both pipelining and parallelism, but often only a fraction of the cells in an array are active in each time slot. One-dimensional arrays for convolution are characterized by the fact that their I/O bandwidth requirement is independent of the size of the convolution kernel. This contrasts with other types of 2-D arrays, for which the I/O bandwidth increases as the kernel increases. For adaptive convolution kernels, the critical factor is the adaption time constant, i.e., the rate at which the kernel can be modified. It may be difficult to modify the kernel without disturbing the ongoing convolution process. Communication bandwidth between the system memory and the systolic array is a limiting factor.

8.8 WAVE FRONT ARRAYS

A *wave front array* is an n-dimensional structural pipeline with asynchronous communication between the PEs. The principle is illustrated in Figure 8.23. The main difference between the systolic array and the wave front array is that communication between PEs in the latter case is maintained by a handshaking protocol.

The program memory issues a sequence of instructions and the memories provide data requested by the top row and leftmost column of PEs. As data and instructions become available to the PEs, the corresponding operations are executed and the results passed on to the neighbors. The operations for a given set of indices propagate diagonally downward like a plane wave. Operations in this architecture are controlled by the availability of data and not by a clock signal. Hence, the wave front array can accommodate more irregular algorithms than the systolic array, such as algorithms where the execution time in the PEs is data dependent.

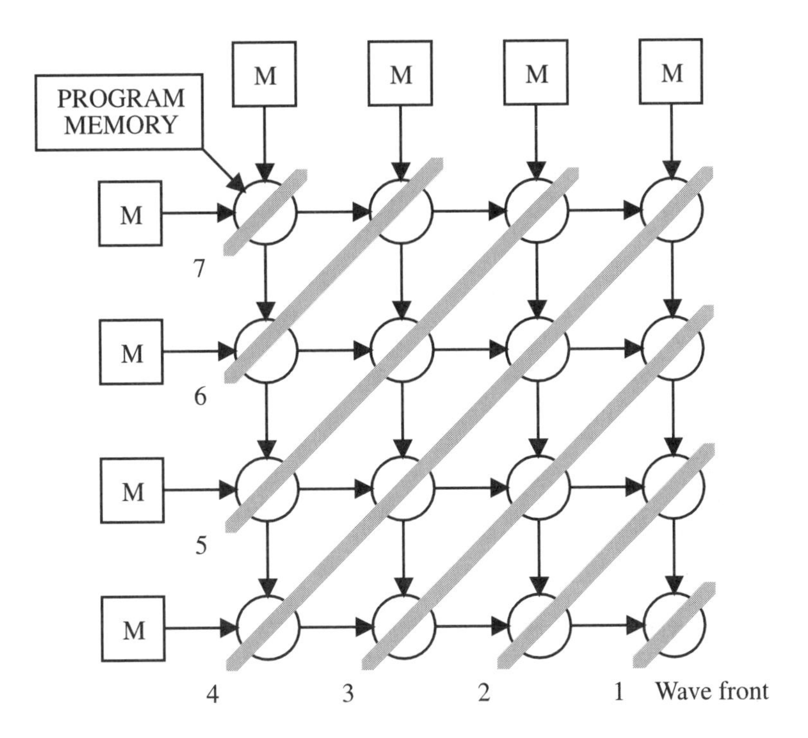

Figure 8.23 Wave front array

Systolic and wave front array implementations of FIR and IIR filters achieve a maximum throughput rate of one output per multiply–accumulate cycle. For example, lattice wave digital filters with branches implemented using Richards structures can be implemented efficiently using systolic and wave front arrays. They use, at most, twice as many PEs as the filter order.

8.8.1 Datawave™

ITT Intermetal has developed a single-chip multiple-instruction multiple-data (MIMD) processor, Datawave [19], for video applications. Figure 8.24 shows the architecture. The PEs are organized in a 4×4 rectangular array and the ICN provides nearest neighbor communication. Several chips can be combined to synthesize larger arrays. The chip has a data word length of only 12 bits, which is sufficient for most video applications. Communication between PEs, as well as with the outside, takes place via two 12–bit wide buses and requires only one clock cycle, except for interchip communication, which is somewhat slower.

Communication between PEs is data-driven. Operations in the algorithm are therefore statically scheduled at compile time, although only the order of operations is scheduled, not the timing. FIFOs have been inserted between the PEs to improve performance by averaging out variations in work load between the PEs. Self-timed handshake protocols are used to synchronize the asynchronous data

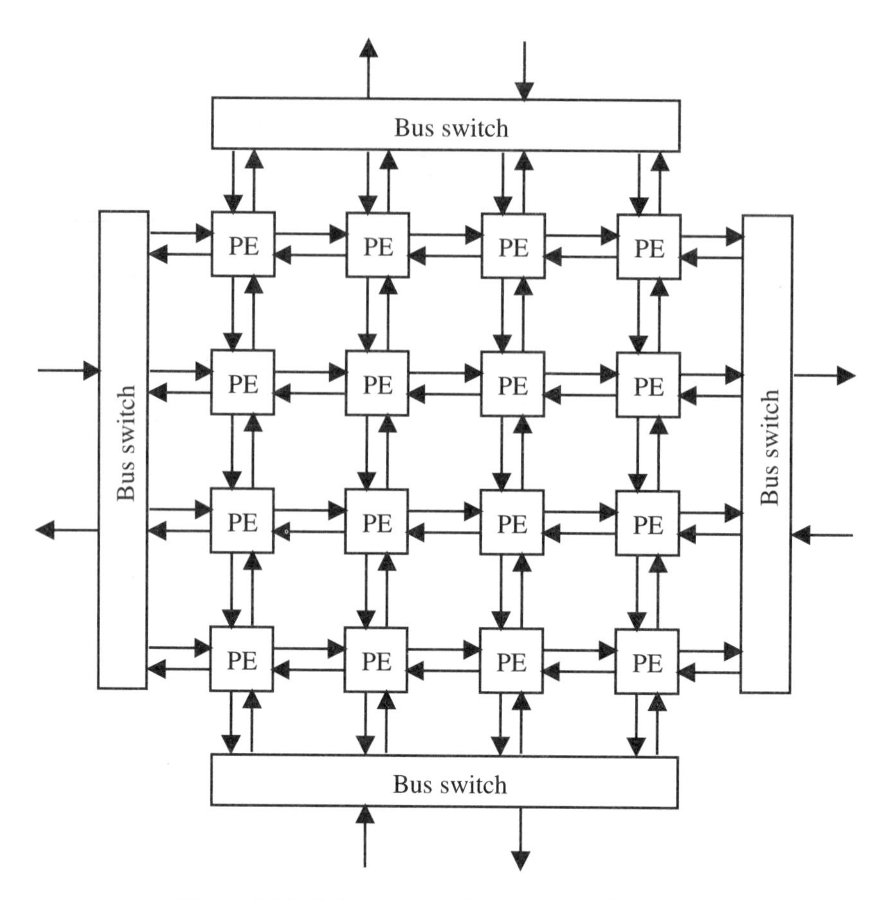

Figure 8.24 Datawave multiprocessor architecture

transfers. The PEs use local clocks derived from the global clock. Figure 8.25 shows the architecture of the PEs. The core of a PE is a 12-bit RISC processor with local program and data stores. Three 12-bit ring buses are used to connect the PE core with adjacent PEs via FIFOs.

The two outer buses are used to deliver data to the MAC (multiplier–accumulator) and ALU while the third bus is used to deliver the results to functional units and the outside world. The MAC has a 12×12-bit multiplier and a 29-bit accumulator. The ALU works in parallel with the MAC. Each PE, which is pipelined, can start a multiply-and-accumulate operation every clock cycle (125 MHz). Hence, a very high peak performance of 4 GOPS is obtained.

The program memory can store only 64 46-bit–wide instructions, but this is usually sufficient since the chip is assumed to run very high sample rate applications. There is time to execute only a few instructions per sample. Program memories are loaded via a serial bus connected to all PEs. The local data memory is a four-port register with 16 words. Unfortunately, current technologies do not allow large on-chip data memories. Such large memories are needed, for example, to store several lines, or even several frames, of a TV image.

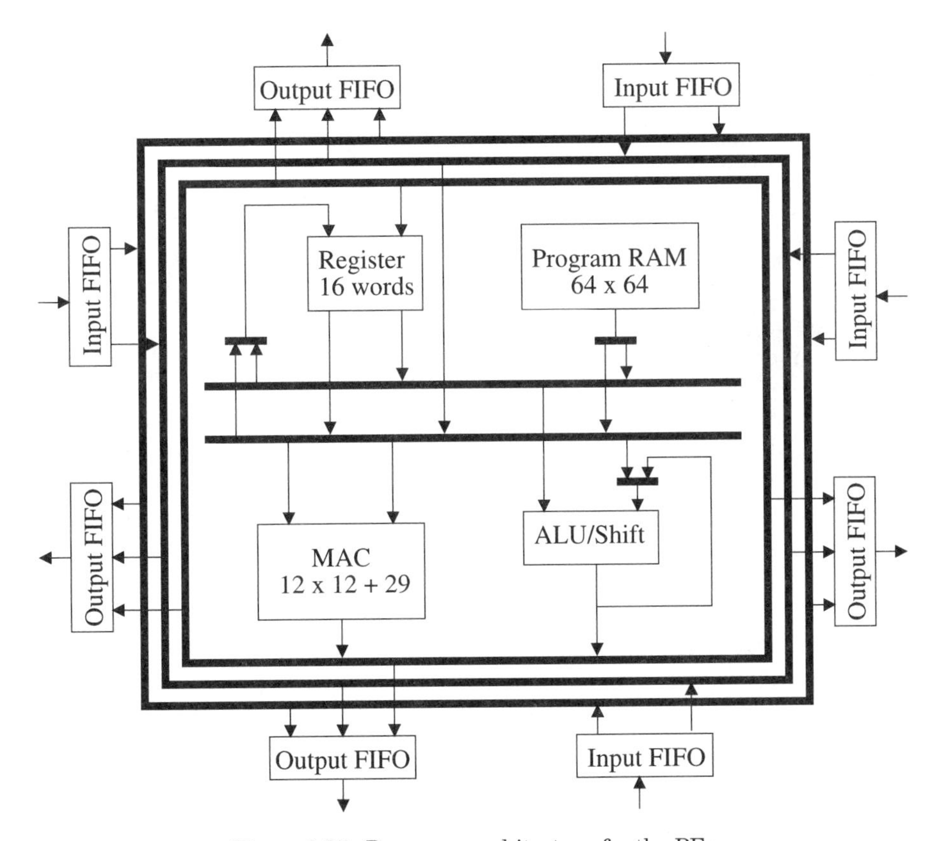

Figure 8.25 Processor architecture for the PEs

The Datawave chip is implemented in a 0.8-μm CMOS process. Domino logic circuits are used for the time critical parts. Each PE requires about 75,000 devices and 8 mm^2 while the whole chip requires about 1.2 million devices and 150 mm^2. The chip is housed in a 124-pin package and consumes 8 W.

8.9 SHARED-MEMORY ARCHITECTURES

Friends share all things.
Pythagoras

A special case of a multiprocessor architecture, henceforth called *shared-memory architecture,* is shown in Figure 8.26. In order to simultaneously provide the processors with several (K) input data, the shared memory is partitioned into K memories. Using a rotating access schedule, each processor gets access to the memories once every N cycles, at which time the processor either writes a result into one or several of the memories, or reads in parallel at most one value from each memory. This simple access scheme guarantees that access conflicts can be completely avoided. The system is tightly coupled, since all processors have the same access time to the memories.

Shared-memory architectures are well suited for tightly coupled DSP algorithms—for example, recursive algorithms with complicated data dependencies. Unfortunately, the shared-memory architecture can accommodate only a small number of processors due to the memory bandwidth bottleneck. In many DSP applications the required work load is too large for a single shared-memory architecture based on current circuit technologies. Fortunately, top-down synthesis techniques tend to produce systems composed of loosely coupled subsystems that are tightly coupled internally. Typically, a signal processing system consists of a mix of subsystems in parallel and cascade. The system is often implemented using a message-based architecture since the subsystems usually have relatively low intercommunication requirements, while the tightly coupled subsystems are implemented using shared-memory architectures. Generally, the subsystems are fully synchronous, while global communication may be asynchronous.

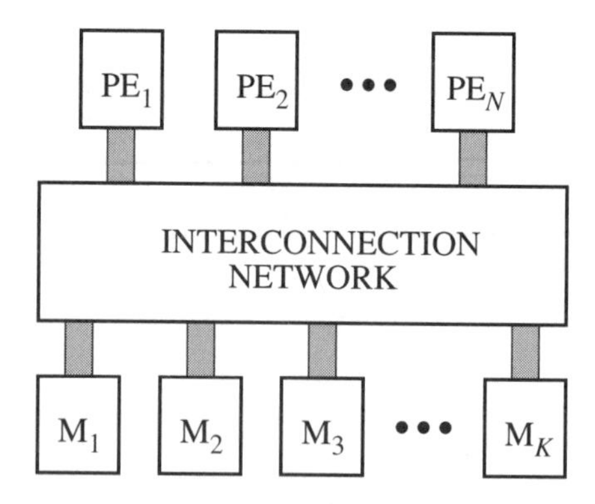

Figure 8.26 Multiprocessor architecture

8.9.1 Memory Bandwidth Bottleneck

The major limitation of shared-memory architecture is the well-known memory bandwidth bottleneck. Each processor must be allocated two memory time slots: one for receiving inputs and the other for storing the output value into the memories. To fully utilize N processors with execution time T_{PE}, the following inequality must hold:

$$T_{PE} \geq 2\,N\,T_M \tag{8.1}$$

where T_M is the cycle time for the memories. However, T_{PE} and T_M are of the same order. Hence, very few processors can be kept busy because of this *memory bandwidth bottleneck*. Thus, there is a fundamental imbalance between computational capacity and communication bandwidth in shared-memory architecture. In the following sections we will discuss some methods to counteract this imbalance and reduce the implementation cost.

For simplicity, we assume that the PEs perform only constant-time operations. $T_M = T_R = T_W$ are the read and write times for the memories. According to inequality (8.1), there are only three factors that can be modified by the design. In section 8.9.2, we will discuss methods of reducing the cycle time of the memories, and in section 8.9.3, we will discuss methods of reducing the number of memory–PE transactions. Finally, in Chapter 9, we will propose an efficient method based on slow PEs.

8.9.2 Reducing the Memory Cycle Time

The effective cycle time can be reduced by interleaving memories [9]. Each memory in the original architecture is substituted with K memories, as illustrated in

Figure 8.27. Operation of the memories is skewed in time. The general idea is to arrange the memories so that K sequential memory accesses fall in K distinct memories, allowing K accesses to be under way simultaneously. In fact, this arrangement is equivalent to pipelining the memory accesses. The effective memory cycle time is reduced by a factor K.

By choosing $K = 2N$, a good balance is obtained. Interleaving leads to expensive overhead because of the necessary duplication of decoders, sense amplifiers, etc. Furthermore, losses in processing efficiency may be incurred by memory access conflicts.

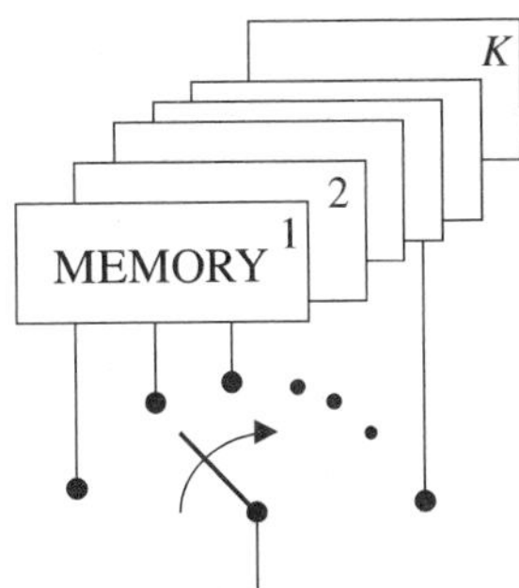

Figure 8.27 Interleaving of K memories

Another method of reducing the right-hand side of inequality (8.1), by a factor of two, is to use two separate sets of memories. Results from the PEs are written into one set while the other set is used for reading values which will become inputs to the PEs. The role of these two sets of memories alternates every other PE cycle. In this way memory conflicts are avoided.

Vector PEs usually employ interleaving of memories for accessing sequences of vector elements placed in consecutive memories. The memory elements do not need to be consecutive. It is enough that the access pattern is known in advance.

8.9.3 Reducing Communications

Most schemes to reduce the memory bandwidth requirement exploit some inherent property of the DSP algorithm to reduce the number of memory accesses. We will discuss only the most common techniques.

Broadcasting

The number of memory read cycles can be reduced if all PEs operate on the same input data. Figure 8.28 shows an array of processing elements supported by K memories [1]. The outputs from the memories are broadcast to the PEs. Hence, only one memory read cycle is needed. The required number of write cycles depends on the algorithm. Only one write cycle is needed if the results from the PEs are written into different memories such that access conflicts do not occur ($N \leq K$), but N cycles are needed if all results are written into the same memory. Hence, we have

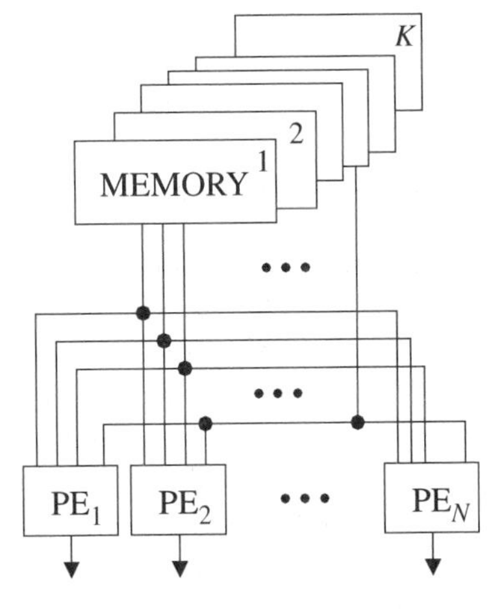

Figure 8.28 Broadcasting of data

$$T_{PE} \geq n\, T_M, \quad 2 \leq n \leq N + 1$$

Thus, a good balance between processing capability and communication bandwidth can be obtained for certain types of algorithms. This type of architecture can easily be time-shared among several input channels. However, it has too large a processing capacity for many applications.

In the case of homogeneous PEs capacity can be adjusted to the application by multiplexing the PEs. Other advantages are that the PEs can be optimized to their individual work load and that the interconnection network is simple and fixed. A typical example of broadcasting data can be found in Chapter 11 where an implementation of a discrete cosine transform using several PEs is proposed. All the PEs operate on the same set of input values. Each PE computes one of the DCT components.

Interprocessor Communication

A common method of reducing the number of memory transactions is to provide communication channels between the PEs, as illustrated in Figure 8.29. In general, some memory cells must be placed between the PEs (pipelining). Popular uses for this type of architectures are systolic and wave front arrays, which combine pipelining and parallel processing. This important class of architectures will be discussed in Chapter 11.

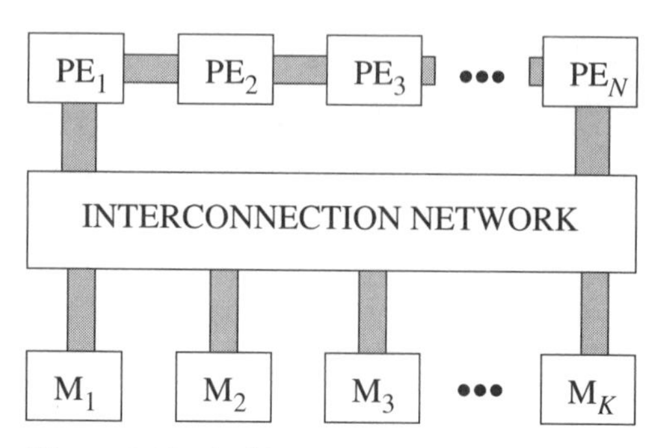

Figure 8.29 Architecture with direct interprocessor communication

---

Example 8.2

Figure 8.30 shows a shared-memory architecture for an FFT using a single, but composite, butterfly PE. The architecture has direct interprocessor communication and uses complex data. The real and imaginary parts of a complex value are stored as one word in the memory—i.e., at the same address. Discuss balance between computation and memory bandwidth.

Since only one memory location is needed for each complex value, the inequality (8.1) reduces to

$$T_{PE} \geq 2T_M$$

where

$$T_{PE} = T_{CMult} + T_{CAdd}$$

Thus, the computation versus communication balance is reasonably good if we assume that

$$T_M \approx T_{CMult} \approx T_{CAdd}$$

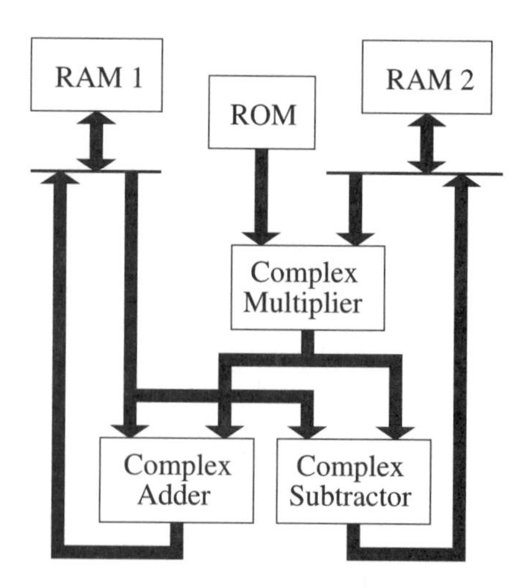

Figure 8.30 FFT architecture with a single butterfly

A perfectly balanced architecture is obtained if the number of memories is doubled—i.e., we have an interleaving factor of two. An even faster implementation is obtained if the PE is pipelined by introducing a set of flip-flops between the multiplier and the adder and subtractor. This architecture can again be balanced by increasing the interleaving of the memories by a further factor of two. This is possible since the FFT is a nonrecursive algorithm.

Cache Memories

A common technique to reduce the communication demand is to provide the processors with fast private memories. The processor can therefore access its cache memory without interference from the other processors. This scheme works well if the relevant data are kept in the cache memories and if the communication demands between the cache memories and the main memory are relatively small. This scheme allows the use of slower and less expensive main memories. We will, in the examples used as case studies, use cache memories to obtain balanced architectures.

8.9.4 Large Basic Operations

The third factor in inequality (8.1) affecting architectural balance is execution time for the PEs. Obviously, if we use PEs with a large granularity, execution time will be longer, however more useful work will be done. For example, a butterfly PE is preferred over separate PEs that perform simpler operations such as add, subtract, and multiply. Further, fewer memory transactions may be needed if direct interprocessor communications are allowed. It is interesting to note that, conversely to the current trend of RISC technology, it is advantageous to use basic operations that are as large as possible, since large basic operations tend to do more useful work on a given set of input data and thereby tend to reduce the communication requirement.

In Chapter 9 we will show that it is efficient to use more but slower PEs to obtain a balanced architecture [24, 25].

References

[1] Almasi G.S. and Gottlieb A.: *Highly Parallel Computing,* Benjamin/Cummings, Redwood City, CA, 1989.

[2] Anceau F.: *The Architecture of Microprocessors,* Addison-Wesley, Wokingham, England, 1986.

[3] Bhuyan L.N., Yang Q., and Agrawal D.P.: Performance of Multiprocessor Interconnection Networks, *IEEE Computers,* Vol. 22, No. 2, pp. 25–37, Feb. 1989.

[4] Chen C.H. (Ed.): *Signal Processing Handbook,* Marcel Dekker, New York, 1988.

[5] DeCegama A.L.: *The Technology of Parallel Processing: Parallel Architectures and VLSI Hardware,* Vol. 1, Prentice Hall, Englewoood Cliffs, New Jersey, 1989.

[6] Fettweis A.: Modified Wave Digital Filters for Improved Implementation by Commercial Digital Signal Processors, *Signal Processing,* Vol. 16, No. 3, pp. 193–207, March 1989.

[7] Fountain T.: *Processor Arrays, Architecture and Applications,* Academic Press, London, 1987.

[8] Glasser L.A. and Dobberpuhl D.W.: *The Design and Analysis of VLSI Circuits,* Addison-Wesley, Reading, MA, 1985.

[9] Goor A.J. van de *Computer Architecture and Design,* Addison-Wesley, Wokingham, England, 1989.

[10] Hwang K. and Briggs F.A.: *Computer Architecture and Parallel Processing,* McGraw-Hill Co. New York, 1984.

[11] Inmos: *Digital Signal Processing,* Prentice Hall, 1988.

[12] Kung S.Y.: *VLSI Array Processors,* Prentice Hall, Englewood Cliffs, New Jersey, 1988.

[13] Lawson H.W., Svensson B., and Wanhammar L.: *Parallel Processing in Industrial Real-Time Applications,* Prentice Hall, Englewood Cliffs, New Jersey, 1992.

[14] Lin Kun-Shan (ed.): *Digital Signal Processing Applications with the TMS320 Family,* Vol. 1, Prentice Hall, 1987.

[15] Nilsson P., Torkelson M., Vesterbacka M., Wanhammar L.: CMOS On-Chip Clock for Digital Signal Processors. *Elect. Letters,* Vol. 29, No. 8, pp. 669–670, April 1993.

[16] McCanny J., McWhirter J., and Swartzlander Jr. E.E. (eds.): *Systolic Array Processors,* Prentice Hall, New York, 1989.

[17] Motorola Inc.: *DSP56000/DSP56001 Digital Signal Processor, Users Manual,* Phoenix, AZ,1990.

[18] Motorola Inc.: *DSP96002 IEEE Floating-Point Dual-Port Processor, Users Manual,* Phoenix, AZ, 1989.

[19] Prisch P.: *Architectures for Digital Signal Processing,* John Wiley & Sons, Chichester, 1998.

[20] Schmidt U. and Caesar K.: Datawave: A Single-Chip Multiprocessor for Video Applications, *IEEE Micro,* Vol. 11, No. 3, pp. 22–25 and 88–94, June 1991.

[21] Seitz C.L.: Concurrent VLSI Architectures, *IEEE Trans. on Computers,* Vol. C-33, No. 12, pp. 1247–1265, Dec. 1984.

[22] Smith M.R.: How RISCy is DSP?, *IEEE Micro,* Vol. 12, No. 6, pp. 10–23, Dec. 1992.

[23] Swartzlander Jr. E.E.: *VLSI Signal Processing Systems,* Kluwer Academic Pub., Boston, 1986.

[24] Wanhammar L., Sikström B., Afghahi M., and Pencz J.: A Systematic Bit-Serial Approach to Implement Digital Signal Processing Algorithms, *Proc. 2nd Nordic Symp. on VLSI in Computers and Communications,* Linköping, Sweden, June 2-4, 1986.

[25] Wanhammar L.: On Algorithms and Architecture Suitable for Digital Signal Processing, *Proc. The European Signal Processing Conf., EUSIPCO–86,* The Hague, The Netherlands, Sept. 1986.

Problems

8.1 Define in what sense an architecture can be ideal. State suitable criteria for an ideal DSP architecture

8.2 What are the main limitations in shared memory architectures? Discuss different approaches in overcoming or reducing these limitations?

8.3 Determine the peak performance and I/O data rate for the Datawave™ chip if I/O takes two clock cycles.

8.4 Compare the various standard fixed-point DSPs with respect to power consumption per sample for some typical algorithms—for example, FIR filters and second-order sections.

8.5 Determine maximum throughput for the autonomous PE shown in Figure 8.12. Assume that the PE is a multiplier with a coefficient word length of W_c and that the data word length is $W_d > W_c$.

8.6 Estimate the required chip area and achievable clock frequency if the processor TMS320C25 would be implemented using a 0.35-μm CMOS process. Assume for sake of simplicity that the original and new CMOS processes have the same number of metal layers.

SYNTHESIS OF DSP ARCHITECTURES

9.1 INTRODUCTION

Digital signal processing has, as a result of the development of robust and efficient algorithms and advances in VLSI technology, attained widespread use. A major factor in this development is the independence of element sensitivity inherent in digital systems. The advent of commercially available digital signal processors further increased the competitiveness of digital over analog signal processing techniques. However, DSP-based implementations are generally not competitive for low- to medium-performance and low-cost systems. SC techniques are often more suitable for these types of applications. Other difficult DSP applications are systems with very high sampling rates and large computational work loads.

We believe that the somewhat limited success in designing competitive, cost-effective DSP systems is largely due to the lack of general and systematic procedures for exploring the computational properties of the algorithm and synthesis procedures for the corresponding optimal circuit architectures. We will therefore focus on design techniques that match the architecture to the algorithm and not vice versa.

In previous chapters we have discussed the problems of scheduling operations and subsequent resource allocation and assignment. As a result of these design steps a class of ideal architectures was defined. The type and number of PEs, and the size and number of memories, including the required communication channels, were determined. The required control signals, component performance requirements (e.g., T_{PE}, T_M, and communication channel bandwidths) were also derived in these design steps. In this chapter we will present methods of synthesizing optimal architectures based upon these results. The synthesis process leads to a wide class of shared-memory architectures that can be designed to have a good balance between communication bandwidth and processing capacity. We will also discuss a few simple cases where the scheduling, resource allocation, and assignment steps are trivial. In the first case all of the processes are mapped onto a single processor. This approach leads to power- and area-efficient implementations useful in applications with small work loads. In the second case each process is assigned to a dedicated PE. This approach trades large power consumption and chip area for high throughput.

9.2 MAPPING OF DSP ALGORITHMS ONTO HARDWARE

The majority of conventional architectures proposed for digital signal processing represent extreme points in the resource-time domain as illustrated in Figure 9.1. At one extreme, only one processor is used to execute all processes. The processor must be able to execute all types of operations and the processing time is equal to the sum of all execution times.

At the other extreme, one or several dedicated PEs are assigned to each process as discussed in section 7.5.4. The PEs can therefore be customized (optimized) to execute only one particular operation. The usable resources are limited by the parallelism in the algorithm and the maximum sample rate is determined by the critical loop. In both cases, the scheduling and PE assignment problems become trivial. However, the latter approach often leads to low utilization of the computational resources since most applications do not need the large processing power provided by a fully parallel architecture. The challenge is therefore to find architectures that provide just enough processing power using a minimum amount of hardware resources.

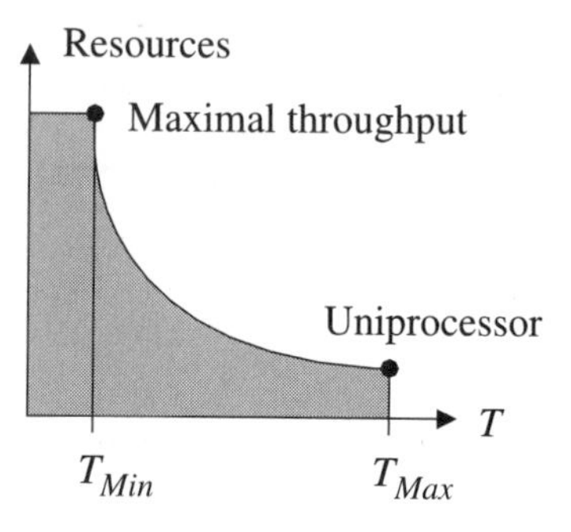

Figure 9.1 Resource-time domain

Battery-powered systems with stringent power budgets are becoming more and more common. Surprisingly, the maximally fast implementation can achieve reduced power consumption. The power consumption of a CMOS circuit is proportional to the square of the power supply voltage, while the speed is approximately inversely proportional. Hence, if the required system speed is much lower than the speed of the maximally fast implementation, the power supply voltage can be reduced with a significant reduction in power consumption. For example, the digital filters discussed in Examples 7.3 and 7.4 are maximally fast implementations—i.e., they require a minimum number of clock cycles per sample. These implementations can be used to achieve low power consumption in applications that require much lower sample rates than 130 MHz. See Problem 9.20.

To summarize: Low power consumption can be achieved by increasing the amount of parallelism and hardware resources (chip area). See Problem 2.7.

9.2.1 Design Strategy

Generally, relationships between operations in the algorithm and the processing elements are established by directly mapping operations executed in each time slot onto the processing elements. Thus the schedule implicitly determines the number of processing elements, memories, communication channels, and the control for a class of ideal architectures.

In the next design step, the best candidate among these architectures is chosen and optimized subject to the system requirements. However, scheduling assumes an underlying hardware architecture—for example, types of processing elements and number of memories. Therefore we assume that the scheduling of the processes is fixed when we explore the architectural candidates for a given schedule.

Another schedule leads to another set of architectural alternatives. Hence, the synthesis process is essentially downward, as illustrated in Figure 9.2, but the design steps are highly interdependent. If the requirements can not be met in a certain design step, the synthesis process has to be restarted at an earlier stage. Possibly a better algorithm has to be chosen. In some cases it may be necessary to continue upward in the hierarchy and redesign the basic building blocks, choose a better process technology, etc.

The choice of a particular algorithm determines the class of usable architectures. Therefore, the synthesis process must in practice be followed by an evaluation phase and possibly by a subsequent redesign iteration involving the selection of another algorithm. The design process may have to be iterated several times in order to arrive at a satisfactory solution. It may also have to be extended to a lower level of abstraction—for example, to the logic and electrical circuit levels.

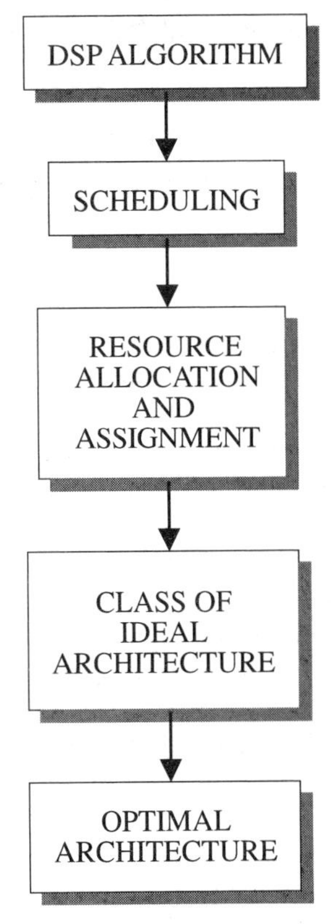

Figure 9.2 Synthesis path for optimal architectures

9.3 UNIPROCESSOR ARCHITECTURES

Conceptually the simplest approach to implementing a DSP algorithm is to use a standard digital signal processor. The use of only one processor represents an extreme point in the resource-time domain. Figure 9.3 illustrates the major steps in the implementation process. Any sequential schedule for the DSP algorithm is acceptable. Programming standard digital signal processors is therefore simple, but the long execution time is a severe drawback. Advantages with this approach are that decision making and irregular algorithms can be easily accommodated. A common drawback of standard digital signal processors is that

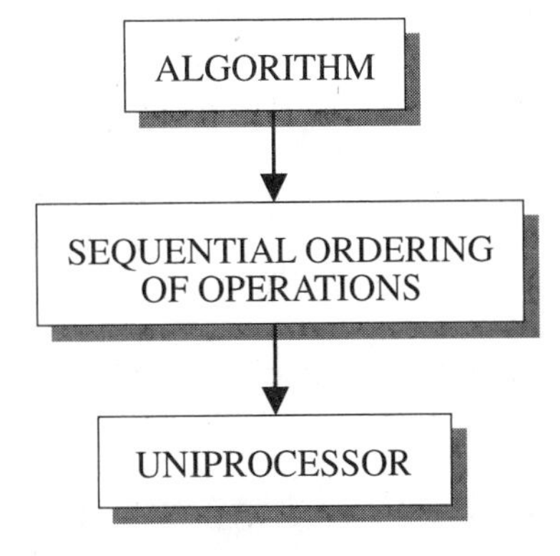

Figure 9.3 Mapping of the operations onto a single processing element.

data and coefficient word lengths are too short. Typical word lengths are in the range of 18 to 24 bits.

Generally, the power consumption of standard signal processors is large. Note that floating-point arithmetic may also provide too low an accuracy—i.e., the mantissa is too short. In fact, digital signal processing generally requires high accuracy, but the dynamic range requirement is comparatively low.

Application-specific architectures using only one processor are of great importance when the application has a relatively low work load and stringent requirements on low power dissipation—for example, in battery powered applications. A number of such architectures are discussed in this section.

EXAMPLE 9.1

Figure 9.4 shows a block diagram for a single PE (multiplier–accumulator) architecture for FIR filters in direct form [4] that implements the convolution:

$$y(n) = \sum_{i=0}^{N-1} a_i x(n-i)$$

$$= [[...[[0 + a_{N-1}\,x(n-N+1)] + a_{N-2}\,x(n-N+2)] +...+ a_1\,x(n-1)] + a_0\,x(n)]$$

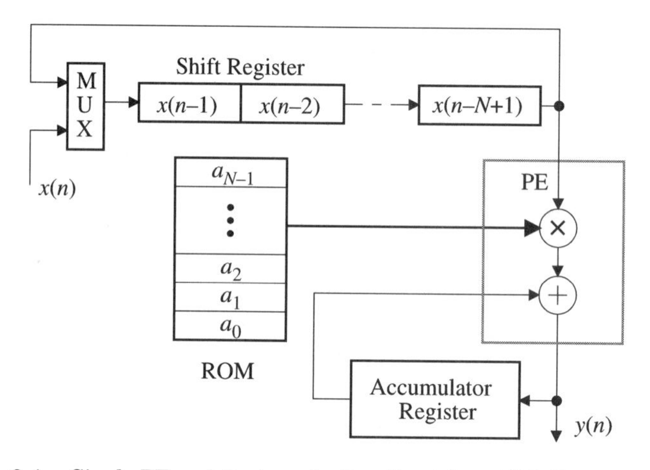

Figure 9.4 Single PE architecture for the direct form FIR filter structure

Data are processed bit-serially and stored in a long shift register. The multiplier accepts data in bit-serial form and coefficients in bit-parallel form. The design of such serial/parallel multipliers will be discussed in Chapter 11. The filter coefficients are therefore stored bit-parallel in a ROM. Determine how the output values are computed.

The operation of the FIR filter is illustrated in Table 9.1. The computations start with a cleared accumulator register. In the first time step the product $a_{N-1}\,x(n-N+1)$ is computed and added to the cleared accumulator. In the next time step the shift register is shifted, $x(n)$ is loaded into the first shift register position, and, simultaneously, the product $a_{N-2}\,x(n-N+2)$ is computed and added to the accumulator.

Input		$x(n)$				$x(n+1)$		
Mult.	$x(n-3)$	$x(n-2)$	$x(n-1)$	$x(n)$	$x(n-2)$	$x(n-1)$	$x(n)$	$x(n+1)$
Coeff.	a_3	a_2	a_1	a_0	a_3	a_2	a_1	a_0
Part. Prod.	p_3	p_2	p_1	p_0	p_3	p_2	p_1	p_0
Output	$y(n-1)$				$y(n)$			

Table 9.1 Operation of the FIR filter in Figure 9.4 with $N = 4$

Next, the remaining products in the convolution are successively computed and accumulated. In the last time step $y(n)$ is obtained by adding the product a_0 $x(n)$ to the accumulator register. Note that the input samples are written into the memory position of the oldest input sample no longer needed.

EXAMPLE 9.2

Find a block diagram for a single PE architecture to implement a transposed, direct form FIR filter structure.

The transposed, direct form FIR filter structure computes the convolution according to the following algorithm:

$$
\begin{aligned}
y(n) &:= a_0\, x(n) + PS_1 \\
PS_1 &:= a_1\, x(n) + PS_2 \\
PS_2 &:= a_2\, x(n) + PS_3 \\
PS_3 &:= a_3\, x(n) + PS_4 \\
&\vdots \\
PS_{N-2} &:= a_{N-2}\, x(n) + PS_{N-1} \\
PS_{N-1} &:= a_{N-1}\, x(n)
\end{aligned}
$$

Each input sample is multiplied by all coefficients and each product is accumulated as the partial sum of products, PS_i. The partial sums are stored in a long shift register, as illustrated in Figure 9.5. Long shift registers have been used in Figures 9.4 and 9.5 to illustrate the basic principle. In practice it is more efficient, in terms of chip area and power consumption, to implement the shift registers using RAMs.

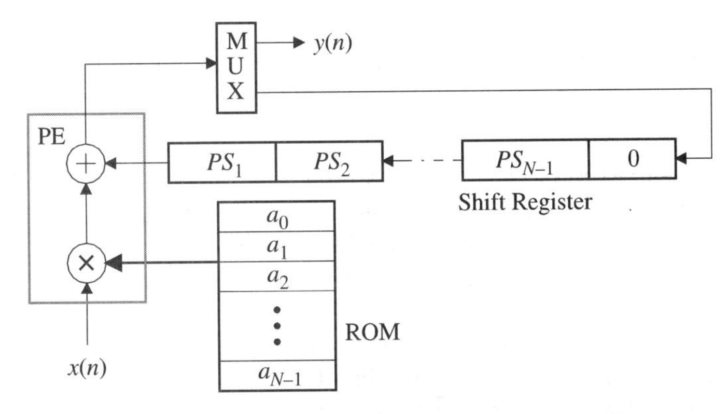

Figure 9.5 Single PE architecture for the direct form FIR filter structure

EXAMPLE 9.3

Find a block diagram for a single PE architecture to implement transposed, direct form FIR filter structures, but the generation of the partial sums should be generated in reverse order compared to the order used in Example 9.2.

The transposed, direct form FIR filter structure with partial sums computed in reverse order is realized according to the following algorithm:

$$
\begin{aligned}
PS_{N-1}(n+1) &:= a_{N-1}\,x(n)\\
PS_{N-2}(n+1) &:= a_{N-2}\,x(n) + PS_{N-1}(n)\\
PS_{N-3}(n+1) &:= a_{N-3}\,x(n) + PS_{N-2}(n)\\
&\vdots\\
PS_{2}(n+1) &:= a_{2}\,x(n) + PS_{3}(n)\\
PS_{1}(n+1) &:= a_{1}\,x(n) + PS_{2}(n)\\
y(n) &:= a_{0}\,x(n) + PS_{1}(n)
\end{aligned}
$$

Figure 9.6 shows the corresponding implementation with a single PE

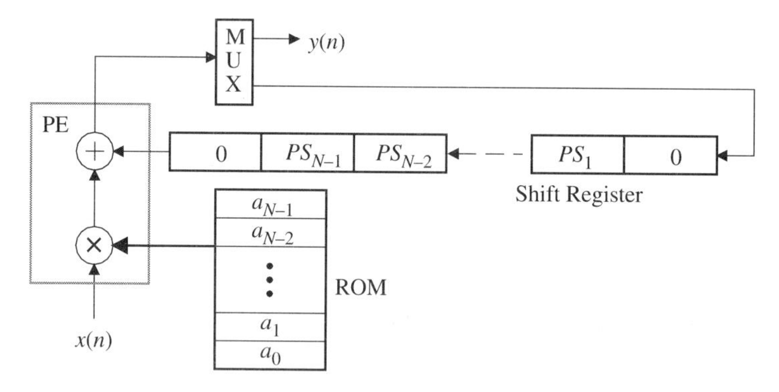

Figure 9.6 Single PE architecture for the transposed, direct form FIR filter structure. The partial sums are generated in reverse order

It is possible to use one of the unused time slots to increase the delay. The output value, instead of a zero, is fed back into the rightmost shift register stage, as shown in Figure 9.7

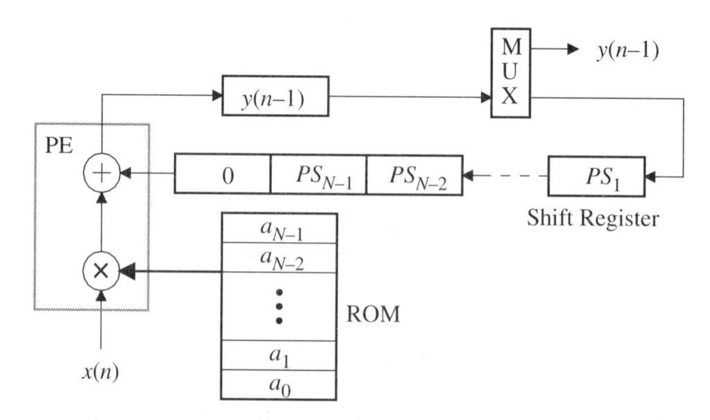

Figure 9.7 Implementation with longer delay than the implementation in Figure 9.6

EXAMPLE 9.4

Show that the two implementations, shown in Figures 9.6 and 9.7, can be combined to explore coefficient symmetry in a transposed linear-phase FIR filter structure.

Assume, for the sake of simplicity, that the filter order is odd ($N - 1$ = odd) and $N = 2Q$. Then the linear-phase FIR filter is described by

$$y(n) = \sum_{i=0}^{Q-1} a_i[x(n-i) + x(n-2Q-1+1+i)$$

The transposed FIR structure, shown in Figure 9.5, implements the following equations

$$\begin{aligned}
y(n) &:= a_0\, x(n) + PS_1(n) \\
PS_1(n+1) &:= a_1\, x(n) + PS_2(n) \\
PS_2(n+1) &:= a_2\, x(n) + PS_3(n) \\
&\vdots \\
PS_Q(n+1) &:= a_Q\, x(n) + PS_{Q+1}(n)
\end{aligned}$$

$$\begin{aligned}
PS_{Q+1}(n+1) &:= a_{Q+1}\, x(n) + PS_{Q+2}(n) \\
PS_{Q+2}(n+1) &:= a_{Q+2}\, x(n) + PS_{Q+3}(n) \\
&\vdots \\
PS_{N-2}(n+1) &:= a_{N-2}\, x(n) + PS_{N-1}(n) \\
PS_{N-1}(n+1) &:= a_{N-1}\, x(n)
\end{aligned}$$

The first half of these equations can be implemented by the architecture illustrated in Figure 9.5. Now, the order of the equations in the second half of equations can be changed

$$\begin{aligned}
PS_{N-1}(n+1) &:= a_{N-1}\, x(n) \\
PS_{N-2}(n+1) &:= a_{N-2}\, x(n) + PS_{N-1}(n) \\
PS_{N-3}(n+1) &:= a_{N-3}\, x(n) + PS_{N-2}(n) \\
&\vdots \\
PS_{Q+1}(n+1) &:= a_{Q+1}\, x(n) + PS_{Q+2}(n)
\end{aligned}$$

Further, symmetry in the impulse response yields

$$a_i = a_{N-1-i}$$

Hence, we get

$$\begin{aligned}
PS_{N-1}(n+1) &:= a_0\, x(n) \\
PS_{N-1}(n+1) &:= a_1\, x(n) + PS_{N-1}(n) \\
PS_{N-3}(n+1) &:= a_2\, x(n) + PS_{N-2}(n) \\
&\vdots \\
PS_{Q+1}(n+1) &:= a_Q\, x(n) + PS_{Q+2}(n)
\end{aligned}$$

The set of equations can be implemented by the architecture shown in Figure 9.7. Only one multiplier is needed, since the coefficients and the input value for each row are the same for the two halves. The resulting single multiplier

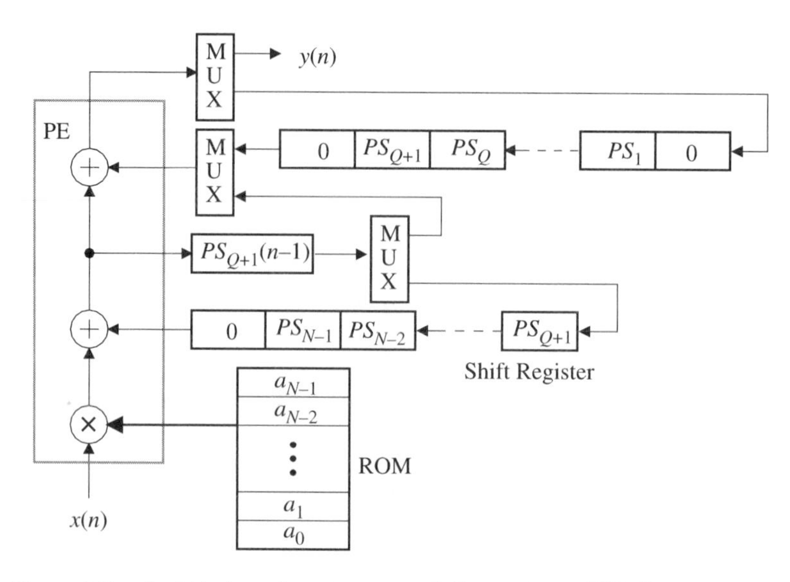

Figure 9.8 Architecture for a transposed, linear-phase FIR filter structure

architecture is shown in Figure 9.8, which represents the following set of computations

$P := a_0\, x(n)$	$y(n) := P + PS_1(n)$	$PS_{N-1}(n+1) := P$
$P := a_1\, x(n)$	$PS_1(n+1) := P + PS_2(n)$	$PS_{N-2}(n+1) := P + PS_{N-1}(n)$
$P := a_2\, x(n)$	$PS_2(n+1) := P + PS_3(n)$	$PS_{N-3}(n+1) := P + PS_{N-2}(n)$
$\vdots$		
$P := a_Q\, x(n)$	$PS_Q(n+1) := P + PS_{Q+1}(n)$	$PS_{Q+1}(n+1) := P + PS_{Q+2}(n)$

These types of FIR filter implementations can be easily multiplexed among several input signals, multiplexed to implement several filter functions simultaneously, and used for decimation and interpolation of the sample frequency [4]. Of particular interest are half-band FIR filters and complementary FIR filters for which large savings in the computational work load can be obtained. Note that recursive filter structures can also be implemented by incorporating appropriate feedback paths.

9.4 ISOMORPHIC MAPPING OF SFGs

A common implementation approach is to allocate one PE to each operation in the algorithm [7–9, 14, 17, 21, 29]. The PEs are interconnected according to the signal-flow graph illustrated in Figure 9.9. There is thus an isomorphy between the interconnection network and the signal-flow graph. The main advantages of this approach are that it is easy to understand, ease of multiplexing the hardware among many channels, and each PE can be optimized for its type of operation and computational work load. However, it does not generally lead to maximally fast implementations or a high resource utilization unless the scheduling of the operations are performed over multiple sample intervals, as discussed in Chapter 7.

The usefulness of this approach depends to a high degree on the type of arithmetic used in the PEs, since there are many PEs. Parallel arithmetic requires a large chip area for the PEs. A more interesting case is when bit-serial arithmetic is used in the PEs. The advantage is that PEs can easily be optimized for chip area, speed, and work load. Single-chip implementation is often possible.

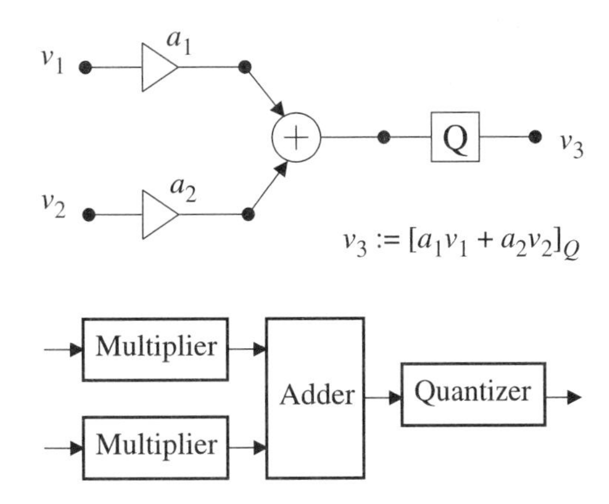

Figure 9.9 Isomorphic mapping of operations onto the processing elements

The major disadvantage of this approach results from the fixed binding of PEs to operations, which makes it difficult or impossible to exploit the processing power fully by multiplexing. This drawback is particularly pronounced in low sample rate applications, where the low utilization of the PEs results in wasted chip area. However, several development tools have been created along these lines—e.g., *Cathedral I*, *Next* [5] and *Parsifal I* [16]. The main drawback of these tools is that only operations within a single sample interval are considered.

9.4.1 Cathedral I

Cathedral I was developed at Catholic University of Leuven and IMEC, Leuven, Belgium [17]. A later commercial version, called *Mistral I*™, is marketed by Mentor Graphics in the tools set *DSP Station*™. This system also uses an isomorphic mapping of the DSP algorithm as well as bit-serial PEs. Multiplication is implemented by a shift-and-add and shift-and-subtract approach using canonic signed digit coded coefficients. Only a few basic building blocks are required, and the methodlogy is efficient from a chip-area point of view [14, 22].

Cathedral I is a design system for weakly programmable digital filters with sample rates in the range of 1 to 5 MHz. It is supported by digital filter synthesis tools (Falcon) as well as simulation and validation tools [13, 17, 25]. Figure 9.10 illustrates the main steps in the design process used in Cathedral I.

In the first design step, the digital filter is designed. Design of wave digital filters is supported. The filter coefficients are optimized in order to minimize the number of nonzero digits—i.e., the coefficients are represented by signed digit code (see Chapter 11). The aim is to implement the multiplications by shifts, additions, and subtractions, as illustrated in Figure 9.10. The control circuitry is based on bit-serial building blocks.

In the next step the high-level description is translated into a bit-serial architecture, according to the principle just described. The design system automatically introduces shimming delays and performs retiming of the signal-flow graph to minimize the memory required.

Cathedral I is based on about 20 bit-serial building blocks. Due to the bit-serial processing, only a small number of memory and logic blocks such as flip-flops, shift registers, adders, and sign-extenders are needed. Sign-extenders, which will be further discussed in Chapter 11, are special circuits required to perform multiplication with signed numbers. These blocks can be designed in CMOS using a standard-cell approach, as shown in Figure 9.10, or by a more efficient layout style using tiles, which was used in Cathedral I. Further, the design system provides automatic placement and wire routing.

The main advantages with this approach are

- ❑ Complete design path, from specification to working circuits
- ❑ Short design time
- ❑ High probability of first-time working circuitry
- ❑ No need to understand the VLSI technology
- ❑ High speed due to the inherent pipelining in bit-serial arithmetic
- ❑ Automatic layout, because of the standard-cell approach
- ❑ Supports of scan-testing, because of the bit-serial processing

The main disadvantages are

- ❑ It does not lead to a maximally fast implementation unless the periodic scheduling formulation is used instead of the single interval formulation
- ❑ It is difficult to multiplex the hardware components, hence a low resource utilization
- ❑ Nonlinear or variable functions are difficult to implement
- ❑ It is difficult to use different data word lengths
- ❑ Only operations with constant execution time are possible
- ❑ The continuous movement of data leads to high power consumption

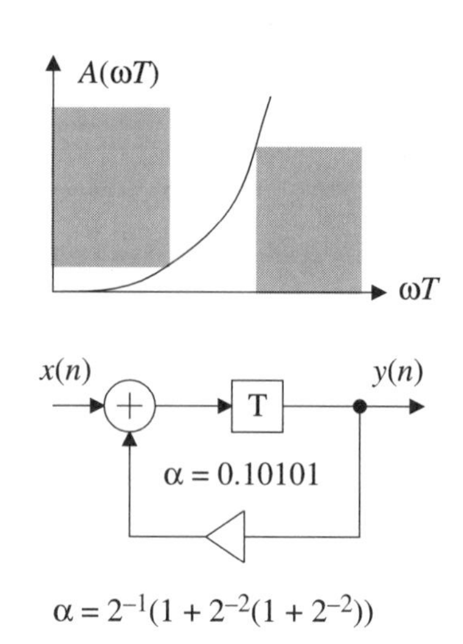

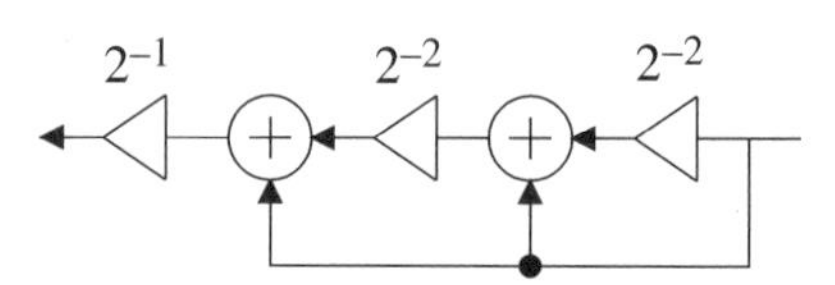

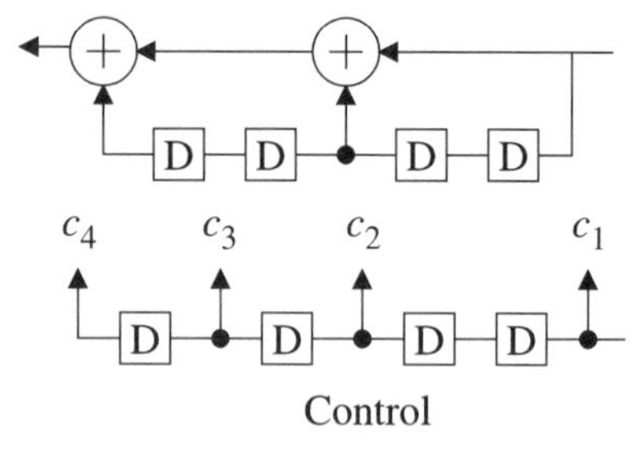

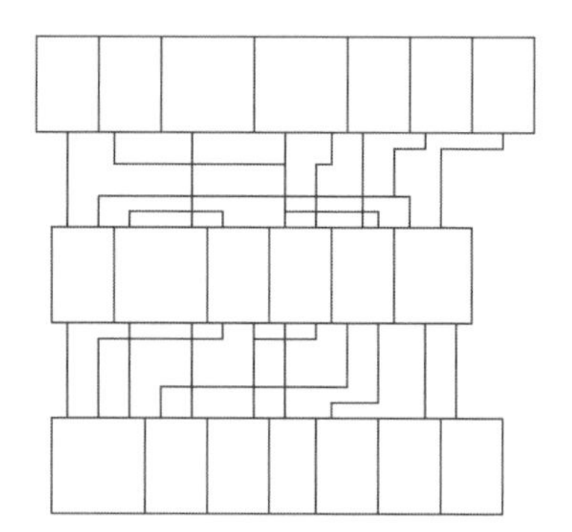

Figure 9.10 Design path for Cathedral I

9.5 IMPLEMENTATIONS BASED ON COMPLEX PEs

Arithmetic operations in most DSP algorithms generally involve several inputs and operations with only one or a few outputs [31, 34–38]. Typical examples are butterflies, two-port adaptors, and sum-of-products. Hence, basic PEs should be of the type illustrated in Figure 9.11.

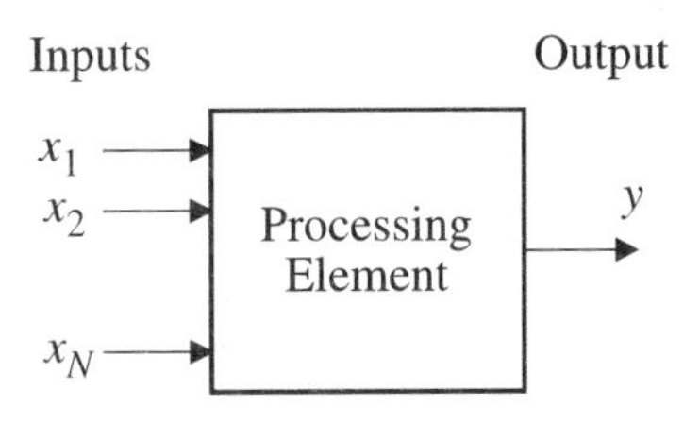

Figure 9.11 Typical DSP processing element

The basic approach is to use a dedicated PE for each value in the algorithm that has to be computed explicitly. We will discuss the selection of these values later. It is often advantageous from a chip area point of view to use high-speed bit-serial communication between the PEs and broadcast the outputs from the multiple-input PEs via a bit-serial network, as shown in Figure 9.12. A bit-serial network requires less area and power.

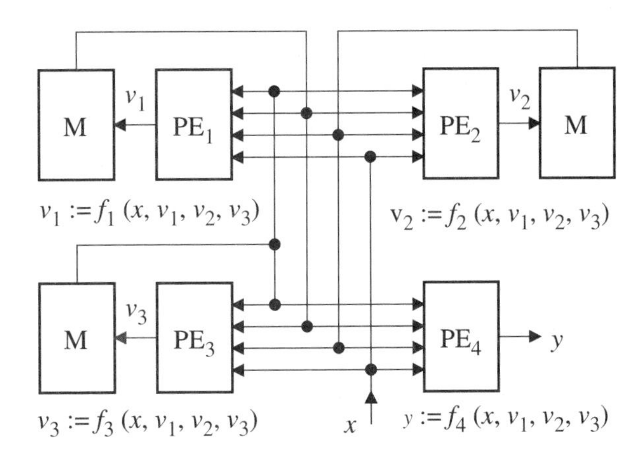

Figure 9.12 Fully parallel architecture with a minimum number of PEs

Such an architecture has a minimum number of PEs and a large granularity. It can easily be time-shared between many input signals by expanding the memory. However, in many applications the required number of PEs from a computational point of view is less than the number of expressions. The processing elements are therefore not fully utilized. In such cases a more elaborate scheme having a few multiplexed PEs may be considered. This approach is efficient when the DSP task fully exploits the processing power of the PEs. In fact, when the PEs are time-shared among many input signals, this is one of the most efficient approaches available for implementation of high-performance digital filters.

9.5.1 Vector-Multiplier–Based Implementations

The basic operation used in digital filters and many other DSP algorithms is the sum-of-products

$$y = \sum_{i=1}^{N} a_i x_i = \mathbb{a} \cdot \mathbb{x} \tag{9.1}$$

which also can be viewed as a vector multiplication, or inner product, or dot product, between a constant vector $\mathbb{a}$ and a data vector $\mathbb{x}$. The latter type of vector multiplication can be implemented efficiently using distributed arithmetic, which will be discussed in detail in Chapter 11. The cost in terms of execution time and power consumption for such a vector multiplication is the same as for a scalar multiplication; the area though is somewhat larger. A vector multiplier with less than eight terms requires only slightly larger chip area than an ordinary multiplier. The input and output values are bit-serial in distributed arithmetic. Hence, implementations based on vector-multipliers are generally highly efficient. Further, the design cost is low since the required circuits are highly regular, and modularity allows automatic layout.

The direct form FIR filter shown in Figure 4.4 can be implemented directly by a single vector-multiplier that computes the output and a set of shift registers. However, the required chip area for a distributed arithmetic-based implementation increases exponentially with the length of the filter and will become excessively large for N larger than about 12 to 14. In Chapter 11 we will discuss various techniques to reduce the area.

The increase in chip area, compared to an ordinary multiplier, will generally be small for recursive digital filters since N is typically less than 5 to 8. We demonstrate by an example how vector-multipliers can be used to implement recursive digital filters.

EXAMPLE 9.5

Determine the number of vector-multipliers and a block diagram for the corresponding implementation of the bandpass filter discussed in Examples 4.4 and 4.5. Use pipelined second-order sections.

The bandpass filter was implemented in cascade form with four second-order sections in direct form I. The numbers of scalar multipliers and adders for a classical implementation are 20 and 16, respectively. The number of shift registers is 14.

Now, observe that the output of each section is a vector-multiplication between constant coefficient vectors and data vectors of the form

$$y = \mathbb{a} \cdot \mathbb{x} \tag{9.2}$$

where $\mathbb{a} = (a_{0i}, a_{1i}, a_{2i}, b_{1i}, b_{2i})$. The data vectors, with $v_0(n-1) = x(n)$ and $v_4(n) = y(n)$, are

$$\mathbb{x} = [v_i(n-1), v_i(n-2), v_i(n-3), v_{i+1}(n-1), v_{i+1}(n-2)]^T \text{ for } i = 0, \ldots, 3$$

Only four vector-multipliers are needed, one per second-order section. The cost in terms of PEs has been reduced significantly. The block diagram for the implementation is shown in Figure 9.13. Note that the vector-multipliers can work in parallel and that this approach results in a modular and regular hardware that can be implemented with a small amount of design work.

The maximal sample frequency is proportional to the data word length. Clock frequencies of 400 MHz or more, can be achieved in a 0.8-μm CMOS process. Hence, with a typical data word length of 20 bits, sample frequencies up to $f_{CL}/W_d = 400/20 = 20$ Msamples/s can be achieved.

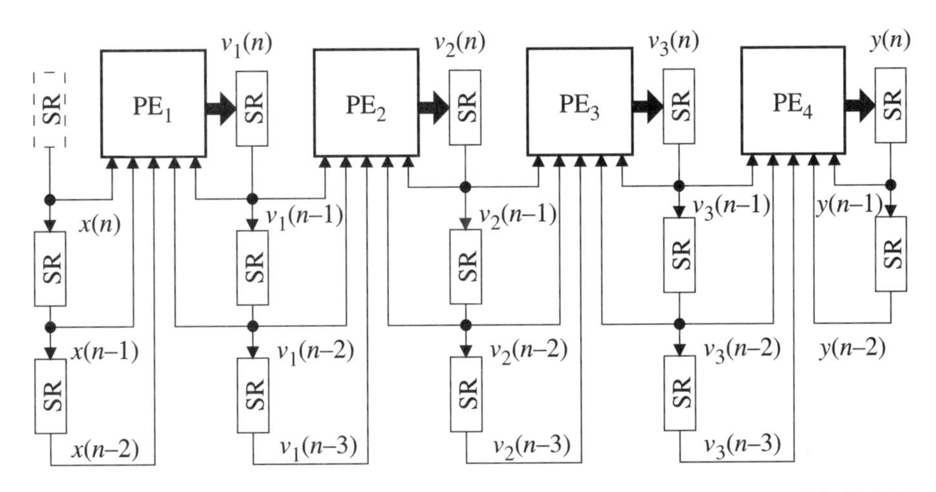

Figure 9.13 Implementation of eighth-order bandpass filter using vector-multiplier PEs

9.5.2 Numerically Equivalent Implementation

In this section we will show how to derive a numerically equivalent algorithm that is equivalent to, or in some respects even better than, numerical properties compared to the original algorithm. The new algorithm is suitable for implementation using vector-multipliers [31, 37].

Consider the recursive digital filter depicted in Figure 9.14. Arithmetic operations are represented by the network N. We assume that all values are computed with full precision and that the quantizations are done only in front of the delay elements in the original algorithm. Hence, no overflow or rounding errors occur, and the outputs and $v_i(n)$, $i = 1,..., M$ are computed exactly before they are quantized. The node values, which must be computed explicitly, are the outputs and the values that shall be stored in the delay elements—i.e., $v_i(n)$, $i = 1,..., M$.

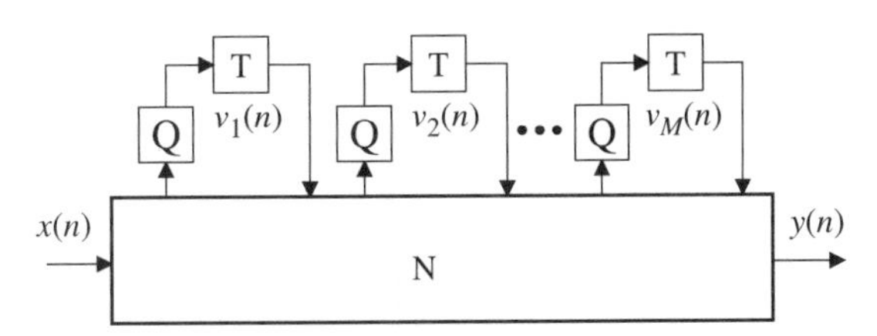

Figure 9.14 Recursive digital filter

These values can be computed in either of two numerically equivalent ways: First, as just mentioned, by doing the multiplications using extended precision so that no errors occur; and second, by precomputing new equivalent coefficients we get a set of expressions with fewer multiplications, but usually more than in the original algorithm. Thus, the new algorithm is numerically equivalent to the original since the quantization errors that occur in the two algorithms are the same.

The new algorithm can be described by the state-space equations

$$\begin{cases} \mathbb{v}(n+1) = \mathbb{A}\ \mathbb{v}(n) + \mathbb{B}\ x(n) \\ \mathbb{y}(n) = \mathbb{C}\ \mathbb{v}(n) + \mathbb{D}\ x(n) \end{cases} \tag{9.3}$$

where $\mathbb{A}$, $\mathbb{B}$, $\mathbb{C}$, $\mathbb{D}$, and $\mathbb{v}$ are matrices and vectors with dimensions $M \times M$, $M \times 1$, $1 \times M$, 1×1, and $M \times 1$, respectively. In order to demonstrate this implementation technique and some possible trade-offs, we explicitly write the matrix equations for the special case $M = 5$, as shown in Equation (9.4):

$$\begin{pmatrix} v_1(n+1) \\ v_2(n+1) \\ v_3(n+1) \\ v_4(n+1) \\ v_5(n+1) \\ y(n) \end{pmatrix} = \begin{pmatrix} a_{11} & a_{12} & a_{13} & a_{14} & a_{15} & b_{11} \\ a_{21} & a_{22} & a_{23} & a_{24} & a_{25} & b_{21} \\ a_{31} & a_{32} & a_{33} & a_{34} & a_{35} & b_{31} \\ a_{41} & a_{42} & a_{43} & a_{44} & a_{45} & b_{41} \\ a_{51} & a_{52} & a_{53} & a_{54} & a_{55} & b_{51} \\ c_{11} & c_{12} & c_{13} & c_{14} & c_{15} & d_{11} \end{pmatrix} = \begin{pmatrix} v_1(n) \\ v_2(n) \\ v_3(n) \\ v_4(n) \\ v_5(n) \\ x(n) \end{pmatrix} \tag{9.4}$$

Each new coefficient will essentially consist of a product of some of the original coefficients. The new coefficient word length will therefore be proportional to the sum of the word lengths of the coefficients in the products. Hence, the word length will be excessively large for highly sequential algorithms since there will be many more factors in the products. Highly parallel algorithms, on the other hand, will have fewer factors and therefore a shorter word length.

Each of the left-hand variables can be computed using six vector-multipliers and broadcasting the right-hand variables, as illustrated in Figure 9.15. Note that the PEs work in parallel and the hardware structure is regular and modular. The hardware structure can easily be multiplexed among several input signals. Such applications are common in telephone systems where the cost for the multiplexed

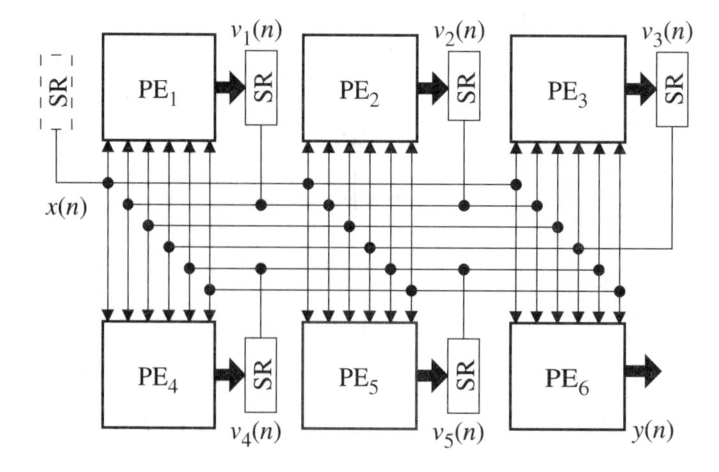

Figure 9.15 Numerically equivalent implementation using vector-multiplier PEs

system is shared by many speech channels. The system can also be used to process the same input signal several times—for example, to implement a digital filter bank. In applications with low work loads, one vector-multiplier can be multiplexed to reduce the hardware cost.

EXAMPLE 9.6

Show that filter structures such as the one in Figure 9.16 yield significant hardware reductions.

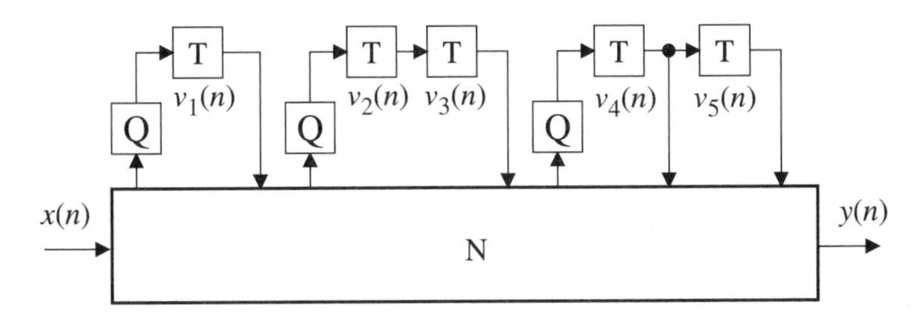

Figure 9.16 Recursive digital filter suitable for implementation using the numerically equivalent technique

Nodes $v_3(n)$ and $v_5(n)$ need not be computed by PEs since the values are already available in the preceding delay elements. The corresponding simplified matrix equation is shown in Equation (9.5). Further, node value $v_2(n)$ does not have a path to the network N. Hence, the elements in the corresponding column in Equation (9.5) will have only zeros except for the third row.

The simplified hardware structure is shown in Figure 9.17. Note that the number of signals entering the vector-multipliers has been decreased from six to

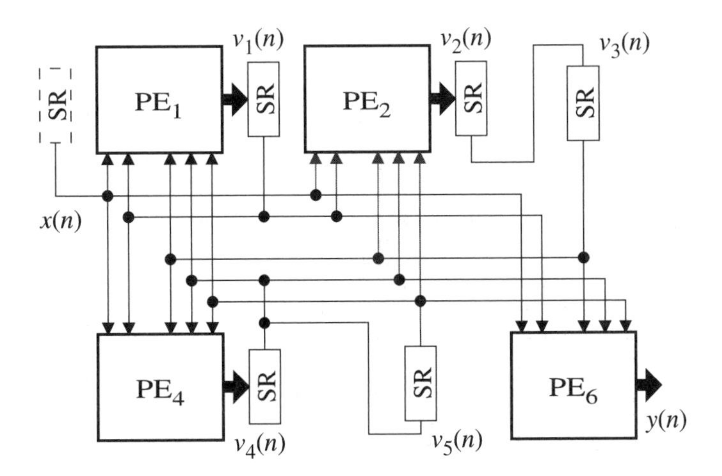

Figure 9.17 Simplified hardware structure corresponding to Equation (9.5)

five. This will only reduce the implementation cost slightly since the number of inputs is small.

$$\begin{pmatrix} v_1(n+1) \\ v_2(n+1) \\ v_3(n+1) \\ v_4(n+1) \\ v_5(n+1) \\ y(n) \end{pmatrix} = \begin{pmatrix} a_{11} & 0 & a_{13} & a_{14} & a_{15} & b_{11} \\ a_{21} & 0 & a_{23} & a_{24} & a_{25} & b_{21} \\ 0 & 1 & 0 & 0 & 0 & 0 \\ a_{41} & 0 & a_{43} & a_{44} & a_{45} & b_{41} \\ 0 & 0 & 0 & 1 & 0 & 0 \\ c_{11} & c_{12} & c_{13} & c_{14} & c_{15} & d_{11} \end{pmatrix} = \begin{pmatrix} v_1(n) \\ v_2(n) \\ v_3(n) \\ v_4(n) \\ v_5(n) \\ x(n) \end{pmatrix} \quad (9.5)$$

9.5.3 Numerically Equivalent Implementations of WDFs

Wave digital filters can also be implemented efficiently in a numerically equivalent form [31, 34 – 37]. Sensitivity properties as well as the more important pseudo-passivity is retained since proper magnitude quantizations are done in front of the delay elements. Pseudo-passivity is required in order to suppress parasitic oscillations. The word length for the coefficients in the numerically equivalent algorithm can be made short by exploiting the low sensitivity of WDFs to quantize the adaptor coefficients properly. However, the most important factor is selection of a good filter structure.

Generally, algorithms with a high degree of parallelism are better suited to this approach, because the coefficient word length will be lower. Word length is proportional to the number of adaptors through which the different computational paths pass. Therefore, wave digital lattice filters with branches implemented using Richards' structures are preferable. Such a filter is shown in Figure 9.18. The coefficient word length will be determined by only two adaptor coefficients.

EXAMPLE 9.7

Determine the elements in the numerically equivalent form for the lattice wave digital filter shown in Figure 9.18. Scaling multipliers corresponding to transformers in the reference filter have been inserted in order to scale the signal levels using L_2-norms. Assume the following coefficient values which have been optimized for a conventional implementation using canonic signed digit code (see section 11.4):

$$\alpha_1 = \frac{13}{16} \qquad \alpha_2 = -\frac{63}{64} \qquad \alpha_3 = \frac{61}{64} \qquad \alpha_4 = -\frac{105}{128} \qquad \alpha_5 = -\frac{31}{32}$$

$$c_1 = 2 \qquad c_2 = \frac{1}{8} \qquad c_3 = 8 \quad cy = 1 \quad c_5 = \frac{1}{8} \qquad c_6 = \frac{1}{2}$$

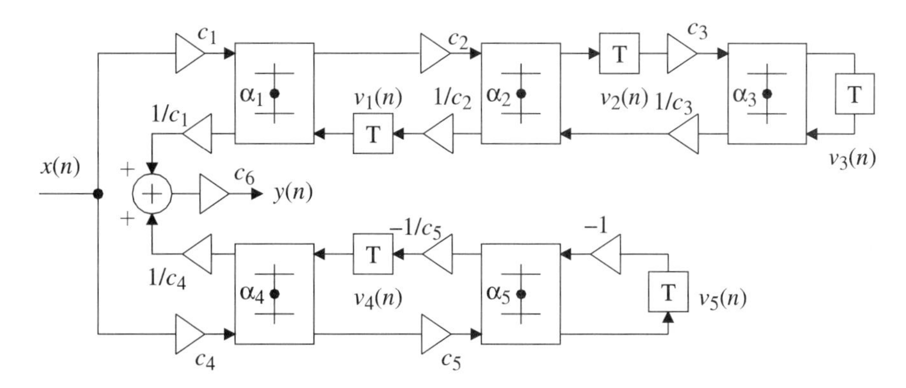

Figure 9.18 Lattice WDF with the branches implemented using Richards structures

Obviously, the two branches of the lattice filter introduce zeros in the corresponding matrix, as shown in Equation (9.6).

$$\begin{pmatrix} v_1(n+1) \\ v_2(n+1) \\ v_3(n+1) \\ v_4(n+1) \\ v_5(n+1) \\ y(n) \end{pmatrix} = \begin{pmatrix} a_{11} & a_{12} & a_{13} & 0 & 0 & b_{11} \\ a_{21} & a_{22} & a_{23} & 0 & 0 & b_{21} \\ 0 & a_{32} & a_{33} & 0 & 0 & b_{31} \\ 0 & 0 & 0 & a_{44} & a_{45} & b_{41} \\ 0 & 0 & 0 & a_{54} & a_{55} & b_{51} \\ c_{11} & 0 & 0 & c_{14} & 0 & d_{11} \end{pmatrix} = \begin{pmatrix} v_1(n) \\ v_2(n) \\ v_3(n) \\ v_4(n) \\ v_5(n) \\ x(n) \end{pmatrix} \tag{9.6}$$

The number of delay elements that contribute to the output is minimized by placing delay elements closest to the output as shown in Figure 9.18. The remaining delay elements are placed alternately in the upper and lower branches. The longest path from the output of a delay element to the input of a delay element or to the output of the filter is at most through two adaptors. Hence, the coefficient a_{ij} is a product of at most two adaptor coefficients; independent of the filter order. The number of nonzero coefficients in Equation (9.6) is 19; but only six vector-multipliers are required for implementation. In generally, $N + 1$ vector-multipliers are required. By simple calculation we get

$$\begin{pmatrix} -\alpha_1\alpha_2 & -(\alpha_2+1)\dfrac{\alpha_3}{c_2} & \dfrac{(\alpha_2+1)(\alpha_3+1)}{c_2c_3} & 0 & 0 & (\alpha_1-1)\alpha_2c_1 \\ \alpha_1(1-\alpha_2)c_2 & -\alpha_2\alpha_3 & \dfrac{\alpha_2(\alpha_3+1)}{c_3} & 0 & 0 & (1+\alpha_1)(1+\alpha_2)c_2c_3 \\ 0 & (1-\alpha_3)c_3 & \alpha_3 & 0 & 0 & 0 \\ 0 & 0 & 0 & \alpha_4\alpha_5 & -\dfrac{(\alpha_5+1)}{c_5} & (1-\alpha_4)\alpha_4\alpha_5 \\ 0 & 0 & 0 & \alpha_4(\alpha_5-1)c_5 & -\alpha_5 & (1-\alpha_4)(\alpha_5-1)c_4c_5 \\ \dfrac{(\alpha_1+1)}{c_1} & 0 & 0 & \dfrac{\alpha_4+1}{c_4} & 0 & -(\alpha_1+\alpha_4) \end{pmatrix}$$

$$= \begin{pmatrix} \frac{819}{1024} & -\frac{61}{512} & \frac{125}{4096} & 0 & 0 & \frac{189}{512} \\ \frac{1651}{8192} & \frac{3843}{4096} & -\frac{7875}{32768} & 0 & 0 & \frac{381}{4096} \\ 0 & \frac{3}{8} & \frac{61}{64} & 0 & 0 & 0 \\ 0 & 0 & 0 & \frac{3255}{4096} & \frac{1}{4} & -\frac{7223}{16384} \\ 0 & 0 & 0 & \frac{-6615}{32768} & \frac{31}{32} & \frac{14679}{131072} \\ \frac{29}{64} & 0 & 0 & \frac{23}{64} & 0 & \frac{1}{256} \end{pmatrix}$$

Note that the adaptor coefficients have been optimized such that the number of nonzero digits is minimized. However, in this approach the word lengths of the coefficients appearing in the matrix should be minimized.

The numerically equivalent implementation has a minimum number of quantization nodes and can be optimally scaled. If a low-sensitivity filter structure is used, the roundoff noise will also be low. Hence, the total roundoff noise at the output of the filter will be low and a short data word length can be used.

The coefficient word length can also be reduced by pipelining the original filter structure. For example, lattice WDFs of the type shown in Figure 4.48 can be pipelined by placing delay elements between the two sections. This reduces the coefficient word length as well as the number of input signals to the vector-multipliers at the cost of two additional vector-multipliers to compute the values in the delay element between the sections.

9.6 SHARED-MEMORY ARCHITECTURES WITH BIT-SERIAL PEs

The interconnection network for the shared-memory architecture shown in Figure 9.19 has been split into two parts. The first part provides communication channels from the PEs to the memories while the second part provides channels from the memories to the PEs. The PEs receive their inputs bit-serially, and their results are also produced bit-serially. Hence, only a single wire is required for each signal. The cost in terms of chip area and power consumption is therefore low. However, bit-serial communication leads to high clock frequencies.

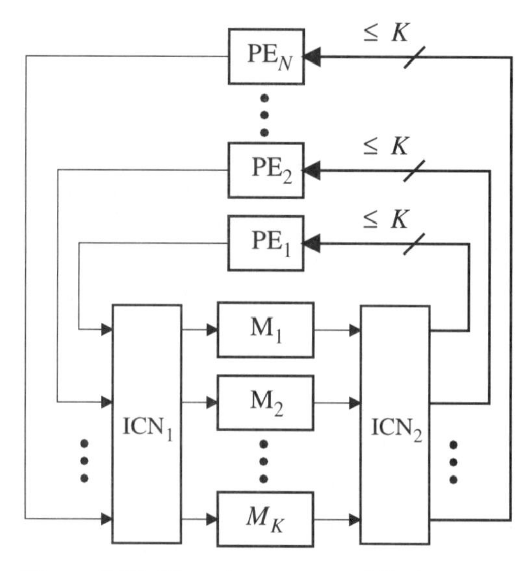

Figure 9.19 Shared-memory architecture with bit-serial PEs

The shared-memory space has been split into K parts (logical memories) so that K values can be accessed simultaneously, as illustrated in Figure 9.19. Memory has therefore to be assigned such that access conflicts are avoided.

As discussed in Chapter 8, a shared-memory architecture can only accommodate a few PEs, since the cycle time for the memories and the execution time for the processing elements are of the same order.

The imbalance becomes even more severe for DSP algorithms, since the processing elements often require many input values from the memories for each operation. Obviously, it is not efficient to use PEs so fast that the memories become the bottleneck. Generally, PEs that are faster than necessary consume more power and chip area. An efficient way to obtain a balanced architecture is to increase the execution time of the PEs as much as allowed by the application requirements. The cost of the system can thereby be reduced.

9.6.1 Minimizing the Cost

The throughput per unit chip area is $1/(A\ T_{PE})$. As a measure of cost we use the reciprocal,

$$\text{Cost} \propto A\ T_{PE}$$

Many different PE implementations are possible at the same cost, $A\ T_{PE}$. Hence, the designer can make a trade-off between chip area and execution time for PEs at the same cost. For example, we can use bit-serial instead of bit-parallel PEs. In Chapter 11, we will discuss the implementation of bit-serial PEs. A bit-serial PE typically uses less chip area:

$$A_s < \frac{A_p}{W_d}$$

than the corresponding bit-parallel PEs, but has longer execution time. For the sake of argument, we make the highly pessimistic assumption that

$$T_{PEs} \approx W_d\ T_{PEp}$$

Thus, the cost of executing an algorithm is

$$A_p\ T_{PEp} \approx A_s\ T_{PEs} \approx \text{constant}$$

If the speed of the PEs is reduced, then the number of PEs has to be increased proportionally to perform the same amount of work. Thus, this simplified analysis indicates that the total chip area for the PEs is almost the same for bit-serial and bit-parallel arithmetic. A more accurate comparison also involving other important factors (e.g., power consumption) is very difficult. In practice, the number of PEs can be more closely adjusted to the actual throughput requirement in the bit-serial case. Further, the chip area required for routing wires etc. can be significantly reduced. However, it is necessary that parallelism in the algorithm is sufficiently high to enable enough bit-serial PEs to be employed.

9.6.2 Uniform Memory Access Rate

To obtain a uniform memory access pattern we propose a shared-memory architecture with a multibus ICN, with each logical memory connected to its own bus. Further, the PEs are provided with a set of cache memories connected to the buses, as illustrated in Figure 9.20. Each cache memory is split into two parts, one of which is connected to the PE and the other to the memory. The two parts change role every PE cycle. The cache memories provide a uniform access pattern to the RAMs. The cache memories are simple shift registers that can be loaded and read bit-parallel. In section 9.6.4 we will show that a perfectly balanced architecture can be obtained by a proper design.

The complex and irregular part of the two interconnection networks, shown in Figure 9.19, is bit-serial and requires little chip area, while the interface between the memories and the serial/parallel converters is simple and regular. In practice, the interconnection networks can be significantly simplified. ICN_1 often has only a one bit-serial wire from each PE. A control signal is used to load the value in the shift register onto the memory bus. This control signal can be generated by a control unit placed close to the memory. Network ICN_2 is also bit-serial, but it may contain both switched and shimming delays. It is efficient to use bit-serial rather than bit-parallel communication for the main part of the system since the former consumes less chip area for wiring and drivers.

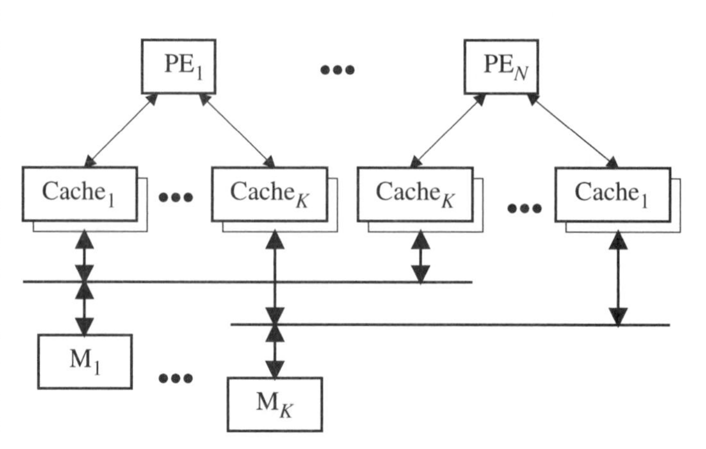

Figure 9.20 Multibus architecture

The bit-parallel buses can be made very short by placing the shift registers close to the RAMs. The longer distance, between shift register and PE, uses bit-serial communication and requires only two wires. Moreover, shimming delays are often needed between the shift registers and the PEs to synchronize different bit-streams. A bit-serial shimming delay (shift register stage) is essentially two cascaded clocked inverters. These inverters can also be used to provide the necessary drive capability to transmit data over long distances.

The minimum number of memories, K, depends on the number of input values that must be provided simultaneously to the PEs. Hence,

$$K = \max_t \left\{ \sum_{i=1}^{N} n_i(t) \right\} \tag{9.7}$$

where $n_i(t)$ is the number of inputs to PE_i at time instant t. Thus, the number of memories may be excessively large. However, the PEs in Figure 9.20 may be skewed in time so that only one PE requests inputs or delivers results at a time. The number of memories is therefore equal to, or less than, the largest number of inputs to any of the PEs:

$$K \leq \max_t \{n_i(t)\} \tag{9.8}$$

This scheme allows the PEs to be supplied with inputs using only a single memory or a few memories. Often, a single memory is enough to support several PEs.

9.6.3 Fast Bit-Serial Memories

PEs must be supported by very fast memories, since the clock frequency of bit-serial PEs can be made very high in modern CMOS technology—e.g., 100 to 400 MHz. At this stage, it is interesting to note that shift registers are the fastest circuits possible, since they are the most trivial clocked logic circuits possible. On the other hand, RAM is comparatively slow, 30 to 100 MHz, but is area-efficient and consumes much less power than shift registers. Since data can not be moved without power consumption, it is desirable to make a trade-off between shift registers and RAM. A long, fast bit-serial memory can be implemented by combining a shift register with a RAM, as shown in Figure 9.21.

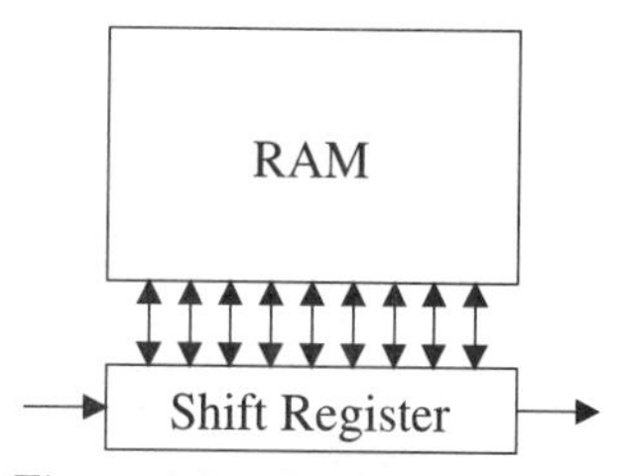

Figure 9.21 Realization of a long, fast bit-serial memory

Data are read bit-parallel from the RAM and stored into the shift register. The data are then shifted out, bit-serially, at high speed. A data word can be written into the RAM by first shifting it into the shift register and then writing it, bit-parallel, into the RAM. The RAM addressing is only required to be cyclic. It is often advantageous to use a pseudo-random generator (i.e., a shift register with feedback paths) for this purpose.

The difference in speed between the shift register and the RAM can be adjusted by using an appropriate word length in the RAM. Using this technique with shift registers as serial-parallel converters, fast and efficient bit-serial memories can be constructed.

The effective speed can be doubled by using a dual-port RAM, which can perform a read and a write operation simultaneously.

9.6.4 Balancing the Architecture

As discussed in Chapter 8, shared-memory architecture can only accommodate a few PEs, but in practice this bottleneck can be avoided by a proper partitioning of the system. However, the communication bit-rates differ considerably in the shared-memory architecture shown in Figure 9.20. Serial/parallel converters, previously called cache memories, have therefore been placed between the RAMs and the interconnection network, shown in Figure 9.22, to emulate a high-speed bit-serial memory. This architecture can be designed to have a perfect balance between processing capacity and communication bit-rates.

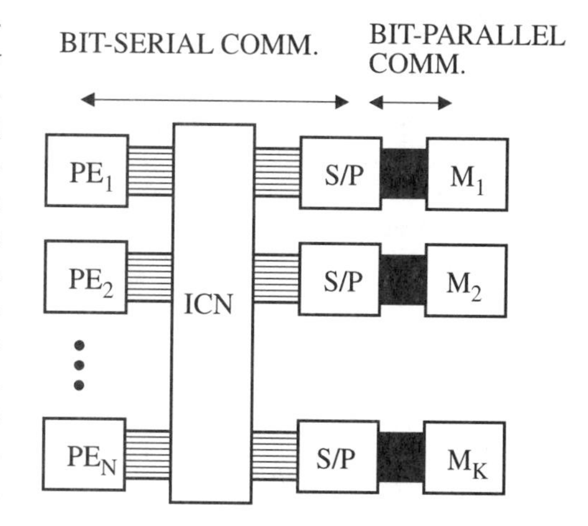

Figure 9.22 Multiprocessor architecture with serial/parallel converters

To support the bit-serial PEs with data so they can remain busy, the following inequality must hold between the bit-rates in the bit-serial and bit-parallel parts:

$$T_{PEs} \geq 2N_s \frac{T_M}{W_M} \tag{9.9}$$

where N_s is the number of bit-serial PEs, and T_{PEs} is their clock frequency. Thus, by choosing an appropriate word length, W_M, in the memories, a sufficiently high bit-rate can be obtained. To avoid fractions of data words at the same address, the memory word length, W_M, must be chosen so that

$$n\, W_M = W_d, \quad n = \text{integer}$$

The maximum number of PEs in one system is limited either by the parallelism in the algorithm or by the data word length. We have

$$N_{smax} \leq \frac{T_{PEs}}{T_M} \frac{W_d}{2} \tag{9.10}$$

Typically, we have $4T_{PEs} \approx T_M$. Thus, one system can typically support 2 to 3 PEs, since normal data word lengths are in the range 16 to 24 bits. In general, it is possible to use different data word lengths and to use multiple word lengths in different PEs and subsystems. Note that the factor 2, in Equation (9.10), can be removed by using two sets of memories. Hence, the number of PEs in one system can be increased to about $W_d/4$, or 4 to 6 PEs. Note that this represents a significant amount of processing capacity.

9.6.5 Mode of Operation

A more detailed example of a shared-memory architecture is shown in Figure 9.23. Each RAM has been augmented with a set of input and output shift registers. The PEs have their own shift registers connected to the RAMs from which they import and export values. The number of shift registers is equal to the number of PEs, while the number of memories corresponds to the largest number of inputs to any one of the PEs. In this case, only two RAMs are needed, since the PEs in Figure 9.23 have only two inputs. Each RAM can read and write a word to the shift registers through the bit-parallel buses.

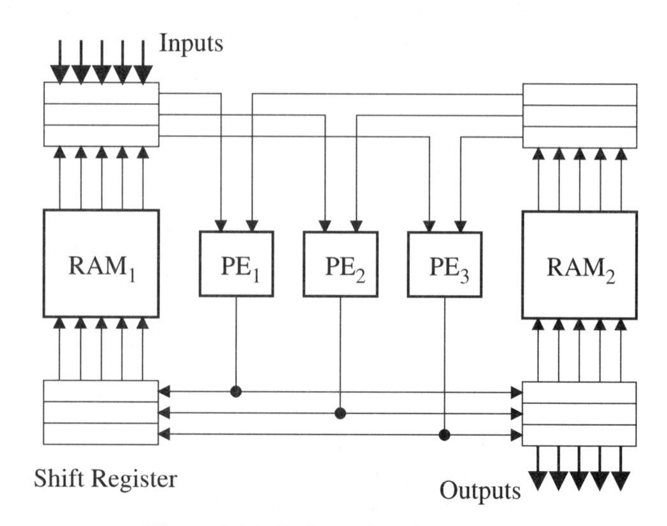

Figure 9.23 Balanced architecture

Evaluation of the first expression, using PE_1, starts with writing the appropriate values into the shift registers associated with PE_1. These values are then shifted bit-serially into PE_1 where the processing takes place. Simultaneously, a result from the previous operation will appear at the output of PE_1, since pipelining is inherent in the bit-serial PEs. The result is shifted into the shift registers, at the bottom of Figure 9.23, and finally written into the appropriate RAM. These two operations take two RAM cycles: one read and one write. In the next RAM cycle the input values to PE_2 can be written into its shift registers and so on. The operational scheme for all PEs is the same, but the PEs are skewed in time by two clock cycles, as shown in Figure 9.24. The operation of the system is fully synchronous.

The shift registers connected to the inputs of the PEs are emptied as the values are shifted out. At the same time, the results are shifted out of the PEs. Clearly, the outputs of the PEs can be shifted directly into the former set of shift registers, as shown in Figure 9.25. The inputs and outputs of the whole system can be either bit-serial or bit-parallel.

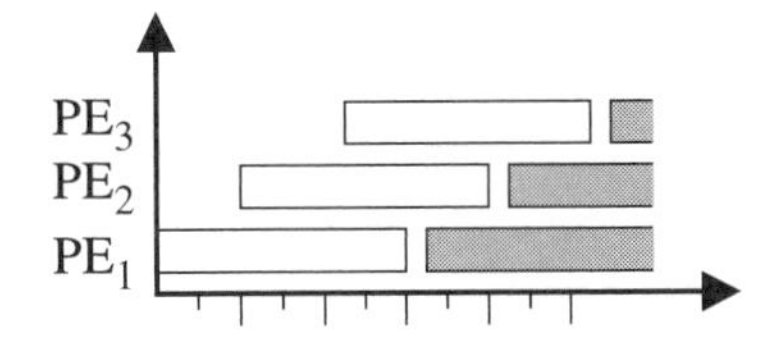

Figure 9.24 Scheduling of the PEs

In practice, this type of architecture is almost always free from memory access conflicts and is inherently free from bus contention.

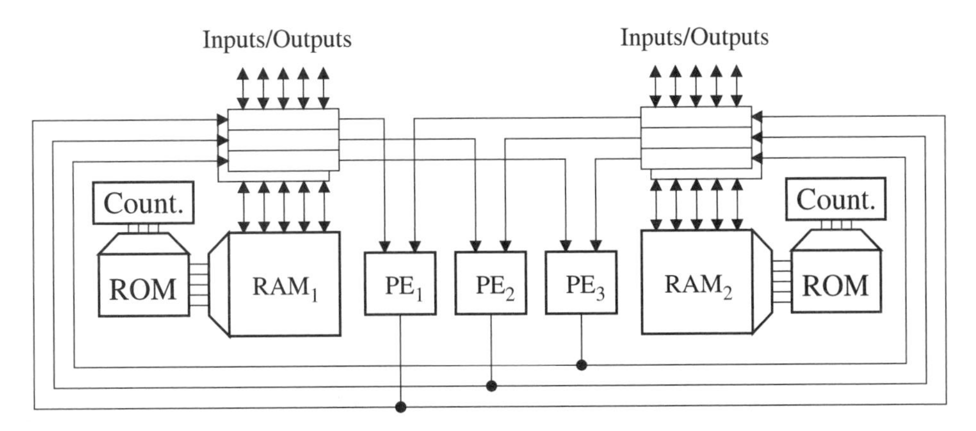

Figure 9.25 Architecture with common input/output shift registers

9.6.6 Control

Control signals are of two types: clocks and signals that set up communication channels and control the flow of data. Clock signals will be discussed later; only the latter type of control signals is discussed in this section. Since this class of architectures is strongly coordinated, control circuits can be distributed to the PEs and RAMs. Figure 9.25 shows how the control circuits can be distributed.

Each RAM receives its addresses from an associated ROM which contains successive read and write addresses. In practice, several ROMs can share the same address decoder. The ROMs are addressed by a counter which sequences through all values. The RAMs also receive a read and write signal every other cycle. The ROMs also contain information about which RAM the outputs of the PEs shall be stored into.

9.7 BUILDING LARGE DSP SYSTEMS

Several systems of the type just discussed can be connected in cascade, parallel, or in a nested form, as illustrated in Figures 9.26 through 9.28. Obviously, this kind of system can be easily interfaced to general-purpose host computers or standard digital signal processors. Bit-serial subsystems, conventional bit-parallel architectures, and standard digital signal processors can be readily mixed, as illustrated in Figure 9.26.

Several subsystems can be implemented on a single chip, since a bit-serial PE requires very little chip area. For example, a bit-serial multiplier typically requires much less than 1 mm^2 of chip area. Hence, a complete signal processing system with many PEs and memories can be implemented on a single chip.

Functional testing of these architectures is simplified since the different parts can be tested separately. Furthermore, ample means for supplying and observing test vectors are provided, since few external pins are needed when bit-serial communication is used.

EXAMPLE 9.8

Determine a shared-memory architecture with bit-serial PEs that implements the schedule derived in Example 7.2 for a pipelined second-order section in direct form II.

The schedule has been redrawn in Figure 9.29 on page 412. The scheduling period is 20 time units and the sample period is T_{sample} = 10 time units. Multiplication takes four time units and addition takes one time unit.

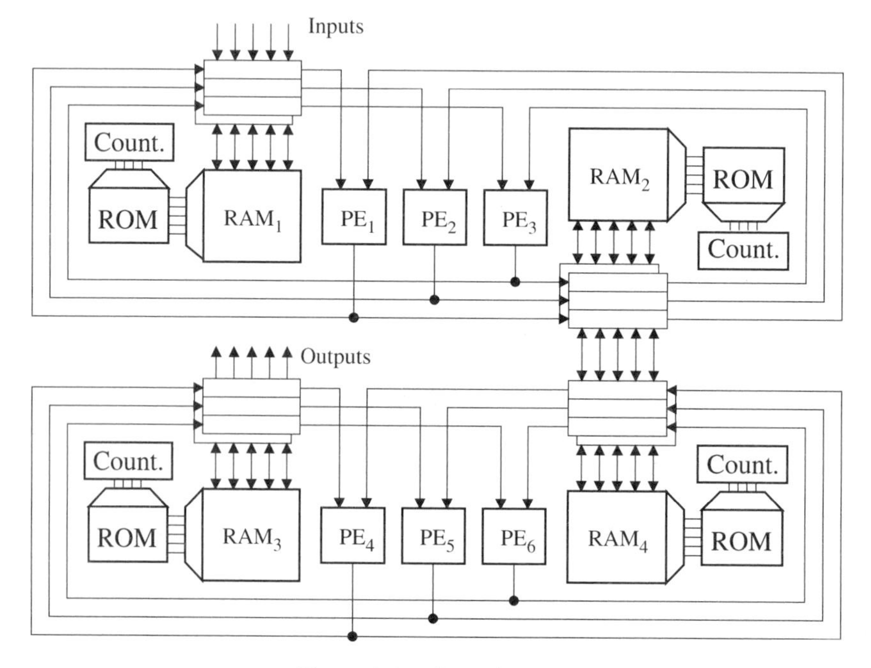

Figure 9.26 Cascade form

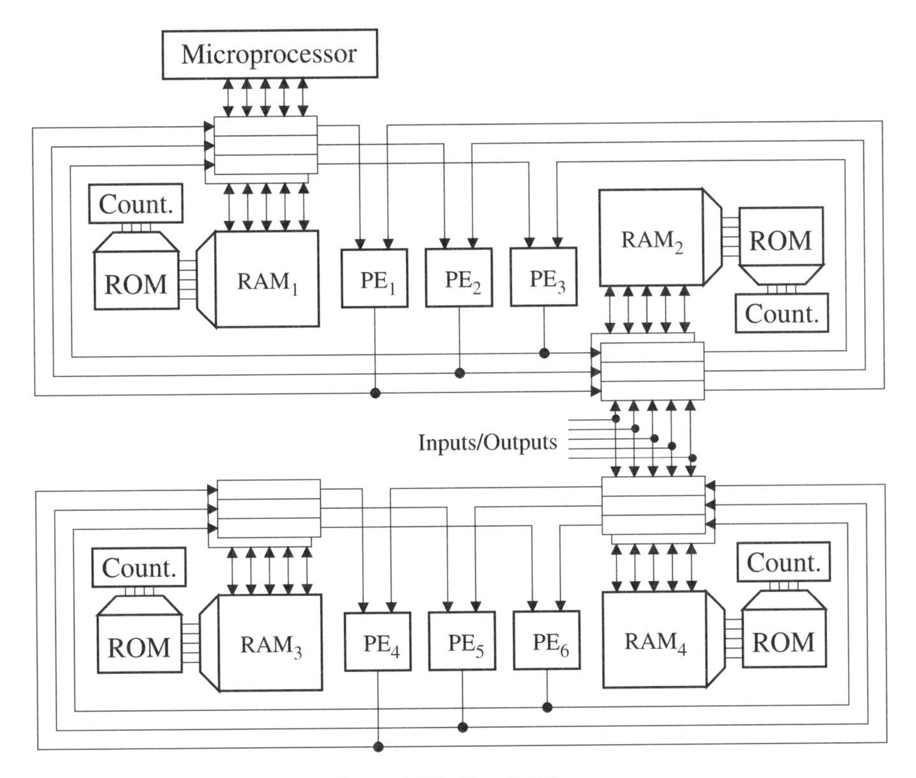

Figure 9.27 Parallel form

Scheduling was done with the aim of minimizing the number of PEs. Sharing of delays has been explicitly shown in Figure 9.29. From the schedule it is evident that 76 time units of storage are required. In a general shared-memory architecture the required number of memories is five, since five memory transactions take place at time slot 0 (20). However, the number of memories may be reduced to four if the inputs are scheduled to take place in time slots 5 and 13. Then four memories are required, since four memory transactions take place at both time slots 8 and 16. Storage is reduced to 68 units. In this case, even more efficient schedules are possible.

The class of ideal architectures for this schedule has three processing elements (two multipliers and one adder) and four memories. The communication pattern and required channels can be deduced from the schedule. In the next step we make trade-offs within this class of ideal architectures.

The shared-memory architecture shown in Figure 9.30 allows several memory–PE transactions to take place within a single time slot. The number of memories is thereby reduced according to Equation (9.9). Hence, only two memories are required, since the maximum number of inputs to the adder is two. The balanced architecture, with three PEs and two memories, each with three shift registers, is shown in Figure 9.30. Two switches have been provided to simplify the memory assignment. In practice, of course, it is not efficient to use a too simple PE such as the bit-serial adder. Instead, the multipliers and the adder should be combined into a multiplier–accumulator PE.

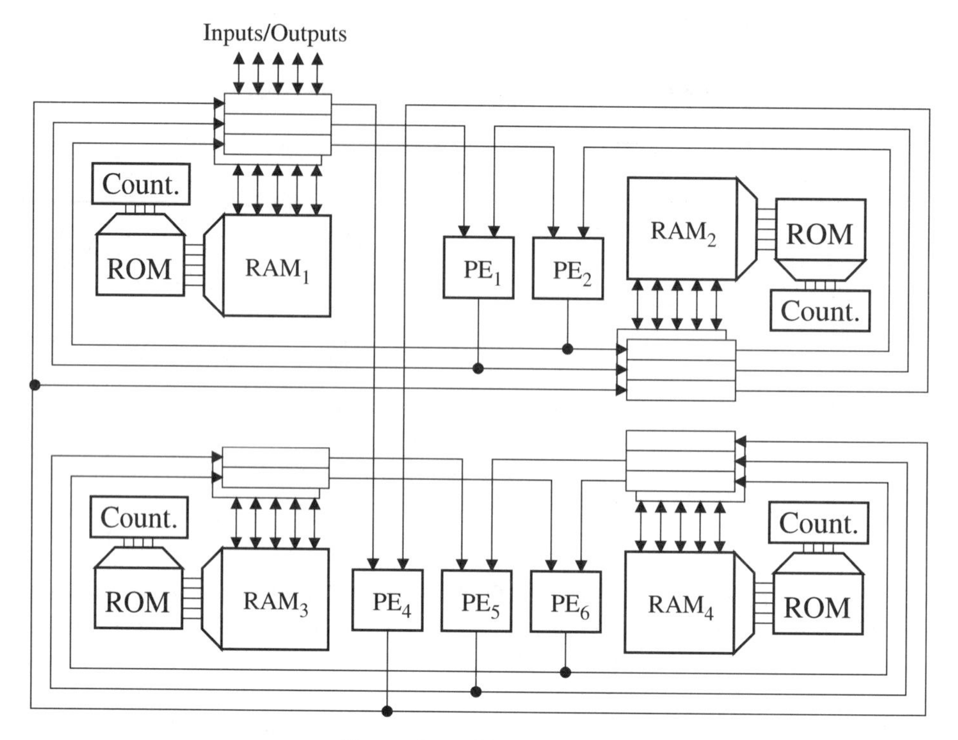

Figure 9.28 Nested form

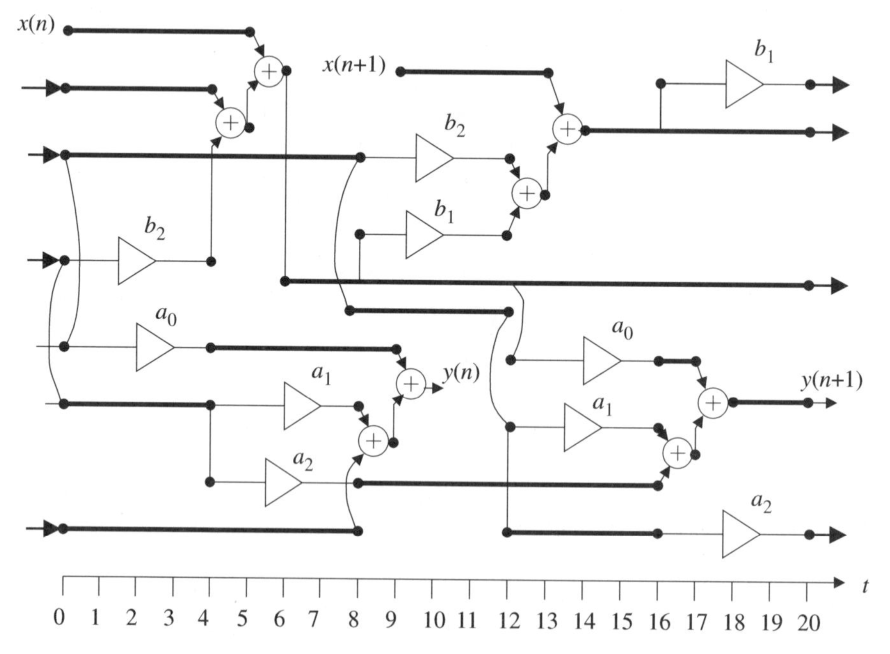

Figure 9.29 Schedule for the pipelined second-order section in direct form II

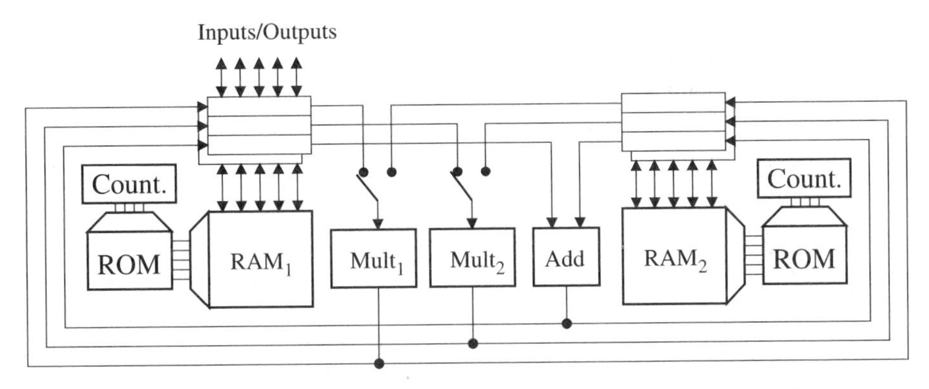

Figure 9.30 Balanced architecture with three PEs and two memories

9.8 INTERPOLATOR, CONT.

The wave digital filter used for interpolation was scheduled in 6 time units, using the adaptors as basic operations. Four PEs each with two inputs and two outputs were needed. Hence, to support four PEs we need $2 \cdot 4 = 8$ memories in the ideal architecture. A single memory can support several PEs. The schedule implies that 24 adaptor operations are executed per sample interval and each adaptor operation requires two read and two write operations. Hence, a memory with access rate

$$f_{RAM} = 4 \cdot 24 \cdot f_{sample} = 153.6 \text{ MHz}$$

would be enough. It is possible to design such fast memories, but it is difficult. Therefore we choose to use two memories working at only 76.8 MHz. Each memory has 22 21-bit words. The resulting balanced architecture is shown in Figure 9.31.

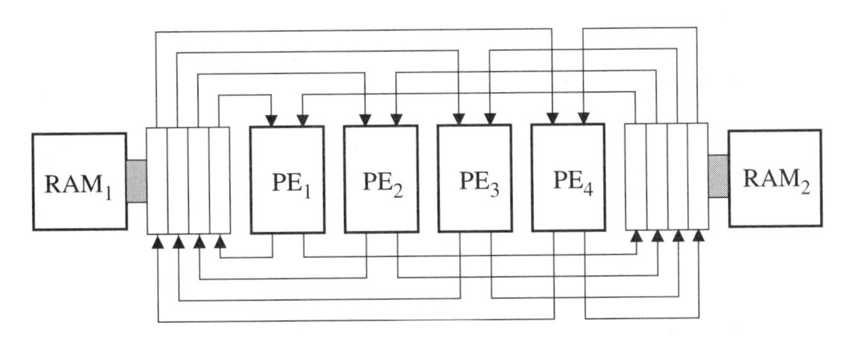

Figure 9.31 Architecture for the interpolator

9.9 FFT PROCESSOR, CONT.

The schedule for the FFT processor used two butterfly PEs and we elect to use the shared-memory architecture with bit-serial PEs and bit-parallel RAMs. The ideal architecture needs only two memories, if the real and imaginary parts of a complex value are stored as a single word. The memories shall support 5120 butterfly operations in 0.32 ms using an access rate for the memories of 32 MHz. A set of shift registers, one for each input/output of the PEs, is used as converters between the bit-parallel and bit-serial parts. The shift registers can be viewed as cache memories used to obtain uniform access patterns for the RAMs.

9.9.1 Selecting the Interconnection Network

In this section we will discuss the effect on the interconnection network (ICN) resulting from different choices of RAM and PE assignments. In general, the assignments, discussed in section 7.12, lead to four different architectures, since we have two alternative assignments each for RAMs and PEs. We summarize the resulting architectures next. Detailed derivation of the architectures is left as an exercise.

Architecture 1

The first architectural alternative is obtained by selecting the first assignment alternative for the PEs and the second alternative for the RAMs. This architectural alternative is described by the following constraints:

- A butterfly PE shall always use data from two different RAMs.
- Butterflies 0 to $N/4-1$ are assigned to PE_0 and the rest to PE_1.

The resulting architecture shown in Figure 9.32 has a simple RAM assignment and an XOR pattern for the PE assignment. Its main advantage is simple address generation for the RAMs. However, the ICN contains switches on the bit-serial side of the architecture. The control structure would become simpler if they were removed.

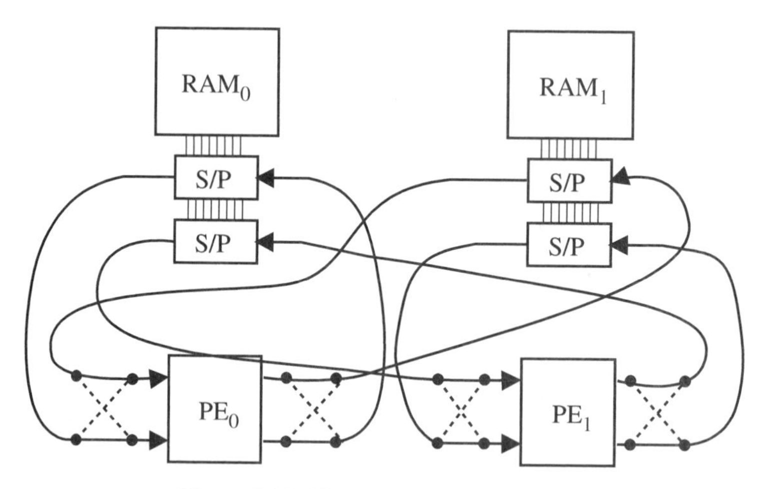

Figure 9.32 First architectural alternative

Architecture 2

The second architectural alternative is obtained by selecting the first assignment alternatives for both the PEs and RAMs. The resulting architecture is shown in Figure 9.33. In this case too, switches are needed on the bit-serial side.

This selection is described by the following constraints:

- The variables with indices 0 to $N/2 - 1$ are assigned to RAM_0 and the rest to RAM_1.
- Butterflies 0 to $N/4 - 1$ are assigned to PE_0 and the rest to PE_1.

Architecture 3

The third architecture is obtained by using the second alternative for both PEs and RAMs—i.e., assignment according to an XOR pattern. This interesting architecture is shown in Figure 9.34. This architecture is characterized by the following constraints:

- A butterfly PE always takes data from the two different RAMs.
- A RAM is not connected to both inputs of a butterfly PE.

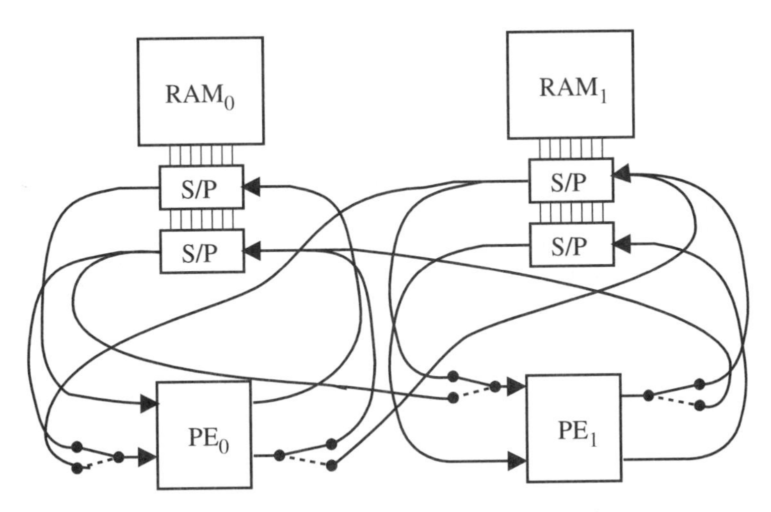

Figure 9.33 Second architectural alternative

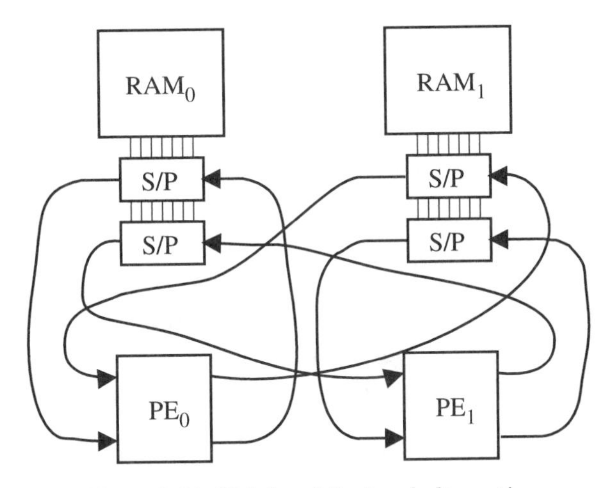

Figure 9.34 Third architectural alternative

An advantage is that the high-speed, bit-serial interconnection network is fixed and does not require any control signals. However, a switching function is required on the bit-parallel side. We must decide which of the two serial/parallel (S/P) converters the RAM should be using. Further, address generation for the RAM can be designed so that this control becomes simple.

Architecture 4

The fourth architecture is obtained by using the second alternative for PEs and the first alternative for the RAMs. This architecture is characterized by the following constraint:

- The variables with indices 0 to $N/2 - 1$ are assigned to RAM_0 and the rest to RAM_1.
- A RAM is not connected to both inputs of a butterfly PE.

It is not possible to find an architecture with these properties since they are contradictory. This can be explained from Figure 7.71. For example, in the second stage a butterfly uses data with indices $i = 0$ and $i+N/4 = 4$. According to the first constraint just listed, these data are assigned to the same RAM, but according to the second constraint, this is not allowed.

The main advantage of having no, or few, switches is that control becomes simpler. In order to evaluate the three feasible architectures properly, we must investigate the complexity of control of the architectures. We have elected to evaluate the third architecture since it lacks switches and since the shift registers have fixed connections to the butterfly PE.

9.9.2 Re-Partitioning the FFT

In the previous sections we have performed resource allocation and assignment for the processes that perform the arithmetic operations involved in the FFT. Also, we have chosen to continue the design on the third architecture alternative. This means we have decided which butterfly PE will perform a certain butterfly process, and which memory cell will store a particular data item. We must now design the control processes that control the architecture according to this assignment. Data must be read from the memories, communication channels must be set up, coefficients must be sent to the butterfly PEs, etc. The control processes consists of the address generators for RAM, ROM, cache memories, and the two loops. In this section we will repartition the design to find suitable processes to control the architecture. It will also be shown that it is possible to map the resulting processes directly onto hardware structures.

Address Generation for the FFT

In the section about RAM allocation and assignment we described two different assignment schemes. In the first scheme, the first half of the FFT variables were assigned to the first RAM and the second half to the second. In the second, the variables were assigned to RAM using an XOR pattern. The assignment of specific data can be found by evaluating the P function in the first assignment alternative. However, we need not necessarily evaluate this function in the address generation. The address in a RAM is not necessarily equal to the index of a variable.

In this section we will find a mapping from data index to RAM address. Address generation for the RAMs could then be split into several processes, one for each memory. The goal of this section is to find which function these processes must have to get the desired address sequences for the RAMs. We will describe the control and address generation for the third architectural alternative only. This architecture does not have switches on the bit-serial side, instead the ICN contains a switching function on the bit-parallel side to select which of the registers in the S/P converters to use.

The index generation for data was described in Box 7.4. The index generation is described by the following indices for the variables:

$$k_1 = 4N_s \left\lfloor \frac{m}{N_s} \right\rfloor + m \bmod(N_s) \qquad \text{(integer division)} \tag{9.11}$$

$$k_{1N_s} = k_1 + N_s \tag{9.12}$$

$$k_2 = \begin{cases} k_1 + N_s/2 & Stage = 1 \\ k_1 + 2N_s & Stage \geq 2 \end{cases} \tag{9.13}$$

$$k_{2N_s} = k_2 + N_s \tag{9.14}$$

$$p = k_1 \, 2^{Stage-1} \bmod(N/2) \tag{9.15}$$

These are the four indices for data to be used in the butterflies and the exponent, p, to compute W^p. In the following index k1 will be denoted the base index. The other indices, k_{1N_s}, k_2, and k_{2N_s} can be computed by setting bits in the binary representation of the base index. In the equation for k_2 there is a term $N_s/2$ or $2N_s$ that is added to the base index. This term is hereafter called BF, denoting that it is used to derive the index for the second butterfly process.

We must now find a way to transform this into a description with two RAMs, two butterfly PEs, cache memories, and some address generation for the RAMs and cache memories. First we will study how to transform from the indices to the selection of RAMs and PEs in the chosen assignment. Variables are assigned to RAMs through the P function defined in Equation (7.8). We can derive the following assignments of the various indices:

$$RAM_P(k_1) = P(k_1)$$

$$RAM_P(k_1N_s) = RAM_{\overline{P(k_1)}}$$

$$RAM_P(k_2) = RAM_{\overline{P(k_1)}}$$

$$RAM_P(k_2N_s) = RAM_{P(k_1)}$$

For example, $P(k_{1Ns})$ is always the inverse of $P(k_1)$ since k_{1Ns} is computed by setting one, initially reset, bit in k_1. Also, the main property of this RAM assignment was that a butterfly PE always uses data from both RAMs. Now, since k_1 is equal to the rearranged bits of m, we can also state that

$$P(k_1) = P(m)$$

A mapping between the index and the assigned butterfly has also to be found. In general, the PE that has a variable with index i as input can be derived from the following two equations:

$$S_{BF} = N_s \left\lfloor \frac{i}{N_s} \right\rfloor + i \bmod(N_s) \tag{9.16}$$

$$S_{PE} = P(S_{BF})$$

The first equation, Equation (9.15), describes the transformation from index to butterfly in a stage. The second equation describes the remaining transformation to a butterfly PE. In the following we will derive the assignment for the base index

$$S_{BF,k_1} = N_s \left\lfloor \frac{k_1}{2N_s} \right\rfloor + k_1 \bmod(N_s) =$$

$$= N_s \left(\frac{4\left\lfloor \frac{m}{N_s} \right\rfloor + m \bmod(N_s)}{2N_s} \right) + \left(4N_s \left\lfloor \frac{m}{N_s} \right\rfloor + m \bmod(N_s) \right) \bmod(N_s)$$

$$= 2N_s \left\lfloor \frac{m}{N_s} \right\rfloor + m \bmod(N_s)$$

Hence, since S_{BF,k_1} is equal to the rearranged bits of m

$$S_{PF,k_1} = P(S_{BF,k_1}) = P(m)$$

Assignment of the other indices follows from the fact that the k_1N_s index must be assigned to the same PE as k_1, and indices k_2 and k_2N_s to the other. We can conclude that

$$S_{PE,k_1N_s} = S_{PE,k_1}$$
$$S_{PE,k_2} = S_{PE,k_1}$$
$$S_{PE,k_2N_s} = S_{PE,k_2}$$

These equations will form the basis of an analysis of the requirements for address generation and control of the ICN.

We have previously mentioned that m need not to be incremented in binary order. If it is, we must compute the function $P(m)$ for every m. A better solution is to let m be incremented in Gray code order. Gray code has the property that only one bit at a time changes. Hence, $P(m)$ will switch every time. It is possible to implement a Gray counter as a binary counter followed by a Gray encoder. Then, the function $P(m)$ is equal to the least significant bit of the binary counter. Next we summarize the transactions that must take place between RAMs and PEs.

When $P(m) = 0$,

k_1 from RAM_0; address $k_1 \bmod(N/2)$, to PE_0

k_{1N_s} from RAM_1; address $(k_1 + N_s) \bmod(N/2)$, to PE_0

k_2 from RAM_1; address $(k_1 + BF) \bmod(N/2)$, to PE_1

k_{2N_s} from RAM_0; address $(k_1 + N_s + BF) \bmod(N/2)$, to PE_1

When $P(m) = 1$,

k_1 from RAM_1; address $k_1 \bmod(N/2)$, to PE_1

k_{1N_s} from RAM_0; address $(k_1 + N_s) \bmod(N/2)$, to PE_1

k_2 from RAM_0; address $(k_1 + BF) \bmod(N/2)$, to PE_0

k_{2N_s} from RAM_1; address $(k_1 + N_s + BF) \bmod(N/2)$, to PE_0

At this point, we find that the partitioning into cooperating processes must be refined. This is because the present address generation should be split into several processes—for example, one for each memory. Hence, we split the "*addresses*" process into several parts, corresponding to the address generation of the different RAMs, ROMs, and cache memories. Also, the "*Stage*" and "*m*" processes are replaced by processes with a different but similar function. The variable *Stage* itself is not used at all, since N_s can replace *Stage* as a means of control. We partition the address generation into seven blocks: a process for the base index and *P*(*m*), an address process for each RAM, a control process for each cache, and a process for each W^p process, as shown in Figure 9.35. Also, we introduce processes for the cache memories.

This partitioning can be mapped to a hardware structure with RAMs, PEs, and control, as shown in Figure 9.36.

The "base index generator" generates k_1 and *P*(*m*). It consists of a binary counter followed by a Gray encoder. The function *P*(*m*) is equal to the least significant bit of the binary counter. The address generators increase the base index by N_s and *BF*. "Address Generator 0" generates two addresses, A_{00} and A_{10}, for RAM_0 from k_1 and *P*(*m*). Its architecture is described in Box 9.1. "Address Generator 1" described inf Box 9.2, generates two addresses, A_{01} and A_{11}, for RAM_1 in a similar way.

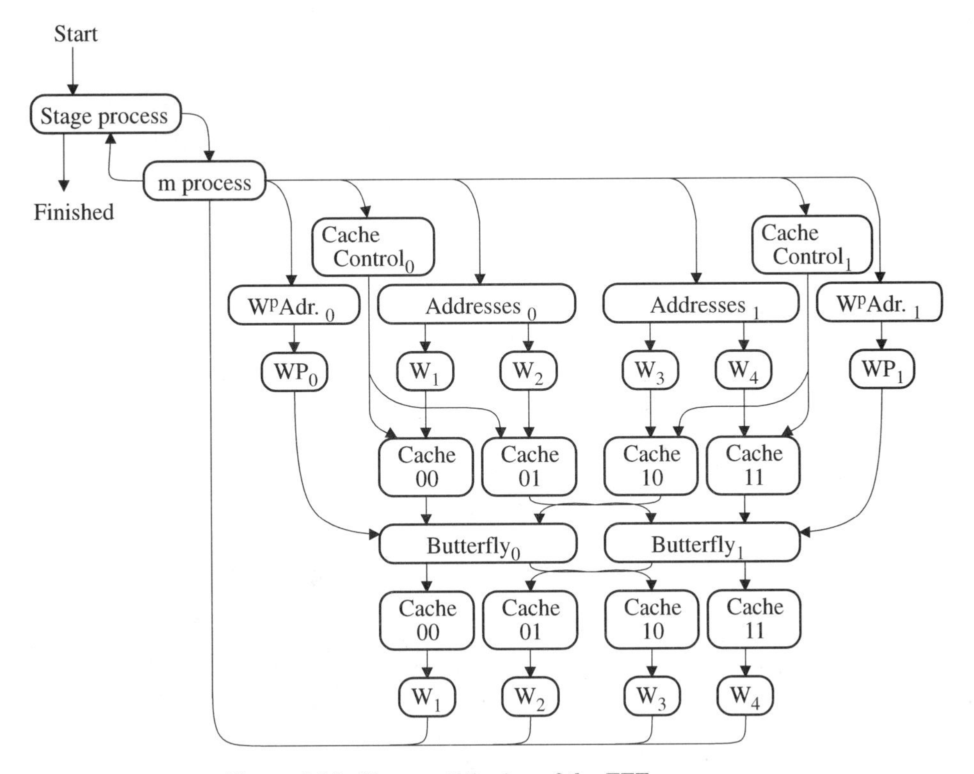

Figure 9.35 New partitioning of the FFT processor

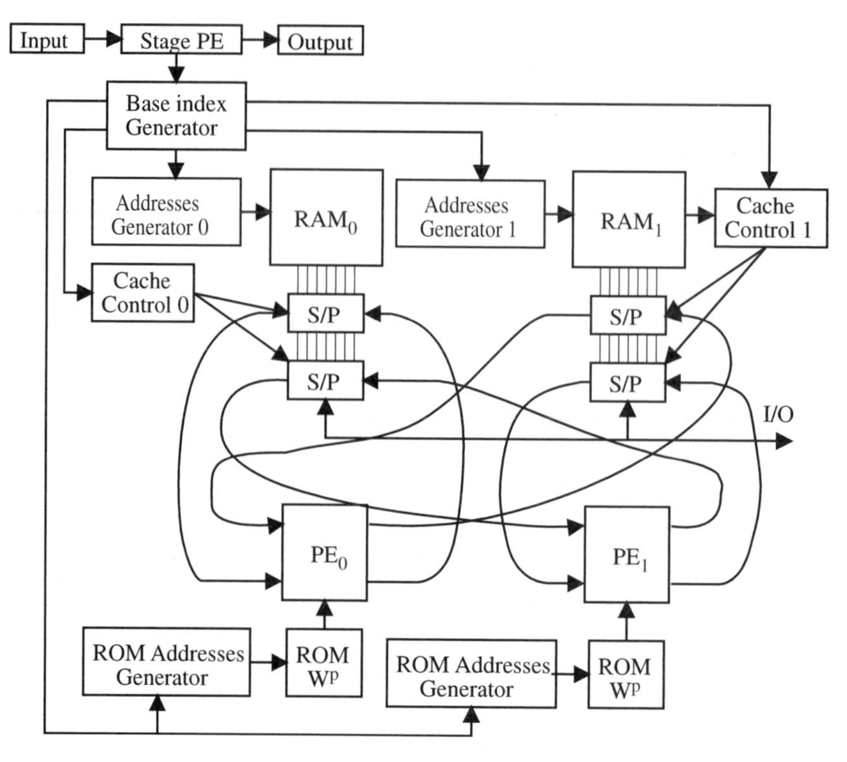

Figure 9.36 Architecture with address generation and control

```
architecture Beh_Address_Gen_0 of Address_Gen_0 is
...
  begin
    if P(m) = 0 then
      A00 = k1 mod(N/2);
      A10 = (k1 + Ns + BF) mod(N/2);
    else
      A00 = (k1 + BF) mod(N/2);
      A10 = (k1 + Ns) mod(N/2);
    end if;
  end Address_Gen_0;
```

Box 9.1 Address generator 0

The "Cache control 0" and "Cache control 1" blocks have the same behavior. We always write first into register 0, then into register 1. Hence, their function is in fact independent of $P(m)$. Implementation of these blocks is trivial.

In addition to the read addresses for the RAMs, the address generators must also provide write addresses. Due to in-place use of the memory cells, these addresses are the same as the read addresses, though suitably delayed.

```
architecture Beh_Address_Gen_1 of Address_Gen_1 is
...
  begin
    if P(m) = 0 then
      A01 = (k1 + Ns) mod(N/2);
      A11 = (k1 + BF) mod(N/2);
    else
      A01 = (k1 + Ns + BF) mod(N/2);
      A11 = k1 mod(N/2);
    end if;
  end Address_Gen_1;
```

Box 9.2 Address generator 1

Address Generation for the Input and Output

During the input and output phases we transfer N complex data words sequentially in the order defined by the data index. Data are distributed in the two RAMs according to the XOR pattern assignment. The data index is controlled by the i-loop process in the input and output loops. Inside the loops a data word is either read from or written into a memory cell in one of the memories. Hence, only one memory is active at a time. The I/O rate is 16 MHz. Thus the data rate is only half the data rate of the FFT computation.

From the index i we must first compute which RAM this data will be, or are, stored in. We can not avoid computing $P(i)$ during the input and output since data arrive in index order and must be distributed to the RAMs correctly. During the output phase we must reverse the digits in the index to unscramble the data. This is not done during the input phase. Finally, the RAM address inside the RAM is equal to the last 9 bits of the index, or digit-reversed index if we are in the output phase. Notice that in some applications the digit reverse operation is not used.

The processes that compute addresses during the input and output phase can, with one exception, be mapped to the same hardware modules that are used during the FFT computation. The basic function required is the same—namely, counting. This can be performed by the base index generator, without the Gray encoder. However, it must also be able to compute $P(i)$ to find the correct RAM for a data word. The function of the other modules must be modified slightly. For example, the address generators for RAMs must be able to inhibit read or write. This must be controlled by $P(i)$. Also, they must never add bits to the base index. Altogether, the extension of the function is trivial. One extra module must be included for the digit reversal operation.

9.9.3 The Final FFT Architecture

We are now able to present the final architecture, as shown in Figure 9.37. The base index generator will have three modes of operation: input, FFT computation, and output. It will operate in similar fashion in three modes. The basic function is

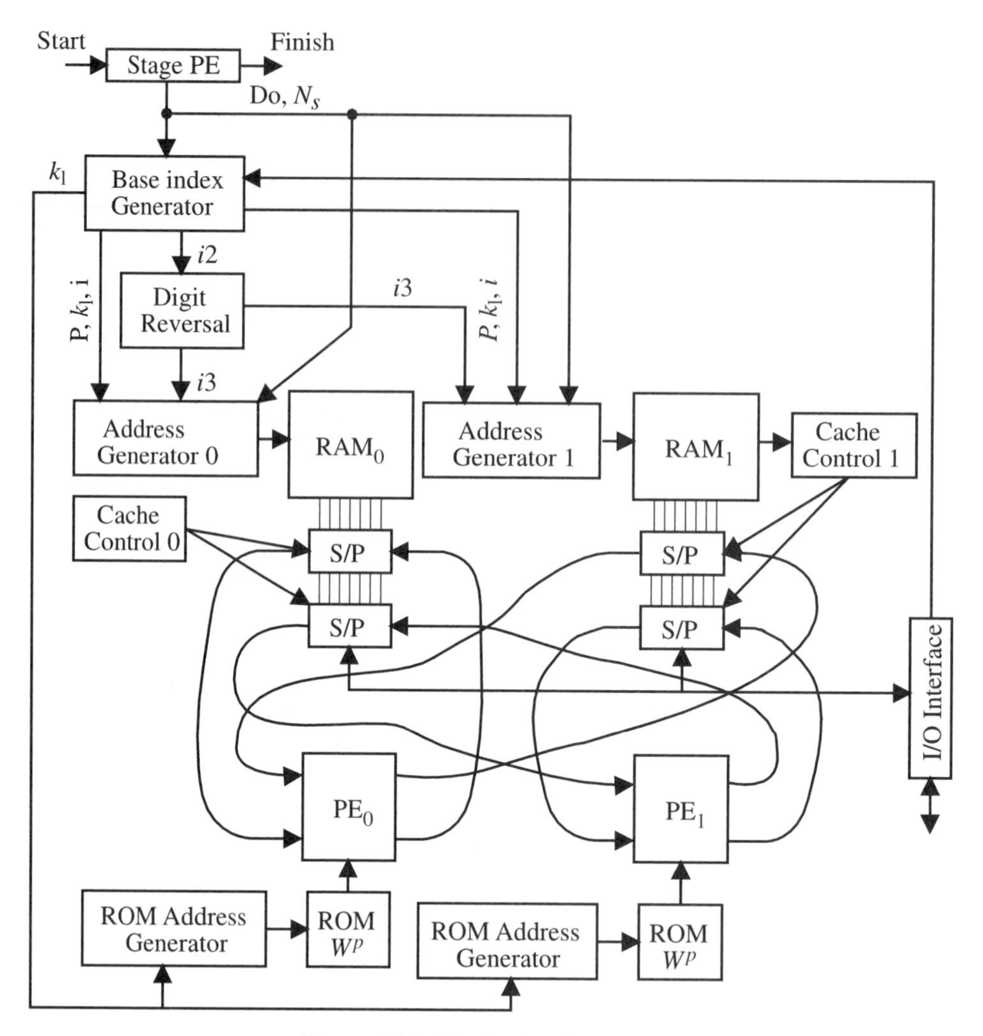

Figure 9.37 The final architecture

a binary counter. In the input and output phases, the base index is used almost directly as an address to a RAM.

The base index generator must evaluate which RAM a data word belongs to and send this RAM the base index. In the output phase, the index must be digit reversed before being used as an address. Also, during the input and output phases the I/O interface must be controlled. In the FFT computation phase the base index is first Gray encoded and then sent to the address generators, which extend it with the appropriate bits.

Next summarize the behavioral descriptions of the address generation processes. The behavior of the stage process is described in Box 9.3 while the behavior of base index generator is described in Box 9.4. The behaviors of Address Generator 0 and Address Generator 1 are described in Boxes 9.5 and 9.6, respectively.

```
architecture Behv_Stage_process of Stage_process is
...
  begin
    if Start_in = True then
      Do_out := Input;
      N_S := N;
      while N_S > 1 loop
        N_S := N_S/2;
        N_S_out := N_S;
        Do_out := FFTStage;
      end loop;
      Do_out := Output:
      Finished_out := True;
    end if;
  end Stage_process;
```

Box 9.3 Behavior of the Stage process

```
architecture Behv_Bas_Index_Gen of Base_Index_Gen is
...
  begin
    if Do_in = Input then
      for m in 0 to N – 1 loop
        i_out := m mod(N/2);
        P_out := P(m);
      end loop;
    end if;
    if Do_in = FFTStage then
      for m_binary in 0 to ((N/4)–1) loop
        m := GrayCode(m_binary)
        k1_out := 4*N_S_in*(m/N_S_in) + m mod(Ns_in);
        P_out := P(m);
      end loop;
    end if;
    if Do_in = Output then

      for m in 0 to N – 1 loop
        i2_out := m mod(N/2);
        P_out := P(m);
      end loop;
    end;
  end Base_Index_Gen;
```

Box 9.4 Behavior of the base index generator

```
architecture  Beh_Address_Gen_0 of Address_Gen_0 is
...
  begin
    if Do_in = Input then
      if P_in = 0 then
        A_out := i_in;
      end if;
```

```
        end if;
        if Do_in = FFTStage then
              if Ns_in = (N/2) then
                 BF := Ns_in/2;
           else
           BF := 2 * Ns_in;
           end if;
        if P_in = 0 then
              A00_out = k1_in mod(N/2);
              A10_out = (k1_in + Ns_in + BF) mod(N/2);
        else
              A00_out = (k1_in + BF) mod(N/2);
              A10_out = (k1_in + Ns_in) mod(N/2);
           end if;
        if Do = Output then
           if P_in = 0 then
              A_out := i3_in;
           end if;
        end if;
     end Address_Gen_0;
```

Box 9.5 Behavior of address generator 0

```
architecture Beh_Address_Gen_1 of Address_Gen_1 is
...
   begin
      if Do = Input then
         if P_in = 1 then
            A_out := i_in;

         end if;
      end if;
      if Do = FFTStage then
         if Ns_in = (N/2) then
            BF := Ns_in/2
         else
            BF := 2 * Ns_in;
         end if;
         if P_in = 0 then
            A01_out = (k1_in + Ns_in) mod(N/2);
            A11_out = (k1_in + BF) mod(N/2);
         else
            A01_out = (k1_in + Ns_in + BF) mod(N/2);
            A11_out = k1_in mod(N/2);
         end if;
      end if;
      if Do = Output then
         if P_in = 0 then
            A_out := i3_in;
         end if;
      end if;
   end Address_Gen_1;
```

Box 9.6 Behavior of address generator 1

9.10 DCT PROCESSOR, CONT.

It was determined in Chapter 7 that only two 1-D DCT processing elements are required. Each PE requires 16 input values and delivers 16 output values. According to the schedule data are written into the memories in the first phase (16 time units), while during the last phase data are read from the memories. Hence, from a memory point of view, a 1-D DCT computation involves either a write or a read operation. Thus, the ideal architecture, with two PEs, should have $2 \cdot 16 = 32$ memories. The schedule length of 34 time units implies an execution time for the DCT PEs, including communication time and either a read or a write cycle, of

$$T_{PE} = \frac{2}{34 \cdot 486 \cdot 10^3} = 121.0 \text{ ns}$$

Hence, $f_{PE} = 8.262$ MHz. Using only one memory to support both PEs the required access rate is

$$f_{RAM} = 8.262 \cdot 10^6 \cdot 2 \cdot 16 = 264.4 \text{ MHz}$$

which is much too high. We therefore choose to use four memories, two per PE. The required access rate becomes a more reasonable 66.1 MHz. Each memory has $8 \cdot 16 = 128$ 12-bit words. The resulting architecture for the 2-D DCT is shown in Figure 9.38.

The external FIFOs feed one of the PEs while the other PE produces outputs. After 16 time units the PEs exchange roles. The two RAMs should provide the PEs with 16 values in bit-serial form—i.e., eight values from each RAM. Each RAM has therefore been provided with eight shift registers (serial/parallel converters).

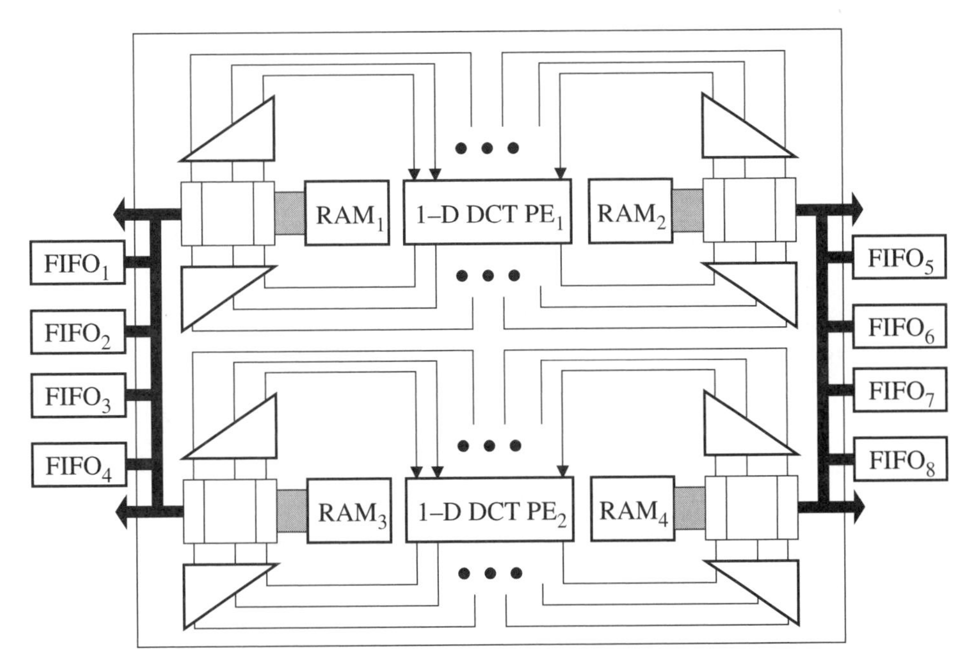

Figure 9.38 Architecture for the DCT processor with external memories

The inputs should arrive at the PEs synchronously, but the memory delivers the values sequentially. Hence, the first delivered values must be delayed in order to arrive at the PE at the same time as the last value delivered by the memory. Shimming delays have therefore been inserted on both sides of the serial/parallel converters, as shown in Figure 9.38.

To avoid access conflicts the intermediate array must be stored in the memories properly. Further, transposition of the intermediate array can be obtained free by storing the array, as illustrated in Figure 9.39 [26, 27].

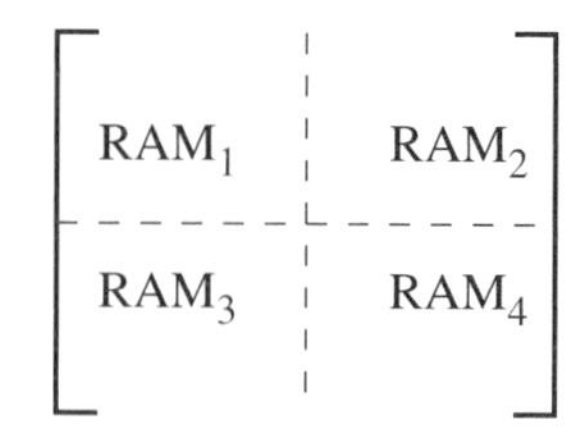

Figure 9.39 Storage of intermediate array in RAM_1 and RAM_2

In the first computational phase, the PE delivers eight values to each RAM corresponding to one row in the intermediate array. In the second phase, values corresponding to columns should be accessed. This scheme could be implemented by putting a multiplexer after the PEs, but can be accomplished more easily by exploring the symmetry of the basis function.

9.11 SIC (SINGLE-INSTRUCTION COMPUTER)

Shared-memory address space can be partitioned into several sections associated with special attributes, as illustrated in Figure 9.40 [3, 19]. The memory may be used for interprocess communication, possibly between asynchronous processing elements. Some sections of the memory may be implemented using external memories while other sections represent on-chip memory. The data word length may vary and the different memory sections may be of type ROM, RAM, or cyclic buffer. Certain memory cells are associated with processing elements (PEs), DMA channels, standard digital signal processors, RISC processors, or other computational systems. The shared memory can be implemented physically using a mixture of external and internal memories. Hence, advanced standard memories can be used in applications that require the storage of large amounts of data.

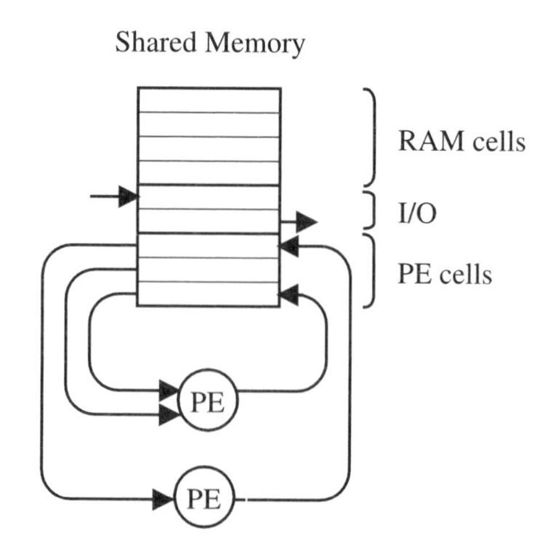

Figure 9.40 Realization of interprocess communication using a SIC processor

One memory cell is associated with each input and output of those PEs connected to the shared memory. The processors only task is to move data between PEs and memory cells. The only instruction needed therefore is the Move instruction.

Communication between PEs is via a conditional move instruction. That is, a value in a memory cell is placed in the input register of the PEs and the result is returned to the output cell of the PE. The SIC can therefore be viewed as an intercommunication processor.

The PEs are asynchronous and work only when they have valid input data. Power is thereby conserved. We assume that most of the PEs perform large computational tasks, so they are slow compared with shared memory. In practice, bit-serial PEs are efficient in terms of power consumption, chip area, and throughput.

New functions can easily be added by providing new PEs. No new op-codes are required. This simplifies the designer's trade-off between hardware and software. The system can be updated as new and improved PEs become available, thereby prolonging its lifetime.

This approach places emphasis on communication issues in the real-time system. We will later describe how conventional programming constructs can be implemented, if needed, in the processor.

9.11.1 Partitioning of Large DSP Systems

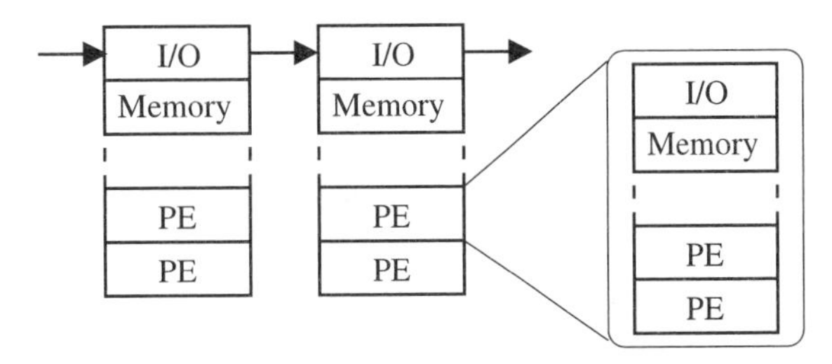

Figure 9.41 Large DSP system that is implemented as co-operating SIC processors

A large DSP system can usually be partitioned into a set of subsystems. Each of these subsystems can be implemented by an SIC processor having its own shared-memory address space, as illustrated in Figure 9.41. Typically each SIC has only a few PEs to achieve balance between communication bandwidth and computational capacity. It is possible to connect SIC processors that implement the subsystems according to the organization of the large DSP system. Hence, several of these SIC processors can be used in parallel, cascade, or hierarchical fashion as illustrated in Figure 9.41. Compare also with Figures 9.26 through 9.28.

It is possible to use a mix of different types of processors within the same memory address space. A standard DSP may, for example, be substituted for one of the PEs in the SIC processor. This nonhomogeneous multiprocessor architecture allows the same framework to be used for a wide range of applications.

Notice that the use of high-level DSP operations will significantly reduce the complexity of programming the highly pipelined, nonhomogeneous multiprocessor system. This would otherwise become a major problem.

9.11.2 Implementation of Various SIC Items

Programming concepts such as subroutines, interrupts, and local variables, can be easily implemented as illustrated in Figure 9.42.

The Single Instruction Since there is only one move instruction, no instruction decoding is necessary. The program memory (*PM*) need only contain the source address and the destination address of the moves.

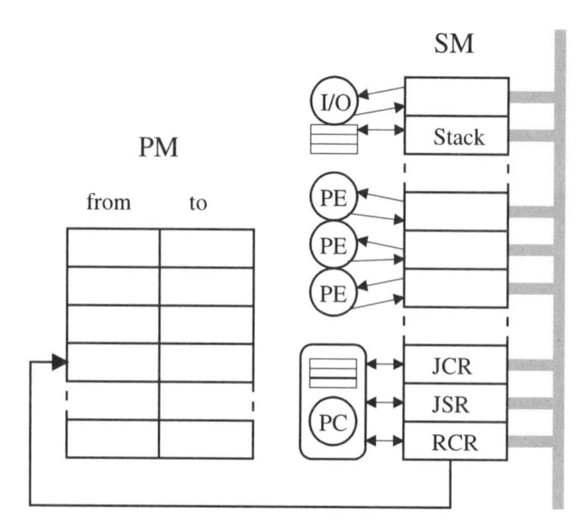

Figure 9.42 The SIC architecture

Constants Program constants are stored in ROM cells in the shared memory (*SM*).

The Program Counter The program counter (*PC*) works as the other processes in the architecture with its interfacing registers in the SM area. However, the program counter register (*PCR*) has an additional output for pointing to the source and destination addresses in the PM of the next move to be performed. Unconditional jumps are facilitated by an immediate move of the next program address into the PCR. To facilitate conditional jumps, a jump condition register (*JCR*) is set up to contain the jump address. The JCR is moved into the PCR when a process signals the PC that the jump condition has been met; for example, a comparator process could signal when its two operands are equal.

To issue a subroutine call the start address of the subroutine is moved to a jump subroutine register (JSR). This in turn causes the value of the PCR to be put on a stack local to the PC-process before writing the new address into the PCR from the JSR. When returning from a subroutine call, the JSR is moved into the PCR, then the top value of the stack is popped and transferred to the JSR displaying the next return value. A similar operation could be implemented for external interrupts by causing the PC to load an interrupt register into the PCR whenever an interrupt occurs.

Stack Implementation A stack can be implemented as another process with an input/output register in the SM. Writing a value to this register would then cause the process to put the register value onto a stack. Reading the register value would pop the next value off the stack and into the register, making it available for any subsequent reading of the register.

Indexing Indexing can be implemented by an index register pointing to a cell within a memory frame together with an input/output register that interfaces the selected cell with the SM. For sequential accesses of the indexed memory, automatic increment/decrement of the index register can be performed post-prereading or writing to the frame, thus eliminating the need to update the index register.

Circular Buffers A circular buffer can easily be implemented by using the concept of the indexed buffer just mentioned, together with automatic increment/decrement of the index register.

References

[1] Afghahi M., Matsumura S., Pencz J., Sikström B., Sjöström U., and Wanhammar L.: An Array Processor for 2-D Discrete Cosine Transforms, *Proc. The European Signal Processing Conf.,* EUSIPCO-86, The Hague, The Netherlands, Sept. 1986.

[2] Almasi G.S. and Gottlieb A.: *Highly Parallel Computing,* Benjamin/Cummings, Redwood City, CA, 1989.

[3] Azaria H. and Tabak D.: Design Consideration of a Single Instruction Microcomputer A Case Study, *Microprocessing and Microprogramming,* Vol. 1, pp. 187–194, 1983.

[4] Bellanger M.G. and Bonnerot G.: Premultiplication Scheme for Digital FIR Filters with Application to Multirate Filtering, *IEEE Trans. on Acoustics, Speech, and Signal Processing,* Vol. ASSP-26, No. 1, pp. 50–55, Feb. 1978.

[5] Cheung Y.S. and Leung S.C.: A Second Generation Compiler for Bit-Serial Signal Processing Architecture, *Proc. IEEE Intern. Conf. on Acoustics, Speech, and Signal Processing,* pp. 487–490, 1987.

[6] Denyer P.B.: An Introduction to Bit-Serial Architectures for VLSI Signal Processing, in Randell R., and Treleaven P. C. (eds.): *VLSI Architecture,* Prentice Hall, pp. 225–241, 1983.

[7] Denyer P.B. and Renshaw D.: Case Studies in VLSI Signal Processing Using a Silicon Compiler, *Proc. IEEE Intern. Conf. on Acoustics, Speech, and Signal Processing,* ICASSP-83, Boston, pp. 939–942, 1983.

[8] Denyer P.B., Murray A.F., and Renshaw D.: FIRST—Prospect and Retrospect, in Lyon R. (ed.): *VLSI Signal Processing,* IEEE Press, pp. 252–263, 1984.

[9] Denyer P.B. and Renshaw D.: *VLSI Signal Processing: A Bit-Serial Approach,* Addison-Wesley, Reading, MA 1985.

[10] Dinha F., Sikström B., Sjöström U., and Wanhammar L.: A Multi-Processor Approach to Implement Digital Filters, *Nordic Symp. on VLSI in Computers and Communications,* Tampere, Finland, June 13–16, 1984.

[11] Dinha F., Sikström B., Sjöström U., and Wanhammar L.: LSI Implementation of Digital Filters—A Multi-Processor Approach, *Intern. Conf. on Computers, Systems and Signal Processing,* Bangalore, India, Dec. 10–12, 1984.

[12] Fountain T.: *Processor Arrays, Architecture and Applications,* Academic Press, London, 1988.

[13] Gajski D.D. (ed.): *Silicon Compilation,* Addison-Wesley, Reading, MA, 1988.

[14] Ginderdeuren van J.K.J., De Man H.J., De Loore B.J.S., Vanden Wijngaert H., Dearuelle A., and van den Audenaerde G.G.J: A High-Quality Digital Audio Filter Set Designed by Silicon Compiler CATHEDRAL-1, *IEEE J. of Solid-State Circuits,* Vol. SC-21, No. 6, pp. 1067–1075, Dec. 1986.

[15] Goor van de A.J.: *Computer Architecture and Design*, Addison-Wesley, Wookingham, England, 1989.

[16] Hartley R. and Jasica J.: Behavioral to Structural Translation in a Bit-Serial Silicon Compiler, *IEEE Trans. on Computer-Aided Design,* Vol. CAD-7, No. 8, pp. 877–886, Aug. 1988.

[17] Jain R., Catthoor F., Vanhoof J., De Loore B.J.S., Goossens G., Goncalvez N.F., Claesen L.J.M., Ginderdeuren van J.K.J., Vandewalle J., and De Man H.J.: Custom Design of a VLSI PCM-FDM Transmultiplexer from System Specifications to Circuit Layout Using a Computer-Aided Design System, *IEEE Circuits and Systems,* Vol. CAS-33, No. 2, pp. 183–195, Feb. 1986.

[18] Kung S.Y.: *VLSI Array Processors,* Prentice Hall, Englewood Cliffs, New Jersey, 1988.

[19] Lipovski G.J.: Architecture of a Simple, Effective Control Processor, Second Symposium on Micro Architecture, *EUROMICRO,* pp. 187–194, 1976.

[20] McCanny J., McWhirter J., and Swartzlander Jr. E.E. (eds.): *Systolic Array Processors,* Prentice Hall, London, 1989.

[21] Murray A.F. and Denyer P.B.: A CMOS Design Strategy for Bit-Serial Signal Processing, *IEEE J. on Solid-State Circuits,* Vol. SC-20, No. 3, pp. 746–753, June 1985.

[22] Nordhamn E., Sikström B., and Wanhammar L.: Design of an FFT Processor, *Fourth Swedish Workshop on Computer Architecture,* DSA-92, Linköping, Sweden, Jan. 1992.

[23] Nordhamn E.: *Design of an Application-Specific FFT Processor,* Linköping Studies in Science and Technology, Thesis, No. 324, Linköping University, Sweden, June 1992.

[24] Note S., Meerbergen van J., Catthoor F., and De Man H.J.: Automated Synthesis of a High Speed Cordic Algorithm with CATHEDRAL III Compilation System, IEEE Inter. symmp. on circuts and systems, *ISCAS-88,* Espoo, Finland, June 6–9, 1988.

[25] Rabaey J., De Man H.J., Vanhoof J., Goossens G., and Catthoor F.: CATHEDRAL II: A Synthesis System for Multi-Processor DSP Systems, in Gajski (ed.), *Silicon Compilation,* pp. 311–360, 1988.

[26] Sikström B., Wanhammar L., Afghahi M., and Pencz J.: A High Speed 2-D Discrete Cosine Transform Chip, *Proc. 2nd Nordic Symp. on VLSI in Computers and Communications,* Linköping, Sweden, June 2–4, 1986.

[27] Sikström B., Wanhammar L., Afghahi M., and Pencz J.: A High Speed 2-D Discrete Cosine Transform Chip, *Integration, the VLSI Journal,* Vol. 5, No. 2, pp. 159–169, June 1987.

[28] Smith S.G., Denyer P.B., Renshaw D., Asada K., Coplan K.P., Keightley M., and Mhar J.I.: Full-Span Structural Compilation of DSP Hardware, *Proc. ICASSP-87,* pp. 495–499, 1987.

[29] Swartzlander Jr. E.E.: *VLSI Signal Processing Systems,* Kluwer Academic Pub., Boston, 1986.

[30] Torkelson M.: *Design of Applications Specific Digital Signal Processors,* Diss. 158, Lund University, Sweden, 1990.

[31] Wanhammar L.: Implementation of Wave Digital Filters with Distributed Arithmetic, *Proc. 4th Intern. Symp. on Network Theory,* Ljubljana, Yugoslavia, pp. 385–397, Sept. 1979.

[32] Wanhammar L., Sikström B., Afghahi M., and Pencz J.: A Systematic Bit-Serial Approach to Implement Digital Signal Processing Algorithms, *Proc. 2nd Nordic Symp. on VLSI in Computers and Communications,* Linköping, Sweden, June 2–4, 1986.
[33] Wanhammar L.: On Algorithms and Architecture Suitable for Digital Signal Processing, *Proc. The European Signal Processing Conf.,* EUSIPCO – 86, The Hague, The Netherlands, Sept. 1986.
[34] Wanhammar L.: Implementation of Wave Digital Filters Using Distributed Arithmetic, *Signal Processing,* Vol. 2, No. 3, pp. 253–260, July 1980.
[35] Wanhammar L.: Implementation of Wave Digital Lattice Filters, *Proc. 1980 European Conf. on Circuit Theory and Design,* Warsaw, Poland, Sept. 1980.
[36] Wanhammar L.: Implementation of Wave Digital Filters Using Vector-Multipliers, *Proc. First European Signal Processing Conf.,* EUSIPCO-80, Lausanne, Switzerland, pp. 21–26, Sept. 1980.
[37] Wanhammar L.: *An Approach to LSI Implementation of Wave Digital Filters,* Linköping Studies in Science and Technology, Diss. No. 62, Sweden, April 1981.
[38] Wanhammar L., Afghahi M., and Sikström B.: On Mapping of Algorithms onto Hardware, *IEEE Intern. Symp. on Circuits and Systems,* ISCAS-88, Espoo, Finland, pp. 1967–1970, June 1988.

PROBLEMS

9.1 Sketch an architecture for implementation of a complementary FIR filter that uses only one PE of the multiplier–accumulator type. Determine also the relevant system parameters when $N = 128$.

9.2 Modify the architectures shown in Figure 9.5 to compute
(a) The correlation function between two sequences
(b) The autocorrelation function

9.3 Describe the major steps in a systematic approach to implementing DSP algorithms into an ASIC. In what sense is the approach optimal?

9.4 Show by an example that an isomorphic mapping of the signal-flow graph onto hardware resources does not always yield a maximally fast implementation. Also discuss resource utilization.

9.5 Propose a scheme based on a single, multiplexed vector-multiplier that implements the bandpass filter discussed in Example 9.5.

9.6 Implement the wave digital lattice filter shown in Figure P9.6 using a minimum number of vector-processors.
(a) Determine the appropriate set of difference equations using the following adaptor coefficients:

$$\alpha_0 = \frac{1}{4}, \alpha_1 = \frac{3}{8}, \alpha_2 = \frac{5}{16}, \alpha_3 = \frac{3}{16}, \alpha_4 = \frac{5}{32}$$

For simplicity do not scale the filter.

(b) Determine the coefficient word lengths.

(c) How many vector-processors are required?

(d) How many input signals are there to the PEs?

9.7 Implement the wave digital lattice filter discussed in Example 6.5 using a minimum number of vector-processors. Can the implementation be made maximally fast?

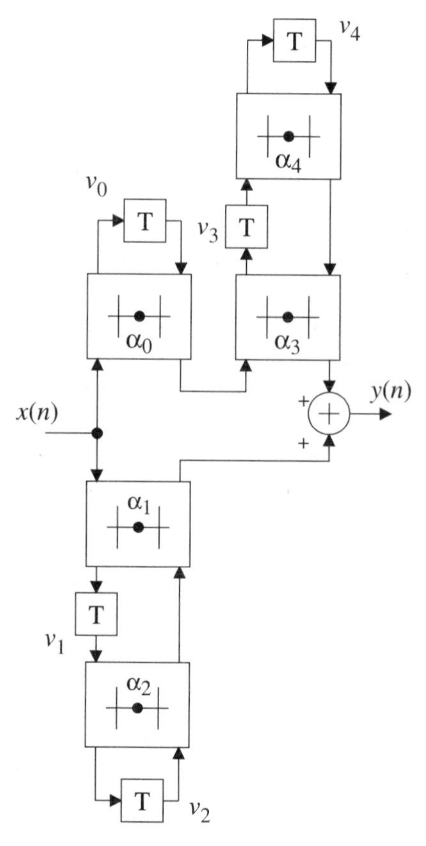

Figure P9.6 Lattice WDF

9.8 (a) Determine the system of difference equations in numerically equivalent form for the lattice wave digital filter shown in Figure. P9.8. Symmetric adaptors are used. The adaptor coefficients are

$$\alpha_1 = \frac{3}{16},\ \alpha_3 = \frac{3}{16},\ \alpha_5 = \frac{3}{16}$$

(b) Determine the number of vector-multipliers, the number of terms in each difference equation, and the coefficient word length needed to implement the filter.

9.9 Show that a significant simplification can be obtained for the implementation in Problem 9.8 by explicitly computing outputs u_{34} and u_{54} of the adaptors instead of the outputs of the filter shown in Figure P9.8.

9.10 A DSP algorithm is implemented using N_{PE} bit-serial PEs with data word length W_{PE}, The number of memories is N_{RAM} each with word length

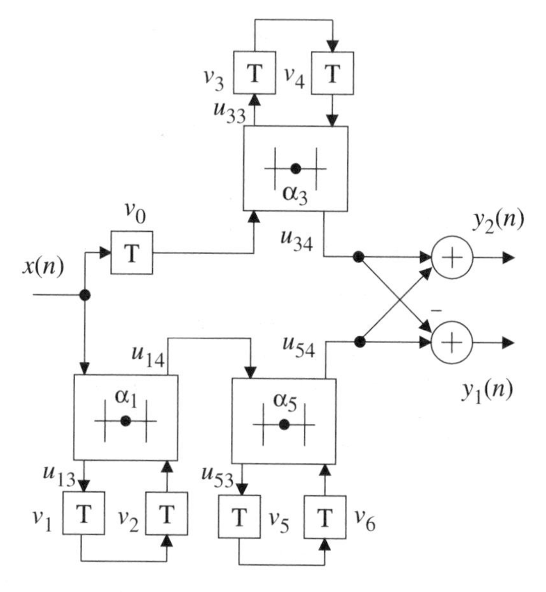

Figure P9.8 Lattice wave digital filter

W_{RAM}. The clock period for the processors is T_{PE} and read and write time for the memories is $T_R = T_W = T_{RAM}$.

(a) Suggest a suitable architecture.

(b) How many memories are required, given N_{PE}, W_{PE}, T_{PE}, W_{RAM}, and T_{RAM}, in order to obtain balance between processing capability and communication bandwidth?

9.11 The architecture in Figure P9.11 is used to implement a 1024-point complex FFT. The FFT is performed in less than 1 ms and the internal data word length, W_d, is 21 bits for both the real and complex parts.

The input word length is 16 bits. Estimate the necessary clock rate at section A–A and the clock rate for the RAM memory. The memory word length is chosen to be 21 bits. What will happen if the memory word length is doubled? Is one RAM enough to support the butterfly PE?

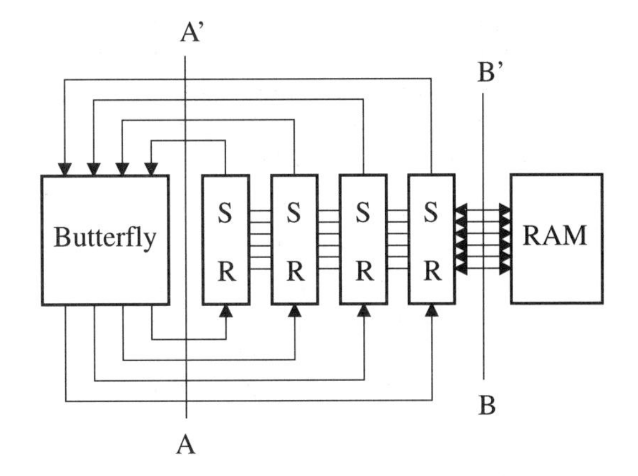

Figure P9.11

9.12 The shared memory architecture in Figure P9.12 with bit-serial PEs is used to implement a digital filter that performs an interpolation of the sample rate from 44.1 to 705.6 kHz.

Two such interpolators are implemented using the same hardware. The digital filter consists of four stages, each of which interpolates with a factor of two. The filter stages are bireciprocal lattice wave digital filters. The orders of the four stages are 17, 9, 5, and 5, respectively. The data word length has been estimated to be 20 bits. Estimate the clock frequency at section A–A and choose an appropriate memory word length at section B–B such that the memory clock frequency becomes less than 20 MHz. Can this architecture be used for an interpolator with two channels?

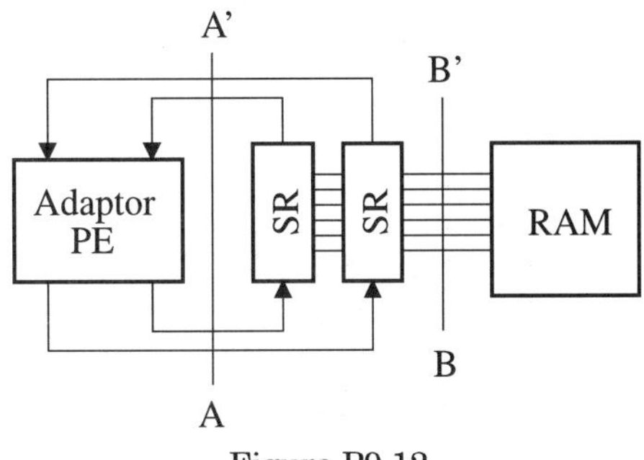

Figure P9.12

9.13 A 2-D DCT meeting the following requirements shall be implemented:

- 16·16 pixels
- Data word length of W_d = 12 bits
- 240,000 2-D transforms/s
- 1-D DCT PEs based on distributed arithmetic

- ❑ Processing time of 14 clock cycles
- ❑ Maximum clock frequency for CMOS, f_{CL} = 120 MHz
- ❑ Read–write time for the RAMs, $T_{read} = T_{write}$ = 17 ns

(a) Determine the number of PEs and a suitable PE schedule.
(b) Suggest an architecture. Specify the PEs, memories, I/O, communication channels, and control.
(c) Determine the required clock frequency for the PEs.
(d) Determine the required number of memories when the memory word length is the same as the data word length.
(e) Determine the size of the RAMs.
(f) Select the number of RAMs when the form factor ≤ 1:4.
(g) Determine the communication network.

9.14 An FFT processor meeting the following specification shall be implemented:

- ❑ 1024 complex points
- ❑ Processing time ≤ 1 ms
- ❑ Data word length of W_d = 21 bits
- ❑ The PEs implementing a butterfly
- ❑ Processing time of 23 clock cycles
- ❑ Maximum clock frequency for CMOS, f_{CL} = 120 MHz
- ❑ Read–write time for the RAMs, $T_{read} = T_{write}$ = 17 ns

(a) Determine the number of PEs and a suitable PE schedule.
(b) Suggest a suitable architecture. Specify the PEs, memories, I/O, communication channels, and control.
(c) Determine the required clock frequency for the PEs.
(d) Determine the required number of memories when the memory word length is the same as the data word length.
(e) Determine the size of the RAMs.
(f) Select the number of RAMs when the form factor ≤ 1:4.
(g) Determine the communication network.

9.15 An interpolating wave digital filter meeting the following specification shall be implemented:

- ❑ Interpolation with a factor 16 in 4 stages
- ❑ Two signal channels
- ❑ Sampling frequency (input) = 44.1 kHz
- ❑ Sampling frequency (output) = 16 · 44.1 = 705.6 kHz
- ❑ Filter orders of 17, 9, 5, and 5
- ❑ The adaptors in each stage of operating at the lower sample rate
- ❑ Data word length of W_d = 22 bits
- ❑ Coefficient word length of W_c = 14 bits
- ❑ PE implementing a two-port adaptor
- ❑ Processing time of 24 clock cycles
- ❑ Maximum clock frequency for CMOS, f_{CL} = 120 MHz
- ❑ Read–write time for the RAMs, $T_{read} = T_{write}$ = 17 ns

(a) Determine the number of PEs and a suitable PE schedule.
(b) Suggest an architecture. Specify the PEs, memories, I/O, communication channels, and control.

(c) Determine the required clock frequency for the PEs.
(d) Determine the required number of memories when the memory word length is the same as the data word length.
(e) Estimate the size of the RAMs.
(f) Select the number of RAMs when the form factor ≤ 1:4.
(g) Determine the communication network.

9.16 Determine the number of shimming delays required in the 2-D DCT processor shown in Figure 9.39.

9.17 Show how transposition is accomplished in the 2-D DCT processor. See Figure 9.40.

9.18 Verify the four architectural alternatives discussed in section 9.9.

9.19 Complete the behavioral descriptions in Boxes 9.4 through 9.6.

9.20 Estimate the reduction in power consumption for the implementation discussed in Example 7.3 if the power supply voltage is reduced to a minimum. The required sample rate is only 10 MHz. Assume that $|V_{Tn}| \approx |V_{Tp}| \approx 0.7$ volt.

9.21 Develop an SIC architecture suitable for the adaptive FIR filters discussed in section 3.14.1.

10

DIGITAL SYSTEMS

10.1 INTRODUCTION

As discussed in Chapter 1, large systems are generally, for reasons of complexity, partitioned into a hierarchy of subsystems. The lower levels of the hierarchy are typically mapped directly onto hardware structures while higher-level functions are implemented as virtual machines built of combinations of software and hardware. Most digital systems are designed to have a hierarchical organization that mirrors the system hierarchy, as illustrated in Figure 10.1.

The architectural components used to build such virtual machines, as well as the digital hardware modules, are PEs, ICNs, memories, and control units. Complex hardware modules can normally be decomposed into a set of simpler modules. For example, a vector-multiplier may be decomposed successively into multipliers, adders, buses, registers, and finite-state machines (FSMs) down to full-adders, flip-flops, and even wires. At the lowest-level gates, transistors, wires, simple memory elements, and clocks are used as basic entities. Notice also that memories can be similarly decomposed into PEs, ICNs, memories, and control units (decoders, I/O circuitry, basic memory elements, clocks). The communication parts may differ slightly since, at the lowest level, they are decomposed into switches, wires, and clocks.

A logic function is implemented by execution of some logic algorithm which consists of a sequence of elementary logic functions operating on binary data. The algorithm may be fully sequential or consist of a combination of sequential and recursive operations.

The task of specifying the gates and their interconnections to implement a given behavior generally involves a large amount of work. Often, this work equals or exceeds the work required in the VLSI design phase. However, the use of automatic synthesis tools may change this in the near future.

In this chapter, we will discuss some general aspects of digital systems. In particular, we will discuss different approaches to decompose a logic function into manageable pieces that can more easily be designed to meet specified area and speed requirements. Another important issue discussed will be the various clocking regimes used in digital systems.

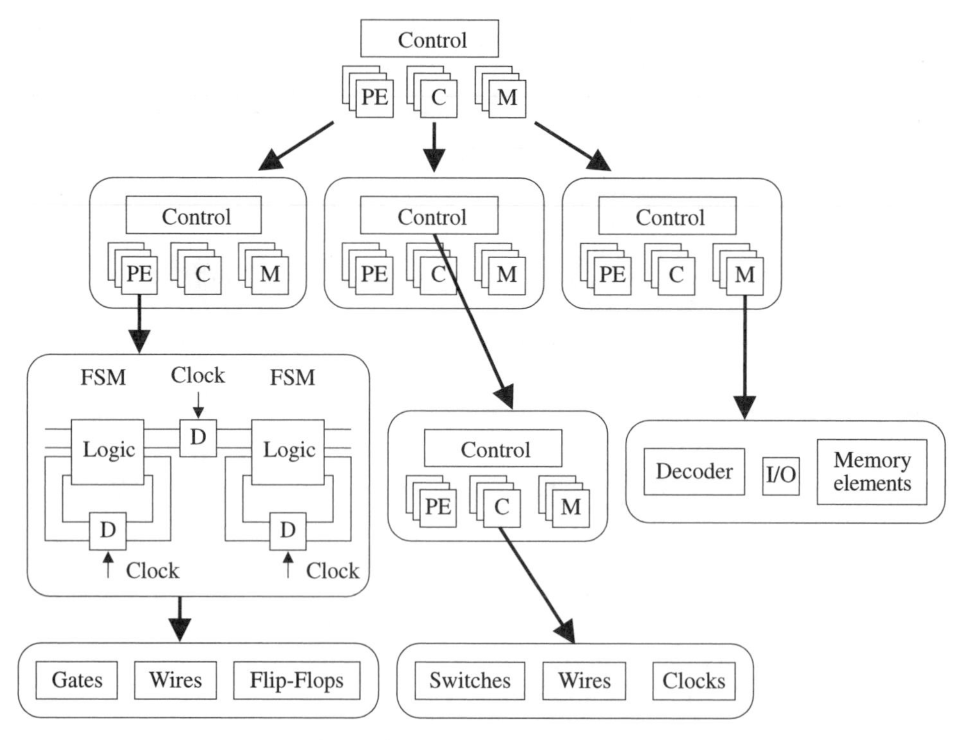

Figure 10.1 Hierarchical decomposition of a digital system

10.2 COMBINATIONAL NETWORKS

A combinational network for implementing a general Boolean function requires at least two levels as illustrated in Figure 10.2. Traditionally such networks are implemented using NAND–NAND or NOR–NOR structures, but other more general structures are becoming increasingly common. Various heuristics are used because of the high complexity involved in the design of optimal two-level networks. The aim is to make a suitable trade-off between speed, chip area (time-space), and design time. Notice that the required chip area depends mainly on the amount and regularity of the wiring, whereas the number of transistors and contacts is often less important.

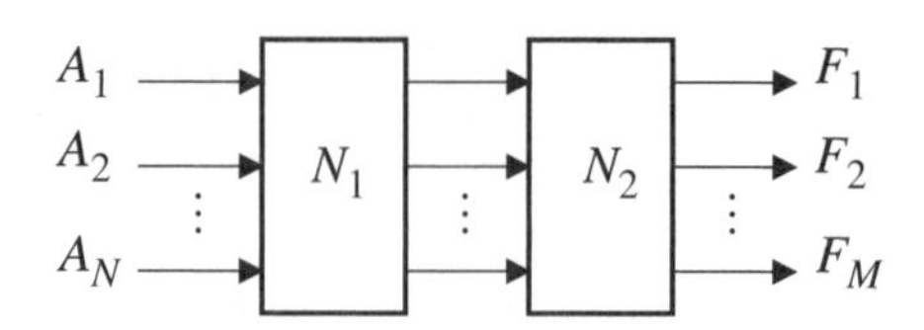

Figure 10.2 Two-level network

Various optimization strategies may be chosen depending on the numbers of inputs, and outputs as well as the Boolean functions to be implemented. An important issue in optimization is the handling of dont care values in the inputs and outputs.

The first case is to, instead of optimizing N_1 and N_2 simultaneously, select N_1 fixed and optimize N_2 only. An important example of this type of logic circuit is a ROM. A ROM characteristic is that all input combinations are valid. The network N_1 is therefore designed as an address decoder and N_2 as a set of NOR gates that select the proper decoder output.

The second case is to choose N_2 fixed and optimize only N_1. If a NOR–NOR structure is chosen for N_1 and N_2, the Boolean function should be written in POS (products-of-sums) form. If, instead, a NAND–NAND structure is chosen, the appropriate form is SOP (sum-of-products). Which of these two forms is the more efficient depends on the properties of the Boolean functions to be implemented.

The third case, where both N_1 and N_2 are optimized, occurs in the implementation of FSM (finite-state machines). Two-level logic functions with many common Boolean terms can be implemented effectively using PLAs (programmable logic arrays). Boolean functions with many don't care input and output values are particularly suitable for PLA implementation. Figure 10.3 shows the structure of a PLA that has an AND and an OR plane that generate the product and sum terms, respectively. Drawbacks of PLAs are low operation speed for large logic networks due to large capacitance in the wiring.

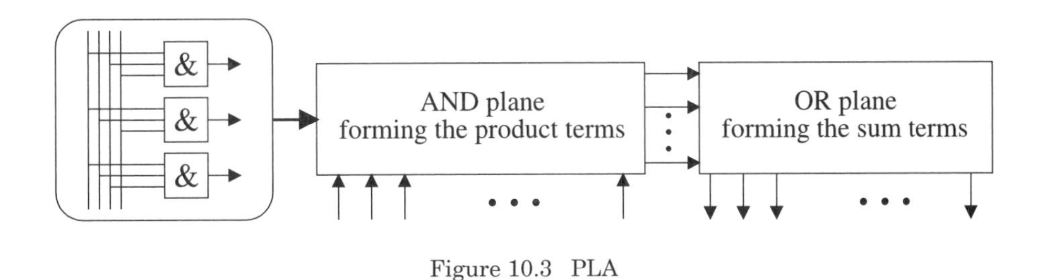

Figure 10.3 PLA

10.3 SEQUENTIAL NETWORKS

Any reasonably complex digital system is based on sequential evaluation of a set of logic expressions. Each expression is evaluated using combinational logic. The expressions must, however, be evaluated in the proper order. Therefore, a control signal, usually referred to as a clock, must be introduced in order to control the operation of the combinational circuitry. Such logic networks are called sequential networks or finite-state machines.

Finite-state machines can be implemented in two generic forms, Mealy's and Moore's, shown in Figures 10.4 and 10.5, respectively. The Mealy form has a direct path from the inputs to the outputs, while the Moore form has a delay. The inputs have an immediate effect on the outputs in the Mealy from. This may cause problems in some cases since the outputs may change asynchronously. The Moore form is therefore preferred when several finite-state machines are cascaded.

A PLA of the type shown in Figure 10.3 can be used to implement an FSM by connecting some of the outputs to the inputs via a register. FSMs are usually generated and optimized by computer programs because of their high computational complexity [19].

Throughput of the FSM is limited by the iteration period bound which may be relatively long for large PLAs, because of large capacitive loads. Large FSMs are therefore often partitioned into several smaller FSMs because of their higher speed. In time-critical applications, more efficient implementations with a lower iteration period bound must be used. In section 10.6 we will discuss methods to derive fast implementations of FSMs. PLA-based finite-state machines

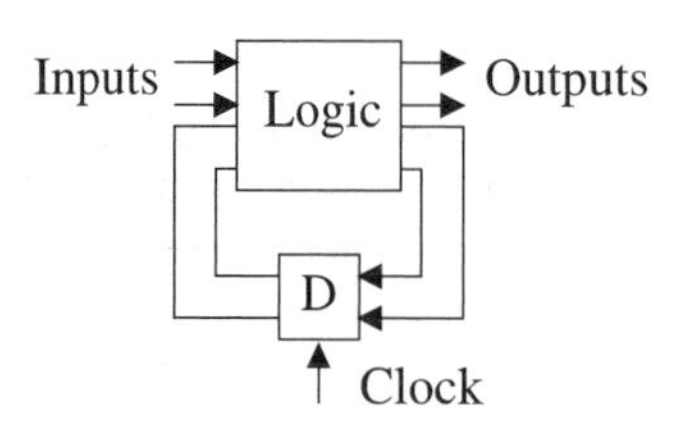

Figure 10.4 Mealys form of FSM.

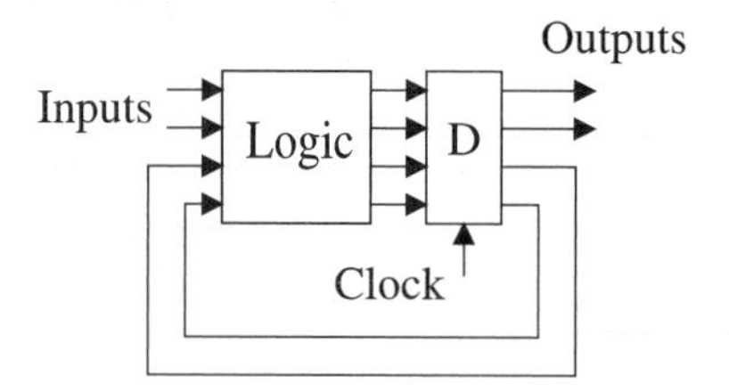

Figure 10.5 Moores form of FSM

are often used to implement noncritical (time) functions. Fast FSMs are generally implemented using so-called random logic in which the critical paths have been minimized.

10.4 STORAGE ELEMENTS

Flip-flop: The song on the other side of a hit record

A storage element is a device that can store a value applied to the input of the element. The storage element is controlled by a clock signal that periodically changes between logic 0 and logic 1. These two states are referred to as clock phases. Depending on their behavior during these phases, storage elements are classed as either latches or flip-flops.

A *latch* holds its value during part of the clock cycle. It is transparent and propagates the input to the output during another part of the clock cycle. In Figure 10.6, a latch and its corresponding timing diagram are shown. The latch in Figure 10.6 is transparent with the input being fed directly to the output during the interval when the clock is high (shaded in the diagram). The input signal latches when the clock goes low and becomes the output until the next time the clock goes high. Of course, latches that perform the latching operation when the clock goes high also exist.

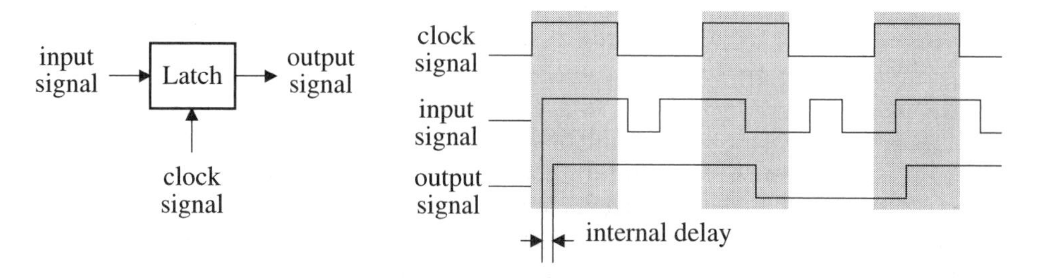

Figure 10.6 Latch with corresponding timing diagram

The other type of storage element, the *flip-flop*, is characterized by the fact that there is never any transparency between input and output;. The input signal is stored into the edge-triggered D flip-flop at positive (or negative) transitions of the clock signal (shaded in Figure 10.7). The input signal to a practical flip-flop must be valid before and during the positive clock edge to ensure proper operation. A small delay between input and output is incurred.

Flip-flops are edge-triggered circuits—that is, the data value is captured and the output(s) may only change in response to a transition at the clock input. Latches are level-sensitive circuits—that is, changes at the data input will be observed at the latch output(s) as long as the clock signal remains at its active level.

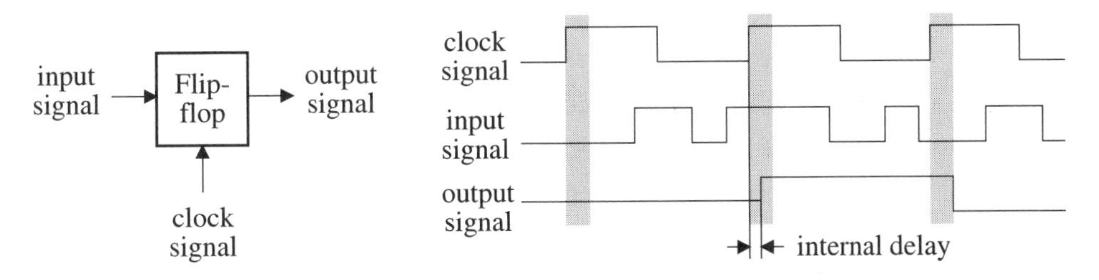

Figure 10.7 Positive edge-triggered D flip-flop with corresponding timing diagram

10.4.1 Static Storage Elements

The state (value) of a memory element is generally stored as a charge in a capacitor. Charge leakage will in practical circuits cause the storage node to degrade and the charge must therefore either be periodically refreshed or retained using positive feedback. Circuits based on the former technique are therefore called *dynamic* memory elements. A *static* memory element, using positive feedback, need not be refreshed, while a *semistatic* memory element is a dynamic memory element in which the charge is automatically refreshed periodically.

The cross-coupled inverters shown in Figure 10.8 can be used as the basic storage element of a static memory cell. The circuit has a stable state in which the input and output are complementary. The inverter driven by the output node is designed to have a low drive capability so that an input signal driving the memory cell can force node V_1 to any desired value, irrespective of the state of the cell. The memory will remain in its stable state if the storage element is left without a driving input. Hence, the circuit does not need refreshing. The memory element can be turned into a latch by putting a switch between the driving source and the element as shown in Figure 10.8. A data value is loaded into the latch by the clock signal that operates the switch. Switches can be implemented as a pair of MOS transistors. However, there is a potential charge-sharing problem with this circuit, which might drive the latch into an undefined state [21].

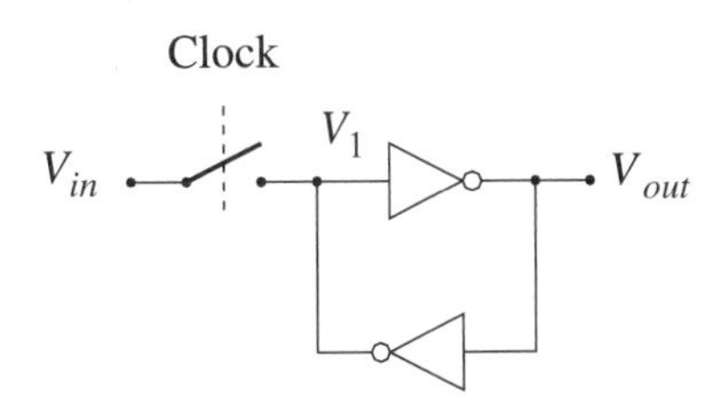

Figure 10.8 Memory element, (inverting latch) consisting of two cross-coupled inverters

An asynchronous SR memory element can be implemented as the cross-coupled NOR gates shown in Figure 10.9, or by cross-coupled NAND gates as shown in Figure 10.10. Notice that the inputs S and R are inverted for the NAND implementation compared to the NOR implementation.

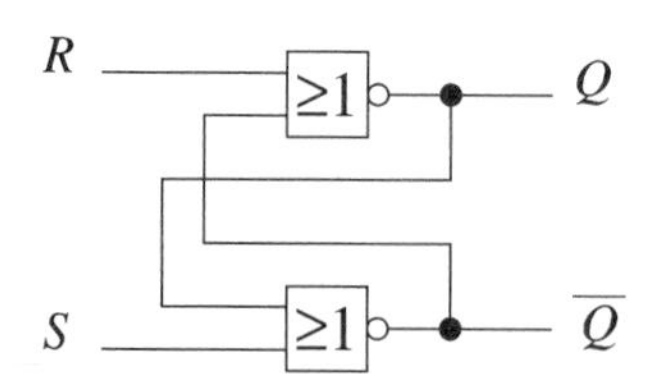

Figure 10.9 Cross-coupled NOR gate implementation of an SR memory element

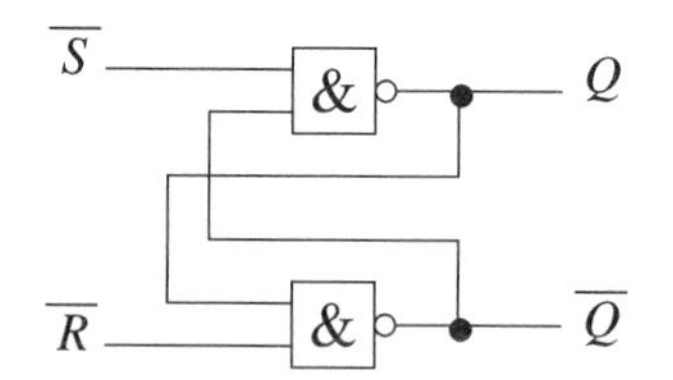

Figure 10.10 Cross-coupled NAND gate implementation of an SR memory element

The state of the memory element is defined by the two complementary outputs Q and $\overline{Q}$. If input $S = 1$ and input $R = 0$, the next state becomes $Q = 1$ and $\overline{Q} = 0$. If $S = 0$ and $R = 1$, then the next state becomes $Q = 0$ and $\overline{Q} = 1$. When both $S = 0$ and $R = 0$, the memory element remains in the state it had when one of the two inputs was high. The input combination $S = R = 1$ is not allowed. The behavior of the SR memory element can be expressed as

$$Q^+ = S + \overline{R}\,Q \tag{10.1}$$

where Q and Q^+ are the current and new state, respectively, and with the design limitation that $R \cdot S = 0$. The operation of the SR memory element is summarized in Table 10.1.

$S(t)$	$R(t)$	$Q(t+1)$	$\overline{Q}(t+1)$
0	0	$Q(t)$	$\overline{Q}(t)$
0	1	0	1
1	0	1	0
1	1	Not allowed	Not allowed

Table 10.1 States of the SR memory element

For the memory element to change its state, inputs S and R must both be stable at a high level for a period of time at least equal to the total delay through the feedback loop in order to propagate the effect of an input change to the output. This time is called the *latching time*.

The SR memory elements can be turned into a clocked SR latch as shown in Figure 10.11. The major difference between this circuit and the clocked dynamic latch, which is discussed in section 10.4.2, is that if both S and R are set to 0, the circuit retains its state indefinitely. The clock Φ acts as an enable signal for the level-sensitive latch. If $\Phi = 0$, the latch holds its state and is unaffected by changes in the two inputs. If $\Phi = 1$, however, the circuit reverts effectively to the asynchronous latch circuit of Figure 10.9, with changes in S and R affecting Q and $\overline{Q}$ directly.

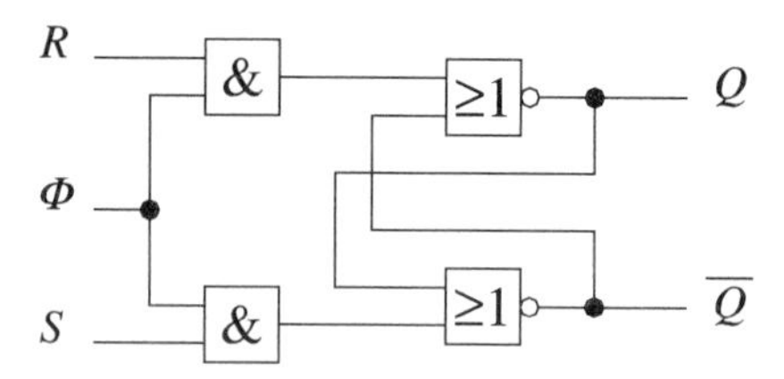

Figure 10.11 Clocked static SR latch

Clocked SR latches are rarely encountered in practice. Instead, clocked JK or D latches are more commonly used. A JK latch behaves in the same way as an SR latch (with $J = S$ and $K = R$) except that the response to $J = K = 1$ is that the latch changes state. This is expressed as

$$Q^+ = J\overline{R} + \overline{K}Q \tag{10.2}$$

The next state of a D latch is the same as the current input. A JK flip-flop can be changed into a D flip-flop by letting $D = J = \overline{K}$.

The edge-triggered flip-flop shown in Figure 10.12 may only change state when the clock signal changes. Changes in the input signal during the clock cycle have no effect on the stored state, which depends solely on conditions prevailing immediately before the active clock edge.

A master–slave flip-flop can be implemented by cascading two latches controlled by two different clock signals. Both the type of latch and the relationship between clock signals vary in different implementations. In its simplest form, two latches are driven by a single clock signal, Φ_2 being derived from Φ_1 by passing it through an inverter. A common variation is to use two independent clock signals, Φ_1 and Φ_2, that are nonoverlapping.

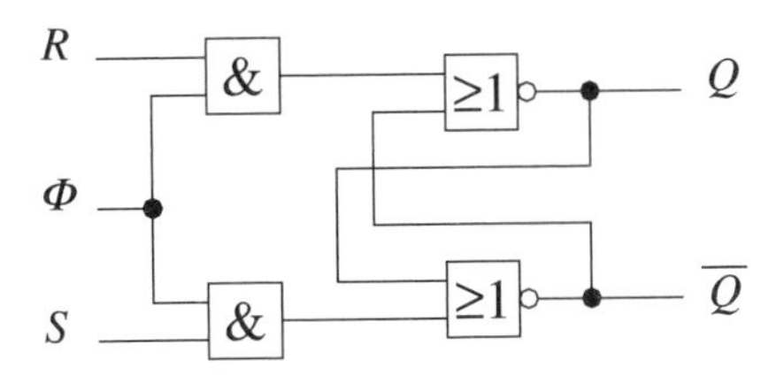

Figure 10.12 Edge-triggered flip-flop.

Inputs to an edge-triggered flip-flop must be stable before the arrival of the clock edge. The required time interval is called the setup time for the inputs, which includes a delay due to possible clock skew.

10.4.2 Dynamic Storage Elements

The input capacitance to a gate built of MOS transistors represents a significant capacitance that can be exploited as a dynamic memory element. Dynamic logic circuits rely on temporary storage of charges in gate capacitances to simplify and speed up the circuits. However, the state (charged or uncharged node) of a dynamic node is retained only for a short time (ms), because of leakage currents. The node must therefore be recharged periodically.

The basic principle of dynamic storage is illustrated in Figure 10.13. A control signal, called a clock signal, controls a switch. When the switch is closed, input signal V_{in} is applied to the input of a logic circuit, in this case an inverter. During this phase, neglecting any delays in the switch and the inverter, output signal V_{out} is equal to the inverse of V_{in}. After the switch has been opened by the control signal, the output signal will remain at the same value, since the input to the inverter will remain the same due to the charge stored in the gate capacitance, C_{gate}.

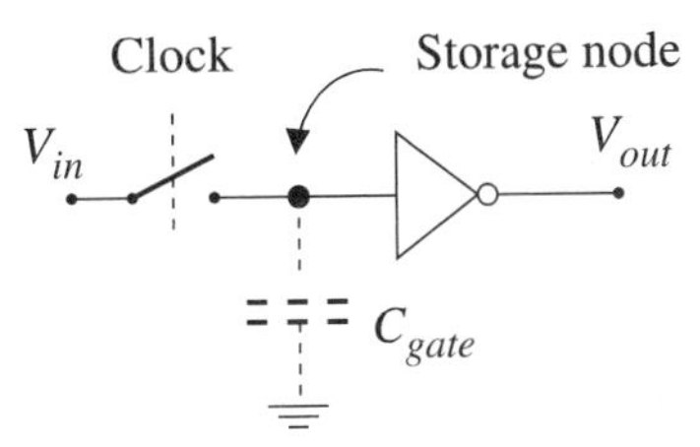

Figure 10.13 Principle of a dynamic storage element

10.4.3 Metastability

A bistable device may enter into what is referred to as the metastable state where the output is undefined—i.e., it is not properly defined in terms of the inputs [3, 9]. Two types of metastability can occur: analog and oscillatory [23]. The former causes the output of the device to stay at an electrical level near the input threshold voltage, while the latter causes the output to toggle repeatedly between the two logic levels. In both cases, the bistable device settles to either valid state, within a resolving time that is a function of the input timing and certain circuit parameters.

This anomalous behavior, called *metastability*, is unavoidable in any type of bistable device, and the resolving time needed for the device to settle to either state is, in principle, unbounded. However, several design measures can be taken to keep undesired effects within certain limits [11]. The main issue is to use circuits that have a high gain-bandwidth product. Degradation of the power supply voltage and increased chip temperature are highly detrimental.

10.5 CLOCKING OF SYNCHRONOUS SYSTEMS

Digital (logic) systems are in principle built using combinational and sequential networks. The inputs and outputs of the networks are interconnected via storage elements. These storage elements carry information from one evaluation phase to the next phase, in a similar way to delay elements in digital filter algorithms. Systems in which the information flow is fully controlled by the clock signals are called *synchronous systems*. The storage elements can be implemented as latches or flip-flops, as just discussed. Several different clocking schemes have been developed to control the information flow [18, 25]. The most common clocking schemes will be discussed next.

10.5.1 Single-Phase Clock

By using storage elements (SEs) controlled by one or more clock signals, a variety of clocking schemes can be obtained. A straightforward clocking scheme is the single-phase clocking illustrated in Figure 10.14. For simplicity, the lines in the figures may represent one or several signals. Obviously, this prevents the circuit from working properly if the storage elements are implemented with latches that become transparent during a part of the clock cycle. If the transparent phase is shorter than the internal delay in the combinational logic, the circuit may work, but this is a very risky situation and should be avoided. Notice that the basic idea with digital circuits is that they should be robust and insensitive to circuit and timing variations.

Signal delays incurred in the logic circuitry may create race problems. Another source of problems results from delay difference incurred by clock signals that have propagated through different paths. This type of delay is called *clock skew*. Both of these problems will ultimately limit the maximum clock frequency. Race problems can be reduced by using circuits with short critical paths between the clocked elements. This can be accomplished by introducing pipelining at the circuit level. The clock skew problem can be reduced by using proper clocking strategies. The choice of logic style is also important to minimize the effect of clock skew.

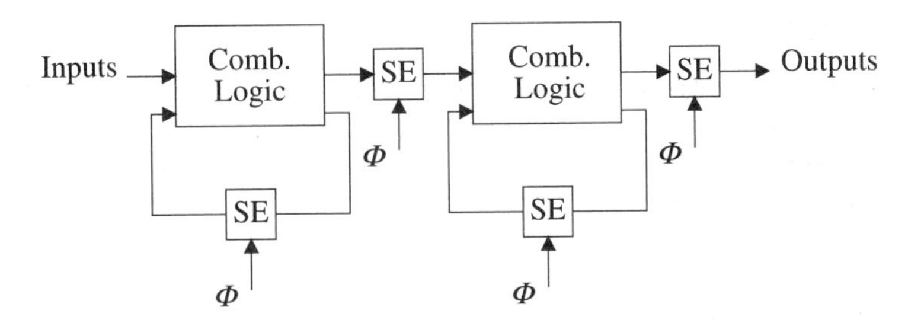

Figure 10.14 Logic circuit with single-phase clocking

Edge-triggered D flip-flops can be used to obtain working logic systems by cascading several single-phase clocked blocks, as shown in Figure 10.14. However, the latches separating the blocks may not be transparent in the same clock phase, since the logic signals may not flow directly through the blocks. Two successive latches must therefore be operated so that the transparent phases do not overlap. This is done by letting every other dynamic circuit be controlled by the inverse of the clock signal. The clock signals Φ and are shown in Figure 10.15.

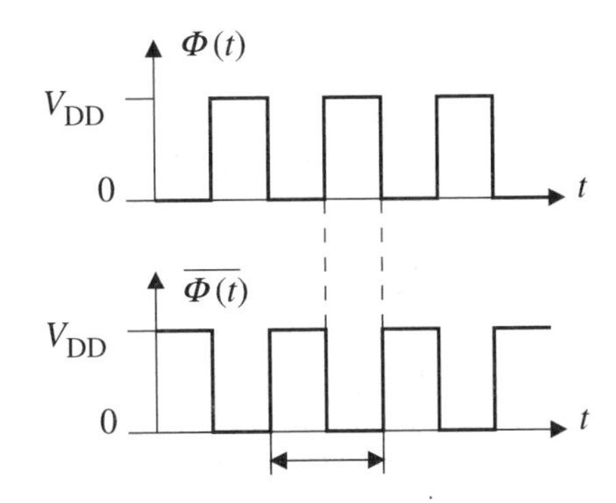

Figure 10.15 Single-phase clock, $\Phi(t)$ and its inverse

10.5.2 Single-Phase Logic

Single-phase logic is a high-speed CMOS circuit style with the with the advantage that only one clock signal is needed. Thus, the inverse of the clock signal is not required. The principle of single-phase logic is illustrated in Figure 10.16.

The circuit must have alternating n- and p-logic blocks. Outputs of n-blocks are used as inputs to p-blocks, and vice versa. A single clock signal, Φ, is used for the entire circuit. The clock signal controls circuit operation as follows:

The clock is low, $\Phi = 0$.

n-block: The precharge node P_n in the n-block is precharged to 1. This ensures that transistors n_2 and p_2 that enclose node F_n are both turned off. The second stage of the n-block will therefore function as a dynamic memory and the value of F_n will be stored in its stray capacitance.

p-block: The precharge node P_p was in the previous clock phase ($\Phi = 1$) precharged to 0. Now, when the clock goes low, transistor p_3 will turn on and the logic function of the p-block will be evaluated by the pMOS network. If the p-network conducts, the node P_p will be charged to 1, otherwise it will remain at 0. The p-block will evaluate correctly, since transistors p_2 and n_2 guarantee a stable output value from the n-block during $\Phi = 0$, and transistor p_4 will

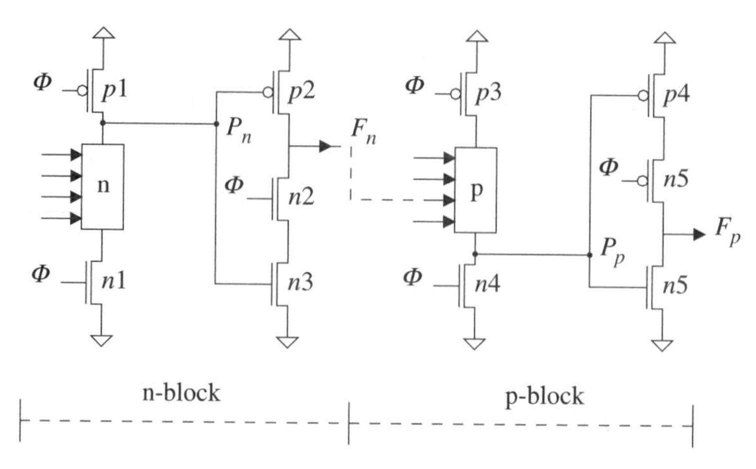

Figure 10.16 Single-phase logic

force the evaluation to take place only when $\Phi = 0$. Since $\Phi = 0$, p_5 will be on and the second stage of the p-block will function as an inverter. Hence, the inverted value of P_p will appear at F_p.

The clock is high, $\Phi = 1$.

n-block: Transistor n_1 switches on and the n-block will evaluate its logic function. The resulting value of node P_n will become low if the n-network conducts, otherwise it will remain high. The second stage of the n-block will now function as an inverter so that the inverted value of P_n will appear at F_n.

p-block: P_p is precharged to 0 in this phase, and the value at node F_p will be stored dynamically in its stray capacitance.

The clock signal and possible signal transitions are shown in Figure 10.17. Obviously, the p- and n-blocks evaluate on positive-and negative-going clocks, respectively. The output from one type of block is stable while the other type evaluates.

This circuit technique leads to high speed since logic evaluation takes place at every clock transition. However, problems may arise because of charge redistribution. This may occur if the capacitive load of a block is altered during the precharge phase—e.g., if the output is connected to a transmission gate that opens. If this happens, the total charge is redistributed among the capacitances and the signal level of the output is changed. To avoid this problem, such connections should only be made during the evaluation phase of the block.

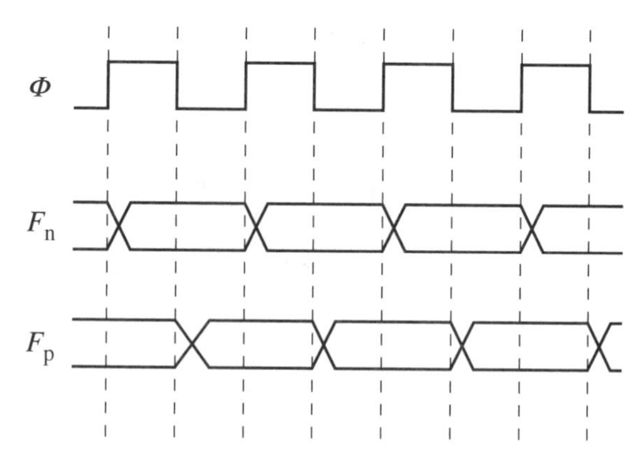

Figure 10.17 Signals in single-phase logic

Another disadvantage is the voltage variations that may occur at the output of a block due to evaluation time in the block. This results in peaks in the output voltage of a p-block, when it is low for two consecutive clock periods, on a negative clock transition. For an n-block the opposite situation occurs with voltage drops on positive clock transitions when the output is high. Neither of these cases cause problems in single-phase logic, but connections to other kinds of logic have to be made with caution.

An advantage is that only one clock wire needs to be distributed over the logic circuitry, resulting in fewer clock drivers and no clock skew. However, the requirement on sharp clock edges is strict.

A dynamic shift register using a single-phase clock is shown in Figure 10.18. When $\Phi = 0$, the first inverter is in its precharge mode and the second inverter is off. When $\Phi = 1$, the first inverter is in its evaluate mode and the second inverter is turned on. The third inverter is in its precharge mode and the fourth inverter is off. When $\Phi = 0$ the next time, the third inverter is in its evaluate mode and the fourth inverter is on. At the end of the clock period, the output becomes equal to the input.

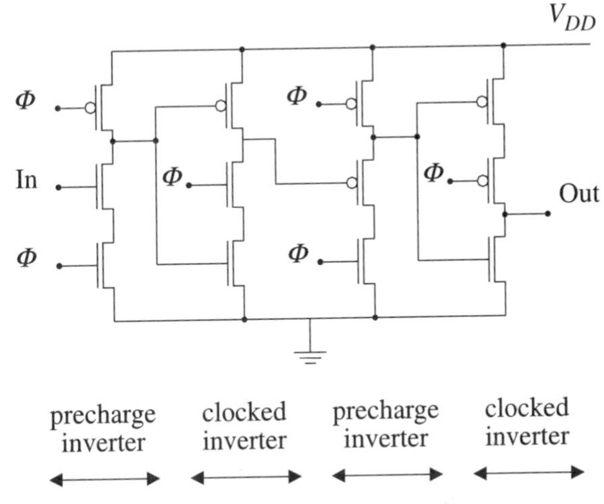

Figure 10.18 CMOS shift register; using a single-phase clock

10.5.3 Two-Phase Clock

The logic circuit shown in Figure 10.14 may still malfunction if the clock signals, Φ and $\overline{\Phi}$, are skewed. If the clock skew is large enough, both Φ and $\overline{\Phi}$ are high for a short time, and the signal may propagate through two adjacent logic blocks. To prevent this happening, a nonoverlapped two-phase clock, Φ_1 and Φ_2, can be used. The nonoverlapping parts of the clock interval allow for some skew of Φ_1 and Φ_2. In Figure 10.19, the two-phase clock, Φ_1 and Φ_2, is shown.

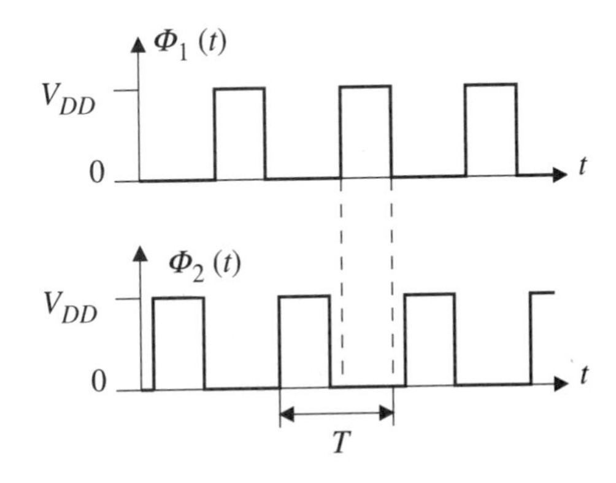

Figure 10.19 Two-phase non-overlapping clock, $\Phi_1(t)$ and $\Phi_2(t)$

The switches needed in a dynamic latch can be implemented by using either pass-transistors or transmission gates. A dynamic shift register can easily be implemented by alternately cascading transmission gates (TG) and inverters. A shift register stage consists of two cascaded pairs of TG-inverters. Figure 10.20 shows part of a shift register controlled by the two clock signals Φ_1 and Φ_2. The operation of the shift register during one full clock cycle is illustrated in Figure 10.21.

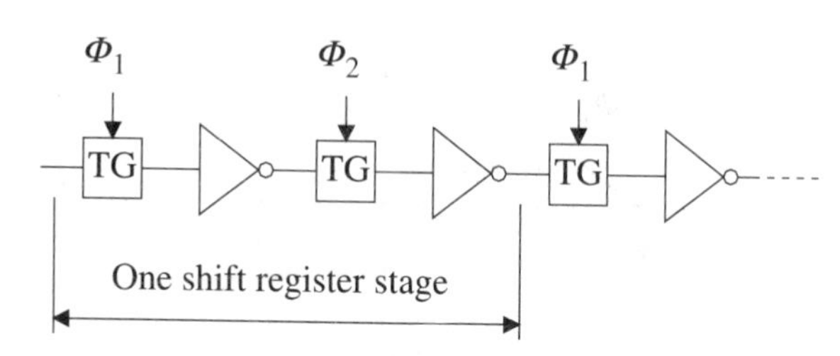

Figure 10.20 One and a half stages of a dynamic shift register

The initial state is $\Phi_1 = 1$ and a logic signal x is applied at the input. The status of the shift register is illustrated for more than one full clock period. Notice the nonoverlapping intervals where $\Phi_1 = \Phi_2 = 0$.

This basic mechanism for data transportation is essentially valid for all dynamic clocking strategies. Logic can be incorporated into this scheme by using combinational logic circuits instead of inverters.

Dynamic logic occupies significantly less chip area compared to static logic that uses static storage elements for the registers. However, dynamic circuits must ensure that the capacitance at the storage node is sufficiently large.

Two-Phase Clock With Dual Latches

The transparency problem with latches can be circumvented by using two latches in each storage node, as shown in Figure 10.22. Each pair of latches is controlled by a two-phase nonoverlapping clock.

The nonoverlapping clock guarantees that each pair of latches can not be transparent simultaneously. However, to ensure correct operation, the skew between the clock signals must be carefully designed. This circuit works in a way similar to the previously described circuit. This type of circuit is referred to as master–slave.

Two-Phase Clock With Distributed Latches

All computation done by the circuit shown in Figure 10.22 is performed during one clock phase. This clocking scheme is used often even though the circuit is asymmetric and unbalanced. However, this drawback can easily be avoided by distributing the combinational logic between both clock phases, as shown in Figure 10.23.

Latches are simple to implement in CMOS as compared to edge-triggered flip-flops. Dynamic latches are implemented by using stray capacitances. Clocking techniques using latches are therefore preferred.

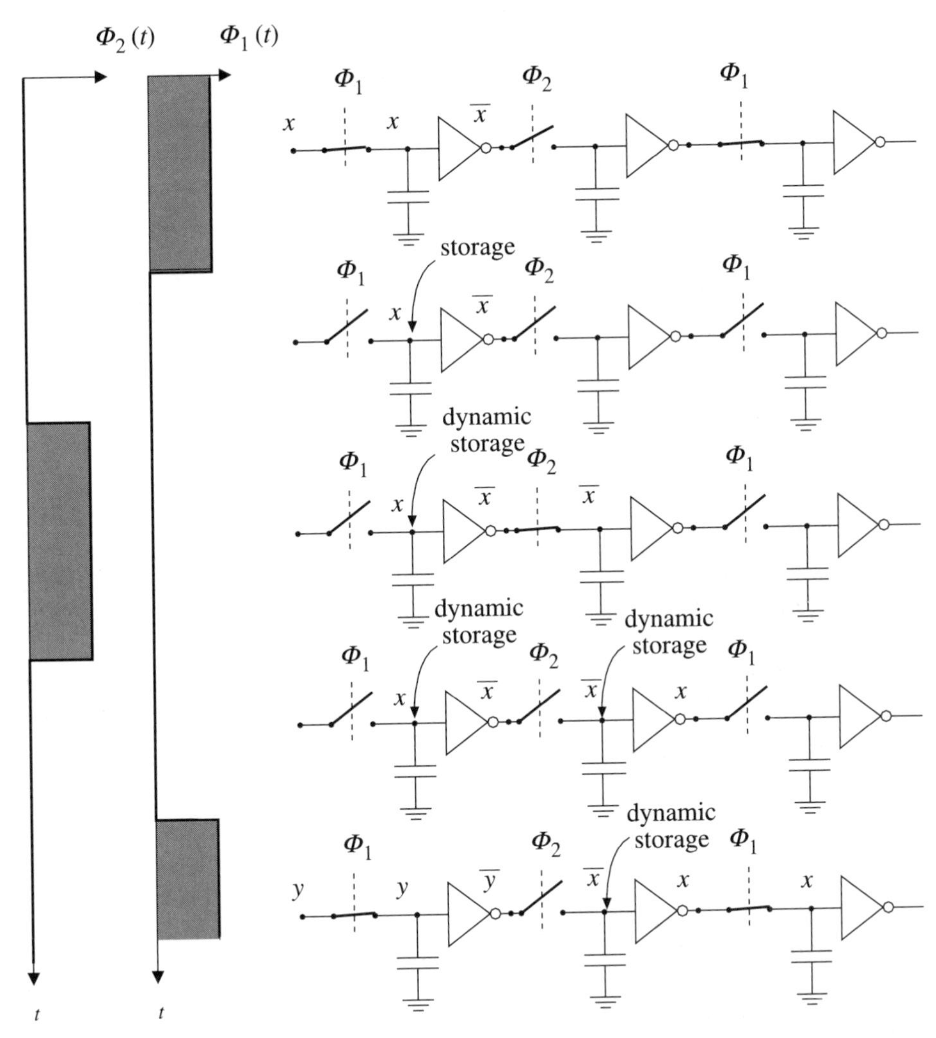

Figure 10.21 Signal transitions in a dynamic shift register during one clock phase

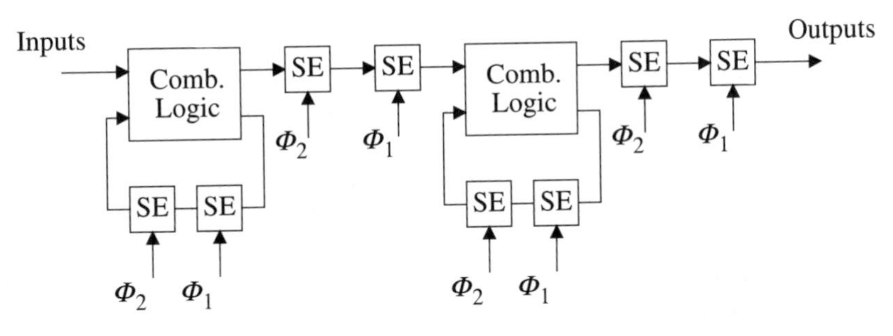

Figure 10.22 Two-phase clock structure using dual latches

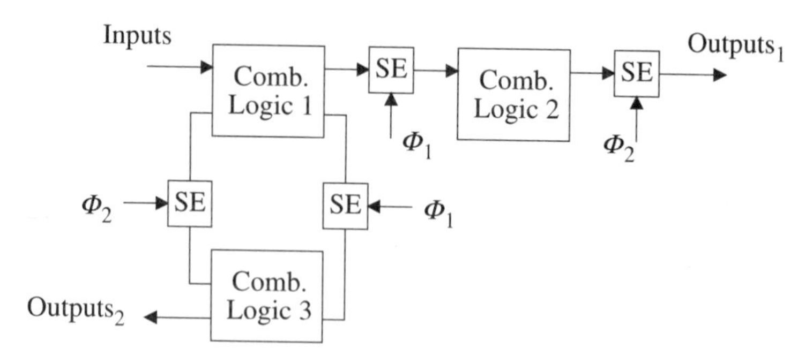

Figure 10.23 Two-phase clock structure using distributed latches

10.5.4 Clock Skew

If more than one clock signal is used in the system, problems can arise because of the different clock signal delays. These originate partly from internal delays in the clocking circuits and drivers and partly from the clock wires on the chip. Clock signals are global signals that result in long wires. Clock skew and delay are limiting factors in high-speed logic circuits and become more severe as chip size increases.

Clock skew is the maximum time difference between the active clock edges for any two clocked elements. Clock skew is viewed as an uncertainty on the position of the active clock edge, and as the skew increases larger margins must be allocated for the setup and hold times of the clocked registers. Clock delay is the propagation delay from the system clock generator to the clocked element. Chip-to-chip interfaces must take clock delay into account. A large fraction of the cycle time may be wasted when two or more chips have a clock delay imbalance.

10.6 ASYNCHRONOUS SYSTEMS

In synchronous system design, the sequences of data and time are associated to each other by a global clock signal. As a consequence, a fixed time is used for each computation. In asynchronous systems computations start whenever preceding computations are finished. Hence, the data signals themselves are used to control the data flow. A handshaking protocol with request and acknowledge signals is used for local communication [12, 17, 18], while unacknowledged communication is used between chips. Asynchronous systems can achieve speeds comparable with their synchronous counterparts, and they are potentially more reliable and easier to design [2, 10, 17, 19, 21].

There are two aspects of a system that can be either asynchronous or synchronous: the computational steps and communication. A collection of subsystems within a system is synchronous with respect to computations if computation step n cannot occur unless step $n1$ has been completed in all of the subsystems.

A collection of subsystems is synchronous with respect to communication if two processes, executed in two different subsystems, are required to rendezvous—i.e., wait for each other and communicate through a memoryless channel. A synchronous system is one in which transfer of information between combinational blocks is performed in synchrony with a global clock signal.

In an asynchronous system the subsystems are not required to wait on each other, and the transfer of information between combinational blocks is not performed in synchrony with a global clock signal, but rather at times is determined by the latencies of the blocks themselves. In the asynchronous scheme, processing time approaches an average for all tasks rather than being set by a clock period in the worst-case completion time of all tasks.

A self-timed circuit is a circuit that performs computational steps whose initiation is caused by signal evens at its inputs and whose completion is indicated by signal events at its outputs. Self-timed systems, (a subclass of asynchronous systems) are legal interconnections of self-timed circuits.

A circuit is *delay-insensitive* when its correct operation is independent of any assumptions on delays of operators and wires except that delays are finite [22]. Such circuits do not use a clock signal or knowledge about delays. It has been proved in [16] that the class of delay-insensitive circuits is very limited. Different asynchronous techniques distinguish themselves in the choice of the compromises to delay-insensitivity.

A *speed-independent system*, with zero-delay wires, implements a logic behavior independent of delay in the logic modules, but there are no delays in wires. Self-timed techniques assume that a circuit can be decomposed into equipotential regions inside which wire delays are negligible [24].

Communication between two asynchronous systems is subject to a communication protocol, as illustrated in Figure 10.24. Four-phase signaling is one such protocol that will be discussed later.

Self-timed circuits are a means of synchronizing different blocks at a local level, thereby alleviating problems associated with distributing global clock signals over the entire face of the chip. Most designers agree that asynchronous transfer of information is required at the higher levels in a complex system, but may disagree on how small the appropriate block size should be. However, it seems unreasonable to use blocks consisting of a single gate, due to the overhead associated with asynchronous transfers. As the speed of the technology has increased, block size has shrunk. Self-timed circuits, in addition to performing some logic function, provide a completion signal to indicate when the computation is finished.

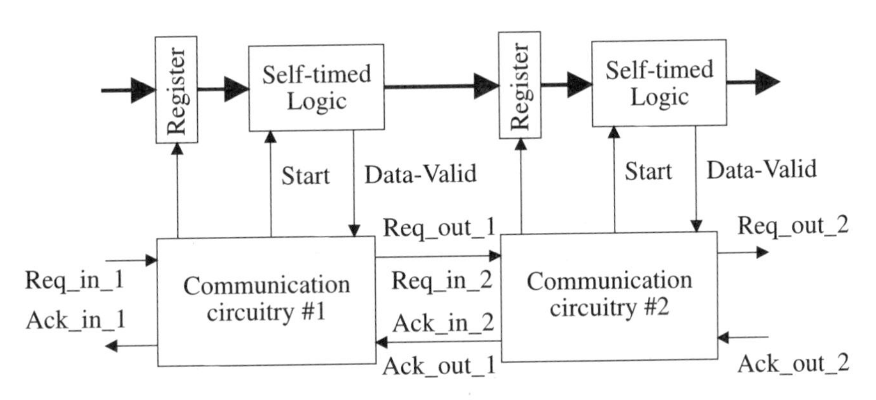

Figure 10.24 Communication between two asynchronous systems

For an asynchronous combinational circuit to operate properly, it must be free of hazards—i.e., a signal should not make temporary transitions when it is required to be stable, or change more than once when it is required to change only once. Figure 10.25 shows a basic building block, the so-called Müller C element, often used in asynchronous systems. A Müller C element has the property that the output will not change until all inputs have changed.

A problem often associated with asynchronous systems is metastability. Asynchronous communication or computation is not the cause of metastability problem, it is rather the architecture in which the circuits are employed in. Hence, the metastability problem can in practice be reduced at the architectural level.

Figure 10.26 shows a possible logic implementation of a four-phase signaling protocol and Figure 10.27 shows the corresponding state graph.

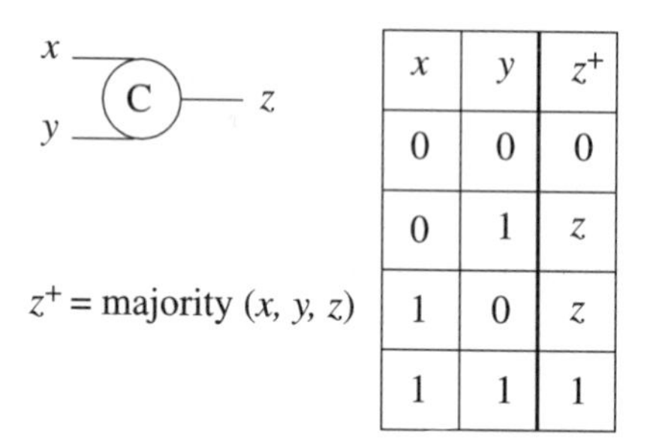

x	y	z^+
0	0	0
0	1	z
1	0	z
1	1	1

Figure 10.25 Müller C element

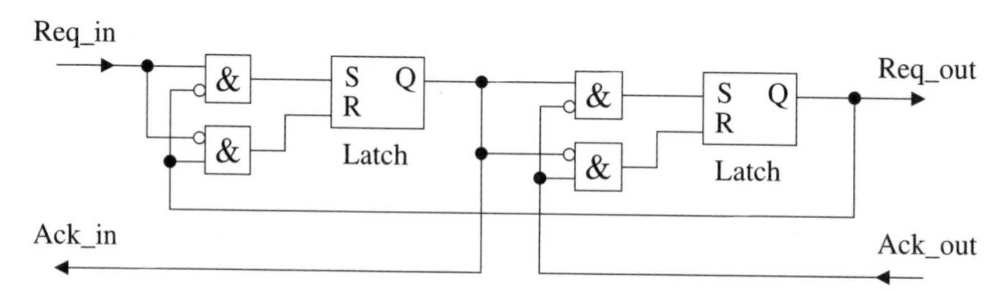

Figure 10.26 Implementation of the four-phase signaling protocol

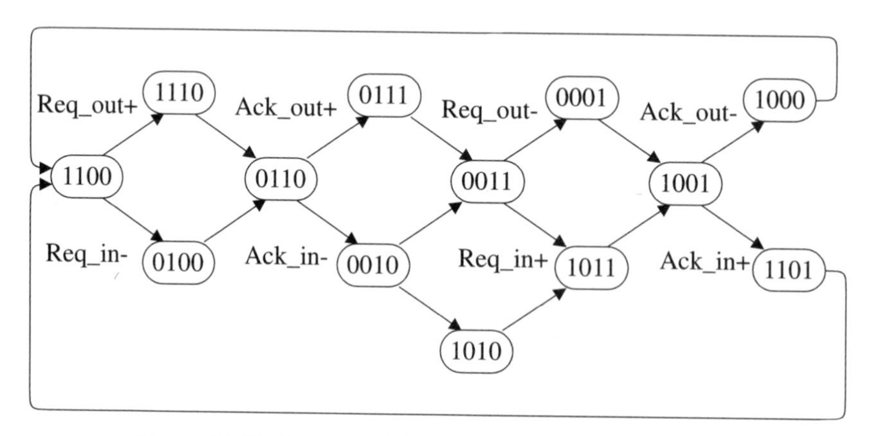

Figure 10.27 State graph for four-phase signaling protocol

The design of such protocols and circuits is beyond the scope of this book [10, 12, 17, 19–21]. The signals marked + and – indicate transitions low-to-high and high-to-low, respectively. Figure 10.28 shows how the four-phase protocol can be implemented by using Müller C elements to obtain a self-timed processing element.

The self-timed PE is provided with a local ring counter that generates a clock signal guaranteed to be lower than the speed of the PE. This can be accomplished, in spite of speed variations between individual chips due to process variations, if the ring counter is placed close to the PE. The ring counter is started by the signal *Start* and stopped after a given number of clock cycles. In this way, the circuit design problem is simplified significantly. The circuit has only to be sufficiently fast.

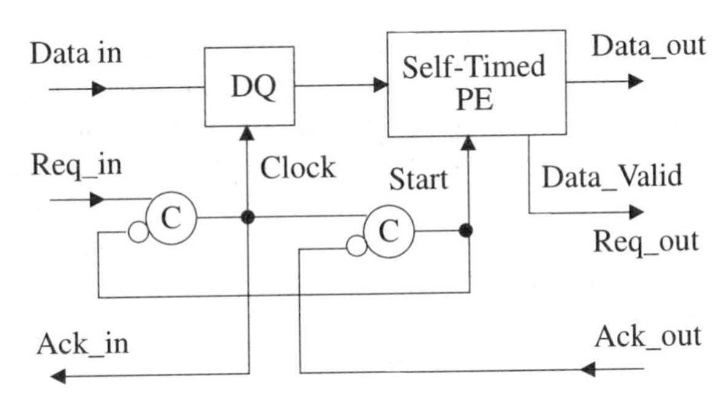

Figure 10.28 Self-timed processing element.

10.7 FINITE STATE MACHINES (FSMs)

Finite state machines (FSMs) are used in many digital systems. An FSM can be described by

$$s(n+1) = T(s(n), x(n)) \tag{10.4}$$

$$y(n) = U(s(n), x(n)) \tag{10.5}$$

where $s(n)$, $x(n)$, and $y(n)$ are the state, input, and output at time instant n, while T and U are the state transition function and output function, respectively.

An FSM has an infinite memory span if the present state is determined by an infinite number of past states. Compare this with an IIR filter. The throughput for such an FSM is therefore limited by the iteration period bound, obtained from Equation (10.4). Techniques similar to the ones discussed in Chapter 6 have also been developed for decreasing the iteration period bound for this class of FSM [14].

An FSM has a finite memory span if the current state is determined by a finite number of past states. Hence, such FSMs have no iteration period bounds, and can therefore achieve arbitrarily large throughputs by using pipelining and interleaving in the same way as nonrecursive FIR filters. We will not therefore discuss this class of FSM further.

There are two basic ways to increase the throughput of a recursive algorithm. The first way is to reduce the computation time for the critical loop—for example, by removing unnecessary operations from the critical loop. This approach is obviously application dependent and relaxes only the iteration period bound. The second way is to use various forms of look-ahead and block processing techniques, as was done in section 6.9.

10.7.1 Look-Ahead FSMs

The basic principle of look-ahead techniques is to iterate Equations (10.4) and (10.5) N times so that N inputs are used to compute N output values in each iteration step. This approach effectively increases the number of delay elements in the critical loop.

The throughput is therefore increased if the recursive calculation in the modified FSM can be performed faster than N times the time required by the original FSM.

An FSM can be described by a state vector $\mathbb{S}$ that has an element $s_i(n)$ for each state in the FSM. The element $s_i(n) = 1$ if, and only if, the FSM is in state i at time instant n. The next state is computed by multiplying the binary state transition matrix $\mathbb{T}(n)$ by the current state. For simplicity we neglect possible input signals. We get

$$\mathbb{S}(n+1) = \mathbb{T}(n) \cdot \mathbb{S}(n) \tag{10.6}$$

The scalar multiplications and additions are defined as logic AND and OR, respectively. Iterating N steps forward we get

$$\mathbb{S}(n+N+1) = \mathbb{D}(n+N) \cdot \mathbb{S}(n) \tag{10.7}$$

where

$$\mathbb{D}(n+N) = \mathbb{T}(n+N1) \cdot \mathbb{T}(n+N2) \cdots \mathbb{T}(n) \tag{10.8}$$

is the new state transition matrix. Now, since the matrix $\mathbb{D}(n+N)$ can be computed outside the recursive loop, the iteration period bound has been reduced by a factor up to N.

Figure 10.29 illustrates the new faster algorithm where a modified output function is used to compute the output values from a block of input values and the current state. The output function does not belong to the critical loop of the FSM. Both the precomputation and postcomputation parts can therefore be pipelined to any degree since they are outside the recursive loop.

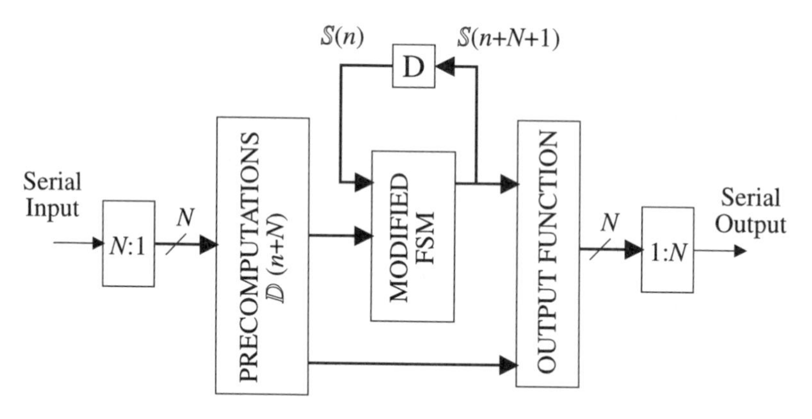

Figure 10.29 Structure of the look-ahead FSM

The hardware complexity of the look-ahead approach depends on the state transition function, which is often nonlinear and does not have a simple closed form.

A simple, but expensive, way to implement the look-ahead FSM is to use a look-up table. The required amount of resources increases exponentially with the throughput improvement. This limits the method to cases where the number of states is moderate.

Other, more efficient, methods are the matrix multiplier and the linearized look-ahead method [14], which differ in the way the computation of $\mathbb{D}(n+N)$ is implemented.

EXAMPLE 10.1

Use the look-ahead method to double the throughput of the bit-serial adder shown in Figure 10.30. The addition begins with addition of the LSB (least-significant bit) of binary inputs a, b, and *carryin* which is obtained from the previous addition step. Initially, the D flip-flop is reset to 0. Bit-serial adders are discussed in detail in Chapter 11.

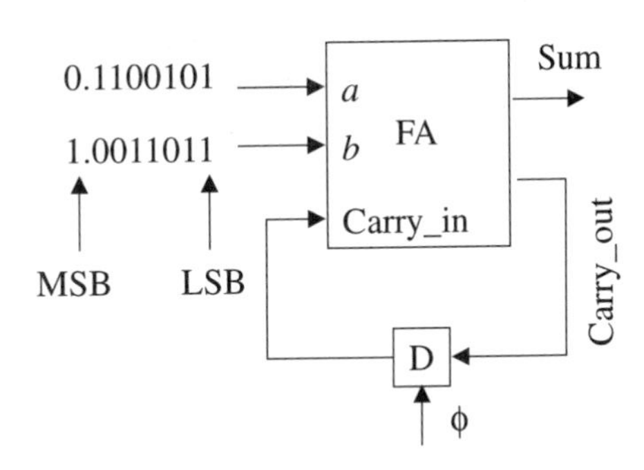

Figure 10.30 Bit-serial adder

The bit-serial adder is described by

$$Sum = a \oplus b \oplus c$$

$$Carry\text{-}out = a \cdot b + a \cdot c + b \cdot c$$

The state transition diagram is shown in Figure 10.31. The two states are s_0: carry = 0 and s_1: carry = 1. The matrix representation is

$$\mathbb{T}(n) = \begin{pmatrix} \overline{a(n) \cdot b(n)}, \bar{a}(n) \cdot \bar{b}(n) \\ a(n) \cdot b(n), a(n) + b(n) \end{pmatrix}$$

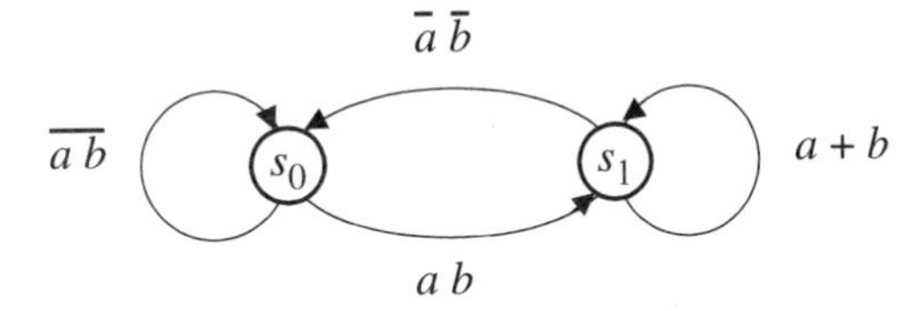

Figure 10.31 State transition diagram for a bit-serial adder

where $a(n)$ and $b(n)$ are the input bits at time n. To double the throughput an expansion is made. The expanded matrix multiplication gives:

$$\mathbb{D}(n+2) = \mathbb{T}(n+1) \cdot \mathbb{T}(n)$$

$$= \begin{pmatrix} \overline{a(n+1) \cdot b(n+1)}, \bar{a}(n+1) \cdot \bar{b}(n+1) \\ a(n+1) \cdot b(n+1), a(n+1) \cdot b(n+1) \end{pmatrix} \cdot \begin{pmatrix} \overline{a(n) \cdot b(n)}, \bar{a}(n) \cdot \bar{b}(n) \\ a(n) \cdot b(n), a(n) \cdot b(n) \end{pmatrix}$$

$$= \begin{pmatrix} d_{11}, d_{12} \\ d_{21}, d_{22} \end{pmatrix}$$

where

$$d_{11} = \overline{a(n+1) \cdot b(n+1)} \cdot \overline{a(n) \cdot b(n)} + \bar{a}(n+1) \cdot \bar{b}(n+1) \cdot a(n) \cdot b(n)$$

$$d_{12} = \overline{a(n+1) \cdot b(n+1)} \cdot \bar{a}(n) \cdot \bar{b}(n) + \bar{a}(n+1) \cdot \bar{b}(n+1) \cdot [a(n) + b(n)]$$

$$d_{21} = a(n+1) \cdot b(n+1) \cdot \overline{a(n) \cdot b(n)} + [a(n+1) + b(n+1)] \cdot a(n) \cdot b(n)$$

$$d_{22} = a(n+1) \cdot b(n+1) \cdot \bar{a}(n) \cdot \bar{b}(n) + [a(n+1) + b(n+1)] \cdot [a(n) + b(n)]$$

Matrix $\mathbb{D}(n+2)$ can be calculated outside the critical loop of the FSM. The critical loop consists only of two levels of gates to compute the new state. We assign a variable c to store the two states s_0 and s_1. This gives $\mathbb{S}(n) = (\bar{c}(n), c(n))^T$.

By multiplying by $\mathbb{D}(n+2)$, we get two equations: one for $c(n)$ and one for $\bar{c}(n)$. These equations are complementary—i.e., one and only one equation evaluates to true. The equation for $c(n)$ is the same as the equation for the carry term of a 2-bit adder. This method is equivalent to using the periodic scheduling formulation over two sample periods discussed in section 7.5.4.

10.7.2 Concurrent Block Processing

Another method with lower hardware complexity is based on concurrent calculation of blocks. By buffering and dividing the input into concurrently computed blocks, the throughput can be increased. Each block computation must, however, be made independent of the others. There are two ways to achieve this independence. Consider the state trellis diagram in Figure 10.32.

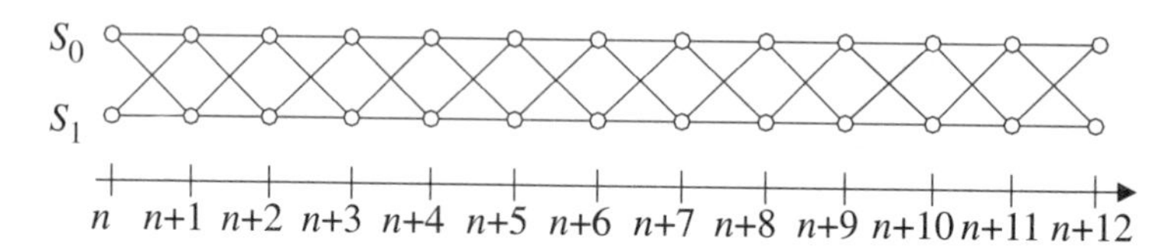

Figure 10.32 State trellis diagram

Block precomputation, as illustrated in Figure 10.33, uses look-ahead techniques to calculate the initial state for each block before the output value is calculated. Concurrent calculations using the original FSM algorithm can then be performed in each separate block.

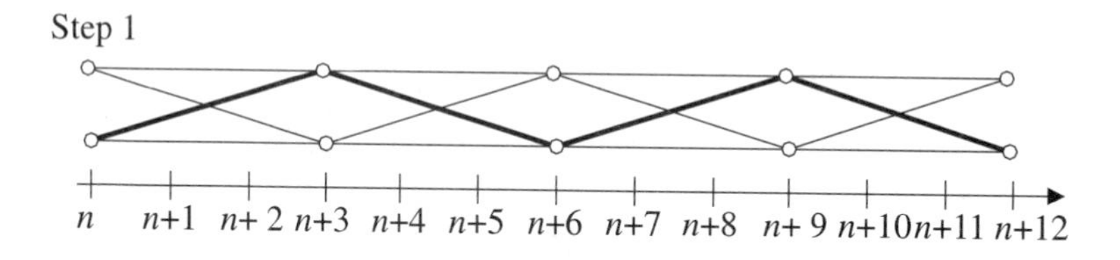

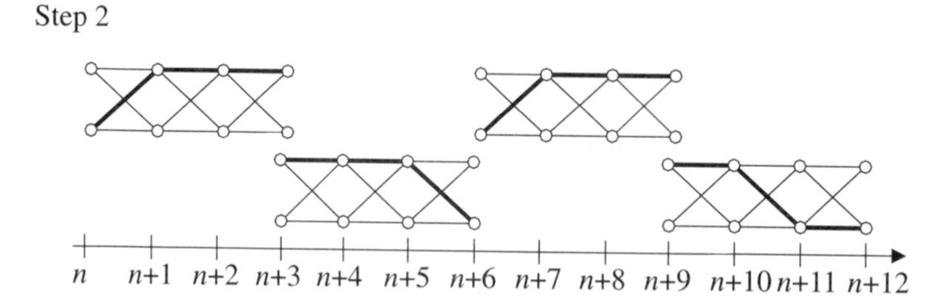

Figure 10.33 Block precomputation

Block postcomputation, as illustrated in Figure 10.34, is useful if the FSM has a moderate state size. For every possible initial state the resulting output sequence is calculated. This is done concurrently in every block. Only the starting and ending states are then considered in step 2. The next step is to use the initial state. The idea is to calculate the state sequence of every initial state.

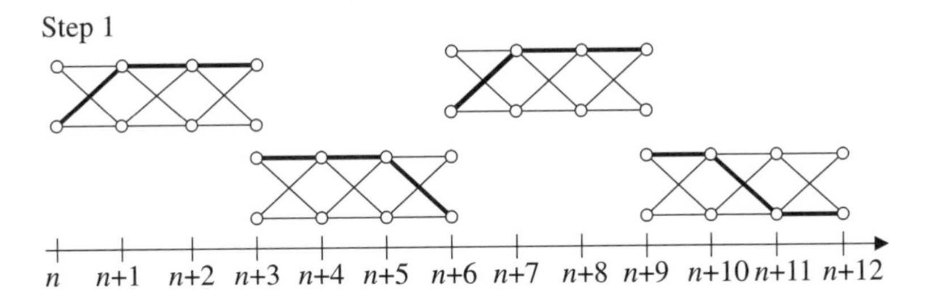

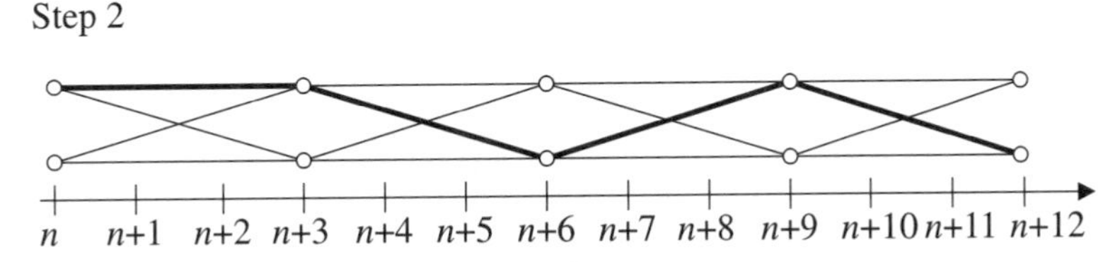

Figure 10.34 Block postcomputation

EXAMPLE 10.2

We use the bit-serial adder to illustrate block postcomputation. The state trellis of the adder is the state diagram expanded over time, as illustrated in Figure 10.35.

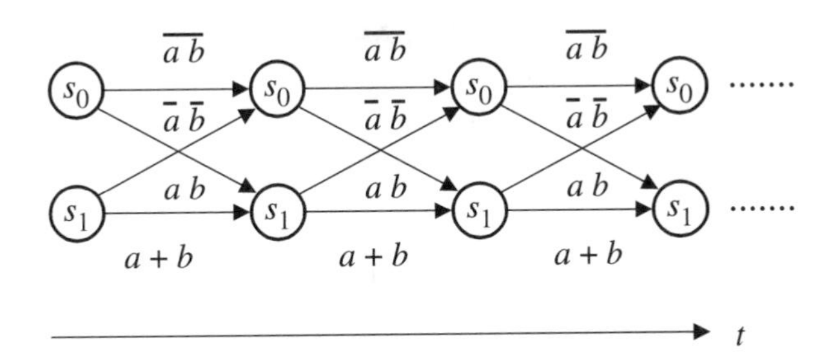

Figure 10.35 State trellis for the carry in a bit-serial adder

The input sequence is divided into blocks, as illustrated in Figure 10.36. In each block the path matches the corresponding input for each possible initial state found (layer 1). The inputs are discarded and each initial state of a block is paired with an ending state and an associated output subsequence. The path traversal at layer 2 can be considered as a new FSM. It can be traversed sequentially, given the initial state. Each branch in layer 2 then corresponds, in this example, to a path of length four.

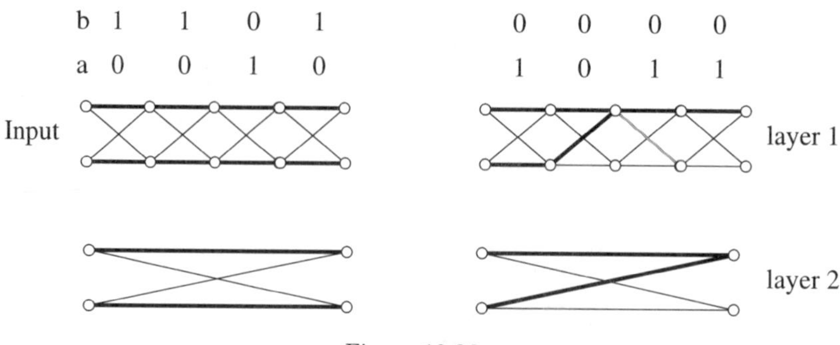

Figure 10.36

References

[1] Afghahi M. and Svensson C.: A Unified Single-Phase Clocking Scheme for VLSI Systems, *IEEE J. Solid-State Circuits,* Vol. SC-25, No. 1, pp. 225–233, Feb. 1990.

[2] Afghahi M. and Svensson C.: Performance of Synchronous and Asynchronous Schemes for VLSI Systems, *IEEE Trans. on Computers,* Vol. C-41, No. 7, pp. 858–872, July 1992.

[3] Anceau F.: *The Architecture of Microprocessors,* Addison-Wesley, Wokingham, England, 1986.

[4] Bakoglu H.B.: *Circuits, Interconnections, and Packaging for VLSI,* Addison-Wesley, Reading, MA, 1990.

[5] Brayton R., Hachtel G., McMullen C., and Sangiovanni-Vincentelli A.: *Logic Minimization Algorithms for VLSI Synthesis,* Kluwer Academic Pub., 1984.

[6] Day P. and Woods J.V.: Investigation into Micropipeline Latch Design Styles, *IEEE Trans. on VLSI Systems,* Vol. 3, No. 2, pp. 264–272, June 1995.

[7] Glasser L. and Dobberpuhl D.W.: *The Design and Analysis of VLSI Circuits,* Addison-Wesley, Reading, MA 1985.

[8] Hartly R.I. and Parhi K.K.: *Digit-Serial Computation,* Kluwer Academic Pub., Boston, 1995.

[9] Horstmann J.U., Eichel H.W., and Coates R.L.: Metastability Behavior of CMOS ASIC Flip-Flops in Theory and Test, *IEEE J. Solid-State Circuits,* Vol. SC-24, No. 1, pp. 146–157, Feb. 1989.

[10] Jacobs G.M. and Brodersen R.W.: A Fully Asynchronous Digital Signal Processor Using Self-Timed Circuits, *IEEE J. on Solid-State Circuits,* Vol. SC-25, No. 6, pp. 1526–1537, Dec. 1990.

[11] Kim L.S. and Dutton R.W.: Metastability of CMOS Latch/Flip-Flop, *IEEE J. on Solid-State Circuits,* Vol. SC-25, No. 4, pp. 942–951, Aug. 1990.

[12] Lau C.H., Renshaw D., and Mavor J.: Data Flow Approach to Self-Timed Logic in VLSI, *IEEE Intern. Symp. on Circuits and Systems,* ISCAS-88 Espoo, Finland, pp. 479–482, June 1988.

[13] Liao J. and Ni L.M.: A New CMOS Circuit Representation for Timing Verification, *IEEE Intern. Symp. on Circuits and Systems,* ISCA-88 Espoo, Finland, pp. 483–486, June 1988.

[14] Lin H.D. and Messerschmitt D.G.: Finite State Machine has Unlimited Concurrency, *IEEE Trans. on Circuits and Systems,* Vol. CAS-38, No. 5, pp. 465–5, May 1991.

[15] Marino L.: General Theory of Metastable Operation, *IEEE Trans. on Computers,* Vol. C-30, No. 2, pp. 107–115, Feb. 1981.

[16] Martin A.: Synthesis of Asynchronous VLSI Circuits, in *Formal Methods for VLSI Design,* ed. J. Staunstrup, North-Holland, 1990.

[17] McAuley A.J.: Dynamic Asynchronous Logic for High-Speed CMOS Systems, *IEEE J. Solid-State Circuits,* Vol. SC-27, No. 3, pp. 382–388, March 1992.

[18] Mead C. and Conway L.: *Introduction to VLSI Systems,* Addison-Wesley, Reading, MA, 1980.

[19] Meng T.H.: *Synchronization Design for Digital Systems,* Kluwer Academic Pub., 1991.

[20] Meng T.H., Brodersen R.W., and Messerschmitt D.G.: Automatic Synthesis of Asynchronous Circuits from High-Level Specifications, *IEEE Trans. on Computer-Aided Design,* Vol. CAD-8, No. 11, pp. 1185–1205, Nov. 1989.

[21] Meng T.H., Brodersen R.W., and Messerschmitt D.G.: A Clock-Free Chip Set for High-Sampling Rate Adaptive Filters, *J. of VLSI Signal Processing,* Vol. 1, No. 4, pp. 345–365, April 1990.

[22] Muller D.E. and Bartky W.S.: A Theory of Asynchronous Circuits, *Annals of the Computation Lab. of Harvard University.,* Vol. 29, Harvard University Press, Cambridge, MA, pp. 204–243, 1959.

[23] Reyneri L.M., Del Corso D., and Sacco B.: Oscillatory Metastability in Homogeneous and Inhomogeneous Flip-Flops, *IEEE J. on Solid-State Circuits,* Vol. SC-25, No. 1, pp. 254–264, Feb. 1990.

[24] Staunstrup J. and Greenstreet M.R.: Designing Delay-Insensitive Circuits using Synchronized Transitions, IMEC, *IFIP International workshop on Applied Formal Methods for Correct VLSI Design,* 1989.

[25] Weste N. and Eshraghian K.: *Principles of CMOS VLSI Design: A System Perspective,* Addison-Wesley, Reading, MA, 1985.

PROBLEMS

10.1 Identify different hierarchical levels, as illustrated in Figure 10.1, for the FFT processor.

10.2 What is the difference between latches and flip-flops? Also show that a D flip-flop can be realized as two cascaded latches and determine their clocking.

10.3 Determine the type of flip-flop shown in Figure 10.12. Which edge of the clock signal is active?

10.4 Suggest a logic circuit that implements a nonoverlapping two-phase clock.

10.5 Derive the complete 2-bit serial: adder in Example 10.1 with the appropriate input and output signals.

10.6 Use the look-ahead method to double the throughput of a bit-serial subtractor.

10.7 (a) Determine the maximal throughput for the algorithm shown in Figure P10.7.

(b) Use the look-ahead method to double the throughput.

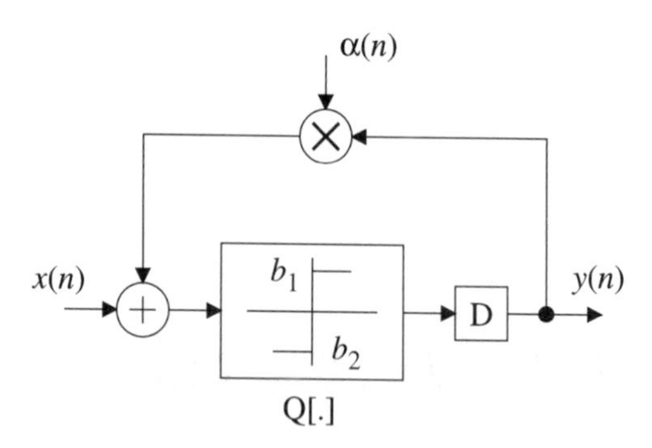

Figure P10.7 Quantization loop

11

PROCESSING ELEMENTS

11.1 INTRODUCTION

In this chapter we will discuss the implementation of basic arithmetic operations such as addition, subtraction, and multiplication using fixedpoint arithmetic, as well as compound operations such as multiple-constant multiplication, vector-multiplication, and multiplication involving complex values. The two latter types of arithmetic operations can be implemented using so-called distributed arithmetic, which will be discussed in detail in section 11.14. Arithmetic operations used in DSP algorithms can be implemented with many different techniques. For example, the data involved can be either bit-serial or bit-parallel processed. Arithmetic operations can also be done by using different binary representations for the negative numbers. These representations differ with respect to their implementation cost (space–time–power). In this chapter we will therefore first review the types of number representations and arithmetic most commonly used in DSP algorithms.

11.2 CONVENTIONAL NUMBER SYSTEMS

Not all numbers were created equal.

A number system is defined by the set of values that each digit can assume and by an interpretation rule that defines the mapping between the sequence of digits and their numerical values. Conventional number systems are nonredundant, weighted, positional number systems. In such number systems every number,[1] $\boldsymbol{x}$, represented by a sequence of digits $(x_0 x_1 x_2, \ldots, x_{W_d-1})$ has a unique representation.

The value of a number $\boldsymbol{x}$, is

$$x = \sum_{i=0}^{W_d-1} w_i x_i = 1 \quad (11.1)$$

1. We will use bold face to denote the number and plain face to denote the digits in the corresponding sequence of digits whenever both are used in the same context.

where W_d is the word length and w_i is a weight associated with each digit. In a positional number system, the weight w_i depends only on the position of the digit x_i. In conventional number systems the weight w_i is the ith power of a fixed integer r— i.e., $w_i = r^i$. Such number systems are called fixed-radix systems. The digits have to satisfy $0 \le x_i \le r - 1$.

Generally, a number can have an integer and a fractional part. This is done by letting the k leftmost digits represent the integer part and the W_d–k remaining digits represent the fractional part. For example, the binary representation of the number $\boldsymbol{x}$ with $k = 2$ is $x_0x_1{\bullet}x_2...x_{W_d - 1}$. This corresponds to instead using the weights $w_i = r^{k - i - 1}$. The weight of the least significant digit is $Q = w_{W_d - 1} = r^{k - W_d}$. Notice that the radix point is not stored; its position is only understood. Such numbers are called fixed-point numbers. Generally, a fixed-point number is written from left to right with the most significant bit (*MSB*) at the far left, and the least-significant bit (*LSB*) to the far right.

There are two distinct forms of fixed-point arithmetic: integer and fractional. In integer arithmetic the right-hand part is considered to be most "important." For example, multiplication of two integer values yields a product with twice as long a word length. If the word length needs to be reduced to the original length, the right-hand part is kept and the left-hand part is dropped. An error occurs only if the product is too large, but the error will be very large if it occurs.

In fractional fixed-point arithmetic, the numbers may be considered to be in the range $1 < \boldsymbol{x} < 1$. Hence, the product will be in the same range. The word length is reduced by keeping the left- and dropping the right-hand part. This will result in a small error in the quantized product.

Conversely, a number in floating-point arithmetic is described by a signed mantissa and a signed exponent,—e.g., $\pm m\ 2^{\pm exp}$. The magnitude of the mantissa is usually normalized to the interval [0.5, 1[. Floating-point arithmetic is used in general-purpose computers to handle values that vary over a large range. In most DSP algorithms, however, the necessary signal range can be made small by appropriate scaling of the signal levels. Moreover, good signal processing algorithms tend to generate low round-off noise and use the available signal range well.

An advantage of fractional fixed-point arithmetic is that parasitic oscillations are more easily suppressed than they are in floating-point arithmetic. Fixed-point arithmetic also requires less chip area and is much faster than floating-point arithmetic. Hence, in most VLSI circuits for dedicated DSP applications calculations are done using fractional fixed-point arithmetic.

Performance of the processing elements with respect to speed, chip area, and power dissipation depends on the number representation used. In this section we will discuss the following fixed-point representations:

- Signed-magnitude representation
- One's-complement representation
- Two's-complement representation
- Binary offset representation

11.2.1 Signed-Magnitude Representation

In signed-magnitude representation the magnitude and sign are represented separately. The first digit represents the sign and the remaining digits the magnitude. A normalized W_d-bit binary word using signed-magnitude representation represents the value

$$\boldsymbol{x} = (1 - 2x_0) \sum_{i=1}^{W_d - 1} x_i 2^{-i} \tag{11.2}$$

A normalized value is in the range $1 \le \boldsymbol{x} \le 1$. Hence, the binary point is placed immediately to the right of x_0. If the need arises, the number range can be changed by moving the binary point. In the signed-magnitude representation numbers having the same magnitude are represented by the same binary word, except for the sign bit. For example, we have

$$(+0.828125)_{10} = (0.110101)_{SM}$$
$$(-0.828125)_{10} = (0.110101)_{SM}$$
$$(0)_{10} = (0.000000)_{SM} \text{ or } (1.000000)_{SM}$$

All values lie in the range $1 + Q \le \boldsymbol{x} \le 1 \;\; Q$ where $Q = 2^{-W_d + 1}$. There are as many negative numbers as positive numbers. Notice that zero can be represented as either +0 or 0. This may complicate the implementation of an ALU since two checks for two different zeros have to be made. The values +1 and 1 can not be represented. A major disadvantage of the signed-magnitude representation is that addition and subtraction depend on the signs of the operands. These operations are more complicated to implement compared to one's-complement, two's-complement, and binary offset representations. Multiplication, and division by a power of two, can easily be done by shifting the fractional part of the binary word.

11.2.2 Complement Representation

One of the main advantages of a complement representation is that addition and subtraction can be performed without regard of the sign of the operands. Positive numbers are in one's-complement, two's-complement, and binary offset representations represented by the same number as in the signed-magnitude representation. Negative numbers are represented by

$$-\boldsymbol{x} = \boldsymbol{R} - \boldsymbol{x} \tag{11.3}$$

where $\boldsymbol{R}$ is to be determined shortly.

If a positive, $\boldsymbol{x}$, and a negative number, $\boldsymbol{y}$, are to be added, the negative number is represented by the complement, $(\boldsymbol{R} - \boldsymbol{y})$. Hence,

$$\boldsymbol{x} + (\boldsymbol{R} - \boldsymbol{y}) = \boldsymbol{R} - (\boldsymbol{y} - \boldsymbol{x}) \tag{11.4}$$

If $\boldsymbol{y} > \boldsymbol{x}$, then the negative result $-(\boldsymbol{y} - \boldsymbol{x})$ is in the correct complemented form $\boldsymbol{R} - (\boldsymbol{y} - \boldsymbol{x})$ and there is no need to make any corrections. However, if $\boldsymbol{x} > \boldsymbol{y}$, then the correct result should be $(\boldsymbol{x} - \boldsymbol{y})$ but the actual result is $\boldsymbol{R} - (\boldsymbol{y} - \boldsymbol{x}) = \boldsymbol{R} + (\boldsymbol{x} - \boldsymbol{y})$. Hence, the error term $\boldsymbol{R}$ must be eliminated. This can be done without any costs in terms of extra operations if the value of $\boldsymbol{R}$ is selected properly.

It is in practice important that it is easy to compute the complement of a value. We define the complement of a single digit x_i as

$$\bar{x}_i = (r - 1) - x_i \tag{11.5}$$

Let $\bar{x}$ denote the number obtained by complementing each digit in the corresponding number x. It is easy to show that

$$x + \bar{x} + Q = r^k \tag{11.6}$$

If this number is stored in a register holding k integer digits and $W_d - k$ fractional digits, the register overflows. The most significant digit is discarded and the result is zero. This is equivalent to taking the remainder after dividing by r^k.

Now, if we select $R = r^k$ we obtain the so-called radix complement, or two's-complement representation in the binary case

$$R - x = r^k - x = \bar{x} + Q \tag{11.7}$$

No correction of Equation (11.4) is needed since the error term R is discarded when computing $R + (x - y)$. Another possible choice is

$$R = r_k - Q \tag{11.8}$$

This choice is called diminished-radix complement, or one's-complement representation in the binary case. We obtain the complement

$$R - x = r^k - Q - x = \bar{x} \tag{11.9}$$

The computation of the complement is much simpler in this case. All digits can be complemented in parallel. However, a correction of the obtained result is required if $x > y$ [22].

11.2.3 One's-Complement Representation

One's-complement representation as just discussed is a diminished-radix complement representation with $R = r^k - Q = 1 - Q$ for $r = 2$ and $k = 0$. A normalized W_d-bit binary word in one's-complement representation is interpreted as

$$x = -x_0(1 - Q) + \sum_{i=1}^{W_d - 1} x_i 2^{-i} \tag{11.10}$$

The values lie in the range $1 + Q \le x \le 1 \;\; Q$ where $Q = 2^{-W_d + 1}$. Also, in one's-complement representation, the zero value has a redundant binary representation which reduces the value range and complicates the checking for zero. The values +1 and 1 can not be represented. Numbers having the same magnitude, but different signs, are represented by different binary words. For example, we have

$$(+0.828125)_{10} = (0.110101)_{1C}$$

$$(-0.828125)_{10} = (0.110101)_{1C}$$

$$(0)_{10} = (0.000000)_{1C} \text{ or } (1.111111)_{1C}$$

For $x > 0$, one's-complement representation has the same binary words as, for example, signed-magnitude representation. For $x < 0$, the values are the bit-complement of the corresponding positive values. Addition and subtraction in one's-complement representation are simpler to implement than in binary offset and two's-complement representations, which will be discussed later.

EXAMPLE 11.1

Show that the negative value of a number, $\boldsymbol{x}$, in one's-complement representation can be obtained by inverting all bits in the binary word.

We have from Equation (11.10)

$$\boldsymbol{y} = -\boldsymbol{x} = -\left[-x_0(1-Q) + \sum_{i=1}^{W_d-1} x_i 2^{-i}\right]$$

$$= x_0(1-Q) - \sum_{i=1}^{W_d-1} x_i 2^{-i} + \sum_{i=1}^{W_d-1} (1-x_i) 2^{-i}$$

$$= -(1-x_0)(1-Q) + \sum_{i=1}^{W_d-1} (1-x_i) 2^{-i} = -\overline{x}_0(1-Q) + \sum_{i=1}^{W_d-1} \overline{x}_i 2^{-i}$$

Hence, the sign of a number in one's-complement representation can be changed by just taking the bit-complement. Thus, subtraction can be accomplished by addition of the bit-complemented word.

A change of sign is easier in signed-magnitude representation compared to binary offset and two's-complement representations, since using these representations, changing the sign of a number is done by taking the bit-complement and adding 1 to the least-significant position. Multiplication using one's-complement is more difficult to implement than with two's-complement or binary offset since an end-around-carry is required.

11.2.4 Two's-Complement Representation

"Two's complement, three's a crowd"
Tony Platt, SAAB Military Aircraft

Two's-complement representation is a so-called radix complement representation with $\boldsymbol{R} = r^k = 1$ for $r = 2$ and $k = 0$. Two's-complement representation is the most common type of arithmetic used in digital signal processing. The value of a normalized W_d-bit binary word in two's-complement representation is

$$\boldsymbol{x} = -x_0 + \sum_{i=1}^{W_d-1} x_i 2^{-i} \tag{11.11}$$

The values lie in the range $1 \le \boldsymbol{x} \le 1 \;\; Q$, where $Q = 2^{-W_d-1}$. There is one more negative number than positive numbers and the value +1 can not be represented. For $\boldsymbol{x} > 0$ two's-complement has the same binary word as signed-magnitude representation. The negative value of a number in two's-complement representation can be obtained from the corresponding positive number by adding Q to the bit-complement. For example,

$$(+0.828125)_{10} = (0.110101)_{2C}$$
$$(-0.828125)_{10} = (1.001010)_{2C} + (0.000001)_{2C} = (1.001011)_{2C}$$
$$(0)_{10} = (0.000000)_{2C}$$

EXAMPLE 11.2

Show that the negative value of a number, $\boldsymbol{x}$, in two's-complement representation can be obtained by inverting all bits in the binary word and adding a 1 in the least significant position.

We have

$$\boldsymbol{y} = -\boldsymbol{x} = -\left[-x_0 + \sum_{i=1}^{W_d-1} x_i 2^{-i}\right] = x_0 - \sum_{i=1}^{W_d-1} 2^{-i} + \sum_{i=1}^{W_d-1} (1-x_i)2^{-i}$$
$$= -(1-x_0) + \sum_{i=1}^{W_d-1} (1-x_i)2^{-i} + 2^{-W_d+1} = -\bar{x}_0 + \sum_{i=1}^{W_d-1} \bar{x}_i 2^{-i} + Q$$

A useful property of two's-complement representation is that, if the sum lies in the proper range, several two's-complement numbers can be added even though the partial sums may temporarily overflow the available number range. Thus, the numbers can be added in arbitrary order without considering possible overflow as long as the final sum lies within the proper range.

EXAMPLE 11.3

Show by an example that the sum of n ($n \geq 3$) numbers in two's-complement representation can be added without regard to overflows of the partial sums if the final sum is within the number range. Assume that the numbers are

$6/8 = (0.110)_2$, $4/8 = (0.100)_2$, and $7/8 = (1.001)_2$.

We first add

$$S_2 = 6/8 + 4/8$$
$$= (1.110)_2 + (0.100)_2$$
$$= \{\text{overflow}\} = (1.010)_2 = -6/8$$

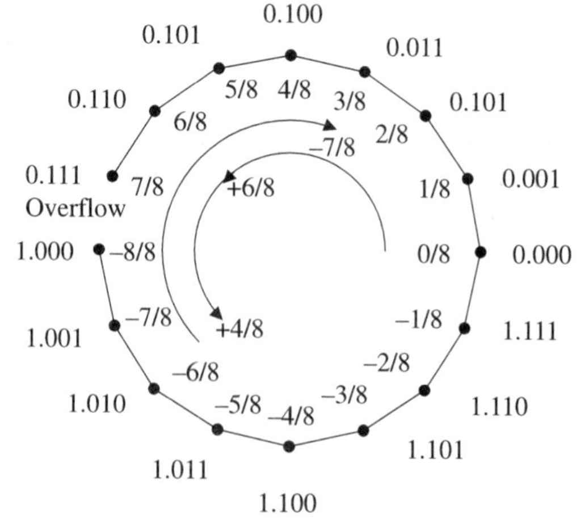

Figure 11.1 Additions with overflow using two's-complement numbers

and then we add the third number (neglecting the temporary overflow that occurred and neglecting the second overflow)

$$\begin{aligned} S_2 &= -6/8 + (-7/8) \\ &= (1.010)_2 + (1.001)_2 \\ &= \{\text{overflow}\} = (0.011)_2 = 3/8 \end{aligned}$$

yielding the correct result. Figure 11.1 illustrates the cyclic property of two's-complement representation. A correct result is obtained as long as the amounts of left and right "rotations" cancel.

11.2.5 Binary Offset Representation

The value of a normalized W_d-bit binary word in binary offset representation is

$$\boldsymbol{x} = (x_0 - 1) + \sum_{i=1}^{W_d - 1} x_i 2^{-i} \qquad (11.12)$$

The values lie in the range $1 \le \boldsymbol{x} \le 1 \;\; Q$, where $Q = 2^{-W_d - 1}$. Binary offset representation is also a nonredundant representation. The sequence of digits is equal to the two's-complement representation, except for the sign bit which is complemented. For example, we have

$$\begin{aligned} (+0.828125)_{10} &= (0.110101)_{BO} \\ (-0.828125)_{10} &= (1.001011)_{BO} \\ (0)_{10} &= (1.000000)_{BO} \end{aligned}$$

In the same way as for the two's-complement representation, it can be shown that the sign of a number can be changed by first taking the bit-complement and then adding Q.

11.3 REDUNDANT NUMBER SYSTEMS

Conventional fixed-radix number systems where the radix is a positive (usually even) integer are commonly used in arithmetic units in general-purpose computers and standard digital signal processors. In application-specific arithmetic units it is often advantageous to use more unconventional, redundant number systems— for example, negative radix systems [22] or signed digit representations. By using redundant number systems it is possible to simplify and speed up certain arithmetic operations. In fact, addition and subtraction can be performed without long carry (borrow) paths. Typically this is achieved at the expense of increased complexity for other arithmetic and nonarithmetic operations and larger registers. Zero and sign detection and conversion to and from conventional number representations are time consuming. In this section we will discuss three such representations:

- Signed-digit code
- Canonic signed digit code
- On-line arithmetic

Signed-digit codes (*SDCs*) differ from the binary representations, just discussed, since each digit is allowed to have a sign. Typically digits can take one of three values: 1, 0, or +1.

One advantage of these number systems is that long carry propagations can be eliminated in additions and subtractions. Another advantage is that the required number of addition/subtraction cycles can be reduced in some multiplication algorithms.

On-line arithmetic refers to redundant number systems that allow arithmetic operations to be performed by using the most significant bit first. These properties can be exploited to reduce the latency inherent in long sequential computations and operations involving very long word lengths.

11.3.1 Signed-Digit Code

A number, $\boldsymbol{x}$, in the range $2 + Q \leq \boldsymbol{x} \leq 2 \quad Q$, where $Q = 2^{-W_d}$, is represented in signed digit code by

$$\boldsymbol{x} = \sum_{i=1}^{W_d - 1} x_i 2^{-i} \text{ where } x_i = -1, 0, \text{ or } +1 \qquad (11.13)$$

For example, the decimal number $(15/32)_{10} = (0.01111)_{2C}$ can be represented by $(0.10001)_{SDC}$ or by $(15/32)_{10} = (0.01111)_{SDC}$. The decimal number $(15/32)_{10} = (1.10001)_{2C}$ can be represented by $(0.10001)_{SDC}$ or $(0.01111)_{SDC}$.

A minimal SDC number has a minimum number of nonzero digits. Obviously, the signed digit code is not unique. This causes problems when a number shall be quantized (rounded or truncated). In practice, a signed digit code is therefore converted to a "conventional" representation before it is quantized. In a conventional number system the numbers are inherently ordered according to the values. It is therefore simple to compare the magnitude of two numbers or to round a number to the "closest" number with appropriate word length. In practice it may be too costly to convert SDC numbers for the necessary quantization and overflow checks in every recursive loop. However, the SDC has an advantage that the time taken for addition of two numbers is independent of the word length.

Notice also that the quantization step for SDC is not equidistant. The resolution is higher for small numbers. The maximum and minimum step size depend on the number of nonzero bits and the word length.

EXAMPLE 11.4

Show that carry propagation can be avoided when adding two numbers in SDC representation using the algorithm described in Table 11.1 [40]. Assume for simplicity that the two numbers are $(1111)_{SDC} = (5)_{10}$ and $(0111)_{SDC} = (1)_{10}$. The sum is

$$s_i = z_i + c_{i+1}$$

$x_i\, y_i$ or $y_i\, x_i$	0 0	0 1	0 1	0 1	0 1	1 1	1 1	1 1
$x_{i+1}\, y_{i+1}$	—	Neither is –1	At least one is –1	Neither is –1	At least one is –1	—	—	—
c_i	0	1	0	0	1	0	1	1
z_i	0	1	1	1	1	0	0	0

Table 11.1 Rules for adding SDC numbers

Apply the rules in Table 11.1 from right to left. The carries have been shifted one step to the left in order to align the digits for the final summation. The result of the addition is $(0100)_{SDC} = (4)_{10}$; see Table 11.2. Notice that all of the intermediate terms z_i and carries can be generated in parallel by considering two successive sets of digits.

i		0	1	2	3
x_i		1	–1	1	–1
y_i		0	–1	1	1
c_{i+1}	0	–1	1	0	—
z_i		1	0	0	0
s_i		0	1	0	0

Table 11.2 Obtained result

11.3.2 Canonic Signed Digit Code

A canonic signed digit code (CSDC) is a special case of a signed digit code in that each number has a unique number representation. CSDC has some interesting properties that make it useful in the mechanization of multiplication. A number, $\boldsymbol{x}$, in the range $-4/3 + Q \le \boldsymbol{x} \le 4/3-Q$, where $Q = 2^{-W}$, $W = W_d-1$ for W_d = odd, and $W = W_d-2$ for W_d = even, is represented in canonic signed digit code by

$$\boldsymbol{x} = \sum_{i=1}^{W_d-1} x_i 2^{-i} \text{ where } x_i = -1, 0, \text{or } +1 \tag{11.14}$$

where no two consecutive digits are nonzero— i.e.,

$$x_i \cdot x_{i+1} = 0,\ 0 \le i \le W_d - 1$$

For example, the number $(15/32)_{10}$ is represented by $(0.10001)_{CSDC}$ and $(15/32)_{10}$ by $(0.10001)_{CSDC}$. Furthermore, the CSDC has a minimum number of nonzero digits. It can be shown that the average number of nonzero digits is

$$\frac{W_d}{3} + \frac{1 + 2^{-W_d}}{9}$$

Hence, for moderately large W_d the average number of nonzero bits is about $W_d/3$ as opposed to $W_d/2$ in the usual (nonsigned) binary representations.

Conversion of Two's-Complement to CSDC Numbers

The conversion from two's-complement representation to CSDC is straightforward. It is based on the identity

$$2^{k+n+1} - 2^k = 2^{k+n} + 2^{k+n-1} + 2^{k+n-2} + \ldots + 2^k$$

A string of 1s in the two's-complement representation of a number can be replaced by a 1, followed by 0s, followed by a 1. For example, $(0.011111)_{2C} = (31/64)_{10}$ corresponds to $(0.100001)_{CSDC} = (32\ \ 1/64)_{10}$.

A two's-complement number can be converted to CSDC in an iterative manner by the following algorithm. Strings of consecutive 1s separated by 0s are first identified. A string of two or more 1s,— for example, $(...00111100...)_{2C}$, is converted into $(...01000100...)_{CSDC}$. Isolated 1s are left unchanged. After this first step, the entire string is reexamined and pairs of type (1, 1) are changed to (0, 1) and triplets of type (0, 1, 1) are changed to (1, 0, 1). This process is repeated until no two adjacent nonzero digits remain. For example, $(0.110101101101)_{2C} = (1.001010010101)_{CSDC}$. An alternative algorithm for converting two's-complement numbers into CSDC that is suitable for hardware implementation is given in [17, 33].

Conversion of SDC to Two's-Complement Numbers

A conversion of an SDC number into a representation where the numbers are ordered according to size is often necessary. For example, in a recursive algorithm the word length increases and must be quantized (rounded or truncated) and checked for overflow. The standard method of converting an SDC number into two's-complement representation is to separate the SDC number into two parts. One part holds the digits that are either 0 or 1 and the other part the 1 digits. Finally, these two numbers are subtracted. This carry propagation in this operation causes a significant delay. Notice that adding (subtracting) a two's-complement number to an SDC number can be performed directly and does not require any conversion.

11.3.3 On-Line Arithmetic

On-line arithmetic refers to number systems with the property that it is possible to compute the ith digit of the result, using only the first $(i+\delta)$th digits, where δ is a small positive constant [42]. Thus, after that the first δ digits have become available, the first digit of the result can be computed and the following digits can successively be computed for each new digits of the operands become available. Thus, the latency corresponds to δ digits. This property can be favorable in recursive algorithms using numbers with very long word lengths.

Obviously, the signed digit code, just discussed can be used for on-line addition and subtraction since the carry propagation is only one position, $\delta = 1$, regardless of the word length. Multiplication and division can also be computed on-line using SDC. A more efficient and faster method of converting an SDC number into two's-complement representation that even allows on-line operations is described in [15].

11.4 RESIDUE NUMBER SYSTEMS

Another way of performing arithmetic operations that avoids carry propagation is based on RNSs (residue number systems). This old Chinese technique has gained renewed interest as the VLSI technique has become available. In fixed-point RNS a number is represented by a set of residues obtained after an integer division by mutual prime factors (moduli), m_i, $i = 1, 2, ..., p$.

For a given integer number x and moduli set $\{m_i\}$, $i = 1, 2, \ldots, p$, we may find the elements $(r_1, r_2, \ldots, r_p)$ of the RNS by the relation

$$x = q_i m_i + r_i \tag{11.15}$$

where q_i, m_i, and r_i are integers. x can be written as

$$x = (r_1, r_2, \ldots, r_p) \tag{11.16}$$

The advantage of RNS is that the arithmetic operations (+, –, *) can be performed for each residue independently of all the other residues. Individual arithmetic operations can thus be performed by independent operations on the residue. The Chinese remainder theorem[2] is of fundamental importance in residue arithmetic.

Theorem 11.1

Let $A = (a_1, a_2, \ldots, a_p)$ and $B = (b_1, b_2, \ldots, b_p)$

Then $A \oplus B = [(a_1 \oplus b_1)m_1, (a_2 \oplus b_2)m_2, \ldots, (a_p \oplus b_p)m_p]$ (11.17)

where $\oplus$ is a general arithmatic operation, module m_i, such as addition, subtraction, or multiplication.

As can be seen, each residue digit is dependent only on the corresponding residue digits of the operands.

Example 11.5

Use a residue number system with the modules 5, 3, and 2 to add the numbers 9 and 19 and multiply the numbers 8 and 3. Also determine the number range.

The number range is $D = 5 \cdot 3 \cdot 2 = 30$

$$9 + 19 = (4, 0, 1)_{RNS} + (4, 1, 1)_{RNS} = \big((4 + 4)_5, (0 + 1)_3, (1 + 1)_2\big)_{RNS}$$

$$= (3, 1, 0)_{RNS} = 28$$

$$8 \cdot 3 = (3, 2, 0)_{RNS} \cdot (3, 0, 1)_{RNS} = \big((3 \cdot 3)_5, (2 \cdot 0)_3, (0 \cdot 1)_2\big)_{RNS}$$

$$= (4, 0, 0)_{RNS} = 24$$

A number range corresponding to more than 23 bits can be obtained by choosing the following mutual prime factors 29, 27, 25, 23, 19. Then $D = m_1 \cdot m_2 \cdot \ldots \cdot m_5 = 8,554,275 > 2^{23}$. Each residue is represented in the common weighted binary number system as a 5-bit number.

RNS is restricted to fixed-point operations and gives no easily obtainable information about the size of the numbers. For example, it is not possible to determine directly which is the closest number— i.e., comparison of numbers, overflow detection, and quantization operations are difficult. Hence, in order to round off an

2. After the Chinese mathematician Sun-Tsü, AD 100.

RNS number it has to be converted to a normal number representation. RNS is therefore not suitable for use in recursive loops. Division is possible only in special cases and would seriously complicate the system. Further, the dynamic range must then be kept under strict control. Overflow would spoil the results. The RNS method is effective only for special applications— e.g., nonrecursive algorithms.

11.5 BIT-PARALLEL ARITHMETIC

"Heigh-ho, Heigh-ho, it's off to work we go...
Larry Morey and Frank Churchill

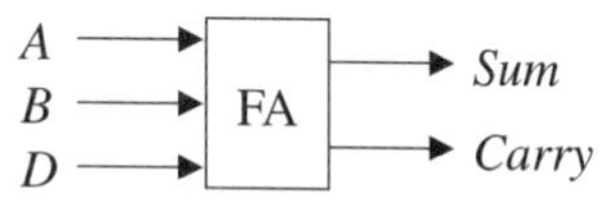

Figure 11.2 Full-adder

In this section we will discuss methods to implement arithmetic operations such as addition, subtraction, and multiplication using bit-parallel methods. These operations are simple to implement in two's-complement representation, since they are independent of the signs of the numbers involved. Hence, two's-complement representation is used predominantly. However, in many cases it is advantageous to exploit other number systems to improve the speed and reduce the required chip area and power consumption.

The basic computation element for addition, subtraction, and multiplication is a full-adder (FA). It accepts three binary inputs: A, B, and D, called addend, augend, and carry-in, respectively. The two outputs are the sum, S, and the carry-out, C. A symbol for the full-adder is shown in Figure 11.2. We have for the full-adder

$$\begin{aligned} S &= A \oplus B \oplus D = \{\text{Parity}\} \\ &= A \cdot B \cdot D + A \cdot \overline{B} \cdot \overline{D} + \overline{A} \cdot \overline{B} \cdot D + \overline{A} \cdot B \cdot \overline{D} \end{aligned} \tag{11.18}$$

$$C = A \cdot B + A \cdot D + B \cdot D = A \cdot B + D \cdot (A + B) \tag{11.19}$$

These expressions can be modified to be more suitable for implementation in CMOS with XOR gates:

$$Sum = \begin{cases} \overline{D} \text{ if } A \oplus B = 1 \\ D \text{ if } A \oplus B = 0 \end{cases} \tag{11.20}$$

$$Carry = \begin{cases} D \text{ if } A \oplus B = 1 \\ A \text{ if } A \oplus B = 0 \end{cases} \tag{11.21}$$

11.5.1 Addition and Subtraction

Two binary numbers, $\boldsymbol{x} = (x_0 \bullet x_1 x_2 \dots x_{Wd-1})$ and $\boldsymbol{y} = (y_0 \bullet y_1 y_2 \dots y_{Wd-1})$, can be added bit-serial in $O(W_d)$ steps and, by using more advanced techniques and bit-parallel operation, the number of steps, or gate levels, can be reduced to only $O(\log(W_d))$. Notice, however, that the times required for a computational step may differ between different addition algorithms, since the switching time also depends on gate loads. In practice the addition time for one algorithm with more steps may therefore be shorter than for one with fewer steps. Thus, to accurately estimate the addition time a detailed analysis and circuit realization are required.

Ripple-Carry Adder

The ripple-carry adder (RCA) is the simplest form of adder [22]. Two numbers using two's-complement representation can be added by using the circut shown in Figure 11.3. A W_d-bit RCA is built by connecting W_d full-adders so that the carry-out from each full-adder is the carry-in to the next stage. The sum and carry bits are generated sequentially, starting from the LSB. The carry-in bit into the rightmost full-adder, corresponding to the LSB, is set to zero, i.e., (c_{Wd} = 0). The speed of the RCA is determined by the carry propagation time which is of order $O(W_d)$. Special circuit realization of the full-adders with fast carry generation are often employed to speed the operation. Pipelining can also be used.

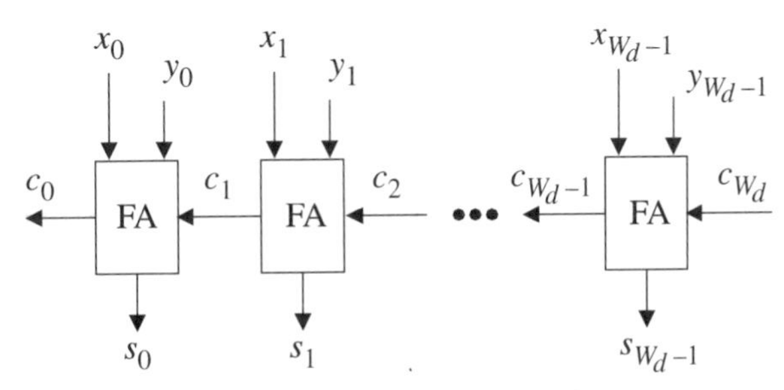

Figure 11.3 Ripple-carry adder

Figure 11.4 shows how a 4-bit RCA can be used to obtain an adder/subtractor. Subtraction is performed by adding the negative value. The subtrahend is bit-complemented and added to the minuend while the carry into the LSB is set to 1 (c_{Wd} = 1).

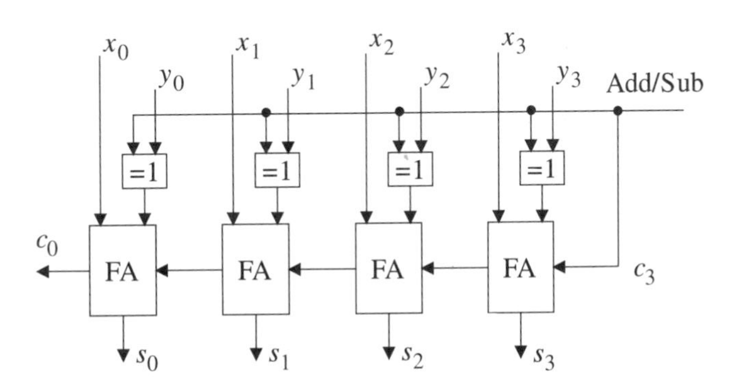

Figure 11.4 Ripple-carry adder/subtractor

Addition time is essentially determined by the carry path in the full-adders. Notice that even though the adder is said to be bit-parallel, computation of the output bits is done sequentially. In fact, at each time instant only one of the full-adders performs useful work. The others have already completed their computations or may not yet have received a correct carry input, hence they may perform erroneous computations. Because of these wasted switch operations, power consumption is higher than necessary.

Carry–Look–Ahead Adder

In the carry–look–ahead adder (*CLA*), the carry propagation time is reduced to $O(\log_2(W_d))$ by using a treelike circuit to compute the carry rapidly [5, 18, 22, 44]. The area requirement is $O(W_d \log_2(W_d))$. The CLA algorithm was first introduced by Weinberger and Smith [44], and several variants have since been developed. Brent and Kung [5] have derived an adder with an addition time and area proportional to $2 \log_2(W_d)$ and $2W_d \log_2(W_d)$, respectively.

The CLA exploits the fact that the carry generated by a bit-position depends on the three inputs to that position. If $x_i = y_i = 1$, a carry is generated independently of the carry from the previous bit-position and if $x_i = y_i = 0$, no carry is generated. If $x_i \neq y_i$, a carry is generated if and only if the previous bit-position generates a carry. It is possible to compute all the carries with only two gate delays (although this would require gates with excessive fan-in). The fan-in of logic gates increases linearly with the number of bits in the adder. The high fan-in forces the CLA to be partitioned into blocks with carry–look–ahead. The block size is usually selected in the range three to five.

A 1992 comparison among adders [8] shows that the CLA adder, which also is amenable to pipelining, is one of the fastest adders. The implementation of a CLA adder, using dynamic CMOS logic, was reported in 1991 [41].

Carry–Save Adder

Carry–save adders are suitable when three or more operands are to be added, as in some multiplication schemes [22]. In this adder a separate sum and carry bits are generated for the partial results, except when the last operand is added. For example, if three numbers are to be added, the first two are added using a carry–save adder. The partial result is two numbers corresponding to the sum and the carry. The last operand is added using a second carry–save adder stage. The result becomes a sum and carry number. Thus, a carry-save adder reduces the number of operands by one for each adder stage. Finally, the sum and carry numbers are added using an adder with carry propagation—for example, carry-look-ahead adder.

Carry–Select Adder

Carry–select adders (*CSAs*) and carry–skip adders provide a compromise between RCA and CLA adders. The carry–select adder is based on precomputation of some partial results for the two possible values of the carry bits and using the carry from the previous bit-position to select the proper result [3].

The CSA adder is for complexity reasons also divided into blocks of RCAs. The carry into the first block is known, but the carries into the rest of the blocks are unknown. Now, 0 and 1 are the only possible carries into a block and the two possible sums and carries out from each block can therefore be computed. Next, the carry from the previous block selects the correct sum and carry-out, which in turn select the next blocks sum and carry-out, until the last sum and is selected.

Carry–Skip Adder

The carry–skip adder is also divided into RCA blocks [22]. The first block has no input carry while the carry into a subsequent block is generated by the skip circuit of the previous block. As in the CSA, all blocks receive the inputs simultaneously. The basic principle is that a carry will be propagated, by the blocks skip circuit, only if all of the inputs to the block are in the propagate state; otherwise a 0 or 1 carry will generated [9, 43]. If a block is in the propagate state, then the carry-out is the same as the carry into the block. Hence, the carry skips over the block. On the other hand, if the block is in a nonpropagate state, then the carry-out of the block is determined by the blocks inputs only.

The sum bits in a block must be computed with the proper input carry. These computations, for all but the last block, can be overlapped with the computation of the sum bits of the last block. The carry propagation path is through the carry skip circuitry and the final RCA. The addition time is $O(\sqrt{W_d})$. The *Manchester adder* is a special case of carry–skip adder. A carry-skip adder usually requires less chip area than the corresponding CSA; however, it can not be pipelined.

Conditional–Sum Adder

The conditional–sum adder generates all possible sums and carries in a manner similar to the carry–select adder [22]. The conditional–sum adder uses a modified half-adder to generate sums and carries in the first phase. The second phase uses $\log_2(W_d)$ levels of multiplexers to conditionally combine neighboring bits into the final sum [36]. This adder can easily be pipelined by placing latches after the first phase and after every multiplexer stage. The conditional–sum adder, implemented using dynamic CMOS circuits [25], is usually faster than a carry–look–ahead adder, but both adders have complex interconnections that require large chip areas. The conditional–sum adder should be considered as a candidate for high-speed CMOS adders.

11.5.2 Bit-Parallel Multiplication

In order to multiply two two's-complement numbers, $\boldsymbol{a}$ and $\boldsymbol{x}$, we form the partial bit-products $a_i \cdot x_k$, as shown next in Figure 11.5. High-speed, bit-parallel multipliers can be divided in three classes [19, 27]. The first type, so-called *shift-and-add multipliers,* generates partial bit-products sequentially and accumulates them successively as they are generated. This type is therefore the slowest multiplier, but the required chip area is low.

The second type generates all bit-products in parallel and uses a multi-operand adder (i.e., an adder tree) for their accumulation. This multiplier type is also known as *parallel multiplier.* A parallel multiplier structure can be partitioned into three parts: partial product generation, carry-free addition, and carry-propagation addition. These three parts can be implemented using different schemes. For example, the partial product generation can be implemented by AND gates or by using the Booth's algorithm, as described in section 11.5.4. The carry-free addition part is often implemented by using a Wallace or redundant binary addition tree. The last part can employ one of the addition schemes just described.

Finally, the third type of multipliers uses an array of almost identical cells for generation of the bit-products and accumulation. This type of multipliers is called

$a_0 \cdot x_{W_d-1} \bullet\bullet\bullet \; a_{W_c-2} \cdot x_{W_d-1} \; a_{W_c-1} \cdot x_{W_d-1}$

$a_0 \cdot x_{W_d-2} \; a_1 \cdot x_{W_d-2} \quad a_{W_c-1} \cdot x_{W_d-2}$

$\vdots$

$-a_0 \cdot x_2$

$-a_0 \cdot x_1 \quad -a_1 \cdot x_1$

$-a_0 \cdot x_0 \quad -a_1 \cdot x_0 \quad -a_2 \cdot x_0 \quad \bullet\bullet\bullet \quad -a_{W_c-1} \cdot x_0$

$y_{-1} \quad y_0 \bullet \; y_1 \quad \bullet\bullet\bullet \quad y_{W_d+W_c-3} \; y_{W_d+W_c-2}$

Figure 11.5 Bit-products used in multiplication

array multipliers. Of these types, the array multiplier takes up the least amount of area, but it is also the slowest with a latency proportional to $O(W_d)$ where W_d is the word length of the operands. The parallel (tree-based) multipliers have latencies of $O(\log_2(W_d))$, but they take up more area due to more irregular wiring.

11.5.3 Shift-and-Add Multiplication

A sequential multiplier is obtained by generating the bit-products row-wise. Consider the multiplication of two numbers, $\boldsymbol{y} = \boldsymbol{a} \cdot \boldsymbol{x}$. Using two's-complement representation we have

$$\boldsymbol{y} = \boldsymbol{a}\left(-x_0 + \sum_{i=1}^{W_d-1} x_i 2^{-i}\right) = -\boldsymbol{a}x_0 + \sum_{i=1}^{W_d-1} \boldsymbol{a}x_i 2^{-i} \tag{11.22}$$

The multiplication can be performed by generating the partial products (for example, rowwise) as illustrated in Figure 11.6. A *shift-and-add multiplier* is obtained by using only one bit-parallel adder and successively adding the partial products row or columnwise. Figure 11.7 shows the block diagram of a shift-and-add multiplier in which the partial products are implemented using a multiplexer instead of AND gates.

Addition starts with the partial products corresponding to the LSB. If $x_i = 0$, zero is accumulated via the multiplexer, while if $x_i = 1$, then the partial product, $\boldsymbol{a} \cdot x_i$, is accumulated. Starting with bit x_{W_d-1}, the bit-products $(a \cdot x_{W_d-1})$ is added to the initially cleared accumulator register and the sum is shifted one step to the right, which corresponds to division by 2.

Next, the row of bit-products $(\boldsymbol{a} \cdot x_{W_d-2})$ is added to the accumulator register. Upon reaching x_0, the bit-products $(\boldsymbol{a} \cdot x_0)$ are subtracted from the value in the accumulator. In practice it is more common to add the bit-products $(-\boldsymbol{a} \cdot x_0)$ instead.

A multiplication between two W_d-bit numbers will require W_d^2 AND operations and W_d-1 add and shift operations. The result will be a (W_d+W_c-1)-bit number.

Notice that division by 2, accomplished by shifting a two's-complement number one step to the right, requires that the sign bit be copied. For example, $(0.5)_{10} = (1.100)_{2C}$. Shifting one step to the right and copying the sign bit gives $(1.110)_{2C} = (0.25)_{10}$.

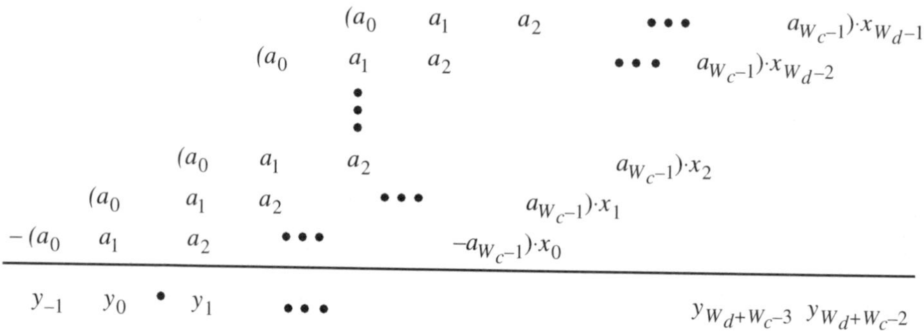

$(a_0 \quad a_1 \quad a_2 \quad \cdots \quad a_{W_c-1})\cdot x_{W_d-1}$
$(a_0 \quad a_1 \quad a_2 \quad \cdots \quad a_{W_c-1})\cdot x_{W_d-2}$
$\vdots$
$(a_0 \quad a_1 \quad a_2 \quad a_{W_c-1})\cdot x_2$
$(a_0 \quad a_1 \quad a_2 \quad \cdots \quad a_{W_c-1})\cdot x_1$
$-(a_0 \quad a_1 \quad a_2 \quad \cdots \quad -a_{W_c-1})\cdot x_0$

$y_{-1} \quad y_0 \; \bullet \; y_1 \quad \cdots \quad y_{W_d+W_c-3} \quad y_{W_d+W_c-2}$

Figure 11.6 Sequential forming of partial bit-products

The number of shift-and-add cycles can be reduced for fixed coefficients if the coefficients are represented by CSDC. On average the number of cycles reduced to $W_d/3$, with worst case is $W_d/2$. However, it may be difficult to use PEs with variable execution time.

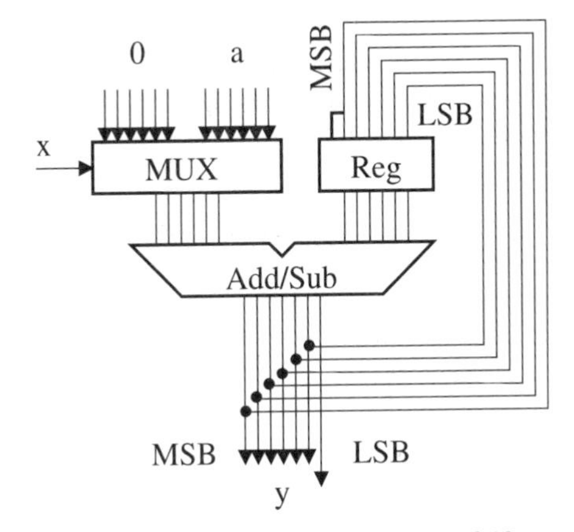

Figure 11.7 Block diagram for a shift-and-add-multiplier

11.5.4 Booth's Algorithm

Most modern general-purpose processors—for example, the MIPS R4000™—contain a dedicated processor for handling integer arithmetic operations. Multiplication in such integer processors, as well as in floating-point processors, usually uses some version of Booth's modified recoding algorithm.

Consider the multiplication involving a 16-bit two's-complement number. The multiplicand can be rewritten

$$x = \sum_{i=1}^{15} x_i 2^{-i} - x_0 2^0 = \sum_{i=1}^{8} x_{2i-1} 2^{-2i+1} + \sum_{i=1}^{7} x_{2i} 2^{-2i-1} - x_0 2^0$$

$$= \sum_{i=1}^{8} x_{2i-1} 2^{-2i+1} + \sum_{i=1}^{7} x_{2i} 2^{-2i-1} - 2\sum_{i=1}^{7} x_{2i} 2^{-2i-1} - x_0 2^0$$

Since $x_{16} = 0$, the second and third terms can be rewritten

$$x = \sum_{i=1}^{8} x_{2i-1} 2^{-2i+1} + \sum_{i=1}^{7} x_{2i} 2^{-2i-1} - 2\sum_{i=2}^{8} x_{2(i-1)} 2^{-2i+1} - x_0 2^0$$

Finally, we get

$$x = \sum_{i=1}^{8} \left[x_{2i-1} + x_{2i} - 2x_{2(i-1)}\right] 2^{-2i+1}$$

The multiplication $x \cdot y$ can now be written

$$x \cdot y = \sum_{i=1}^{8} \left[x_{2i-1} + x_{2i} - 2x_{2(i-1)}\right] y 2^{-2i+1} \tag{11.23}$$

Thus, three bits at a time from one of the input operands, x, is used to recode the other input operand, y, into partial products that are added (subtracted) to form the result. The value of the term in parentheses is either 0, ±1, or ±2. Hence, the number of partial products generated is reduced and they are simple multiples of the input operand ($-2y$, $-y$, 0, y, $2y$). A partial product of the type $\pm 2y$ (i.e., a multiplication by a factor two) is achieved with a left, or right, shift.

Booth's algorithm reduces the number of additions/subtractions cycles to $W_d/2 = 8$, which is only half that required in the ordinary shift-and-add algorithm. Notice that the reduction of the number of cycles directly corresponds to a reduction in the power consumption. Typically, decoding, shifting, and addition are pipelined in order to achieve higher throughput. Typically, an 11×16-bit multiplier can be operated with a cycle time of about 20 ns and requiring about 4200 devices and about 0.6 mm^2 using a standard 0.8-μm CMOS process. Booth's algorithm is also suitable for bit-serial or digit-serial implementation.

11.5.5 Tree-Based Multipliers

In order to simplify the illustration of a multiplication algorithm we will sometimes use the shorthand version of the bit-products shown in Figure 11.8 instead of the one shown in Figure 11.5.

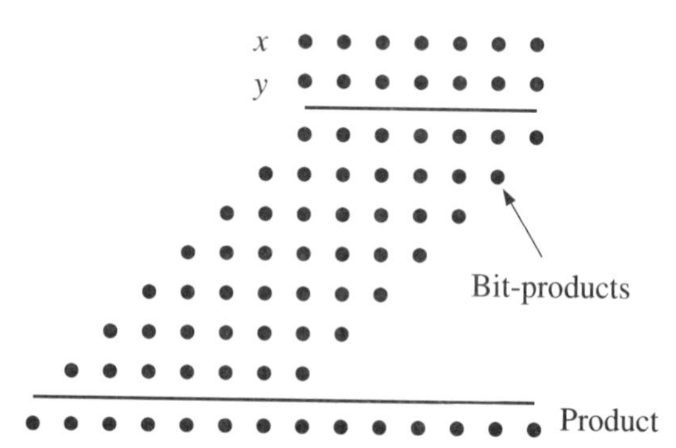

Figure 11.8 Simplified notation for multiplication

In tree-based multipliers the bit-products are added using full-adders. A full-adder can be considered as a counter, denoted (3, 2)-counter, that adds three inputs and forms a two-bit result. The output equals the number of 1s in the inputs. A half-adder is denoted (2, 2)-counter.

In 1964 Wallace showed that a tree structure of such counters is an efficient method, $O(\log_2(W_d))$, to add the bit-products. Figure 11.9 illustrates how the number of bit-products can be accumulated successively.

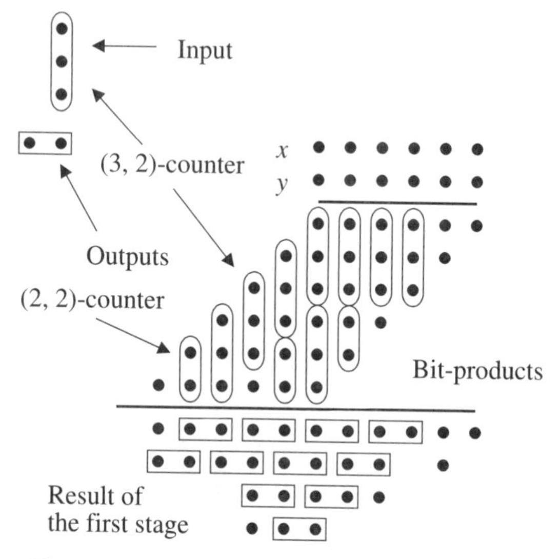

Figure 11.9 Reduction of the number of partial products

In the first stage, eight (3, 2)-counters and three (2, 2)-counters are used. The largest number of partial products in any column is reduced to four. In the second stage, we need, as shown in Figure 11.10, five (3, 2)-counters and three (2, 2)-counters to reduce the largest number of products to three, as shown in Figure 11.10. In the third stage, we need three (3, 2)-counters and three (2, 2)-counters to reduce the largest number of products to two.

Finally in the fourth stage, an RCA or a CLA can be used to obtain the product. For a 6×6 multiplier 13 (3, 2)-counters and nine (2, 2)-counters are needed, not counting the final carry–look–ahead adder.

Several alternative addition schemes are possible. For example, Dadda has derived a scheme where all bit-products with the same weight are collected and added using a Wallace tree with a minimum number of counters and minimal critical paths. However, the multiplier is irregular and difficult to implement efficiently since a large wiring area is required. Wallace tree multipliers should therefore only be used for large word length and where the performance is critical.

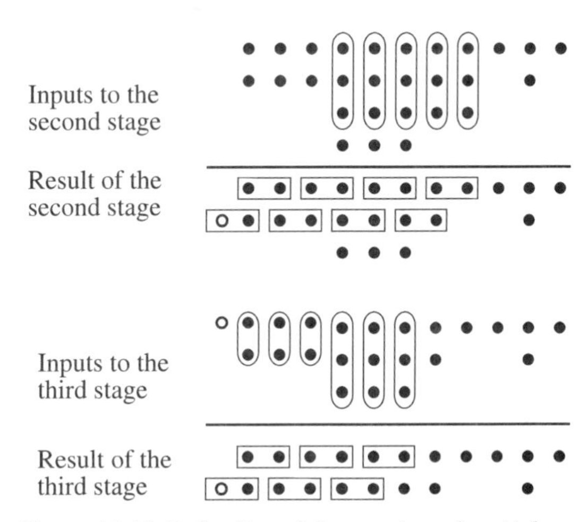

Figure 11.10 Reduction of the number of partial products

11.5.6 Array Multipliers

Many array multiplier schemes have been proposed. Varying degrees of pipelining are used, ranging from nonpipelined to fully systolic or wave front arrays. In this section we will only discuss a typical array multiplier—the *Baugh–Wooley's multiplier* with a multiplication time proportional to $2W_d$.

The Baugh–Wooley's multiplier scheme can be derived as follows:

$$\boldsymbol{P} = \boldsymbol{x} \cdot \boldsymbol{y} = \left(-x_0 + \sum_{i=1}^{W_d-1} x_i 2^{-i}\right)\left(-y_0 + \sum_{i=1}^{W_d-1} y_i 2^{-i}\right)$$

$$= x_0 \cdot y_0 + \sum_{i=1}^{W_d-1}\sum_{j=1}^{W_d-1} x_i \cdot y_j 2^{-i-j} - x_0 \sum_{i=1}^{W_d-1} y_i 2^{-i} - y_0 \sum_{i=1}^{W_d-1} x_i 2^{-i}$$

Each of the two negative terms may be rewritten

$$-\sum_{i=1}^{W_d-1} x_0 \cdot y_i 2^{-i} = -1 + 2^{-W_d+1} + \sum_{i=1}^{W_d-1} (1 - x_0 \cdot y_i) 2^{-i}$$

and by using the overflow property of two's-complement representation we get

$$-\sum_{i=1}^{W_d-1} x_0 \cdot y_i 2^{-i} = 1 + 2^{-W_d+1} + \sum_{i=1}^{W_d-1} \overline{x_0 \cdot y_i} 2^{-i}$$

We get

$$\boldsymbol{P} = 2 + 2^{-W_d+2} + x_0 \cdot y_0 + \sum_{i=1}^{W_d-1}\sum_{j=1}^{W_d-1} x_i \cdot y_j 2^{-i-j}$$
$$+ \sum_{i=1}^{W_d-1} \overline{x_0 \cdot y_i} 2^{-i} + \sum_{i=1}^{W_d-1} \overline{y_0 \cdot x_i} 2^{-i}$$

The partial products for a 4×4 multiplication (W_d = 4 bits) are shown in Figure 11.11. Notice that some slight variations in the way the partial products are formed occur in the literature. From the preceding expression and Figure 11.11 the logic realization of the Baugh–Wooley's multiplier can easily be identified.

				x_0	x_1	x_2	x_3
				y_0	y_1	y_2	y_3
			1	$\overline{x_0 \cdot y_3}$	$x_1 \cdot y_3$	$x_2 \cdot y_3$	$x_3 \cdot y_3$
			$\overline{x_0 \cdot y_2}$	$x_1 \cdot y_2$	$x_2 \cdot y_2$	$x_3 \cdot y_2$	
		$\overline{x_0 \cdot y_1}$	$x_1 \cdot y_1$	$x_2 \cdot y_1$	$x_3 \cdot y_1$		
1	$x_0 \cdot y_0$	$\overline{x_1 \cdot y_0}$	$\overline{x_2 \cdot y_0}$	$\overline{x_3 \cdot y_0}$			
p_{-1}	$p_0 \bullet$	p_1	p_2	p_3	p_4	p_5	p_6

Figure 11.11 Partial products for a 4×4 Baugh–Wooley's multiplier

A logic realization of the 4×4 array multiplier is shown in Figure 11.12. All bit-products are generated in parallel and collected through an array of full-adders and a final RCA. The first term (2) is taken into account by adding a 1 in the stage (p_0) that corresponds to the sign-bit. However, if the number range is reduced by not allowing the number 1, the multiplier can be simplified by removing the last stage (p_1) and p_0 is used as the sign-bit. Also the first level of full-adders can be simplified to half-adders. Notice that the multiplier has a regular structure that

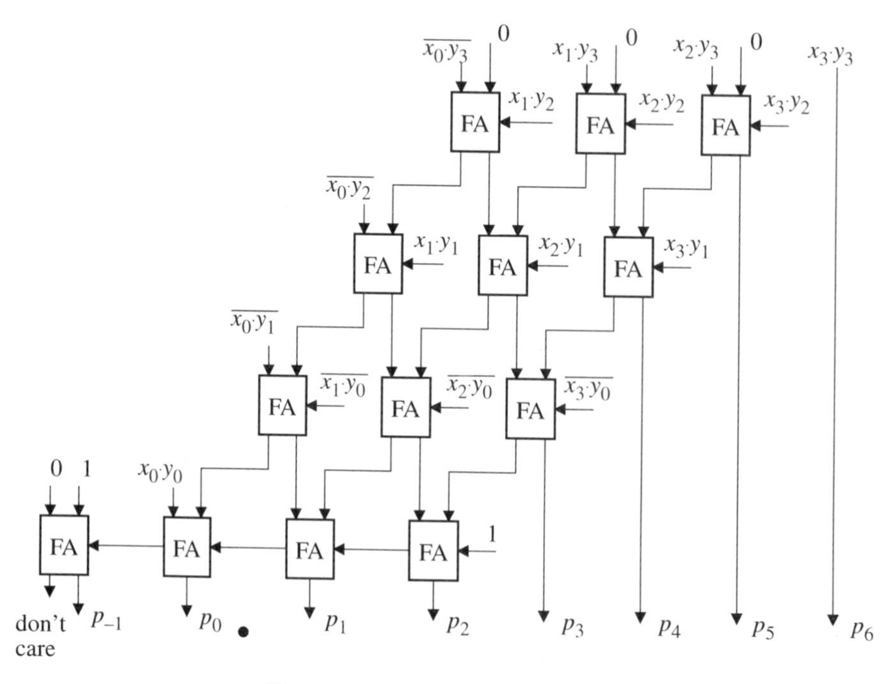

Figure 11.12 Baugh–Wooley's multiplier

simplifies the wiring and the layout. The Baugh–Wooley's multiplier algorithm can also be used to realize serial/parallel multipliers. The power consumption is of the order $O(W_d^3)$.

11.5.7 Look-Up Table Techniques

Multiplication can also be done by using look-up tables. A straightforward implementation would require a table with $2^{W_d+W_c}$ words. A ROM of this size would require too much chip area and also be too slow for the typical word length used in digital filters. A more efficient approach is based on the identity

$$x \cdot y = \frac{(x+y)^2}{4} - \frac{(x-y)^2}{4}$$

The multiplication can be carried out by only one addition, two subtractions, and two table look-up operations. The size of the lookup table, which stores squares, is reduced to only 2^{W_d} words. Hence, this technique can be used for numbers with small word lengths—for example, up to eight to nine bits.

11.6 BIT-SERIAL ARITHMETIC

Bit-serial arithmetic is a viable alternative in digital signal processing applications to traditional, bit-parallel arithmetic. A major advantage of bit-serial over bit-parallel arithmetic is that it significantly reduces chip area. This is done in two ways. First, it eliminates wide buses and simplifies wire routing. Second, by using small processing elements, the chip itself will be smaller and require shorter wiring. A small chip can support higher clock frequencies and is therefore faster. Two's-complement and binary offset representations are suitable for DSP algorithms implemented with bit-serial arithmetic, since the bit-serial operations can be done without knowing the sign of the numbers involved. Since two's-complement and binary offset representation use similar algorithms, we will discuss only bit-serial arithmetic with two's-complement representation.

A major issue in the design of the building blocks is to decide how data shall be transmitted and processed. There are two basic possibilities: bit-serial or bit-parallel transmission. Usually, this choice also governs how operations are performed. When making this choice, it is easy to draw the wrong conclusion given the following argument: "Since bit-serial arithmetic processes only one bit at each time instance, bit-parallel arithmetic must be about W_d times faster, assuming the data word length is W_d." In reality, the ratio in speed will be much smaller due to the long carry propagation paths present in parallel arithmetic. Furthermore, bit-parallel arithmetic uses more than W_d times as large a chip area as does bit-serial arithmetic. In fact, the computational throughput per unit chip area is higher than for parallel arithmetic. On the whole, bit-serial arithmetic is often superior.

The issue of comparing power consumption is, however, more complicated. Bit-parallel arithmetic suffers from energy losses in glitches that occur when the carry propagates, but the glitches will be few if successive data are strongly correlated. Driving long and wide buses consumes large amounts of power. Bit-serial arithmetic, on the other hand, will only perform useful computations without any glitches, but require more clocked elements that will consume significant amounts of power. Power-efficient realization of the clocked elements is therefore important. To summarize, bit-parallel and bit-serial arithmetics each have their advantages and disadvantages.

11.6.1 Bit-Serial Addition and Subtraction

In bit-serial arithmetic the numbers are normally processed with the least-significant bit first. Bit-serial numbers in a two's-complement representation can be added or subtracted with the circuits shown in Figure 11.13.

Since the carries are saved from one bit position to the next the circuits are called *carry–save adder* and *carry–save subtractor*, respectively. At the start of the computation the D flip-flop is reset (set) for the adder (subtractor), respectively. We have for addition

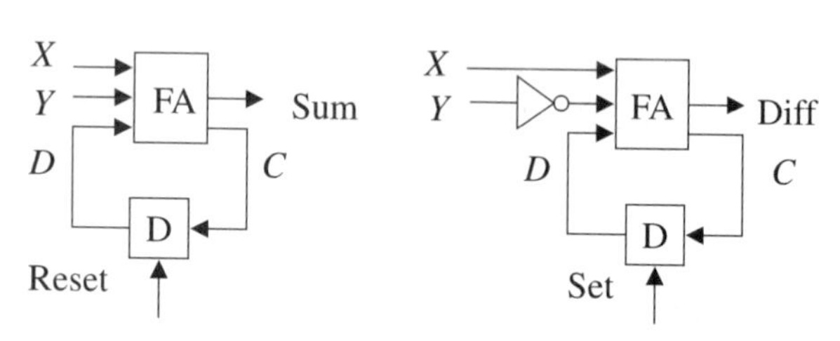

Figure 11.13 Bit-serial adder and subtractor

$$S_i = XOR\{X_i, Y_i, D_i\} = X_i \oplus Y_i \oplus D_i$$
$$C_i = \text{Majority}\{X_i, Y_i, D_i\} = X_i \cdot Y_i + X_i \cdot D_i + Y_i \cdot D_i$$
$$D_i = C_{i-1} \quad \text{for } i = W_d - 1, W_d - 2, \ldots, 1, 0$$
$$D_{W_d - 1} = 0$$

and for subtraction:

$$S_i = XOR\{X_i, \overline{Y_i}, D_i\}$$
$$C_i = \text{Majority}\{X_i, \overline{Y_i}, D_i\}$$
$$D_i = C_{i-1} \quad \text{for } i = W_d - 1, W_d - 2, \ldots, 1, 0$$
$$D_{W_d - 1} = 1$$

11.6.2 Bit-Serial Multiplication

According to Figure 11.5, the multiplication of two two's-complement numbers, $\boldsymbol{y} = \boldsymbol{a} \cdot \boldsymbol{x}$, requires the formation of the partial bit-products by multiplying the coefficient, $\boldsymbol{a}$, by the bits of $\boldsymbol{x}$. The partial products can then be added bit-serially with the proper weighting. This addition can be performed by a variety of schemes. The slowest bit-serial multipliers are obtained if only one bit-product is generated and added in each time slot. Most bit-serial multipliers, however, are in practice based on the shift-and-add algorithm where several bit-products are added in each time slot. We will describe several such bit-serial multipliers in the next sections.

11.6.3 Serial/Parallel Multiplier

Many forms of so-called serial/parallel multipliers are have been proposed [24,37]. In a serial/parallel multiplier the multiplicand, $\boldsymbol{x}$, arrives bit-serially while the multiplier, $\boldsymbol{a}$, is applied in a bit-parallel format. Many different schemes for bit-serial multipliers have been proposed. They differ mainly in which order bit-products are generated and added and in the way subtraction is handled. A common approach is to generate a row, or diagonal, of bit-products in each time slot (see Figure 11.8) and perform the additions of the bit-products concurrently. We will in this and the following sections describe several alternative serial/parallel multiplier algorithms and their implementations.

First, lets consider the special case when data is positive, $\boldsymbol{x} \geq 0$. Here the shift-and-add algorithm can be implemented by the circuit shown in Figure 11.14, which uses carry-save adders. The coefficient word length is five bits. Since $\boldsymbol{x}$ is processed bit-serially and coefficient $\boldsymbol{a}$ is processed bit-parallel, this type of multiplier is called a *serial/parallel multiplier.* Henceforth we do not explicitly indicate that the D flip-flops are clocked and reset at the beginning of a computation.

Addition of the first set of partial bit-products starts with the products corresponding to the LSB of $\boldsymbol{x}$. Thus, in the first time slot, at bit x_{W_d-1}, we simply add $\boldsymbol{a} \cdot x_{W_d-1}$ to the initially cleared accumulator.

Next, the D flip-flops are clocked and the sum-bits from the FAs are shifted one bit to the right, each carry-bit is saved and added to the FA in the same stage, the sign-bit is copied, and one bit of the product is produced at the output of the accumulator. These operations correspond to multiplying the accumulator contents by 2^{-1}. In the following clock cycle the next bit of $\boldsymbol{x}$ is used to form the next set of bit-products which are added to the value in the accumulator, and the value in the accumulator is again divided by 2.

This process continues for W_d–1 clock cycles, until the sign bit of $\boldsymbol{x}$, x_0, is reached, whereupon a subtraction must be done instead of an addition. At this point, the accumulator has to be modified to perform this subtraction of the bit-products, $\boldsymbol{a} \cdot x_0$. We will present an efficient method to perform the subtraction in Example 11.5. Recall that we assumed that the data here are positive. Hence, $x_0 = 0$ and the subtraction is not necessary, but a clock cycle is still required. The highest clock frequency is determined by the propagation time through one AND gate and one full-adder.

During the first W_d clock cycles, the least significant part of the product is computed and the most significant is stored in the D flip-flops. In the next W_c–1 clock cycles, zeros are therefore applied to the input so that the most significant part of the product is shifted out of the multiplier. Hence, the multiplication requires W_d+W_c–1 clock cycles. Two successive multiplications must therefore be separated by W_d+W_c clock cycles since one clock cycle is required to clear the accumulator.

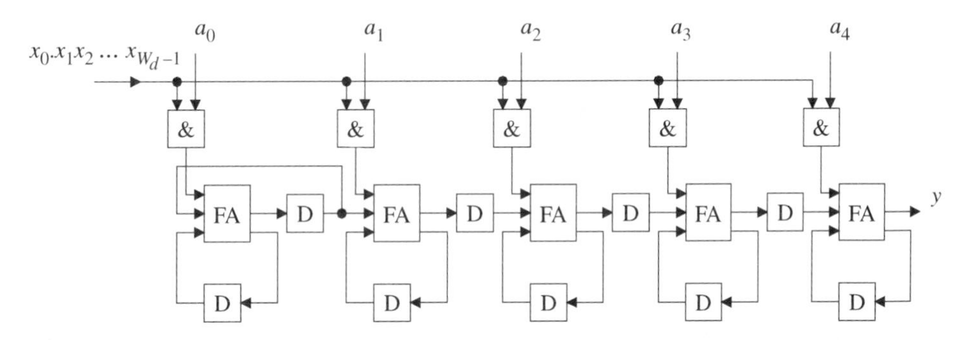

Figure 11.14 Serial/parallel multiplier based on carry-save adders

EXAMPLE 11.6

Show that subtraction of the bit-products required for the sign-bit in the serial/parallel multiplier can be avoided by extending the input by W_c–1 copies of the sign-bit.

After W_d–1 clock cycles the most significant part of the product is stored in the D flip-flops. In the next W_c clock cycles the sign bit of $\boldsymbol{x}$ is applied to the multipliers input. This is accomplished by the sign extension-circuit shown in Figure 11.15. The sign extension-circuit consists of a latch that transmits all bits up to the sign-bit and thereafter latches the sign-bit. For simplicity, we assume that W_d = 6 bits and W_c = 5 bits.

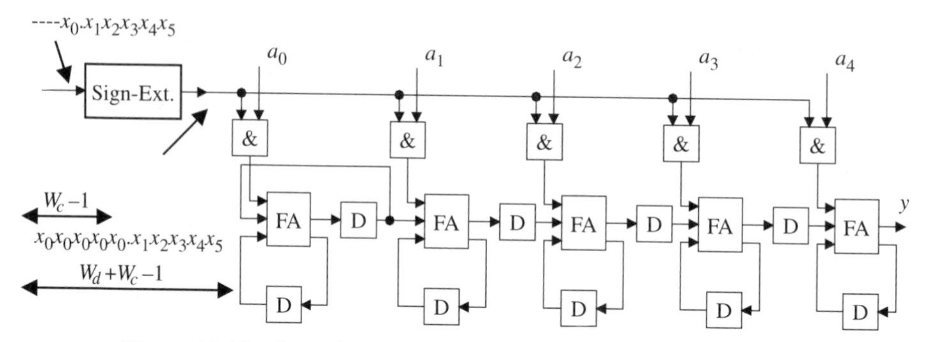

Figure 11.15 Serial/parallel multiplier with sign-extension circuit

The product is

$$\boldsymbol{y} = \boldsymbol{a} \cdot \boldsymbol{x} = \boldsymbol{a} \cdot \left\{-x_0 + \sum_{i=1}^{5} x_i 2^{-i}\right\}$$

but the multiplier computes

$$\begin{aligned} y_{Mult} &= \boldsymbol{a} \cdot \left\{x_0 2^4 + x_0 2^3 + x_0 2^2 + x_0 2^1 + x_0 2^0 + \sum_{i=1}^{5} x_i 2^{-i}\right\} \\ &= \boldsymbol{a} \cdot \left\{x_0 2^4 + x_0 2^3 + x_0 2^2 + x_0 2^1 + x_0 2^1 + \sum_{i=1}^{5} x_i 2^{-i}\right\} \\ &= \boldsymbol{a} \cdot \boldsymbol{x}_0 \{2^4 + 2^3 + 2^2 + 2^1 + 2^1\} + \boldsymbol{a} \cdot \boldsymbol{x} \\ &= \boldsymbol{a} \cdot x_0 2^5 + \boldsymbol{a} \cdot \boldsymbol{x} \end{aligned} \quad (11.24)$$

The first term here contributes an error in the desired product. However, as shown next, there will not be an error in the W_d+W_c–1 least-significant bits since the error term only contributes to the bit positions with higher significance.

A bit-serial multiplication takes at least W_d+W_c–1 clock cycles. In section 11.15, we will present a technique that uses partially overlapping of subsequent multiplications to increase the throughput. These serial/parallel multipliers, using this technique, can be designed to perform one multiplication every $max\{W_d, W_c\}$ clock cycles. A 16-bit serial/parallel multiplier implemented using two-phase logic in a 0.8-μm CMOS process requires an area of only 90 μm × 550 μm ≈ 0.050 mm^2.

An alternative solution to copying the sign-bit in the first multiplier stage is shown in Figure 11.16. The first stage, corresponding to the sign-bit in the coefficient, is replaced by a subtractor. In fact, only a half-adder is needed since one of the inputs is zero. We will later see that this version is often the most favorable one.

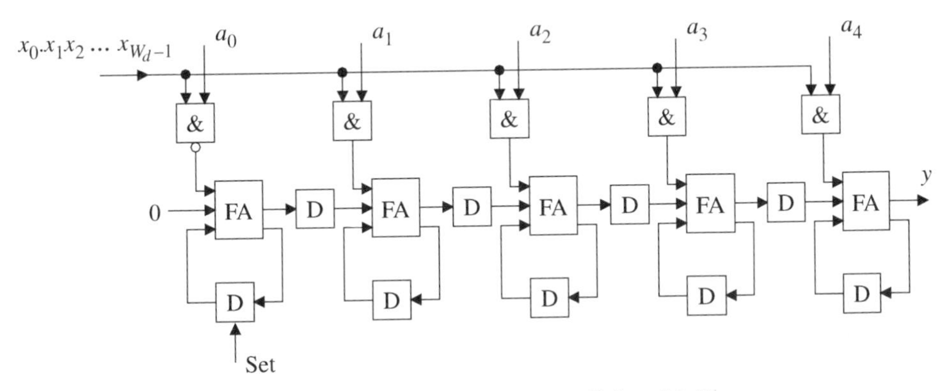

Figure 11.16 Modified serial/parallel multiplier

11.6.4 Transposed Serial/Parallel Multiplier

An alternative realization of the serial/parallell multiplier is shown in Figure 11.17, which adds the bit-products in Figure 11.6 columnwise. This multiplier structure, which can be derived for Figure 11.16 by using the transposition theorem, suffers from a long sum-propagation path. However, this disadvantage can be alleviated by pipelining, at a cost of two D flip-flops per stage. An advantage is that truncation or rounding of the product can be done so that the multiplication time can be reduced to only W_d clock cycles [26]. Further, this multiplier structure can also be modified into a serial/serial multiplier, or squarer, where both the multiplier and the multiplicand arrive bit-serially.

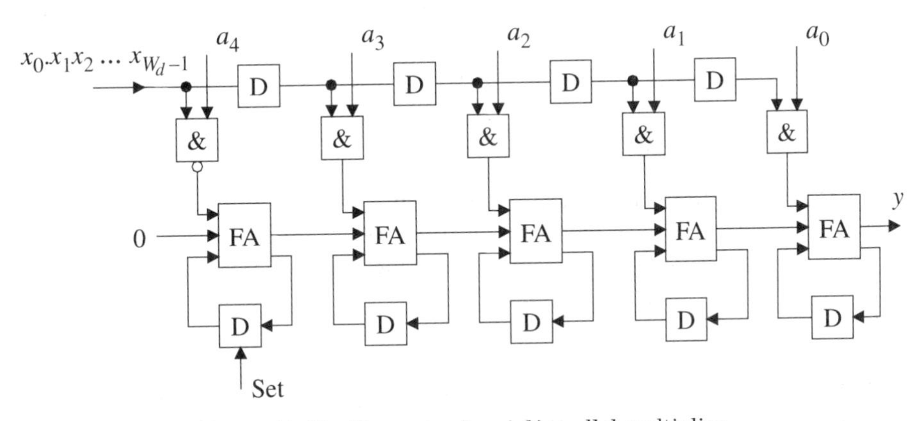

Figure 11.17 Transposed serial/parallel multiplier

11.6.5 S/P Multiplier–Accumulator

An alternative realization of the serial/parallel SP multiplier with an additional input that allows computations of the type $\boldsymbol{y} = \boldsymbol{a} \cdot \boldsymbol{x} + \boldsymbol{z}$ is shown in Figure 11.18. The extra input allows a value $\boldsymbol{z}$ to be added at the same level of significance as $\boldsymbol{x}$. A multiplier–accumulator is obtained if the output $\boldsymbol{y}$ is truncated or rounded to the same word length as $\boldsymbol{x}$ and added to the subsequent multiplication. A full precision multiplier–accumulator is obtained if the part of $\boldsymbol{y}$ that is truncated is saved and used to set the sum D flip-flops instead of resetting them at the start of a multiplication.

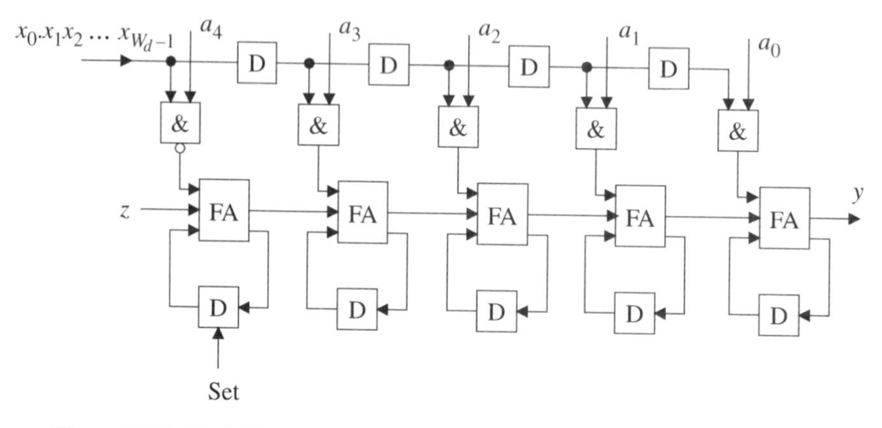

Figure 11.18 Serial/parallell multiplier with an inherent input for addition

11.7 BIT-SERIAL TWO-PORT ADAPTOR

The PEs must be able to handle overflow and quantization of data. In order to detect overflow, extra guard bits must be introduced by extending the sign bit to the left. If the output from a PE is correct, the sign bit and the guard bits are equal, otherwise they differ. Hence, overflow detection is simple. However, due to the bit-serial processing, the correction of overflow must be delayed until the result is available. This may be a problem in recursive loops.

Quantization of data must also be performed in the recursive loops. In the case of truncation, quantization is simple and does not incur any extra clock cycles. The penalty for using more complicated quantization schemes may be additional clock cycles. In nonrecursive algorithms, such as the FFT and the DCT, extra pipeline stages can be used for overflow and quantization. Thus, the throughput will not be decreased.

Bit-serial adders and multipliers implement adaptors in a wave digital filter efficiently. A block diagram for the bit-serial implementation of a symmetric two-port adaptor based on a serial/parallel multiplier is shown in Figure 11.19. For simplicity the word length of the adaptor coefficient is selected to be only 5 bits. Pipelining has been applied such that the longest path is a full-adder and a few gates. Execution time for the adaptor is

$$T_{PE} = (W_c + W_d + 2)\, T_{CL}$$

and latency is $2T_{CL}$ where T_{CL} is the clock period.

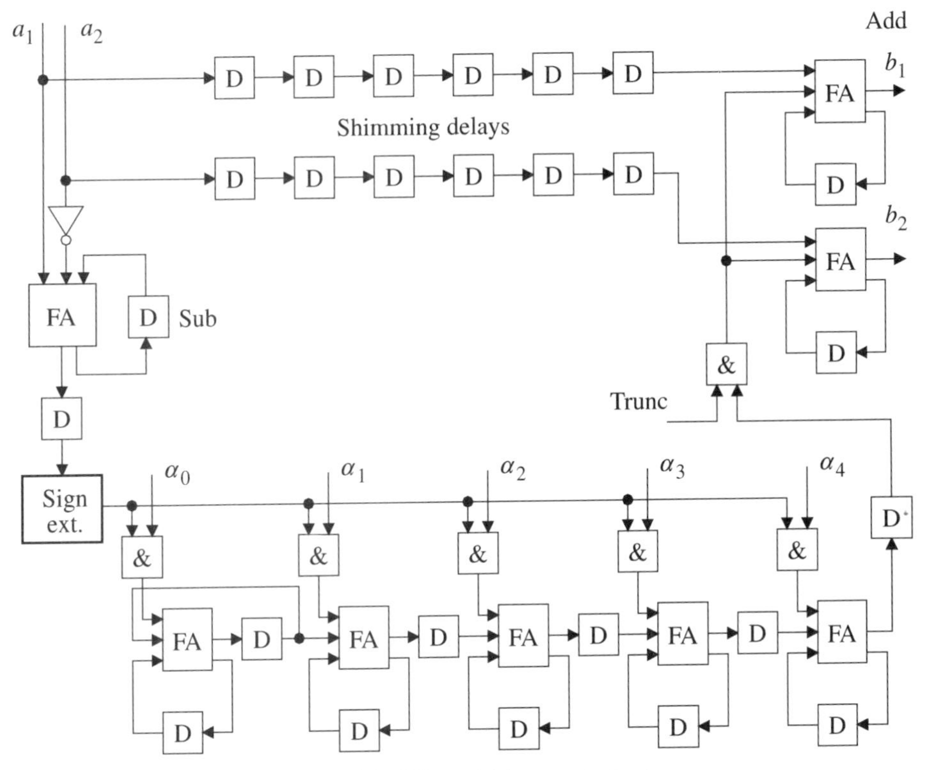

Figure 11.19 Bit-serial two-port adaptor

All the D flip-flops are reset except the subtractor, which is set, when LSBs are applied to the inputs. Shimming delays are needed to synchronize the data streams from the two inputs with the output of the multiplier. Control signal *Trunc* performs a truncation of the two's-complement product. As discussed in Chapter 5, a sufficient condition to avoid limit cycles in wave digital filters is given by Equation (5.2)

$$|b_i(n)_Q| < |b_i(n)_{exact}|$$

The magnitude of $b_i(n)$ is reduced by either the quantization or the overflow correction. The magnitude of the actual value is less than its correct value. This scheme guarantees that the wave digital filter suppresses parasitic oscillations.

A two's-complement number with a word length of W_d bits has a range [–1, 1[. When two such numbers are added, an overflow may occur. Overflow corrections must be performed in front of the multipliers in the two-port adaptors, as discussed in Chapter 5. To avoid overflow, the numerical range is increased to [–2, 2[using one guard bit. Three cases occur:

- Sign bit = guard bit => no action
- Sign bit = 1, guard bit = 0
  => positive overflow, i.e., the result is in the range [1, 2[
  => set the value to $(0.1111...1)_2$
- Sign bit = 0, guard bit = 1
  => negative overflow, i.e., the result is in the range [–2, –1[
  => set the value to $(1.0000...0)_2$

The serial/parallell converters used between the RAMs and the PEs can be modified to perform overflow correction. Figure 11.20 shows how overflow correction can be implemented.

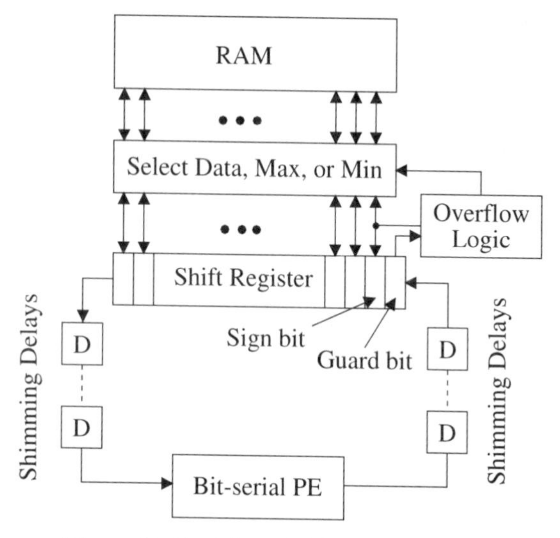

Figure 11.20 Bit-serial PE with overflow correction

When a value is inside the shift register, the sign bit and the guard bit are positioned in the two rightmost memory cells. If an overflow is detected, the largest or smallest value is instead written into the RAM. The necessary circuitry consists of a set of switches to select the proper values. Notice that only one set of switches is needed for each RAM. No extra clock cycles are needed. The quantization procedure in the two-port adaptor is

- Truncate the output from the multiplier to W_d bits.
- Add the sign bit of $b_1(n)$ and $b_2(n)$ to the least significant bit of $b_1(n)$ and $b_2(n)$, respectively.

In order not to waste clock cycles we store the uncorrected values of $b_1(n)$ and $b_2(n)$ in the RAM, and instead add the sign bits when the data are read from the RAM. Addition of the sign bits is performed using two half-adders, as shown in Figure 11.21. This procedure incurs one clock cycle. However, this is no real cost, since we have spare clock cycles in the loop PE–shimming delays–shift register. The length of this loop has to be a multiple of W_d.

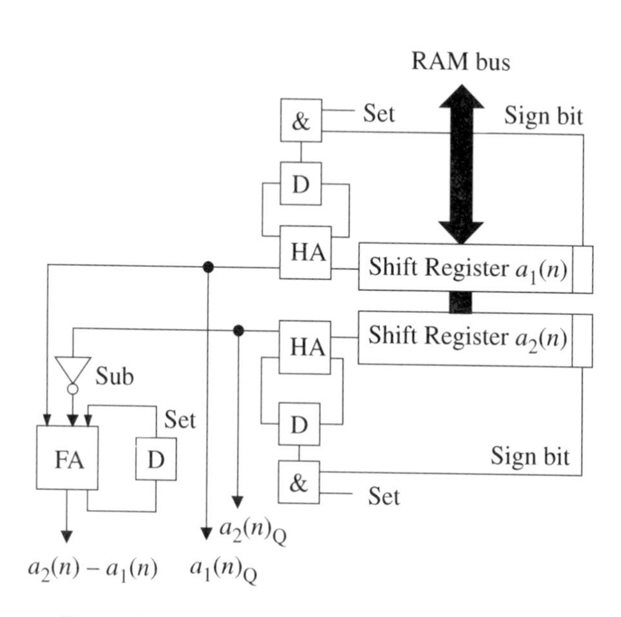

Figure 11.21 Magnitude truncation in two-port adaptors

Note that series and parallel two-port adaptors can be realized more efficiently than the symmetric two-port adaptor using the S/P multiplier shown in Figure 11.18. See problem 11.15.

11.8 S/P MULTIPLIERS WITH FIXED COEFFICIENTS

The multiplier can be simplified significantly if the coefficient is fixed. If a certain bit in the coefficient is 1, then the corresponding AND gate can be replaced by a wire. On the other hand, if a certain bit is 0, then the corresponding AND gate can be removed. We demonstrate further possible simplifications by the means of an example.

EXAMPLE 11.7

Simplify the serial/parallel multiplier shown in Figure 11.22 for the fixed coefficient $\boldsymbol{a} = 0.1011$.

First, we remove unnecessary AND gates as shown in Figure 11.23. Next, we notice that the D flip-flops are cleared at the beginning of the multiplication. The left-most FA therefore has only zeros as inputs.

Hence, it will always produce sum and carry bits that are zero, and can therefore be removed. Notice that the FA can not be removed if the coefficient is negative. We will later show that this case can be avoided.

Inputs to the second FA will be the x-bit, a zero from the preceding stage, and a zero from the initially cleared D flip-flop (carry). This FA will produce a sum bit equal to the x-bit, and the carry will always be zero. The second FA can therefore be removed, leaving only the D flip-flop. The third FA can also be

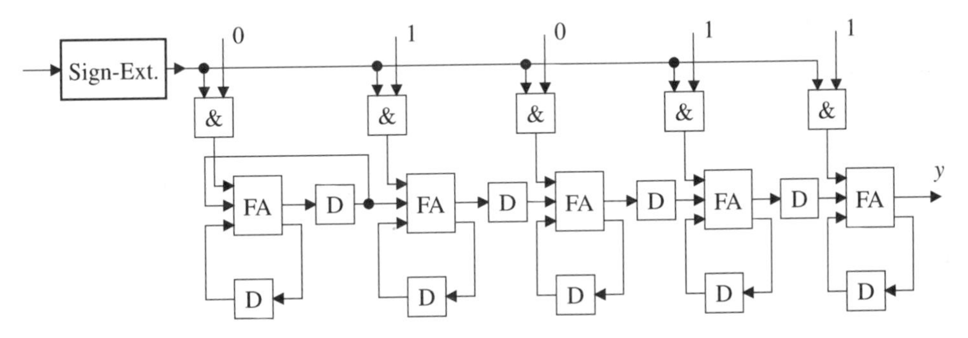

Figure 11.22 Serial/parallel multiplier with coefficient $\boldsymbol{a} = 0.1011$

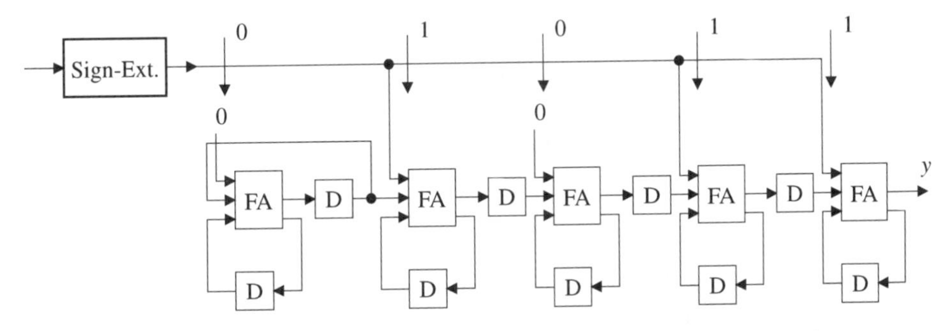

Figure 11.23 Serial/parallell multiplier with AND gates removed

removed since two of its inputs always are zero. The much simplified multiplier is shown in Figure 11.24.

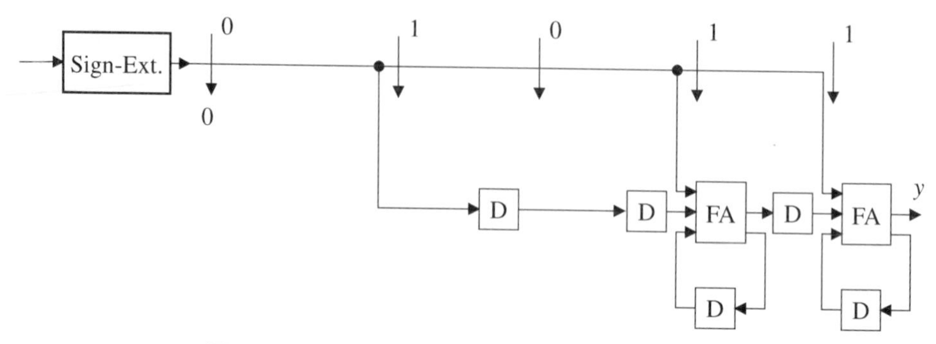

Figure 11.24 Simplified serial/parallell multiplier

Generally, the serial/parallel multiplier can be simplified by using the Box 11.1.

1. Remove all AND gates.
2. Remove all FAs and corresponding D flip-flops, starting with the most significant bit in the coefficient, up to the first 1 in the coefficient.
3. Replace each FA that corresponds to a zero-bit in the coefficient with a feedthrough.

Box. 11.1 Simplification of S/P multipliers with fixed coefficients

The number of FAs is one less that the number of 1's in the coefficient. The number of D flip-flops equals the number of 1-bit positions between the first and last bit positions. Thus, substantial savings can be made for fixed coefficients. Hence, it is common practice in digital filter design to optimize coefficients to have a minimum number of 1's.

11.8.1 S/P Multipliers with CSDC Coefficients

The serial/parallel multiplier can be simplified significantly if the coefficient is fixed. The cost is essentially determined by the number of nonzero bits in the coefficient. In CSDC, which was discussed in section 11.3.2, the average number of nonzero bits is only about $W_c/3$, compared to $W_c/2$ for two's-complement numbers. Further reductions in hardware resources are therefore possible.

A number in CSDC can be written

$$\boldsymbol{c} = \pm c_0 \pm c_1 2^{-1} \pm c_2 2^{-2} \pm c_3 2^{-3} \pm \dots \pm c_{Wc-1} 2^{-W_c+1}$$

where most of the bits are zero. The number $\boldsymbol{c}$ can be rewritten as a difference between two numbers with only positive coefficients,

$$\boldsymbol{c} = \boldsymbol{c}_+ - \boldsymbol{c}_-$$

A multiplication can now be written

$$y = c\,x = (c_+ - c_-)\,x = c_+ x + c_-(-x) = c_+ x + c_-(\bar{x} + 2^{-W_d+1})$$

where represents the original value with all bits inverted. Obviously, the multiplication can be implemented using the technique just discussed, except the x-inputs to the FAs in bit positions with negative coefficient weights are inverted, and the corresponding carry D flip-flops are initially set.

EXAMPLE 11.8

Simplify the implementation of a serial/parallell multiplier with coefficient $a = (0.00111)_2$ using CSDC. Also compare the cost with an implementation with a in two's-complement coefficient representation.

We first rewrite the coefficient, as just discussed, using CSDC. We get

$$a = (0.00111)_2 = (0.0100\text{–}1)_{CSDC} = (0.01000)_2 - (0.00001)_2$$

Only one FA and four D flip-flops are needed, as shown in Figure 11.25, while an implementation using the two's-complement coefficient would require two FAs and five D flip-flops. Hence, the CSDC implementation is better in this case. The D flip-flops— except for the carry D flip-flop, which is set—are cleared at the beginning of the multiplication.

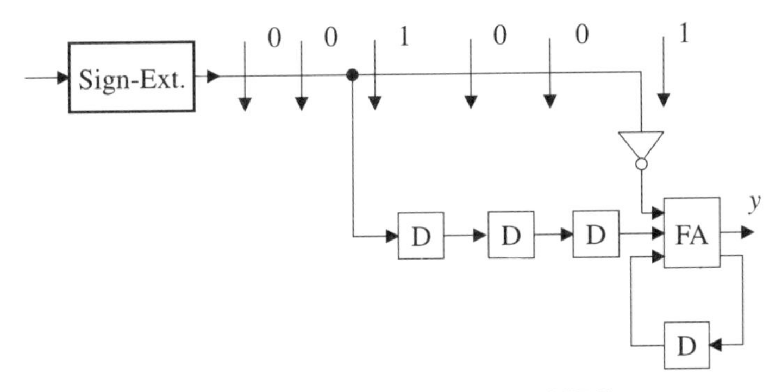

Figure 11.25 Serial/parallell multiplier with CSDC representation of the coefficient $a = (0.01001)_{CSDC}$

A significant implementation problem is the setting and resetting of the D flip-flops. These trivial operations are often costly in terms of power consumption and chip area, because the flip-flops represent a significant capacitive load. However, these operations can often be simplified and sometimes even avoided [13].

11.9 MINIMUM NUMBER OF BASIC OPERATIONS

There are many applications of fixed-point multiplications with fixed coefficients. In such cases the implementation cost can often be reduced if the multiplications are replaced by elementary operations. The most interesting cases are when the multiplication is realized by only using the following operations:

Addition only
Addition and subtraction
Addition and shift
Addition and subtraction and shift

As before, we will not differentiate between addition and subtraction, since their implementation cost is roughly the same. A shift operation corresponds to a multiplication with a power-of-two. A shift operation can be implemented in bit-parallel arithmetic either by a barrel shifter if the number of shifts varies or simply by a skewed wiring if the number of shifts is fixed. In bit-serial arithmetic a shift operation corresponds to a cascade of D flip-flops. Thus, in parallel arithmetic a fixed amount of shift may have almost negligible cost while both the chip area and power consumption are significant in bit-serial arithmetic.

In this section we will discuss various schemes to reduce the number of add/sub-and-shift operations. We will mainly discuss the bit-serial case with integer coefficients, since the case with fractional bits can be obtained by a simple modification of the algorithms. Further, the bit-parallel case is almost the same as the bit-serial case.

11.9.1 Multiplication with a Fixed Coefficient

As discussed before, a multiplication with a fixed coefficient (multiplicand) can be simplified if the latter is expressed in canonic signed digit code (CSDC). The number of add/sub operations equals the number of nonzero digits in the multiplicand minus one. However, the number of adders/subtractors required by this approach is not always a minimum if the multiplicand is larger than 44. For example, the number $(45)_{10} = (1010101)_{CSDC}$ requires three additions (subtractions) using the technique discussed in section 11.8.1. Another more efficient alternative representation is $(45)_{10} = (2^3 + 1)(2^2 + 1)$, which requires only two additions. Hence, it is of great interest to find the best algorithm to perform the multiplication with fewest basic operations.

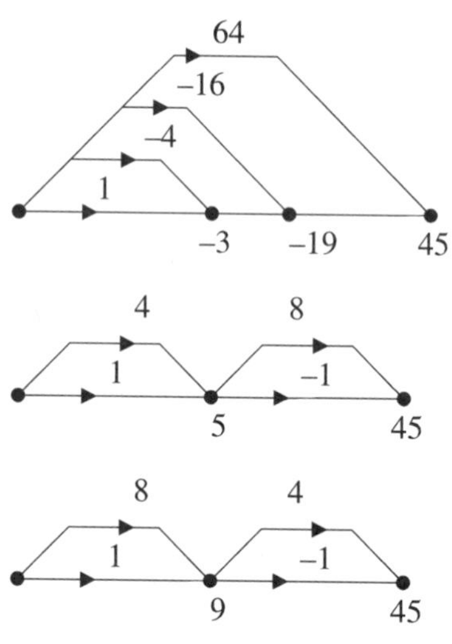

Figure 11.26 Alternative graphs representing

To this purpose it is useful to use the graphs such as those in Figure 11.26 to describe the various ways in which a multiplicand can be represented. The leftmost and rightmost nodes are the initial node and terminal node, respectively. They are connected with branches with associated binary weights—i.e., $\pm 2^n$. These weights correspond to shift operations. Each intermediate node, which represents an addition or subtraction, must have two input branches and at least one output branch. The uppermost graph describes the CSDC representation while the other two describe the more efficient algorithms.

For sake of simplicity we will only discuss multiplication with integers. Hence, $n \geq 0$. Multiplication with the corresponding fractional coefficient is easily obtained from the same graph and we need only to consider positive numbers, since the negative counterparts can be obtained by changing the sign of the branch weights. As mentioned before, we do not differentiate between the two since adders and subtractors are assumed to have equal costs.

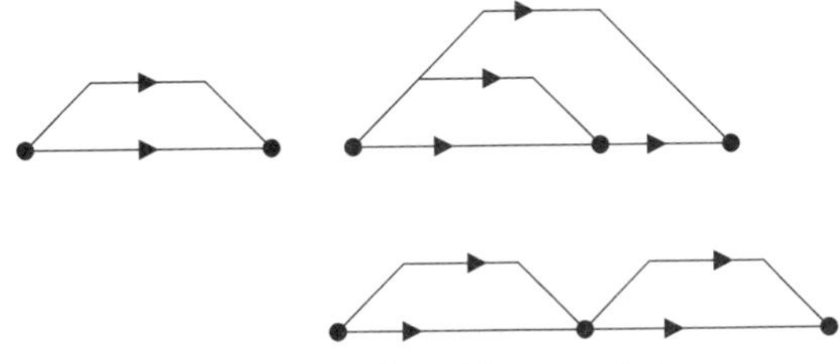

Figure 11.27 Graphs with one and two adders/subtractors multipliers

Figures 11.27 and 11.28 show all possible graphs with up to three adders or subtractors. There is only one graph with a single adder/subtractor and two graphs with two adders/subtractors. There are eight different graphs with three adders/subtractors and 32 different graphs with four adders/subtractors. The number of possible graphs grows rapidly with the number of adders/subtractors.

Obviously there must exist a representation that yields a minimum number of elementary operations (add/subtract-shift), but this number may not be unique. Heuristic algorithms for determining the best representation for a given set of basic operations are given in [6, 14, 23]. Note that for some numbers the CSDC representation is still the best.

All numbers in the range [4096, 4096], which correspond to a 12-bit word length, and, of course, many outside this range can be realized with only four adder/subtractors.

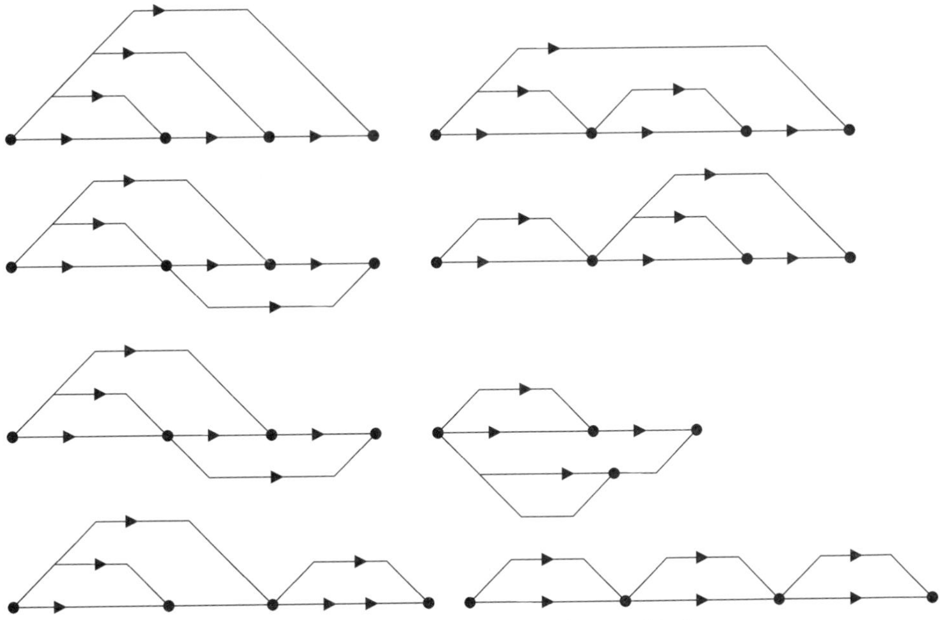

Figure 11.28 Graphs for three adder multipliers

For a 12-bits word length, the optimal multipliers achieve an average reduction of 16% in the number of adders required over CSDC. Note, however, that for a particular multiplicand the reduction may be much larger.

In digital filters, FFTs, etc., it is in practice sufficient to use no more than five additions, which can represent all coefficients with 16-bit word lengths and, of course, many numbers corresponding to the longer word lengths.

In practice it may be efficient to once and for all generate and store a table of the best alternative realizations for word length of interest—for example, up to 12 to 14 bits. When a search for favorable coefficient is performed, the cost can then be obtained from this table.

EXAMPLE 11.9

Find possible graphs for realization of the multiplicand 75 using two or three adders/subtractors. Derive also the corresponding bit-serial multiplication realization with the coefficient 75/128 from these graphs.

Figure 11.29 shows two alternative algorithms for the integer 75. The first, which requires three adders, corresponds to the realization

$$c_1 = 9 \cdot 8 + 3 = (1 + 2^3)\, 2^3 + (1 + 2^1)$$

and the second alternative, which only requires two adders, corresponds to

$$c_2 = 5 \cdot 15 = (1 + 2^2)\,(1 + 2^4)$$

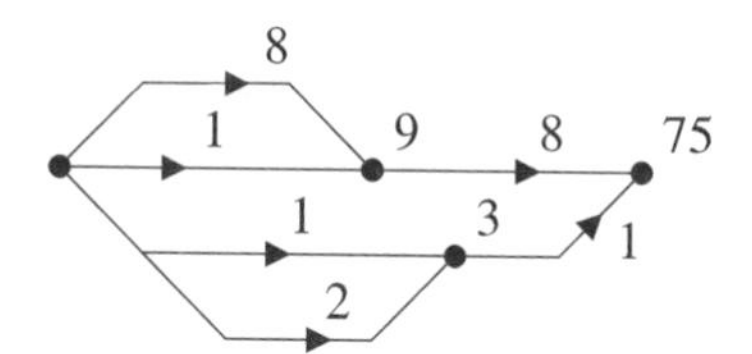

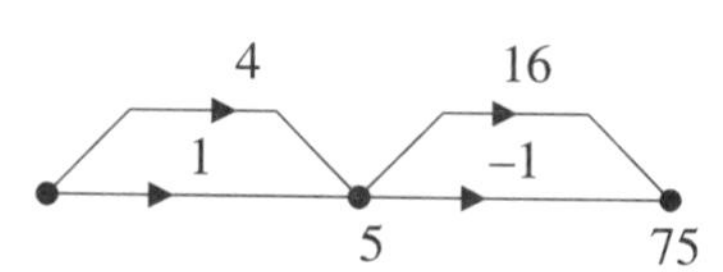

Figure 11.29 Alternative graphs for multiplication with 75

To derive the graphs corresponding to multiplication with 75/128 we insert a D flip-flop for every power of 2 and scale the output with 1/128 by moving the binary point seven positions to the left. The resulting realizations are shown in Figure 11.30.

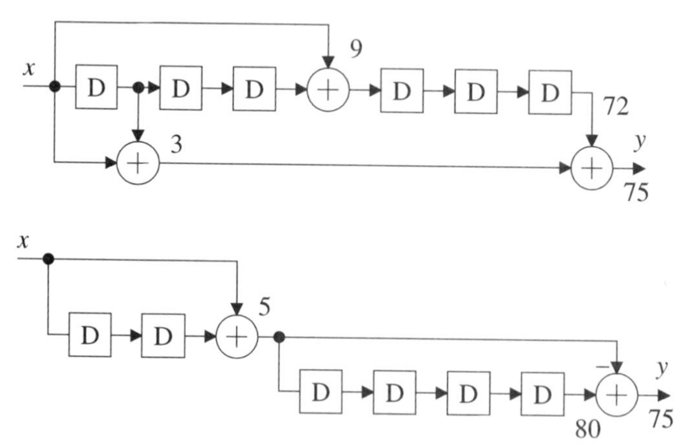

Figure 11.30 Alternative realizations of multiplication with the coefficient 75/128

11.9.2 Multiple-Constant Multiplications

Most DSP algorithms involve multiplication of one variable with several constants as shown in Figure 11.31 or the transposed version shown in Figure 11.32. Such composite operations can be significantly simplified by eliminating common subexpressions as illustrated in Figure 11.33 [32]. The aim is to reduce the number of basic operations (i.e., add/sub-and-shift operations), by factoring out common factors in the coefficients c_i. By first generating a set of common subexpressions that in the next step are multiplied with simple coefficients or added/subtracted to generate new subexpressions, the overall complexity of the implementation can be reduced.

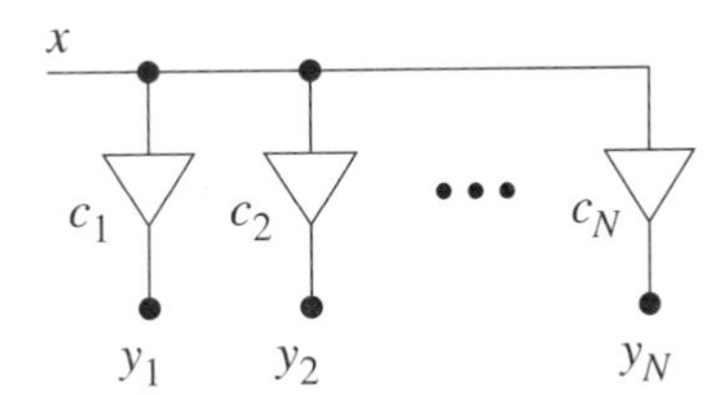

Figure 11.31 Multiple-constant multiplication.

For multiple-constant multiplication with many coefficients (for example, for long FIR filters), the average increase of the number of adders/subtractors approaches one for an increase of the length of the filter by one. This is due to the fact that most subexpressions have already been generated and only a single addition/subtraction is needed to generate another coefficient.

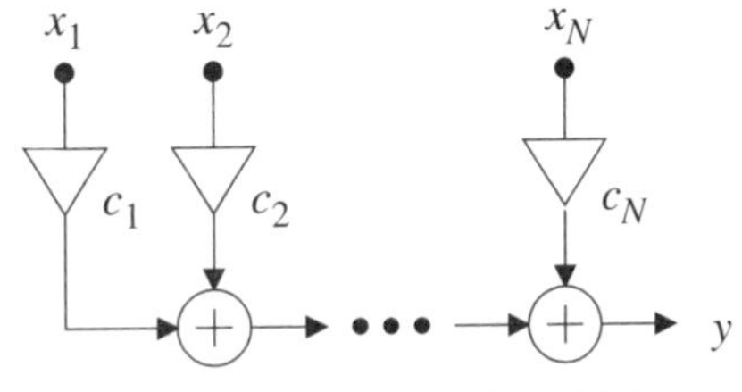

Figure 11.32 Transposed multiple-constant multiplication

In a particular case, a more irregular strategy for generating common subexpressions than the one shown in Figure 11.33 may be more efficient. Note, however, that the wiring may become very complex and irregular and in practice this may significantly offset the savings in adders/subtractors and make the layout and transistor sizing very time consuming.

The resulting composite operation can either be implemented using bit-parallel or bit-serial arithmetic. In the former case, the number of adds/subs are the most important, since shifts can be implemented virtually free of cost by fixed wiring. In the bit-serial case, both the number of adds/subs and shifts are important, since they correspond to hardware that consumes chip area and power.

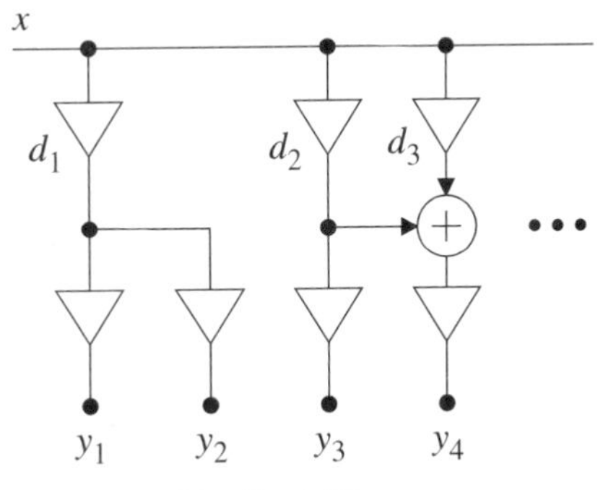

Figure 11.33 Exploiting common subexpressions to simplify a multiple-constant multiplication

The problem of finding optimal solutions to the multiple-constant multiplication problem is NP-complete. Heuristic algorithms have therefore been derived [6, 13, 14]. Note, however, many practical problems are very small and a complete search can be performed with a reasonable execution time.

Example 11.10

Find possible graphs that simultaneously realize the multiplications

$$y_1 = c_1 x$$

and

$$y_2 = c_2 x$$

where $c_1 = 7$ and $c_2 = 106$.

Figure 11.34 shows two alternative realizations. In both cases only three adder/subtractors are required.

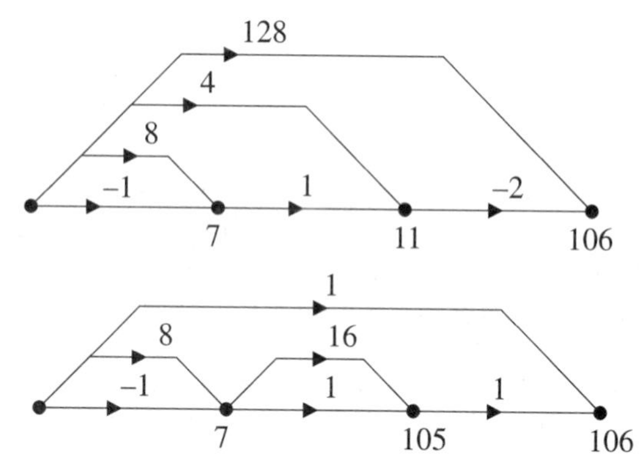

Figure 11.34 Alternative graphs for multiplication with 7 and 106

11.10 BIT-SERIAL SQUARERS

To derive a suitable algorithm for implementing a unit that computes the square $\boldsymbol{x}^2$ of a bit-serial number $\boldsymbol{x}$, we will first discuss the special case where $\boldsymbol{x}$ is a fractional, unsigned binary number [46].

11.10.1 Simple Squarer

Since the square is to be computed in a bit-serial fashion, we need to decompose the square into suitable bit components to find the necessary operations for computing each bit in the wanted result. An n-bit, fractional, unsigned binary number $\boldsymbol{x}$ is written as

$$\boldsymbol{x} = \sum_{i=1}^{n} x_i 2^{-i} \tag{11.25}$$

Now, let the function $\boldsymbol{f}$ represent the square of the number $\boldsymbol{x}$—i.e.,

$$\boldsymbol{f}(\boldsymbol{x}) = \boldsymbol{x}^2 \tag{11.26}$$

Then, the computation of $\boldsymbol{f}$ can be carried out in n iterations by repeatedly squaring the sum of the most significant bit of a number and the other bits of that number. In the first step, $f_1 = \boldsymbol{f}(\boldsymbol{x})$ is decomposed into the square of the most significant bit of $\boldsymbol{x}$ with a rest term and a remaining square f_2.

$$f_1\left(\sum_{i=1}^{n} x_i 2^{-i}\right) = \left(\sum_{i=1}^{n} x_i 2^{-i}\right)^2 = \left(x_1 2^{-1} + \sum_{i=2}^{n} x_i 2^{-i}\right)^2$$

$$= x_1 2^{-2} + x_1 2_o \sum_{i=2}^{n} x_i 2^{-i} + \left(\sum_{i=2}^{n} x_i 2^{-i} \right)^2$$
$$= x_1 2^{-2} + x_1 2_o \sum_{i=2}^{n} x_i 2^{-i} + f_2 \left(\sum_{i=2}^{n} x_i 2^{-i} \right) \quad (11.27)$$

In the next step, $f_2 = \boldsymbol{f}(\boldsymbol{x}-x_1)$ is decomposed in the same manner into the square of the most significant bit of $\boldsymbol{x}-x_1$ with rest term and a remaining square f_3. The scheme is repeated as long as there are bits left to process in the remaining square— i.e., until the square f_n is reached.

Examining this scheme we find that in order to input a bit-serial word $\boldsymbol{x}$ with the least significant bit first, we have to reverse the order of the iterations in the preceding scheme to have a suitable algorithm.

The iterative algorithm then can be written as

$$f_j = \Delta_j + f_{j+1}$$

where

$$j = n, n-1, \ldots, 1$$
$$\Delta_j = 2^{-2j} x_j + 2^1 - j_{xj} \sum_{i=j+1}^{n} x_i 2^{-i} \quad (11.28)$$
$$f_{n+1} = 0$$

In each iteration j we accumulate the previous term f_{j+1} and input the next bit x_j. If $x_j = 1$ then we add the square of the bit weight and the weights of the bits that have arrived prior to bit x_j shifted left $1-j$ positions. Then we store bit x_j for the next iteration. Examination of the bit weights accumulated in each iteration reveals that the sum converges toward the correct square with at least one bit in each step, going from the least significant bit in the result toward more significant bits.

An implementation of the preceding algorithm is shown in Figure 11.35. It uses a *shift-accumulator* to shift the accumulated sum to the right after each iteration. Thus, left-shifting of the stored x_i's are avoided and the addition of the squared bit weight of the incoming x_j is reduced to a shift to the left in each iteration. The implementation consists of n regular bit-slices which makes it suitable for hardware implementation.

The operation of the squarer in Figure 11.35 is as follows: All D flip-flops and SR flip-flops are assumed to be cleared before the first iteration. In the first iteration, the control signal $c(0)$ is high while the remaining control signals are low. This allows the first bit $x(0) = x_n$ to pass the AND gate on top of the rightmost bit-slice. The value x_n is then stored in the SR flip-flop of the same bit-slice for later use. Also, x_n is added to the accumulated sum via the OR gate in the same bit-slice,[3]

3. An OR gate is sufficient to perform the addition.

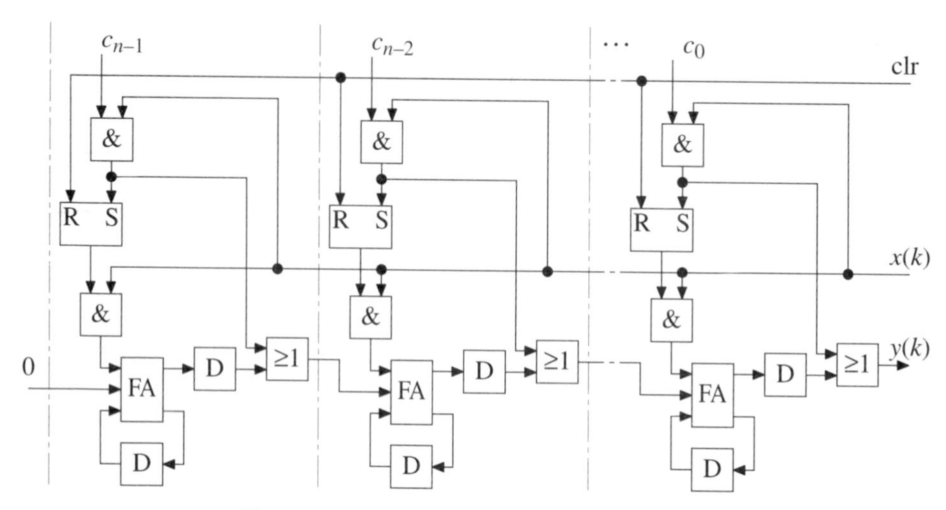

Figure 11.35 Squarer for unsigned numbers

implementing the squared part of the accumulation. At the same time, x_n controls the AND gates present in front of each full adder, implementing the addition of previously stored bits when x_n is high.[4] The least significant output bit $y(0) = y_{2n}$ then becomes available at the output after pure gate delays. Then a shift to the left follows. The following iterations are carried out in the same manner, with the input bits $x(1) = x_{n-1}$, $x(2) = x_{n-2}$, ..., $x(n-1) = x_1$ in sequence along with one of the control signals $c(1)$, $c(2)$, ..., $c(n-1)$ high, respectively. During the iterations, the result bits $y(1) = y_{2n-1}$, $y(2) = y_{2n-2}$, ..., $y(n-1) = y_{n+1}$ will become available at the output. Then, $x(i)$ has to be zero-extended n times to access the bits $y(n) = y_n$, ..., $y(2n-1) = y_0$ stored in the accumulator. The last control signal *init* is used to clear the SR flip-flops before the next computation can take place. A feature of the proposed implementation is that it computes a square without any error. This makes it unnecessary to clear the D flip-flops, since the shift–accumulator will contain the zero extended part on completion, which is zero in all remaining positions.

11.10.2 Improved Squarer

A more efficient implementation of the squaring algorithm is shown in Figure 11.36. Here, the leftmost bit-slice has been simplified into an AND gate. The simplification is based on the observation that the stored value in the leftmost SR flip-flop never will be used, since all incoming $x(i)$'s will be zero after the arrival of the most significant bit.

Moreover, the last active control signal $c(n-1)$ is used to clear the SR flip-flops, removing the need for the *init* signal. Also, zero-extension of the input signal $x(i)$ is no longer necessary, since any input is neglected after the most significant bit has arrived. This feature also makes it possible to compute the square of sign magnitude numbers with the same implementation. Then, the sign bit is input after the most significant bit with no other alteration to the scheme. This sign bit will be neglected by the squarer, which still yields a correct result, since $(-\boldsymbol{x})^2 = \boldsymbol{x}^2$.

---

4. Actually, since no bits have been stored when the first bit arrives, a zero will be added in the first iteration.

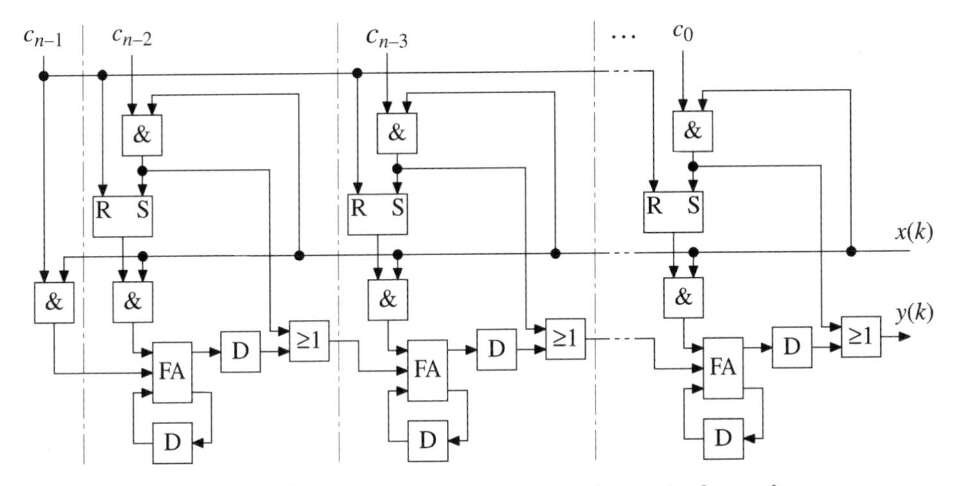

Figure 11.36 Squarer for unsigned or signed-magnitude numbers

The squarer in Figure 11.36 can further be modified to compute the square of a two's-complement number. An n-bit two's-complement number $\boldsymbol{x}$ can be written as

$$\boldsymbol{x} = -x_0 + \sum_{i=1}^{n-1} x_i 2^{-i} \tag{11.29}$$

To adapt the squarer to this case, we sign-extend a two's-complement number to at least $2n$–1 bits and do not clear the SR flip-flops until the next computation is about to take place. Then, the squarer computes

$$\begin{aligned}
\boldsymbol{y} &= \sum_{j=1}^{n-1} x_0 2^{j+1} \sum_{i=1}^{n-1} x_i 2^{-i} + \left(\sum_{i=0}^{n-1} x_i 2^{-i}\right)^2 \\
&= x_0(2^{n+1} - 4) \sum_{i=1}^{n-1} x_i 2^{-i} + \left[x_0 + 2x_0 \sum_{i=1}^{n-1} x_i 2^{-i} + \left(\sum_{i=0}^{n-1} x_i 2^{-i}\right)^2\right] \\
&= x_0 2^{n+1} \sum_{i=1}^{n-1} x_i 2^{-i} + \left[x_0 + 2x_0 \sum_{i=1}^{n-1} x_i 2^{-i} + \left(\sum_{i=0}^{n-1} x_i 2^{-i}\right)^2\right] \\
&= x_0 2^{n+1} \sum_{i=1}^{n-1} x_i 2^{-i} + \left(-x_0 + \sum_{i=0}^{n-1} x_i 2^{-i}\right)^2 \\
&= x_0 2^{n+1} \sum_{i=1}^{n-1} x_i 2^{-i} + \boldsymbol{x}^2
\end{aligned} \tag{11.30}$$

Here we can see that the result will contain an error in the accumulated sum. But, since this error only exists in bits of higher significance than the result, this error will not affect the output result. Further, if we sign extend the input signal with more than $2n-1$ bits, the error will be scaled accordingly, and stay within the squarer. An implementation of a squarer for two's-complement numbers is shown in Figure 11.37.

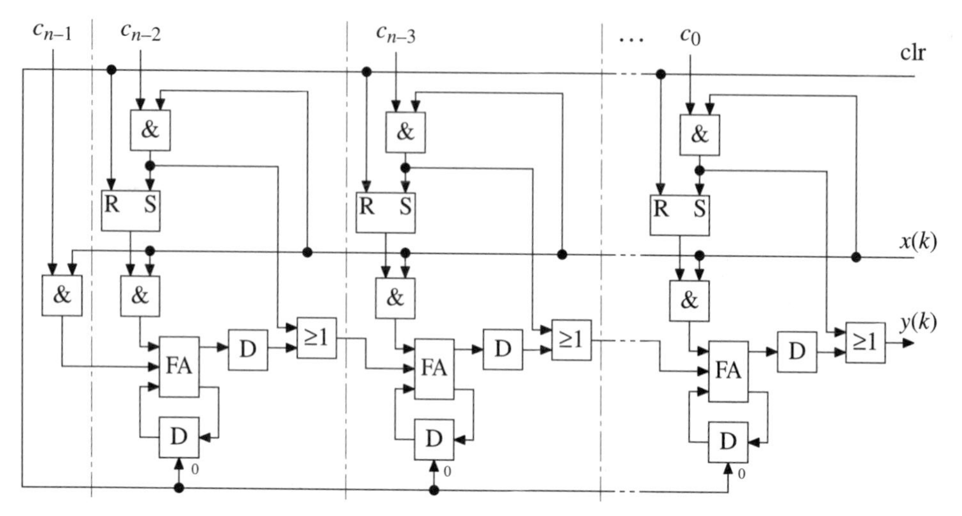

Figure 11.37 Squarer for two's-complement numbers

The only drawback compared to the squarer in Figure 11.36, is that we now need to clear out the error in the accumulator before the next computation can take place.

Notice that it is sufficient to clear the D flip-flops containing the carry bits. When the sign-extension of $\boldsymbol{x}$ takes place, the leftmost bit-slice consisting of an AND gate will only deliver zeroes to the next bit-slice. Since the error in this bit-slice will be $x_0 \cdot x_1$ according to Equation (11.30), and $x_0 \cdot x_1$ is added during the sign-extension phase, the carry will be $x_0 \cdot x_1$ in this phase and the sum will be zero. The zero will propagate to the next bit-slice in the next clock cycle, which will propagate the error $x_0 \cdot x_2$ in the same manner as the previous bit-slice, and so on. Since the minimum period of sign extension lasts as many clock cycles as there are bit-slices to clear, the sum paths are already zero at the moment of clearing.

11.11 SERIAL/SERIAL MULTIPLIERS

A true serial/serial multiplier can be useful in many applications—for example, to compute correlations and in adaptive filters. The approach to derive a squaring algorithm [46] is also applicable to multiplication of two bit-serial numbers $\boldsymbol{x}$ and $\boldsymbol{y}$ by use of the identity

$$(\boldsymbol{x}+\boldsymbol{y})^2 = \boldsymbol{x}^2+\boldsymbol{y}^2+2\boldsymbol{xy} \Rightarrow \boldsymbol{xy} = \frac{1}{2}[(\boldsymbol{x}+\boldsymbol{y})^2-(\boldsymbol{x}^2-\boldsymbol{y}^2)] \tag{11.31}$$

As before, we will first investigate the special case where the numbers are fractional, unsigned binary numbers. First, we let a function $\boldsymbol{f}$ represent our operation—i.e., the product of the numbers

$$f(\boldsymbol{x}, \boldsymbol{y}) = \boldsymbol{x} \cdot \boldsymbol{y} \tag{11.32}$$

Then, the computation of $\boldsymbol{f}$ can be carried out in n iterations by repeatedly squaring the sum of the most significant bit of a number and the other bits of that number. In the first step, $f_1 = \boldsymbol{f}(\boldsymbol{x})$ is decomposed into the square of the most significant bit of $\boldsymbol{x}$ with a rest term and a remaining square f_2.

$$f_1\left(\sum_{i=1}^{n} x_i 2^{-i}\right) = \left(\sum_{i=1}^{n} x_i 2^{-i}\right)^2$$

$$= \left(x_1 2^{-1} + \sum_{i=2}^{n} x_i 2^{-i}\right)^2$$

$$= x_1 2^{-2} + x_1 2^{0} \sum_{i=2}^{n} x_i 2^{-i} + \left(\sum_{i=2}^{n} x_i 2^{-i}\right)^2$$

$$= x_1 2^{-2} + x_1 2^{0} \sum_{i=2}^{n} x_i 2^{-i} + f_2\left(\sum_{i=2}^{n} x_i 2^{-i}\right)$$

The logic realization of the serial/serial multiplier is shown in Figure 11.38.

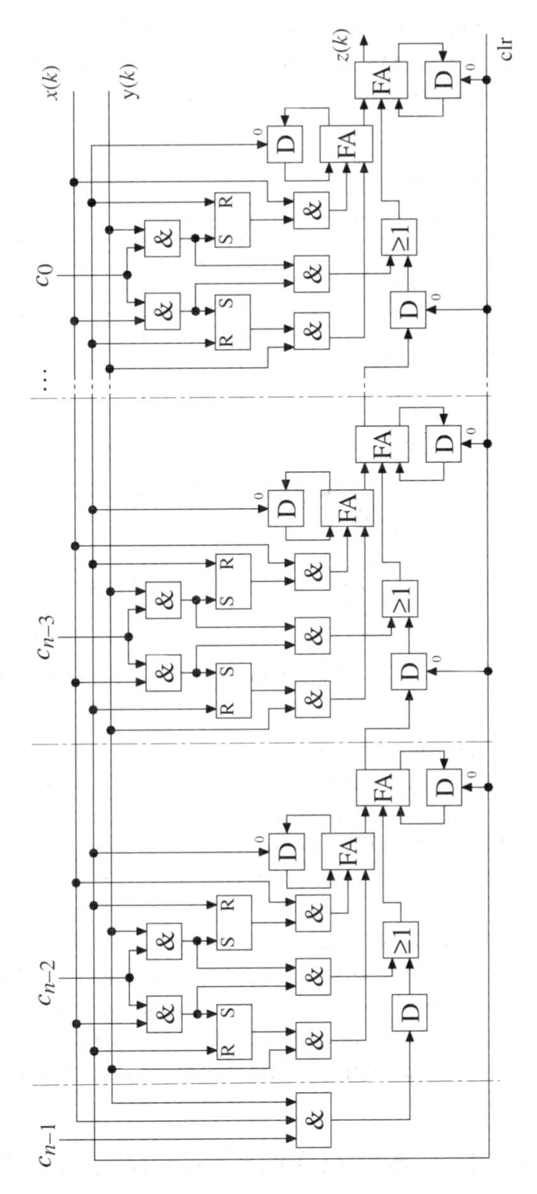

Figure 11.38 Serial/serial multiplier

Several variations of the serial/parallell and serial/serial multipliers just described are possible [26, 37]. For example, multiply and add $\boldsymbol{y} = \boldsymbol{a}\ \boldsymbol{x} + \boldsymbol{b}$ and multipliers that process two or more bits at a time in order to increase the throughput and at the same time reduce the power consumption by reducing the number of D flip-flops [37].

Note that bit-serial arithmetic requires power-efficient D flip-flops to compete with bit-parallel arithmetic. Normal standard-cell libraries contain only power-consuming D flip-flops and are therefore unsuited for bit-serial arithmetic. In some logic styles the logic function and the D flip-flops are merged into one circuit, which reduces the number of transistors and the capacitive load.

11.12 DIGIT-SERIAL ARITHMETIC

From speed and power consumption points of view, it may sometimes be advantageous to process several bits at a time, so-called digit-serial processing. The number of bits processed in a clock cycle is referred to as the digit size. Most of the principles, just discussed, for bit-serial arithmetic can easily be extended to digit-serial arithmetic. A complete design environment named *Parsifal* for digit-serial implementations has been developed by General Electric Research Center [17].

11.13 THE CORDIC ALGORITHM

The well-known CORDIC algorithm can be used to compute, for example, trigonometric functions, square roots, etc. The CORDIC algorithm iteratively computes magnitude and phase, or rotation of a vector in circular, linear, and hyperbolic coordinate systems. The CORDIC algorithm is multiplier-free, using only shift, add/sub, and compare operations. The basic equations to be iterated are

$$x_{n+1} = x_n + \sigma_n 2^{-S(m,n)y_n} \tag{11.34}$$

$$y_{n+1} = y_n + \sigma_n 2^{-S(m,n)x_n} \tag{11.35}$$

$$z_{n+1} = z_n - \sigma_n \alpha(m,n) \tag{11.36}$$

where m denotes the coordinate system, σ_n the rotation direction, $S(m, n)$ the shift sequence, and $\alpha(m, n)$ the rotation angle. The latter are related according to

$$\alpha(m,n) = \frac{1}{\sqrt{m}}\arctan\left(\sqrt{m}2^{-S(m,n)}\right) \tag{11.37}$$

Two operational modes are possible: rotation or vectoring. The rotation direction factor σ_n is determined by the following equations, depending on the specified iteration goal

$$\sigma_n = \begin{cases} \text{Sign}(z_n) & \text{for } z_n \to 0 \text{ (rotation mode)} \quad (11.38) \\ -\text{Sign}(x_n)\text{Sign}(y_n) & \text{for } z_n \to 0 \text{ (vectoring mode)} \quad (11.39) \end{cases}$$

The solution of the iteration is given by Equations.(11.34) through (11.36) with k_m being the scaling factor and x_0, y_0, and z_0 the starting values of the iterations. The final values of x, y, and z are

$$\begin{cases} z_n \to 0 \quad \text{(rotation)} & (11.40) \\ x_n = k_m\left(x_0\cos(\sqrt{m}z_0) - \sqrt{m}y_0\sin(\sqrt{m}z_0)\right) & (11.41) \\ y_n = k_m\left(y_0\cos(\sqrt{m}z_0) - \frac{1}{\sqrt{m}}x_0\sin(\sqrt{m}z_0)\right) & (11.42) \end{cases}$$

and

$$\begin{cases} y_n \to 0 \quad \text{(vectoring)} & (11.43) \\ x_n = k_m \sqrt{x_0^2 + m y_0^2} & (11.44) \\ z_n = z_0 + \dfrac{1}{\sqrt{m}} \arctan\left(\dfrac{\sqrt{m} y_0}{x_0}\right) & (11.45) \end{cases}$$

By selecting the appropriate system (m = 0, 1, –1 means a linear, circular, and hyperbolic coordinate system, respectively) and iteration mode, we obtain a variety of computable functions, unmatched in terms of simplicity by any other unified algorithm. On the other hand, CORDIC belongs to the digit-by-digit algorithms with linear convergence and sequential behavior. This means that for W-bit precision, approximately W iterations are needed. Equations (11.34) through (11.36) indicate that the main speed limiting factor is addition/subtraction. Further, the iterations given by Equations (11.34) through (11.36) do not guarantee that $k_m = 1$— i.e., the final vector magnitude may be expanded or contracted (see Equations (11.40) through (11.45)). Scaling factor compensation is therefore required [41].

11.14 DISTRIBUTED ARITHMETIC

Most digital signal processing algorithms involve sum-of-products (inner products) of the type

$$\boldsymbol{y} = \mathbb{a} \cdot \mathbb{x} = \sum_{i=1}^{N} \boldsymbol{a}_i \boldsymbol{x}_i \tag{11.46}$$

Typically, the coefficients $\boldsymbol{a}_i$ are fixed, but there are several common cases where both $\mathbb{a}$ and $\mathbb{x}$ are variable vectors, for example, computation of correlation, which require a serial/parallell or serial/serial multiplier. Distributed arithmetic is an efficient procedure for computing inner products between a fixed and a variable data vector. The basic principle is owed to Croisier *et al.* [11], and Peled and Liu [30] have independently presented a similar method.

11.14.1 Distributed Arithmetic

Consider the inner product in Equation (11.46). Coefficients, $\boldsymbol{a}_i$, i = 1, 2, ..., N, are assumed to be fixed. A two's-complement representation is used for both coefficients and data. The data are scaled so that $|\boldsymbol{x}_i| \le 1$. The inner product can be rewritten

$$\boldsymbol{y} = \sum_{i=1}^{N} \boldsymbol{a}_i \left[-x_{i0} + \sum_{k=1}^{W_d - 1} x_{ik} 2^{-k} \right]$$

where x_{ik} is the kth bit in $\boldsymbol{x}_i$. By interchanging the order of the two summations we get

$$\boldsymbol{y} = -\sum_{i=1}^{N} \boldsymbol{a}_i x_{i0} + \sum_{k=1}^{W_d-1} \left[\sum_{i=1}^{N} \boldsymbol{a}_i x_{ik} \right] 2^{-k}$$

which can be written

$$\boldsymbol{y} = -F_0(x_{10}, x_{20}, \ldots, x_{N0}) + \sum_{k=1}^{W_d-1} F_k\left(x_{1k}, x_{2k}, \ldots, x_{Nk}\right) 2^{-k} \tag{11.47}$$

where

$$F_k(x_{1k}, x_{2k}, \ldots, x_{Nk}) = \sum_{i=1}^{N} \boldsymbol{a}_i x_{ik}$$

F is a function of N binary variables, the ith variable being the kth bit in the data x_i. Since F_k can take on only a finite number of values, 2^N, it can be computed and stored in a look-up table. This table can be implemented using a ROM (read-only memory). Using Horners method[5] for evaluating a polynomial for $x = 0.5$, Equation (11.47) can be rewritten

$$\boldsymbol{y} = ((\ldots((0 + F_{W_d-1})2^{-1} + F_{W_d-2})2^{-1} + \ldots + F_2)2^{-1} + F_1)2^{-1} - F_0 \tag{11.49}$$

Figure 11.39 shows a block diagram for computing an inner product according to Equation (11.49). Since the output is divided by 2, by the inherent shift, the circuit is called a *shift-accumulator* [20, 35].

Inputs, $\boldsymbol{x}_1$, $\boldsymbol{x}_2, \ldots$, $\boldsymbol{x}_N$ are shifted bit-serially out from the shift registers with the least significant bit first. Bits x_{ik} are used as an address to the ROM storing the look-up table.

Computation of the inner product starts by adding F_{W_d-1} to the initially cleared accumulator register, REG. In the next clock cycle, outputs from the shift registers address F_{W_d-2}, which is added to the value in the accumulator register. After W_{d-1} clock cycles, F_0 is subtracted from the value in the accumulator register. The computational

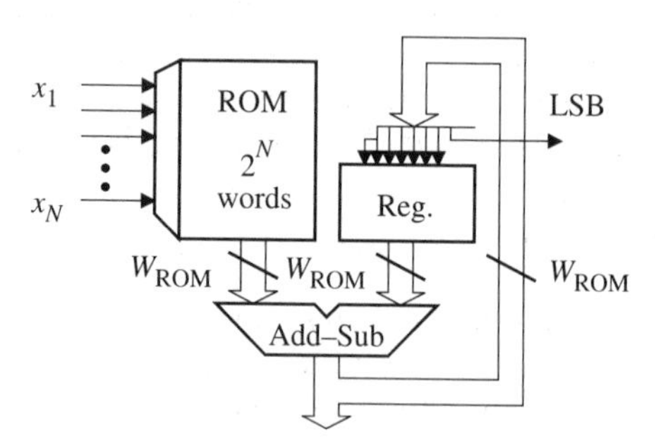

Figure 11.39 Block diagram for distributed arithmetic

5. William Horner, 1819. In fact, this method was first derived by Isaac Newton in 1711.

time is W_d clock cycles. The word length in the ROM, W_{ROM}, depends on the F_k with the largest magnitude and the coefficient word length, W_c, and

$$W_{ROM} \leq W_c + \log_2(N) \tag{11.50}$$

EXAMPLE 11.11

Determine the values that should be stored in ROM for the inner product

$$y = a_1 x_1 + a_2 x_2 + a_3 x_3$$

where $a_1 = 33/128 = (0.0100001)_2$, $a_2 = 85/128 = (0.1010101)_2$, and $a_3 = -11/128 = (1.1110101)_2$. Also determine the necessary word length in the shift-accumulator.

We shall precompute all possible linear combinations of coefficients according to Equation (11.48). These values shall then be stored in ROM at the addresses given by the binary weights in the linear combination. Table 11.3 shows the values both in binary and decimal form with the corresponding addresses.

$x_1\ x_2\ x_3$	F_k	F_k	F_k
0 0 0	0	0.0000000	0.0000000
0 0 1	a_3	1.1110101	0.0859375
0 1 0	a_2	0.1010101	0.6640625
0 1 1	$a_2 + a_3$	0.1001010	0.5781250
1 0 0	a_1	0.0100001	0.2578125
1 0 1	$a_1 + a_3$	0.0010110	0.1718750
1 1 0	$a_1 + a_2$	0.1110110	0.9218750
1 1 1	$a_1 + a_2 + a_3$	0.1101011	0.8359375

Table 11.3 ROM contents

The shift-accumulator must be able to add correctly the largest possible value obtained in the accumulator register and in the ROM. The largest value in the accumulator register is obtained when the largest (magnitude) value stored in the ROM is repeatedly accumulated. Thus, at the last clock cycle, corresponding to the sign bit, the value in REG according to Equation (11.49) is

$$|y| = ((\ldots((0 + F_{max})2^{-1} + F_{max})2^{-1} + \ldots + F_{max})2^{-1} + F_{max})2^{-1} \leq F_{max}$$

Hence, the shift-accumulator must be able to add numbers of magnitude $\leq F_{max}$. The necessary number range is $\pm$ 1. The word length in the shift-accumulator must be 1 guard bit for overflow detection + 1 sign bit + 0 integer bits + 7 fractional bits = 9 bits.

Notice the similarity between Equations (11.22) and (11.47). In Equation (11.22), the partial products are summed while in Equation (11.47) the values of F_k are summed. In both cases, the same type of shift-accumulator can be used. Hence, the distributed arithmetic unit essentially consists of a serial/parallell multiplier augmented by a ROM.

EXAMPLE 11.12

Show that a second-order section in direct form I can be implemented by using only a single PE based on distributed arithmetic. Also show that the PE can be pipelined.

In Figure 11.40 a set of D flip-flops has been placed between the ROM and the shift-accumulator to allow the two operations to overlap in time—i.e., the two operations are pipelined. The number of words in the ROM is only $2^5 = 32$.

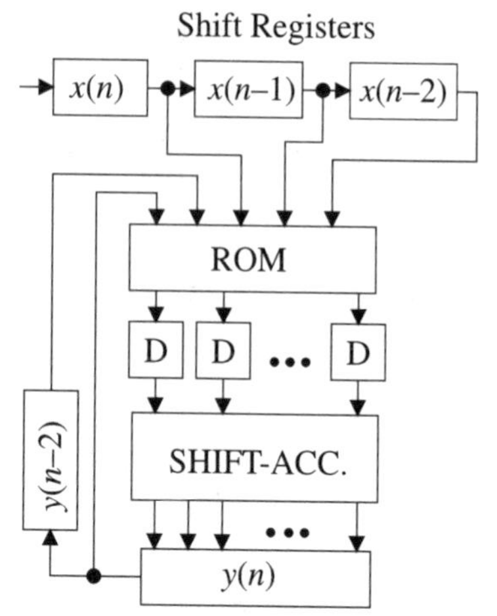

Figure 11.40 Direct form I implemented with distributed arithmetic

EXAMPLE 11.13

Suggest a scheme to implement a linear-phase FIR structure using distributed arithmetic. Assume that N is even.

Figure 11.41 shows an implementation of an eleventh-order linear-phase FIR filter. $N/2$ bit-serial adders (subtractors) are used to sum the symmetrically placed values in the delay line. This reduces the number of terms in the inner product. Only 64 words are required whereas $2^{12} = 4096$ words are required for a nonlinear-phase FIR filter. For higher-order FIR filters the reduction in the number of terms by 50% is essential. Further, the logic circuitry has been pipelined by introducing D flip-flops between the adders (subtractors) and the ROM, and between the ROM and the shift-accumulator.

The number of words in the ROM is 2^N where N is the number of terms in the inner product. The chip area for the ROM is small for inner products with up to five to six terms. The basic approach is useful for up to 10 to 11 terms. However, inner products containing many terms can be partitioned into a number of smaller inner products which can be computed and summed by using either distributed arithmetic or an adder tree. A complete inner product PE and several similar types of PEs (for example, adaptors and butterflies) can be based on distributed arithmetic. The main parts of an inner product PE are the shift-accumulator, the coefficient ROM with decoder, and the control circuits. Overflow and quantization circuitry are also necessary, but it is often advantageous to move parts of this circuitry to the serial/parallell converters used in the interconnection network, as discussed in Chapter 8.

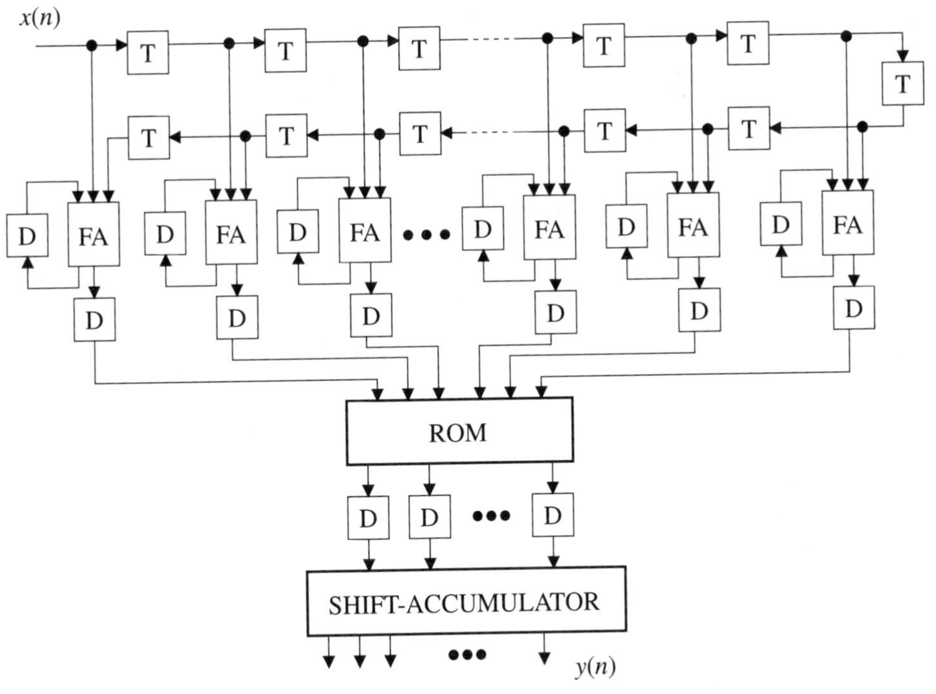

Figure 11.41 Linear-phase FIR filter with $N = 12$

11.14.2 Parallel Implementation of Distributed Arithmetic

Distributed arithmetic can, of course, be implemented in parallel form—i.e., by allocating a ROM to each of the terms in Equation (11.47). The ROMs, which are identical, can be addressed in parallel and their values, appropriately shifted, added using an adder tree as illustrated in Figure 11.42. The critical path is through a ROM and through the adder tree. The critical path can be broken into small pieces to achieve very high speed by pipelining.

11.15 THE BASIC SHIFT-ACCUMULATOR

The shift-accumulator shall perform a shift-and-add operation and a subtraction in the last time slot. Obviously, for typical word lengths in ROM of 8 to 18 bits, a ripple–through adder or a carry–look–ahead adder is unsuitable for speed and complexity reasons. The shift-accumulator, shown in Figure 11.43, uses carry–save adders instead. This yields a regular hardware structure, with short delay paths between the clocking elements. Furthermore, the shift-accumulator can be implemented using a modular (bit-slice) design. The number of bits in the shift-accumulator can be chosen freely.

In the first time slot word F_{W_d-1} from the ROM shall be added to the initially cleared accumulator. In the next time slot F_{W_d-2} shall be added to the previous

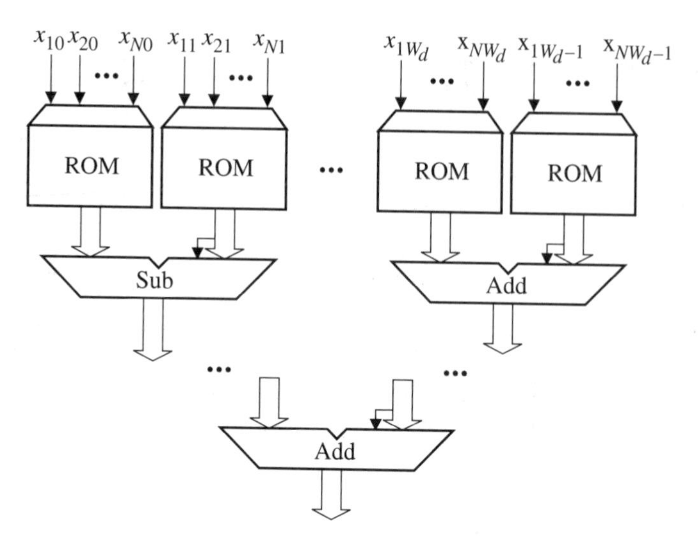

Figure 11.42 Parallel implementation of distributed arithmetic

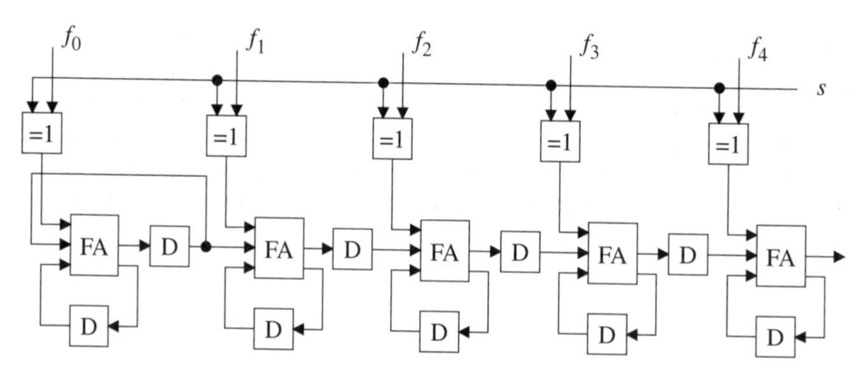

Figure 11.43 Shift-accumulator using carry-save adders

result divided by 2. This division is done by shifting F_{W_d-1} one step to the right and copying the sign bit. One bit of the result is obtained during each clock cycle.

This procedure is continued until F_0, corresponding to the sign bit of the data, is being subtracted. This is done by adding F_0, i.e., inverting all the bits in F_0 using the XOR gates and the signal s, and adding one bit in the least-significant position. We will explain later how this last addition is done. After F_0 has been added, the most significant part of the inner product must be shifted out of the accumulator. This can be done by accumulating zeros. The number of clock cycles for one inner product is W_d+W_{ROM}. A more efficient scheme is to free the carry–save adders in the accumulator by loading the sum and carry bits of the carry–save adders into two shift registers as shown in Figure 11.44 [12, 35]. The outputs from these can be added by a single carry–save adder.

This scheme effectively doubles the throughput since two inner products are computed concurrently for a small increase in chip area.

The result will appear with the least significant part in the output of the shift-accumulator, and the most significant part in the output of the lower carry–save

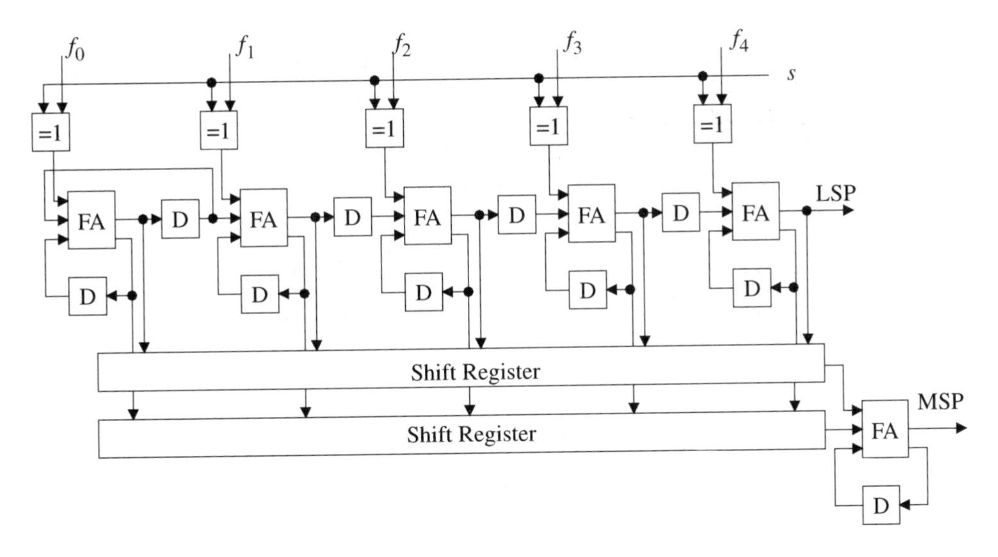

Figure 11.44 Shift-accumulator augmented with two shift registers

adder. Thus, a special end-bit-slice is needed to separate the relevant bits in the outputs. A special first bit-slice is also needed to copy the sign bit.

Figure 11.45 shows how the carry–save adders are connected to two shift registers that are loaded bit-parallel. The loading is accomplished via the multiplexers. The input of the rightmost multiplexer in the lower shift register is used to set the input of the carry–save adder to 1, in order to get a proper subtraction of F_0. The required number of bit-slices is equal to the word length, W_{ROM}.

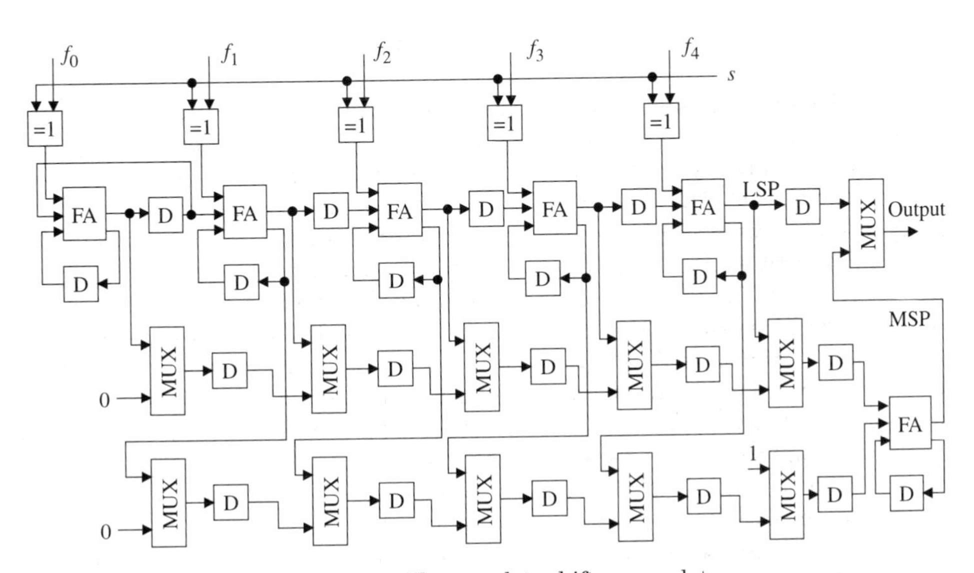

Figure 11.45 The complete shift-accumulator

The first W_d clock cycles are used to accumulate values from the ROM while the last W_{ROM} clock cycles are used to shift the result out of the shift registers. Hence, the required number of clock cycles is

$$N_{CL} = \max\{W_{ROM}, W_d\} \leq \max\{W_c + \log(N)_2, W_d\} \tag{11.51}$$

Notice that these two phases can be overlapped with subsequent operations so that two operations are performed concurrently. In a typical filter implementation W_d = 16 to 22 bits and W_{ROM} = 4 to 16 bits. Hence, the number of clock cycles necessary is W_d in most applications. The latency between the inputs and outputs is W_{ROM} clock cycles, and a new computation can start every W_d clock cycles. The word length of the result will be $W_d + W_{ROM} - 1$ bits. The result is split into two parts; the least significant part comes from the output of the last full-adder in the accumulator and the most significant part is formed as the bit-serial sum of the carry-register and the sum-register. A special end-bit-slice is needed to form the desired output.

A local control unit can be integrated into the shift-accumulator. All local control signals can be generated from a single external synchronization signal that initiates a new computation.

Each bit-slice is provided with a D flip-flop which forms a shift register generating delayed versions of the synchronization signal. The local control signal needed for selection of the least and most significant parts of the output is generated using this shift register. The control is therefore independent of the word length of the shift-accumulator. This simplifies the layout design and decreases the probability of design errors. It also decreases the probability of timing problems that can occur when a signal is distributed over a large distance.

11.16 REDUCING THE MEMORY SIZE

The amount of memory required becomes very large for long inner products. There are mainly two ways to reduce the memory requirements. The two methods can be applied at the same time to obtain a very small amount of memory.

11.16.1 Memory Partitioning

One of several possible ways to reduce the overall memory requirement is to partition the memory into smaller pieces that are added before the shift-accumulator as shown in Figure 11.46. The amount of memory is reduced from 2^N words to $2 \cdot 2^{N/2}$ words if the original memory is partitioned into two parts. For example, for $N = 10$ we get $2^{10} = 1024$ words to $2 \cdot 2^5 = 64$ words. Hence, this approach reduces the memory significantly at the cost of an additional adder.

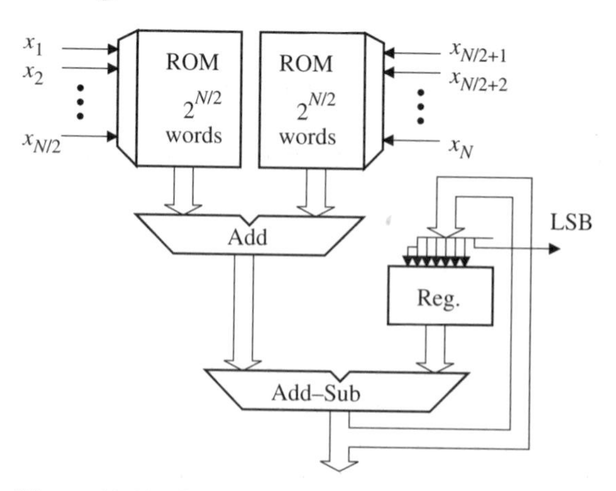

Figure 11.46 Reducing the memory by partitioning

Notice, depending on the values in the ROM, it is often favorable from both speed and area points of view to implement a ROM by PLAs. Large ROMs tend to be slow.

11.16.2 Memory Coding

The second approach is based on a special coding of the ROM content. Memory size can be halved by using the ingenious scheme [7, 11] based on the identity

$$\boldsymbol{x} = \frac{1}{2}[\boldsymbol{x} - (-\boldsymbol{x})] \tag{11.52}$$

In two's-complement representation, a negative number is obtained by inverting all bits and then adding a 1 to the least significant position of the original number. The identity can be rewritten

$$\boldsymbol{x} = \frac{1}{2}\left[-x_0 + \sum_{k=1}^{W_d-1} x_k 2^{-k} - \left(-\overline{x_0} + \sum_{k=1}^{W_d-1} \overline{x_k} 2^{-k} + 2^{-(W_d+1)}\right)\right]$$

$$= -(x_0 - \overline{x_0})2^{-1} + \sum_{k=1}^{W_d-1} (x_k - \overline{x_k})2^{-k-1} - 2^{-W_d}$$

Notice that $x_k - \overline{x_k}$ can only take on the values 1 or +1. Inserting this expression into Equation (11.46) yields

$$\boldsymbol{y} = \sum_{k=1}^{W_d-1} F_k(x_{1k}, \ldots, x_{Nk})2^{-k-1} - F_0(x_{10}, \ldots, x_{N0})2^{-1} + F(0, \ldots, 0)2^{-W_d} \tag{11.53}$$

where

$$F_k(x_{1k}, x_{2k}, \ldots, x_{Nk}) = \sum_{i=1}^{N} \boldsymbol{a}_i(x_k - \overline{x_k}) \tag{11.54}$$

The function F_k is shown in Table 11.4 for $N = 3$. Notice that only half the values are needed, since the other half can be obtained by changing the signs. To explore this redundancy we make the following address modification shown to the right in Table 11.4:

$$u_1 = x_1 \otimes x_2$$

$$u_2 = x_1 \otimes x_3$$

Here, we have selected variable x_1 as the control signal, but any of the variables will do. The add/sub control (i.e., x_1) must also provide the correct addition/subtraction function when the sign-bits are accumulated. Therefore, we form the following control signal to address the ROM:

$$A/S = x_1 \otimes x_{sign\text{-}bit}$$

where the control signal $x_{sign\text{-}bit}$ is zero at all times except when the sign-bit arrives.

$x_1\ x_2\ x_3$	F_k	$u_1\ u_2$	A/S
0 0 0	$\boldsymbol{a}_1\ \boldsymbol{a}_2\ \boldsymbol{a}_3$	0 0	A
0 0 1	$\boldsymbol{a}_1\ \boldsymbol{a}_2 + \boldsymbol{a}_3$	0 1	A
0 1 0	$\boldsymbol{a}_1 + \boldsymbol{a}_2\ \boldsymbol{a}_3$	1 0	A
0 1 1	$\boldsymbol{a}_1 + \boldsymbol{a}_2 + \boldsymbol{a}_3$	1 1	A
1 0 0	$+\boldsymbol{a}_1\ \boldsymbol{a}_2\ \boldsymbol{a}_3$	1 1	S
1 0 1	$+\boldsymbol{a}_1\ \boldsymbol{a}_2 + \boldsymbol{a}_3$	1 0	S
1 1 0	$+\boldsymbol{a}_1 + \boldsymbol{a}_2\ \boldsymbol{a}_3$	0 1	S
1 1 1	$+\boldsymbol{a}_1 + \boldsymbol{a}_2 + \boldsymbol{a}_3$	0 0	S

Table 11.4 ROM content

Figure 11.47 shows the resulting principle for distributed arithmetic with halved ROM. Only $N-1$ variables are used to address the memory. The XOR gates used for halving the memory can be merged with the XOR gates used for inverting F_k in Figure 11.43. In section 11.17 we will show that this technique for reducing the memory size can easily be implemented using a small modification of the shift-accumulator discussed in section 11.12.

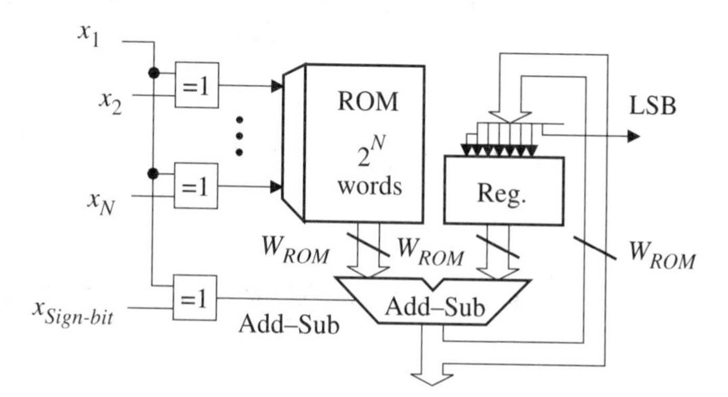

Figure 11.47 Distributed arithmetic with smaller ROM

11.17 COMPLEX MULTIPLIERS

In many algorithms, such as in the FFT, the data are complex-valued. A direct implementation of a complex multiplication requires four real multiplications and two additions. The number of real multiplications can, however, be reduced to three, at the expense of more additions. See Problem 11.24. In this section we will show that two distributed arithmetic PEs are sufficient.

Let

$$\boldsymbol{X} = \boldsymbol{A} + j\boldsymbol{B} \text{ and } \boldsymbol{K} = \boldsymbol{C} + j\boldsymbol{D}$$

where $\boldsymbol{K}$ is the fixed coefficient and $\boldsymbol{X}$ is the data. Once again we use the identity

$$\boldsymbol{x} = \frac{1}{2}[\boldsymbol{x} - (-\boldsymbol{x})] = \frac{1}{2}\left[-x_0 + \sum_{i=1}^{W_d-1} x_i 2^{-i} - \left(-\bar{x}_0 + \sum_{i=1}^{W_d-1} \bar{x}_i 2^{-i} + 2^{-W_d+1}\right)\right]$$

$$= -(x_0 - \bar{x}_0)2^{-1} + \sum_{i=1}^{W_d-1} (x_i - \bar{x}_i)2^{-i-1} - 2^{-W_d}$$

Now, the product of two complex numbers can be written

$$
\begin{aligned}
\boldsymbol{K} \cdot \boldsymbol{X} &= (\boldsymbol{CA} - \boldsymbol{DB}) + j(\boldsymbol{DA} + \boldsymbol{CB}) \\
&= \left\{ -\boldsymbol{C}(a_0 - \overline{a}_0)2^{-1} + \sum_{i=1}^{W_d - 1} \boldsymbol{C}(a_i - \overline{a}_i)2^{-i-1} - \boldsymbol{C}2^{-W_d} \right\} \\
&= \left\{ -\boldsymbol{D}(b_0 - \overline{b}_0)2^{-1} + \sum_{i=1}^{W_d - 1} \boldsymbol{D}(b_i - \overline{b}_i)2^{-i-1} - \boldsymbol{D}2^{-W_d} \right\} \\
&\quad + j\left\{ -\boldsymbol{D}(a_0 - \overline{a}_0)2^{-1} + \sum_{i=1}^{W_d - 1} \boldsymbol{D}(a_i - \overline{a}_i)2^{-i-1} - \boldsymbol{D}2^{-W_d} \right\} \\
&\quad + j\left\{ -\boldsymbol{C}(b_0 - \overline{b}_0)2^{-1} + \sum_{i=1}^{W_d - 1} \boldsymbol{C}(b_i - \overline{b}_i)2^{-i-1} - \boldsymbol{C}2^{-W_d} \right\} \\
&= \left\{ -\boldsymbol{F}_1(a_0, b_0)2^{-1} + \sum_{i=1}^{W_d - 1} \boldsymbol{F}_1(a_i, b_i)2^{-i-1} + \boldsymbol{F}_1(0, 0)2^{-W_d} \right\} \\
&\quad + j\left\{ -\boldsymbol{F}_2(a_0, b_0)2^{-1} + \sum_{i=1}^{W_d - 1} \boldsymbol{F}_2(a_i, b_i)2^{-i-1} + \boldsymbol{F}_2(0, 0)2^{-W_d} \right\}
\end{aligned}
\tag{11.55}
$$

Hence, the real and imaginary parts of the product can be computed using just two distributed arithmetic units. The binary functions F_1 and F_2 can be stored in a ROM, addressed by the bits a_i and b_i. The ROM content is shown in Table 11.5.

a_i	b_i	F_1	F_2
0	0	$(\boldsymbol{C}-\boldsymbol{D})$	$(\boldsymbol{C}+\boldsymbol{D})$
0	1	$(\boldsymbol{C}+\boldsymbol{D})$	$(\boldsymbol{C}-\boldsymbol{D})$
1	0	$(\boldsymbol{C}+\boldsymbol{D})$	$(\boldsymbol{C}-\boldsymbol{D})$
1	1	$(\boldsymbol{C}-\boldsymbol{D})$	$(\boldsymbol{C}+\boldsymbol{D})$

Table 11.5 ROM contents for the complex multiplier

It is obvious from Table 11.4 that only two coefficients are needed: $(\boldsymbol{C}+\boldsymbol{D})$ and $(\boldsymbol{C}-\boldsymbol{D})$. The appropriate coefficients can be directed to the accumulators via a 2:2-multiplexer. If $a_i \oplus b_i = 1$ the values are applied directly to the accumulators, and if $a_i \oplus b_i = 0$ the values are interchanged. Further, the coefficients are either added to, or subtracted from, the accumulators registers depending on the data bits a_i and b_i.

Generally, the number of clock cycles is W_d+1, because of the last term in both the real and imaginary parts in Equation (11.55). However, the number can be reduced to only W_d, since these values are data independent and they can therefore be loaded into the accumulators instead of being cleared between two computations. A block diagram of the implementation is shown in Figure 11.48.

To summarize, only two shift-accumulators are required to compute a complex multiplication. A shift-accumulator is from area, speed, and power consumption points of view comparable to a real serial/parallell multiplier.

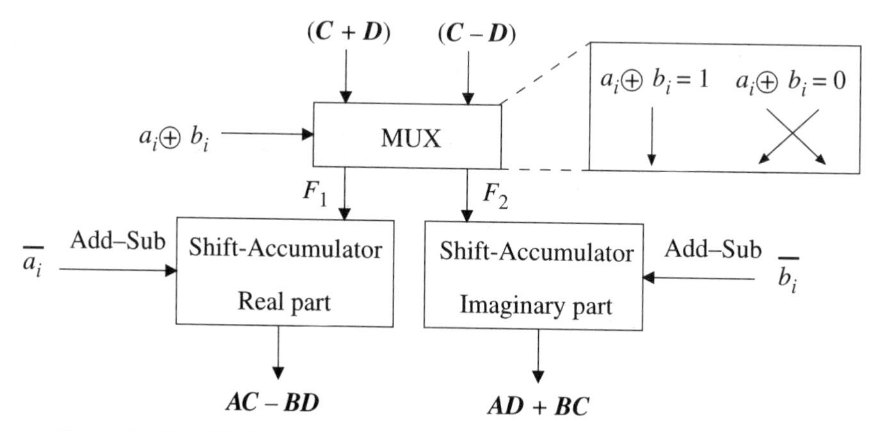

Figure 11.48 Block diagram for a complex multiplier using only two shift-accumulators and a multiplexer

11.18 IMPROVED SHIFT-ACCUMULATOR

The last term in Equation (11.53) shall be added to the first term in the sum, F_{W_d-1}, at the same level of significance. This can be accomplished by initially setting carry D flip-flops to $F(0, 0,..., 0)$, as illustrated in Figure 11.49 where only the upper part of the shift-accumulator part is shown. Note that one of the inputs, S, to the first bit-slice is free. This input can be used to add a value to the inner product

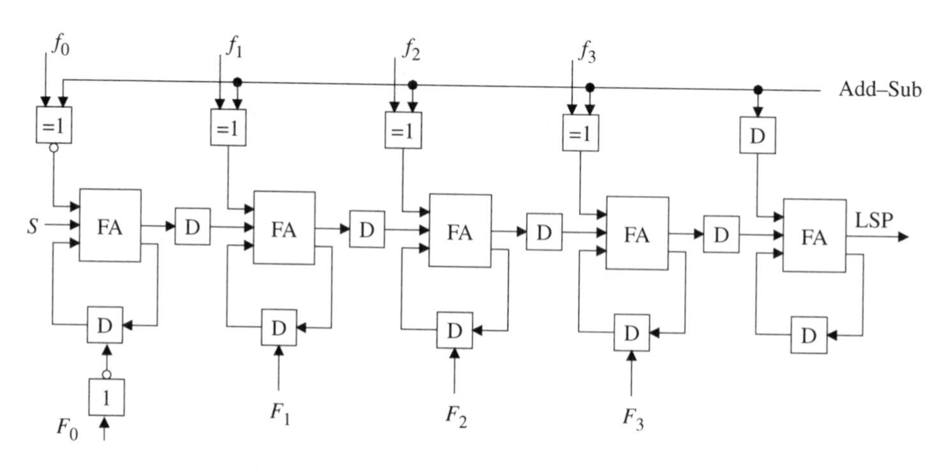

Figure 11.49 Improved shift-accumulator

11.18.1 Complex Multiplier Using Two-Phase Logic

Figure 11.50 shows the layout of one half of a complex multiplier based on the improved shift-accumulator using two-phase clocking[47]. The coefficient word length is 8 bits. A 0.8-μm double metal CMOS process was used. The shift-accumulator include the multiplexer circuitry and register needed for use as a complex multiplier. However, these parts are small compared to the shift-accumulator itself. The maximal clock frequency is about 175 MHz at 5 V. The chip area is 440 μm × 200 μm = 0.088 mm^2.

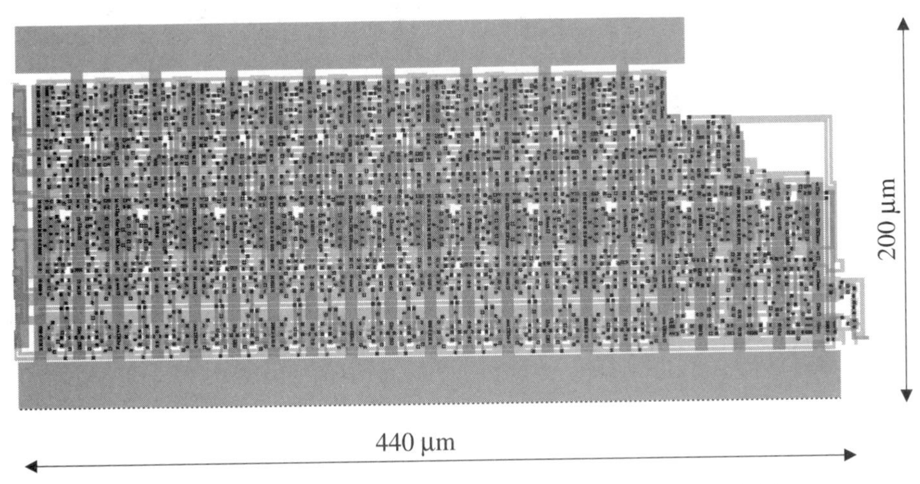

Figure 11.50 Half of a complex multiplier using two-phase clocking

Drivers for the clock are estimated to require an additional 0.016 mm^2. The power consumption is estimated to be

$$P_{tot} \approx P_{clock+control} + P_{shift\text{-}acc.} = 23.9 \text{ mW} + 10.2 \text{ mW} = 34.1 \text{ mW}$$

at 5 V and 175 MHz. Notice that the power consumption due to the clocking and control circuitry is more than twice as large as the power consumption in the logic circuitry itself. The energy consumed per complex multiplication (16-bit sample and 8-bit coefficient) is

$$E = 2P_{tot}\frac{16}{(175 \cdot 10^6)} = 6.24 \text{ nJ/sample}$$

11.18.2 Complex Multiplier Using TSPC Logic

Figure 11.51 shows the layouts of one-half of a complex multiplier [47] based on the improved shift-accumulator using TSPC (true single-phase clocking) [20, 21, 48]. Also this shift-accumulator includes the multiplexer circuitry and register needed for use as a complex multiplier. The maximal clock frequency is about 400 MHz at 5 V. The chip area is 540 μm × 250 μm = 0.135 mm^2. Drivers for the clock are estimated to require an additional 0.052 mm^2.

The power consumption is estimated to

$$P_{tot} \approx P_{clock+control} + P_{shift\text{-}acc.} = 117 \text{ mW} + 57 \text{ mW} = 174 \text{ mW}$$

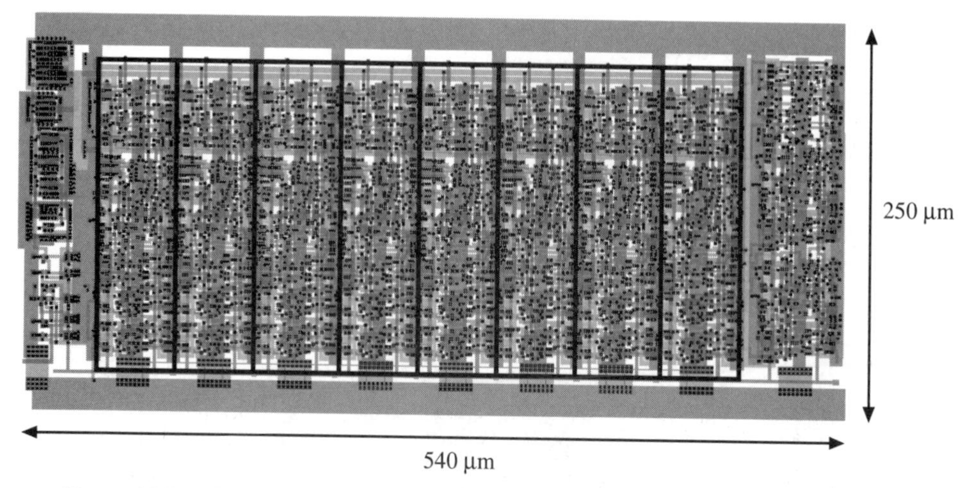

Figure 11.51 Half of a complex multiplier using TSPC (true single-phase clocking)

at 5 V and 400 MHz. Also in this case the power consumption due to the clocking and control circuitry is much larger than the power consumption in the logic circuitry. However, the clock load can be reduced significantly by using D flip-flops with fewer clocked transistors. The energy consumed per complex multiplication (16-bit sample and 8-bit coefficient) is

$$E = 2P_{tot}\frac{16}{(400 \cdot 10^6)} = 13.9 \text{ nJ/sample}$$

Hence, the higher speed of this multiplier compared to the one realized with two-phase logic is obtained at the expense of increased chip area and, more importantly, much higher power consumption per sample. It is therefore too early in the design process to determine the number of processing elements and the required clock frequency. It may often, from a power consumption point of view, be efficient to increase the number of processing elements and thereby allow the use of a more power-effective but slower logic style. Further, any excess speed may be converted to a reduction in the power consumption by reducing the power supply voltage until the speed is just adequate.

11.19 FFT PROCESSOR, Cont.

The decimation-in-frequency radix-2 bit-serial butterfly PE has been implemented in a 0.8-μm standard CMOS process. The PE has a built-in coefficient generator that can generate all twiddle factors in the range 0 to 128, which is sufficient for a 1024-point FFT. The butterfly PE can be implemented using the complex multiplier and the two complex adders (subtractors), as shown in Figure 11.52.

A complex adder requires only two real adders while the multiplier requires two distributed arithmetic units. The maximum clock frequency at 3 V supply voltage is 133 MHz with a power consumption of 30 mW (excluding the power consumed by the clock). In order to estimate the power consumption in the different parts of the butterfly, simulations were made in *Lsim Power Analyst*™. The simu-

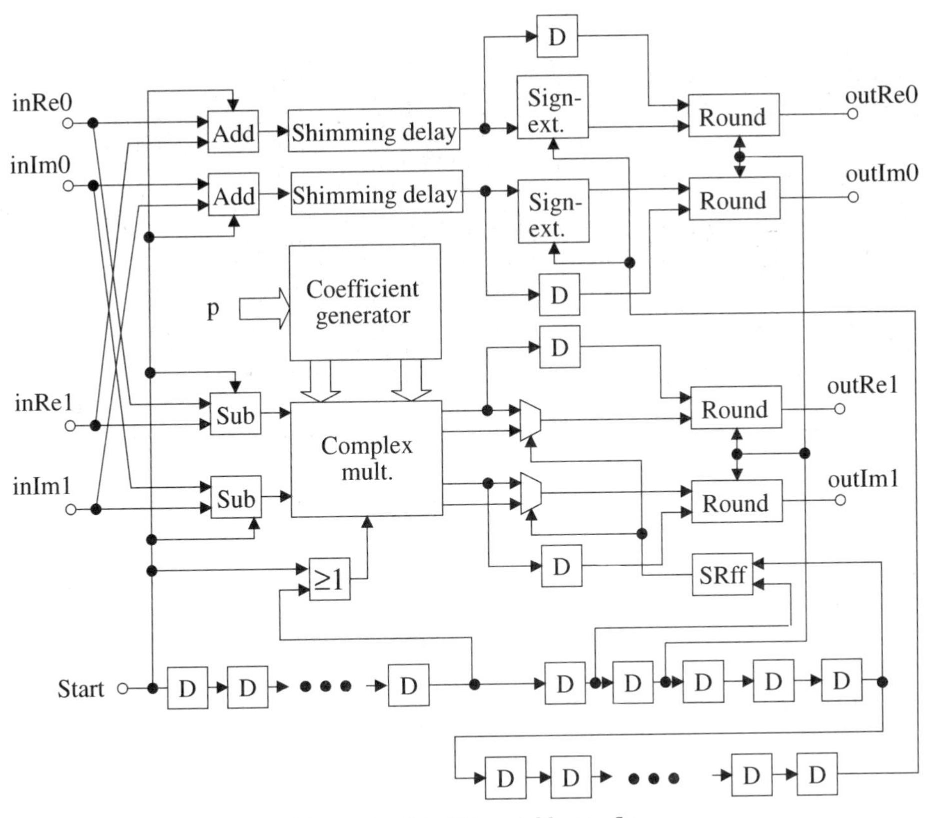

Figure 11.52 Bit-serial butterfly

lations indicate that 80% of the power is consumed in the complex multiplier and 5% in the coefficient generator. The rest (15%) is evenly distributed in the rest of the butterfly.

The D flip-flops and the gates at the bottom of the block diagram are the control. The computation is initiated by a high start signal which then travels through the D flip-flops. The coefficient generator is also included in the PE. To get a high maximum clock frequency at the expense of high power consumption, TSPC logic style was chosen for the implementation. Other benefits of TSPC are easy distribution of the clock (only one clock phase) and its suitability for heavily pipelined structures like bit-serial arithmetic units. The full custom layout of the butterfly is made using an AMS 0.8-μm double metal CMOS process. The layout is shown in Figure 11.53. It is clear that the coefficient generator and the complex multiplier occupy most of the area. The area is 1.47 mm^2.

11.19.1 Twiddle Factor PE

Twiddle factors can be generated in several ways: by using a CORDIC PE, via trigonometric formulas, or by reading from a precomputed table. Here we will use the latter method—that is, a PE that essentially consists of a ROM. We have previously shown that it is possible to use only one W^p PE. However, here it is better to use one for each butterfly PE, because the required chip area for a ROM is

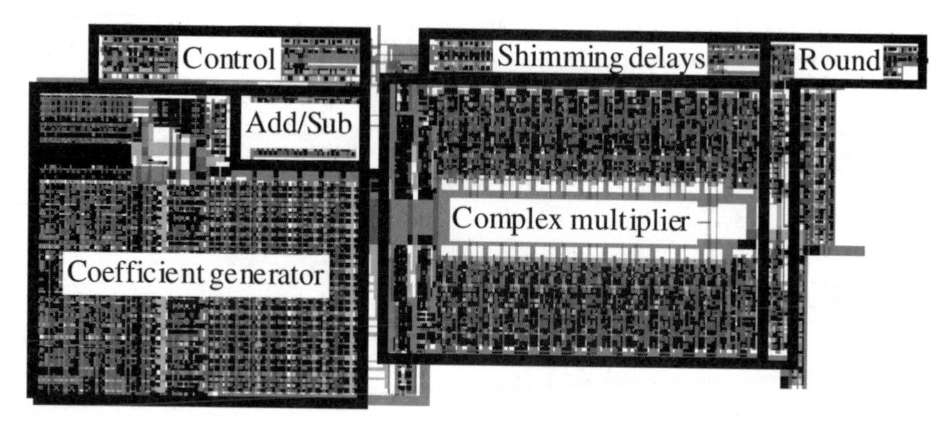

Figure 11.53 Layout of the butterfly PE

relatively small. If only one ROM were used, it would have been necessary to use long bit-parallel buses, which are costly in terms of area, to distribute the twiddle factors to the butterfly PEs.

The values of the twiddle factors, $\boldsymbol{W}$, are spaced uniformly around the unit circle. Generally, there are N twiddle factors, but it is possible to reduce the number of unique values by exploring symmetries in the trigonometric functions. In fact, it can be shown that only $N/8 + 1$ coefficient values need be stored in a table [1].

Instead of storing $\boldsymbol{W}^p$, we will store the values

$$\frac{\boldsymbol{C}(\boldsymbol{a}) + \boldsymbol{S}(\boldsymbol{a})}{2} = \frac{1}{2}(\cos(\boldsymbol{a}) - \sin(\boldsymbol{a})) = \frac{1}{\sqrt{2}}\sin\left(\boldsymbol{a} - \frac{\pi}{4}\right)$$

$$\frac{\boldsymbol{C}(\boldsymbol{a}) + \boldsymbol{S}(\boldsymbol{a})}{2} = \frac{1}{2}(\cos(\boldsymbol{a}) + \sin(\boldsymbol{a})) = \frac{1}{\sqrt{2}}\sin\left(\frac{\pi}{4} + \boldsymbol{a}\right)$$

where, $\boldsymbol{a} = 2\pi \boldsymbol{p}/N$. Table 11.6 shows that the twiddle factors in the eight octants can be expressed in terms of the twiddle factors in the range 0 to $\pi/4$.

Notice that the three most significant bits in the binary representation of $\boldsymbol{a}$ (i.e., $a_0 a_1 a_2$) can be used to select the appropriate entry in the table. For example, a_2 is used to select the argument—i.e., either $\boldsymbol{b}$ or $\pi/4 - \boldsymbol{b}$. Let the seven least significant bits in the binary representation of $\boldsymbol{a}$ be $a_3 a_4 a_5 a_6 a_7 a_8 a_9$ then the binary representation for $\pi/4 - \boldsymbol{b}$ is

$$\overline{a_3 a_4 a_5 a_6 a_7 a_8 a_9}$$

Octant	a	$a_0 a_1 a_2$	b		
0	$0 \le \boldsymbol{a} < \pi/4$	000	a	$\sin(\boldsymbol{b})$	$\cos(\boldsymbol{b})$
1	$\pi/4 \le \boldsymbol{a} < 2\pi/4$	001	$\boldsymbol{a} - \pi/4$	$\sin(\boldsymbol{b})$	$\cos(\boldsymbol{b})$
2	$2\pi/4 \le \boldsymbol{a} < 3\pi/4$	010	$\boldsymbol{a} - 2\pi/4$	$\cos(\boldsymbol{b})$	$\sin(\boldsymbol{b})$
3	$3\pi/4 \le \boldsymbol{a} < 4\pi/4$	011	$\boldsymbol{a} - 3\pi/4$	$\cos(\boldsymbol{b})$	$\sin(\boldsymbol{b})$

Table 11.6 Address generation for the modified twiddle factors

Hence, a_2 can be used to control a set of XOR gates to invert the address to the memory in which the first octants of $\sin(\boldsymbol{b})$ and $\cos(\boldsymbol{b})$ are stored.

Notice that the argument of the sine and cosine functions, $\boldsymbol{b}$ or $\pi/4-\boldsymbol{b}$, are always in the range 0 to $\pi/4$. This means that to get all values of $(\boldsymbol{C}+\boldsymbol{S})/2$ and $(\boldsymbol{C}-\boldsymbol{S})/2$ we need only to store $\sin(\boldsymbol{b})/\sqrt{2}$/and $\cos(\boldsymbol{b})/\sqrt{2}$/for $0 \le \boldsymbol{b} \le \pi/4$. Bit a_1 can be used to select if the $\sin(\boldsymbol{b})$-part, or the $\cos(\boldsymbol{b})$-part of the memory shall be used. We choose always to first read $\sin(\boldsymbol{a}+\pi/4)$ and then $\sin(\boldsymbol{a}-\pi/4)$. Hence, $a_1 = 0$ is used to select the $\sin(\boldsymbol{b})$-part for the first value and the $\cos(\boldsymbol{b})$-part for the second value, and vice versa for $a_1 = 1$.

The sign of the twiddle factor has to be derived by using the three most significant bits, $a_0a_1a_2$. The change of sign of the twiddle factors can be incorporated into the control signals for the multipliers. The ROM requires only $2(N/8+1) = 2 \cdot 128 + 2 = 258$ 14-bits words. However, there are only $N/4+1$ different twiddle factors since $\sin(\pi/4) = \cos(\pi/4)$. Hence, we can store all coefficients in a ROM with only 129 words. The principle of this W^p processor is illustrated in Figure 11.54.

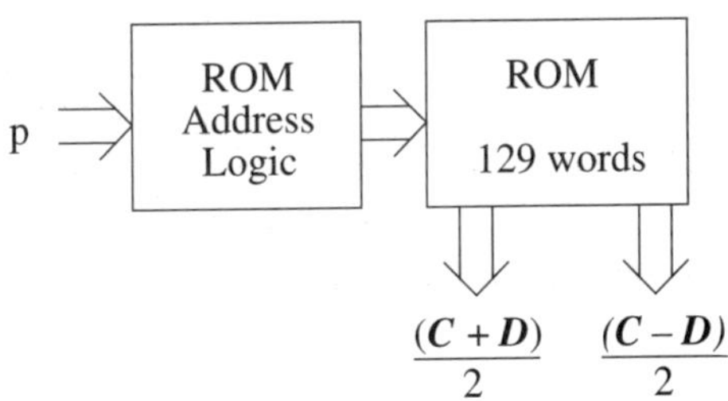

Figure 11.54 The W^p processor

The ROM address logic must transform $\boldsymbol{p}$ to the right address in the ROM. Since $\boldsymbol{a} = 2\pi \boldsymbol{p}/N$, the following applies to $\boldsymbol{b}$:

$$\boldsymbol{b} = \frac{2\pi}{N}(\boldsymbol{p} \bmod(N/8))$$

$$\frac{\pi}{4} - \boldsymbol{b} = \frac{2\pi}{N}(N/8 - \boldsymbol{p} \bmod(N/8))$$

The square parentheses can be interpreted as an address in the coefficient ROM. We can use addresses 00000000_2 to 10000000_2 (0_{10} to 128_{10}) in the ROM. If $\boldsymbol{p} = p_8p_7p_6p_5p_4p_3p_2p_1p_0$ in binary, then $\boldsymbol{p}$ **mod**$(N/8) = 0p_6p_5p_4p_3p_2p_1p_0$. This is used as the address to the ROM in octants 1 and 3. In octants 0 and 2 we will instead use $N/8-\boldsymbol{p}$ **mod**$(N/8)$ as the address. Since subtracting is equivalent to adding the two's-complement, we have

$$\begin{aligned} N/8 - \boldsymbol{p}\ \mathbf{mod}(N/8) &= 10000000_2 - 0p_6p_5p_4p_3p_2p_1p_0 \\ &= 10000000_2 + 1\overline{p_6p_5p_4p_3p_2p_1p_0} + 00000001_2 \\ &= 0\overline{p_6p_5p_4p_3p_2p_1p_0} + 00000001_2 \end{aligned}$$

This is, in turn, equal to the two's-complement of $1p_6p_5p_4p_3p_2p_1p_0$. Further design of the W^p processor will be postponed until after we have designed the address processors.

We will store $\sin(\boldsymbol{b})$ and $\cos(\boldsymbol{b})$ for $0 \le \boldsymbol{b} \le \pi/4$—i.e., we store only $2(N/8+1)$ coefficients. These are accessible through the addresses $(00000000)_2$ to $(10000000)_2$—i.e., 0 to 128 in a ROM.

Let $\boldsymbol{a} = \boldsymbol{a_b}\, 2\pi/N$, where $\boldsymbol{a_b} = a_9a_8a_7a_6a_5a_4a_3a_2a_1a_0$. Then $\boldsymbol{a_b}$ will be in the range $0 \le \boldsymbol{a_b} < 1024$, i.e., $(0000000000)_2 \le \boldsymbol{a_b} < (10000000000)_2$. We use bit a_7 to select the proper octant. Thus, if $a_7 = 1$, we use $a_6a_5a_4a_3a_2a_1a_0$ as the address to the ROM and when $a_7 = 0$, we use the two's-complement $\overline{a_7a_6a_5a_4a_3a_2a_1a_0} = (a_7\overline{a_6}\overline{a_5}\overline{a_4}\overline{a_3}\overline{a_2}\overline{a_1}\overline{a_0} + 00000001)_2$ as the address.

11.19.2 Control PEs

It is possible to implement the loop counters i and k with an 8-bit counter [29]. The indices of $x(i)$ can be interpreted as addresses in memories which can be extracted from this counter. For the coefficients, $\boldsymbol{W}^p$, we need the exponent p. Notice that $\boldsymbol{W}$ = exp($2\pi j/N$) and hence the binary representation of $\boldsymbol{p}$ can be interpreted as $\boldsymbol{a_b}$ as just discussed. It is also possible to extract $\boldsymbol{p}$ from the same 8-bit counter [29].

A drawback of implementing the control units using a bit-serial style is the high clock frequency that is required. It may therefore be better to use a bit-parallel or digit-serial implementation to reduce the required clock frequency and possibly also the power consumption.

11.19.3 Address PEs

Control of the architecture is captured in the address processors for the RAMs. A behavioral description of the address processors was derived in Section 9.9. The base index generator has three modes of operation: input, FFT computation, and output. During the input and output phases it must compute the data index, i, and which RAM this data belongs to. The index is in the interval $0 \le i < 1024$. During the FFT computation, the base index generator computes m, k_1, and the Gray code equivalent of k_1.

During the input and output phases, index i can be almost directly used as an address to the RAMs. The exception is that the index must be bit-reversed during the output phase and must be truncated to 9 bits. Hence, the function of the address generators for the RAMs is trivial.

During the FFT computation, index m must be extended to become an address. The location of the extra bits is controlled by the *Stage*. The positions of these bits have the same weight as N_s and BF:

$$N_s = N2^{-Stage}$$

$$BF = \begin{cases} N_s/2 \text{ if } Stage = 1 \\ 2N_s \text{ if } Stage \ge 2 \end{cases}$$

The address generators compute two addresses each. These are computed by adding N_s and/or BF to the base index and truncating the result to 9 bits. Address generator 0 computes addresses A_{00} and A_{10} according to the algorithm shown in Box 9.1. Address generator 1 computes addresses A_{01} and A_{11} according to the algorithm shown in Box 9.2.

It is instructive to study how a RAM address is built from the base index k_1. This is shown in Figure 11.55. The least and the most significant bits in k_1 are denoted as k_{1L} and k_{1H}, respec-

Stage	Addresses
1	BF, k_{1L}
2	Ns, k_{1L}
3	BF, Ns, k_{1L}
4	k_{1H}, BF, Ns, k_{1L}
5	k_{1H}, BF, Ns, k_{1L}
6	k_{1H}, BF, Ns, k_{1L}
7	k_{1H}, BF, Ns, k_{1L}
8	k_{1H}, BF, Ns, k_{1L}
9	k_{1H}, BF, Ns, k_{1L}
10	k_{1H}, BF, Ns

Figure 11.55 Derivation of the RAM addresses

tively. In k_1 the bits denoted *BF* and N_s are set to 0. Hence, the addresses are computed by setting these bits to 1.

11.19.4 Base Index Generator

The base index generator must have three basic functions: a binary counter, a Gray encoder, and a unit to compute *P*(*m*). The counter can be implemented as a bit-serial counter. This implementation requires a set of D flip-flops (a shift register) and a bit-serial half-adder.

The translation of a binary number, $\boldsymbol{x} = x_n x_{n-1} \cdots x_1 x_0$, to its Gray coded equivalent, $\boldsymbol{y} = y_n y_{n-1} y_1 y_0$, can be described by

$$y_i = \begin{cases} x_i \oplus x_{i+1} & \text{if } 0 \le i < n \\ x_i & \text{if } 0 \le i < n \end{cases}$$

This operation can also be performed bit-serially. The computation of *P*(*m*) can be described as

$$P(m) = (((\ldots((m_0) \oplus m_1) \oplus m_2) \oplus m_3) \oplus \ldots) \oplus m_7$$

This computation can also be done bit-serially. The implementation of a device that increments *m*, computes the Gray code representation of *m*, computes *P*(*m*), and extends this by two bits to k_1 is shown in Figure 11.56.

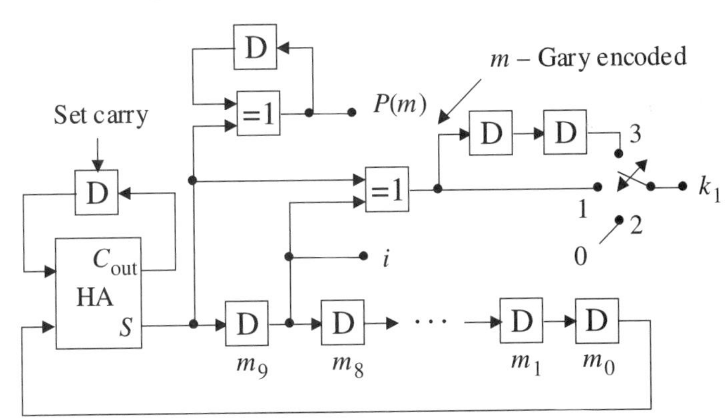

Figure 11.56 The base index generator

The generator consists of four parts: the counter, the Gray encoder, the unit that computes *P*(*m*), and the unit that extends *m* by two bits. The counter is formed around the *m* register and the half-adder. To increment *m* by one's we first use "Set carry" to set the carry D flip-flop of the bit-serial half-adder. The half-adder adds 1 to the LSB of *m*, and thus the LSB of *m*+1 is present at the output. Next we let 10 (the *m* register length) clock periods pass. Simultaneously with the incrementation of *m* we compute the Gray code representation of *m* using an XOR gate. The *i* and *P*(*m*) outputs are used during the input and output phases.

To extend *m* by two bits, the switch is initially set to position 1. The register is clocked to the position where the extra bits should be inserted. The switch is then set to position 2 and two zeros are entered during two clock cycles. After this, the switch is set to 3 and the remaining part of the base index is clocked out.

After 10 clock periods in the FFT computation phase, m_8 is checked. If $m_8 = 1$, it means that *m* has reached $2^8 = 256$, and we have computed all butterflies in a stage. When this happens, *m* is reset, and a done signal is sent to the Stage PE.

11.19.5 RAM Address PEs

The address processors must compute the RAM addresses from the base index by conditional setting of the extended bits. This can be accomplished by the unit shown in Figure 11.57.

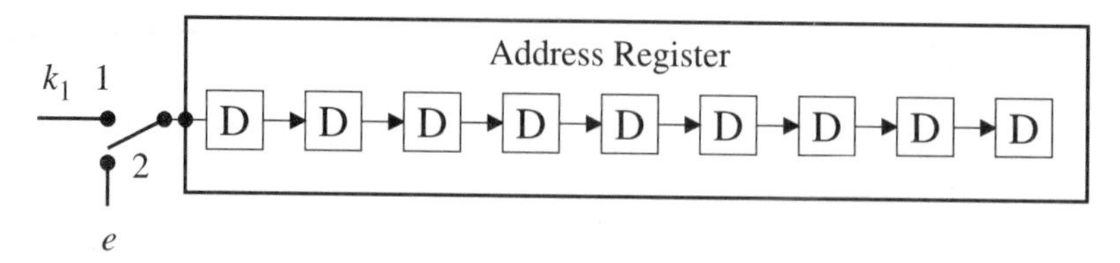

Figure 11.57 Logic implementation of the address generation

The e input is used for setting the extended bits. To save area we can share the switch in this address register with the switch in the base index generator. However, this sharing leads to a small modification of the partitioning of the address generation. The Stage PE can be implemented as a shift register with $\log_2(N/2) = 9$ bits. In fact, we use N_s instead of *Stage*. The input to the computation of the coefficients is p. The exponent p relates to k_1 by

$$p = k_1 2^{Stage-1} \bmod (N/2)$$

Hence, p is equal to k_1 shifted *Stage*−1 steps to the right. This can be performed in a shift register. We need both p and its two's-complement value, which can be computed using a bit-serial subtractor.

11.20 DCT PROCESSOR, Cont.

The chip area needed to implement a vector-multiplier using distributed arithmetic grows as $O(2^N)$ where N is the number of terms in the inner product. The chip area required for implementing a one-dimensional MSDCT can be reduced by exploiting the symmetry (antisymmetry) of the basis functions. A detailed study shows that the basis functions exhibit both symmetry and antisymmetry in their coefficients. Using the same principle that was used in the linear-phase FIR filter

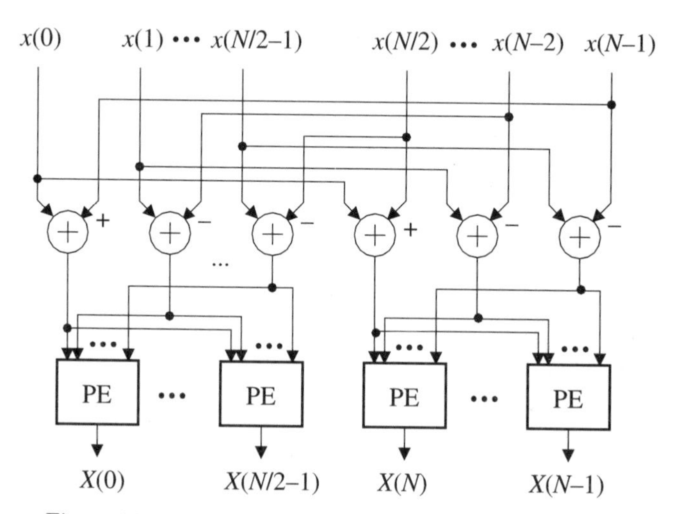

Figure 11.58 DCT processor with direct interconnection between the processing elements

structure, the pixels that are multiplied by the same coefficient are added (or subtracted). This reduces the number of terms in the remaining inner products by 50%. The chip area is thereby reduced from $O(2^N)$ to $O(2^{N/2})$, which is a significant reduction. In comparison, the area for the bit-serial adders is insignificant. Figure 11.58 shows a block diagram for the DCT PE.

A 2-D DCT for 16 × 16 pixels can be built using only one 1-D DCT PE which itself consists of 16 distributed arithmetic units with $N = 8$. The TSPC-based shift-accumulator in Figure 11.51 can be used to implement a distributed arithmetic unit. The length of the shift-accumulator depends on the word length, W_{ROM}, which depends on the coefficients in the vector–products. In this case we assume that $W_{ROM} = W_c + 1 = 12$ bits. The ROM corresponding to each bit-slice is organized to have eight rows and 2^{N-3} columns in order to have about the same width as the bit-slices.

The area for a 1-D DCT PE is estimated to

$$A_{DCT} \approx 16\,A_{DA} + A_{Wire} \approx 16 \cdot 0.246 \cdot 1.3 \text{ mm}^2 \approx 5.2 \text{ mm}^2$$

where we have assumed that the area reserved for wiring is about 30%.

REFERENCES

[1] Agrawal J.P. and Ninan J.: Hardware Modification in Radix-2 Cascade FFT Processors, *IEEE Acoust., Speech, Signal Processing*, Vol. ASSP-26, No. 2, pp. 171–172, April 1978.

[2] Akl S.G. and Meijer H.: On the Bit Complexity of Parallel Computations, Integration, *The VLSI Journal*, Vol. 6, pp. 201–212, 1988.

[3] Bedrij O.J.: Carry-Select Adder, *IRE Trans. Elect. Comp.* EC-11, pp. 340–346, 1962.

[4] Bickerstaff K.C., Schulte M.J., and Swartzlander Jr. E.E.: Parallel Reduced Area Multipliers, *J. of VLSI Signal Processing*, Vol. 9, No. 3, pp. 181–191, April 1995.

[5] Brent R.P. and Kung H.T.: A regular Layout for Parallel Adders, *IEEE Trans. on Computers*, Vol. C-31, pp. 280–284, 1982.

[6] Bull D.R. and Horrocks D.H.: Primitive Operator Digital Filters, *IEE Proc. G*, Vol. 138, No. 3, pp. 401–411, June 1991.

[7] Büttner M. and Schüßler H.W.: On Structures for the Implementation of the Distributed Arithmetic, *Nachrichtentechn. Z.*, Vol. 29, No. 6, pp. 472–477, 1976.

[8] Callaway T.K. and Swartzlander E.E.: Optimizing Arithmetic Elements for Signal Processing, in *VLSI Signal Processing* V, ed. K. Yao et al., IEEE Pub., New York, pp. 91–100, 1992.

[9] Chan P.K. and Schlag M.D.F.: Analysis and design of CMOS Manchester Adders with Variable Carry-Skip, *IEEE Trans. on Computers*, Vol. C-39, pp. 983–992, 1990.

[10] Chatterjee A., Roy R.K., and dAbreu M.: Greedy Hardware Optimization for Linear Digital Circuits Using Number Splitting and Refactorization, *IEEE Trans. on Very Large Scale Integration*, Vol. 1, No. 4, pp. 423–431, Dec. 1993.

[11] Croisier A., Esteban D.J., Levilion M.E., and Rizo V.: Digital Filter for PCM Encoded Signals, *U. S. Patent 3777130*, Dec. 4, 1973.

[12] De Man H.J., Vandenbulcke C.J., and van Cappellen M.M.: High-Speed NMOS Circuits for ROM-Accumulator and Multiplier Type Digital Filters, *IEEE J. on Solid-State Circuits*, Vol. SC-13, No. 5, pp. 565–572, Oct. 1978.

[13] Dempster A.G. and Macleod M.D.: Constant Integer Multiplication Using Minimum Adders, *IEE Proc. Circuits Devices Systems*, Vol. 141, No. 5, pp. 407–413, Oct. 1994.

[14] Dempster A.G. and Macleod M.D.: Use of Minimum-Adder Multiplier Blocks in FIR Digital Filters, *IEEE Trans. on Analog and Digital Signal Processing*, Vol. 42, No. 9, pp. 569–577, Sept. 1995.

[15] Ercegovac M.D. and Lang T.: On-the-Fly Conversion of Redundant into Conventional Representations, *IEEE Trans. on Computers*, Vol. C-36, No. 7, pp. 895–897, July 1987.

[16] Ginderdeuren van J.J., De Man H.J., Goncalves N.F., and van Noije W.A.M.: Compact NMOS Building Blocks and a Methodology for Dedicated Digital Filter Applications, *IEEE J. on Solid-State Circuits*, Vol. SC-18, No. 3, pp. 306–316, June 1983.

[17] Hartly R.I. and Parhi K.K.: *Digit-Serial Computation*, Kluwer Academic Pub., Boston, 1995.

[18] Hwang I.S. and Fisher A.L.: Ultra Fast Compact 32-Bit CMOS Adder in Multi-Output Domino Logic, *IEEE J. Solid-State Circuits*, Vol. SC-24, pp. 358–369, 1989.

[19] Hwang K.: *Computer Arithmetic: Principles, Architecture, and Design*, New York, John Wiley & Sons, 1979.

[20] Ingelhag P., Jonsson B., Sikström B., and Wanhammar L.: A High-Speed Bit-Serial Processing Element, *ECCTD-89*, pp. 162–165, Brighton, UK, Sept. 1989.

[21] Karlsson I.: True Single Phase Dynamic CMOS Circuit Technique, *IEEE Intern. Symp. on Circuits and Systems*, ISCAS-88, pp. 475–478, Espoo, Finland, June 1988.

[22] Koren I.: *Computer Arithmetic Algorithms*, Prentice Hall, Englewood Cliffs, New Jersey, 1993.

[23] Li D.: Minimum Number of Adders for Implementing a Multiplier and Its Application to the Design of Multiplierless Digital Filters, *IEEE Trans. on Circuits and Systems II*, Vol. CAS-II-42, No. 7, pp. 453–460, July 1995.

[24] Li W.: A Fully Parallel Implementation of Distributed Arithmetic, *IEEE Intern. Symp. on Circuits and Systems*, ISCAS-88, pp. 1511–1515, Espoo, Finland, June 1988.

[25] Lindkvist H. and Andersson P.: Techniques for Fast CMOS-Based Conditional Sum Adders, *IEEE Intern. Conf. on Computer Design*, Cambridge, MA, pp. 626–635, 1994.

[26] Lyon R.F.: Two's Complement Pipeline Multipliers, *IEEE Trans. on Communication*, Vol. Com-24, No. 4, pp. 418–424, April 1976.

[27] Ma G.K. and Taylor F.J.: Multiplier Policies for Digital Signal Processing, *IEEE ASSP Magazine*, Vol. 7, No. 1, pp. 6–20, Jan. 1990.

[28] Melander J., Widhe T., Sandberg P., Palmkvist K., Vesterbacka M., and Wanhammar L.: Implementation of a Bit-Serial FFT Processor with a Hierarchical Control Structure, *Proc. European Conf. on Circuit Theory and Design*, ECCTD '95, Istanbul, Turkey, Aug. 1995.

[29] Nordhamn E.: *Design of an Application-Specific FFT Processor*, Linköping Studies in Science and Technology, Thesis, No. 324, Linköping University, Sweden, 1992.

[30] Peled A. and Liu B.: A New Hardware Realization of Digital Filters, *IEEE Trans. on Acoust., Speech, Signal Processing*, Vol. ASSP-22, No. 6, pp. 456–462, Dec. 1974.

[31] Peled A. and Liu B.: *Digital Signal Processing*, John Wiley & Sons, New York, 1976.

[32] Potkonjak M., Srivastava M.B., and Chandrakasan A.P.: Multiple Constant Multiplications: Efficient and Versatile Framework and Algorithms for Exploring Common Subexpression Elimination, *IEEE Trans. on Computer-Aided Design of Integrated Circuits and Systems*, Vol. VLSI-15, No. 2, pp. 151–165, Feb. 1996.

[33] Reitwiesner G.W.: *Binary Arithmetic, Advances in Computers,* Vol. 1, pp. 232–308, 1960.

[34] Shanbhag N.R. and Siferd R.E.: A Single-Chip Pipelined 2-D FIR Filter Using Residue Arithmetic, *IEEE J. Solid-State Circuits*, Vol. SC-26, No. 5, pp. 796–805, May 1991.

[35] Sikström B. and Wanhammar L.: A Shift-Accumulator for Signal Processing Applications, *Proc. European Conf. on Circuit Theory and Design*, ECCTD81, The Hague, The Netherlands, pp. 919–924, 1981.

[36] Sklansky J.: Conditional-Sum Addition Logic, *IRE Trans. Elect. Comp.* EC-9, pp. 226–231, 1960.

[37] Smith S.G. and Denyer P.B.: *Serial-Data Computation,* Kluwer Academic Pub., Boston, 1988.

[38] Soderstrand M.A., Jenkins W.K., Jullien G.A., and Taylor F.J.: *Residue Number System Arithmetic: Modern Applications in Digital Signal Processing*, IEEE Press, 1986.

[39] Swartzlander Jr. E.E. (ed.): *Computer Arithmetic*, Benchmark Papers in Electrical Engineering and Computer Science, Vol. 21, Dowden, Hutchinson & Ross, Inc., 1980.

[40] Takagi N., Yassuura H., and Yajima S.: High Speed VLSI Multiplication Algorithm with a Redundant Binary Addition Tree, *IEEE Trans. on Computers*, Vol. C-34, pp. 789–796, Sept. 1985.

[41] Timmermann D., Hahn H., Hosticka B.J., and Rix B.: A New Addition Scheme and Fast Scaling Factor Compensation Methods for CORDIC Algorithms, *Integration, the VLSI Journal*, Vol. 11, pp. 85–100, 1991.

[42] Trevedi K.S. and Ercegovac M.D.: On-Line Algorithms for Division and Multiplication, *IEEE Trans. on Computers*, Vol. C-26, No. 7, pp. 681–687, July 1977.

[43] Turrini S.: Optimal Group Distribution in Carry-Skip Adders, *Proc. 9th Symp. Computer Arithmetic*, pp. 96–103, Sept. 1990.

[44] Weinberger A. and Smith J.L.: A Logic for High Speed Addition, *Nat. Bur. Stand. Circ.,* Vol. 591, pp. 3–12, 1958.

[45] Vesterbacka M.: *Implementation of Maximally Fast Wave Digital Filters*, Thesis 495, LiU-Tek-Lic-1995:27, Linköping University, June 1995.

[46] Vesterbacka M., Palmkvist K., and Wanhammar L.: Serial Squarers and Serial/Serial Multipliers, *Nat. Conf. on Radio Science—RVK96, Luleå Institute of Technology*, Luleå, Sweden, June 3–6, 1996.

[47] Widhe T., Melander J., and Wanhammar L.: Implementation of a Bit-Serial Butterfly PE, *Natn. Conf. on Radio Science—RVK96,* Luleå Institute of Technology, Luleå, Sweden, June 3–6, 1996.

[48] Yuan J. and Svensson C.: High-Speed CMOS Circuit Technique, *IEEE J. on Solid-State Circuits*, Vol. SC-24, No. 1, pp. 62–70, Feb. 1989.

Problems

11.1 Two cascaded bit-parallel adders are used to add three values. The data word length is 20 bits. Assume that the propagation times for the full-adders are $t_{carry} = t_{sum} = 2$ ns.

(a) Determine the addition time.
(b) Introduce one level of pipelining and determine the new throughput and latency.

11.2 Multiply the binary integers corresponding to the decimal numbers $(15)_{10}$ and $(13)_{10}$ using Booth's algorithm.

11.3 Determine the number range if the following moduli set is used in a residue number system:

(a) {13, 11, 9, 7, 5, 4}
(b) {32, 31, 29, 27, 25, 23, 19}

11.4 (a) Show by an example that the sign-extension method used in the serial/parallel multiplier in Figure 11.14 yields correct products. Use, for example, the coefficient $\boldsymbol{a} = 1.1101$ and $\boldsymbol{x} = 1.001$.
(b) How much would the throughput increase if the serial/parallell multiplier in Figure 11.14 is replaced by the one in Figure 11.44 when the data word length is 16 bits.

11.5 Simplify the basic serial/parallell multiplier in terms of the number of D flip-flops and full-adders for the coefficient $(0.0101)_2$.

(a) Draw the logic diagram for the minimized multiplier.
(b) Determine the required control signals.
(c) Determine the throughput when the data word length is 12 bits.
(d) Validate the function of the minimized multiplier by performing a multiplication with the value $\boldsymbol{x} = (1.010)_2$.

11.6 Simplify the basic serial/parallell multiplier in terms of the number of D flip-flops and full-adders, and draw its logic diagram when the coefficient is represented in canonic signed digit code. The coefficient is $\boldsymbol{a} = (0.1001)_{CSDC}$.

11.7 A multiplication by a factor $\boldsymbol{a} = (1.11011)_2$ shall be implemented by using a serial/parallell multiplier.

(a) Determine a schematic diagram for the multiplier with a general coefficient.
(b) Simplify the multiplier in terms of number of D flip-flops and full-adders and draw its logic diagram.

11.8 Find the simplest implementation of a serial/parallel multiplier with fixed coefficient when the coefficient is

(a) $(0.011001)_2$
(b) $(0.111011)_2$
(c) $(1.011001)_2$
(d) $(1.011011)_2$
(e) $(0.000001)_2$
(f) $(1.000001)_2$

11.9 (a) Determine for the filter shown in Figure P11.9 the number of D flip-flops, excluding the D flip-flops used in the bit-serial adders (subtractors) required in a bit-serial isomorphic implementation. $W_d = 20$ bits and $\alpha = 0.375$.

(b) Determine the number of D flip-flops and bit-serial adders (subtractors) required to implement the multiplication using a simplified S/P multiplier.

(c) Determine the maximal sampling frequency when the full-adders can be clocked at 400 MHz.

(d) Sketch the attenuation for the filter.

11.10 (a) Modify the two-port adaptor in order to implement the multiplications by 1 that may appear in lattice wave digital filters.

(b) Show that it is sufficient to invert all bits, including the sign bit, in the reflected waves that exceed the signal range, in order to suppress overflow oscillations in wave digital filters. Two's-complement representation is used for the signals.

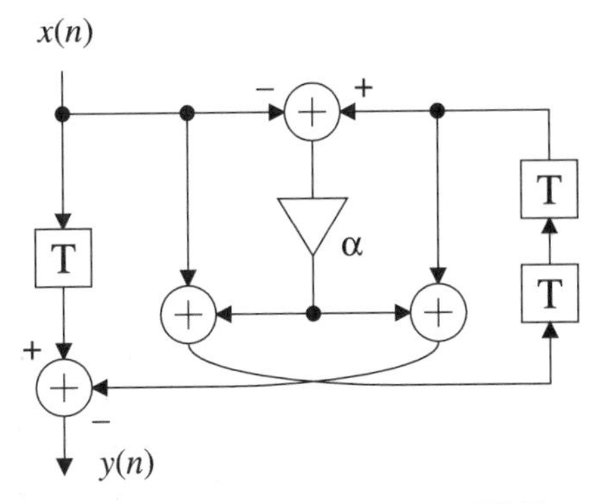

Fig P11.9 Third-order lattice WDF

11.11 Suggest a bit-serial PE for implementation of a product, $x\,y$, of two variables. The external interface should be bit-parallel.

11.12 Suggest a bit-serial PE based on serial/parallell multipliers for implementation of a

(a) Direct form FIR filter with fixed coefficients.

(b) Transposed direct form FIR filter with fixed coefficients.

11.13 Derive a bit-serial realization of both a series and parallel two-port adaptor.

11.14 Consider the sum of products

$$y = a_1 x_1 + a_2 x_2 + a_3 x_3$$

which shall be evaluated by a PE based on distributed arithmetic. The data word length is $W_d = 17$ bits, and the coefficients are

$$a_1 = (0.5)_{10} = (0.1)_2$$
$$a_2 = (0.375)_{10} = (1.101)_2$$
$$a_3 = (0.5625)_{10} = (0.1001)_2$$

Suggest a block diagram for the PE and describe how it works. How many bit-slices are required? Determine also the contents in the ROM.

11.15 Distributed arithmetic is used to compute: $y = \sum_{i=1}^{N} a_i x_i$

where x_i are variables and a_i are fixed coefficients. The data word length is W_d and the word length in the ROM is W_{ROM}. Further we have $W_d > W_{ROM}$.

(a) Express the word length in the W_{ROM} in terms of the coefficients a_i.
(b) How is the hardware cost and throughput related to W_{ROM} and W_d?

11.16 A fourth-order digital filter is realized in cascade form using second-order sections in direct form II. The second-order sections are implemented using distributed arithmetic. How many PEs are required? How many words are required in the ROMs? Sketch the major parts of the implementation.

11.17 The filter structure in Figure P11.17 shall be implemented using distributed arithmetic.

(a) Sketch the major parts of the implementation.
(b) How many PEs are required? How many words are needed in the ROMs?
(c) Determine the contents in the ROMs.
(d) Show that all computations can be performed in parallel if the structure is pipelined. How is the transfer function affected?

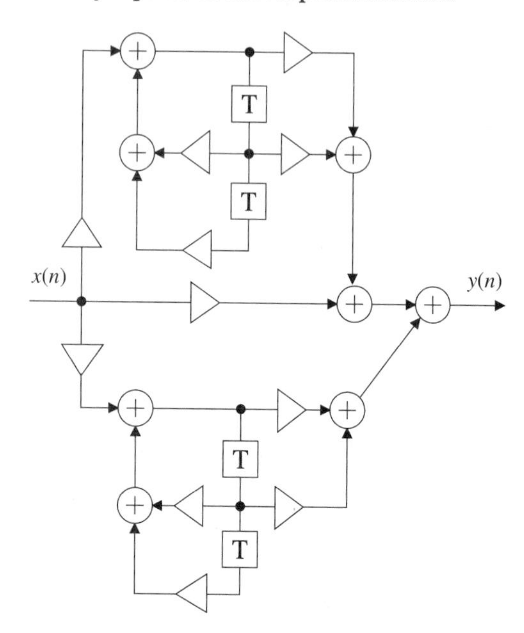

Figure P11.17 Parallel form

11.18 The filter structure shown in Problem 6.6 shall be implemented using vector-multipliers (distributed arithmetic). Determine first the difference equations in computable order. Sketch the most important parts in such an implementation. Determine the number of words in the ROMs.

11.19 (a) Describe how the so-called interpolated FIR filter (IFIR) in Figure P11.19 can be implemented using distributed arithmetic. How many PEs are required?
(b) Determine the contents in the ROMs.

11.20 The lattice WDF shown in Figure P11.20 shall be implemented using PEs (distributed arithmetic) that compute inner products with fixed coefficients. The adaptor coefficients α_i have a word length of 6 bits while the data have 18 bits (two's-complement representation).

(a) Determine the number of inner product operations required per sample.
(b) Determine the best placement of the delay elements—i.e., minimize the number of words in the ROMs.
(c) Determine the maximal sample frequency when the PEs and the shift registers can be clocked at least at 400 MHz.
(d) Sketch a block diagram for an implementation where the memories are implemented as circulating shift registers.

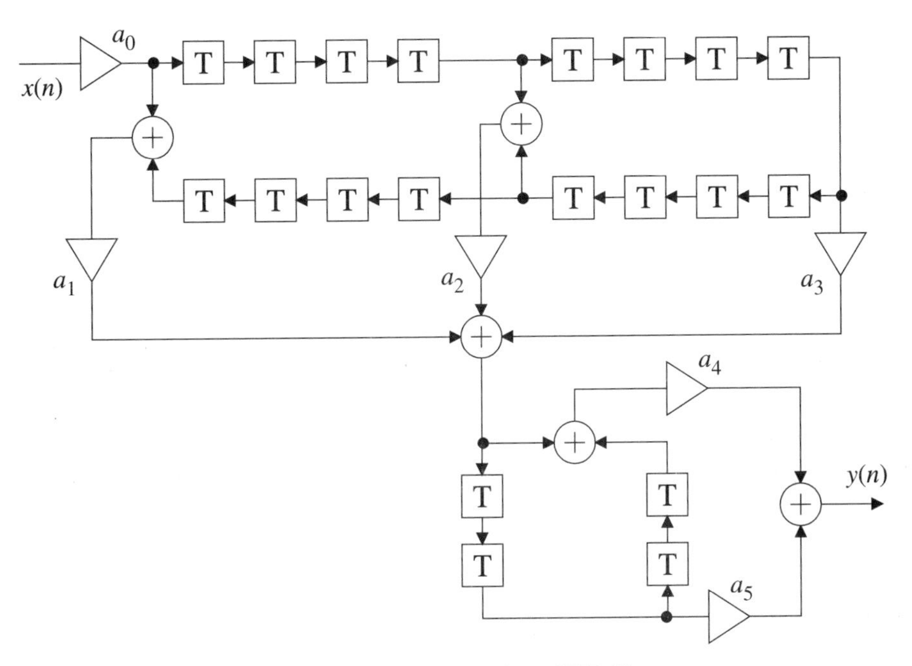

Figure P11.19 Linear-phase IFIR filter

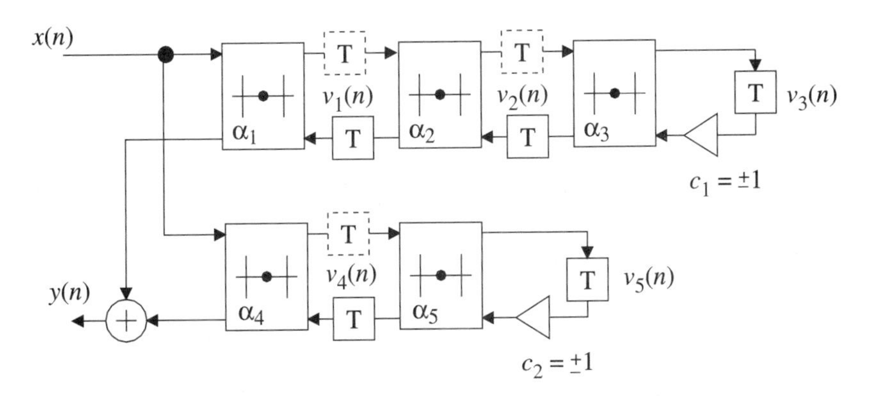

Figure P11.20 Lattice WDF with the branches implemented by Richards' structures

11.21 Implement the first-order section with scattered look-ahead pipelining that was discussed in Example 6.13 using distributed arithmetic.

11.22 Show that a butterfly can be implemented using only three real multiplications.

11.23 Suggest a multi-input, multi-output PE for a butterfly of type (1) decimation-in-time and (2) decimation-in-frequency. Use both serial/parallell multipliers and distributed arithmetic.

11.24 Determine the difference in ROM size if the MDCT is not used. Which of the other DCTs can be used?

11.25 (a) Find a bit-serial multiplier that implements the coefficient

$$\boldsymbol{a} = 2^1(1 + 2^2(1 + 2^2))$$

(b) Determine the normal serial/parallell multiplier for the same coefficient value.

(c) Determine the relationship between the two multiplier structures.

11.26 Redo Example 6.7, but include appropriate control signals as well.

11.27 Derive bit-parallel implementations of the control processes used in the FFT processor. Discuss trade-offs between the bit-serial and bit-parallel versions.

11.28 Conversion between RGB and $YCbCr$ digital color video image formats can be performed by the following transformations:

$$R = Y + 350\ Cr/256 - 175$$
$$G = Y - 86\ Cb/256 - 178\ Cr/256 + 132$$
$$B = Y + 44\ Cb/256 - 222$$

and

$$Y = (77\ R + 150\ G + 29\ B)/256$$
$$Cb = (-44\ R - 87\ G + 131\ B)/256 + 128$$
$$Cr = (131\ R - 110\ G - 21\ B)/256 + 128$$

The color components are quantized to 8 bits. Derive an implementation based on

(a) Bit-serial multipliers
(b) Distributed arithmetic
(c) Compare the two implementations.

11.29 Suggest a logic realization of error-feedback for both a serial/parallell multiplier and a distributed arithmetic PE. Can the input, S, in Figure 11.49 be used? What are the potential advantages of such an approach?

11.30 Derive a VHDL description of a serial/parallell multiplier.

11.31 Derive an implementation of the fast 1-D DCT using serial/parallell multipliers and compare the result with the implementation using distributed arithmetic.

11.32 Derive an implementation of the 1-D DCT using column symmetry and distributed arithmetic.

11.33 Suggest a bit-parallel implementation of the complex multiplier, which is based on distributed arithmetic, and compare it to the version used in the case study.

12

INTEGRATED CIRCUIT DESIGN

12.1 INTRODUCTION

The difficulty involved in designing large integrated circuits has grown far beyond the capabilities of a purely manual approach. The successful completion of a large circuit necessitates both a highly structured design methodology and computer-aided support [4, 12, 14, 17].

The VLSI design process can be viewed as a sequence of transformations, starting with a functional or behavioral description of the system at the logic level and ending with a geometric description that can be used for manufacturing the circuit. This process is inherently very complex and involves large amounts of design information. A structured approach is mandatory not only to cope with the high complexity, but also to reduce the amount of design data. Generally, a structured design approach aims at a highly modular and regular design that is partitioned into small, noninteracting parts that are easy to design. Design complexity is reduced if the modules can successively be partitioned (assembled) into smaller (larger) modules without too much effort. The complexity and the amount of design data are further reduced in a regular design since such a design has only a few types of modules.

Obviously, the approach used in the system design phase has a major impact on the complexity incurred in the VLSI design phase. Experience indicates that design decisions taken at the higher levels of abstractions are more important than optimization at lower levels. As discussed in Chapter 1, the complexity at the VLSI design level is essentially reduced by various ways of restricting the design space. The restrictions are introduced in several ways and at different levels of abstraction. It is often appropriate to refer to these kinds of design restrictions as *standardization*. Standardization reduces the probability of errors and provides a basis for the design tools. Moreover, a hierarchy of well-defined modules provides a necessary framework for automatic design tools.

12.2 LAYOUT OF VLSI CIRCUITS

The layout of an integrated circuit is divided into a sequence of steps, in order to reduce the complexity of the problem, as illustrated in Figure 12.1. Floor planning is an important first step in the layout of an integrated circuit. It is the process of determining the shapes and relative positions of the modules so that the chip size becomes small and a desired aspect ratio of the chip is obtained. I/O pin locations, timing constraints, power dissipation, temperature gradients, and wire density estimates are other factors guiding the relative placement.

Most of the early systems developed for layout focused on minimizing the chip area. However, performance issues—in particular, speed and power consumption have recently become a major concern. Notice that a smaller chip represents a smaller capacitive load and may therefore be potentially faster and consume less power than a larger chip. Floor planning is used to verify the feasibility of a design without performing the detailed layout and design of all the blocks and functions.

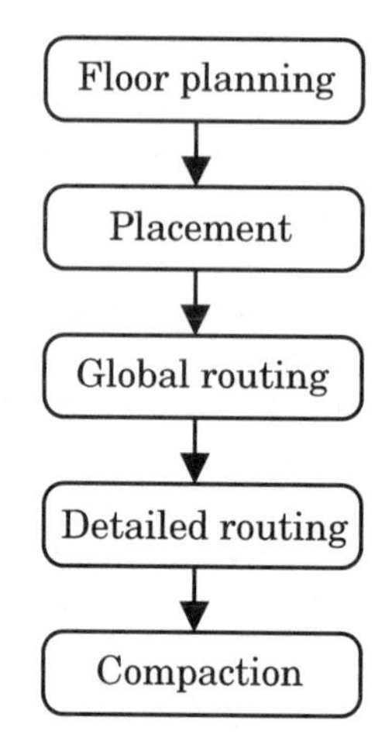

Figure 12.1 Main steps in the layout of VLSI circuits

The next step, referred to as placement, is determining the precise positions and orientations of the modules. The goal is to complete the placement and interconnections in the smallest possible area consistent with a set of technological and performance constraints.

Wire routing is usually partitioned into two steps: global and detailed. Global routing of the nets involves selecting the particular routing regions each net will go through. Thereafter, in the detailed routing step, the actual or relative locations of all interconnections are determined. Often the entire routing process is carried out in a symbolic form which is technology-independent. In the subsequent steps, the wires and contacts are given their actual sizes and any surplus area is removed by compaction. Often, detailed routing and compaction are combined into a single step. Naturally, all these steps are mutually dependent.

Unfortunately, most problems in floor planning, placement, and routing are NP-complete problems (i.e., large problems can not be solved in reasonable time). Thus, in practice various heuristic algorithms are used. A heuristic algorithm explores only a small region of the configuration space by sequentially generating a new configuration according to a set of rules. Hence, a heuristic algorithm can not be guaranteed to find the global optimum.

12.2.1 Floor Planning and Placement

The floor planning problem is defined as the problem of determining the geometry, locations, aspect ratios, and pin positions for all modules, so that all the constraints, listed shortly, are satisfied and so that a weighted sum of the total chip area, net length, delay, and power dissipation is minimized. Placement is a special, more restricted, case of floor planning.

The floor planning constraints are

- A set of modules with variable geometry, pin positions, estimated delays, and power dissipation
- Constraints on the positions of some of the modules

- Constraints on the estimated area and aspect ratio of the chip
- Constraints on the aspect ratios of the modules
- Constraints on the positions of the pins
- A net list specifying which pins have to be interconnected
- Constraints on power dissipation and maximum temperature
- Estimated delay per unit length of interconnection
- Constraints on delays at the chip level

The algorithms needed for solving this problem are NP-complete. Computation time therefore increases exponentially with the number of modules. This forces us to revert to heuristic methods. In practice heuristic or statistical algorithms may be efficiently combined with hints and guidance by the designer.

Ultimately, the goal is to automatically generate a good floor plan, do the placement, as well as automatically route the wires. The placement problem is defined as the problem of determining the locations of all modules so that all the constraints, given shortly, are satisfied and so that a weighted sum of the total area, net length, delay, and power dissipation is minimized.

The placement constraints are

- A set of modules with fixed geometry, pin positions, delays, and power consumption
- Constraints on the positions of some of the modules
- Constraints on area and on the aspect ratio of the chip
- A net list specifying which pins have to be interconnected
- Constraints on the power dissipation and maximum temperature
- Estimated delay per unit length of interconnection
- Constraints on delays at the chip level

12.2.2 Floor Plans

The problem of determining the optimal positions of the modules so that the total area of the chip is minimized, is an NP-complete problem. Of similar complexity is the problem of determining the aspect ratios of the modules such as the ones involved in the floor planning phase. However, there are some constrained placements that yield a simple and elegant algorithm to determine the optimal area and corresponding orientations and aspect ratios of the components: the so-called slicing structures. It is useful to classify floor planning and placement algorithms according to the structure of the layout generated by the algorithm—i.e., slicing structures or unconstrained placement. In order to simplify the problem, the modules are often assumed to have rectangular shapes. The area of all modules is fixed while the dimensions of some of the modules may be either fixed or flexible within a range of possible aspect ratios. Some of the modules may have fixed positions and orientations.

A floor plan is a slicing floor plan if it only contains a single rectangle or if there exists a vertical or a horizontal cut that divides the slicing structure into two parts, each of which is a slicing floor plan. A slicing floor plan can be recursively partitioned by repeated use of the following rule:

> Cutting a rectangle is defined as dividing it into two rectangles by either a horizontal or a vertical line.

A slicing floor plan is defined as a floor plan with n basic rectangles obtained by cutting the rectangular floor plan into smaller and smaller rectangles.

Figures 12.2 and 12.3 show examples of slicing and nonslicing floor plans, respectively. An advantage with the sliced floor plan is that an optimal order in which the channels can be routed is easily determined. The slicing order is indicated in Figure 12.2 and the routing order is the reverse of the slicing order.

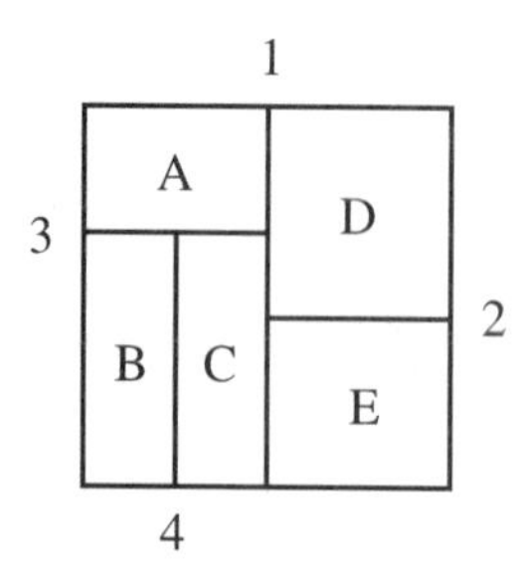

Figure 12.2 Sliced floor plan

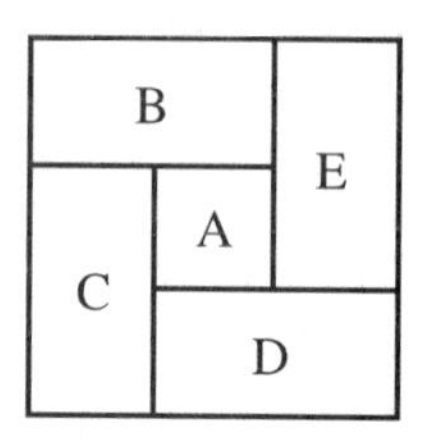

Figure 12.3 A nonslicing floor plan

12.2.3 Global Routing

The routing problem is to connect all the pins in each net while using minimum routing area, number of vias, and wire length. Other factors, such as preference of one layer over another or consideration of coupling with neighboring nets, can be included by setting cost parameters appropriately. The problem of interconnecting a large number of circuit elements in a chip takes up an increasing amount of design time and chip area. Typically, 30% of total design time and about 90% of the chip area are expended merely to interconnect the circuit elements.

Global routing is an important part of the VLSI layout, since it is a preliminary planning stage for the final routing. The conventional method for global routing is to first define a channel graph and assign the nets to the channels, as shown in Figure 12.4. The global routing phase can be partitioned into several subphases:

1. Channel definition
2. Channel ordering
3. Channel assignment

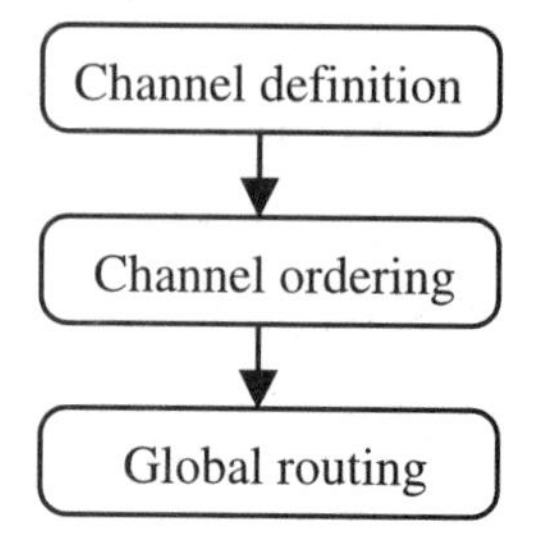

Figure 12.4 Overview of the global routing phase

Generally, the aim is to define channels only as routing areas, since channels are simpler than the more irregular regions. Further, for slicing floor plans, the channel routing order is the reverse of the slicing order. This simplifies automation of these steps.

12.2.4 Detailed Routing

General routers attempt to find the shortest Manhattan path between two adjacent points in a net, around any intervening obstructions. Since the routing problem is in general NP-complete, different solutions have been derived for special cases. The most common routers can be classified into one of the following three groups:

1. *Area routers*. This is the most general case with an arbitrary shape of the routing area and pin positions.
2. *Switchbox routers*. The routing area is restricted to be a rectangle.
3. *Channel routers*. The routing area is two-sided.

Area Routers

> "Would you tell me please, which way I ought to go from here?"
> "That depends a good deal on where you want to get to," said the cat.
> "I don't much care where," said Alice.
> "Then it doesn't matter which way you go," said the cat.
> —Lewis Carroll

An example of area routers is the Lee–Moore's algorithm, sometimes referred to as a wavefront or maze-running router. The path to be routed follows a rectangular grid, moving from one cell on the grid to an immediately adjacent cell. For example, consider the situation illustrated at the top of Figure 12.5, where the white cells indicate space available for interconnection routing, and the black cells are obstructions due to prior routing or other items. If we require to route from source cell position S to target position T, we first number the available cells as follows:

1. All cells immediately surrounding S are numbered "1."
2. All cells immediately surrounding the cells numbered "1" are numbered "2."
3. Continue numbering cells surrounding every already numbered cell with the next higher integer number until the target T is encountered.

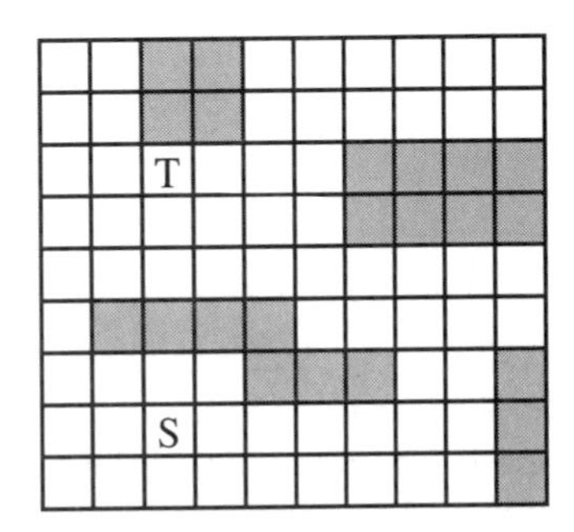

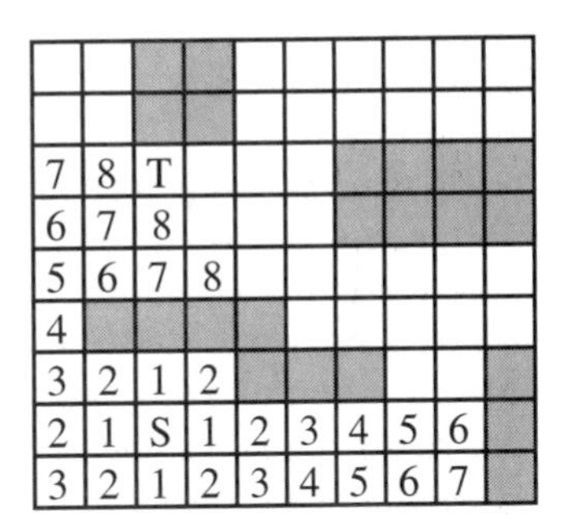

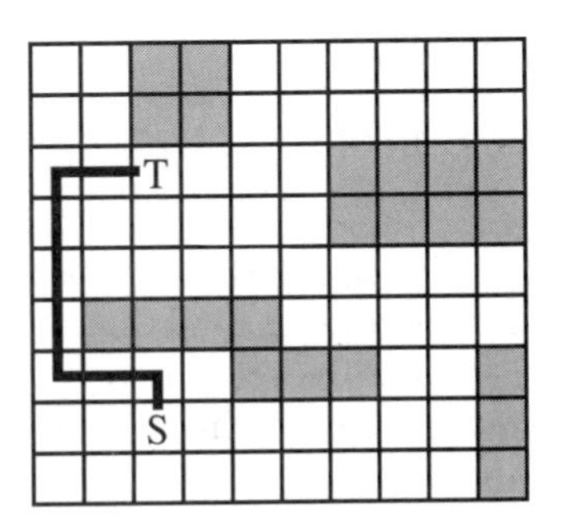

Figure 12.5 Lee–Moore's algorithm

The Lee–Moore's routing algorithm then selects a routing path between S and T by backtracking from T toward S, always moving from one grid to a next lowernumbered cell, such as shown in Figure 12.5. The advantages of Lee–Moore's algorithm are

1. Provided that the initial cell numbering can be extended from S to T, then a path between S and T is guaranteed.
2. The path chosen will always be the shortest Manhattan distance between S and T.
3. It can readily cope with nets with more than two points S and T.

Disadvantages are

1. Complications occur where there is a choice of cells in the backtracking, since in theory any cell numbered (x–1) may be chosen to follow cell x.
2. The choice of a particular route between S and T may preclude subsequent net list connections from being routed, with deletion and rerouting (by hand) of the previously completed nets necessary for 100% completion.
3. A considerable and unnecessary area of the available wiring space may be covered and enumerated by the grid radiating outward from S, with high computer memory requirements and long central processing unit running time.

Various techniques to refine the basic procedure have been advanced, particularly to avoid the labeling of matrix squares which are generally running in an inappropriate direction from the target cell T. The Lee–Moore's algorithm always finds the optimal path with respect to the metrics used, if it exists, between any two points in a gridded routing region with any number of obstacles. The drawback of the Lee–Moore's algorithm is that it is memory and time consuming.

Switchbox Routers

Switchbox routers are used to route wires in a region that is confined on all four sides. The switchbox routing problem is much more difficult than the channel routing problem. In fact, a solution to a given switchbox problem may not even exist.

Channel Routers

The channel routing problem can be defined as follows: A rectangular region, called a channel, with pins located on the top and bottom edges aligned on the grid, is given. Horizontal and vertical grid lines are referred to as tracks. The objective of channel routing is to connect all the nets using a minimum number of horizontal tracks. Typically, two different layers are used for horizontal and vertical tracks. Efficient algorithms exist for solving the channel routing problem. The left-edge algorithm is one such algorithm. However, it does not always lead to an optimal solution with a minimum number of horizontal tracks.

12.2.5 Compaction by Zone Refining

The compaction problem is inherently connected to the floor planning, placement, and routing problems. Several compaction approaches exist. However, most methods are only one-dimensional. The resulting compaction of one-dimensional compactors is usually poor.

An interesting approach, based on simulated annealing, is the compaction by zone refining algorithm. The method is based on an analogy with refinement of silicon rods used for integrated circuits. The modules are initially placed at the top of the placement area. They are then moved downwards, according to the simulated annealing algorithm, to the bottom of the placement area where they are placed properly. This process is repeated several times in all four directions.

12.3 LAYOUT STYLES

There are several major design styles that are identified by the standardization constraints imposed in the design at the geometrical level. Each style imposes its unique set of constraints on the design method as well as on the tools used for the design and testing.

The standard-cell, gate array, and sea-of-gates styles are called *semicustom* approaches by the silicon vendors, since the circuit design involves only some of the mask layers or uses a vendor-supplied cell library. The latter two approaches are based on prefabricated wafers that can be personalized by using a few interconnection layers. In other layout styles all mask layers are specified by the designer and these styles are therefore called *full-custom* styles by the silicon vendors. These terms are somewhat confusing, are irrelevant from the system designer's point of view, and are therefore not recommended. A more useful classification divides the layout styles along the imposed standardization constraints placed on the building blocks.

12.3.1 The Standard-Cell Design Approach

In the standard-cell approach the designer builds the system by using cells that are stored in a cell library that often is provided by the silicon vendor. Most cell libraries contain only simple gates and flip-flops, but there is a trend that the vendors also include more complex logic and even analog building blocks. The basic idea was originally to provide the designer with a route to design VLSI circuits in the same way as was used for design with SSI and MSI circuits. The standard-cell approach is suitable for mapping existing designs, built with now outdated SSI and MSI components, onto silicon. Of course, it is also used for new designs. The approach is simple and therefore an economically important approach to VLSI design.

The simplicity comes from the similarity to conventional (older) design techniques and the standardized floor plan. A typical floor plan for a standard-cell design approach is shown in Figure 12.6 [4, 14]. The cells implementing the logic circuits are placed in rows, and routing areas, also called *routing channels*, are provided between the rows. Notice that the heights of the routing channels, which are often referred to as the channel widths, can be selected according to the requirements. The height of the cells is fixed while the cell length varies from cell to cell.

In this approach the designer enters the design by selecting the appropriate circuits (logic primitives) from a library and by providing a *netlist* with their interconnections. The cells have output buffers that can drive a few other cells so the designer is only required to check a maximum fan-out design rule. The design is typically validated by logic and simple timing simulations. Automatic CAD tools are then used to map the design onto silicon. The tool selects the proper positioning of the cells and performs the wiring. The placement of the cells is restricted in order to simplify placement and to reduce the required number of geometrical variations of a cell.

In a standard cell the power and grounds are distributed using horizontal wires inside the cells so that abutting cells automatically make the necessary connections. The power is fed from the end of each row of cells. Inputs and outputs of the cells appear at the top and bottom. The wiring between the cells is

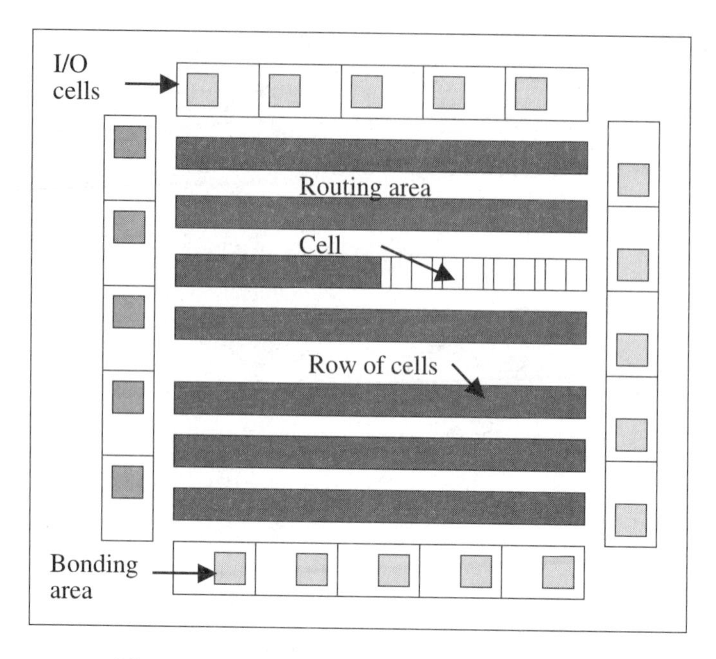

Figure 12.6 Floor plan for a standard-cell chip

done automatically by computer programs, which are called *autorouters*. The routing can be done by using only one level of metal and one level of polysilicon, or polysilicide, but usually for performance reasons two levels of metal are used.

Placement of the cells (i.e., selecting a particular row and position for a cell) is simplified because of the standardized floor plan. The routing of wires is also simplified because of the shape of the routing area. The routing area can be kept small since the height of the routing area can be adjusted to the requirements. Efficient tools for the automatic layout of standard-cell chips exist today.

Characteristics for the standard-cell design approach are:

- The cells are placed in rows. The wire routing between cells is done between the rows in routing channels. The channel widths can be adjusted according to the needs. A significant part of the rotated chip is occupied by the routing channels [12].
- The design involves mainly schematic or netlist entry followed by logic verification–validation. The placement and routing includes simple simulation, fan-out, and wire length checking, which are done by special tools. Normally, the vendor does the detailed timing analysis and checking of the layout for design rule violations.
- Device density and switching speed are higher than those of gate arrays, since the cells are internally optimized and less routing is required. The standard-cell approach has the potential to attain device densities in the range 5000 to 8000 transistors per mm^2 for a double metal, 0.8-μm CMOS process.
- Manufacturing cost is higher than for the gate array approach since all mask layers are circuit specific.
- Changing vendor or VLSI process is simple. Often, only the timing of the circuit needs to be checked by simulation.

- Custom cells (for example, RAM and ROM) as well as analog cells can be incorporated. Appropriate I/O drivers can be selected. A significant percentage of the power is often dissipated in the I/O drivers.
- The cell library is fully dependent on a specific VLSI process.

12.3.2 The Gate Array Design Approach

A simple layout style that significantly reduces processing costs is the *gate array design approach* [7]. In this approach, all but the last few mask layers are predefined by the silicon vendor and the designer is presented with an array of gates (transistors), as shown in Figure 12.7.

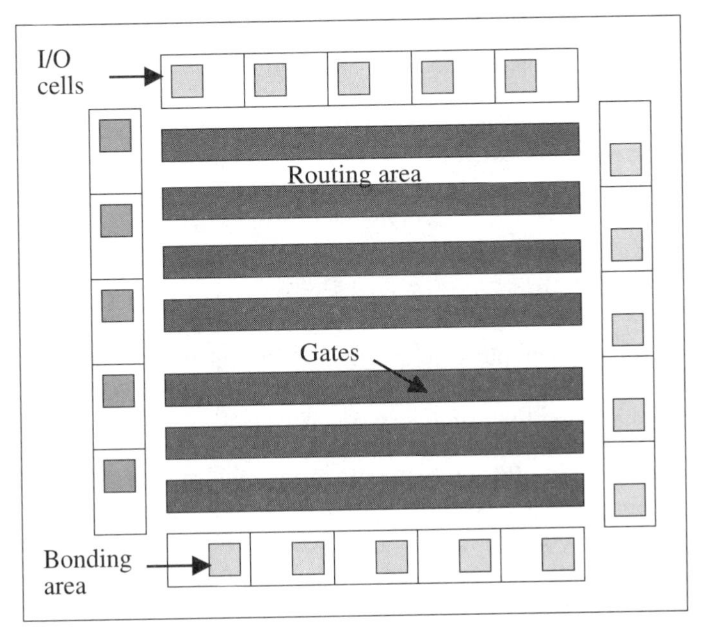

Figure 12.7 Floor plan of a gate array chip

Placement of the transistors, contacts, and routing areas is predefined by the vendor. Typically, four transistors are placed in a group that can be wired into different types of gates and flip-flops by the designer. The design process is similar to the standard-cell design approach. The design starts with schematic entry of the digital circuit—i.e., a list of gates and a *netlist* describing the interconnections from which the automatic layout tool can produce a complete layout of the gate array. The tool performs placement of circuit functions (gates), customization of devices into gates, and interconnection of the gates. The layout task is significantly simplified because of the standardization of the layout.

Today, customization of the gate array is usually done with two layers of metal or one metal and one polysilicon layer. Typically, one layer is used predominantly for vertical runs and the other for horizontal runs of wires. If a third or fourth metal layer is available, it is usually reserved for power distribution.

As an example of a typical gate array, the Fujitsu CE61 series accommodates 132 to 2000 kgates in a 0.35-μm CMOS process with a gate delay on 85 ps. The power consumption is typically 0.64 μW/MHz/gate at 3.3 V.

Gate arrays are relatively inefficient in utilization of chip area for two main reasons. Firstly, gate arrays come only in standard sizes. It is necessary to select a chip that contains more gates than needed. Typically, less than 80% of the available gates can be used. Secondly, the routing areas are made wide in order to accommodate a large number of wires. This means that the routing areas are wider than necessary.

Gate arrays provide the means by which a large portion of the general-purpose logic used in most large electronic systems can be combined into a single-package solution. Both the standard-cell and the gate array approaches reduce the designer's layout work by unloading it onto a computer.

The customization of the gate array requires only a few processing steps. Hence, the turnaround time can be very short, since the silicon vendor can stock prefabricated gate array wafers. The total manufacturing cost for a gate array design depends mainly on the chip size, packaging, and processing cost—and the processing cost is very low.

For large-volume applications, the relative inefficiency with respect to chip area may become significant. Larger chips result in higher system costs due to higher chip cost since there are fewer working chips on each wafer and a larger and more expensive package must be used. A more area-efficient design, using a more expensive design approach, may therefore more than balance the higher design cost. The gate array approach is therefore an economical alternative for low- and medium-volume systems.

BiCMOS technologies are particularly suitable for gate arrays because of their ability to drive nodes representing large capacitances, which are characteristic for gate arrays. The bipolar devices are used for I/O drivers, clock drivers, and time critical circuits.

Gate arrays are inefficient for realization of large memories. Special gate arrays are therefore provided with a RAM. Gate arrays with analog building blocks are also available. Fast gate arrays based on GaAs can accommodate more that 30,000 equivalent two-input NAND gates while dissipating 8 W. Silicon-based gate arrays achieve much higher gate counts. For example, Texas Instruments has a BiCMOS gate array with 150 kgates in a 0.8-μm process.

Characteristics for the gate array design approach are:

- The layout pattern of the transistors is fixed. Only the placements of cells, customization of transistors into cells, and routing are required.
- The design involves mainly schematic or netlist entry followed by logic verification–validation. The placements (position and rotation) of cells and wire routing including simple simulation, fan-out, and wire length checking are done by special tools. Normally, the vendor does the detailed timing analysis and checking of the layout for design rule violations.
- The device density and the switching speed are lower than for standard-cell designs, but this is often compensated for by the availability of more advanced processes with smaller geometries.
- The turnaround time is shorter and the processing cost is much lower than for standard-cell designs.
- Changing vendor or VLSI process is simple. Often, only the timing needs to be checked by simulation.

12.3.3 The Sea-of-Gates Design Approach

The sea-of-gates design approach is a variation of the gate array theme. The areas dedicated for wiring have been removed and the gate pattern has been made more uniform. The device density is therefore much higher than for ordinary gate arrays. The wiring is done over unused devices, and contacts to active devices are created whenever needed. Performance approaching that of full-custom design is obtained because of the high device density and ease of wiring. Figure 12.8 illustrates the basic idea for a cell that implements a three-input NAND gate. The second metal layer is used to customize the array. A drawback of both gate arrays and sea-of-gate arrays is that all devices have the same *W/L* ratio. Some CMOS logic circuit styles (for example, dynamic logic, transmission gate flip-flops, ROMs, and RAMs) require transistors of different sizes.

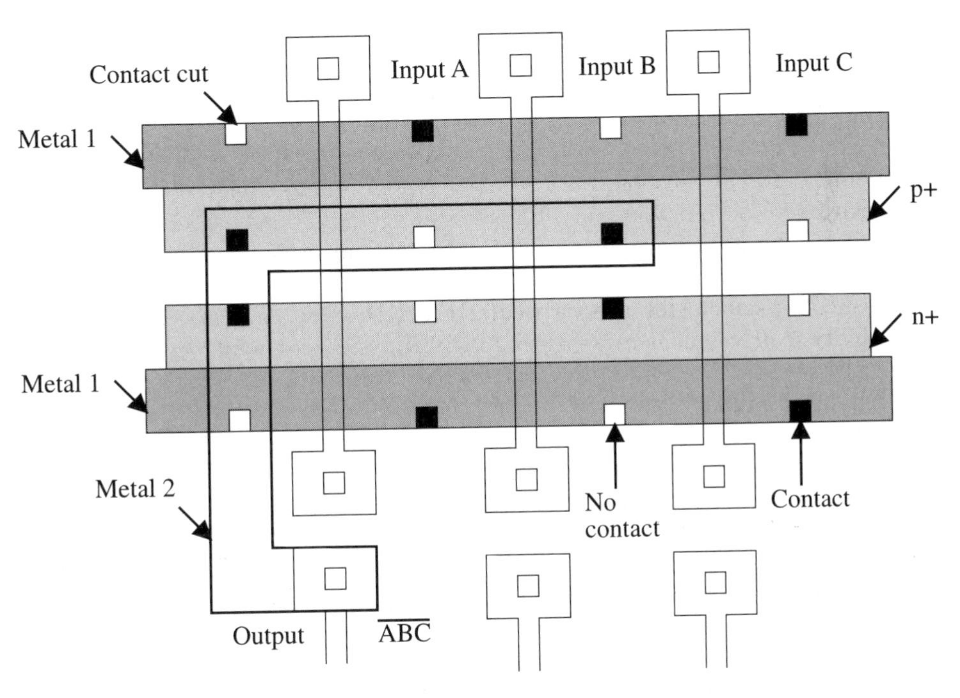

Figure 12.8 Sea-of-gates layout of a three-input NAND gate

High-speed, sea-of-gates arrays with more gates than the corresponding gate arrays are available. The number of equivalent gates on both gate and sea-of-gate arrays increases very rapidly with the shrinking of the geometries. Sea-of-gates arrays have achieved up to 1600 two-input NAND gates per mm^2, ROMs with 17 kbit/mm^2, and static RAMs with 1.1 kbit/mm^2 using a 0.8-µm process with three metal layers.

12.3.4 The Unconstrained-Cell Design Approach

The unconstrained-cell design or macrocell design approach allows arbitrary shapes and placement of the building blocks. Typically, the layout of the building blocks is synthesized from high-level functional or behavioral descriptions and the designer has only limited control of their shapes. A bottom-up assembly of the building blocks therefore becomes an important design step. Significant amounts of chip area may therefore appear between the blocks. However, this area is not wasted, because it can be used to implement decoupling capacitors between the power lines. Such capacitors are necessary to reduce switching noise. Figure 12.9 illustrates a floor plan with routed wires for a chip that is designed using the unconstrained-cell approach. Only the highest level in the cell hierarchy is shown.

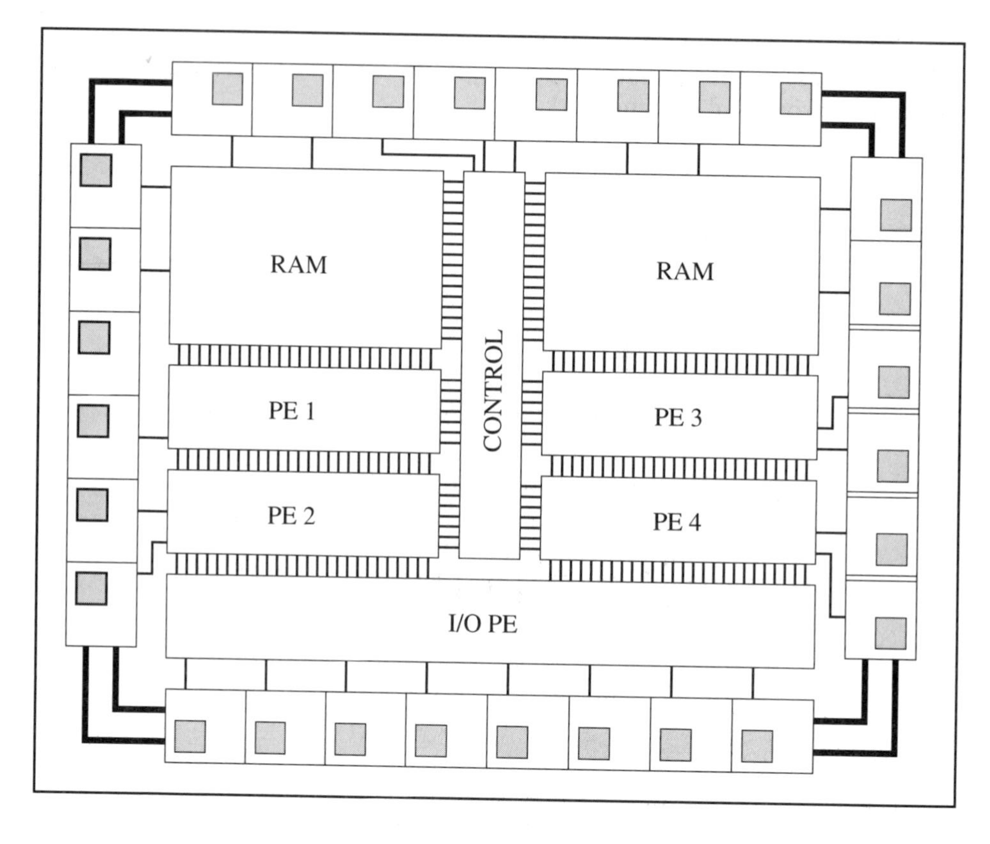

Figure 12.9 Floor plan of a chip designed with the unconstrained-cell approach

Of course, it is also possible to combine the unconstrained-cell design approach with a top-down design process. The first step is to estimate the shape and chip area needed for the different functions—e.g., processors, memories, and control units. These blocks and the necessary I/O modules are given a placement in the next step, which is called *floor planning*, and the wires between the blocks are routed. The goal of the placement and routing steps is to minimize the required chip area while satisfying the timing restrictions. This process is repeated for all levels in the design hierarchy until the remaining modules are simple cells that can easily be designed or are available in the cell library. Noncritical parts of a chip are often implemented using a standard-cell approach.

EXAMPLE 12.1

Figure 12.10 show a layout of a 128-point FFT/IFFT processor with two RAMs and a single butterfly processing element.

An unconstrained-cell approach is used for all blocks. The processor computes a 128-point FFT, or IFFT in 64 ms (including I/O). The architecture is analogous to the FFT in the case study. The processing element runs at 128 MHz, the bit-serial control unit runs at 256 MHz, and the memory runs at 32 MHz. The core area is 8.5 mm^2 and the total chip area including pads is 16.7 mm^2. The number of devices is 37,000. The chip was fabricated in AMS 0.8-μm double metal CMOS process. The power consumption is about 400 mW at 3.0 V. The power consumption is high in this case, mainly because of the use of the TSPC logic style and an overdesigned clock driver and clock network.

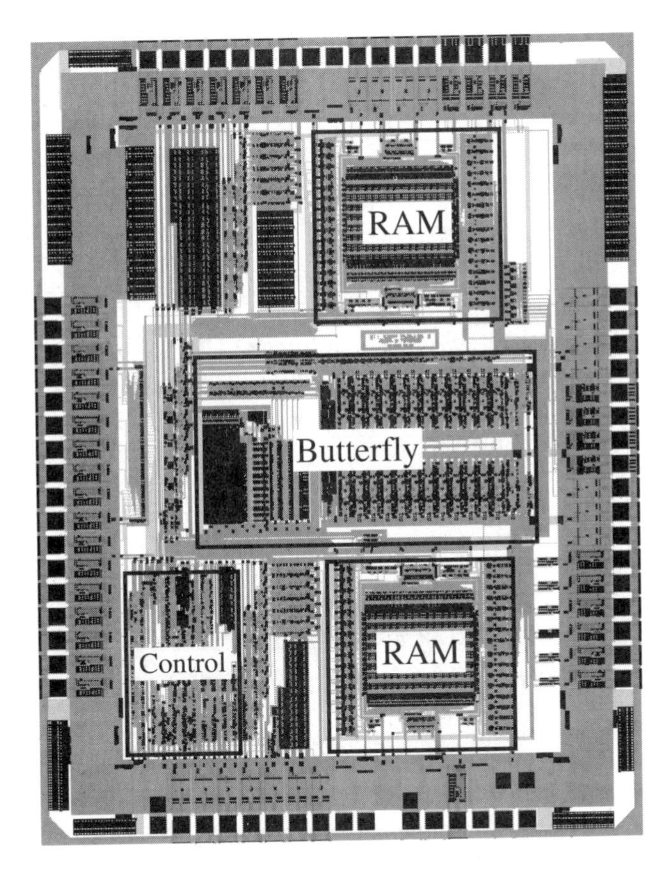

Figure 12.10 A 128-point FFT/IFFT processor using the unconstrained-cell approach

The aim of the unconstrained-cell design approach is to achieve high performance circuits and at the same time reduce the amount of design work by using previously designed building blocks or blocks that can be generated by a synthesis tool. Standardization is only used at the lowest layout level to reduce the design space. It takes the form of *design rules* that impose geometrical and topological constraints on the layout. These constraints originate from two sources:

- Limitations in the manufacturing process—for example, misalignment among mask layers, minimum width, and spacing of features.
- Physical limitations of circuits such as electromigration, current densities, junction breakdown, punch through, and latch-up.

Often, a *Manhattan geometry* is imposed on the layout. In this standardization of the layout, the wires are required to run only horizontally or vertically. The Manhattan layout style significantly simplifies the checking of the design rules. Many design rule checkers only accept Manhattan layouts.

Characteristics for the unconstrained-cell approach are:

- Generally, no restrictions are put on the shape, area, and placement of modules or the wire routing. The layout of the low-level cells is done in detail and allows all aspects of the cells to be optimized, but it is costly in terms of design time. This cost may be reduced by using automatic design tools for the cell design.
- The design involves not only schematic and netlist entry, but also detailed layout, design rule checking, logic and electrical simulation, etc. As opposed to the semicustom approaches, the designer is responsible for the whole design.
- Sophisticated software is required to generate the building blocks.
- The potential device density and switching speed are very high. Typical device densities for a double metal, 0.8-μm CMOS process are in the range of 6000 to 11000 transistors per mm^2. However, device density, normalized with respect to the shrinking geometry, which are achieved in microprocessor chips are continuing to decrease [12]. The reason for this decline in device density is to be found in the inefficient design methodologies and design tools that are being used.
- Turnaround time and cost are the same as for standard-cell designs.
- The design time may be long and less well controlled.
- Changing vendor or VLSI process may be difficult, but may be somewhat simpler if automatic or semi-automatic layout tools are used.
- Digital and analog circuits can be mixed. Medium- to relatively large-size memories can be implemented in a standard process for digital circuits.

12.3.5 The Unconstrained Design Approach

This is the most powerful and flexible design method. The designer has full freedom to specify all design parameters, transistor sizes, placement, routing, etc. This freedom gives the fullest control over area, speed, power consumption, etc., both for the devices and for the whole chip. However, because of the large flexibility, the design effort becomes very large and costly and the lack of standardization makes many design tools unusable. Therefore, the unconstrained design approach in practice is used only for critical circuits where performance is of major concern—for example, RAM and ROM cells.

Characteristics for the unconstrained design approach are:

- The layout is done in detail and allows all aspects of the circuits to be optimized, but it is costly in terms of design time. The design involves not only schematic or netlist entry, but also detailed layout, design rule checking, logic, and electrical simulation.
- The potential device density and switching frequency are very high.
- Turnaround time and cost are the same as for standard- and unconstrained-cell designs.
- Changing vendor or VLSI process may be very difficult. Often, the basic cells need to be completely redesigned.
- Digital and analog circuits can be mixed. Medium- to relatively large-size memories can be implemented using standard processes for digital circuits.

12.4 FFT PROCESSOR, Cont.

In this section we estimate the required chip area of the FFT processor assuming a 0.8-µm CMOS process. The FFT processor requires a large memory with a capacity of 1024 × (23+23)-bit words. For the FFT processor we have chosen to partition each of the two logic memories into four physical memories. This means that we will use eight memories with 128 × (23+23)-bit words. The partitioning is done to save chip area since the RAMs are a dominant part of the FFT processor. To save further chip area, it is possible to let several memory arrays share a row decoder. A decoder to select the right memory array is also needed. The floor plan for a 128 × (23+23)-bit memory is shown in Figure 12.11. The required read and write frequency is 31 MHz.

A lower bound for the chip area required for the eight memories in the FFT processor is

$$A_{\text{Memory}} = 8\,A_{\text{RAM}} = 8 \cdot 1.1 \cdot 1.3 = 11.4 \text{ mm}^2$$

Notice that a significant area is required for wiring and a large part of the chip area is wasted. Figure 12.12 shows the floor plan for the butterfly processor. A preliminary floor plan for the complete FFT processor is shown in Figure 12.13. The RAMs will consume most of the chip area.

In practice, it is recommended to use at least one power pin, for both V_{DD} and *Gnd*, for every eight I/O pins. The FFT processor requires only 32 input/output pins, and at least three pins for communication and clocking; only four V_{DD} and four *Gnd* pins are required. Further, we assume that five pins are needed for

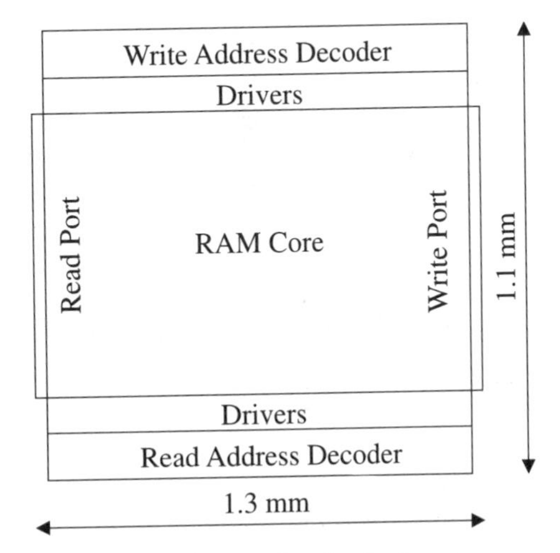

Figure 12.11 Floor plan for dual-port, single-phase RAM with 128 × (23 + 23)-bits

testing purposes. Thus, about 48 pins are required. The pad frame goes around the whole chip and the width of the pad frame is about 0.26 mm. Large empty areas are found between the butterfly processing elements. These areas will be used for address generation, control, decoupling capacitances, etc. The total chip area for the FFT processor is estimated to be

$$A_{\text{FFT}} \approx 6 \times 5 \approx 30 \text{ mm}^2$$

The power consumption is estimated to be about 150 mW for each RAM and 300 mW for each butterfly processor at 5 V. Allowing 400 mW for control and I/O yields a total power consumption of about

$$P \approx 2\, P_{\text{Butterfly}} + 8\, P_{\text{RAM}} + P_{\text{Control}} + P_{\text{I/O}} \approx 2.2 \text{ W}$$

Note that this estimate is a very crude estimate and the actual power consumption may deviate significantly. If a more power-efficient logic style [8, 9] is used, the power consumption may be reduced to only 60%—i.e., only 1.3 W is it required. Further, it should be noted that the power consumption for state-of-the-art CMOS processes, which today have geometries in the range of 0.25 to 0.35 μm, would be significantly lower. The chip area would also be lower due to the smaller geometry, as well as to the fact that few processing elements would be required sine the inherent circuit speed would be higher.

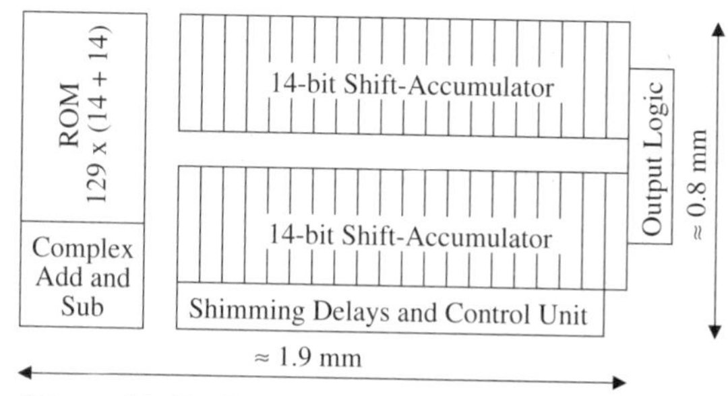

Figure 12.12 Butterfly processor with distributed arithmetic and TSPC logic

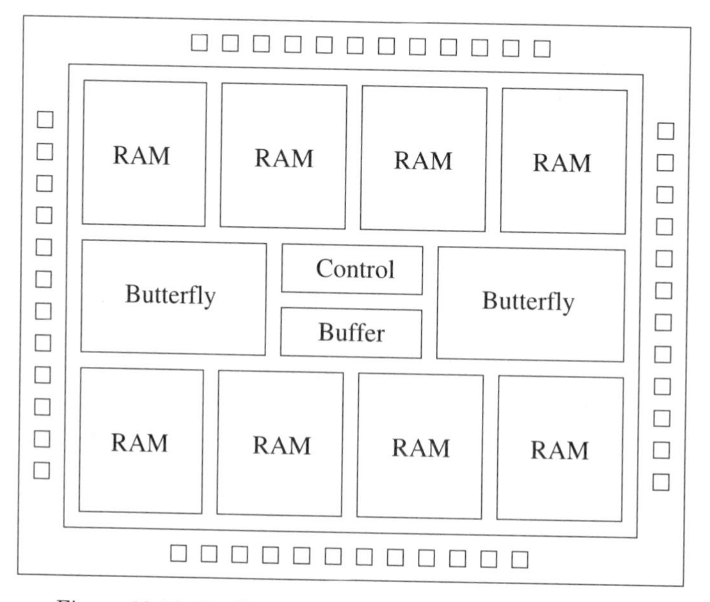

Figure 12.13 Preliminary floor plan of the FFT processor

12.5 DCT PROCESSOR, CONT.

Two sets of memories are required for the DCT processor, since two images are processed concurrently. Each memory is implemented as two dual-port RAMs with 128 × 16-bits. The size of one such RAM is

$$A_{RAM} = 0.7 \times 1.3 \approx 0.9 \text{ mm}^2$$

A floor plan for the RAM is shown in Figure 12.14. The area required for the memories is estimated to

$$A_{Memory} = 4\,A_{RAM} \approx 3.6 \text{ mm}^2$$

Figure 12.15 shows the floor plan for the complete DCT processor. The required chip area is estimated as

$$A_{DCT} = 4.4 \times 2.7 \approx 12 \text{ mm}^2$$

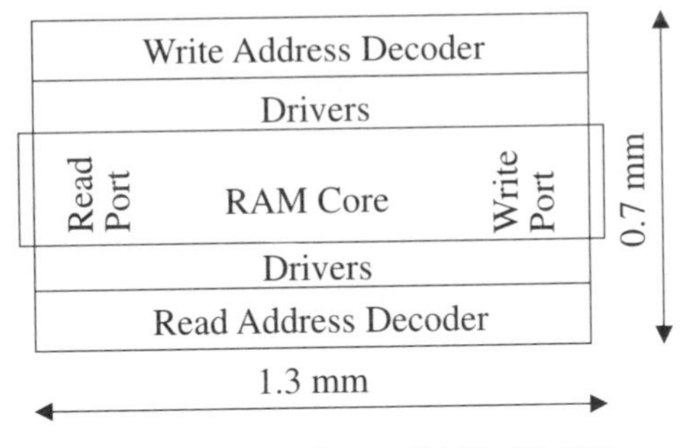

Figure 12.14 Dual-port RAM with 128 × 16-bits

The DCT processor requires 72 input and 72 output pins, and at least 3 pins for communication and clocking. Here it is necessary to use about 18 V_{DD} and 18 *Gnd* pins. We assume that 5 pins are needed for testing purposes. Thus, altogether we need about 185 pins. The active circuitry is about 4.4 × 2.7 mm = 12 mm^2. A die of this size can accommodate about 100 pads, so the chip is pad limited. The die

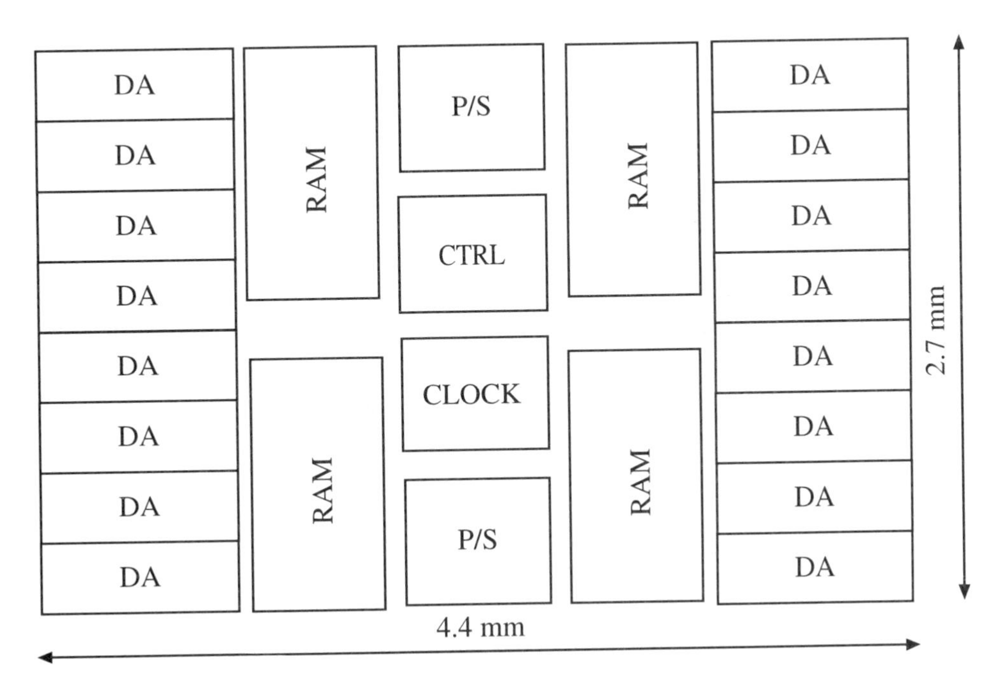

Figure 12.15 Floor plan for the 2-D DCT using TSPC logic

size is therefore increased to 7.5 × 7.5 mm to accommodate this large number of pads. The required die size, including pads and the necessary scribe margin between the dice, is

$$A_{DCT} \approx 7.5 \times 7.5 = 56 \text{ mm}^2$$

Hence, about 20% of the die is used for active circuitry. A more efficient design would involve either multiplexing I/O pins or possibly including more of the system functions on the chip. Often special, small pads are used for pin-limited chips to reduce the chip area.

The power consumption at 5 V is estimated to be 50 mW for the control unit, 60 mW for each RAM, 50 mW for each parallel/serial converter, 30 mW for the clock, and 170 mW for each distributed arithmetic unit. The total power consumption is estimated to be

$$P = 50 + 4 \cdot 60 + 2 \cdot 50 + 30 + 16 \cdot 170 \approx 3.2 \text{ W}$$

12.6 INTERPOLATOR, CONT.

The adaptor coefficients are 9 bits and the data word length is 21 bits. The floor plan for the adaptor is shown in Figure 12.16. The main area is occupied by the multiplier. The serial/parallel multiplier is based on the circuit shown in Figure 11.34. The execution time is

$$T_{PE} = \left(\max\{(W_c + 2), W_d\} + 3\right) T_{CL}$$

Also a ROM is needed for the six coefficients in each adaptor. However, the area for this ROM is insignificant. Allowing only 30% overhead for routing, etc., a 9-bit adaptor requires

$$A_{adaptor} \approx 0.9 \times 0.3 \approx 0.27 \text{ mm}^2$$

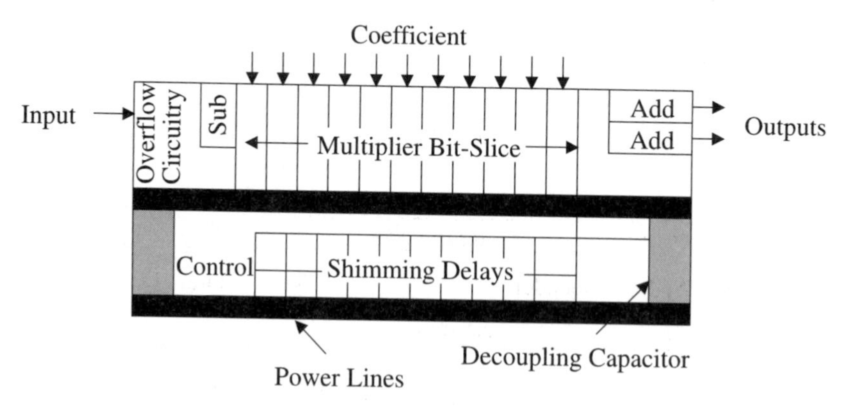

Figure 12.16 Floor plan for a 9-bit two-port adaptor using TSPC logic

Thus, less than 2 mm^2 is needed for the four PEs in the interpolator.

Two memories are required for the interpolator: one with five 21-bit words and the other with ten 21-bit words. Figure 12.17 shows the floor plan for the first

RAM. The area of the first RAM, with eight words of which only five are used, is estimated as

$$A_{RAM1} = 0.5 \cdot 0.14 \approx 0.07 \text{ mm}^2$$

The area for the second RAM, with 16 words of which only 10 are used, is estimated as

$$A_{RAM2} = 0.5 \cdot 0.2 = 0.1 \text{ mm}^2$$

Of course, the unused cells in the memories and the corresponding parts of the address decoders are never implemented. The corresponding floor plan is shown in Figure 12.18. The interpolator requires 16 input pins, 16 output pins, and at least 3 pins for communication and clocking. We will use 4 V_{DD} and 4 *Gnd* pins. Further, we assume that 5 pins are needed for testing purposes. Thus, altogether we need about 48 pins. The floor plan for the interpolator is shown in Figure 12.19.

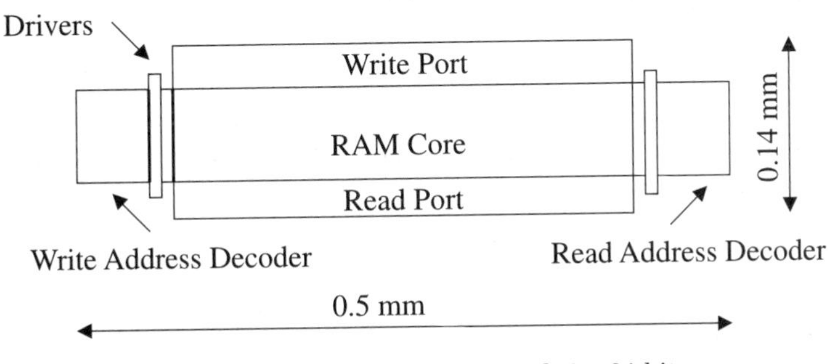

Figure 12.17 Dual-port RAM with 8 × 21 bits

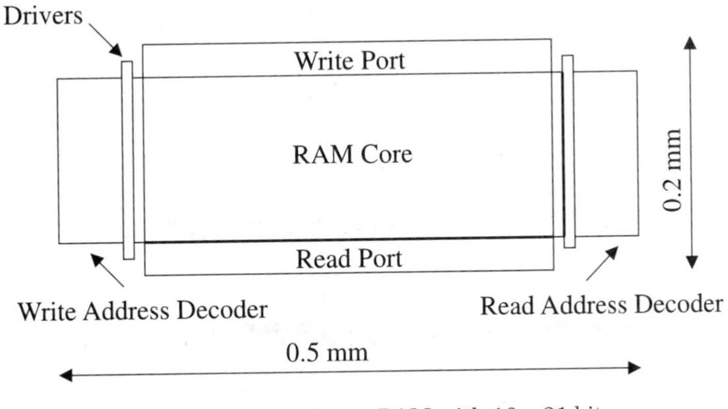

Figure 12.18 Dual-port RAM with 16 × 21 bits

The active circuitry is about 1.1 × 1.9 mm ≈ 2.1 mm^2. The pads can not be placed with a spacing of less than 135 µm. Hence, the die can accommodate only

$$2\frac{1.1 + 1.9}{0.135} \approx 44 \text{ pins}$$

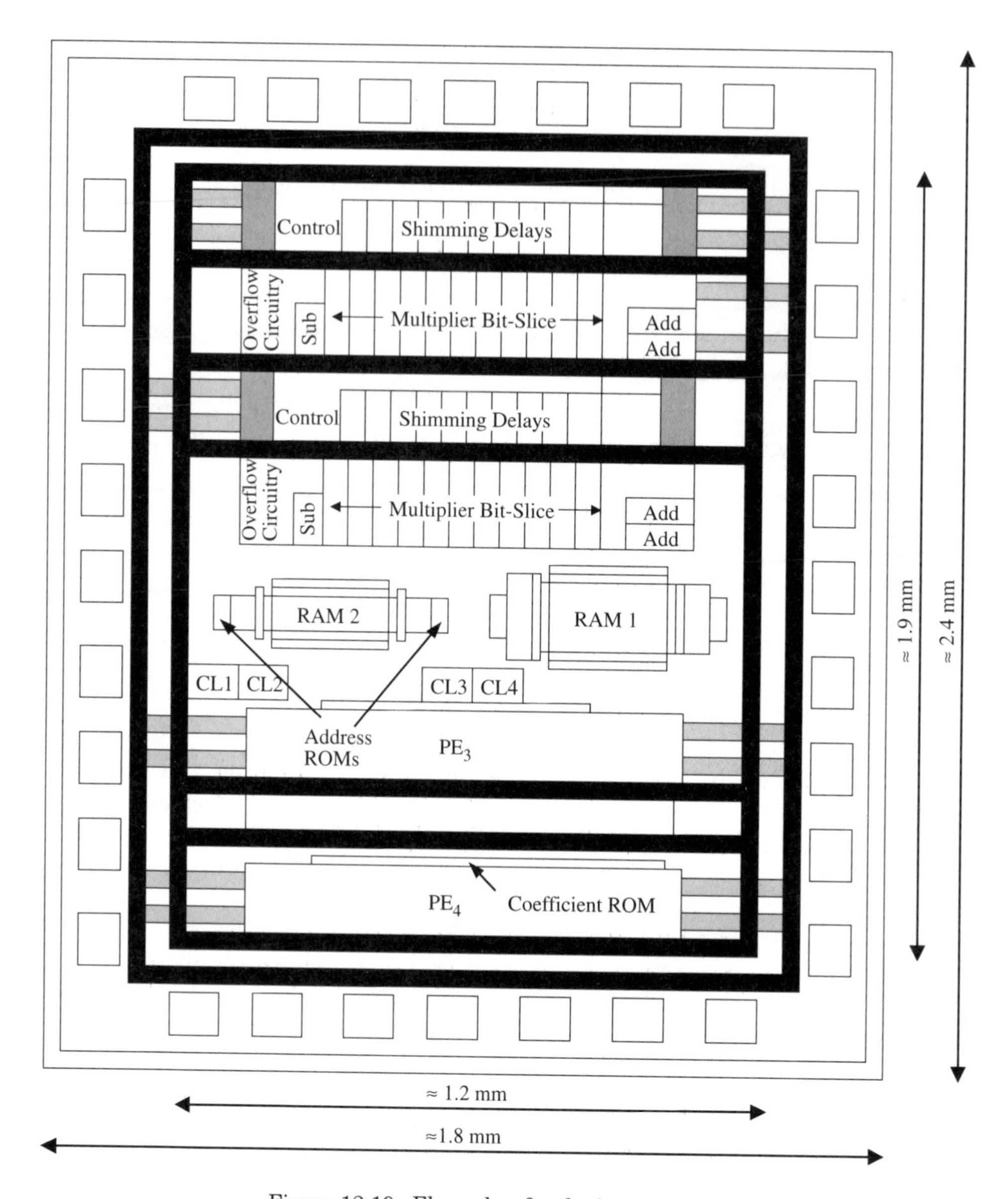

Figure 12.19 Floor plan for the interpolator

The die size must therefore be increased so that the circumference of the active circuitry becomes at least $48 \cdot 0.135 \approx 6.5$ mm. We may choose to increase the size to $1.9 \times 1.4 \approx 2.7$ mm^2. The pads and the necessary scribe margin add another 0.5 mm on each side. The required die size is $2.9 \times 2.4 \approx 7.0$ mm^2. Hence, only 32% of the die is used for active circuitry. The interpolator is obviously a pad-limited circuit.

The power consumption for each adaptor is about 180 mW. The power consumption for the complete interpolator is estimated as

$$P \approx 0.9 \text{ W}$$

Also in this case, the power consumption is unnecessarily high, mainly because of an overdesigned clock driver and clock network. Another reason is the use of a logic style—i.e., TSPC that requires a lot of power. More power-efficient logic styles [8, 9] are available that require only about 60% of the power of TSPC—i.e., only 0.55 W would be required. Yet a more modern CMOS process would reduce both the chip area and power consumption significantly.

Since only a small fraction of the available area is used, it may be advantageous to instead use an architecture with more processing elements with a reduction of the speed requirements. The excess speed may then be traded for lower power consumption by using voltage scaling—i.e., reducing the power supply voltage until the circuit just meets the requirements.

12.7 ECONOMIC ASPECTS

Today, application-specific VLSI circuits are costly and require a long design time, but this is to a large extent also true for conventional implementation techniques. It is therefore necessary to address the economic constraints as well as the designer's productivity and the risks and uncertainties associated with development of an integrated circuit. A manufacturable, reliable design must be developed on an aggressive time schedule with a minimum effort to keep pace with the rapidly changing market.

Most DSP applications require such high performance with respect to power dissipation, computational throughput, and size that it can not be met with conventional implementation technologies. The required number of chips is often small, which means that the development cost must be shared among just a few systems. The unit cost is therefore dominated by the development cost. However, it can be expected that the large investments currently being made in computer-aided and automatic design tools for DSP and integrated circuits will significantly reduce the development costs as well as the design and manufacturing time.

A VLSI package may cost $20 or more in low-volume quantities. The cost of developing and running test programs on expensive testers is also significant. An unprocessed wafer costs about $10 and the cost after processing is in the range of $200 to $800, but a wafer contains one to several hundreds of potentially good dies. Thus, silicon real-estate is relatively cheap even if large chip size means fewers chips per wafer and a higher probability of defective chips.

12.7.1 Yield

Integrated circuits are fabricated by batch processing several wafers simultaneously. Typical lot sizes may vary between 20 to 200 wafers. The number of square dice per wafer is

$$N_{Dice} \approx \pi \left(\frac{D_w}{2L_c} - 2 \right)^2 \tag{12.1}$$

where L_c = die edge length and D_w = wafer diameter. Today, 6- and 8-inch wafers are common.

The two major factors that influence the yield of an IC process line are the maturity of the line and the size of the dice. New process lines tend to have low yields in their infancy, but the yields tend to improve with processing experience and fine-tuning of the processing steps. A measure of performance for a process line is the number of defects/cm^2. It is assumed that any defects falling within the active area of a given chip will cause that chip to fail. Typical defect densities are in the range $D \approx 0.5$ to 5 defects/cm^2. Hence, the yield will be low for large chips. The simplest model of the yield is (Poisson distribution)

$$Y = e^{-dAD}, dAD < 1 \tag{12.2}$$

where d = device density (percent active area), A = die area, and D = defect density. The Poisson model assumes that defects are uniformly distributed across the area of the wafer. In practice, however, actual yield results show a somewhat slower descent of yield with increasing chip area than predicted by the Poisson model. A more accurate model for large dice ($d\,AD > 1$) is the composite Murphy–Moore's model:

$$Y = 0.5\left[\frac{1-e^{-dAD}}{dAD}\right] + 0.5e^{-\sqrt{dAD}} \tag{12.3}$$

In practice the yield is somewhat lower for dice close to the wafer periphery—i.e., the outermost 10 mm. Further, a number of circuits may be damaged during testing, bonding, etc.

EXAMPLE 12.2

Estimate the number of working chips from a batch of 50 wafers of diameter 6 inches ≈ 150 mm when the dice are square with an area of 120 mm^2. The defect density is 2 defects/cm^2. Assume that the device density—i.e., the area susceptible to defects, is $d = 0.7$.

We have $d\,A\,D = 0.7 \cdot 120 \cdot 0.02 = 1.68$

$$Y = 0.5\left[\frac{1-e^{-1.68}}{1.68}\right]^2 + 0.5e^{-\sqrt{1.68}} \approx 0.117 + 0.137 \approx 31\%$$

The estimated number of working chips is

$$N \approx 50 \cdot \pi\left(\frac{150}{2\sqrt{120}} - 1\right)^2 Y \approx 50 \cdot 107 \cdot 0.31 \approx 1658$$

REFERENCES

[1] Bakoglu H.B.: *Circuits, Interconnections, and Packaging for VLSI*, Addison-Wesley, Reading, MA, 1990.

[2] Cai H. and Hegge J.J.A.: Comparison of Floorplanning Algorithms for Full Custom ICs, *IEEE 1988 Custom Integrated Circuit Conf.*, pp. 7.2.1–7.2.4, Rochester, NY, May 1988.

[3] Cunning J.A.: The Use and Evaluation of Yield Models in Integrated Circuit Manufacturing, *IEEE Trans. on Semiconductor Manufacturing*, Vol. 3, No. 2, pp. 60–71, 1990.

[4] De Micheli G., Sangiovanni-Vincentelli A., and Antognetti P. (eds.): *Design Systems for VLSI Circuits, Logic Synthesis and Silicon Compilation*, Kluwer Academic Pub., Boston, 1987.

[5] Dillinger T.E.: *VLSI Engineering*, Prentice-Hall, Englewood Cliffs, NJ, 1988.

[6] Glasser L.A. and Dobberpuhl D.W.: *The Design and Analysis of VLSI Circuits*, Addison-Wesley, Reading, MA, 1985.

[7] Hollis E.E.: *Design of VLSI Gate Array ICs*, Prentice-Hall, Englewood Cliffs, NJ, 1987.

[8] Karlsson M., Vesterbacka M., and Wanhammar L.: A Robust Differential Logic Style with NMOS Logic Nets, *Proc. IEE IWSSIP*, Poznan, Poland, May 1997.

[9] Karlsson M., Vesterbacka M., and Wanhammar L.: Design and Implementation of a Complex Multiplier Using Distributed Arithmetic, *IEEE Workshop on Signal Processing*, SiPS, Leicester, UK, Nov. 1997.

[10] Lengauer T.: *Combinatorial Algorithms for Integrated Circuits*, J. Wiley & Sons, 1990.

[11] Ohtsuki T. (ed.): Layout Design and Verification, *Advances in CAD for VLSI*, Vol. 4, North-Holland, Amsterdam, 1986.

[12] Preas B. and Lorenzetti M. (eds.): *Physical Design Automation of VLSI Systems*, Benjamin/Cummings, Menlo Park, CA, 1988.

[13] Revett M.C.: Custom CMOS Design Using Hierarchical Floor Planning and Symbolic Cell Layout, *Proc. IEEE Intern. Conf. on Computer-Aided Design*, ICCAD-85, pp. 146–148, Santa Clara, CA, Nov. 1985.

[14] Sangiovanni-Vincentelli A.: Automatic Layout of Integrated Circuits, in De Micheli G., Sangiovanni-Vincentelli A., and Antognetti P. (eds.): *Design Systems for VLSI Circuits, Logic Synthesis and Silicon Compilation*, Kluwer Academic Pub., Boston, 1987.

[15] Sharman D.: SPLAT—Symbolic Cell Placement and Routing Tool, *1986 Canadian Conf. on VLSI*, pp. 343–347, Montreal, Oct. 1986.

[16] Sechen C.: *VLSI Placement and Global Routing Using Simulated Annealing*, Kluwer Academic Pub., Boston, 1988.

[17] Sherwani N.: *Algorithms for VLSI Physical Design Automation*, Kluwer Academic Pub., Boston, 1993.

[18] Shoji M.: *CMOS Digital Circuit Technology*, Prentice-Hall, Englewood Cliffs, NJ, 1988.

[19] Wang N.: *Digital MOS Integrated Circuits*, Prentice-Hall, Englewood Cliffs, NJ, 1989.

[20] Watanabe T. and Baba H.: A Floor plan Design System for LSI Layout, *Proc. Intern. Symp. on Circuits and Systems*, ISCAS-85, Kyoto, Japan, pp. 9–12, June 1985.

[21] Wong D.F., Leong H.W., and Liu C.L.: *Simulated Annealing for VLSI Design*, Kluwer Academic Pub., Boston, 1988.

[22] Wong D.F., and Liu C.L.: Floorplan Design of VLSI Circuits, *Algorithmica*, Vol. 4, pp. 263–291, 1989.

PROBLEMS

12.1 Estimate the unit cost, excluding development costs, for the FFT processor in the case study when
(a) 500 (b) 5000 (c) 500,000
units are to be produced. Make the appropriate assumptions.

12.2 What is a slicing floor plan? What are its advantages and disadvantages?

12.3 Estimate the power consumption for the FFT processor if the two RAMs are placed off-chip.

12.4 The required clock frequency of a commercial CMOS circuit is 50 MHz. The chip is estimated to dissipate 0.4 W and is housed in a plastic chip carrier that has an effective thermal resistance of 172 °C/W. The process spread is ±35% and the variation in power supply is ±5%. Estimate the required design margin for the clock frequency.

12.5 Is it possible to significantly reduce the power consumption for the FFT processor by using another logic style for the shift-accumulators?

12.6 Estimate the power consumption for the FFT processor if the execution time is increased to 2 µs.

12.7 Estimate the required chip area and power consumption if an FIR filter solution is used for the interpolator.

12.8 Estimate the potential reduction in power consumption that is possible by using only power supply voltage scaling for the case studies involving a
(a) DCT processor
(b) Interpolator

Assume that the maximal clock frequency is

$$f_{CLmax} \approx k \frac{(V_{DD} - V_T)^{\alpha}}{V_{DD}}$$

where $\alpha \approx 1.55$, $V_T = 0.7$ V, and a 12-bit S/P multiplier in the logic style [8, 9] typically requires 0.25 mm^2 and 35 mW at 250 MHz and 3 V.

12.9 Estimate the potential reduction in power consumption that is possible by using both power supply voltage scaling and a modified architecture for the case studies involving a
(a) DCT processor
(b) Interpolator

INDEX

D